Spon's Architects' and Builders' Price Book

2017

Spon's Architects' and Builders' Price Book

Edited by

AECOM

2017

One hundred and forty-second edition

CRC Press
Taylor & Francis Group

First edition 1873
One hundred and forty-second edition published 2017
by CRC Press
2 Park Square, Milton Park, Abingdon, Oxon, OX14 4RN

and by CRC Press
Taylor & Francis, 6000 Broken Sound Parkway, NW, Suite 300, Boca Raton, FL 33487

CRC Press is an imprint of the Taylor & Francis Group, an informa business

© 2017 Taylor & Francis

British Library Cataloguing in Publication Data
A catalogue record for this book is available from the British Library

ISBN: 978-1-4987-8611-9
Ebook: 978-1-3153-1742-7

ISSN: 0306-3046

Typeset in Arial by Taylor & Francis Books

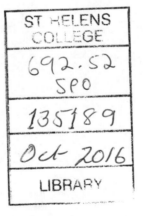

MIX
Paper from
responsible sources
FSC
www.fsc.org   FSC® C013056

Printed and bound in Great Britain by
TJ International Ltd, Padstow, Cornwall

Contents

PART 4: PRICES FOR MEASURED WORKS

PART 5: FEES FOR PROFESSIONAL SERVICES

PART 6: DAYWORK AND PRIME COST

PART 7: USEFUL ADDRESSES

PART 8: TABLES AND MEMORANDA

Preface to the One Hundred and Forty-Second Edition

Recent Construction Activity Trends

Despite moderating nationwide construction industry output capacity constraints are unmistakeable. Wages and salaries, skills surveys, constructor views and sentiment all indicates an extension of the purple patch for UK construction until very recently. Some clouding of the outlook are the discussions around the European Union, underlying fundamentals for the UK economy and global economic issues. These converging challenges are likely to affect UK construction in the medium term.

Around the regions there are mixed messages from main contractors. Many firms are operating at or near full capacity. Yet others still have significant gaps in their order books. Certainly there are good prospects for UK regions, with some large projects waiting for the green light.

Headline UK output data published by the Office for National Statistics (ONS) suggests that a weaker rate of construction growth is apparent. This is a continuing trend seen since the second half of 2015. Between January and March 2016, output in the construction industry was estimated to have decreased by 1.1% compared with 2015Q4. Downward pressure on the quarter came from all new work which decreased by 0.6% and repair and maintenance (R&M) which decreased by 1.9%. Between 2015Q1 and 2016Q1 output was estimated to have decreased by 1.9%.

Materials

Construction materials experiencing the largest price increases in the 12 months to March 2016 in the UK are shown in the table below:

Construction materials	Change between March 2015 and March 2016
Insulating materials	+4.8%
Concrete blocks, bricks, tiles and flagstones	+4.1%
Sanitary ware	+3%
Precast concrete products	+2.9%
Builders woodwork	+2.6%
Sand and gravel	+2.6%
Imported sawn or planed wood	−6.5%
Fabricated structural steel	−10%
Imported plywood	−13.2%
Concrete reinforcing bars	−16.5%

From March 2015 to March 2016 the average rise for the BIS recorded materials show a slight fall of 0.7%. This figure does however hide some dramatic differences for individual materials.

Labour – Wage Agreements

All wage agreements have yet to be settled. We have assumed a 3.5% increase to last year's rates. We will publish the agreed rates in our *Update* in November/December.

Construction industry pay levels generally follow a similar trend to whole economy pay levels. More recently however they have been exceeding the trend and show higher levels of pay.

Book Price Level

The price level of *Spon's A&B 2017* has been indexed at **572**. Readers of *Spon's A&B* are reminded that Spon's is the only known price book in which key rates are checked against current tender prices. Users should note that this number is based on book prices arrived at by pricing our usual tender price models with prices taken from the book. It does not reflect a particular date which would appear in our normal published indices.

Profits and Overheads

The 2017 edition includes a 5% allowance main contractor's overheads and profit.

Preliminaries

There are continuing signs that preliminaries costs are beginning to harden, but they still typically range from 10% to 15%, sometimes lower, sometimes higher. Procurement route, location and a host of other things effect the preliminaries allowances. We have set our example provision for preliminaries this year at 13.2%.

- Preliminaries are ***not included*** within the main *Prices for Measured Works* or in the *Approximate Estimating Rates* sections of the book.
- Preliminaries are ***included*** in the rates within *Building Prices per Functional Unit* and *Building Prices per Square Metre* sections.

Prices included within this edition do not include for VAT, professional fees etc. which must be added if appropriate.

PARTS OF THIS BOOK

Part 1: General

This section contains advice on various construction specialisms; capital allowances; legislation; taxes; insurances; building cost and tender price indices and regional price variations.

Part 2 Rates of Wages

Shows current industry level wage agreement rates and how we have built up the gang labour rates being used in the book.

Part 3: Approximate Estimating

This section contains distinct areas:

- *Building Prices per Functional Unit; Building Prices per Square Metre* and *Building Cost Models*. It should be noted that these sections all include *site preliminaries*. The only occasion this happens within the book.
- *Approximate Estimating Rates* shows typical composite built-up rates organized by building elements. Please note these rates *do not include for any site preliminaries*.

There is also a section where we show typical preliminaries build up for a project valued at approximately £4,000,000. This is intended for guidance only and should not be used as part of any tender submission.

Part 4: Prices for Measured Works

These sections contain Prices for Measured Works organized using the NRM2 Work Sections for building works.
NOTE: All prices in Part 4 exclude the main contractor's preliminaries costs.

Part 5: Fees for Professional Services

This section contains guidance on fee levels for professional services; Quantity Surveyors; Architects' and Consulting Engineers. Readers should always obtain fee proposals for their project prior to commencement as there are many factors that influence fee submissions.

NOTE: Professional fees are not included in any rates in the book.

Part 6: Daywork and Prime Cost

This section contains Daywork and Prime Cost allowances issued by the Royal Institute of Chartered Surveyors.

Part 7: Useful Addresses for Further Information
A list of useful trade associations, professional bodies contact details.

Part 8: Tables and Memoranda

This section contains general formulae, weights and quantities of materials, other design criteria and useful memoranda associated with each trade.

While every effort is made to ensure the accuracy of the information given in this publication, neither the Editors nor Publishers in any way accept liability for loss of any kind resulting from the use of such information.

AECOM Ltd
Aldgate Tower
2 Leman Street
London
E1 8FA

Building Construction Handbook, 11th Edition
CHUDLEY & GREENO

The Building Construction Handbook is THE authoritative reference for all construction students and professionals. Its detailed drawings clearly illustrate the construction of building elements, and have been an invaluable guide for builders since 1988. The principles and processes of construction are explained with the concepts of design included where appropriate. Extensive coverage of building construction practice, techniques, and regulations representing both traditional procedures and modern developments are included to provide the most comprehensive and easy to understand guide to building construction.

This new edition has been updated to reflect recent changes to the building regulations, as well as new material on the latest technologies used in domestic construction.

Building Construction Handbook is the essential, easy-to-use resource for undergraduate and vocational students on a wide range of courses including NVQ and BTEC National, through to Higher National Certificate and Diploma, to Foundation and three-year Degree level. It is also a useful practical reference for building designers, contractors and others engaged in the construction industry.

April 2016: 234 x 156 mm: 1012 pp
Pb: 978-1-138-90709-6: £28.99

Special Acknowledgements

Acodrain
ACO Business Park
Hitchin Road
Shefford
Bedfordshire
SG17 5TE
Tel: 01462 816666
buildingdrainage@aco.co.uk
www.aco.co.uk
Drainage channels

the future is safer with altro

Altro Floors
Works Road
Letchworth
Hertfordshire
SG6 1NW
Tel: 0870 6065432/01462 707600
Fax: 0870 5113388/01462 707504
info@altro.co.uk
www.altro.co.uk
Floor coverings

Andrews Marble Tiles
324–330, Meanwood Road
Leeds
Yorkshire
LS7 2JE
Tel: 0113 262 4751
Fax: 0113 239 2184
sales@andrews-tiles.co.uk
www.andrews-tiles.co.uk
Floor and wall tiles

ASSA ABLOY

Assa Abloy Ltd
The Meadows
Cannock Road
Wolverhampton
West Midlands
WV10 0RR
Tel: 01902 364 648
Fax: 01902 364 666
info@assaabloy.com
www.assaabloy.co.uk
Doorsets & furniture

Building Innovation
Unit 30
Berrington Road
Sydenham Industrial Estate
Leamington Spa
Warwickshire
CV31 1NB
Tel: 01926 888808
info@building-innovation.co.uk
www.building-innovation.co.uk
Tapered insulation

HALFEN
YOUR BEST CONNECTIONS

Halfen Ltd
Unit 2
Humphreys Road
Woodside Estate
Dunstable
Bedfordshire
LU5 4TP
Tel: 01582 470 300
brick@halfen.co.uk
www.halfen.co.uk
Brick accessories – channels and special products

Hare Structural Engineers
Brandlesholme House
Brandlesholme Road
Bury
Lancashire
BL81JJ
Tel: 0161 609 0000
info@hare.co.uk
www.hare.co.uk
Structural steelwork

Kingspan Insulated Panels
Greenfield Business Park No 2
Holywell
Flintshire
CH8 7GJ
Tel: 01358 716100
info@kingspanpanels.com
www.kingspanels.co.uk
Insulated wall and roof panels

Parker & Highland Joinery Ltd
14A Chartwell Road
Lancing Business Park
Lancing
West Sussex
BN15 8TU
Tel: 01903 756 283
sales@parker-joinery.com
www.parker-joinery.com
Purpose-made joinery

Profile 22
Stafford Park 6
Telford
Shropshire
TF3 3AT
Tel: 01952 290910
mail@profile22.co.uk
www.profile22.co.uk
uPVC windows

NDM Metal Roofing & Cladding Ltd
Mettalum House
Unit 3, 89, Manor Farm Road
Alperton
Wembley
HA0 1BA
Tel: 0208 991 7310
enquiries@ndmltd.com
www.ndmltd.com
Metal cladding and roofing

Severfield Limited
Dalton Airfield Industrial Estate
Dalton
Thirsk
North Yorkshire
YO7 3JN
Tel: 01845 577896
www.severfield.com
Structural steel

Acknowledgements

Acodrain
ACO Business Park
Hitchin Road
Shefford
Bedfordshire
SG17 5TE
Tel: 01462 816666
buildingdrainage@aco.co.uk
www.aco.co.uk
Drainage channels

Allgood PLC
297, Euston Road
London
NW1 3AQ
Tel: 020 7387 9951
info@allgood.co.uk
www.allgood.co.uk
Ironmongery

Altro Floors
Works Road
Letchworth
Hertfordshire
SG6 1NW
Tel: 0870 6065432/01462 707600
info@altro.co.uk
www.altro.co.uk
Floor coverings

Alumasc Exterior Building Products Ltd
White House Works
Bold Road
Sutton
St Helens
WA9 4JG
Tel: 01744 648400
info@alumasc-exteriors.co.uk
www.alumasc-exteriors.co.uk
Aluminium rainwater goods

Alumasc Interior Building Products Ltd
Halesfield 19
Telford
Shropshire
TF7 4QT
Tel: 01952 580590
sales@pendock.co.uk
www.pendock.co.uk
Skirting trunking & casings

Amwell Systems Ltd
Buntingford Business Park
Baldock Road
Buntingford
Hertfordshire
SG9 9ER
Tel: 01763 276200
sales@amwell-systems.com
www.amwell-systems.com
Toilet cubicles

Andrews Marble Tiles
324–330, Meanwood Road
Leeds
Yorkshire
LS7 2JE
Tel: 0113 262 4751
sales@andrews-tiles.co.uk
www.andrews-tiles.co.uk
Floor and wall tiles

Armitage Shanks Group
Armitage
Rugeley
Staffordshire
WS15 4BT
Tel: 01543 490253
arm-idealinfo@idealstandard.com
www.armitage-shanks.co.uk
Sanitary fittings

Armstrong Floor Products UK Ltd
Customer Service Centre
Fleck Way
Teeside Industrial Estate
Thornaby on Tees
Cleveland
TS17 9JT
Tel: 01642 768660
commercial_uk@armstrong.com
www.armstrong-flooring.co.uk
Flooring products

Assa Abloy Ltd
The Meadows
Cannock Road
Wolverhampton
West Midlands
WV10 0RR
Tel: 01902 364 648
Fax: 01902 364 666
info@assaabloy.com
www.assaabloy.co.uk
Ironmongery

Bison Concrete Products Ltd
Millenium Court
First Avenue Centrum 100
Burton-upon-Trent
DE14 2WR
Tel: 01283 495000
concrete@bison.co.uk
www.bison.co.uk
Precast concrete floors

Bolton Gate Co
Waterloo Street
Bolton
Lancashire
BL1 2SP
Tel: 01204 871000
sales@boltongate.co.uk
www.boltongate.co.uk
Insulated shutter doors

Bradstone Structural Sales
North End Farm Works
Ashton Keynes
Wiltshire
SN6 6QX
Tel: 01285 646 884
bradstone-structural@aggregate.com
www.bradstone-structural.com
Walling, dressings and other products

British Gypsum Ltd
Pandy Lane
Barrow-upon-Soar
Loughborough
Leicestershire
LE12 8GB
Tel: 01509 817 200
bgtechnical.enquiries@bpb.com
www.british-gypsum.com
Plasterboard and plaster products

Building Innovation
Unit 30
Berrington Road
Sydenham Industrial Estate
Leamington Spa
Warwickshire
CV31 1NB
Tel: 01926 888808
info@building-innovation.co.uk
www.building-innovation.co.uk
Tapered insulation

Burlington Slate Ltd
Cavendish House
Kirby-in-Furness
Cumbria
LA17 7UN
Tel: 01229 889 681
sales@burlingtonstone.co.uk
www.burlingtonstone.co.uk
Westmoreland slating

Buttles
Soothouse Spring
Valley Road Industrial Estate
St Albans
Hertfordshire
AL3 6NX
Tel: 01727 834 242
beeweb@buttle.co.uk
www.buttles.co.uk
Timber products

Catnic Ltd
Pontypandy Industrial Estate
Caerphilly
Mid Glamorgan
CF83 3GL
Tel: 02920 337900
www.catnic.com
Steel lintels

Cavity Trays Ltd
Administration Centre
Lufton Trading Estate
Yeovil
Somerset
BA22 8HU
Tel: 01935 474769
sales@cavitytrays.co.uk
www.cavitytrays.com
Cavity trays, closers and associated products

Cementation Foundations Skansa Ltd
Maple Cross House
Denham Way
Maple Cross
Rickmansworth
Hertfordshire
WD3 2SW
Tel: 01923 423100
cementation.foundations@skanska.co.uk
www.skanska.co.uk
Piling systems

Colour Centre
29a, Offord Road
London
N1 1AE
Tel: 020 7609 1164
www.colourcentre.com
Paints, varnishes and stains

Concrete Canvas Ltd (UK)
Unit 3
Block A22
Treforest Industrial Estate
CF37 5SP
Tel: 0845 680 1908
info@concretecanvas.co.uk
www.concretecanvas.co.uk
Concrete canvas

Cox Building Products
Unit 1
Shaw Road
Bushbury
Wolverhampton
WV10 9LA
Tel: 01902 371800
sales@coxspan.co.uk
www.coxbp.com
Rooflights

CPM-Group
Mells Road
Mells
Frome
BA11 3PD
Tel: 0117 981 2791
sales@cpm-group.co.uk
www.cpm-group.co.uk
Concrete pipes etc.

Custom Metal Fabrications
Central Way
Feltham
Middlesex
TW14 0XJ
Tel: 020 8844 0940
dgibbs@cmf.co.uk
www.cmf.co.uk
Metalwork and balustrading

Decra Roof Systems (UK) Ltd
Unit 3
Faraday Centre
Faraday Road
Crawley
West Sussex
RH10 9PX
Tel: 01293 545 058
sales@decra.co.uk
www.decra.co.uk
Roofing system and accessories

Dow Construction Products
2 Heathrow Boulevard
284 Bath Road
West Drayton
Middlesex
UB7 0DQ
Tel: 020 8917 5050
styrofoam-uk@dow.com
www.styrofoameurope.com
Insulation products

Dreadnought Clay Roof Tiles
Dreadnought Works
Pensnett
Brierley Hill
Staffordshire
DY5 4TH
Tel: 01384 77405
sales@dreadnought-tiles.co.uk
www.dreadnought-tiles.co.uk
Clay roof tiling

Dufaylite Developments Ltd
Cromwell Road
St Neots
Cambridgeshire
PE19 1QW
Tel: 01480 215000
enquiries@dufaylite.com
www.dufaylite.com
Clayboard void former

Envirodoor Ltd
Viking Close
Great Gutter Lane East
Willerby
Hull
East Yorkshire
HU10 6BS
Tel: 01482 659375
sales@envirodoor.com
www.envirodoor.com
Sliding and folding doors

Expamet Building Products
Greatham Street
Longhill Industrial Estate (North)
Hartlepool
TS25 1PR
Tel: 01429 866611
sales@expamet.net
www.expamet.co.uk
Builder's metalwork products

Forbo Flooring UK Ltd
High Holborn Road
Ripley
Derbyshire
DE5 3NT
Tel: 01773 744 121
www.forbo-flooring.co.uk
Mats and matwells

Forterra Building Products Ltd
5 Grange Park Court
Roman Way
Northampton
NN4 5EA
Tel: 01604 707600
www.forterra.co.uk
Facing bricks; aircrete blocks; aggregate blocks

Forticrete Ltd
Boss Avenue
off Grovebury Road
Leighton Buzzard
Beds
LU7 4SD
Tel: 01525 244 900
masonry@forticrete.com;
roofing@forticrete.com
nwblocksales@forticrete.com
www.forticrete.co.uk
Blocks and roof tiles

Garador Ltd
Bunford Lane
Yeovil
Somerset
BA20 2YA
Tel: 01935 443700
www.garador.co.uk
Garage doors

Grace Construction Products
Expansion Jointing and Waterproofing Division
Ajax Avenue
Slough
Berkshire
SL1 4BH
Tel: 01753 692929
uksales@grace.com
www.uk.graceconstruction.com
Expansion joint fillers and waterbars

Gradus Wall Protection
Park Green
Macclesfield
Cheshire
SK11 7LZ
Tel: 01625 428 922
sales@gradusworld.com
www.gradusworld.com
Stair nosings

GRP Tanks
33 Rye Road
Hoddesdon
Hertfordshire
EN11 0JE
Tel: 0871 200 2082
www.grptanks.net
GRP water tanks

H+H Celcon UK Ltd
Celcon House
Ightham
Sevenoaks
Kent
TN15 9HZ
Tel: 01732 886333
info@hhcelcon.co.uk
www.hhcelcon.co.uk
Concrete blocks

Halfen Ltd
Unit 2
Humphreys Road
Woodside Estate
Dunstable
Bedfordshire
LU5 4TP
Tel: 01582 470 300
brick@halfen.co.uk
www.halfen.co.uk
Brick accessories – channels and special products

Hare Structural Engineers
Brandlesholme House
Brandlesholme Road
Bury
Lancashire
BL81JJ
Tel: 0161 609 0000
info@hare.co.uk
www.hare.co.uk
Structural steelwork

Hathaway Roofing Ltd
Tindal Crescent
Bishop Auckland
County Dunham
DL14 9TL
Tel: 01388 605 636
steven.price@hathaway-roofing.co.uk
www.hathaway-roofing.co.uk
Sheet wall and roof claddings

Hillaldam Coburn Ltd
Unit 16
Merton Industrial Park
Lee Road
London
SW19 3HX
Tel: 020 8545 6680
sales@hillaldam.co.uk
www.coburn.co.uk
Sliding and folding door gear

HSS Hire Shops
Group Office
25 Willow Lane
Mitcham
Surrey
CR4 4TS
Tel: 020 8260 3100
www.hss.co.uk
Tool hire

Hudevad
Unit 5
Cyan Way
Phoenix Way (A444)
Coventry
CV2 4QP
Tel: 02476 881200
sales@hudevad.co.uk
www.hudevad.co.uk
Radiators

Hunter Plastics Ltd
Nathan Way
London
SE28 0AE
Tel: 020 8855 9851
info@hunterplastics.co.uk
www.hunterplastics.co.uks
Plastic rainwater goods

Ibstock Building Products
Leicester Road
Ibstock
Leicestershire
LE67 6HS
Tel: 01530 261999
marketing@ibstock.co.uk
www.ibstock.co.uk
Facing bricks; tilebricks

Icopal Ltd
Barton Dock Road
Stretford
Manchester
M32 0YL
Tel: 0843 224 7400
info.uksales@icopal.com
www.icopal.co.uk
Damp-proof products

James Latham
Badminton Road
Yate
Bristol
Avon
BS37 5JX
Tel: 01454 315 421
marketing@lathams.co.uk
www.lathamtimber.co.uk
Hardwood and panel products

Jeld-Wen UK Ltd
Watch House Lane
Doncaster
South Yorkshire
DN5 9LR
Tel: 0870 126 0000
marketing@jeld-wen.co.uk
www.jeld-wen.co.uk
Doors and windows

John Brash and Co Ltd
The Old Shipyard
Gainsborough
Lincolnshire
DN21 1NG
Tel: 01427 613858
info@johnbrash.co.uk
www.johnbrash.co.uk
Roofing shingles

John Guest Speedfit Ltd
Horton Road
West Drayton
Middlesex
UB7 8JL
Tel: 01895 449 233
www.speedfit.co.uk
Speedfit product range

Junkers Ltd
Unit 1a
1 Wheaton Road
Witham
Essex
CM8 3UJ
Tel: 01376 534 700
sales@junkers.co.uk
www.junkers.co.uk
Hardwood flooring

Kalzip Ltd
Haydock Lane
Haydock
St Helens
Merseyside
WA11 9TY
Tel: 01942 295500
kalzip-uk@corusgroup.com
www.kalzip.com
Kalzip roofing

KB Rebar Ltd
Unit 5 Dobson Park Industrial Estate
Dobson Park Way
Ince
Wigan
WN2 2DY
Tel: 0161 790 8635
www.kbrebar.co.uk
Reinforcement bar and mesh

Kingspan Access Floors
Burma Drive
Marfleet
Hull
HU9 5SG
Tel: 01482 781 710
enquiries@kingspanaccessfloors.co.uk
www.kingspanaccessfloors.co.uk
Raised access floors

Kingspan Environmental
College Road North
Aston Clinton
Aylesbury
Bucks
HP22 5EW
Tel: 01296 633000
pollutiongb@kingspanenv.com
www.kingspanenv.com
Interceptors and septic tanks

Kingspan Insulated Panels
Greenfield Business Park No 2
Holywell
Flintshire
CH8 7GJ
Tel: 01358 716100
info@kingspanpanels.com
www.kingspanels.co.uk
Insulated wall and roof panels

Kingspan Insulation Ltd
Pembridge
Leominster
Herefordshire
HR6 9LA
Tel: 0870 850 8555
info.uk@insulation.kingspan.com
www.insulation.kingspan.com
Insulation products

Kingspan Structural Products
Sherburn
Malton
YO17 8PQ
Tel: 01944 712 000
sales@kingspanstructural.com
www.kingspan.com
Multibeam purlins

Knauf Insulation Ltd
PO Box 10
Stafford Road
St Helens
WA10 3NS
Tel: 01744 766 666
sales@knaufinsulation.com
www.knaufinsulation.co.uk
Insulation products

Landpro Ltd
14 Upper Bourne Lane
Farnham
Surrey
GU10 4RQ
Tel: 01252 795030
info@landpro.co.uk
www.landpro.co.uk
Landscaping consultants

Lignacite
Meadgate Works
Nazeing
Essex
EN9 2PD
Tel: 01992 464 441
info@lignacite.co.uk
www.lignacite.co.uk
Concrete blocks

Maccaferri
7400 The Quorum
Oxford Business Park North
Garsington Road
Oxford
OX4 2JTZ
Tel: 01865 770 555
marketing@maccaferri.co.uk
www.maccaferri.co.uk
Gabions

Magrini Ltd
Unit 5
Maybrook Industrial Estate
Brownhills
Walsall
West Midlands
WS8 7DG
Tel: 01543 375311
sales@magrini.co.uk
www.magrini.co.uk
Baby equipment

Manhole Covers Ltd
Airfield Industrial Estate
Cheddington Lane
Long Marston
Bucks
HP23 4QR
Tel: 01296 668850
sales@manholecovers
www.manholecovers.com
Manhole covers

Marshalls Mono Ltd (drainage)
Landscape House
Premier Way
Lowfields Business Park
Elland
HX5 9HT
Tel: 01422 312000
drainage@marshalls.co.uk
www.marshalls.co.uk
Drainage channels

Marshalls Mono Ltd (pavings)
Landscape House
Premier Way
Lowfields Business Park
Elland
HX5 9HT
Tel: 01422 312000
marshallspaving@marshalls.co.uk
www.marshalls.co.uk
Pavings

Metsec Lattice Beams Ltd
Rolls Royce Estate
Spring Road
Ettingshall
Wolverhampton
WV4 6JX
technical@metseclatticebeams.com
www.metseclatticebeams.com
Lattice beams

Monier Ltd
Sussex Manor Business Park
Gatwick Road
Crawley
West Sussex
RH10 9NZ
Tel: 01293 618418
roofing.redland@monier.com
www.redland.co.uk
Redland roof tiles

NDM Metal Roofing & Cladding Ltd
Mettalum House
Unit 3, 89, Manor Farm Road
Alperton
Wembley
HA0 1BA
Tel: 020 8991 7310
enquiries@ndmltd.com
www.ndmltd.com
Metal cladding and roofing

Parker & Highland Joinery Ltd
14A Chartwell Road
Lancing Business Park
Lancing
West Sussex
BN15 8TU
Tel: 01903 756 283
sales@parker-joinery.com
www.parker-joinery.com
Purpose-made joinery

Pegler Yorkshire
St Catherine's Avenue
Doncaster
South Yorkshire
DN4 8DF
Tel: 0844 243 4400
info@yorkshirefittings.co.uk
www.pegleryorkshire.co.uk
Yorkshire & Kuterlite fittings

Plumb Centre
The Wolseley Centre
Harrison Way
Leamington Spa
Warwickshire
CV31 3HH
Tel: 0870 1622 557
www.plumbcentre.co.uk
Cylinders & general plumbing

Polyflor Ltd
PO Box 3
Radcliffe New Road
Whitefield
Manchester
M45 7NR
Tel: 0161 767 1111
www.polyflor.com
Polyfloor contract flooring

Polypipe Terrain
New Hythe Business Park
New Hythe Lane
Aylesford
Kent
ME20 7PJ
Tel: 01622 717811
commercialenquiries@polypipe.com
www.terraindrainage.com
Drainage goods

Premdor Ltd
Gemini House
Hargreaves Road
Groundwell Industrial Estate
Swindon
Wiltshrie
SN25 5AJ
Tel: 01793 708200
enquiries@premdor.com
www.premdor.co.uk
Doors and windows

Premier Loft Ladders
2 Dawson Drive
Trimley
St Mary
Felixstowe
Suffolk
IP11 0YW
Tel: 0845 9000 195
sales@premierloftladders.com
www.premierloftladders.com
Loft ladders

Pressalit Care plc
100 Longwater Avenue
Green Park
Reading
Berkshire
RG2 6GP
Tel: 0844 880 6950
www.pressalitcare.com
Bathroom equipment

Profile 22
Stafford Park 6
Telford
Shropshire
TF3 3AT
Tel: 01952 290910
mail@profile22.co.uk
www.profile22.co.uk
uPVC windows

Promat UK
The Stirling Centre
Easton Road
Bracknell
Berkshire
RG12 2ST
Tel: 01344 381 301
promat@promat.co.uk
www.promat.co.uk
Fireproofing materials

Protim Solignum Ltd
Fieldhouse Lane
Marlow
Buckinghamshire
SC7 1LS
Tel: 01628 486644
info@osmose.co.uk
www.osmose.co.uk
Paints and timber treatment

Radius Systems Ltd
PO Box 1
Hillcote Plant
Blackwell
Alfreton
Derbyshire
DE55 5JD
Tel: 01773 811 112
www.radius-systems.co.uk
MDPE pipes and fittings

Rawlplug Ltd
Skibo Drive
Thronliebank Industrial Estate
Glascow
Scotland
G46 8JR
Tel: 0141 638 225
info@rawlplug.co.uk
www.rawlplug.co.uk
Anchoring and fixing systems

Richard Potter Timber Merchants
Millstone Lane
Nantwich
Cheshire
CW5 5PN
Tel: 01270 625791
richardpotter@fortimber.demon.co.uk
www.fortimber.demon.co.uk
Carcasssing softwood

Rockwool Ltd
Pencoed
Bridgend
Glamorganshire
CF35 6NY
Tel: 01656 862 261
info@rockwool.co.uk
www.rockwool.co.uk
Pipe and other insulation products

Ryton's Building Products
Design House
Orion Way
Kettering Business Park
Kettering
Northamptonshire
NN15 6NL
Tel: 01536 511874
admin@rytons.com
www.vents.co.uk
Roof ventilation products

Safeguard Europe Ltd
Redkin Close
Horsham
West Sussex
RH13 5QL
Tel: 01403 212004
www.safeguardeurope.com
Damp-proofing and waterproofing

Saint Gobain PAM UK
Lows Lane
Stanton-by-Dale
Illkeston
Derbyshire
DE7 4QU
Tel: 0115 930 5000
sales.uk.pam@saint-gobain.com
www.saint-gobain-pam.co.uk/
Cast iron soil, water and rainwater pipes and fittings

Sandtoft Roof Tiles Ltd
Sandtoft
Doncaster
South Yorkshire
DN8 5SY
Tel: 01427 871200
info@sandtoft.co.uk
www.sandtoft.co.uk
Clay roof tiling

Schiedel Rite-Vent
Crowther Road
Washington
NE38 0AQ
Tel: 0191 416 1150
sales@schiedel.co.uk
www.isokern.co.uk
Flue pipes and gas blocks

Screeduct Ltd
Unit 29
Alderminster
Startford - Upon- Avon
Warwickshire
CV37 8NY
Tel: 01789 459 211
sales@screeduct.com
www.screeduct.com
Trunking systems and conduits

Severfield Limited
Dalton Airfield Industrial Estate
Dalton
Thirsk
North Yorkshire
YO7 3JN
Tel: 01845 577896
www.severfield.com
Structural steel

Sheet Piling UK Ltd
Oakfield House
Rough Hey Road
Grimsargh
Preston
PR2 5AR
Tel: 01772 794 141
enquiries@sheetpilinguk.com
www.sheetpilinguk.com
Sheet piling

Sheffield Insulation
Unit 1
New England Estate
Gascoigne Road
Barking
Essex
IG11 7NZ
Tel: 020 8477 9500
barking@sheffins.co.uk
www.SheffieldInsulation.co.uk
Insulation products

Siderise Insulation Ltd
Wales Office
Forge Industrial Estate
Maestag
Bridgend
CF34 0AZ
Tel: 01656 730833
sales@siderise.com
www.siderise.com
Fire barriers

Slate UK David Wallace International Ltd
Unit 6, Airfield Approach Business Park
Flookburgh
Grange-over-Sands
Lake District
Cumbria
Tel: 015395 59289
www.slate.uk.com
Spanish roof slates

Stainless UK Ltd
Newhall Road Works
Sheffield
S9 2QL
Tel: 0114 244 1333
sales@stainless-uk.co.uk
www.stainless-uk.co.uk
Stainless steel rebar

Sterling Hydraulics (Huntley & Sparks) Ltd
Building Products Division
Sterling House
Blacknell Lane
Crewkerne
Somerset
TA18 8LL
Tel: 01460 722 22
Rigifix column guards

Stirling Lloyd Polychem Ltd
Union Bank
King Street
Knutsford
Chieshire
WA16 6EF
Tel: 01565 633 111
www.stirlinglloyd.com
Integritank products

Stressline Ltd
Foxbank Industrial Estate
Stoney Stanton
Leicester
LE9 4LX
Tel: 0870 7503167
sales@stressline.ltd.uk
www.stressline.ltd.uk
Concrete lintels

Swish Building Products
Pioneer House
Lichfield Road Industrial Estate
Tamworth
Staffs
B29 7TF
Tel: 01827 317200
marketing@swishbp.co.uk
www.swishbp.co.uk
Swish celuka

Szerelmey Ltd
369, Kennington Lane
Vauxhall
London
SE11 5QY
Tel: 020 7735 9995
info@szerelmey.com
www.szerelmey.com
Stonework

Tarkett-Marley Floors Ltd
Dickley Lane
Lenham
Maidstone
Kent
ME17 2QX
Tel: 01622 854000
uksales@tarkett.com
www.tarkett-floors.com
Sheet and tile flooring

Tarmac Ltd
Pudding Mill Lane
Bow
London
E15 2PJ
Tel: 020 8555 2415
enquiries@tarmac.co.uk
www.tarmac.co.uk
Ready-mixed concrete

Tarmac Mortar and Screeds
Tunstead House
Buxton
Derbyshire
SK17 8TG
Tel: 08701 116 116
mortar@tarmac.co.uk
www.tarmac.co.uk
Readymix screeds and mortar

Tarmac Topblock Ltd
Wergs Hall
Wergs Hall Road
Wolverhampton
West Midlands
WV8 2HZ
Tel: 01902 754 131
enquiries@tarmac.co.uk
www.topblock.co.uk
Concrete blocks

TATA Steel Ltd
PO Box 1
Brigg Road
Scunthorpe
North Lincolnshire
DN16 1BP
Tel: 01724 405 060
www.tatasteeleurope.com
Steel

Timbmet
Kemp House
Chawley Park
Cumnor Hill
Oxford
OX2 9PH
Tel: 01865 860351
marketing@timbmet.com
www.timbmet.com
Hardwood

Travis Perkins Trading Company
Lodge Way House
Lodge Way
Harlestone Road
Northampton
Northamptonshire
NN5 7UG
Tel: 01604 752484
www.travisperkins.co.uk
Builders merchant

UK Glass Force
32–34 Eldon Way Industrial Estate
Spa Road
Hockley
SS5 4AD
Tel: 0800 393 827
support@ukglassforce.co.uk
www.ukglassforce.co.uk
Glass and glazing

VA Hutchison Flooring Ltd
Units 1–3, Building NA
Beeding Close
Southern Cross Trading Estate
Bognor Regis
West Sussex
PO22 9TS
Tel: 01243 841 175
julia@hutchisonflooring.co.uk
www.hutchisonflooring.co.uk
Hardwood flooring

Vandex
Safeguard Europe Ltd
Redkiln Close
Horsham
Sussex
RH13 5QL
Tel: 01403 210204
info@safeguardeurope.com
www.vandex.com
Vandex super and premix products

Velfac Ltd
The Old Livery
Hildersham
Cambridge
CB21 6DR
Tel: 01223 897 100
post@velfac.co.uk
www.velfac.co.uk
Composite windows

Velux Company Ltd
Wellington Road
Kettering Parkway
Kettering
Northants
NN15 6XR
Tel: 0870 166 7676
enquires@velux.co.uk
www.velux.co.uk
Velux roof windows & flashings

Visqueen Building Products
Maerdy Industrial Estate
Rhymney
Tredegar
NP22 5PY
Tel: 01685 840 672
enquiries@visqueenbuilding.co.uk
www.visqueenbuilding.co.uk
Visqueen products

Wavin Plastics Ltd
Parsonage Way
Chippenham
Wiltshire
SN15 5PN
Tel: 01249 766600
info@wavin.co.uk
www.wavin.co.uk
uPVC drainage goods

Web Dynamics Ltd
Moss Lane
Station Road
Blackrod
Lancs BL6 5JB
Tel: 01204 695666
www.webdynamics.co.uk
Breather membranes

Welco
Woodgate Business Park
Kettles Wood Drive
Birmingham
B32 3GH
Tel: 0121 4219000
enquiries@welconstruct.co.uk
www.welconstruct.co.uk
Lockers and shelving systems

Welsh Slate Ltd
Penrhyn Quarry
Bethesda
Bangor
Gwynedd
LL57 4YG
Tel: 01248 600 656
enquiries@welshslate.com
www.welshslate.com
Natural Welsh slates

Wilde Contracts Ltd
Chareau House
1, Miles Street
Oldham
OL1 3NW
Tel: 0161 624 6824
www.rogerwilde.com
Glass blocks

Yeoman Aggregates Ltd
Stone Terminal
Horn Lane
Acton
London
W3 9EH
Tel: 020 8896 6800
debra.ward@yeoman-aggregates.co.uk
www.yeoman-aggregates.co.uk
Hardcore, gravels, sand

Yorkshire Copper Tube Ltd
East Lancashire Road
Kirkby
Liverpool
Merseyside
L33 7TU
Tel: 0151 546 2700
info@yct.com
www.yorkshirecopper.com
Copper tube

Spon's Asia Pacific Construction Costs Handbook, Fifth Edition

LANGDON & SEAH

In the last few years, the global economic outlook has continued to be shrouded in uncertainty and volatility following the financial crisis in the Euro zone. While the US and Europe are going through a difficult period, investors are focusing more keenly on Asia. This fifth edition provides overarching construction cost data for 16 countries: Brunei, Cambodia, China, Hong Kong, India, Indonesia, Japan, Malaysia, Myanmar, Philippines, Singapore, South Korea, Sri Lanka, Taiwan, Thailand and Vietnam.

May 2015: 234X156 mm: 452 pp
Hb: 978-1-4822-4358-1: £160.00

How to use this Book

First time users of *Spon's Architects' and Builders' Price Book* (*Spon's A&B*) and others who may not be familiar with the way in which prices are compiled may find it helpful to read this section before starting to calculate the costs of building work. The level of information on a scheme and availability of detailed specifications will determine which section of the book and which level of prices users should refer to.

Rates in the book include overhead and recovery margins but do not include main contractors' preliminaries: except for two sections:

- *Building Prices per Functional Units* and
- *Building Prices per Square Metre* which are both within *Part 3 Approximating Estimating*.

Prices in the book do not necessarily, and are not intended to represent the lowest possible prices achievable, but are intended as a guide to expected price levels for the items described. AECOM cost a series of building models using current Spon's rates to calculate a book Tender Price Index (TPI). For this edition of the book TPI has been calculated at **572**.

NEW RULES OF MEASUREMENT(NRM)

The NRM suite covers the life cycle of cost management and means that, at any point in a building's life, quantity surveyors will have a set of rules for measuring and capturing cost data. In addition, the BCIS *Standard Form of Cost Analysis* (SFCA) 4th edition has been updated so that it is aligned with the NRM suite. This book is intended for use with NRM1 and NRM2.

The three volumes of the NRM are as follows.

- **NRM1** – Order of cost estimating and cost planning for capital building works.
- **NRM2** – Detailed measurement for building works.
- **NRM3** – Order of cost estimating and cost planning for building maintenance works.

APPROXIMATE ESTIMATING

For preliminary estimates/indicative costs before drawings are prepared or very little information is available, users are advised to refer to the average overall *Building Prices per Functional Units* and multiply this by the proposed number of units to be contained within the building (i.e. number of bedrooms etc.) or *Building Prices per Square Metre* rates and multiply this by the gross internal floor area of the building (the sum of all floor areas measured within external walls) to arrive at an overall initial *Order of Cost Estimate*. These rates include preliminaries, but make no allowance for the cost of external works, VAT, or fees for provisional services.

Where preliminary drawings are available, one should be able to measure approximate quantities for all the major components of a building and multiply these by individual rates contained in the *Building Cost Models* or *Approximate Estimating Rates* sections to produce an *Elemental Cost Plan*. This should produce a more accurate estimate of cost than simply using overall prices per square metre. Labour and other incidental associated items, although normally measured separately within Bills of Quantities, are deemed included within approximate estimating rates. These rates do not include preliminaries or fees for professional services.

MEASURED WORKS

For more detailed estimates or documents such as Bills of Quantities (Quantities of supplied and fixed components in a building, measured from drawings), use rates from *Prices for Measured Work*. Depending upon the overall value of the contract reader may want to adjust the value and we have added a simple chart which shows typical adjustments that could be applied as shown later in this chapter. Items within the Measured Works sections are made up of many components: the cost of the material or product; any additional materials needed to carry out the work; plant required and the labour involved in unloading and fixing, etc.

Measured Works Rates

These components are usually broken down into:

Prime Cost

Commonly known as the PC; Prime Cost is the actual price of the material such as bricks, blocks, tiles or paint, as sold by suppliers. Prime Cost may be given as *per square metre*, *per 100 bags* or *each* according to the way the supplier sells the product. Unless otherwise stated, prices in *Spon's A&B* are deemed to be delivered to site (in which case transport costs will be included) and also take account of trade and quantity discounts. Part loads generally cost more than whole loads but, unless otherwise stated, Prime Cost figures are based on average prices for full loads delivered to a hypothetical site in Acton, West London. Actual prices for live projects will vary depending on the contractor, supplier, the distance from the supplier to the site, the accessibility of the site, whether the whole quantity ordered is to be supplied in one delivery or at specified dates and market conditions prevailing at the time. Prime Cost figures for commonly used alternative materials are supplied in listed form at the beginning of some work sections.

Where a PC rate is entered alongside an item rate then the cost allowed for that item is in the overall material cost. For instance, bricks need mortar; paving needs sand bedding, so the PC cost is simply for the cost of bricks or paving, thus allowing the user to simply substitute an alternative product cost if desired.

Labour

This figure covers the cost of the operation and is calculated on the gang wage rate (skilled or unskilled) and the time needed for the job. A full explanation and build-up is provided. Large regular or continuous areas of work are more economical to install than smaller complex areas.

Plant

Plant covers the use of machinery ranging from excavators and dumpers to static plant and includes running costs such as fuel, water supply, electricity and waste disposal.

For this year's book we are showing Plant costs separately. Previously it has been included in the Materials column. We believe this will make the rates more transparent and understandable to users who did not always realize that Plant costs were included. Small hand held plant is not included.

Materials

Material prices include the cost of any ancillary materials, nails, screws, waste, etc., which may be needed in association with the main material product/s. If the material being priced varies from a standard measured rate, then identify the difference between the original PC price and the material price and add this to your alternative material price before adding to the labour cost to produce a new overall total rate. Alternative material prices, where given, are largely based upon list prices, before the deduction of quantity discounts etc., and therefore require discount adjustment before they can be substituted in place of PC figures given for Measured Work items.

Example:

Item *(NB example data only)*	PC £	Labour £	Material £	Unit	Total Rate £
Half brick wall in common bricks	380.00	18.05	27.14	m²	**45.19**
Forterra Brick; Brecken Grey	476.40				
Calculation: (PC) £476.40 – £380.00 = £96.40 + 5% (OHP) = £4.82 add to materials rate of £27.14 = £31.96					
Therefore, Brecken Grey bricks =	476.40	18.05	31.96	m²	**50.01**

Unit

The Unit is generally based upon measurement guidelines laid out in the New Rules of Measurement – Detailed measurement for building works (NRM2).

Total Rate

Prices in the Total Rate column generally include for the supply and fix of items, unless otherwise described.

Overheads and Profit

The general overheads of the Main Contractor's business – the head office overheads and any profit sought on capital and turnover employed, is usually covered under a general item of overheads and profit which is applied either to all measured rates as a percentage, or alternatively added to the tender summary or included within Preliminaries for site specific overhead costs.

Within this edition we are including an allowance of 5% for overheads and profit on built-up labour rates and material prices.

Preliminaries

Site specific Main Contractor's overheads on a contract, such as insurances, site accommodation, security, temporary roads and the statutory health and welfare of the labour force, are not directly assignable to individual items so they are generally added as a percentage or calculated allowances after all building component items have been costed and totalled. Preliminaries will vary from project to project according to the type of construction, difficulties of the site, labour shortage, or involvement with other contractors, etc. The overall addition for a scheme should be adjusted to allow for these factors. For this edition we have shown a calculated typical Preliminaries cost example of approximately 13%.

Sub/Specialist-Contractor's Costs

For the purpose of this book, these are deemed to include all the above costs, and assume a 2.5% main contractor's discount.

With the exclusion of main contractor's preliminaries, the above items combine to form item rates in the Prices for Measured Works sections. It should be appreciated that a variation in any one item in any group will affect the final measured work price. Any cost variation must be weighed against the total cost of the contract, and a small variation in Prime Cost where the items are ordered in thousands may have more effect on the total cost than a large variation on a few items, while a change in design which introduces the need to use, for example earth moving equipment, which must be brought to the site for that one task, will cause a dramatic rise in the contract cost. Similarly, a small saving on multiple items could provide a useful reserve to cover unforeseen extras.

COST PLANNING

Order of Cost Estimate

The purpose of an Order of Cost Estimate is to establish if the proposed building project is affordable and, if affordable, to establish a realistic cost limit for the project. The cost limit is the maximum expenditure that the employer is prepared to make in relation to the completed building project, which will be managed by the project team.

An Order of Cost Estimate is produced as an intrinsic part of RIBA Work Stages A: Appraisal and B: Design Brief or OGC Gateways 1 (Business Justification) and 2 (Delivery Strategy).

There are comprehensive guidelines within the NRM documentation and readers are recommended to read the relevant sections of the NRM where more detailed explanations and examples can be found.

At this early stage, in order for the estimate to be representative of the proposed design solution, the key variables that a designer needs to have developed to an appropriate degree of certainty are:

* The floor areas upon which the estimate is based
* Proposed elevations
* The implied level of specification

Rates will need to be updated to current estimate base date by the amount of inflation occurring from the base date of the cost data to the current estimate base date. The percentage addition can be calculated using published indices (i.e. tender price indices [TPI]).

Example 1:

New secondary school

Note: example data only
Gross Internal Floor Area (GIFA) = 15,000 m^2
Cost plan prepared with a TPI = 556
Start on site Q4 2017 TPI = 586
Location: North West adjustment = 0.89
From *Building Prices per Square Metre*

Assume rate of say £1,800 per m^2

		Cost (£)
School rate, say	£1,800 /m^2 × 15,000 m^2 =	27,000,000
Adjust for inflation to start date	(586/560) say +4.5%	1,215,000
	subtotal	28,215,000
Adjust for location	−10%, say	−2,822,000
	subtotal	25,393,000
Allow for contingencies	say 10%	250,000
Total Order of Cost Estimate		**25,643,000**

Main contractor's preliminaries, overheads and profit need not be added to the cost of building works as they are included within the Spon's building prices per square metre rates, but you will need to add on professional fees and other enabling works costs such as site clearance, demolition, external works, car parking, bringing services to site etc.

Elemental Cost Planning

Elemental cost plans are produced as an intrinsic part of RIBA Work Stages C: Concept, D: Design Development, E: Technical Design and F: Production Information; or when the OGC Gateway Process is used, Gateways 3A (Design Brief and Concept Approval) and 3B (Detailed Design Approval).

Cost Models can be used to quickly extract £/m^2 of GIFA:

Example 2:

Health centre

Note: example data only
Gross Internal Floor Area = 1,000 m^2
Cost plan prepared with TPI = 560
Start on site Q4 2017 TPI = 586
Location: South West (adjustment = 0.91)

Element	Rate (£/m^2)	Cost (£)
Substructures	106.53	106,530
Frame and upper floors	160	160,000
Roof	24	24,000
And so on for each element to give a total of	**1,300.56**	**1,300,560**
Contractors preliminaries, overheads and profit	say 15%	195,000
	subtotal	1,495,560
Adjust for inflation to start date	(580/560) say +4.5%	67,000
	subtotal	1,615,560
Adjust for location	say −10%	−162,000
	subtotal	1,453,560
Allow for contingencies	say 5%	73,000
Total Elemental Cost Plan		**1,526,560**

Other allowances such as consultants fees, design fee, VAT, risk allowance, client costs, fixed price adjustment may need to be added to each of the examples above.

Formal Cost Planning Stages

The NRM schedules a number of formal cost planning stages, which are comparable with the RIBA Design and Pre-Construction Work Stages and OGC Gateways 3A (Design Brief and Concept Approval) and 3B (Detailed Design Approval) for a building project. The employer is required to 'approve' the cost plan on completion of each RIBA Work Stage before authorising commencement of the next RIBA Work Stage.

Formal Cost Plan	RIBA Work Stage
1	C: Concept
2	D: Design Development
3 {	E: Technical Design
	F: Production Information

Formal Cost Plan 1 is prepared at a point where the scope of work is fully defined and key criteria are specified but no detailed design has begun. Formal Cost Plan 1 will provide the frame of reference for Formal Cost Plan 2. Likewise, Formal Cost Plan 2 will provide the frame of reference for Formal Cost Plan 3. Neither Formal Cost Plans 2 nor 3 involve the preparation of a completely new Elemental Cost Plan; they are progressions of the previous cost plans, which are developed through the cost checking of price significant components and cost targets as more design information and further information about the site becomes available.

Cost plans can be developed using from Elemental Cost Plans using both Approximate Estimating Rates and/or Prices for Measured Works depending upon the level of information available.

The cost targets within each formal cost plan approved by the employer will be used as the baseline for future cost comparisons. Each subsequent cost plan will require reconciliation with the preceding cost plan and explanations relating to changes made. In view of this, it is essential that records of any transfers made to or from the risk allowances and any adjustments made to cost targets are maintained, so that explanations concerning changes can be provided to both the employer and the project team.

Adjustment According to Contract Sum Cost

The construction costs for a project will depend on the size, type of building, standard of finish required, and location, the economic climate of the construction industry i.e. if there is a shortage of construction work available firms will reduce their tender in order to try and attract work. If the opposite is the case and there is a lot of work available, firms will increase their tenders, as they will not be too keen to obtain the contract which will stretch their resources unless it is worth their while financially. In a recession, construction firms can literally buy work in order to keep their workforce and to ensure some cash flow.

Building Costs can vary between builders/developers. This can be due to the size or purchasing abilities of a company or the discount that it receives from suppliers.

Adjustments can be made to reflect the value of a project by using the following table.

Contract Sum	% Adjustment
£5,000,000	−1%
£4,000,000	0%
£3,000,000	1%
£2,000,000	2%
£1,000,000	5%

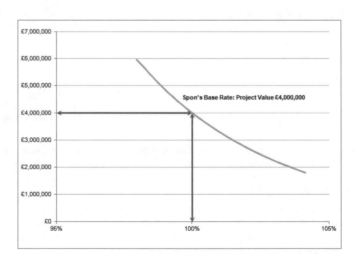

Spon's Architects' & Builders' Price Book is targeted at new build projects with a value range of approximately £3,000,000–£5,000,000.

Users should not simply apply percentage adjustments to any project regardless of size. We recommend project values between £2,000,000 and £5,000,000 for rates found in *Spon's A&B* could be adjusted according to the above table. This is given only as an indication and users should always remember that there are many factors that affect the overall project costs.

These numbers are only intended as a guide and are not explicit and should be applied to the total project value only, not individual rates.

It is one thing to
imagine a better world.
It's another to deliver it.

**Understanding change,
unlocking potential, creating
brilliant new communities.**

The Tate Modern extension
takes an iconic building and
adds to it. Cost management
provided by AECOM.

AECOM
Built to deliver a better world

aecom.com

BIM and Quantity Surveying
PITTARD & SELL

The sudden arrival of Building Information Modelling (BIM) as a key part of the building industry is redefining the roles and working practices of its stakeholders. Many clients, designers, contractors, quantity surveyors, and building managers are still finding their feet in an industry where BIM compliance can bring great rewards.

This guide is designed to help quantity surveying practitioners and students understand what BIM means for them, and how they should prepare to work successfully on BIM compliant projects. The case studies show how firms at the forefront of this technology have integrated core quantity surveying responsibilities like cost estimating, tendering, and development appraisal into high profile BIM projects. In addition to this, the implications for project management, facilities management, contract administration and dispute resolution are also explored through case studies, making this a highly valuable guide for those in a range of construction project management roles.

Featuring a chapter describing how the role of the quantity surveyor is likely to permanently shift as a result of this development, as well as descriptions of tools used, this covers both the organisational and practical aspects of a crucial topic.

December 2015: 234 x 156 mm: 258 pp
Pb: 978-0-415-87043-6; £24.99

To Order: Tel: +44 (0) 1235 400524 Fax: +44 (0) 1235 400525
or Post: Taylor and Francis Customer Services,
Bookpoint Ltd, Unit T1, 200 Milton Park, Abingdon, Oxon, OX14 4TA UK
Email: book.orders@tandf.co.uk

For a complete listing of all our titles visit:
www.tandf.co.uk

Estimator's Pocket Book

Duncan Cartlidge

The Estimator's Pocket Book is a concise and practical reference covering the main pricing approaches, as well as useful information such as how to process sub-contractor quotations, tender settlement and adjudication. It is fully up-to-date with NRM2 throughout, features a look ahead to NRM3 and describes the implications of BIM for estimators.

It includes instructions on how to handle:

- the NRM order of cost estimate;
- unit-rate pricing for different trades;
- pro rata pricing and dayworks
- builders' quantities;
- approximate quantities.

Worked examples show how each of these techniques should be carried out in clear, easy-to-follow steps. This is the indispensible estimating reference for all quantity surveyors, cost managers, project managers and anybody else with estimating responsibilities. Particular attention is given to NRM2, but the overall focus is on the core estimating skills needed in practice.

May 2013 186x123: 310pp
Pb: 978-0-415-52711-8: £21.99

To Order: Tel: +44 (0) 1235 400524 Fax: +44 (0) 1235 400525
or Post: Taylor and Francis Customer Services,
Bookpoint Ltd, Unit T1, 200 Milton Park, Abingdon, Oxon, OX14 4TA UK
Email: book.orders@tandf.co.uk

For a complete listing of all our titles visit:
www.tandf.co.uk

PART 1

General

This part contains the following sections:

Metric Handbook
Planning and Design
Data, 5th Edition

Pamela Buxton

The Metric Handbook is the major handbook of planning and design information for architects and architecture students. Covering basic design data for all the major building types, it is the ideal starting point for any project. For each building type, the book gives the basic design requirements and all the principal dimensional data, and succinct guidance on how to use the information and what regulations the designer needs to be aware of.

As well as building types, the Metric Handbook deals with broader aspects of design such as materials, acoustics and lighting, and general design data on human dimensions and space requirements. The Metric Handbook provides an invaluable resource for solving everyday design and planning problems.

Mar 2015: 203 x 292: 858 pp
Pb: 978-0-415-72542-2 £39.99

To Order: Tel: +44 (0) 1235 400524 Fax: +44 (0) 1235 400525
or Post: Taylor and Francis Customer Services,
Bookpoint Ltd, Unit T1, 200 Milton Park, Abingdon, Oxon, OX14 4TA UK
Email: book.orders@tandf.co.uk

For a complete listing of all our titles visit:
www.tandf.co.uk

Capital Allowances

Introduction

Capital Allowances provide tax relief by prescribing a statutory rate of depreciation for tax purposes in place of that used for accounting purposes. They are utilized by government to provide an incentive to invest in capital equipment, including assets within commercial property, by allowing the majority of taxpayers a deduction from taxable profits for certain types of capital expenditure, thereby reducing or deferring tax liabilities.

The capital allowances most commonly applicable to real estate are those given for capital expenditure on existing commercial buildings in disadvantaged areas, and plant and machinery in all buildings other than residential dwellings. Relief for certain expenditure on industrial buildings and hotels was withdrawn from April 2011, although the ability to claim plant and machinery remains.

Enterprise Zone Allowances are also available for capital expenditure within designated areas only where there is a focus on high value manufacturing. Enhanced rates of allowances are available on certain types of energy and water saving plant and machinery assets.

The Act

The primary legislation is contained in the Capital Allowances Act 2001. Major changes to the system were introduced in 2008 and 2014 affecting the treatment of plant and machinery allowances.

Plant and Machinery

Various legislative changes and case law precedents in recent years have introduced major changes to the availability of Capital Allowances for property expenditure. The Capital Allowances Act 2001 precludes expenditure on the provision of a building from qualifying for plant and machinery, with prescribed exceptions.

List A in Section 21 of the 2001 Act sets out those assets treated as parts of buildings:

- *Walls, floors, ceilings, doors, gates, shutters, windows and stairs.*
- *Mains services, and systems, for water, electricity and gas.*
- *Waste disposal systems.*
- *Sewerage and drainage systems.*
- *Shafts or other structures in which lifts, hoists, escalators and moving walkways are installed.*
- *Fire safety systems.*

Similarly, List B in Section 22 identifies excluded structures and other assets.

Both sections are, however, subject to Section 23. This section sets out expenditure, which although being part of a building, may still be expenditure on the provision of Plant and Machinery.

List C in Section 23 is reproduced below:

Sections 21 and 22 do not affect the question whether expenditure on any item in List C is expenditure on the provision of Plant or Machinery

1.   Machinery (including devices for providing motive power) not within any other item in this list.
2.   Gas and sewerage systems provided mainly –
       a.   to meet the particular requirements of the qualifying activity, or
       b.   to serve particular plant or machinery used for the purposes of the qualifying activity.
3.   Omitted.
4.   Manufacturing or processing equipment; storage equipment (including cold rooms); display equipment; and counters, checkouts and similar equipment.
5.   Cookers, washing machines, dishwashers, refrigerators and similar equipment; washbasins, sinks, baths, showers, sanitary ware and similar equipment; and furniture and furnishings.
6.   Hoists.
7.   Sound insulation provided mainly to meet the particular requirements of the qualifying activity.
8.   Computer, telecommunication and surveillance systems (including their wiring or other links).
9.   Refrigeration or cooling equipment.
10.   Fire alarm systems; sprinkler and other equipment for extinguishing or containing fires.
11.   Burglar alarm systems.
12.   Strong rooms in bank or building society premises; safes.
13.   Partition walls, where moveable and intended to be moved in the course of the qualifying activity.
14.   Decorative assets provided for the enjoyment of the public in hotel, restaurant or similar trades.
15.   Advertising hoardings; signs, displays and similar assets.
16.   Swimming pools (including diving boards, slides & structures on which such boards or slides are mounted).
17.   Any glasshouse constructed so that the required environment (namely, air, heat, light, irrigation and temperature) for the growing of plants is provided automatically by means of devices forming an integral part of its structure.
18.   Cold stores.
19.   Caravans provided mainly for holiday lettings.
20.   Buildings provided for testing aircraft engines run within the buildings.
21.   Moveable buildings intended to be moved in the course of the qualifying activity.
22.   The alteration of land for the purpose only of installing Plant or Machinery.
23.   The provision of dry docks.
24.   The provision of any jetty or similar structure provided mainly to carry Plant or Machinery.
25.   The provision of pipelines or underground ducts or tunnels with a primary purpose of carrying utility conduits.
26.   The provision of towers to support floodlights.
27.   The provision of –
       a.   any reservoir incorporated into a water treatment works, or
       b.   any service reservoir of treated water for supply within any housing estate or other particular locality.
28.   The provision of –
       a.   silos provided for temporary storage, or
       b.   storage tanks.
29.   The provision of slurry pits or silage clamps.
30.   The provision of fish tanks or fish ponds.
31.   The provision of rails, sleepers and ballast for a railway or tramway.
32.   The provision of structures and other assets for providing the setting for any ride at an amusement park or exhibition.
33.   The provision of fixed zoo cages.

Capital Allowances on plant and machinery are given in the form of writing down allowances at the rate of 18% per annum on a reducing balance basis. For every £100 of qualifying expenditure £18 is claimable in year 1, £14.76 in year 2 and so on until either all the allowances have been claimed or the asset is sold.

Integral Features

The category of qualifying expenditure on 'integral features' was introduced with effect from April 2008. The following items are integral features:

- An electrical system (including a lighting system)
- A cold water system
- A space or water heating system, a powered system of ventilation, air cooling or air purification, and any floor or ceiling comprised in such a system
- A lift, an escalator or a moving walkway
- External solar shading

The Integral Features legislation introduced certain assets which were not usually previously allowable as plant and machinery such as electrical, lighting and cold water systems.

Integral Features are given in the form of writing down allowances at the rate of 8% per annum on a reducing balance basis.

Thermal Insulation

From April 2008, expenditure incurred on the installation of thermal insulation to existing buildings qualifies as plant and machinery which is available on a reducing balance basis at 8% per annum.

Long-Life Assets

A reduced writing down allowance of 8% per annum is available on long-life assets.

A long-life asset is defined as plant and machinery that has an expected useful economic life of at least 25 years. The useful economic life is taken as the period from first use until it is likely to cease to be used as a fixed asset of any business. It is important to note that this likely to be a shorter period than an item's physical life.

Plant and machinery provided for use in a building used wholly or mainly as dwelling house, showroom, hotel, office or retail shop or similar premises, or for purposes ancillary to such use, cannot be long-life assets.

In contrast certain plant and machinery assets in buildings such as factories, cinemas, hospitals and so on are all potentially long-life assets.

Case Law

The fact that an item appears in List C does not automatically mean that it will qualify for capital allowances. It only means that it may potentially qualify.

Guidance about what can qualify as plant is found in case law since 1887. The case of Wimpy International Ltd and Associated Restaurants Ltd v Warland in the late 1980s is one of the most important case law references for determining what can qualify as plant.

The Judge in that case said that there were three tests to be applied when considering whether or not an item is plant.

1. Is the item stock in trade? If the answer yes, then the item is not plant.
2. Is the item used for carrying on the business? In order to pass the business use test the item must be employed in carrying on the business; it is not enough for the asset to be simply used in the business. For example, product display lighting in a retail store may be plant but general lighting in a warehouse would fail the test.
3. Is the item the business premises or part of the business premises? An item cannot be plant if it fails the premises test, i.e. if the business use is as the premises (or part of the premises) or place on which the business is conducted. The meaning of part of the premises in this context should not be confused with the law of real property. The Inland Revenue's internal manuals suggest there are four general factors to be considered, each of which is a question of fact and degree:

- Does the item appear visually to retain a separate identity
- With what degree of permanence has it been attached to the building
- To what extent is the structure complete without it
- To what extent is it intended to be permanent or alternatively is it likely to be replaced within a short period

Certain assets will qualify as plant in most cases. However, many others need to be considered on a case-by-case basis. For example, decorative assets in a hotel restaurant may be plant but similar assets in an office reception area may not be.

Refurbishment Schemes

Building refurbishment projects will typically be a mixture of capital costs and revenue expenses, unless the works are so extensive that they are more appropriately classified a redevelopment. A straightforward repair or a 'like for like' replacement of part of an asset would be a revenue expense, meaning that the entire amount can be deducted from taxable profits in the same year.

Where capital expenditure is incurred that is incidental to the installation of plant or machinery then Section 25 of the Capital Allowances Act 2001 allows it to be treated as part of the expenditure on the qualifying item. Incidental expenditure will often include parts of the building that would be otherwise disallowed, as shown in the Lists reproduced above. For example, the cost of forming a lift shaft inside an existing building would be deemed to be part of the expenditure on the provision of the lift.

The extent of the application of section 25 was reviewed for the first time by the Special Commissioners in December 2007 and by the First Tier Tribunal (Tax Chamber) in December 2009, in the case of JD Wetherspoon. The key areas of expenditure considered were overheads and preliminaries where it was held that such costs could be allocated on a pro-rata basis; decorative timber panelling which was found to be part of the premises and so ineligible for allowances; toilet lighting which was considered to provide an attractive ambience and qualified for allowances; and incidental building alterations of which enclosing walls to toilets and kitchens and floor finishes did not qualify but tiled splash backs, toilet cubicles and drainage did qualify along with the related sanitary fittings and kitchen equipment.

Annual Investment Allowance

The annual investment allowance is available to all businesses of any size and allows a deduction for the whole of the first £500,000 from 19 March 2014 (£250,000 before 19 March 2014) of qualifying expenditure on plant and machinery, including integral features and long-life assets. The current Annual Investment Allowances rates as at May 2016 are:

- 1 April 2014 to 31 December 2015 – £500,000
- 1 January 2016 onwards – £200,000

For accounting periods less or greater than 12 months or if claiming in periods where the rates have changed, time apportionment rules will apply to calculate hybrid rates applicable to the period of claim.

The Enhanced Capital Allowances Scheme

The scheme is one of a series of measures introduced to ensure that the UK meets its target for reducing greenhouse gases under the Kyoto Protocol. 100% first year allowances are available on products included on the Energy Technology List published on the website at www.eca.gov.uk and other technologies supported by the scheme. All businesses will be able to claim the enhanced allowances, but only investments in new and unused Machinery and Plant can qualify.

There are currently 15 technologies with multiple sub-technologies currently covered by the scheme:

- Air-to-air energy recovery
- Automatic monitoring and targeting (AMT)
- Boiler equipment
- Combined heat and power (CHP)
- Compressed air equipment
- Heat pumps
- Heating ventilation and air conditioning (HVAC) equipment
- High speed hand air dryers

- Lighting
- Motors and drives
- Pipework insulation
- Radiant and warm air heaters
- Refrigeration equipment
- Solar thermal systems
- Uninterruptible power supplies

Finance Act 2003 introduced a new category of environmentally beneficial plant and machinery qualifying for 100% first-year allowances. The Water Technology List includes 14 technologies:

- Cleaning in place equipment
- Efficient showers
- Efficient taps
- Efficient toilets
- Efficient washing machines
- Flow controllers
- Greywater recovery and reuse equipment
- Leakage detection equipment
- Meters and monitoring equipment
- Rainwater harvesting equipment
- Small scale slurry and sludge dewatering equipment
- Vehicle wash water reclaim units
- Water efficient industrial cleaning equipment
- Water management equipment for mechanical seals

Buildings and structures and long-life assets as defined above cannot qualify under the scheme. However, following the introduction of the integral features rules, lighting in any non-residential building may potentially qualify for enhanced capital allowances if it meets the relevant criteria.

A limited payable ECA tax credit equal to 19% of the loss surrendered was also introduced for UK companies in April 2008.

From April 2012 expenditure on plant and machinery for which tariff payments are received under the renewable energy schemes introduced by the Department of Energy and Climate Change (Feed-in Tariffs or Renewable Heat Incentives) will not be entitled to enhanced capital allowances.

Enterprise Zones

The creation of 11 Enterprise Zones was announced in the 2011 Budget. Additional zones have since been added bringing the number to 24 in total. Originally introduced in the early 1980s as a stimulus to commercial development and investment, they had virtually faded from the real estate psyche.

Enterprise zones benefit from a number of reliefs, including a 100% first year allowance for new and unused non-leased plant and machinery assets, where there is a focus on high-value manufacturing.

Flat Conversion Allowances

Tax relief was available on capital expenditure incurred on or after 11 May 2001 on the renovation or conversion of vacant or underused space above shops and other commercial premises to provide flats for rent.

This relief has been abolished for expenditure incurred after April 2013.

Business Premises Renovation Allowance

The Business Premises Renovation Allowance (BPRA) was first announced in December 2003. The idea behind the scheme is to bring long-term vacant properties back into productive use by providing 100% capital allowances for the cost of renovating and converting unused premises in disadvantaged areas. The legislation was included in Finance Act 2005 and was finally implemented on 11 April 2007 following EU state aid approval.

The scheme will apply to properties within the areas specified in the Assisted Areas Order 2007 and Northern Ireland.

BPRA is available to both individuals and companies who own or lease business property that has been unused for 12 months or more. Allowances will be available to a person who incurs qualifying capital expenditure on the renovation of business premises.

An announcement to extend the scheme by a further five years to 2017 was made within the 2011 Budget, along with a further 11 new designated Enterprise Zones.

Legislation was introduced in Finance Bill 2014 to clarify the scope of expenditure qualifying for relief to actual costs of construction and building work, and for certain specified activities such as architectural and surveying services. The changes will have effect for qualifying expenditure incurred on or after 1 April 2014 for businesses within the charge to corporation tax, and 6 April 2014 for businesses within the charge to income tax.

Other Capital Allowances

Other types of allowances include those available for capital expenditure on Mineral Extraction, Research and Development, Know-How, Patents, Dredging and Assured Tenancy.

International Tax Depreciation

The UK is not the only tax regime offering investors, owners and occupiers valuable incentives to invest in plant and machinery and environmentally friendly equipment. Ireland, Australia, Malaysia and Singapore also have capital allowances regimes that are similar broadly similar to the UK and provide comparable levels of tax relief to businesses.

Many other overseas countries have tax depreciation regimes based on accounting treatment, instead of Capital Allowances. Some use a systematic basis over the useful life of the asset and others have prescribed methods spreading the cost over a statutory period, not always equating to the asset's useful life. Some regimes have prescribed statutory rates, whilst others have rates which have become acceptable to the tax authorities through practice.

Value Added Tax

Introduction

Value Added Tax (VAT) is a tax on the consumption of goods and services. The UK introduced a domestic VAT regime when it joined the European Community in 1973. The principal source of European law in relation to VAT is Council Directive 2006/112/EC, a recast of Directive 77/388/EEC which is currently restated and consolidated in the UK through the VAT Act 1994 and various Statutory Instruments, as amended by subsequent Finance Acts.

VAT Notice 708: Buildings and construction (August 2014) provides HMRC's interpretation of the VAT law in connection with construction works, however the UK VAT legislation should always be referred to in conjunction with the publication. Recent VAT tribunals and court decisions since the date of this publication will affect the application of the VAT law in certain instances. The Notice is available on HM Revenue & Customs website at www.hmrc.gov.uk.

The Scope of VAT

VAT is payable on:

- Supplies of goods and services made in the UK
- By a taxable person
- In the course or furtherance of business; and
- Which are not specifically exempted or zero-rated.

Rates of VAT

There are three rates of VAT:

- A standard rate, currently 20% since January 2011
- A reduced rate, currently 5%; and
- A zero rate of 0%.

Additionally some supplies are exempt from VAT and others are considered outside the scope of VAT.

Recovery of VAT

When a taxpayer makes taxable supplies he must account for VAT, known as output VAT at the appropriate rate of 20%, 5% or 0%. Any VAT due then has to be declared and submitted on a VAT submission to HM Revenue & Customs and will normally be charged to the taxpayer's customers.

As a VAT registered person, the taxpayer is entitled to reclaim from HM Revenue & Customs, commonly referred to as input VAT the VAT incurred on their purchases and expenses directly related to its business activities in respect of a standard-rated, reduced-rated and zero-rated supplies. A taxable person cannot however reclaim VAT that relates to any non-business activities (but see below) or depending on the amount of exempt supplies they made input VAT may be restricted or not recoverable.

At predetermined intervals the taxpayer will pay to HM Revenue & Customs the excess of VAT collected over the VAT they can reclaim. However if the VAT reclaimed is more than the VAT collected, the taxpayer who will be a net repayment position can reclaim the difference from HM Revenue & Customs.

Example

X Ltd constructs a block of flats. It sells long leases to buyers for a premium. X Ltd has constructed a new building designed as a dwelling and will have granted a long lease. This first sale of a long lease is VAT zero-rated supply. This means any VAT incurred in connection with the development which X Ltd will have paid (e.g. payments for consultants and certain preliminary services) will be recoverable. For reasons detailed below the contractor employed by X Ltd will not have charged VAT on his construction services as these should be zero-rated.

Use for Business and Non-Business Activities

Where a supply relates partly to business use and partly to non-business use then the basic rule is that it must be apportioned on a fair and reasonable basis so that only the business element is potentially recoverable. In some cases VAT on land, buildings and certain construction services purchased for both business and non-business use could be recovered in full by applying what is known as 'Lennartz' accounting to reclaim VAT relating to the non-business use and account for VAT on the non-business use over a maximum period of 10 years. Following an ECJ case restricting the scope of this approach, its application to immovable property was removed completely in January 2011 by HMRC (business brief 53/10) when UK VAT law was amended to comply with EU Directive 2009/162/EU.

Taxable Persons

A taxable person is an individual, firm, company etc. who is required to be registered for VAT. A person who makes taxable supplies above certain turnover limits is compulsory required to be VAT registered. From 1 April 2016, the current registration limit known as the VAT threshold is £83,000 for 2015–16. If the threshold is exceeded in any 12 month rolling period, or there is an expectation that the value of the taxable supplies in a single 30 day period, or you receive goods into the UK from the EU worth more than the £83,000, then you must register for UK VAT.

A person who makes taxable supplies below the limit is still entitled to be registered on a voluntary basis if they wish, for example in order to recover input VAT incurred in relation to those taxable supplies, however output VAT will then become due on the sales.

In addition, a person who is not registered for VAT in the UK but acquires goods from another EC member state, or make distance sales in the UK, above certain value limits may be required to register for VAT in the UK.

VAT Exempt Supplies

Where a supply is exempt from VAT this means that no output VAT is payable – but equally the person making the exempt supply cannot normally recover any of the input VAT on their own costs relating to that exempt supply.

Generally commercial property transactions such as leasing of land and buildings are exempt unless a landlord chooses to standard-rate its interest in the property by a applying for an option to tax. This means that VAT is added to rental income and also that VAT incurred, on say, an expensive refurbishment, is recoverable.

Supplies outside the scope of VAT

Supplies are outside the scope of VAT if they are:

- Made by someone who is not a taxable person
- Made outside the UK; or
- Not made in the course or furtherance of business.

In course or furtherance of business

VAT must be accounted for on all taxable supplies made in the course or furtherance of business with the corresponding recovery of VAT on expenditure incurred.

If a taxpayer also carries out non-business activities then VAT incurred in relation to such supplies is generally not recoverable.

In VAT terms, business means any activity continuously performed which is mainly concerned with making supplies for a consideration. This includes:

- Any one carrying on a trade, vocation or profession;
- The provision of membership benefits by clubs, associations and similar bodies in return for a subscription or other consideration; and
- Admission to premises for a charge.

It may also include the activities of other bodies including charities and non-profit making organizations.

Examples of non-business activities are:

- Providing free services or information;
- Maintaining some museums or particular historic sites;
- Publishing religious or political views.

Construction Services

In general the provision of construction services by a contractor will be VAT standard rated at 20%, however, there are a number of exceptions for construction services provided in relation to certain relevant residential properties and charitable buildings.

The supply of building materials is VAT standard rated at 20%, however, where these materials are supplied and installed as part of the construction services the VAT liability of those materials follows that of the construction services supplied.

Zero-rated construction services

The following construction services are VAT zero-rated including the supply of related building materials.

The construction of new dwellings

The supply of services in the course of the construction of a new building designed for use as a dwelling or number of dwellings is zero-rated other than the services of an architect, surveyor or any other person acting as a consultant or in a supervisory capacity.

The following basic conditions must ALL be satisfied in order for the works to qualify for zero-rating:

1. A qualifying building has been, is being or will be constructed;
2. Services are made 'in the course of the construction' of that building;
3. Where necessary, you hold a valid certificate;
4. Your services are not specifically excluded from zero-rating.

The construction of a new building for 'relevant residential or charitable' use

The supply of services in the course of the construction of a building designed for use as a relevant residential Purpose (RRP) or relevant charitable purpose (RCP) is zero-rated other than the services of an architect, surveyor or any other person acting as a consultant or in a supervisory capacity.

A 'relevant residential' use building means:

1. A home or other institution providing residential accommodation for children;
2. A home or other institution providing residential accommodation with personal care for persons in need of personal care by reason of old age, disablement, past or present dependence on alcohol or drugs or past or present mental disorder;
3. A hospice;
4. Residential accommodation for students or school pupils;
5. Residential accommodation for members of any of the armed forces;
6. A monastery, nunnery, or similar establishment; or
7. An institution which is the sole or main residence of at least 90% of its residents.

A 'relevant residential' purpose building does not include use as a hospital, a prison or similar institution or as a hotel, inn or similar establishment.

A 'relevant charitable' purpose means use by a charity in either or both of the following ways:

1.   Otherwise than in the course or furtherance of a business; or
2.   As a village hall or similarly in providing social or recreational facilities for a local community.

Non-qualifying use which is not expected to exceed 10% of the time the building is normally available for use can be ignored. The calculation of business use can be based on time, floor area or head count subject to approval being acquired from HM Revenue & Customs.

The construction services can only be zero-rated if a certificate is given by the end user to the contractor carrying out the works confirming that the building is to be used for a qualifying purpose i.e. for a 'relevant residential or charitable' purpose. It follows that such services can only be zero-rated when supplied to the end user and, unlike supplies relating to dwellings, supplies by subcontractors cannot be zero-rated.

The construction of an annex used for a 'relevant charitable' purpose

Construction services provided in the course of construction of an annexe for use entirely or partly for a 'relevant charitable' purpose can be zero-rated.

In order to qualify the annexe must:

1.   Be capable of functioning independently from the existing building;
2.   Have its own main entrance; and
3.   Be covered by a qualifying use certificate.

The conversion of a non-residential building into dwellings or the conversion of a building from non-residential use to 'relevant residential' use where the supply is to a 'relevant' housing association.

The supply to a 'relevant' housing association in the course of conversion of a non-residential building or non-residential part of a building into:

1.   A new eligible dwelling designed as a dwelling or number of dwellings; or
2.   A building or part of a building for use solely for a relevant residential purpose, of any services related to the conversion other than the services of an architect, surveyor or any person acting as a consultant or in a supervisory capacity are zero-rated.

A 'relevant' housing association is defined as:

1.   A private registered provider of social housing;
2.   A registered social landlord within the meaning of Part I of the Housing Act 1996 (Welsh registered social landlords);
3.   A registered social landlord within the meaning of the Housing (Scotland) Act 2001 (Scottish registered social landlords); or
4.   A registered housing association within the meaning of Part II of the Housing (Northern Ireland) Order 1992 (Northern Irish registered housing associations).

If the building is to be used for a 'relevant residential' purpose the housing association should issue a qualifying use certificate to the contractor completing the works. Subcontractors services that are not made directly to a relevant housing association are standard-rated.

The development of a residential caravan parks

The supply in the course of the construction of any civil engineering work 'necessary for' the development of a permanent park for residential caravans of any services related to the construction are zero-rated when a new permanent park is being developed, the civil engineering works are necessary for the development of the park and the services are not specifically excluded from zero-rating. This includes access roads, paths, drainage, sewerage and the installation of mains water, power and gas supplies.

Certain building alterations for disabled persons

Certain goods and services supplied to a 'disabled' person, or a charity making these items and services available to disabled persons can be zero-rated. The recipient of these goods or services needs to give the supplier an appropriate written declaration that they are entitled to benefit from zero rating.

The following services (amongst others) are zero-rated:

1. The installation of specialist lifts and hoists and their repair and maintenance;
2. The construction of ramps, widening doorways or passageways including any preparatory work and making good work;
3. The provision, extension and adaptation of a bathroom, washroom or lavatory; and
4. Emergency alarm call systems.

Approved alterations to protected buildings

The zero rate for approved alterations to protected buildings was withdrawn from 1 October 2012, other than for projects where a contract was entered into or where listed building consent (or equivalent approval for listed places of worship) had been applied for before 21 March 2012.

Provided the application was in place before 21 March 2012, zero rating will continue under the transitional rules until 30 September 2015.

All other projects will be subject to the standard rate of VAT on or after 1 October 2012.

Sale of Reconstructed Buildings

Since 1 October 2012 a protected building shall not be regarded as substantially reconstructed unless, when the reconstruction is completed, the reconstructed building incorporates no more of the original building than the external walls, together with other external features of architectural or historical interest. Transitional arrangements protect contracts entered into before 21 March 2012 for the first grant of a major interest in the protected building made on or before 20 March 2013.

DIY Builders and Converters

Private individuals who decide to construct their own home are able to reclaim VAT they pay on goods they use to construct their home by use of a special refund mechanism made by way of an application to HM Revenue & Customs. This also applies to services provided in the conversion of an existing non-residential building to form a new dwelling.

The scheme is meant to ensure that private individuals do not suffer the burden of VAT if they decide to construct their own home.

Charities may also qualify for a refund on the purchase of materials incorporated into a building used for non-business purposes where they provide their own free labour for the construction of a 'relevant charitable' use building.

Reduced-rated Construction Services

The following construction services are subject to the reduced rate of VAT of 5%, including the supply of related building materials.

Conversion – changing the number of dwellings

In order to qualify for the 5% rate there must be a different number of 'single household dwellings' within a building than there were before commencement of the conversion works. A 'single household dwelling' is defined as a dwelling that is designed for occupation by a single household.

These conversions can be from 'relevant residential' purpose buildings, non-residential buildings and houses in multiple occupation.

A house in multiple occupation conversion

This relates to construction services provided in the course of converting a 'single household dwelling', a number of 'single household dwellings', a non-residential building or a 'relevant residential' purpose building into a house for multiple occupation such as a bed sit accommodation.

A special residential conversion

A special residential conversion involves the conversion of a 'single household dwelling', a house in multiple occupation or a non-residential building into a 'relevant residential' purpose building such as student accommodation or a care home.

Renovation of derelict dwellings

The provision of renovation services in connection with a dwelling or 'relevant residential' purpose building that has been empty for two or more years prior to the date of commencement of construction works can be carried out at a reduced rate of VAT of 5%.

Installation of energy saving materials

A reduced rate of VAT of 5% is paid on the supply and installation of certain energy saving materials including insulation, draught stripping, central heating, hot water controls and solar panels in a residential building or a building used for a relevant charitable purpose.

Buildings that are used by charities for non-business purposes, and/or as village halls, were removed from the scope of the reduced rate for the supply of energy saving materials under legislation introduced in Finance Bill 2013.

Grant-funded installation of heating equipment or connection of a gas supply

The grant-funded supply and installation of heating appliances, connection of a mains gas supply, supply, installation, maintenance and repair of central heating systems, and supply and installation of renewable source heating systems, to qualifying persons. A qualifying person is someone aged 60 or over or is in receipt of various specified benefits.

Grant-funded installation of security goods

The grant-funded supply and installation of security goods to a qualifying person.

Housing alterations for the elderly

Certain home adaptations that support the needs of elderly people were reduced rated with effect from 1 July 2007.

Building Contracts

Design and build contracts

If a contractor provides a design and build service relating to works to which the reduced or zero rate of VAT is applicable then any design costs incurred by the contractor will follow the VAT liability of the principal supply of construction services.

Management contracts

A management contractor acts as a main contractor for VAT purposes and the VAT liability of his services will follow that of the construction services provided. If the management contractor only provides advice without engaging trade contractors his services will be VAT standard rated.

Construction Management and Project Management

The project manager or construction manager is appointed by the client to plan, manage and coordinate a construction project. This will involve establishing competitive bids for all the elements of the work and the appointment of trade contractors. The trade contractors are engaged directly by the client for their services.

The VAT liability of the trade contractors will be determined by the nature of the construction services they provide and the building being constructed.

The fees of the construction manager or project manager will be VAT standard rated. If the construction manager also provides some construction services these works may be zero or reduced rated if the works qualify.

Liquidated and Ascertained Damages

Liquidated damages are outside of the scope of VAT as compensation. The employer should not reduce the VAT amount due on a payment under a building contract on account of a deduction of damages. In contrast an agreed reduction in the contract price will reduce the VAT amount.

Similarly, in certain circumstances HM Revenue & Customs may agree that a claim by a contractor under a JCT or other form of contract is also compensation payment and outside the scope of VAT.

Architect's Legal Pocket Book
Second Edition

Matthew Cousins

A little book that's big on information. The Architect's Legal Pocket Book is the definite reference on legal issues for architects and architectural students. This handy pocket guide covers key legal principles which will help you to quickly understand the law and where to go for further information.

Now in a fully updated new edition, this bestselling book covers a wide range of subjects focused on the UK including building legislation and the Localism Bill, negligence, liability, planning policy and development, listed buildings, party wall legislation, and rights of light. This edition also contains greater coverage of contracts including the RIBA contracts, dispute resolution and legal issues in professional practice.

Illustrated with clear diagrams and featuring key cases, this is an invaluable source of practical information and a comprehensive guide of the current law for architects. It is a book no architect should be without.

August 2015 186x123: 402pp
Pb: 978-1-138-82144-6: £24.99

To Order: Tel: +44 (0) 1235 400524 Fax: +44 (0) 1235 400525
or Post: Taylor and Francis Customer Services,
Bookpoint Ltd, Unit T1, 200 Milton Park, Abingdon, Oxon, OX14 4TA UK
Email: book.orders@tandf.co.uk

For a complete listing of all our titles visit:
www.tandf.co.uk

The Aggregates Levy

The Aggregates Levy came into operation on 1 April 2002 in the UK, except for Northern Ireland where it has been phased in over five years from 2003.

It was introduced to ensure that the external costs associated with the exploitation of aggregates are reflected in the price of aggregate, and to encourage the use of recycled aggregate. There continues to be strong evidence that the levy is achieving its environmental objectives, with sales of primary aggregate down and production of recycled aggregate up. The Government expects that the rates of the levy will at least keep pace with inflation over time, although it accepts that the levy is still bedding in.

The rate of the levy remains at £2.00 per tonne from 1 April 2014 and is levied on anyone considered to be responsible for commercially exploiting 'virgin' aggregates in the UK and should naturally be passed by price increase to the ultimate user.

All materials falling within the definition of 'Aggregates' are subject to the levy unless specifically exempted.

It does not apply to clay, soil, vegetable or other organic matter.

The intention is that it will:

- Encourage the use of alternative materials that would otherwise be disposed of to landfill sites
- Promote development of new recycling processes, such as using waste tyres and glass
- Promote greater efficiency in the use of virgin aggregates
- Reduce noise and vibration, dust and other emissions to air, visual intrusion, loss of amenity and damage to wildlife habitats

Definitions

'Aggregates' means any rock, gravel or sand which is extracted or dredged in the UK for aggregates use. It includes whatever substances are for the time being incorporated in it or naturally occur mixed with it.

'Exploitation' is defined as involving any one or a combination of any of the following:

- Being removed from its original site, a connected site which is registered under the same name as the originating site or a site where it had been intended to apply an exempt process to it, but this process was not applied
- Becoming subject to a contract or other agreement to supply to any person
- Being used for construction purposes
- Being mixed with any material or substance other than water, except in permitted circumstances

The definition of 'aggregate being used for construction purposes' is when it is:

- Used as material or support in the construction or improvement of any structure
- Mixed with anything as part of a process of producing mortar, concrete, tarmacadam, coated roadstone or any similar construction material

Incidence

It is a tax on primary aggregates production – i.e. 'virgin' aggregates won from a source and used in a location within the UK territorial boundaries (land or sea). The tax is not levied on aggregates which are exported or on aggregates imported from outside the UK territorial boundaries.

It is levied at the point of sale.

Exemption from Tax

An 'aggregate' is exempt from the levy if it is:

- Material which has previously been used for construction purposes
- Aggregate that has already been subject to a charge to the Aggregates Levy
- Aggregate which was previously removed from its originating site before the start date of the levy
- Aggregate which is moved between sites under the same Aggregates Levy Registration
- Aggregate which is removed to a registered site to have an exempt process applied to it
- Aggregate which is removed to any premises where china clay or ball clay will be extracted from the aggregate
- Aggregate which is being returned to the land from which it was won provided that it is not mixed with any material other than water
- Aggregate won from a farm land or forest where used on that farm or forest
- Rock which has not been subjected to an industrial crushing process
- Aggregate won by being removed from the ground on the site of any building or proposed building in the course of excavations carried out in connection with the modification or erection of the building and exclusively for the purpose of laying foundations or of laying any pipe or cable
- Aggregate won by being removed from the bed of any river, canal or watercourse or channel in or approach to any port or harbour (natural or artificial), in the course of carrying out any dredging exclusively for the purpose of creating, restoring, improving or maintaining that body of water
- Aggregate won by being removed from the ground along the line of any highway or proposed highway in the course of excavations for improving, maintaining or constructing the highway otherwise than purely to extract the aggregate
- Drill cuttings from petroleum operations on land and on the seabed
- Aggregate resulting from works carried out in exercise of powers under the New Road and Street Works Act 1991, the Roads (Northern Ireland) Order 1993 or the Street Works (Northern Ireland) Order 1995
- Aggregate removed for the purpose of cutting of rock to produce dimension stone, or the production of lime or cement from limestone
- Aggregate arising as a waste material during the processing of the following industrial minerals:
  - anhydrite
  - ball clay
  - barytes
  - calcite
  - china clay
  - clay, coal, lignite and slate
  - feldspar
  - flint
  - fluorspar
  - fuller's earth
  - gems and semi-precious stones
  - gypsum
  - any metal or the ore of any metal
  - muscovite
  - perlite
  - potash
  - pumice
  - rock phosphates
  - sodium chloride
  - talc
  - vermiculite
  - spoil from the separation of the above industrial minerals from other rock after extraction
  - material that is mainly but not wholly the spoil, waste or other by-product of any industrial combustion process or the smelting or refining of metal

Anything that consists 'wholly or mainly' of the following is exempt from the levy (note that 'wholly' is defined as 100% but 'mainly' as more than 50%), thus exempting any contained aggregates amounting to less than 50% of the original volumes:

- clay, soil, vegetable or other organic matter
- drill cuttings from oil exploration in UK waters
- material arising from utility works, if carried out under the New Roads and Street Works Act 1991

However, when ground that is more than half clay is mixed with any substance (for example, cement or lime) for the purpose of creating a firm base for construction, the clay becomes liable to Aggregates Levy because it has been mixed with another substance for the purpose of construction.

Anything that consists completely of the following substances is exempt from the levy:

- Spoil, waste or other by-products from any industrial combustion process or the smelting or refining of metal – for example, industrial slag, pulverized fuel ash and used foundry sand. If the material consists completely of these substances at the time it is produced it is exempt from the levy, regardless of any subsequent mixing
- Aggregate necessarily arising from the footprint of any building for the purpose of laying its foundations, pipes or cables. It must be lawfully extracted within the terms of any planning consent
- Aggregate necessarily arising from navigation dredging
- Aggregate necessarily arising from the ground in the course of excavations to improve, maintain or construct a highway or a proposed highway
- Aggregate necessarily arising from the ground in the course of excavations to improve, maintain or construct a railway, monorail or tramway

Relief from the levy either in the form of credit or repayment is obtainable where:

- it is subsequently exported from the UK in the form of aggregate
- it is used in an exempt process
- where it is used in a prescribed industrial or agricultural process
- it is waste aggregate disposed of by dumping or otherwise, e.g. sent to landfill or returned to the originating site

The Aggregates Levy Credit Scheme (ALCS) for Northern Ireland was suspended with effect from 1 December 2010 following a ruling by the European General Court.

An exemption for aggregate obtained as a by-product of railway, tramway and monorail improvement, maintenance and construction was introduced in 2007.

Exemptions to the levy were suspended on 1 April 2014, following an investigation by the European Commission into whether they were lawful under State aid rules. The Commission announced its decision on 27 March 2015 that all but part of one exemption (for shale) were lawful and all the exemptions have, therefore, been reinstated apart from the exemption for shale. The effective date of reinstatement is 1 April 2014, which means that businesses which paid Aggregates Levy on materials for which the exemption has been confirmed as lawful may claim back the tax they paid while the exemption was suspended.

However, under EU law the UK government is required to recover unlawful State aid with interest from businesses that benefited from it. HM Revenue & Customs therefore initiated a process in 2015 of clawing back the levy, plus compound interest, for the deliberate extraction of shale aggregate for commercial exploitation from businesses it believes may have benefited between 1 April 2002 and 31 March 2014.

Discounts

From 1 July 2005 the standard added water percentage discounts listed below can be used. Alternatively a more exact percentage can be agreed and this must be done for dust dampening of aggregates.

- washed sand 7%
- washed gravel 3.5%
- washed rock/aggregate 4%

Impact

The British Aggregates Association suggested that the additional cost imposed by quarries is more likely to be in the order of £3.40 per tonne on mainstream products, applying an above average rate on these in order that by-products and low grade waste products can be held at competitive rates, as well as making some allowance for administration and increased finance charges.

With many gravel aggregates costing in the region of £20.00 per tonne, there is a significant impact on construction costs.

Avoidance

An alternative to using new aggregates in filling operations is to crush and screen rubble which may become available during the process of demolition and site clearance as well as removal of obstacles during the excavation processes.

Example: Assuming that the material would be suitable for fill material under buildings or roads, a simple cost comparison would be as follows (note that for the purpose of the exercise, the material is taken to be 1.80 tonne per m³ and the total quantity involved less than 1,000 m³):

	£/m³	£/tonne
Importing fill material:		
Cost of 'new' aggregates delivered to site	37.10	20.16
Addition for Aggregates Tax	3.60	2.00
Total cost of importing fill materials	**40.70**	**22.61**
Disposing of site material:		
Cost of removing materials from site	26.63	14.79
Crushing site materials:		
Transportation of material from excavations or demolition to stockpiles	0.88	0.49
Transportation of material from temporary stockpiles to the crushing plant	2.36	1.31
Establishing plant and equipment on site; removing on completion	2.36	1.31
Maintain and operate plant	10.62	5.90
Crushing hard materials on site	15.34	8.52
Screening material on site	2.36	1.31
Total cost of crushing site materials	**33.92**	**18.84**

From the above it can be seen that potentially there is a great benefit in crushing site materials for filling rather than importing fill materials.

Setting the cost of crushing against the import price would produce a saving of £6.78 per m³. If the site materials were otherwise intended to be removed from the site, then the cost benefit increases by the saved disposal cost to £33.41 per m³.

Even if there is no call for any or all of the crushed material on site, it ought to be regarded as a useful asset and either sold on in crushed form or else sold with the prospects of crushing elsewhere.

Specimen Unit rates	Unit³	£
Establishing plant and equipment on site; removing on completion		
crushing plant	trip	1,400.00
screening plant	trip	700.00
Maintain and operate plant		
crushing plant	week	8,500.00
screening plant	week	2,100.00
Transportation of material from excavations or demolition places to temporary stockpiles	m³	3.50
Transportation of material from temporary stockpiles to the crushing plant	m³	2.80
Breaking up material on site using impact breakers		
mass concrete	m³	16.50
reinforced concrete	m³	19.00
brickwork	m³	7.00
Crushing material on site		
mass concrete not exceeding 1000m³	m³	15.00
mass concrete 1000–5000m³	m³	14.00
mass concrete over 5000m³	m³	13.00
reinforced concrete not exceeding 1000m³	m³	18.00
reinforced concrete 1000–5000m³	m³	16.00
reinforced concrete over 5000m³	m³	15.00
brickwork not exceeding 1000m³	m³	14.00
brickwork 1000–5000m³	m³	13.00
brickwork over 5000m³	m³	12.00
Screening material on site	m³	2.50

More detailed information can be found on the HMRC website (www.hmrc.gov.uk) in Notice AGL 1 Aggregates Levy published August 2015.

NRM1 Cost Management Handbook

David P Benge

The definitive guide to measurement and estimating using NRM1, written by the author of NRM1

The 'RICS New rules of measurement: Order of cost estimating and cost planning of capital building works' (referred to as NRM1) is the cornerstone of good cost management of capital building works projects - enabling more effective and accurate cost advice to be given to clients and other project team members, while facilitating better cost control.

The NRM1 Cost Management Handbook is the essential guide to how to successfully interpret and apply these rules, including explanations of how to:

- quantify building works and prepare order of cost estimates and cost plans
- use the rules as a toolkit for risk management and procurement
- analyse actual costs for the purpose of collecting benchmark data and preparing cost analyses
- capture historical cost data for future order of cost estimates and elemental cost plans
- employ the rules to aid communication
- manage the complete 'cost management cycle'
- use the elemental breakdown and cost structures, together with the coding system developed for NRM1, to effectively integrate cost management with Building Information Modelling (BIM).

March 2014: 246 x 174: 640pp
Pb: 978-0-415-72077-9: £41.99

To Order: Tel: +44 (0) 1235 400524 Fax: +44 (0) 1235 400525
or Post: Taylor and Francis Customer Services,
Bookpoint Ltd, Unit T1, 200 Milton Park, Abingdon, Oxon, OX14 4TA UK
Email: book.orders@tandf.co.uk

For a complete listing of all our titles visit:
www.tandf.co.uk

Land Remediation

The purpose of this section is to review the general background of ground contamination, the cost implications of current legislation and to consider the various remedial measures and to present helpful guidance on the cost of Land Remediation.

It must be emphasized that the cost advice given is an average and that costs can vary considerably from contract to contract depending on individual Contractors, site conditions, type and extent of contamination, methods of working and various other factors as diverse as difficulty of site access and distance from approved tips.

We have structured this Unit Cost section to cover as many aspects of Land Remediation works as possible.

The introduction of the Landfill Directive in July 2004 has had a considerable impact on the cost of Remediation works in general and particularly on the practice of Dig and Dump. The number of Landfill sites licensed to accept Hazardous Waste has drastically reduced and inevitably this has led to increased costs.

Market forces will determine future increases in cost resulting from the introduction of the Landfill Directive and the cost guidance given within this section will require review in light of these factors.

Statutory Framework

In July 1999 new contaminated land provisions, contained in Part IIA of the Environmental Protection Act 1990 were introduced. Primary objectives of the measures included a legal definition of Contaminated Land and a framework for identifying liability, underpinned by a 'polluter pays' principle meaning that remediation should be paid for by the party (or parties) responsible for the contamination. A secondary, and indirect, objective of Part IIA is to provide the legislative context for remediation carried out as part of development activity which is controlled through the planning system. This is the domain where other related objectives, such as encouraging the recycling of brownfield land, are relevant.

Under the Act action to remediate land is required only where there are unacceptable actual or 'significant possibility of significant harm' to health, controlled waters or the environment. Only Local Authorities have the power to determine a site as Contaminated Land and enforce remediation. Sites that have been polluted from previous land use may not need remediating until the land use is changed; this is referred to as 'land affected by contamination'. This is a risk-based assessment on the site specifics in the context of future end uses. As part of planning controls, the aim is to ensure that a site is incapable of meeting the legal definition of Contaminated Land post-development activity. In addition, it may be necessary to take action only where there are appropriate, cost-effective remediation processes that take the use of the site into account.

The Environment Act 1995 amended the Environment Protection Act 1990 by introducing a new regime designed to deal with the remediation of sites which have been seriously contaminated by historic activities. The regime became operational on 1 April 2000. Local authorities and/or the Environment Agency regulate seriously contaminated sites which are known as 'special sites'. The risks involved in the purchase of potentially contaminated sites are high, particularly considering that a transaction can result in the transfer of liability for historic contamination from the vendor to the purchaser.

The contaminated land provisions of the Environmental Protection Act 1990 are only one element of a series of statutory measures dealing with pollution and land remediation that have been and are to be introduced. Others include:

- Groundwater regulations, including pollution prevention measures
- An integrated prevention and control regime for pollution
- Sections of the Water Resources Act 1991, which deals with works notices for site controls, restoration and clean up.

April 2012 saw the first revision of the accompanying Part IIa Statutory Guidance. This has introduced a new categorization scheme for assessing sites under Part IIA. Category 1 is land which definitely is Contaminated Land and Category 4 is for land which definitely is not Contaminated Land. This is intended to assist prioritization of sites which pose the greatest risk. Still included in the statutory guidance are matters of inspection, definition, remediation, apportionment of liabilities and recovery of costs of remediation. The measures are to be applied in accordance with the following criteria:

- The planning system
- The standard of remediation should relate to the present use
- The costs of remediation should be reasonable in relation to the seriousness of the potential harm
- The proposals should be practical in relation to the availability of remediation technology, impact of site constraints and the effectiveness of the proposed clean-up method

Liability for the costs of remediation rests with either the party that 'caused or knowingly permitted' contamination, or with the current owners or occupiers of the land.

Apportionment of liability, where shared, is determined by the local authority. Although owners or occupiers become liable only if the polluter cannot be identified, the liability for contamination is commonly passed on when land is sold.

If neither the polluter nor owner can be found, the cleanup is funded from public resources.

The ability to forecast the extent and cost of remedial measures is essential for both parties, so that they can be accurately reflected in the price of the land.

At the end of March 2012, the National Planning Policy Framework replaced relevant planning guidance relating to remediation, most significantly PPS 23 Planning and Pollution Control. This has been replaced by a need to investigate and assess land contamination, which must be carried out by a competent person.

The EU Landfill Directive

The Landfill (England and Wales) Regulations 2002 came into force on 15 June 2002 followed by Amendments in 2004 and 2005. These new regulations implement the Landfill Directive (Council Directive 1999/31/EC), which aims to prevent, or to reduce as far as possible, the negative environmental effects of landfill. These regulations have had a major impact on waste regulation and the waste management industry in the UK.

The Scottish Executive and the Northern Ireland Assembly will be bringing forward separate legislation to implement the Directive within their regions.

In summary, the Directive requires that:

- Sites are to be classified into one of three categories: hazardous, non-hazardous or inert, according to the type of waste they will receive
- Higher engineering and operating standards will be followed
- Biodegradable waste will be progressively diverted away from landfills
- Certain hazardous and other wastes, including liquids, explosive waste and tyres will be prohibited from landfills
- Pre-treatment of wastes prior to landfilling will become a requirement

On 15 July 2004 the co-disposal of hazardous and non-hazardous waste in the same landfill site ended and in July 2005 new waste acceptance criteria (WAC) were introduced which also prevents the disposal of materials contaminated by coal tar.

The effect of this Directive has been to dramatically reduce the hazardous disposal capacity post July 2004, resulting in a **SIGNIFICANT** increase in remediating costs. This has significantly increased travelling distance and cost for disposal to landfill. The increase in operating expenses incurred by the landfill operators has also resulted in higher tipping costs.

However, there are now a growing number of opportunities to dispose of hazardous waste to other facilities such as soil treatment centres, often associated with registered landfills potentially eliminating landfill tax. Equally, improvements in on-site treatment technologies are helping to reduce the costs of disposal by reducing the hazardous properties of materials going offsite.

All hazardous materials designated for disposal off-site are subject to WAC tests. Samples of these materials are taken from site to laboratories in order to classify the nature of the contaminants. These tests, which cost approximately £200 each, have resulted in increased costs for site investigations and as the results may take up to 3 weeks this can have a detrimental effect on programme.

As from 1 July 2008 the WAC derogations which have allowed oil contaminated wastes to be disposed in landfills with other inert substances were withdrawn. As a result the cost of disposing oil contaminated solids has increased.

There has been a marked slowdown in brownfield development in the UK with higher remediation costs, longer clean-up programmes and a lack of viable treatment options for some wastes.

The UK Government established the Hazardous Waste Forum in December 2002 to bring together key stakeholders to advise on the way forward on the management of hazardous waste.

Effect on Disposal Costs

Although most landfills are reluctant to commit to future tipping prices, tipping costs have generally stabilized. However, there are significant geographical variances, with landfill tip costs in the North of England typically being less than their counterparts in the Southern regions.

For most projects to remain viable there is an increasing need to treat soil in situ by bioremediation, soil washing or other alternative long-term remediation measures. Waste untreatable on-site such as coal tar remains a problem. Development costs and programmes need to reflect this change in methodology.

Types of hazardous waste

- Sludges, acids and contaminated wastes from the oil and gas industry
- Acids and toxic chemicals from chemical and electronics industries
- Pesticides from the agrochemical industry
- Solvents, dyes and sludges from leather and textile industries
- Hazardous compounds from metal industries
- Oil, oil filters and brake fluids from vehicles and machines
- Mercury-contaminated waste from crematoria
- Explosives from old ammunition, fireworks and airbags
- Lead, nickel, cadmium and mercury from batteries
- Asbestos from the building industry
- Amalgam from dentists
- Veterinary medicines

[Source: Sepa]

Foam insulation materials containing ODP (Ozone Depletant Potential) are also considered as hazardous waste under the EC Regulation 2037/2000.

LAND REMEDIATION TECHNIQUES

There are two principal approaches to remediation – dealing with the contamination in situ or ex situ. The selection of the approach will be influenced by factors such as: initial and long term cost, timeframe for remediation, types of contamination present, depth and distribution of contamination, the existing and planned topography, adjacent land uses, patterns of surface drainage, the location of existing on-site services, depth of excavation necessary for foundations and below-ground services, environmental impact and safety, interaction with geotechnical performance, prospects for future changes in land use and long-term monitoring and maintenance of in situ treatment.

On most sites, contamination can be restricted to the top couple of metres, although gasholder foundations for example can go down 10 to 15 metres. Underground structures can interfere with the normal water regime and trap water pockets.

There could be a problem if contaminants get into fissures in bedrock.

In situ techniques

A range of in situ techniques is available for dealing with contaminants, including:

- Clean cover – a layer of clean soil is used to segregate contamination from receptor. This technique is best suited to sites with widely dispersed contamination. Costs will vary according to the need for barrier layers to prevent migration of the contaminant.
- On-site encapsulation – the physical containment of contaminants using barriers such as slurry trench cut-off walls. The cost of on-site encapsulation varies in relation to the type and extent of barriers required, the costs of which range from £50/m² to more than £175/m².

There are also in situ techniques for treating more specific contaminants, including:

- Bioremediation – for removal of oily, organic contaminants through natural digestion by microorganisms. Most bioremediation is ex situ, i.e. it is dug out and then treated on site in bio-piles. The process can be slow, taking up to three years depending upon the scale of the problem, but is particularly effective for the long-term improvement of a site, prior to a change of use.
- Phytoremediation – the use of plants that mitigate the environmental problem without the need to excavate the contaminant material and dispose of it elsewhere. Phytoremediation consists of mitigating pollutant concentrations in contaminated soils, water or air, with plants able to contain, degrade or eliminate metals, pesticides, solvents, explosives, crude oil and its derivatives and various other contaminants from the media that contain them.
- Vacuum extraction – involving the extraction of volatile organic compounds (e.g. benzene) from soil and groundwater by vacuum.
- Thermal treatment – the incineration of contaminated soils on site. Thermal processes use heat to increase the volatility to burn, decompose, destroy or melt the contaminants. Cleaning soil with thermal methods may take only a few months to several years.
- Stabilization – cement or lime, is used to physically or chemically bind oily or metal contaminants to prevent leaching or migration. Stabilization can be used in both in situ and ex situ conditions.
- Aeration – if the ground contamination is highly volatile, e.g. fuel oils, then the ground can be ploughed and rotovated to allow the substance to vaporize.
- Air sparging – the injection of contaminant-free air into the subsurface enabling a phase transfer of hydrocarbons from a dissolved state to a vapour phase.
- Chemical oxidization – the injection of reactive chemical oxidants directly into the soil for the rapid destruction of contaminants.
- Pumping – to remove liquid contaminants from boreholes or excavations. Contaminated water can be pumped into holding tanks and allowed to settle; testing may well prove it to be suitable for discharging into the foul sewer subject to payment of a discharge fee to the local authority. It may be necessary to process the water through an approved water treatment system to render it suitable for discharge.

Ex situ techniques

Removal for landfill disposal has, historically, been the most common and cost-effective approach to remediation in the UK, providing a broad spectrum solution by dealing with all contaminants. As mentioned above, the implementation of the Landfill Directive has resulted in other techniques becoming more competitive for the disposal of hazardous waste.

If used in combination with material-handling techniques such as soil washing, the volume of material disposed at landfill sites can be significantly reduced. The disadvantages of these techniques include the fact that the contamination is not destroyed, there are risks of pollution during excavation and transfer; road haulage may also cause a local nuisance. Ex situ techniques include:

- Soil washing – involving the separation of a contaminated soil fraction or oily residue through a washing process. This also involves the excavation of the material for washing ex situ. The de-watered contaminant still requires disposal to landfill. In order to be cost effective, 70–90% of soil mass needs to be recovered. It will involve constructing a hard area for the washing, intercepting the now-contaminated water and taking it away in tankers.
- Thermal treatment – the incineration of contaminated soils ex situ. The uncontaminated soil residue can be recycled. By-products of incineration can create air pollution and exhaust air treatment may be necessary.

Soil treatment centres are now beginning to be established. These use a combination of treatment technologies to maximize the potential recovery of soils and aggregates and render them suitable for disposal to the landfill. The technologies include:

- Physicochemical treatment – a method which uses the difference in grain size and density of the materials to separate the different fractions by means of screens, hydrocyclones and upstream classification.
- Bioremediation – the aerobic biodegradation of contaminants by naturally occurring microorganisms placed into stockpiles/windrows.
- Stabilisation/solidification – a cement or chemical stabilisation unit capable of immobilising persistent leachable components.

COST CONSIDERATIONS

Cost drivers

Cost drivers relate to the selected remediation technique, site conditions and the size and location of a project.

The wide variation of indicative costs of land remediation techniques shown below is largely because of differing site conditions.

Indicative costs of land remediation techniques for 2016 (excluding general items, testing, landfill tax and backfilling)		
Remediation technique	**Unit**	**Rate (£/unit)**
Removal – non-hazardous	disposed material (m³)	40–100
Removal – hazardous		
Note: excluding any pre-treatment		
of material	disposed material (m³)	75–200
Clean cover	surface area of site (m²)	20–45
On-site encapsulation	encapsulated material (m³)	30–95
Bioremediation (in situ)	treated material (m³)	15–40
Bioremediation (ex situ)	treated material (m³)	20–45
Chemical oxidation	treated material (m³)	30–80
Stabilization/solidification	treated material (m³)	20–65
Vacuum extraction	treated material (m³)	25–75
Soil washing	treated material (m³)	40–95
Thermal treatment	treated material (m³)	100–400

Many other on-site techniques deal with the removal of the contaminant from the soil particles and not the wholesale treatment of bulk volumes. Costs for these alternative techniques are very much Engineer designed and site specific.

Factors that need to be considered include:

- waste classification of the material
- underground obstructions, pockets of contamination and live services
- ground water flows and the requirement for barriers to prevent the migration of contaminants
- health and safety requirements and environmental protection measures
- location, ownership and land use of adjoining sites
- distance from landfill tips, capacity of the tip to accept contaminated materials, and transport restrictions
- the cost of diesel fuel, currently approximately £1.08 per litre (at April 2016 prices)

Other project related variables include size, access to disposal sites and tipping charges; the interaction of these factors can have a substantial impact on overall unit rates.

The tables below set out the costs of remediation using dig-and-dump methods for different sizes of project, differentiated by the disposal of non-hazardous and hazardous material. Variation in site establishment and disposal cost accounts for 60–70% of the range in cost.

Variation in the costs of land remediation by removal: Non-hazardous waste

Item	Disposal Volume (less than 3,000 m³) (£/m³)	Disposal Volume (3,000 – 10,000 m³) (£/m³)	Disposal Volume (more than 10,000 m³) (£/m³)
General items and site organization costs	55–90	25–40	7–20
Site investigation and testing	5–12	2–7	2–6
Excavation and backfill	18–35	12–25	10–20
Disposal costs (including tipping charges but not landfill tax)	20–35	20–35	20–35
Haulage	15–35	15–35	15–35
Total (£/m³)	**113–207**	**74–142**	**54–116**
Allowance for site abnormals	*0–10 +*	*0–15 +*	*0–10 +*

Variation in the costs of land remediation by removal: Hazardous waste

Item	Disposal Volume (less than 3,000 m³) (£/m³)	Disposal Volume (3,000 – 10,000 m³) (£/m³)	Disposal Volume (more than 10,000 m³) (£/m³)
General items and site organization costs	55–90	25–40	7–20
Site investigation and testing	10–18	5–12	5–12
Excavation and backfill	18–35	12–25	10–20
Disposal costs (including tipping charges but not landfill tax)	80–170	80–170	80–170
Haulage	25–120	25–120	25–120
Total (£/m³)	**188–433**	**147–367**	**127–342**
Allowance for site abnormals	*0–10 +*	*0–15 +*	*0–10 +*

The strict health and safety requirements of remediation can push up the overall costs of site organization to as much as 50% of the overall project cost (see the above tables). A high proportion of these costs are fixed and, as a result, the unit costs of site organization increase disproportionally on smaller projects.

Haulage costs are largely determined by the distances to a licensed tip. Current average haulage rates, based on a return journey, range from £1.95 to £4.50 per mile. Short journeys to tips, which involve proportionally longer standing times, typically incur higher mileage rates, up to £9.00 per mile.

A further source of cost variation relates to tipping charges. The table below summarizes typical tipping charges for 2016, exclusive of landfill tax:

Typical 2016 tipping charges (excluding landfill tax) Waste classification	Charges (£/tonne)
Non-hazardous wastes	15–45
Hazardous wastes	35–90
Contaminated liquid	40–75
Contaminated sludge	125–400

Tipping charges fluctuate in relation to the grades of material a tip can accept at any point in time. This fluctuation is a further source of cost risk. Furthermore, tipping charges in the North of England are generally less than other areas of the country.

Prices at licensed tips can vary by as much as 50%. In addition, landfill tips generally charge a tip administration fee of approximately £25 per load, equivalent to £1.25 per tonne. This charge does not apply to non-hazardous wastes.

Landfill Tax, which increased on 1 April 2016 to £84.40 a tonne for active waste, is also payable. Exemptions are no longer available for the disposal of historically contaminated material (refer also to *Landfill Tax* section).

Tax Relief for Remediation of Contaminated Land

The Finance Act 2001 included provisions that allow companies (but not individuals or partnerships) to claim tax relief on capital and revenue expenditure on the 'remediation of contaminated land' in the United Kingdom. The relief is available for expenditure incurred on or after 11 May 2001.

From 1 April 2009 there was an increase in the scope of costs that qualify for Land Remediation Relief where they are incurred on long-term derelict land. The list includes costs that the Treasury believe to be primarily responsible for causing dereliction, such as additional costs for removing building foundations and machine bases. However, while there is provision for the list to be extended, the additional condition for the site to have remained derelict since 1998 is likely to render this relief redundant in all but a handful of cases. The other positive change is the fact that Japanese Knotweed removal and treatment (on-site only) will now qualify for the relief under the existing legislation, thereby allowing companies to make retrospective claims for any costs incurred since May 2001 – provided all other entitlement conditions are met.

A company is able to claim an additional 50% deduction for 'qualifying land remediation expenditure' allowed as a deduction in computing taxable profits, and may elect for the same treatment to be applied to qualifying capital expenditure.

With Landfill Tax exemption (LTE) now phased out, Land Remediation Relief (LRR) for contaminated and derelict land is the Government's primary tool to create incentives for brownfield development. LRR is available to companies engaged in land remediation that are not responsible for the original contamination.

Over 7 million tonnes of waste each year were being exempted from Landfill Tax in England alone, so this change could have a major impact on the remediation industry. The modified LRR scheme, which provides Corporation Tax relief on any costs incurred on qualifying land remediation expenditure, is in the long run designed to yield benefits roughly equal to those lost through the withdrawal of LTE, although there is some doubt about this stated equity in reality.

However, with much remediation undertaken by polluters or public authorities, who cannot benefit from tax relief benefits, the change could result in a net withdrawal of Treasury support to a vital sector. Lobbying and consultation continues to ensure the Treasury maintains its support for remediation.

While there are no financial penalties for not carrying out remediation, a steep escalator affected the rate of landfill tax for waste material other than inert or inactive wastes, which has increased to £84.40/tonne from 1 April 2016. This means that for schemes where there is no alternative to dig and dump and no pre-existing LTE, the cost of remediation has risen to prohibitive levels should contaminated material be disposed off-site to licensed landfills.

Looking forward, tax-relief benefits under LRR could provide a significant cash contribution to remediation. Careful planning is the key to ensure that maximum benefits are realized, with actions taken at the points of purchase, formation of JV arrangements, procurement of the works and formulation of the Final Account (including apportionment of risk premium) all influencing the final value of the claim agreed with HM Revenue & Customs.

The Relief

Qualifying expenditure may be deducted at 150% of the actual amount expended in computing profits for the year in which it is incurred.

For example, a property trading company may buy contaminated land for redevelopment and incurs £250,000 on qualifying land remediation expenditure that is an allowable for tax purposes. It can claim an additional deduction of £125,000, making a total deduction of £375,000. Similarly, a company incurring qualifying capital expenditure on a fixed asset of the business is able to claim the same deduction provided it makes the relevant election within 2 years.

What is Remediation?

Land remediation is defined as the doing of works including preparatory activities such as condition surveys, to the land in question, any controlled waters affected by the land, or adjoining or adjacent land for the purpose of:

- Preventing or minimising, or remedying or mitigating the effects of, any relevant harm, or any pollution of controlled waters, by reason of which the land is in a contaminated state

Definitions

Contaminated land is defined as land that, because of substances on or under it, is in such a condition that relevant harm is or has the significant possibility of relevant harm being caused to:

- The health of living organisms
- Ecological systems
- Quality of controlled waters
- Property

Relevant harm is defined as meaning:

- Death of living organisms or significant injury or damage to living organisms
- Significant pollution of controlled waters
- A significant adverse impact on the ecosystem or
- Structural or other significant damage to buildings or other structures or interference with buildings or other structures that significantly compromises their use

Land includes buildings on the land, and expenditure on asbestos removal is expected to qualify for this tax relief. It should be noted that the definition is not the same as that used in the Environmental Protection Act Part IIA.

Sites with a nuclear license are specifically excluded.

Conditions

To be entitled to claim LRR, the general conditions for all sites, which must all be met, are:

- Must be a company
- Must be land in the United Kingdom
- Must acquire an interest in the land
- Must not be the polluter or have a relevant connection to the polluter
- Must not be in receipt of a subsidy
- Must not also qualify for Capital Allowances (particular to capital expenditure only)

Additional conditions introduced since 1 April 2009:

- The interest in land must be major – freehold or leasehold longer than 7 years
- Must not be obligated to carry out remediation under a statutory notice

Additional conditions particular to derelict land:

- Must not be in or have been in productive use at any time since at least 1 April 1998
- Must not be able to be in productive use without the removal of buildings or other structures

In order for expenditure to become qualifying, it must relate to substances present at the point of acquisition.

Furthermore, it must be demonstrated that the expenditure would not have been incurred had those substances not been present.

Cost Studies of Buildings

Allan Ashworth & Srinath Perera

This practical guide to cost studies of buildings has been updated and revised throughout for the 6th edition. New developments in RICS New Rules of Measurement (NRM) are incorporated throughout the book, in addition to new material on e-business, the internet, social media, building information modelling, sustainability, building resilience and carbon estimating.

This trusted and easy to use guide to the cost management role:

Focuses on the importance of costs of constructing projects during the different phases of the construction process

Features learning outcomes and self-assessment questions for each chapter

Addresses the requirements of international readers

From introductory data on the construction industry and the history of construction economics, to recommended methods for cost analysis and post-contract cost control, Cost Studies of Buildings is an ideal companion for anyone learning about cost management.

July 2015: 246x174: 544pp Pb:
978-1-138-01735-1: £42.99

To Order: Tel: +44 (0) 1235 400524 Fax: +44 (0) 1235 400525
or Post: Taylor and Francis Customer Services,
Bookpoint Ltd, Unit T1, 200 Milton Park, Abingdon, Oxon, OX14 4TA UK
Email: book.orders@tandf.co.uk

For a complete listing of all our titles visit:
www.tandf.co.uk

The Landfill Tax

The Tax

The landfill tax came into operation on 1 October 1996. It is levied on operators of licensed landfill sites in England, Wales and Northern Ireland at the following rates with effect from 1 April 2016

- Inactive or inert wastes [Lower Rate] £2.60 per tonne   Included are soil, stones, brick, plain and reinforced concrete, plaster and glass – lower rate

- All other taxable wastes [Higher Rate] £84.40 per tonne   Included are timber, paint and other organic wastes generally found in demolition work and builders skips – standard rate

The standard and lower rates of Landfill Tax were due to increase in line with the RPI, rounded to the nearest 5 pence, from 1 April 2016. However, legislation introduced in the Finance Bill 2016 provided for new rates of Landfill Tax until 31 March 2019 as follows:

- Lower Rate   From 1 April 2016
  £2.65/tonne
- Lower Rate   From 1 April 2017
  £2.70/tonne
- Lower Rate   From 1 April 2018
  £2.80/tonne
- Standard Rate   From 1 April 2016
  £84.40/tonne
- Standard Rate   From 1 April 2017
  £86.10/tonne
- Standard Rate   From 1 April 2018
  £88.95/tonne

Following industry engagement to address compliance, on 1 April 2015 the government introduced a loss on ignition testing regime on fines (residual waste from waste processing) from waste transfer stations by April 2015. Only fines below a 10% threshold will be considered eligible for the lower rate, though there is a 12 month transitional period where the threshold will be 15%. This transitional period came to an end on 31 March 2016. The government intends to provide further longer term certainty about the future level of landfill tax rates once the consultation process on testing regime has concluded, but in the mean time is committed to ensuring that the rates are not eroded in real terms. (Finance Bill 2014)

The Landfill Tax (Qualifying Material) Order 2011 came into force on 1 April 2011. This has amended the qualifying criteria that are eligible for the lower rate of Landfill Tax. The revisions introduced arose primarily from the need to reflect changes in wider environmental policy and legislation since 1996, such as the implementation of the European Landfill Directive.

A waste will be lower rated for Landfill Tax only if it is listed as a qualifying material in the Landfill Tax (Qualifying Material) Order 2011.

The principal changes to qualifying material are:

- Rocks and subsoils that are currently lower rated will remain so
- Topsoil and peat will be removed from the lower rate, as these are natural resources that can always be recycled/reused
- Used foundry sand, which has in practice been lower rated by extra-statutory concession since the tax's introduction in 1996, will now be included in the lower rate Order
- Definitions of qualifying ash arising from the burning of coal and petroleum coke (including when burnt with biomass) will be clarified

- The residue from titanium dioxide manufacture will qualify, rather than titanium dioxide itself, reflecting industry views
- Minor changes will be made to the wording of the calcium sulphate group of wastes to reflect the implementation of the Landfill Directive since 2001
- Water will be removed from the lower rate – water is now banned from landfill so its inclusion in the list of lower rated wastes is unnecessary; where water is used as a waste carrier the water is not waste and therefore not taxable

Calculating the Weight of Waste

There are two options:

- If licensed sites have a weighbridge tax will be levied on the actual weight of waste
- If licensed sites do not have a weighbridge tax will be levied on the permitted weight of the lorry based on an alternative method of calculation based on volume to weight factors for various categories of waste

Effect on Prices

The tax is paid by landfill site operators only. Tipping charges reflect this additional cost.

As an example, *Spon's A&B* rates for mechanical disposal will be affected as follows:

- Inactive waste
  - *Spon's A&B 2017* net rate — £21.69 per m³
  - Tax, 1.9 tonne per m³ (un-bulked) @ £2.65 — £5.04 per m³
  - Spon's rate including tax — £26.63 per m³
- Active waste — Active waste will normally be disposed of by skip and will probably be mixed with inactive waste. The tax levied will depend on the weight of materials in the skip which can vary significantly.

Exemptions

The following disposals are exempt from Landfill Tax subject to meeting certain conditions:

- Dredging's which arise from the maintenance of inland waterways and harbours
- Naturally occurring materials arising from mining or quarrying operations
- Pet cemeteries
- Material from the reclamation of contaminated land (see below)
- Inert waste used to restore landfill sites and to fill working and old quarries where a planning condition or obligation is in existence
- Waste from visiting NATO forces

The exemption for waste from contaminated land has been phased out completely from 1 April 2012 and no new applications for Landfill Tax exemption are now accepted.

Devolution of Landfill Tax to Scotland from 1 April 2015

The Scotland Act 2012 provided for Landfill Tax to be devolved to Scotland. From 1 April 2015, operators of landfill sites in Scotland are no longer liable to pay UK Landfill Tax for waste disposed at their Scottish sites. Instead, they will be liable to register and account for Scottish Landfill Tax (SLfT).

Operators of landfill sites only in Scotland were deregistered from UK Landfill Tax with effect from 31 March 2015.

The Standard and Lower rates of the Scottish Landfill Tax will mirror the rates applied to the current UK landfill tax for the time being.

Devolution of Landfill Tax to Wales from April 2018

The Wales Act 2014 provides for Landfill Tax to be devolved to Wales. This is expected to take effect in April 2018. Further information will be provided on developments by HM Revenue & Customs at a later date.

For further information contact the HMRC VAT and Excise Helpline, telephone 0300 200 3700.

Property Insurance

The problem of adequately covering by insurance the loss and damage caused to buildings by fire and other perils has been highlighted in recent years by the increasing rate of inflation.

There are a number of schemes available to the building owner wishing to insure his property against the usual risk. Traditionally the insured value must be sufficient to cover the actual cost of reinstating the building. This means that in addition to assessing the current value an estimate has also to be made of the increases likely to occur during the period of the policy and of rebuilding which, for a moderate size building, could amount to a total of three years. Obviously such an estimate is difficult to make with any degree of accuracy; if it is too low the insured may be penalized under the terms of the policy and if too high will result in the payment of unnecessary premiums.

There are variations on the traditional method of insuring which aim to reduce the effects of over estimating and details of these are available from the appropriate offices. For the convenience of readers who may wish to make use of the information contained in this publication in calculating insurance cover required the following may be of interest.

1 Present Cost

The current rebuilding costs may be ascertained in a number of ways:

- Where the actual building cost is known this may be updated by reference to tender price changes
- By reference to average published prices per square metre of floor area. In this case it is important to understand clearly the method of measurement used to calculate the total floor area on which the rates have been based
- By professional valuation
- By comparison with the known cost of another recently built similar building

Whichever of these methods is adopted regard must be paid to any special conditions that may apply, e.g. a confined site; complexity of design; any demolition works and site clearance that may be required.

2 Allowances for Inflation

The Present Cost when established will usually, under the conditions of the policy, be the rebuilding cost on the first day of the policy period. To this must be added a sum to cover future price increases over the period of the policy. For this purpose, using the historical indices as a base and taking account of the likely change in future building costs and the tender climate, an anticipated re-build cost can be arrived at.

3 Fees

To the total of 1 and 2 above must be added an allowance for any design and other consultancy fees.

4 Value Added Tax (VAT)

To the total of 1 to 3 above must be added Value Added Tax.

Example:

An assessment for insurance cover is required from December 2016 for a property which cost £500,000 when initially built in 2010. This calculation uses AECOM Tender Price Indices. There are various methods for determining insurance cover needed. This is simply an example of one method and users would need to allow for any items that would apply to their own project. Indices number would need to be amended to reflect the most recent published data which is in the free Spon's *Updates* and also in the *Market Forecast* and *Cost Update* articles, published quarterly in Building magazine

A) Update original build costs (2010Q4) to present day costs (2015Q4)

Known cost to build at end of 2010			£500,000
Calculation for present day build cost:			
Tender price index at 2010Q4 =		448	
Forecast tender index 2016Q4 =		559	
Increase in tender index (559/448)	24.78%		£123,900
Estimated present build cost (2016Q4)			**£623,900**

B) Calculate allowance for tender inflation to end of current policy: 31 December 2016

Estimated present build cost at end of current policy: 31 December 2016			£623,900
Inflation calculation to end of new policy:			
Forecast tender index 2016Q4		559	
Forecast tender index 2017Q4		586	
Increase in tender index over period (586/559)	4.8%		£30,100
Estimated build cost at end of new policy (2017Q4)			**£654,000**

Assuming that total damage is suffered on the last day of the policy: 31 December 2017, and that planning and documentation and general project set up would require a period of twelve months before re-building could commence on 31 December 2018, then a further allowance must be made to cover any inflation for that period. Again use the tender price index.

C) Calculate allowance for inflation to start of re-construction: 31 December 2017

Anticipated cost at end of policy: 31 December 2017			£654,000
Inflation calculation from end of policy to tender:			
Forecast tender index 2017Q4		586	
Forecast tender index 2018Q4		608	
Increase in tender index (608/586)	3.8%		£24,600
applied to cost at expiry of policy			
Estimated build cost at tender date (2018Q4)			**£678,600**

D) Summary

You now need to add on any other costs such as fees, demolition, VAT as applicable etc.

Estimated cost of reinstatement		£678,600
Demolitions and site clearance		£30,000
Add professional fees	15%	£106,290
		£814,890
Add for VAT	20%	£162,978
Total insurance cover required		**£977,868**
	say	**£980,000**

Please note that this is just a simple example of one method to calculate the cost of re-building for insurance purposes.

You always need to work with the most recent set of indices available if you are calculating the equivalent for your own property which are free Spon's *Updates* and also in the *Market Forecast* and *Cost Update* articles, published quarterly in *Building* magazine.

JCT Contract Administration Pocket Book

Andy Atkinson

Successfully managing your JCT contracts is a must, and this handy reference is the swiftest way to doing just that. Making reference to best practice throughout, the JCT Standard Building Contracts SBC/Q, DB and MW are used as examples to take you through all the essential contract administration tasks, including:

- Procurement
- Payment
- Final accounts
- Progress, completion and delay
- Subcontracting
- Defects and quality control

In addition to the day to day tasks, this also gives you an overview of what to expect from common sorts of dispute resolution under the JCT, as well as a look at how to administer contracts for BIM-compliant projects. This is an essential starting point for all students of construction contract administration, as well as practitioners needing a handy reference to working with JCT contracts.

June 2015 186x123: 144pp
Pb: 978-1-138-78192-4: £24.99

To Order: Tel: +44 (0) 1235 400524 Fax: +44 (0) 1235 400525
or Post: Taylor and Francis Customer Services,
Bookpoint Ltd, Unit T1, 200 Milton Park, Abingdon, Oxon, OX14 4TA UK
Email: book.orders@tandf.co.uk

For a complete listing of all our titles visit:
www.tandf.co.uk

Building Costs Indices, Tender Price Indices and Location Factors

The extract tables below show the changes in building costs and tender prices since 2010 (base date 1976 = 100). To avoid confusion it is essential that the terms building costs and tender prices are clearly defined and understood.

- **Building Costs Indices** are the costs incurred by the contractor in the course of his business, the principal ones being those for labour and materials, i.e. cost to contractor.
- **Tender Price Indices** represents the price for which the contractor offers to do the project for, i.e. cost to client.

Readers are reminded that the following adjustments for time and location should be applied to the project as a whole and not be applied to individual elements or materials.

Building Cost Indices

Building costs are the costs actually incurred by the builder in the course of business i.e. wages, material prices, plant costs, rates, rents, overheads and taxes. AECOM's Building Cost Index measures movement of basic labour and materials costs to the builder. It is a composite index.

This table reflects the fluctuations since 2010 (base date 2010 = 100) in costs to the contractor. No allowance has been made for changes in productivity or for rates or hours worked which may occur in particular conditions and localities.

Year	Annual Average
2010	100
2011	102
2012	105
2013	106
2014	111
2015	110
2016 (F)	112
2017	115

Note: (F) = Forecast

Tender Price Indices

Tender prices represent the cost a client must pay for a building. They include building costs but also take into account market considerations, the tendering climate, constructor sentiment and its assessment of prevailing market conditions amongst other things. They therefore include allowances for profits, risks and other on-costs such as preliminaries. The constructor also has to anticipate cost changes during the lifetime of the contract. This means that, for example, in busier times tender prices may increase at a greater rate than building costs, whilst in a downturn the opposite may apply. The same concepts apply to trade contractors, and there is often a compounding of these trends up through the supply chain. AECOM's Tender Price Index is derived from analysis of project data and tender returns on AECOM's UK projects.

Tender prices are similar to building costs but also take into account market considerations such as the availability of labour and materials, and the prevailing economic situation. This means that in boom periods, when there is a surfeit of building work to be done, tender prices may increase at a greater rate than building costs, whilst in a period when work is scarce, tender prices may actually fall when building costs are rising.

Spon's A&B 2017 has been model calculated to give a TPI of 572, which currently equates to a price level around 2017Q2 in the table below

Year	First quarter	Second quarter	Third quarter	Fourth quarter	Annual average
2010	457	454	452	448	453
2011	446	447	448	444	446
2012	446	446	443	439	444
2013	440	442	446	453	445
2014	457	464	474	482	470
2015	492	505	520	530	512
2016	537 (P)	545 (F)	552	559	548
2017	566	573	579	586	576

Note: (P) = Provisional, (F) = Forecast thereafter

Readers can be kept abreast of tender price movements in the free Spon's *Updates* and also in the *Market Forecast* and *Cost Update* articles, published quarterly in *Building* magazine.

AECOM Tender Price, Building Cost and Retail Price Indices chart

This chart shows the relative movement of the indices since 2010 by rebasing the above data so that 2010 = 100. This makes it much easier to see the relative movement between the indices.

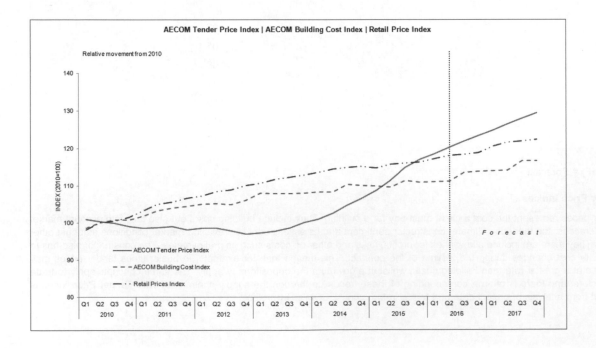

Regional Variations

As well as being aware of inflationary trends when preparing an estimate, it is also important to establish the appropriate price level for the project location.

Prices throughout this book are at a price level index of 572 in Outer London (location factor = 1.00). Regional variations for certain inner London boroughs can be up to 10% higher, while prices in the Yorkshire and Humberside can be as much as 10% lower. Broad regional adjustment factors used to assist with the preparation of initial estimates are shown in the table on the next page.

Over time, price differentials can change depending on regional workloads and local hot spots. Spon's *Updates* together with the *Market Forecast* and *Cost Update* features in *Building* magazine will keep readers informed of the latest regional developments and changes as they occur.

The regional variations shown in the following table are based on our forecast of price differentials in each of the economic planning regions at the time the book was published. Historically the adjustments do not vary very much year-on-year; however there can be wider variations within any region, particularly between urban and rural locations. The following table shows suggested adjustments required to cost plans or estimates prepared using the *Spon's A&B 2017*. Please note: these adjustments need to be applied to the total cost plan value, not to individual material or element items.

Region	Adjustmen to Measured Works Section
Outer London (SPON's 2017)	1.00
Inner London	1.06
South East	0.98
South West	0.91
East of England	0.93
East Midlands	0.90
West Midlans	0.88
Northern	0.86
North West	0.89
Yorkshire and Humberside	0.88
Wales	0.88
Scotland	0.90
Northern Ireland	0.76

Spons Architects' and Builders' Price Book
2017 Edition
(Tender Index 572)

Scotland
0.90

Northern
Ireland
0.76

North East
0.86

Yorkshire and
Humberside
0.88

North West
0.89

East Midlands
0.90

East of England
0.93

West Midlands
0.88

Wales
0.88

South East
(excluding London) 0.98

South West 0.91

Inner London 1.06
Outer London 1

PART 2

Rates of Wages and Labour

This part contains the following sections:

Quantity Surveyor's Pocket Book

2nd Edition

D Cartlidge

This second edition of the Quantity Surveyor's Pocket Book is fully updated in line with NRM1, NRM2 and JCT(11), and remains a must-have guide for students and qualified practitioners. Its focussed coverage of the data, techniques, and skills essential to the quantity surveying role make it an invaluable companion for everything from initial cost advice to the final account stage.

Key features include:

• the structure of the construction industry

• cost forecasting and feasibility studies

• measurement and quantification, with NRM2 and SMM7 examples

• estimating and bidding

• whole life costs

• contract selection

• final account procedure.

June 2012: 186x123: 440pp
Pb: 978-0-415-50110-1: £22.99

To Order: Tel: +44 (0) 1235 400524 Fax: +44 (0) 1235 400525
or Post: Taylor and Francis Customer Services,
Bookpoint Ltd, Unit T1, 200 Milton Park, Abingdon, Oxon, OX14 4TA UK
Email: book.orders@tandf.co.uk

For a complete listing of all our titles visit:
www.tandf.co.uk

Rates of Wages

CIJC Basic Rates of Pay

BUILDING INDUSTRY – ENGLAND, WALES AND SCOTLAND

AUTHORS NOTE: AT THE TIME OF COMPILING THE FOLLOWING CIJC RATES AGREEMENT HAD NOT BEEN ACHIEVED FOR 2016/2017 WAGE LEVELS. WE HAVE ASSUMED AN INCREASE OF 3.5%. WE WILL NOTIFY READERS OF THE ACTUAL RATES IN OUR UPDATE IN NOVEMBER/DECEMBER 2016

The Working Rule Agreement includes a pay structure with general, additional skilled and craft operatives. Plus rates and additional payments will be consolidated into basic pay to provide the following rates (for a normal 39 hour week) which will come into effect from the following dates:

The following basic rates of pay are assumed effective from 27 June 2016:	Rate per 39-hour week (£)	Rate per hour (£)
Craft Rate	457.34	11.73
Skill Rate 1	435.72	11.17
Skill Rate 2	419.09	10.75
Skill Rate 3	392.06	10.05
Skill Rate 4	370.44	9.50
General operative	343.83	8.82

Apprentices/Trainees

The Construction Apprenticeship Scheme (CAS) is aimed mainly at school leavers entering the construction industry for the first time, but is open to all ages and operates throughout England and Wales.

The scheme is well regarded by the industry in providing a measure of protection and commitment to apprenticeship programmes, as well as an important in getting financial support through the Grant Scheme.

CAS is owned by the construction industry and administered by CITB, the Construction Industry Training Board.

For further information telephone CAS helpdesk – 0844 875 2274.

The following basic rates of pay are assumed effective from 27 June 2016:	Rate per 39-hour week (£)	Rate per hour (£)
Year 1	190.00	4.87
Year 2	245.30	6.29
Year 3 without NVQ2	287.29	7.37
Year 3 with NVQ2	365.04	9.36
Year 3 with NVQ3	457.34	11.73
On Completion of Apprenticeship with NVQ2	457.34	11.73

Note: If an apprentice is 22 years and over, and in his/her second year of training, then the National Minimum Wage applies.

BUILDING INDUSTRY – ENGLAND, WALES AND SCOTLAND

BUILDING AND ALLIED TRADES JOINT INDUSTRIAL COUNCIL

The Building and Allied Trades Joint Industrial Council (BATJIC), the partnership between the Federation of Master Builders (FMB) and the Transport and General Workers Union (TGWU), agreed new wage rates which became effective on 27 June 2016 to 25 June 2017.

Subject to the conditions in the Working Rule Agreement the standard weekly rates of wages shall be as follows. Effective from 27 June 2016	Rate per 39-hour week (£)	Rate per hour (£)
Craft Operative (NQV3)	459.81	11.79
Craft Operative (NQV2)	395.85	10.15
Adult General Operative	351.00	9.00

For the latest wage/conditions information, go to www.fmb.org.uk.

PLUMBING AND MECHANICAL ENGINEERING SERVICES INDUSTRY

Authorized rates of wages agreed by the Joint Industry Board for the Plumbing and Mechanical Engineering Services Industry in England and Wales.

The Joint Industry Board for Plumbing and Mechanical Engineering Services in England and Wales
Brook House
Brook Street
St Neots
Huntingdon
Cambridgeshire
PE19 2HW
Tel: 01480 476 925
Fax: 01480 403 081
Email: info@jib-pmes.org.uk

Effective from 2 January 2017	Rate per 37.5-hour week (£)	Rate per hour (£)
Operatives		
Technical Plumber and Gas Service Technician	612.00	16.32
Advanced Plumber and Gas Service Engineer	551.25	14.70
Trained Plumber and Gas Service Fitter	472.88	12.61
Apprentices		
1st Year of Training	229.13	6.11
2nd Year of Training	262.88	7.01
3rd Year of Training	296.63	7.91
3rd Year of Training with NVQ Level 2	360.38	9.61
4th Year of Training	364.88	9.73
4th Year of Training with NVQ Level 3	414.38	11.05
4th Year of Training with NVQ Level 4	457.50	12.20
Adult Trainees		
1st 6 Months of Employment	369.00	9.84
2nd 6 Months of Employment	396.00	10.56
3rd 6 Months of Employment	412.13	10.99

BUILDING INDUSTRY – ENGLAND, WALES AND SCOTLAND

The Joint Industry Board for the Plumbing Industry in Scotland and Northern Ireland
Bellevue House
22 Hopetoun Street
Edinburgh
EH7 4GH
Tel: 0131 556 0600
Email: info@snijib.org

Effective from 4 July 2016	Rate per 37.5-hour week (£)	Rate per hour (£)
Operatives Plumbers & Gas Service Operatives		
Plumber and Gas Service Fitter	450.75	12.02
Advanced Plumber and Gas Service Engineer	513.00	13.68
Technician Plumber and Gas Service Technician	568.13	15.15
Plumbing Labourer	402.00	10.72
Apprentice Plumbers and Fitters		
1st Year Apprentice	130.50	3.48
2nd Year Apprentice	195.00	5.20
3rd Year Apprentice	236.25	6.30
4th Year Apprentice	304.88	8.13

Architect's Pocket Book
Fourth Edition

Ann Ross & Jonathan Hetreed

This handy pocket book brings together a wealth of useful information that architects need on a daily basis – on site or in the studio. It provides guidance on a range of tasks, from complying with the Building Regulations, including the recent revisions to Part L, to helping with planning, use of materials and detailing.

Compact and easy to use, the Architect's Pocket Book has sold well over 65,000 copies to the nation's architects, architecture students, designers and construction professionals who do not have an architectural background but need to understand the basics, fast.

This is the famous little blue book that you can't afford to be without.

February 2011: 186 x 123: 348 pp
Pb: 978-0-08-096959-6: £25.99

To Order: Tel: +44 (0) 1235 400524 Fax: +44 (0) 1235 400525
or Post: Taylor and Francis Customer Services,
Bookpoint Ltd, Unit T1, 200 Milton Park, Abingdon, Oxon, OX14 4TA UK
Email: book.orders@tandf.co.uk

For a complete listing of all our titles visit:
www.tandf.co.uk

Labour Rate Calculations

The format of this section is so arranged that, in the case of work normally undertaken by the Main Contractor, the constituent parts of the total rate are shown enabling the reader to make such adjustments as may be required in particular circumstances. Similar details have also been given for work normally sublet although it has not been possible to provide this in all instances.

As explained in the Preface, there is a facility available to readers, which enables a comparison to be made between the level of prices in this section and current tenders by means of a tender index.

To adjust prices for other regions and times, the reader is recommended to refer to the explanations and examples on how to apply these tender indices.

There follow explanations and definitions of the basis of costs in the Prices for Measured Work section under the following headings:

- Overhead charges and profit
- Labour hours and Labour £ column
- Plant £ column
- Material £ column
- Total rate £ column

Overhead and profit charges

Rates checked against winning tenders include overhead charges and profit at current levels.

Labour Hours and Labour £ columns

Labour rates are based upon typical gang costs divided by the number of primary working operatives for the trade concerned, and for general building work include an allowance for trade supervision (see below). Labour hours multiplied by Labour rate with the appropriate addition for overhead charges and profit gives Labour £. In some instances, due to variations in gangs used, Labour rate figures have not been indicated, but can be calculated by dividing Labour £ by Labour hours.

Building craft operatives and labourers

AUTHORS NOTE: AT THE TIME OF COMPILING THE FOLLOWING CIJC RATES AGREEMENT HAD NOT BEEN ACHIEVED FOR 2016/2017 WAGE LAEVEL. WE HAVE ASSUMED AN INCREASE OF 3.5%. WE WILL NOTIFY READERS OF THE ACTUAL RATES IN OUR UPDATE IN NOVEMBER/DECEMBER 2016

Effective from 27 June 2016 our calculated minimum weekly earnings in the London area for craft operatives and general operatives, rates are £457.34 and £343.83 respectively; to these rates have been added allowances for the items below in accordance with the recommended procedure of the Chartered Institute of Building in its Code of Estimating Practice. The resultant hourly rates on which the Prices for Measured Work have generally been based are £15.55 for craft operative and £11.58 for general operatives.

Calculations take into account the following:

- Lost time
- Construction Industry Training Board levy
- Holidays with pay
- Accidental injury, retirement and death benefits scheme
- Sick pay
- National insurance
- Severance pay and sundry costs
- Employer's liability and third party insurance

The table which follows illustrates how the all-in hourly rates have been calculated based upon CIJC Working Rule Agreement. Productive time has been based on a total of 1727 hours (44.3 weeks) worked per year.

			Craft Operatives		General Operatives	
			£	£	£	£
Wages at standard basic rate						
Productive time	44.3	weeks	457.34	20,259.96	343.83	15,231.81
Lost time allowance, say	0.9	weeks	457.34	411.60	343.83	309.45
Overtime	0	weeks	0	0	0	0
				20,671.56		15,541.26
Extra payments under National Working Rules	45	weeks				
CITB Levy (0.50% of payroll)	0.50	year		116.62		87.68
Holiday pay & Public holiday pay	5.80	weeks	457.34	2,652.55	343.83	1,994.23
Employer's contribution to:						
EasyBuild Stakeholder Pension	52	weeks	7.50	390.00	7.50	390.00
National Insurance (average weekly payment)	52	weeks	38.99	2,115.36	25.24	1312.54
				25,946.09		19,325.71
Severance pay and sundry costs, say	Plus		1.5%	389.19	1.5%	289.89
				26,335.19		19615.59
Employer's Liability and Third Party Insurance	Plus		2.0%	526.71	2.0%	392.31
Total cost per annum				**£26,861.99**		**£20,007.90**
Total cost per hour				**£15.55**		**£11.58**

Similar calculations for other CIJC labour classifications result in the following cost per hour:

Based on basic rates of pay that came into effect 27 June 2016:	**Rate per hour (£)**
Craft Operative	15.55
Skill Rate 1	14.79
Skill Rate 2	14.21
Skill Rate 3	13.27
Skill Rate 4	12.51
General Operative	11.58
Rates for responsibilities:	
Foreman	16.48
Ganger	12.51

Notes:

1.   Absence due to sickness has been assumed to be for periods not exceeding 3 days for which no payment is due (Working Rule 20.7.3).
2.   EasyBuild Stakeholder Pension effective from 1 July 2002. Death and accident benefit cover is provided free of charge. The minimum employer contribution shall be £5.00 per week. Where the operative contributes between £5.01 and £10.0 per week the employer shall increase the minimum contribution to match that of the operative up to a maximum of £10.0 per week.
3.   All NI Payments are at not-contracted out rates applicable from April 2016.
     National Insurance is paid for 52 weeks.
4.   Public holiday pay includes for 8 public holiday days.
5.   This table is an example showing a typical breakdown

The labour rates used in the Measured Work sections have been based on the following gang calculations which generally include an allowance for supervision by a foreman or ganger. Alternative labour rates are given showing the effect of various degrees of bonus.

Gang Build-up	Rate £/hour	Unit rate £/hour	Productive Hours	Labour rates with bonus payment £/hour			
				Normal	+10%	+20%	+30%
Groundwork Gang							
1 Ganger	12.51	12.51					
6 Labourers	11.58	69.49					
		82.00	/ 6.5 =	12.62	13.92	15.23	16.54
Concreting Gang							
1 Foreman	16.48	16.48					
4 Skilled Labourers	12.51	50.05					
		66.53	/ 4.5 =	14.78	16.31	17.84	19.36
Steel Fixing Gang							
1 Foreman	16.48	16.48					
4 Steel Fixers	15.55	62.20					
		78.68	/ 4.5 =	17.48	19.28	21.08	22.87
Formwork Gang							
1 Foreman	16.48	16.48					
10 Carpenters	15.55	155.49					
1 Labourer	11.58	11.58					
		183.55	/ 10.5 =	17.48	19.28	21.08	22.88
Bricklaying/Light Blockwork Gang							
1 Foreman	16.48	16.48					
6 Bricklayers	15.55	93.30					
1 Labourer	11.58	11.58					
		121.36	/ 6.50 =	18.67	20.59	22.51	24.43
Dense Blockwork Gang							
1 Foreman	16.48	16.48					
6 Bricklayers	15.55	93.30					
1 Labourers	11.58	11.58					
		121.36	/ 6.50 =	18.67	20.59	22.51	24.43
Carpentry/Joinery Gang							
1 Foreman	16.48	16.48					
5 Carpenters	15.55	77.75					
1 Labourer	11.58	11.58					
		105.81	/ 5.5 =	19.24	21.22	23.20	25.18
Painter, Slater, etc.							
1 Craft Operative	15.55	15.55	/ 1 =	15.55	17.15	18.75	20.35
1 and 1 Gang							
1 Craft Operative	15.55	15.55					
1 Skilled Labourer	12.51	12.51					
		28.06	/ 1 =	28.06	30.95	33.85	36.74
2 and 1 Gang							
2 Craft Operatives	15.55	31.10					
1 Skilled Labourer	12.51	12.51					
		43.61	/ 2 =	21.81	24.05	26.30	28.54
Small Labouring Gang (making good)							
1 Foreman	16.48	16.48					
4 Skilled Labourers	12.51	50.05					
		66.53	/ 4.5 =	14.78	16.31	17.84	19.36
Drain Laying Gang/Clayware							
2 Skilled Labourers	12.51	25.02	/ 2 =	12.51	13.81	15.10	16.40

Subcontractor's operatives

Similar labour rates are shown in respect of subcontractor trades where applicable.

Plumbing operatives

From 2 January 2017 the hourly earnings for technical and trained plumbers are £15.21 and £11.75 respectively; to these rates have been added allowances similar to those added for building operatives (see below). The resultant average hourly rate on which the Prices for Measured Work have been based is £18.74. The items referred to above for which allowance has been made are:

- Tool allowance
- Plumbers' welding supplement
- Holidays with pay
- Pension and welfare stamp
- National Insurance
- Severance pay and sundry costs
- Employer's liability and third party insurance

No allowance has been made for supervision as we have assumed the use of a team of technical or trained plumbers who are able to undertake such relatively straightforward plumbing works, e.g. on housing schemes, without supervision.

The table which follows shows how the average hourly rate referred to above has been calculated. Productive time has been based on a total of 1,687.50 hours worked per year.

Effective from 2 January 2017		Technical Plumber		Trained Plumber	
		Rate £	Total £	Rate £	Total £
Wages at standard basic rate					
productive time	1,687.5 hrs	16.32	27,540.00	12.61	21,279.38
Plumber's welding supplement (gas and arc)	1,687.5 hrs	0.50	843.75		0.00
			28,383.75		21,279.38
Employer's contribution to:					
Holiday credit/welfare					
Holiday pay for 30 days	30 days	122.40	3,672.00	94.58	2,837.25
Additional holiday pay (maximum)	60 credits	2.25	135.00	2.25	135.00
			32,190.75		24,251.63
Pension, say	7.5%		2,414.31		1,818.87
National Insurance	52 weeks	65.52	3,406.81	43.73	2,273.90
			38,011.86		28,344.39
Severance pay and sundry costs, say	1.5%		570.18		425.17
Liability and third party insurance	2.0%		771.64		575.39
Total cost per annum			**39,353.68**		**29,344.95**
Total cost per hour			**23.32**		**17.39**
Average all-in rate per hour				**20.36**	

Notes:

1. First 37.5 hours paid at normal rate
2. 1.5 overtime rate after 37.5 hrs up to 8 pm, after which overtime rate is at double time
3. National Insurance is payable on top-up funding as this is payable by the employer
4. Absence due to sickness has been assumed to be for periods not exceeding 3 days for which no payment is due. The entitlement is payable from and including the fourth day of illness onwards.
5. This table is an example showing a typical breakdown

LABOUR CATEGORIES

Schedule 1 to the WRA lists specified work establishing entitlement to the Skilled Operative Pay Rate 4, 3, 2, 1 or the Craft Rate as follows:

CIJC WORKING RULE AGREEMENT FOR THE CONSTRUCTION INDUSTRY – SCHEDULE 1 **Specified Work Establishing Entitlement to the Skilled Operative Pay Rate 4,3,2,1 or Craft Rate**	
BAR BENDERS AND REINFORCEMENT FIXERS	
• Bender and fixer of Concrete Reinforcement capable of reading and understanding drawings and bending schedules and able to set out work	Craft Rate
CONCRETE	
• Concrete Leveller or Vibrator Operator	4
• Screeder and Concrete Surface Finisher working from datum such as road-form, edge beam or wire	4
• Operative required to use trowel or float (hand or powered) to produce high quality finished concrete	4
DRILLING AND BLASTING	
• Drills, rotary or percussive: mobile rigs, operator of	3
• Operative attending drill rig	4
• Shotfirer, operative in control of and responsible for explosives including placing, connecting and detonating charges	3
• Operatives attending on shotfirer, including stemming	4
DRYLINERS	
• Operatives undergoing approved training in drylining	4
• Operatives who can produce a certificate of training achievement indicating satisfactory completion of at least one unit of approved drylining training	3
• Dryliners who have successfully completed their training in drylining fixing and finishing	Craft Rate
FORMWORK CARPENTERS	
• 1st year trainee	4
• 2nd year trainee	3
• Formwork Carpenters	Craft Rate
GANGERS AND TRADE CHARGEHANDS (Higher grade payments may be made at the employer's discretion)	4
GAS NETWORK OPERATIONS Operatives who have successfully completed approved training to the standard of:	
• GNO Trainee	4
• GNO Assistant	3
• Team Leader – Services	2
• Team Leader – Mains	1
• Team Leader – Mains and Services	1
LINESMEN – ERECTORS	
• 1st Grade Skilled in all works associated with the assembly, erection, maintenance and dismantling of Overhead Lines Transmission Lines on steel towers, concrete or wood poles, including all overhead lines construction elements.	2
• 2nd Grade As above but lesser degree of skill – or competent and fully skilled to carry out some of the elements of construction listed above.	3
• Linesmen-erector's mate Semi-skilled in works specified above and a general helper	4
MASON PAVIORS	
• Operative assisting a Mason Pavior undertaking kerblaying, block and sett paving, flag laying, in natural stone and precast products	4
• Operative engaged in stone pitching or dry stone walling	3

Labour Rate Calculations

LABOUR CATEGORIES

MECHANICS	
• **Maintenance Mechanic** capable of carrying out field service duties, maintenance activities and minor repairs	2
• **Plant Mechanic** capable of carrying out major repairs and overhauls including welding work, operating metal turning lathe or similar machine and using electronic diagnostic equipment	1
• **Maintenance/Plant Mechanics' Mate** on site or in depot	4
• **Tyre Fitter,** heavy equipment tyres	2

MECHANICAL PLANT DRIVERS AND OPERATORS	
Backhoe Loaders (with rear excavator bucket and front shovel and additional equipment such as blades, hydraulic hammers and patch planers)	
• Backhoe, up to and including 50 kW net engine power; driver of	4
• Backhoe, over 50 kW up to and including 100 kW net engine power; driver of	3
• Backhoe, over 100 kW net engine power; driver of	2
• **Compressors and Generators** Air compressor or generators over 10 kW; operator of	4

Concrete Mixers	
• Operative responsible for operating a concrete mixer or mortar pan up to and including 400 litres drum capacity	4
• Operative responsible for operating a concrete mixer over 400 litres and up to and including 1,500 litres drum capacity	3
• Operative responsible for operating a concrete mixer over 1,500 litres drum capacity	2
• Operative responsible for operating a mobile self-loading and batching concrete mixer up to 2,500 litres drum capacity	2
• Operative responsible for operating a mechanical drag-shovel	4

Concrete Placing Equipment	
• Trailer mounted or static concrete pumps: self-propelled concrete placers: concrete placing booms; operator of	3
• Self-propelled mobile concrete pump, with or without boom, mounted on lorry or lorry chassis; driver/operator of	2

Cranes/Mobile Cranes	
Self-propelled mobile crane on road wheels, rough terrain wheels or caterpillar tracks including lorry mounted:	
• Maximum lifting capacity at minimum radius, up to and including 5 tonne; driver of	4
• Maximum lifting capacity at minimum radius, over 5 tonne and up to and including 10 tonne; driver of	3
• Maximum lifting capacity at minimum radius, over 10 tonne	Craft Rate

Tower Cranes (including static or travelling: standard trolley or luffing jib)	
Up to and including 2 tonne max. Lifting capacity at min. radius; driver of	4
• Over 2 tonne up to and including 10 tonne max. Lifting capacity at min. radius; driver of	3
• Over 10 tonne up to and including 20 tonne max. Lifting capacity at minimum radius; driver of	2
• Over 20 tonne max. Lifting capacity at minimum radius ;driver of	1

Miscellaneous Cranes and Hoists	
• Overhead bridge crane or gantry crane up to and including 10 tonne capacity; driver of	3
• Overhead bridge crane or gantry crane over 10 tonne up to and including 20 tonne capacity; driver of	2
• Power driven hoist or jib crane; operator of	4
• **Slinger/Signaller** appointed to attend Crane or hoist to be responsible for fastening or slinging loads and generally to direct lifting operations	4

Dozers	
• Crawler dozer with standard operating weight up to and including 10 tonne; driver of	3
• Crawler dozer with standard operating weight over 10 tonne and up to and including 50 tonne; driver of	2
• Crawler dozer with standard operating weight over 50 tonne; driver of	1

Dumpers and Dump Trucks	
• Up to and including 10 tonne rated payload; driver of	4
• Over 10 tonne and up to and including 20 tonne rated payload; driver of	3
• Over 20 tonne and up to and including 50 tonne rated payload; driver of	2
• Over 50 tonne and up to and including 100 tonne rated payload; driver of	1
• Over 100 tonne rated payload; driver of	Craft Rate

LABOUR CATEGORIES

Excavators (360° slewing)	
• Excavators with standard operating weight up to and including 10 tonne; driver of	3
• Excavator with standard operating weight over 10 tonne and up to and including 50 tonne; driver of	2
• Excavator with standard operating weight over 50 tonne; driver of	1
• **Banksman** appointed to attend excavator or responsible for positioning vehicles during loading or tipping	4
Fork-Lifts Trucks and Telehandlers	
• Smooth or rough terrain fork lift trucks (including side loaders) and telehandlers up to and including 3 tonne lift capacity; driver of	4
• Over 3 tonne lift capacity; driver of	3
Motor Graders: driver of	2
Motorized Scrapers: driver of	2
Motor Vehicles (Road Licensed Vehicles) Driver and Vehicle Licensing Agency (DVLA)	
• Vehicles requiring a driving licence of category C1; driver of (Goods vehicle with maximum authorized mass (mam) exceeding 3.5 tonne but not exceeding 7.5 tonne and including such a vehicle drawing a trailer with a mam not over 750 kg)	4
• Vehicles requiring a driving licence of category C; driver of (Goods vehicle with a maximum authorized mass (mam) exceeding 3.5 tonne and including such a vehicle drawing a trailer with mam not over 750 kg)	2
• Vehicles requiring a driving licence of category C plus E; driver of (Combination of a vehicle in category C and a trailer with maximum authorized mass over 750 kg)	1
Power Driven Tools	
• Operatives using power-driven tools such as breakers, percussive drills, picks and spades, rammers and tamping machines	4
Power Rollers	
• Roller, up to and including 4 tonne operating weight; driver of	4
• Roller, over 4 tonne operating weight and upwards; driver of	3
Pumps, Power-driven pump(s); attendant of	4
Shovel Loaders, (Wheeled or tracked, including skid steer)	
• Up to and including 2 cubic metre shovel capacity; driver of	4
• Over 2 cubic metre and up to and including 5 cubic metre shovel capacity; driver of	3
• Over 5 cubic metre shovel capacity; driver of	2
Tractors (Wheeled or Tracked)	
• Tractor, when used to tow trailer and/or with mounted compressor, up to and including 100 kW rated engine power; driver of	4
• Tractor, ditto, over 100 kW up to and including 250 kW rated engine power; driver of	3
• Tractor, ditto, over 250 kW rated engine power; driver of	2
Trenchers (Type wheel, chain or saw)	
• Trenching Machine, up to and including 50 kW gross engine power; driver of	4
• Trenching Machine, over 50 kW and up to and including 100 kW gross engine power; driver of	3
• Trenching Machine, over 100 kW gross engine power; driver of	2
Winches, Power driven winch; driver of	4
PILING	
• General Skilled Piling Operative	4
• Piling Chargehand/Ganger	3
• Pile Tripod Frame Winch Driver	3
• CFA or Rotary or Driven Mobile Piling Rig Driver	2
• Concrete Pump Operator	3

LABOUR CATEGORIES

PIPE JOINTERS
- Jointers, pipes up to and including 300 mm diameter — 4
- Jointers, pipes over 300 mm diameter and up to 900 mm diameter — 3
- Jointers, pipes over 900 mm diameter — 2
- **except** in HDPE mains when experienced in butt fusion and/or electrofusion jointing operations — 2

PIPELAYERS
- Operative preparing the bed and laying pipes up to and including 300 mm diameter — 4
- Operative preparing the bed and laying pipes over 300 mm diameter and up to and including 900 mm diameter — 3
- Operative preparing the bed and laying pipes over 900 mm diameter — 2

PRE-STRESSING CONCRETE
- Operative in control of and responsible for hydraulic jacks and other tensioning devices engaged in post-tensioning and/or pre-tensioning concrete elements — 3

ROAD SURFACING WORK (includes rolled asphalt, dense bitumen macadam and surface dressings)
Operatives employed in this category of work to be paid as follows:
- Chipper — 4
- Gritter Operator — 4
- Raker — 3
- Paver Operator — 3
- Leveller on Paver — 3
- Road Planer Operator — 3
- Road Roller Driver, 4 tonne and upwards — 3
- Spray Bar Operator — 4

TIMBERMAN
- Timberman, installing timber supports — 3
- Highly skilled timber man working on complex supports using timbers of size 250 mm by 125 mm and above — 2
- Operative attending — 4

WELDERS
- Grade 4 (Fabrication Assistant)
- Welder able to tack weld using SMAW or MIG welding processes in accordance with verbal instructions and including mechanical preparation such as cutting and grinding — 3
- Grade 3 (Basic Skill Level)
- Welder able to weld carbon and stainless steel using at least one of the following processes SMAW, GTAW, GMAW for plate-plate fillet welding in all major welding positions, including mechanical pre-paration and complying with fabrication drawings — 2
- Grade 2 (Intermediate Skill Level)
- Welder able to weld carbon and stainless steel using manual SMAW, GTAW, semi-automatic MIG or MAG, and FCAW welding processes including mechanical preparation, and complying with welding procedures, specifications and fabrication drawings. — 1
- Grade 1 (Highest Skill Level)
- Welder able to weld carbon and stainless steel using manual SMAW, GTAW, semi-automatic GMAW or MIG or MAG, and FCA w elding processes in all modes and directions in accordance with BSEN287-1 and/or 287-2 Aluminium Fabrications including mechanical preparation and complying with welding procedures, specifications and fabrication drawings. — Craft Rate

YOUNG WORKERS
- Operatives below 18 years of age will receive payment 60% of the General Operative Basic Rate.
- At 18 years of age or over the payment is 100% of the relevant rate.

PART 3

Approximate Estimating

This part contains the following sections:

Construction Materials Reference Book, 2nd Edition

Edited by David Doran & Bob Cather

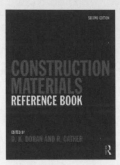

Fully updated to reflect the latest materials and their applications, this second edition of the Construction Materials Reference Book remains the definitive reference source for professionals involved in the conception, design and specification stages of a construction project. The theory and practical aspects of each material are covered in detail, with an emphasis being placed on properties and appropriate use, enabling a deeper understanding of each material and greater confidence in their application.

Containing 38 chapters written by subject specialists, a wide range of construction materials are covered, from traditional materials such as stone through masonry and steel to advanced plastics and composites.

With diagrams, reference tables, chemical and mathematic formulae, and summaries of the appropriate regulations throughout, this is the most authoritative construction materials guide available. This edition features extra material on environmental issues, whole life costing, and sustainability, as well as the health and safety aspects of both use and installation.

July 2013: 246 x 189: 488 pp Hb:
978-0-7506-6376-2: £105.00

Building Prices per Functional Unit

Prices given under this heading are given as a typical range of prices for new work. These should not be interpreted as a minimum or maximum range.

Prices include for Preliminaries, and Overheads and Profit recovery.

Unless otherwise stated, prices do not allow for external works, furniture or special equipment.

On certain types of buildings there exists a close relationship between its cost and the number of functional units that it accommodates. During the very early stages of a project an approximate estimate can be quickly derived by multiplying the proposed unit of accommodation (i.e. hotel bedrooms, car parking spaces etc.) by an appropriate rate.

The following indicative unit areas and costs have been derived from historical data and construction standards. It should be emphasized that the prices must be treated with reserve, as they represent the average of prices from our records and cannot provide more than a rough guide to the cost of a building. There are limitations when using this method of estimating, for example, the functional areas and costs of football stadia, hotels, restaurants etc. are strongly influenced by the extent of front and back of house facilities housed within it, and these areas can vary considerably from scheme to scheme.

The areas may also be used as a 'rule of thumb' in order to check on economy of designs. Where we have chosen not to show indicative areas, this is because either ranges are extensive or such figures may be misleading.

Costs have been expressed within a range, although this is not to suggest that figures outside this range will not be encountered, but simply that the calibre of such a type of building can itself vary significantly.

For assistance with the compilation of an *Order of Cost Estimate*, or of an *Elemental Cost Plan*, the reader is directed to the *Building Prices per Square Metre*, *Approximate Estimating Rates* and *Cost Models* sections of this book.

As elsewhere in this edition, prices do not include Value Added Tax or fees for professional services.

Approximate Estimating Rates

BUILDING PRICES PER FUNCTIONAL UNIT

Building type	Function	Range £		
Utilities, Civil Engineering facilities				
Car Parking				
surface car parking; 20 to 22 m^2/car	car	1800.00	to	2500.00
surface car parking; landscaped 20 to 22 m^2/car	car	2325.00	to	3200.00
multi-storey car parks grade & upper level; 24 to 28 m^2/car	car	10000.00	to	14000.00
multi-storey car parks flat slab; 24 to 28 m^2/car	car	13000.00	to	18000.00
underground car parks partially underground under buildings; natural ventilation; 27 to 30 m^2/car	car	14000.00	to	19500.00
underground car parks completely underground with landscaped roof; mechanical ventilation; 28 to 32 m^2/car	car	32000.00	to	44000.00
underground car parks completely underground under buildings; mechanical ventilation; 28 to 32 m^2/car	car	26000.00	to	36500.00
Offices				
business park; medium quality, non-air conditioned	person	11500.00	to	16000.00
low rise; air conditioned; high quality speculative; 4–8 storey	person	14000.00	to	19500.00
medium rise; air conditioned; high quality speculative; 8–20 storeys	person	21500.00	to	30000.00
Health and Welfare facilities				
district hospitals; 65 to 85 m^2/bed	bed	150000.00	to	210000.00
hospital teaching centres; 120+ m^2/bed	bed	175000.00	to	240000.00
private hospitals; 75 to 100 m^2/bed	bed	185000.00	to	260000.00
residential homes; 40 to 60 m^2/bedroom	bed	61000.00	to	84000.00
nursing homes; 40 to 60 m^2/bedroom	bed	81000.00	to	110000.00
Recreational facilities				
Theatres				
large – over 500 seats	seat	25000.00	to	34500.00
studio/workshop – less than 500 seats	seat	12500.00	to	17500.00
refurbishment	seat	10000.00	to	14000.00
Sports halls				
indoor bowls halls	rink	150000.00	to	210000.00
squash courts	court	85000.00	to	120000.00
Education facilities				
Schools				
nursery schools; 3 to 5 m^2/pupil	pupil	6700.00	to	9400.00
secondary/middle schools; 6 to 10 m^2/pupil	pupil	17000.00	to	23000.00
special schools; 18 to 20 m^2/pupil	pupil	22500.00	to	31000.00
Universities				
arts buildings	student	13500.00	to	19000.00
computer buildings	student	18000.00	to	25000.00
science buildings	student	18000.00	to	25000.00
laboratories	student	35000.00	to	49000.00
Residential facilities				
Housing				
detached; four bedroom; 100 to 140 m^2 gifa	house	150000.00	to	210000.00
semi-detached; three bedroom; 70 to 90 m^2 gifa	house	92000.00	to	130000.00
terraced; 2 bedroom; 55 to 65 m^2 gifa	house	59000.00	to	82000.00
low rise flats; two bedroom; 55 to 65 m^2 gifa	flat	71000.00	to	98000.00
medium rise flats; two bedroom; 55 to 65 m^2 gifa	flat	86000.00	to	120000.00

BUILDING PRICES PER FUNCTIONAL UNIT

Building type	Function	Range £		
Hotels				
luxury; city centre with conference and wet leisure facilities; 70 to 120 m²/bedroom	bed	385000.00	to	530000.00
business; town centre with conference and wet leisure facilities; 70 to 100 m²/bedroom	bed	235000.00	to	325000.00
mid-range; town centre with conference and leisure facilities; 50 to 60 m²/bedroom	bed	110000.00	to	150000.00
budget; city centre with dining and bar facilities; 35 to 45 m²/bedroom	bed	72000.00	to	100000.00
budget; roadside excluding dining facilities; 28 to 35 m²/bedroom	bed	48500.00	to	68000.00
Student residences				
large budget schemes; 200+ units with ensuite accommodation; 18 to 20 m²/bedroom	bed	24500.00	to	34000.00
smaller schemes (40–100 units) with mid range specifications; some with ensuite accommodation; 19 to 24 m²/bedroom	bed	33000.00	to	46000.00
smaller high quality courtyard schemes, college scheme; 24 to 28 m²/bedroom	bed	53000.00	to	73000.00

The Passivhaus Designer's Manual: A technical guide to low and zero energy buildings

Edited by Christina J. Hopfe, Robert S. McLeod

Passivhaus is the fastest growing energy performance standard in the world, with almost 50,000 buildings realised to date. Applicable to both domestic and non-domestic building types, the strength of Passivhaus lies in the simplicity of the concept. As European and global energy directives move ever closer towards Zero (fossil) Energy standards, Passivhaus provides a robust 'fabric first' approach from which to make the next step.

Lavishly illustrated with nearly 200 full colour illustrations, and presented by two highly experienced specialists, this is your one-stop shop for comprehensive practical information on Passivhaus and Zero Energy buildings

Oct 2015: 246X189 mm: 346 pp
Pb: 978-0-415-52269-4 £34.99

Taylor & Francis
Taylor & Francis Group

Building Prices per Square Metre

Prices given under this heading are given as a typical range of prices for new work for a range of typical building types.

Prices in this include for Preliminaries and Overheads and Profit recovery.

Unless otherwise stated, prices do not allow for external works, furniture or special equipment.

Prices are based upon the total gross internal area (GIA) of all storeys measured in accordance with the 6th edition of the *RICS Code of Measuring Practice*, September 2007.

As in previous editions it is emphasized that the prices must be treated with a degree of reserve as they represent the average of prices from historic records and should be used to provide only an approximate order of cost for a building.

Costs have been expressed within a range, although this is not to suggest that figures outside this range will not be encountered, but simply that the calibre of such a type of building can itself vary significantly.

Prices can vary greatly between regions and even between individual sites. Professional interpretation will always need to be applied in the final analysis.

For assistance with the compilation of an *Order of Cost Estimate*, or of an *Elemental Cost Plan*, the reader is directed to the *Cost Models* and *Approximate Estimating Rates* sections in the book.

As elsewhere in this edition, prices do not include Value Added Tax or fees for professional services.

AUTHORS NOTE

RICS Property Measurement

Overview

The Royal Institution of Chartered Surveyors (RICS) has launched 'RICS Property Measurement', which in part, replaces the *Code of Measuring Practice* for office buildings. This came into effect on 1 January 2016.

The RICS Property Measurement incorporates the International Property Measurement Standards (IPMS) for offices, whilst the *Code of Measuring Practice* 6th Edition remains in effect for all other building types.

IPMS was formed in 2013 at the World Bank, with the objective to provide global standardization for measurement of the built environment. It now comprises of more than 70 organizations including RICS. Each member organization is responsible for setting domestic measurement standards. Having successfully launched IPMS office buildings, they are currently working on IPMS residential buildings, followed by industrial, retail and mixed use.

TImetable

RICS Property Measurement was launched in May 2015. The use of IPMS office buildings becomes mandatory for RICS Members from 1 January 2016, unless clients instruct otherwise or local laws specify the use of a different standard.

Proposed Changes from *Code of Measuring Practice* 6th edition: Office Buildings

The first key change is the use of new vocabulary. Gross External Area (GEA), Gross Internal Area (GIA) or Net Internal Area (NIA) are no longer used. This terminology has been replaced with IPMS 1, IPMS 2 and IPMS 3. Whilst these are not direct replacements they can be closely related to GEA, GIA and NIA respectively.

BUILDING PRICES PER SQUARE METRE

The second key change is the concept that whilst IPMS 1, IPMS 2 and IPMS 3 are more inclusive in measurement then GEA, GIA and NIA, these additional measures need to be stated separately. To ensure transparency, marked up plans should accompany measurement reports, which should be detailed to include the breakdown of the measure, this should provide benefit to the end user.

The third key change is the concept of limited use area. Whilst these areas are included in the measurement, they are then stated separately. They include areas with limited height (less than 1.5 metres), areas with limited natural light, internal structural walls/columns, area difference between internal dominant face and the wall/floor junction.

IPMS 1

The IPMS 1 measure includes the area of the building incorporating the external walls. The key differences between IPMS 1 and GEA are the inclusion of balconies under IPMS 1. However covered galleries and balconies are stated separately.

IPMS 2

IPMS 2 is measured to the internal dominant face. The internal dominant face is defined as the 'surface comprising more than 50% of the surface area for each vertical section forming an internal perimeter'. So for example, for an office building with unitized cladding, this would mean the glass surface.

Additional areas which would not be included in a GIA measure but is included under IPMS 2 are the balconies and covered canopies. These would then be stated separately.

IPMS 3

The most significant changes appear in the comparison between IPMS 3 and NIA. IPMS 3 includes area taken by columns, piers/external columns and window reveals. All of these are excluded under NIA. Additionally space taken by balconies, covered galleries, roof terraces are all included but stated separately. Lift lobbies and area with headroom of less than 1.5 metres are also included but stated separately as limited use area.

It will take the wider market some time to wake up to this change, especially with the impact on end user market for valuation purposes, as it will have a direct impact on their comparable databases. There will be a period where the existing GEA, GIA and NIA will remain in favour and these are still being used in this book.

BUILDING PRICES PER SQUARE METRE

Item	Unit	Range £		
UNICLASS D1 UTILITIES, CIVIL ENGINEERING FACILITIES				
Car parking				
surface car parking	m^2	75.00	to	94.00
surface car parking; landscaped	m^2	96.00	to	120.00
Multi-storey car parks				
grade & upper level	m^2	350.00	to	435.00
flat slab	m^2	440.00	to	550.00
Underground car parks				
partially underground under buildings; naturally ventilated	m^2	440.00	to	550.00
completely underground under buildings with mechanical ventilation	m^2	780.00	to	980.00
completely underground with landscaped roof and mechanical ventilation	m^2	950.00	to	1175.00
Transport facilities				
railway stations	m^2	2400.00	to	3000.00
bus and coach stations	m^2	2275.00	to	2900.00
bus garages	m^2	870.00	to	1075.00
petrol stations	m^2	1925.00	to	2400.00
Vehicle showrooms with workshops, garages etc.				
up to 2,000 m^2	m^2	1125.00	to	1400.00
over 2,000 m^2	m^2	960.00	to	1200.00
Vehicle showrooms without workshops, garages etc.				
up to 2,000 m^2	m^2	1125.00	to	1400.00
Vehicle repair and maintenance buildings				
up to 500 m^2	m^2	1650.00	to	2075.00
over 500 m^2 up to 2000 m^2	m^2	1125.00	to	1400.00
car wash buildings	m^2	960.00	to	1200.00
Airport facilities (excluding aprons)				
airport terminals	m^2	2950.00	to	3700.00
airport piers/satellites	m^2	3100.00	to	3900.00
Airport campus facilities				
cargo handling bases	m^2	740.00	to	930.00
distribution centres	m^2	420.00	to	520.00
hangars (type C and D facilities)	m^2	1275.00	to	1600.00
TV, radio and video studios	m^2	1550.00	to	1950.00
Telephone exchanges	m^2	1125.00	to	1400.00
Telephone engineering centres	m^2	760.00	to	950.00
Branch post offices	m^2	1125.00	to	1400.00
Postal delivery offices/sorting offices	m^2	1125.00	to	1400.00
Electricity substations	m^2	1600.00	to	2000.00
Garages, domestic	m^2	680.00	to	860.00
UNICLASS D2 INDUSTRIAL FACILITIES				
B1 Light industrial/offices buildings				
economical shell, and core with heating only	m^2	680.00	to	850.00
medium shell and core with heating and ventilation	m^2	960.00	to	1200.00
high quality shell and core with air conditioning	m^2	1475.00	to	1875.00
developers Category A fit-out	m^2	550.00	to	690.00
tenants Category B fit-out	m^2	360.00	to	455.00
Agricultural storage	m^2	580.00	to	720.00

BUILDING PRICES PER SQUARE METRE

Item	Unit	Range £		
UNICLASS D2 INDUSTRIAL FACILITIES – cont				
Factories				
for letting (incoming services only)	m²	590.00	to	740.00
for letting (including lighting, power and heating)	m²	750.00	to	940.00
nursery units (including lighting, power and heating)	m²	750.00	to	940.00
workshops	m²	700.00	to	870.00
maintenance/motor transport workshops	m²	910.00	to	1125.00
owner occupation for light industrial use	m²	960.00	to	1200.00
owner occupation for heavy industrial use	m²	1600.00	to	2000.00
Factory/office buildings high technology production				
for letting (shell and core only)	m²	630.00	to	790.00
for owner occupation (controlled environment fully finished)	m²	860.00	to	1075.00
High technology laboratory workshop centres, air conditioned	m²	2700.00	to	3400.00
Warehouse and distribution centres				
high bay (10–15m high) for owner occupation (no heating) up to 10,000 m²	m²	305.00	to	380.00
high bay (10–15m high) for owner occupation (no heating) over 10,000 m² up to 20,000 m²	m²	230.00	to	290.00
high bay (16–24m high) for owner occupation (no heating) over 10,000 m² up to 20,000 m²	m²	335.00	to	420.00
high bay (16–24m high) for owner occupation (no heating) over 20,000 m²	m²	280.00	to	350.00
Fit out cold stores, refrigerated stores inside warehouse	m²	610.00	to	760.00
Industrial buildings				
Shell with heating to office areas only				
500–1,000 m²	m²	500.00	to	620.00
1,000–2,000 m²	m²	420.00	to	530.00
greater than 2,000 m²	m²	460.00	to	570.00
Unit including services to production area				
500–1,000 m²	m²	690.00	to	860.00
1,000–2,000 m²	m²	630.00	to	780.00
greater than 2,000 m²	m²	630.00	to	780.00
UNICLASS D3 ADMINISTRATIVE, COMMERCIAL, PROTECTIVE SERVICES FACILITIES				
Embassies	m²	2025.00	to	2500.00
County courts	m²	2450.00	to	3100.00
Magistrates courts	m²	1325.00	to	1675.00
Civic offices				
non-air conditioned	m²	1325.00	to	1675.00
fully air conditioned	m²	1600.00	to	2000.00
Probation/registrar offices	m²	1025.00	to	1275.00
Offices for owner occupation				
low rise, air conditioned	m²	1450.00	to	1800.00
medium rise, air conditioned	m²	1825.00	to	2275.00
high rise, air conditioned	m²	2150.00	to	2700.00
Offices – City and West End				
To developers Cat A finish only				
high quality, speculative 8–20 storeys, air conditioned	m²	2025.00	to	2500.00
high rise, air conditioned, iconic speculative towers	m²	3100.00	to	3900.00

BUILDING PRICES PER SQUARE METRE

Item	Unit	Range £		
Business park offices				
Generally up to 3 storeys only				
functional non-air conditioned less than 2,000 m^2	m^2	1175.00	to	1475.00
functional non-air conditioned more than 2,000 m^2	m^2	1225.00	to	1525.00
medium quality non-air conditioned less than 2,000 m^2	m^2	1275.00	to	1600.00
medium quality non-air conditioned more than 2,000 m^2	m^2	1375.00	to	1700.00
medium quality air conditioned less than 2,000 m^2	m^2	1475.00	to	1875.00
medium quality air conditioned more than 2,000 m^2	m^2	1600.00	to	2000.00
good quality – naturally ventilated to meet BCO specification (exposed soffits, solar shading) less than 2,000 m^2	m^2	1450.00	to	1800.00
good quality – naturally ventilated to meet BCO specification (exposed soffits, solar shading) more than 2,000 m^2	m^2	1475.00	to	1875.00
high quality air conditioned less than 2,000 m^2	m^2	1825.00	to	2275.00
high quality air conditioned more than 2,000 m^2	m^2	1875.00	to	2350.00
Large trading floors in medium sized offices	m^2	2950.00	to	3700.00
Two storey ancillary office accommodation to warehouses/factories	m^2	1175.00	to	1475.00
Fitting out offices (NIA – Net Internal Area)				
City and West End				
basic fitting out including carpets, decorations, partitions and services	m^2	530.00	to	670.00
good quality fitting out including carpets, decorations, partitions and services	m^2	590.00	to	740.00
high quality fitting out including carpets, decorations, partitions, ceilings, furniture, air conditioning and electrical services	m^2	640.00	to	800.00
reception areas	m^2	2700.00	to	3350.00
conference suites	m^2	2550.00	to	3200.00
meeting areas	m^2	3200.00	to	4000.00
back of house/storage	m^2	590.00	to	740.00
sub-equipment room	m^2	1700.00	to	2150.00
kitchen	m^2	3750.00	to	4700.00
Out-of-town (South East England)				
basic fitting out including carpets, decorations, partitions and services	m^2	430.00	to	530.00
good quality fitting out including carpets, decorations, partitions and services	m^2	480.00	to	600.00
high quality fitting out including carpets, decorations, partitions, ceilings, furniture, air conditioning and electrical services	m^2	590.00	to	740.00
reception areas	m^2	1600.00	to	2000.00
conference suites	m^2	2175.00	to	2700.00
meeting areas	m^2	2150.00	to	2700.00
back of house/storage	m^2	480.00	to	600.00
sub-equipment room	m^2	1700.00	to	2150.00
kitchen	m^2	2450.00	to	3100.00
Office refurbishment (including developers finish – (Gross Internal Floor Area – GIFA; Central London)				
minor refurbishment	m^2	500.00	to	630.00
medium refurbishment	m^2	1075.00	to	1325.00
major refurbishment	m^2	1650.00	to	2075.00
Banks and Building Societies				
banks; local	m^2	1550.00	to	1950.00
banks; city centre/head office	m^2	2175.00	to	2700.00
building society branches	m^2	1475.00	to	1875.00
building society; refurbishment	m^2	960.00	to	1200.00

BUILDING PRICES PER SQUARE METRE

Item	Unit	Range £		
UNICLASS D3 ADMINISTRATIVE, COMMERCIAL, PROTECTIVE SERVICES FACILITIES – cont				
Shop shells				
small	m²	650.00	to	810.00
large, including department stores and supermarkets	m²	590.00	to	740.00
Fitting out shell for small shop (including shop fittings)				
simple store	m²	610.00	to	770.00
designer/fashion store	m²	1225.00	to	1525.00
Fitting out shell for department store or supermarket	m²	1775.00	to	2200.00
Retail Warehouses				
shell	m²	450.00	to	560.00
fitting out, including all display and refrigeration units, check outs and IT systems	m²	270.00	to	335.00
Hypermarkets/Supermarkets				
shell	m²	540.00	to	670.00
supermarket fit-out	m²	1075.00	to	1325.00
hypermarket fit-out	m²	750.00	to	940.00
Shopping Centres				
Malls, including fit-out				
comfort cooled	m²	3400.00	to	4300.00
air conditioned	m²	4400.00	to	5500.00
food court	m²	3800.00	to	4750.00
factory outlet centre – enclosed	m²	3100.00	to	3900.00
factory outlet centre – open	m²	620.00	to	780.00
anchor tenants; capped off services	m²	1025.00	to	1275.00
medium/small units; capped off services	m²	960.00	to	1200.00
centre management	m²	2175.00	to	2700.00
enclosed surface level service yard	m²	1550.00	to	1950.00
landlords back of house and service corridors	m²	1550.00	to	1950.00
Refurbishment				
mall; limited scope	m²	1125.00	to	1400.00
mall; comprehensive	m²	1600.00	to	2000.00
Rescue/aid facilities				
ambulance stations	m²	960.00	to	1200.00
ambulance control centres	m²	1475.00	to	1875.00
fire stations	m²	1600.00	to	2000.00
Police stations	m²	1550.00	to	1950.00
Prisons	m²	1925.00	to	2400.00
UNICLASS D4 MEDICAL, HEALTH AND WELFARE FACILITIES				
District hospitals	m²	1700.00	to	2150.00
refurbishment	m²	1275.00	to	1600.00
Hospices	m²	1450.00	to	1800.00
Private hospitals	m²	1775.00	to	2200.00
Pharmacies	m²	1925.00	to	2400.00
Hospital laboratories	m²	2150.00	to	2700.00
Ward blocks	m²	1600.00	to	2000.00
refurbishment	m²	960.00	to	1200.00
Geriatric units	m²	1600.00	to	2000.00
Psychiatric units	m²	1475.00	to	1875.00

BUILDING PRICES PER SQUARE METRE

Item	Unit	Range £		
Psycho-geriatric units	m^2	1450.00	to	1800.00
Maternity units	m^2	1600.00	to	2000.00
Operating theatres	m^2	2150.00	to	2700.00
Outpatients/casualty units	m^2	1600.00	to	2000.00
Hospital teaching centres	m^2	1375.00	to	1725.00
Health centres	m^2	1275.00	to	1600.00
Welfare centres	m^2	1375.00	to	1725.00
Day centres	m^2	1275.00	to	1600.00
Group practice surgeries	m^2	1450.00	to	1800.00
Homes for the mentally handicapped	m^2	1175.00	to	1475.00
Homes for the physically handicapped	m^2	1325.00	to	1675.00
Geriatric day hospital	m^2	1475.00	to	1875.00
Accommodation for the elderly				
residential homes	m^2	960.00	to	1200.00
nursing homes	m^2	1275.00	to	1600.00
Homes for the aged	m^2	1475.00	to	1875.00
refurbishment	m^2	800.00	to	1000.00
Observation and assessment units	m^2	1075.00	to	1325.00
Primary Health Care				
doctors surgery – basic	m^2	1125.00	to	1400.00
doctors surgery/medical centre	m^2	1325.00	to	1675.00
Hospitals				
diagnostic and treatment centres	m^2	2700.00	to	3400.00
acute services hospitals	m^2	2600.00	to	3300.00
radiotherapy and ocology units	m^2	2700.00	to	3400.00
community hospitals	m^2	2175.00	to	2700.00
trauma unit	m^2	2025.00	to	2500.00
UNICLASS D5 RECREATIONAL FACILITIES				
Public houses	m^2	1600.00	to	2000.00
Dining blocks and canteens	m^2	1275.00	to	1600.00
Restaurants	m^2	1450.00	to	1800.00
Community centres	m^2	1375.00	to	1725.00
General purpose halls	m^2	1175.00	to	1475.00
Visitors' centres	m^2	1925.00	to	2400.00
Youth clubs	m^2	1600.00	to	2000.00
Arts and drama centres	m^2	1700.00	to	2150.00
Theatres, including seating and stage equipment				
large – over 500 seats	m^2	3600.00	to	4550.00
studio/workshop – less than 500 seats	m^2	2400.00	to	3000.00
refurbishment	m^2	1925.00	to	2400.00
Concert halls, including seats and stage equipment	m^2	2700.00	to	3400.00
Cinema				
shell	m^2	860.00	to	1075.00
multiplex; shell only	m^2	1700.00	to	2150.00
fitting out, including all equipment, air conditioned	m^2	1075.00	to	1325.00
Exhibition centres	m^2	1600.00	to	2000.00

BUILDING PRICES PER SQUARE METRE

Item	Unit	Range £		
UNICLASS D5 RECREATIONAL FACILITIES – cont				
Swimming pools				
international standard	m²	3600.00	to	4550.00
local authority standard	m²	2400.00	to	3000.00
school standard	m²	1125.00	to	1400.00
leisure pools, including wave making equipment	m²	2800.00	to	3550.00
Ice rinks	m²	1275.00	to	1600.00
Rifle ranges	m²	1075.00	to	1325.00
Leisure centres				
dry	m²	1475.00	to	1875.00
extension to hotels; shell and fit-out, including small pool	m²	2150.00	to	2700.00
wet and dry	m²	2250.00	to	2800.00
Sports halls including changing rooms	m²	1025.00	to	1275.00
School gymnasiums	m²	960.00	to	1200.00
Squash courts	m²	1025.00	to	1275.00
Indoor bowls halls	m²	720.00	to	900.00
Bowls pavilions	m²	880.00	to	1100.00
Health and fitness clubs	m²	1475.00	to	1875.00
Sports pavilions	m²	1075.00	to	1325.00
changing only	m²	1225.00	to	1525.00
social and changing	m²	1225.00	to	1525.00
Clubhouses	m²	1025.00	to	1275.00
UNICLASS D6 RELIGIOUS FACILITIES				
Temples, mosques, synagogues	m²	1375.00	to	1725.00
Churches	m²	1275.00	to	1600.00
Mission halls, meeting houses	m²	1375.00	to	1725.00
Convents	m²	1475.00	to	1875.00
Crematoria	m²	1700.00	to	2150.00
Mortuaries	m²	2350.00	to	2950.00
UNICLASS D7 EDUCATION, SCIENTIFIC AND INFORMATION FACILITIES				
Nursery schools	m²	1275.00	to	1600.00
Primary/junior schools	m²	1325.00	to	1675.00
Secondary/middle schools	m²	1600.00	to	2000.00
Secondary schools and further education colleges				
classrooms	m²	1075.00	to	1325.00
laboratories	m²	1175.00	to	1475.00
craft design and technology	m²	1175.00	to	1475.00
music	m²	1325.00	to	1675.00
Extensions to schools				
classrooms	m²	1275.00	to	1600.00
laboratories	m²	1325.00	to	1675.00
Sixth form colleges	m²	1275.00	to	1600.00
Special schools	m²	1075.00	to	1325.00
Training colleges	m²	1075.00	to	1325.00
Management training centres	m²	1375.00	to	1725.00

BUILDING PRICES PER SQUARE METRE

Item	Unit	Range £		
Universities				
arts buildings	m²	1175.00	to	1475.00
science buildings	m²	1450.00	to	1800.00
College/University Libraries	m²	1175.00	to	1475.00
Laboratories and offices, low level servicing	m²	2250.00	to	2800.00
Computer buildings	m²	1925.00	to	2400.00
Museums and art galleries				
national standard museum	m²	5200.00	to	6500.00
national standard independent specialist museum, excluding fit-out	m²	3600.00	to	4550.00
regional, including full air conditioning	m²	3100.00	to	3900.00
local, including full air conditioning	m²	2275.00	to	2900.00
conversion of existing warehouse to regional standard museum	m²	1450.00	to	1800.00
conversion of existing warehouse to local standard museum	m²	1225.00	to	1525.00
Galleries				
international standard art gallery	m²	3150.00	to	3950.00
national standard art gallery	m²	2400.00	to	3000.00
independent commercial art gallery	m²	1325.00	to	1675.00
Arts and drama centre	m²	1275.00	to	1600.00
Learning resource centre				
economical	m²	1175.00	to	1475.00
high quality	m²	1475.00	to	1875.00
Libraries				
regional; 5,000 m²	m²	2700.00	to	3350.00
national; 15,000 m²	m²	3500.00	to	4350.00
international; 20,000 m²	m²	4550.00	to	5700.00
Conference centres	m²	1875.00	to	2350.00
UNICLASS D8 RESIDENTIAL FACILITIES				
Local Authority and Housing Association schemes				
Housing Asociation Developments (Code for Sustainable Homes Level 3)				
Bungalows				
semi-detached	m²	1025.00	to	1275.00
terraced	m²	980.00	to	1225.00
Two storey housing				
detached	m²	1025.00	to	1275.00
semi-detached	m²	980.00	to	1225.00
terraced	m²	860.00	to	1075.00
Three storey housing				
semi-detached	m²	1025.00	to	1275.00
terraced	m²	860.00	to	1075.00
Apartments/flats				
low rise	m²	1025.00	to	1275.00
medium rise	m²	1275.00	to	1575.00
Sheltered housing with wardens accommodation	m²	980.00	to	1225.00
Private Developments				
single detached houses	m²	1375.00	to	1725.00
two and three storey houses	m²	1425.00	to	1775.00

BUILDING PRICES PER SQUARE METRE

Item	Unit	Range £		
UNICLASS D8 RESIDENTIAL FACILITIES – cont				
Private Developments – cont				
Apartments/flats generally				
standard quality; 3–5 storeys	m^2	1725.00	to	2150.00
warehouse/office conversion to apartments	m^2	1875.00	to	2375.00
high quality apartments in residential tower – Inner London	m^2	2275.00	to	2900.00
Hotels (including fittings, furniture and equipment)				
luxury city-centre with conference and wet leisure facilities	m^2	3050.00	to	3800.00
business town centre with conference and wet leisure facilities	m^2	2250.00	to	2800.00
mid-range with conference and leisure facilities	m^2	1725.00	to	2175.00
budget city-centre with dining and bar facilities	m^2	1525.00	to	1900.00
budget roadside excluding dining facilities	m^2	1325.00	to	1650.00
Hotel accommodation facilities (excluding fittings, furniture and equipment)				
bedroom areas	m^2	920.00	to	1150.00
front of house and reception	m^2	1175.00	to	1475.00
restaurant areas	m^2	1375.00	to	1725.00
bar areas	m^2	1225.00	to	1525.00
function rooms/conference facilities	m^2	1175.00	to	1475.00
Students residences				
large budget schemes with en-suite accommodation	m^2	1175.00	to	1475.00
smaller schemes (40–100 units) with mid range specifications	m^2	1425.00	to	1775.00
smaller high quality courtyard schemes, college style	m^2	1925.00	to	2425.00

Building Cost Models

The authors have been publishing cost models in *Building* magazine for many years. Although the scope and coverage of the cost models has expanded considerably, the objectives have always remained constant. They are:

- To provide detailed elemental cost information derived from a generic building that can be applied to other projects
- To provide a commentary on cost drivers and other design and specification issues
- To compare suitable procurement routes that secure the clients objectives

For this edition we have published updated elemental cost data for 16 building types. All models have been updated to reflect the book tender index stated on the book cover. Locations do vary for each model, so please make a note of the location and location factor for any adjustments that may need to be made.

A summary of UK location factors can be found in *Building Costs Indices, Tender Price Indices and Location Factors* section of this book, together with examples showing how to apply those factors.

Readers should note that these models are based around actual projects and are intended as an indication to build costs only. There are many factors which need to be considered when putting together an early cost plan or estimate and all need to be considered when arriving at a Cost Plan for any particular project.

Readers may refer to the *Approximate Estimating Rates* section of this book to make adjustments to the models for alternative specifications within any of the elements.

As elsewhere in this edition prices do not include Value Added Tax or professional services fees.

Building Cost Models

SMALL INDUSTRIAL UNIT

A single storey new building with a gross internal floor area of 900 m², subdivided into five industrial units. Reinforced concrete ground bearing slab and pads to receive a steel portal frame. Wall and roof cladding is aluminium built up system, with internal blockwork division walls. Each of the five units has a separate entrance door and one roller shutter door, together with a single WC. Units vary in size from 150 m² to 360 m².

Gross internal floor area	900 m²
Model location is South East England	TPI = 561; LF = 0.98

This cost model is copyright of AECOM

Small Industrial Unit	Quantity	Unit	Rate	Total (£)	Cost (£/m²)
Substructure				**79,680**	**88.53**
Excavation and disposal off site	190	m³	28		
Reinforced concrete ground slab, including ground beams and column bases	900	m²	60		
Power floated and hardener	900	m²	8		
Strip foundations for party walls	80	m	160		
Frame and Upper Floors				**66,200**	**73.56**
Steel propped portal frame, cold rolled purlins, surface treatments (@ 40 kg/m²)	36	tonne	1,500		
Intumescent paint fire protection to steelwork	36	tonne	290		
Allowance for miscellaneous works, protecting columns		item	1,800		
Roof				**61,950**	**68.83**
Built up aluminium roof cladding with 180 mm thick insulation, including all labours	950	m²	42		
Extra over for rooflights (10% of total roof area)	95	m²	70		
Mansafe system	80	m	95		
Rainwater drainage, aluminium gutters and downpipes	120	m	65		
External Wall, Windows and Doors				**137,230**	**152.48**
Built up aluminium wall cladding with 130 mm thick insulation	520	m²	46		
2.5 m high inner leaf of 140 thick fairface blockwork	380	m²	37		
3000 × 4600 mm high steel sectional overhead doors	5	nr	3,800		
Aluminium single entrance doors	5	nr	1,350		
Coated aluminium double glazed window system	150	m²	450		
Polycarbonate canopy entrance – approx 1500 × 1000 mm	5	nr	1,200		

SMALL INDUSTRIAL UNIT

Small Industrial Unit	Quantity	Unit	Rate	Total (£)	Cost (£/m²)
Internal Walls and Partitions and Doors				**42,700**	**47.44**
2 hour fire resistant blockwork party walls	450	m²	75		
Fireproofing between blockwork and roof		item	1,800		
Metal stud partitions	50	m²	60		
Laminated faced internal doorset with softwood frames and ironmongery	5	nr	825		
Wall Finishes				**5,200**	**5.78**
Emulsion paint to blockwork wall surfaces generally	1,370	m²	4		
Ceramic wall tiles splashbacks to WC area	5	m²	54		
Floor Finishes				**800**	**0.89**
Screed and non-slip vinyl sheeting to WC areas	15	m²	54		
Ceiling Finishes				**600**	**0.67**
Moisture resistant plasterboard to WC with ceiling grid and paint finish	15	m²	43		
Sanitary Appliances				**6,500**	**7.22**
Disabled WC Suite including all sanitary and fittings	5	nr	1,300		
Disposal Installations				**2,400**	**2.67**
Waste, soil and vent installation; uPVC pipework and fittings	900	m²	3		
Hot and Cold Water Installations				**3,800**	**4.22**
Hot and cold water supplies to WCs	5	nr	750		
Electrical Installations				**31,000**	**34.44**
Small power, basic and emergency lighting	900	m²	24		
Supply to WC for ventilation, heater etc.	5	nr	1,750		
External lighting generally item		item	680		
Incoming Services				**13,500**	**15.00**
Allowance for incoming, electrical, gas and water services		item	13,500		
Protective Installations				**1,100**	**1.22**
Lightning protection	900	m²	1		
Communication Installations				**10,900**	**12.11**
Fire and intruder alarm	900	m²	12		
Builders Work in Connection				**700**	**0.78**
Forming holes and chases etc. @	1	nr	725		
Preliminaries and Contingency				**86,740**	**96.38**
Overheads; profit, site establishment and site supervision @		13.0%			
Contingency @		5.0%			
Construction cost (shell only; rate based on GIFA)				**551,000**	**612.22**

Building Cost Models

INNER LONDON OFFICES

This cost model is based on a typical central London office scheme arranged over 10 floors above ground and one basement floor. it has a gross internal floor area of 25,000 m² with a net-to-gross efficiency of 72% and a wall-to-floor ratio of 0.45. It complies with part L 2014 and achieves a BREEAM 2014 rating of 'Excellent'. Occupational density of the office space has been assumed to be 1:8. It assumes a steel frame building with four-pipe fan coil unit system and a metal tile suspended ceiling. Included are all main contractor preliminaries, fees and risk allowance to complete the building to Category a standard. Demolitions, site preparation, external works and services, fit-out beyond Category a standard, tenant enhancements, professional fees and VAT are all excluded.

Gross internal floor area	25,000 m²
Net internal floor area	18,000 m²
Model location is the Inner London	TPI = 606; LF = 1.06

This cost model is copyright of AECOM

Shell and Core Works	Quantity	Unit	Rate	Total (£)	Cost (£/m²)
Substructure				**4,713,000**	**188.52**
Dewatering		item	270,000		
Break out existing slabs, piles, obstructions and allowance for probing/testing		item	550,000		
Foundations; bored piles with under-ream; ground beams; pile caps	2,150	m²	440		
Piling platform; mini piles and other works to boundary walls		item	330,000		
RC basement slab 300 mm thick, including waterproofing, excavation and disposal	2,150	m²	270		
RC mat slab 1200 mm thick, including waterproofing, excavation and disposal	250	m²	650		
Reinforced concrete retaining walls 300 mm thick	1,100	m²	350		
Reinforced concrete ground floor slab 130 mm thick on profiled metal sheet decking	2,150	m²	160		
Allowance for car park ramp, slab thickenings to stair foundations, lift/escalator pits, drainage channels, concrete transfer walls etc.		item	380,000		
Allowance for crane base including base piles		item	55,000		
Attendance on archaeologists and movement monitoring		item	110,000		
Below slab drainage; other items and sundries		item	600,000		
Frame and Upper Floors				**8,985,100**	**359.40**
Structural steel frame including fittings based on 80 kg/m² overall GIFA including fittings	1,828	tonne	2,200		
Secondary steelwork, based on an extra 5 kg/m²	114	tonne	2,750		
Concrete encased beams at ground floor		item	80,000		
Fire protection to steel frame (90 mins intumescent paint)	1,942	tonne	600		
Reinforced concrete core walls average	3,300	m²	330		

INNER LONDON OFFICES

Shell and Core Works	Quantity	Unit	Rate	Total (£)	Cost (£/m²)
Allowance for other structures (e.g. within plant rooms etc.)		item	160,000		
Allowance for expansion joints and other sundries		item	55,000		
Lightweight reinforced concrete 130 mm thick on profiled steel decking	17,430	m²	105		
Upstands plinths; walkways etc.		item	270,000		
Roof				**1,177,900**	**47.12**
Profiled steel decking with 200 mm lightweight concrete with mesh reinforcement; insulation and acoustics to soffit	2,200	m²	150		
Proprietary roof; paving slabs; insulation and ballast	2,200	m²	220		
Green roof	440	m²	220		
Insulation to exposed soffits and acoustic treatment		item	65,000		
Allowance for upstands to perimeter/terraces	222	m	550		
Upstands/plinths, hatches/ladders, safety hooks and latchways		item	80,000		
Stairs				**237,000**	**9.48**
Steel pan staircases; concrete infills to stair treads; painted mild steel balustrades and handrails (basement to roof)		item	40,000		
Steel pan staircases; concrete infills to stair treads; painted mild steel balustrades and handrails (basement to ground floor – two flights)	2	nr	16,000		
Feature entrance stairs		item	110,000		
Allowance for stairs/cat ladders and safety rails to plant rooms		item	55,000		
External Works				**11,355,800**	**454.23**
Unitized curtain walling; solid spandrel panels; selective high performance glass and some solar control	9,633	m²	1,025		
External structural glazed wall to entrance lobby	200	m²	1,300		
Glazed screen to retail shopfronts	70	m²	1,300		
Aluminium screening to plant enclosures	450	m²	650		
Glass entrance canopies; cantilevered from building	250	m²	1,650		
Extra for louvres		item	80,000		
Blockwork walls at roof level, including wind posts	100	m²	160		
Allowance for visual mock-ups and performance tests		item	330,000		

Building Cost Models

INNER LONDON OFFICES

Shell and Core Works	Quantity	Unit	Rate	Total (£)	Cost (£/m²)
External Windows and Doors				**247,500**	**9.90**
Single and double doors, including disabled pass doors		item	27,500		
Extra over cladding for revolving doors	2	nr	80,000		
Steel roller shutter to loading bay and car park	1	nr	27,500		
Metal doors in service areas		item	32,500		
Internal Walls and Internal Partitions				**2,301,000**	**92.04**
Fairfaced blockwork walls at basement, ground and roof levels	4,000	m²	160		
Curved blockwork entrance feature wall	300	m²	220		
Drylined core walls	6,950	m²	120		
Extra for double thickness drylined core walls	1,000	m²	120		
Other walls/partitions to plant areas, additional walls		item	200,000		
Veneer faced WC cubicles/doors; access panelling	90	nr	4,900		
Internal Doors				**573,000**	**22.92**
Single timber doors	160	nr	2,050		
Double timber doors	30	nr	3,250		
Profilex riser doors	50	nr	1,350		
Other doors: plantrooms; additional access door hatches		item	80,000		
Wall Finishes				**1,329,900**	**53.20**
Stone cladding to main entrance lobby	880	m²	440		
Feature wall panelling on steel frame in main entrance lobby	150	m²	1,200		
Paint to fair face block walls	2,150	m²	11		
Plaster and paint to blockwork/concrete	3,820	m²	22		
Skim coat and paint to drylined walls	1,700	m²	13		
Stone cladding to toilets	450	m²	380		
Granite cladding to lift lobbies	800	m²	440		
Lift architraves		item	110,000		
Floor Finishes				**1,053,600**	**42.14**
Granite/stone tiles to main entrance lobby and lift lobbies	1,250	m²	440		
Stone tiles to toilets including, waterproofing, screed; skirtings	500	m²	440		
Lightweight screed to circulation and core areas	1,280	m²	38		
Sealant/hardener to car park, loading bay and plant rooms	1,140	m²	100		
Vinyl flooring to security areas		item	11,000		
Entrance mats and matwells		item	50,000		

INNER LONDON OFFICES

Shell and Core Works	Quantity	Unit	Rate	Total (£)	Cost (£/m²)
Allowance for white lining to carpark and loading bay		item	27,500		
Allowance for other floor finishes		item	32,500		
Ceiling Finishes				**798,400**	**31.94**
GRG feature ceiling to main entrance lobby	870	m²	440		
Feature drylined ceiling to lift lobbies	380	m²	220		
Metal tile suspended ceilings to toilets	500	m²	110		
Painted plasterboard on metal framing to corridors etc.	840	m²	100		
Insulation to car park/loading bay soffits	1,000	m²	33		
Access panels, bulkheads; detailing; sundry ceiling finishes		item	160,000		
Fittings/Fitting Out (excludes loose furniture)				**764,000**	**30.56**
Main entrance reception desk and security desks		item	110,000		
Stone vanity tops in toilets for basins/taps with mirrors behind	80	m	2,200		
Soap dispensers/tanks, roll holders, paper towels etc.	90	nr	600		
Fittings to disabled toilets	10	nr	1,650		
Rubbish compactor		item	27,500		
Column guards, bollards/crash rails to loading bay/ car park, cycle racks, traffic management, statutory signage		item	380,000		
Sanitary Appliances				**178,200**	**7.13**
WCs, basins, cleaners sinks, urinals (average rate per point)	300	nr	550		
Disabled toilets	12	nr	1,100		
Disposal Installations				**404,500**	**16.18**
Rainwater disposal system	25,000	m²	3		
Soil waste and vent installation	25,000	m²	10		
Drainage to retail areas		item	22,000		
Condensate drainage	25,000	m²	2		
Water Installations				**655,000**	**26.20**
Cold water services: incoming, storage, pumps, distribution	25,000	m²	11		
Hot water heaters and distribution	25,000	m²	9		
Miscellaneous water supplies	25,000	m²	4		
Water services for retail/kitchenettes	25,000	m²	2		

INNER LONDON OFFICES

Shell and Core Works	Quantity	Unit	Rate	Total (£)	Cost (£/m²)
Space Heating and Air Treatment				3,530,000	141.20
Gas installation		item	42,500		
Boilers, flues and primary pipework distribution		item	160,000		
Air handling units and ductwork distribution	25,000	m²	36		
Water cooled chillers	25,000	m²	19		
Heat rejection plant	25,000	m²	14		
LTHW heating including pumps	25,000	m²	19		
CHW installation including pumps and riser pipework	25,000	m²	24		
Condenser water installation	25,000	m²	13		
Metering LTHW/CHW installations	4	nr	27,500		
Reception air conditioning	25,000	m²	4		
Ventilation Installations				652,500	26.10
Toilet and smoke extract ventilation	25,000	m²	8		
Ventilation to plant room, lift motor rooms, refuse area, etc.	25,000	m²	3		
Car park and basement ventilation	25,000	m²	7		
Stair and lobby pressurization	25,000	m²	9		
Electrical Installations				2,677,500	107.10
HV switchgear and transformer	25,000	m²	7		
LV distribution and rising busbars	25,000	m²	46		
Power to mechanical plant	25,000	m²	3		
Small power installation	25,000	m²	7		
Lighting, emergency lighting	25,000	m²	22		
Lighting, emergency lighting to car park and basement	25,000	m²	3		
Enhanced lighting in lobby and other areas		item	55,000		
External building lighting		item	160,000		
Life safety only standby generator installation, including flues, acoustics, fuel installation		item	190,000		
Miscellaneous wireways	25,000	m²	1		
Earthing and bonding	25,000	m²	2		
Lifts				2,280,000	91.20
Passenger lifts, 2.5 m/s, 21 person serving 10 floors with enhanced car finishes	6	nr	230,000		
Goods lift serving 10 floors	1	nr	220,000		
Vehicle park lift	1	nr	170,000		
Firefighting lift	2	nr	170,000		
Builders work, control system etc.		item	170,000		

INNER LONDON OFFICES

Shell and Core Works	Quantity	Unit	Rate	Total (£)	Cost (£/m²)
Protective Installations				**660,000**	**26.40**
Sprinkler Installations; tanks, pumps, risers etc.	25,000	m²	22		
Dry riser installation	25,000	m²	2		
Lightning protection	25,000	m²	2		
Communication Installations				**562,500**	**22.50**
Fire alarm installations	25,000	m²	16		
Landlord security provisions	25,000	m²	4		
Disabled alarms, refuge alarms, induction loops		item	42,500		
Special Installations				**1,252,500**	**50.10**
Building management system	25,000	m²	27		
Leak detection system	25,000	m²	1		
Allowance for façade cleaning equipment		item	550,000		
Builders Work				**600,300**	**24.01**
Builder's work in connection with services installations, including machine bases,	25,000	m²	24		
Preliminaries and Contingencies				**15,430,800**	**617.23**
Contractor's preliminaries and overheads and profit		23.0%			
Risk transfer and Contingency @		8.0%			
Construction cost (shell and core works only)				**62,420,000**	**2,496.80**
Category A Works	**Quantity**	**Unit**	**Rate**	**Total (£)**	**Cost (£/m²)**
Wall Finishes				**149,600**	**5.98**
Emulsion paint finish to office side of core walls	2,150	m²	6		
Column casings, including paint, subframe, etc.	1,450	m²	95		
Floor Finishes				**901,800**	**36.07**
Dust sealer to concrete slabs	18,000	m²	1		
Medium grade fully accessible raised floor, metal faced plycore; 150 nominal depth; including fire barriers	18,000	m²	49		
Ceiling Finishes				**1,080,000**	**43.20**
Concealed grid metal tray suspended ceiling to office areas; acoustic quilt and fire breaks	18,000	m²	60		
Fittings/Fitting Out				**28,800**	**1.15**
Statutory signage	18,000	m²	2		
Space Heating and Air Treatment				**3,295,800**	**131.83**
Four pipe fancoil units	18,000	m²	29		
Distribution ductwork, grilles etc.	18,000	m²	65		
CHW installation; insulation	18,000	m²	43		
LTHW installation; insulation	18,000	m²	33		
Condensate installation; insulation	18,000	m²	13		

INNER LONDON OFFICES

Shell and Core Works	Quantity	Unit	Rate	Total (£)	Cost (£/m²)
Electrical Installations				1,722,600	68.90
Lighting and emergency lighting installation	18,000	m²	65		
Distribution boards	18,000	m²	6		
Small power: supply only floor boxes	18,000	m²	4		
Earthing and bonding	18,000	m²	3		
Mechanical power supplies	18,000	m²	4		
Lighting control	18,000	m²	13		
Protective Installations				396,000	15.84
Sprinkler protection to offices	18,000	m²	22		
Special Installations				432,000	17.28
Building management system	18,000	m²	24		
Builders Work in Connection				145,800	5.83
Builders work in connection with Category A services	18,000	m²	8		
Preliminaries and Contingency				1,605,600	64.22
Contractor's overheads and profit, site establishment and supervision @		14.0%			
Contingency @		5.0%			
Construction cost (Category A only, rate based on GIFA)				9,758,000	390.30

BUSINESS PARK OFFICE

The cost model features a standard office building located in the South-East. The specification includes: GIFA of (6,000 m² (65,000 ft²), two storeys arranged around a central street, steel frame and standard FCUs fitted. The model covers shell-and-core works only. Fit-out costs to category A specification and external works are excluded.

Gross internal floor area   6,000 m²

Model location is South East England   TPI = 561; LF = 0.98

This cost model is copyright of AECOM

Business Park Office	Quantity	Unit	Rate	Total (£)	Cost (£/m²)
Substructure				434,000	71.88
Excavation and disposal off site; reinforced concrete ground slab, including ground beams and column bases	3,100	m³			
Frame, Upper Floors and Stairs				840,100	139.13
Steel propped portal frame, cold rolled purlins, surface treatments (@ 400 kg/m²)	300	tonne	1,550		
Composite slab, metal deck and in situ concrete	2,900	m²	65		
Fire protection to steelwork		item	50,000		
Allowance for miscellaneous works, protecting columns etc.		item	17,500		
Main steel internal staircases; handrails and balustrades (rate per flight)	4	nr	8,800		
Steel external staircases; handrails, balustrades, framing (rate per flight)	4	nr	11,000		
Atrium balustrading	45	m	775		
Miscellaneous metalwork in shafts and risers		item	5,000		
Roof				863,850	113.26
In situ concrete roof slab; metal deck and In situ concrete	3,080	m²	65		
In situ concrete rooflight upstand walls	120	m²	180		
Timber roof trusses		item	60,000		
Monopitch roof, coated aluminium roof cladding, including purlins, trims, cappings, insulation	1,780	m²	110		
Flat roof, asphalt, including insulation, trims, skirtings, flashings, linings to gutters, paving slabs	1,500	m²	100		
Mansafe system		item	1,250		
Rooflights	100	m²	550		
External Wall, Windows and Doors				897,900	148.71
Coated aluminium double glazed curtain walling	150	m²	410		
Fire escape doors	8	nr	1,100		
Coated aluminium double glazed window system	850	m²	390		
Solar shading; fixed aluminium louvres	300	m	330		

BUSINESS PARK OFFICE

Business Park Office	Quantity	Unit	Rate	Total (£)	Cost (£/m²)
Glazed main double entrance doors		item	6,000		
Aluminium wall cladding to roof level plant etc.	610	m²	50		
Render on blockwork, insulation and boarding	550	m²	180		
Facing brickwork in cavity walls	690	m²	170		
Facing brick cladding to columns	675	m²	110		
Miscellaneous masonry		item	70,000		
Internal Walls and Partitions and Doors				**343,100**	**56.82**
Blockwork	1,200	m²	55		
Drywall partition; 1hr fire resistant	570	m²	65		
Hardwood glazed screens; 1 hr fire resistant	180	m²	550		
Hardwood doors and frames; stainless steel ironmongery	80	nr	1,200		
Allowance for fire containment in common areas		item	10,000		
WC cubicles		item	35,000		
Wall Finishes				**69,000**	**11.43**
Emulsion paint to blockwork wall surfaces generally		item	40,000		
Plasterboard to cores and common areas	1,650	m²	18		
Floor Finishes				**82,700**	**13.70**
Screeds to cores and common areas; 150 mm thick to make up levels	640	m²	22		
Carpet tiling to common areas	170	m²	33		
Ceramic tiling to cores and common areas	470	m²	100		
Floor finishes to internal stairs		item	5,000		
Entrance matting	2	nr	5,500		
Ceiling Finishes				**57,800**	**9.57**
Plasterboard and skim to core and common areas	1,050	m²	55		
Furniture and Fittings				**38,000**	**6.29**
Granite vanitary units		item	27,000		
Glass surrounds to vanitary units		item	5,000		
Miscellaneous architectural metalwork		item	6,000		
Sanitary Appliances				**26,600**	**4.41**
WC suite, including disabled facility including all sanitary and fittings	6,050	m²	4		
Disposal Installations				**34,500**	**5.71**
Waste, soil and vent installation; uPVC pipework and fittings	6,050	m²	2		
Rainwater installation	6,050	m²	3		
Water Installations				**3,500**	**5.71**
Hot and cold water supplies to WCs	5	nr	700		

BUSINESS PARK OFFICE

Business Park Office	Quantity	Unit	Rate	Total (£)	Cost (£/m²)
Space Heating and Air Treatment and Ventilation				428,300	70.93
Air handling units, chillers; ductwork, pipework, insulation	6,050	m²	65		
Toilet extract ventilation		item	15,000		
Cooling and ventilation installations to common areas		item	20,000		
Electrical and Gas Installations				294,300	48.74
Mains and sub-mains distribution	6,050	m²	22		
Small power to landlord's areas	975	m²	17		
Electrical supplies to mechanical plant		item	17,500		
Lighting to landlord's areas	975	m²	55		
External feature lighting		item	5,000		
External lighting generally		item	65,000		
Electric trench heating to common areas		item	4,000		
Incoming Services				10,000	1.66
Allowance for incoming electrical, gas and water services		item	10,000		
Protective Installations				8,500	1.41
Lightning protection		item	6,000		
Earthing and bonding		item	2,500		
Communication Installations				39,300	6.51
Fire alarm and smoke detection	6,050	m²	6		
Disabled WC alarm		item	2,000		
Allowance for containment		item	4,000		
Specialist Installations				99,800	16.53
BMS Controls	6,050	m²	17		
Builders Work in Connection				10,000	1.66
Forming holes and chases etc. @		item	10,000		
Preliminaries and Contingency				902,750	149.51
Overheads, profit, site establishment and site supervision @		17.0%			
Contingency @		3.0%			
Construction cost (shell and core only; rate based on GIFA)				5,304,000	878.44

Building Cost Models

DATA CENTRE

This cost model is based on a tier 3 two- storey, new-build data centre for loads up to 1,500 W/m². the scheme comprises two technical spaces, each with a gross internal area of 1,350 m², 850 m² of associated internal plant areas, 600 m² of external plant deck and 950 m² of support facilities. Electrical infrastructure to the site is via A and B supplies with on-site dedicated substations. Further resilience is provided by uninterruptible power supplies and diesel generators at N+1. The technical areas are completed including full fit-out and are capable of delivering up to 20 kW per rack. The m² rate in the cost breakdown is based on the technical area, not gross floor area. Costs of site preparation, external works and external services are included

| Technical Floor area | 2,700 m² | Model location is South East |
| | | TPI = 561; LF = 0.98 |

This cost model is copyright of AECOM

Data Centre	Quantity	Unit	Rate	Total (£)	Cost (£/m²)
Substructure				**1,522,300**	**563.81**
Excavation and disposal off site; reinforced concrete ground slab, including ground beams and column bases	3,500	m²	29		
Reinforced concrete ground slab, hardcore, dpm, ground beams	675	m²	490		
Piled foundations and pile caps 450 mm diameter, 20 m long	490	nr	1,300		
Pile caps	490	nr	650		
Concrete slab to energy centre; 250 mm thick including boundary walls	225	m²	420		
1m high concrete bund to external plant		item	40,000		
Frame, Upper Floors and Stairs				**1,447,200**	**536.00**
Main building steel frame, cold rolled purlins, surface treatments (@ 85 kg/m²):	420	tonne			
Energy centre steel frame, cold rolled purlins, surface treatments (@ 85 kg/m²):	28	tonne			
Additional fire protection	448	tonne			
200 mm in situ reinforced concrete upper floors	2,250	m²			
Internal steel staircase and balustrade with concrete finish		item			
Main entrance reinforced concrete steps		item			
Galvanized steel external escape stairs and balustrades	4	nr			
Galvanized steel external stairs and balustrades	3	nr			
Roof				**202,500**	**75.00**
Kalzip embossed aluminium standing seam roof	2,250	m²	90		
External Wall, Windows and Doors				**669,500**	**247.96**
Kingspan MR panels, 45 mm thick with Fire Safe LC PB approved core including windows, louvres, internal blockwork and Metsec framing	1,700	m²	260		

DATA CENTRE

Data Centre	Quantity	Unit	Rate	Total (£)	Cost (£/m²)
Main entrance security revolving doors		item	70,000		
Coated aluminium double glazed curtain walling		item	17,000		
Energy centre compound frame, cladding and louvres	335	nr	150		
Mechanical centre compound frame and cladding	220	m²	410		
Internal Walls and Partitions and Doors				**415,400**	**153.85**
140 mm thick blockwork walls to plant areas	2,100	m²	47		
Lift shaft walls		item	52,500		
Internal partitions 60/120 min fire rating	1,200	m²	80		
Column encasements and bulkheads		item	52,500		
Bump protection rails to corridors		item	23,000		
Cubicles and vanity units		item	4,650		
Internal single doors 30 min fire rating	40	nr	2,200		
Wall Finishes				**90,000**	**33.33**
Emulsion paint to wall surfaces generally	4,100	m²	11		
Hardwood glazed screens; 1 hr fire resistant	4,200	m²	10		
Wall tiling upstand to kitchen and toilet areas including shower		item	2,450		
Floor Finishes				**432,900**	**160.33**
Raised access floor 850 mm overall height to technical spaces; Kingspan 600 mm square galvanized-steel encapsulated particle board panels	2,700	m²	110		
Raised access floors to other areas including plantrooms	800	m²	105		
Dust sealant to plant and technical areas	3,550	m²	7		
Floor finishes to offices and meeting rooms	400	m²	18		
Floor tiles to reception area	20	m²	180		
Vinyl to stairs		item	1,350		
Epoxy floor paint to plant areas	850	m²	9		
Allowance for protection to raised access floors	2,700	m²	3		
Ceiling Finishes				**131,400**	**48.67**
Suspended 600 × 600 mm grid ceiling to office and technical areas	3,550	m²	37		
Furniture and Fittings				**15,000**	**5.56**
Reception desk and furniture		item	2,900		
Tea kitchens		item	12,000		
Sanitary Appliances				**10,400**	**3.85**
WCs, urinals, wash hand basins; generally	14	nr	470		
Hand dryers	5	nr	410		

DATA CENTRE

Data Centre	Quantity	Unit	Rate	Total (£)	Cost (£/m²)
Shower installations		item	1,750		
Disposal Installations				**11,600**	**4.30**
Above ground drainage installation to support areas	950	m²	12		
Water Installations				**106,400**	**39.41**
Hot and cold water supply to WC, kitchen	950	m²	55		
Hot and cold water installations to plant areas	850		21		
External water installations		Item	8,800		
Water storage installation including ancillary pipework		Item	27,000		
Space Heating and Air Treatment and Ventilation				**3,631,300**	**1,344.93**
Indirect fresh air handling plant (with DX coil) and associated ductwork		item	3,200,000		
water treatment and pipework for adiabatic cooling to air handling plant:		item	105,000		
Supply and extract ventilation system to technical areas	2,700	m²	55		
Ventilation to switch rooms and ancillary areas	850	m²	23		
Office supply and extract ventilation	950	m²	95		
VRF heating and cooling installation to office, meeting rooms and support areas	400	m²	170		
Electrical Installations				**13,143,000**	**4,867.78**
High voltage switchgear and distribution based on a 7 MVA dual supply requirement		item	480,000		
Standby generator installation based on 4nr 2,500 KVA gen sets, providing n+2 complete with flues and 48 hr bulk storage fuel tanks		item	3,200,000		
Low voltage switchgear including, mechanical panels, UPS panels, utility panels, 4 nr 2.5 MVA package transformers including delivery and installation		item	2,800,000		
Static UPS installation consisting of 4nr 1100 KVA modules per system installed in n+1 configuration (IT systems) and 2nr 550 KVA modules installed in n+1 configuration mechanical systems), including distribution		item	3,500,000		
14nr dual input STS PDUs and 14nr single input PDUs including installation		item	1,200,000		
LV distribution including cabling to PDUs, CRAC's, IT and mechanical panels including busbar installation		item	1,300,000		
LV distribution equipment		item	47,500		
General lighting and emergency lighting to all areas complete with PIR controls	4,500	m²	53		

DATA CENTRE

Data Centre	Quantity	Unit	Rate	Total (£)	Cost (£/m²)
External lighting		item	30,000		
Small power installation to plant and support areas	1,800	m²	30		
Primary and secondary containment installation to plant, support and technical areas	4,500	m²	60		
Allowance for trace heating		item	23,000		
Protective and Special Installations				**1,689,300**	**625.67**
Earthing and bonding installation including clean earth to plant	4,500	m²	60		
Lightning protection installation	4,500	m²	8		
Water mist installation to part of plant area and technical areas	3,200	m²	120		
Leak detection installation to plant and technical areas	3,550	m²	20		
Dry riser installation		item	7,100		
Energy management systems/SCADA installation	4,500	m²	130		
Building management system	4,500	m²	75		
Communications Installations				**714,800**	**264.74**
L1 fire and smoke detection and alarm system to plant, support and technical areas	4,500	m²	29		
High sensitivity smoke detection installation to technical areas	2,700	m²	24		
Intruder detection, based on monitored doors, windows and PIR detection	4,500	m²	9		
Access control to technical spaces, plant and support areas including entrances and exits, integrated with intruder detection	4,500	m²	47		
CCTV installation to technical spaces, plant and support areas including entrances and exits	4,500	m²	47		
Interphone video telephony to entrance area		item	21,000		
Public address installation	4,500	m²	4		
Disabled toilet alarm system	2	nr	3,200		
Data installation to office and meeting room areas	400	m²	23		
Lift Installations				**285,000**	**105.56**
Passenger/goods lift; hydraulic 2.5 tonnes; serving two stops	1	nr	150,000		
Passenger lift; hydraulic eight person; serving two stops	1	nr	65,000		
Scissor lifts	2	nr	35,000		

Building Cost Models

DATA CENTRE

Data Centre	Quantity	Unit	Rate	Total (£)	Cost (£/m²)
Testing and Commissioning				271,500	100.56
Testing and commissioning of mechanical services		item	32,500		
Testing and commissioning of electrical services		item	24,000		
IST commissioning		item	160,000		
Allow for electrical heat load test		item	55,000		
Builders Work in Connection				595,900	220.70
Forming holes and chases etc. @		3.0%			
External Works				532,500	197.22
Allowance for vehicle crash barriers		item	32,500		
Allowance for external security fence		item	120,000		
Allowance for soft landscaping works		item	380,000		
Preliminaries and Contingency				4,017,100	1,487.81
Overheads, profit, site establishment and site supervision @		10.0%			
Contingency @		5.0%			
Construction cost (shell and core only; rate based on Technical Floor Area)				29,935,000	11,087.04

OFFICE FIT-OUT

The cost model is based on a notional scheme that is typical for a West End based professional firm. The design is of a high specification, with a mixture of open-planned and cellular office use. It should be noted that the cost of internal finishes can vary considerably and is dependent on a client's aspirations. The cost model assumes that the floor space has been taken from shell and core with a negotiated landlord contribution for the Category A installation.

The office fit-out model is presented as a cost breakdown by functional area rather than in a traditional elemental format. The fit-out is designed to provide the following areas:

Open plan/circulation areas	1,300 m²	Tea points	40 m²
Client meeting rooms	280 m²	Staff dining	150 m²
Client reception	80 m²	Kitchen	60 m²
Cellular offices	1,000 m²	Communications/data rooms	60 m²
Staff meeting rooms	500 m²	Ancillary areas	175 m²
		Total usable area	3,645 m²

Model location is Inner London   TPI = 606; LF = 1.06

This cost model is copyright of AECOM

Office Fit-Out	Quantity	Unit	Rate	Total (£)	Cost (£/m²)
Open Plan/Circulation				**986,400**	**758.77**
Medium grade raised access floor 150 mm high	1,300	m²	51		
Carpet tiles	1,300	m²	32		
Suspended perforated metal ceiling	1,300	m²	90		
Emulsion paint to core walls		item	13,000		
Mechanical and electrical services; including heating, air treatment, ventilation, power, lighting, protective installations, communications, special installations	1,300	m²	490		
Structured cabling	1,300	m²	45		
Roller blinds to perimeter windows	1,300	m²	26		
Statutory signage		item	2,550		
Builders work in connection with services	1,300	m²	13		
Client Meeting Rooms				**1,031,500**	**3,683.93**
Acoustic rated metal stud partitions, 60 minute fire rated, hardwood skirting and ply strengthening	170	m	340		
Double glazed partitions, manifestation	50	m	1,150		
Folding wall; acoustic finish	15	nr	2,550		
American Walnut doors	15	nr	2,850		
Medium grade raised access floor 150 mm high	280	m²	51.00		
Carpet tiles	280	m²	51.00		
Suspended perforated metal ceiling with feature raft	280	m²	130.00		
Fabric wall coverings	900	m²	350.00		

OFFICE FIT-OUT

Office Fit-Out	Quantity	Unit	Rate	Total (£)	Cost (£/m²)
Mechanical and electrical services; including heating, air treatment, ventilation, power, lighting, protective installations, communications, special installations	280	m²	775.00		
Structured cabling	280	m²	38.00		
Black out blinds to perimeter windows	280	m²	65.00		
Statutory signage		item	2,850.00		
Branding and corporate signage		item	13,000.00		
Specialist joinery including credenzas		item	190,000.00		
Builders work in connection with services	280	m²	12.80		
Client Reception				**271,800**	**3,397.50**
Double glazed door	1	nr	5,100.00		
Natural stone flooring on boarded substrate	80	nr	350.00		
Painted plasterboard suspended ceilings including bulkheads, access panels and ceiling feature	80	m²	160.00		
Timber panel and polished plaster wall finishes		item	37,500.00		
Mechanical and electrical services; including heating, air treatment, ventilation, power, lighting, protective installations, communications, special installations	80	m²	775.00		
Structured cabling	80	m²	38.00		
Statutory signage		item	1,300.00		
Branding and corporate signage		item	26,000.00		
Specialist joinery including reception desk		item	95,000.00		
Builders work in connection with services	80	m²	12.80		
Cellular Offices				**1,496,600**	**1,496.60**
Acoustic rated metal stud partitions, 60 minute fire rated, mdf skirting and ply strengthening	600	m	320.00		
Single glazed partitions, manifestation, blinds	150	m	900.00		
Solid core timber veneer doors	120	nr	2,550.00		
Medium grade raised access floor 150 mm high	1,000	m²	51.00		
Carpet tiles	1,000	m²	32.00		
Suspended perforated metal tiled ceiling	1,000	m²	90.00		
Emulsion paint wall finish	3,000	m²	7.70		
Mechanical and electrical services; including heating, air treatment, ventilation, power, lighting, protective installations, communications, special installations	1,000	m²	570.00		
Structured cabling	1,000	m²	45.00		
Roller blinds to perimeter windows	1,000	m²	32.00		
Statutory and corporate signage		item	7,700.00		
Builders work in connection with services	1,000	m²	12.80		

OFFICE FIT-OUT

Office Fit-Out	Quantity	Unit	Rate	Total (£)	Cost (£/m²)
Staff Meeting Rooms				719,100	1,438.20
Acoustic rated metal stud partitions, 60 minute fire rated, mdf skirting and ply strengthening	320	m	320.00		
Double glazed partitions, manifestation	80	m	1,150.00		
Solid core timber veneer doors	15	nr	2,550.00		
Medium grade raised access floor 150 mm high	500	m²	51.00		
Carpet tiles	500	m²	32.00		
Suspended perforated metal tiled ceiling	500	m²	90.00		
Emulsion paint wall finish	1,750	m²	7.70		
Mechanical and electrical services; including heating, air treatment, ventilation, power, lighting, protective installations, communications, special installations	500	m²	570.00		
Structured cabling	500	m²	38.00		
Roller blinds to perimeter windows	500	m²	32.00		
Statutory and corporate signage		item	2,550.00		
Specialist joinery		item	57,500.00		
Builders work in connection with services	500	m²	12.80		
Tea Points				101,100	2,527.50
Standard metal stud partitions, 60 minute fire rated, mdf skirting and ply strengthening	40	m	260.00		
Medium grade raised access floor 150 mm high including tanking to slab	40	m²	75.00		
Slip resistant floor finish on two layers of boarding	40	m²	75.00		
Suspended perforated metal tiled ceiling	40	m²	90.00		
Emulsion paint wall finish and ceramic splashbacks		item	4,450.00		
Mechanical and electrical services; including disposal, water installations, zip taps, heating, air treatment, transfer ventilation, power, lighting, protective installations, communications, leak detection, special installations	40	m²	650.00		
Statutory and signage		item	130.00		
Joinery units, work surfaces and equipment		item	50,000.00		
Builders work in connection with services	40	m²	12.80		
Staff Dining				244,600	1,630.67
Standard metal stud partitions, 60 minute fire rated, mdf skirting and ply strengthening	40	m	260.00		
Medium grade raised access floor 150 mm high including tanking to slab	150	m²	51.00		
Porcelain floor tiles on two layers of boarding and ditra matting	150	m²	190.00		
Suspended perforated metal tiled ceiling	150	m²	130.00		
Emulsion paint wall finish and ceramic splash backs		item	15,000.00		

OFFICE FIT-OUT

Office Fit-Out	Quantity	Unit	Rate	Total (£)	Cost (£/m²)
Mechanical and electrical services; including disposal, water installations, zip taps, heating, air treatment, transfer ventilation, power, lighting, protective installations, communications, leak detection, special installations	150	m²	610.00		
Statutory and signage		item	5,100.00		
Servery and specialist joinery item		item	65,000.00		
Builders work in connection with services	150	m²	12.80		
Kitchen				**258,400**	**4,306.67**
Standard metal stud partitions, 60 minute fire rated, mdf skirting and ply strengthening	50	m	260.00		
Metal pivot door	1	nr	3,850.00		
Medium grade raised access floor 150 mm high including tanking to slab	60	m²	75.00		
Slip resistant floor finish on two layers of boarding and ditra matting	60	m²	75.00		
Hygienic finished ceiling	60	m²	130.00		
Altro Whiterock wall finish	150	m²	160.00		
Mechanical and electrical services; including disposal, water installations, zip taps, heating, air treatment, ventilation, dedicated extract, power, lighting, protective installations, communications, leak detection, special installations	60	m²	1,125.00		
Statutory signage		item	1,300.00		
Stainless steel kitchen units, worktops and equipment		item	130,000.00		
Builders work in connection with services	60	m²	32.00		
Communications/Data Rooms				**220,500**	**3,675.00**
Standard metal stud partitions, 90 minute fire rated, mdf skirting and ply strengthening	40	m	290.00		
Solid timber painted doors; 90 mins. fire rated	1	nr	2,250.00		
Heavy duty raised access floor 150 mm high with anti-static floor finish	60	m²	150.00		
Dust sealant to concrete soffit	60	m²	19.20		
Emulsion paint wall finish	120	m²	7.70		
Mechanical & electrical services; including air treatment, ventilation, stand alone air-conditioning units, electrical, power, lighting, protective installations, communications, special installations, leak detection	60	m²	3,200.00		
Structured cabling	60	m²	45.00		
Statutory signage		item	130.00		
Builders work in connection with services	60	m²	12.80		

OFFICE FIT-OUT

Office Fit-Out	Quantity	Unit	Rate	Total (£)	Cost (£/m²)
Ancillary Areas				253,500	1,448.57
(including stores, reprographics, mail room etc.)					
Standard metal stud partitions, 60 minute fire rated, mdf skirting and ply strengthening	150	m	260.00		
Solid timber painted doors	10	nr	1,900.00		
Medium grade raised access floor 150 mm high	175	m²	45.00		
Sheet vinyl flooring including boarding substrate	175	m²	51.00		
Suspended perforated metal tiled ceiling	175	m²	90.00		
Emulsion paint wall finish	800	m²	7.70		
Mechanical and electrical services; including disposal, water installations, heating, air treatment, ventilation, power, lighting, protective installations, communications, special installations	175	m²	510.00		
Statutory signage		item	320.00		
Fixed fittings and shelving		item	65,000.00		
Builders work in connection with services	175	m²	12.80		
Preliminaries and Contingencies				1,172,500	321.67
Main Contractor's management, site establishment and site supervision @		10.0%			
Main Contractor's overhead & profit @		5.0%			
Design reserve @		5.0%			
Construction cost (rate based on total usable area)				6,756,000	1,853.50

OUT OF TOWN RETAIL UNIT

A 6,500 m² non-food retail unit on an existing retail park in the South East. The unit is comprised of two full height levels which will give the operator the flexibility to trade over both floors without having to compromise on a reduced head height that is often encountered on mezzanine levels of retail park units. The cost model is based on a shell only specification and allows for the construction of the retail unit whilst maintaining continuous trade for all other tenants within the retail park. The overall programme duration for is based on 18 weeks.

Gross internal floor area		6,500 m²
Model location is South East		TPI = 561; LF = 0.98

This cost model is copyright of AECOM

Out of Town Retail Unit	Quantity	Unit	Rate	Total (£)	Cost (£/m²)
Substructure				**620,200**	**95.42**
Excavation, disposal, underground drainage, backfill		item	75,000		
Concrete pad foundations	64	nr	1,850		
Reinforced concrete ground floor slab, 250 mm thick including gas and vapour barrier, edge beams	3,250	m²	120		
Lift and escalator pits	4	nr	9,200		
Frame and Upper Floors				**1,013,500**	**155.92**
Steel propped portal frame, cold rolled purlins, surface treatments (@ 60 kg/m²):	390	tn	1,850		
Secondary steelwork and signage framework		item	25,000		
Holorib and concrete upper floor slab 150 mm thick	3,250	m²	60		
Pre-cast concrete stairs and metal balustrades to cores	4	nr	18,000		
Roof				**546,500**	**84.08**
Standing seam metal roof, insulated roof	2,844	m²	80		
Feature rooflight, clerestory glazed		item	75,000		
Concrete plant deck, inverted roof including waterproofing and ballast	540	m²	100		
Sundry roof works, flashings, forming openings etc.		item	110,000		
Roof parapet screens		item	37,500		
Handrails and GRP designated maintenance walkways		item	42,500		
External Walls, Windows and Doors				**1,639,500**	**252.23**
Structural glazing system to front and side elevations, planar type system, low 'G' value, low iron glass, glazed fins and capless mastic joints, stainless steel trims to perimeter framing	830	m²	775		
Acrylic render system, including insulation and blockwork, to stair cores and soffits of colonnade	740	m²	180		
Punched windows to office areas	9	nr	1,250		
Framed curtain walling to stair cores	130	m²	370		
Half round profiled composite cladding, including block work inner skin	1,270	m²	140		

OUT OF TOWN RETAIL UNIT

Out of Town Retail Unit	Quantity	Unit	Rate	Total (£)	Cost (£/m²)
Granite cladding system to columns	336	m²	490		
Granite cladding system to feature entrance columns	166	m²	700		
Brise soleil, aluminium fins to glazed facades		item	180,000		
Entrance lobby, glazed sides and 2 nr sets of double plus single door and security screens		item	85,000		
Engineering brick base to facades	105	m²	120		
Screens to sprinkler tanks and plant		item	37,500		
Roller shutter doors to service yard		item	12,000		
Fire escape doors		item	18,000		
Internal Walls and Partitions				**37,500**	**5.77**
Sound reducing board and metal stud partitions; generally 2 layers of board each side; various levels of fire and sound insulations; part glazed as appropriate:	500	m²	75		
Internal Doors				**15,000**	**2.31**
Internal doorsets to stair cores; solid core; including vision panels and overpanels; painted finish; ironmongery		item	15,000		
Disposal Installations				**37,500**	**5.77**
Rainwater disposal; downpipes and fittings		item	37,500		
Water Installations					**5.38**
Hot and cold water supply to WC, kitchen		item	37,500		
Electrical and Gas Installations				**37,500**	**5.77**
Incoming electrical services including new substation and termination cubicle		item	100,000		
Incoming gas supply		item	6,100		
Protective, Communications and Special Installations				**146,100**	**22.48**
Lightning protection, earthing and bonding:		item	6,100		
Sprinkler tanks and pump set (internal pipework and sprinkler heads part of fit-out)		item	140,000		
Builders Work				**30,000**	**4.62**
Builders work in connection with services @		item	30,000		
Preliminaries and Contingency				**744,600**	**114.55**
Management costs; site establishment; site supervision @		12.0%			
Contingency @		5.0%			
Construction cost (rate based on GIFA)				**4,974,000**	**765.24**

CAR SHOWROOM

A 1,200 m² building of which 580 m² is workshop, 420 m² is showroom and 200 m² is office space. The level of specification is targeted at a mid-level car dealership built to achieve a BREEAM 'Good' rating. There is a reinforced concrete ground bearing slab, full height glazing to the showroom, microrib wall cladding generally and with a standing seam roof.

Gross internal floor area		1,200 m²
Model location is South East England	TPI = 561; LF = 0.98	

This cost model is copyright of AECOM

Car Showroom	Quantity	Unit	Rate	Total (£)	Cost (£/m²)
Substructure				**122,600**	**102.17**
Excavation and disposal off site	1,200	m³	18		
Reinforced concrete ground slab, hardcore, dpm, ground beams and column bases for steel frame (0.25 W/m².K)	1,000	m²	95		
Power floated and hardener to workshop	580	m²	10		
Frame, Upper Floors and Stairs				**149,100**	**124.25**
Steel propped portal frame, cold rolled purlins, surface treatments (@ 50 kg/m²)	60	tonne	1,850		
Intumescent paint to give 30 minute fire protection to steelwork	60	tonne	370		
Staircase handrails and balustrades	2	nr	6,700		
Allowance for miscellaneous works, protecting columns		item	2,450		
Roof				**150,800**	**125.67**
Built up aluminium insulated roof cladding including all barge boards, trim etc. (0.20 W/m².K)	1,260	m²	100		
Extra for rooflights (10% of total roof area)	95	m²	80		
Mansafe system	80	m	105		
Rainwater drainage, aluminium gutters and downpipes	70	m	90		
External Wall, Windows and Doors				**359,800**	**299.83**
Built up aluminium insulated wall cladding (0.25 W/m².K)	610	m²	100		
2.5 m high inner leaf of 140 mm thick fairface blockwork	400	m²	43		
3000 × 4600 high steel insulated sectional overhead doors (1.5 W/m².K)	2	nr	4,350		
Aluminium single entrance doors	5	nr	1,500		
Coated aluminium double glazed window system to showroom	300	m²	750		
Aluminium louvres to glass frontage 1500 mm projection; including support and brackets:	90	m²	370		
Steel fire escape doors:	3	nr	1,100		

CAR SHOWROOM

Car Showroom	Quantity	Unit	Rate	Total (£)	Cost (£/m²)
Canopy over entrance:	1	nr	3,750		
Internal Walls and Partitions and Doors				**76,000**	**63.33**
Plasterboard and metal stud partitions; generally 2 layers of plasterboard each side; various levels of fire and sound insulations; part glazed as appropriate:	450	m²	90		
140 mm thick fairface blockwork to workshop/showroom wall	245	m²	43		
Fireproofing between partitions and roof:		item	2,450		
Laminated faced internal doorset with softwood frames and ironmongery; including fire resisting where necessary	15	nr	1,500		
Wall Finishes				**5,400**	**4.50**
Emulsion paint to blockwork wall surfaces generally	1,370	m²	4		
Ceramic wall tiles splashbacks to WC area	5	m²	60		
Floor Finishes				**48,000**	**40.00**
Screed and non-slip vinyl sheeting to WC areas	15	m²	60		
Screed and ceramic floor tiles to showroom areas, including skirtings	15	m²	90		
Screed and carpet tiling to office and circulation areas	180	m²	43		
Entrance matwell	5	m²	310		
Ceiling Finishes				**8,800**	**7.33**
Plasterboard and skim to offices, WC etc. with ceiling grid and paint finish	180	m²	49		
Sanitary Appliances				**3,600**	**3.00**
WC suite, including disabled facility including all sanitary and fittings	3	nr	1,200		
Disposal Installations				**800**	**0.67**
Waste, soil and vent installation; uPVC pipework and fittings	3	nr	250		
Water Installations				**1,600**	**1.33**
Hot and cold water supplies to WCs and kitchen	5	nr	310		
Space Heating and Air Treatment and Ventilation				**84,200**	**70.17**
Gas fired VRV comfort cooling system, condensers, air handling unit, ductwork, ceiling void mounted fan coil units	620	m²	90		
Workshop heating; ambirad type heaters	580	m²	49		
Electrical and Gas Installations				**115,700**	**96.42**
Small power, basic and emergency lighting:	1,200	m²	31		
Showroom lighting, including all supply and installation of luminaires:	420	m²	60		
Supply to WC for ventilation, heater etc.:	5	nr	2,000		
Task lighting to workshop to 300 Lux:	580	m²	31		

CAR SHOWROOM

Car Showroom	Quantity	Unit	Rate	Total (£)	Cost (£/m²)
External building lighting generally:	1,200	m²	6		
Allowance for incoming, electrical, gas and water services		item	18,000		
Protective, Communications and Special Installations				**36,200**	**30.17**
Lightning protection, earthing and bonding:	900	m²	2		
Sprinkler installation:	1,200	m²	18		
Fire and intruder alarms, panic alarm buttons:	900	m²	13		
Lift Installation				**18,000**	**15.00**
4 person hydraulic lift		item	18,000		
Builders Work in Connection				**1,900**	**1.58**
Forming holes and chases etc. @		item	1,850		
Preliminaries and Contingency				**220,500**	**183.75**
Overheads, profit, site establishment and site supervision @		13.0%			
Contingency @		5.0%			
Construction cost (shell only; rate based on GIFA)				**1,403,000**	**1,169.17**

PRIVATE PATIENT WARD

The cost model is for a standalone new-build 20-bed private patient ward unit with associated ancillary support services; both internal and external finishes to a good quality. Theatres are provided elsewhere within the hospital estate.

Gross internal floor area, including tiers	1,700 m²
Model location is Outer London	TPI = 572; LF = 1.00

This cost model is copyright of AECOM

Private Patient Ward	Quantity	Unit	Rate	Total (£)	Cost (£/m²)
Substructure				**444,800**	**261.65**
Strip and level site		item	60,000		
Perimeter footings, 0.6 m deep and 0.6 m wide, mass concrete, reinforcement 200 kg/m³	110	m³	625		
Pad foundations, 0.8 m deep and 0.75 × 0.75 m, mass concrete, reinforcement @ 250 kg/m³,	32	m²	750		
150 mm concrete slab; hardcore, dpm, insulation , mesh reinforcement	1,750	m²	120		
Under slab drainage, including excavation		item	70,000		
Allow for trenching of services		item	12,000		
Frame				**86,400**	**50.82**
Precast concrete pad stones bedded on mortar	120	nr	60		
Structural steel beams	20	tonne	2,400		
Structural steel columns	8	tonne	2,400		
Sundry items, additional secondary steel		item	12,000		
Roof				**459,000**	**270.00**
Kalzip aluminium standing seam, 250 mm mineral wool insulation; waterproof membrane and roof structure	700	m²	420		
Single-ply roof membrane, rigid insulation, vapour control layer and roof structure	1,050	m²	140		
Solar shading timber louvered canopy		item	6,000		
Rainwater goods; gutters and downpipes		item	12,000		
External Walls, Windows and Doors				**479,900**	**282.29**
Cavity wall construction, brick external finish, insulation, 100 mm medium-density loadbearing blockwork inner leaf	550	m²	190		
Cavity wall construction, timber cladding system, insulation, 100 mm medium-density load-bearing blockwork inner leaf	450	m²	300		
Cavity wall construction, aluminium cladding panels to external finish, insulation, 100 mm medium-density load-bearing blockwork inner leaf	20	m²	180		
External liquid oxygen enclosure		item	12,000		

PRIVATE PATIENT WARD

Private Patient Ward	Quantity	Unit	Rate	Total (£)	Cost (£/m²)
Generator enclosure		item	6,000		
Curtain walling, double-glazed units, powder coated aluminium, including all necessary ironmongery, fixings, mastic pointing	150	m²	675		
External doors, powder-coated aluminium double-glazed units, including ironmongery and integral cills		item	32,500		
Windows, powder-coated aluminium double-glazed units, including all necessary ironmongery and integral cills,		item	85,000		
Internal Walls, Partitions, Windows and Doors				**285,100**	**167.71**
Load-bearing blockwork partitions, 140 mm thick	1,250	m²	70		
Internal partition, 70 mm studs; 30 min fire rating, 2nr wall board 12.5 mm plasterboard to both sides,	80	m	110		
Internal partition, 70 mm studs with mineral fibre, 90 min fire rating, insulation, 2nr 15 mm plasterboard to both sides	130	m	140		
Internal partition, 70 mm studs, 30 min fire rating, 12.5 mm wallboard plaster board & 1nr 12.5 mm moisture board	180	m	110		
Internal glazed screens, double-glazed partitions	70	m²	800		
IPS panelling, including all necessary fixtures and fittings		item	14,000		
Privacy screens, including all necessary fixtures and fittings	2	nr	600		
Single leaf timber doors and ironmongery	50	nr	1,025		
One and half leaf timber doors and ironmongery	12	nr	1,200		
Double leaf timber doors and ironmongery	8	nr	1,350		
Cupboard/riser timber doors	4	nr	775		
Wall Finishes				**116,100**	**68.29**
2 coat plaster to blockwork; 13 mm thick	2,150	m²	22		
1 coat plaster to blockwork; 3 mm thick	500	m²	6		
Painting to plastered surface, 1 mist and 2 top coats	2,650	m²	10		
Hygiene wall cladding	280	m²	65		
Feature glass, impact resistant, on proprietary carrier system	35	m²	480		
PVC wall protection system		item	6,000		
Floor Finishes				**94,700**	**55.71**
Levelling screed, to insulated floor areas	1,620	m²	6		
Carpet, heavy duty carpet tiles, including timber skirting	120	m²	36		

PRIVATE PATIENT WARD

Private Patient Ward	Quantity	Unit	Rate	Total (£)	Cost (£/m²)
Barrier matting, including aluminium frame	30	m²	330		
Standard vinyl floor finish, including coved skirting	1,050	m²	42		
Non-slip vinyl floor finish, including coved skirting	330	m²	48		
Ceramic floor tiles to entrance and reception	120	m²	80		
Epoxy paint finish	50	m²	24		
Ceiling Finishes				**149,600**	**88.00**
Suspended ceiling, 600 mm × 600 mm, T grid, including suspension system and perimeter edge trim	1,090	m²	48		
Suspended hygiene ceiling, 600 mm × 600 mm, T grid, including suspension system and perimeter edge trim	80	m²	60		
Suspended plasterboard ceiling; MF plasterboard system, including suspension system and perimeter edge trim	300	m²	48		
Slatted timber ceiling, western red cedar, prefabricated timber panels with support profiles and concealed fixings and edge trims	230	m²	180		
Plaster skim coat to plasterboard, 3 mm thick applied in 1 coat	1,700	m²	12		
Painting to plasterboard surfaces, 1 mist coat and 2 top coats	1,700	m²	10		
Fittings and Furnishings				**146,700**	**86.29**
Bespoke reception desk, with built-in storage		item	9,600		
Bespoke reception desk, 2 workstations, with built-in storage		item	4,800		
Touchdown base, 1 workstation, with built-in storage		item	6,000		
Shelving generally		item	6,000		
Mirrors		item	3,600		
Grab rails		item	6,000		
Window blinds and curtains		item	6,000		
Worktops and base units		item	37,500		
Ceiling hoists, including track		item	45,000		
Wall protection generally		item	14,000		
Client and statutory signage		item	6,000		
Baby change unit		item	1,200		
Door stops	50	nr	12		
Coat hooks		item	360		
Mechanical Installations				**638,200**	**375.41**
Clinical wash hand basin	20	nr	540		
Wash hand basins	20	nr	360		

PRIVATE PATIENT WARD

Private Patient Ward	Quantity	Unit	Rate	Total (£)	Cost (£/m²)
Stainless steel sink	4	nr	180		
WC suite and concealed cistern	7	nr	420		
WC suite, Doc M pack	7	nr	1,025		
Macerator pump	1	nr	420		
Bath, to be HTM64 compliant	1	nr	1,800		
Shower, with thermostatic mixer	8	nr	1,200		
Disposal installations		item	32,500		
Water installations – cold water systems		item	37,500		
Water installations – hot water systems		item	130,000		
Heat source		item	85,000		
Space heating and air treatment		item	280,000		
Ventilating systems		item	32,500		
Electrical Installations				**1,111,700**	**653.94**
LV installation including 200 kVa standby generator		item	190,000		
Lighting		item	130,000		
Small power		item	75,000		
Gas installations		item	7,200		
Protective installations		item	12,000		
Communications installations		item	120,000		
Special installations		item	220,000		
Testing and commissioning of building services installations		item	37,500		
Builders work in connection with services		item	80,000		
Incoming infrastructure, gas, statutory water, sewer and telecoms connection		item	240,000		
Preliminaries and Contingency				**599,800**	**352.82**
Management costs, site establishment and site supervision. Contractor's preliminaries, overheads and profit and contingencies @		15.0%			
Construction cost (rate based on GIFA)				**4,612,000**	**2,712.93**

EXTRA CARE HOME

The cost model is for a 77 one and two bedroom apartments housed in a three-storey building of 6,855 m² gross internal floor area, traditional construction.

Gross internal floor area, including tiers   6,855 m²

Model location is North West   TPI = 509; LF = 0.89

This cost model is copyright of AECOM

Extra Care Home	Quantity	Unit	Rate	Total (£)	Cost (£/m²)
Substructure				**679,000**	**99.05**
Site clearance, excavation & disposal	6,210	m²	7		
Site remediation, includes: removal of contaminated material, vaporized membrane to footprint of the building and 600 mm cover to all soft landscaped areas		item	150,000		
Trench fill foundations, beam and block floor, floor insulation and screed	1,970	m²	170		
Piled foundations, beam and block floor, insulation and screed		item	150,000		
Frame and Upper Floors				**489,000**	**71.33**
Steel framing and supporting columns to ground floor main entrance		item	25,000		
Concrete beam and block upper floors	4,884	m²	95		
Roof				**326,600**	**47.64**
Concrete pitched tiled roof, timber structure batterns and felt incl overhang	1,380	m²	140		
Single ply membrane roof, timber structure	750	m²	170		
Front entrance steel and roof structure		item	5,900		
Stairs				**51,000**	**7.44**
Precast concrete stairs over four storeys including balustrade/handrail; over four storeys; half landing	3	nr	17,000		
External Walls, Windows and Doors				**927,500**	**135.30**
Traditional external wall construction comprising facing brick, cavity and internal skin of external wall block work, including dpc, cavity trays and insulation, joints and lintels	916	m²	150		
Traditional external wall construction comprising block work and through colour render, cavity and internal skin of external wall block work including dpc, cavity trays and insulation, joints and lintels	1,131	m²	150		
Traditional external wall construction comprising blockwork and through dressed metal colour cladding, cavity and internal skin of external wall blockwork including dpc, cavity trays and insulation, joints and lintels	39	m²	230		

EXTRA CARE HOME

Extra Care Home	Quantity	Unit	Rate	Total (£)	Cost (£/m²)
Traditional external wall construction comprising blockwork and through accent colour render, cavity and internal skin of external wall blockwork including dpc, cavity trays and insulation, joints and lintels	32	m²	150		
Juliette balconies including steel structure posts	32	m²	440		
Allowance for balconies including steel structure posts, foundations and non-slip decking	26	nr	4,350		
Composite windows	708	m²	370		
Curtain walling and cladding to resident stairwells	110	m²	440		
External doors, single and double	89	nr	1,900		
Internal Walls, Partitions, Windows and Doors				**714,600**	**104.25**
Metal stud partition to internal walls	3,767	m²	38		
Blockwork internal walls – 300 mm	4,249	m²	75		
Internal timber six panel painted apartment entrance door, 60 min fire rated, ironmongery	77	nr	750		
Internal apartment door 30 min fire rated; ironmongery	287	nr	290		
Internal apartment store door, single 30 min fire rated; ironmongery	111	nr	250		
Internal apartment store door; double 30 min fire rated; ironmongery	9	nr	370		
Internal lobby entrance door, powder coated aluminium	2	nr	1,850		
Ancillary communal doors; 30 min fire rated where required, ironmongery	62	nr	390		
Communal circulation door; double veneered 60 min fire rated including vision panel	23	nr	825		
Internal windows to apartments, 60 min fire rated	77	nr	440		
Wall Finishes				**449,400**	**65.56**
Plasterboard and skim finish to all internal walls and the internal face of the external wall face incl emulsion	18,152	m²	21		
ceramic wall tiles to apartment bathrooms & kitchens, craft room kitchenette, staff room kitchen, communals, assisted bathrooms, including trim and sealant	1,427	m²	44		
Floor Finishes				**146,100**	**21.31**
Dust sealer to apartments, halls, bedrooms, lounges and stores	3,375	m²	6		
Safety vinyl flooring with coved skirtings	1,269	m²	41		
Non-slip ceramic tiles	203	m²	70		
Carpet to communal areas	1,206	m²	47		

EXTRA CARE HOME

Extra Care Home	Quantity	Unit	Rate	Total (£)	Cost (£/m²)
High quality carpet	54	m²	53		
Matting (grime buster)	4	m²	95		
Ceiling Finishes				209,400	30.55
Plasterboard and skim suspended ceiling to apartments and communal areas with emulsion finish	5,610	m²	33		
Plasterboard and skim suspended ceiling to wet apartments and communal areas with emulsion finish	501	m²	36		
Allowance for access hatches and reflected suspended ceiling details		item	6,200		
Fittings and Furnishings				408,300	59.56
Standard kitchen units including oven, hob and extract	77	nr	2,500		
Bespoke reception desk allowance		item	6,200		
Allowance for main kitchen and servery including store and cold store		item	95,000		
Shelving generally		item	1,250		
Recycle bins	77	nr	38		
Hair dressers salon and nail bar		item	6,200		
Staff laundry		item	25,000		
Painted curtain battens, window boards, handrails, pipe boxing, etc.		item	25,000		
Internal and external signage		item	12,000		
Lockers and benches to staff room		item	3,750		
Kitchenettes	2	nr	950		
Bathroom cupboards and shelves	77	nr	440		
Stores and meeting room	4	nr	675		
Sanitary Fittings				179,200	26.14
Apartment ensuite incl wash basin, WC, shower and curtain rail	77	nr	2,050		
Assisted bathroom incl argo huntleigh malibu automatic disinfectant bath, wash basin, WC	1	nr	12,000		
Other WC and wash basin	5	nr	1,500		
Staff WC and shower	1	nr	1,800		
Mechanical Installations and Lifts				2,684,900	391.67
Mechanical and electrical installation comprising hot and cold water installation; CHP installation; sprinkler system; lighting and power; TV and telephone; fire/smoke alarm; cctv/intruder alarm; door entry system; external lighting; incoming supply and distribution; heating and ventilation; allowance for BWic with services	6,855	m²	380		

EXTRA CARE HOME

Extra Care Home	Quantity	Unit	Rate	Total (£)	Cost (£/m²)
Lift installation for 13 persons		item	50,000		
Lift installation for 8 persons		item	30,000		
Preliminaries and Contingency				**1,613,000**	**235.30**
Management costs; site establishment; site supervision @		13.0%			
Main contractor's overheads and profit @		5.0%			
Contingency @		3.0%			
Construction cost (rate based on GIFA)				**8,878,000**	**1,295.10**

PRIMARY SCHOOL EXTENSION

A single storey, three-classroom extension to a primary school. Constructed using traditional masonry cavity walls on concrete strip foundations, with a pitched tiled roof over. Individual classrooms are formed by load bearing blockwork partitions

Gross internal floor area including tiers	310 m²	
Model location is South East England	TPI = 561; LF = 0.98	

This cost model is copyright of AECOM

Three Classroom Extension	Quantity	Unit	Rate	Total (£)	Cost (£/m²)
Substructure				**44,600**	**143.87**
Excavation & disposal	140	m³	32		
Concrete strip foundations, masonry work below DPC; blockwork and facing brickwork	90	m	170		
Reinforced in situ concrete ground slab, including service trench; vapour barrier; hardcore, excavation and disposal	310	m²	80		
Roof				**60,800**	**196.13**
Softwood roof trusses	380	m²	48		
Board insulation to roof	310	m²	22		
Cement slate roofing including all eaves, ridge tiles and labours; measured on plan	380	m²	65		
Aluminium rainwater down pipes	40	m	85		
Aluminium gutters	90	m	75		
Fire barriers	40	m	22		
External Walls				**37,400**	**120.65**
Brick cavity wall, facing brick outer skin with cavity, 140 mm inner blockwork leaf	220	m²	170		
Windows and External Doors				**36,500**	**117.74**
Proprietary aluminium framed, double glazed windows, doors and solid aluminium faced panels; powder coated finish	55	m²	460		
Double door steel security doors, including all ironmongery	1	nr	1,850		
Double door aluminium framed glazed doors and screens to paved areas, including all ironmongery	15	m²	625		
Internal Walls and Partitions				**14,700**	**47.42**
Partitions; 100/140 mm blockwork	285	m²	43		
WC cubicle partitions; laminated plastics, including all ironmongery	4	nr	600		

PRIMARY SCHOOL EXTENSION

Three Classroom Extension	Quantity	Unit	Rate	Total (£)	Cost (£/m²)
Internal Doors				**11,900**	**38.39**
Internal fire doors, Georgian wired glass vision panel, stainless steel ironmongery to classrooms	5	nr	1,100		
Double fire door, Georgian wired glass, stainless steel ironmongery to corridor	1	nr	1,500		
Wrot softwood storage cupboard door, stainless steel ironmongery	5	nr	550		
Non-fire-rated doors, stainless steel ironmongery to WC	3	nr	700		
Wall Finishes				**17,300**	**55.81**
Plaster and 1 mist coat, 3 coats of emulsion paint	760	m²	19		
Ceramic wall tiles, full height in toilets and selected classroom areas	40	m²	65		
Floor Finishes				**18,800**	**60.65**
Sand cement screed	310	m²	22		
Carpet tiles	150	m²	27		
Safety vinyl to WC areas & practical areas, including skirtings	90	m²	43		
Heavy duty vinyl to circulation areas, including skirtings	68	m²	48		
Entrance matting with aluminium matwell	2	m²	380		
Ceiling Finishes				**11,900**	**38.39**
Plasterboard ceiling, plaster skim and emulsion paint finish	280	m²	38		
Moisture resistance plasterboard ceiling, plaster skim and emulsion paint finish	30	m²	43		
Furniture and Fittings				**22,900**	**73.87**
Storage trays containers	3	nr	2,900		
Storage units – allowance of 1 double cupboard per classroom	3	nr	600		
Worktops, including cut out for sink, 3000 × 600 mm	3	nr	150		
Coat hooks, fixed to masonry	100	nr	32		
Pinboards, 1000 × 2000 mm	6	nr	85		
Pinboards, 1000 × 1200 mm	2	nr	60		
Whiteboards: Interactive	3	nr	1,700		
Whiteboards: Magnetic	3	nr	120		
Signage	1	item	2,350		
Mirrors 640 × 460 mm	7	nr	48		
Sanitary Fittings				**7,600**	**24.52**
WCs	6	nr	370		
Urinals, including side panel	2	nr	250		

PRIMARY SCHOOL EXTENSION

Three Classroom Extension	Quantity	Unit	Rate	Total (£)	Cost (£/m²)
Hand basins	8	nr	230		
Disabled toilet, including WC, wash hand basin, grab rails and other fittings	1	nr	1,400		
Single stainless steel sinks to classrooms, 1200 × 600 mm	3	nr	270		
Cleaners sink, 510 × 380 mm	1	nr	850		
Disposal Installations				**6,800**	**21.94**
Waste, soil and vent installation; uPVC pipework and fittings	310	m²	22		
Water Installations				**11,700**	**37.74**
Cold water points	21	nr	320		
Hot water points	13	nr	380		
Space Heating and Air Treatment				**43,400**	**140.00**
Space heating, all costs associated with the supply & installation of the heating system, temperature control and distribution pipework	310	m²	140		
Ventilation Installations				**5,400**	**17.42**
Ventilation extraction to toilet areas		item	5,400		
Electrical and Gas Installations				**42,300**	**136.45**
Mains and sub-mains installation	310	m²	27		
Small power installation	310	m²	38		
Lighting and general luminaires; emergency lighting	310	m²	65		
Gas installations, all costs associated with the supply and installation of gas	310	m²	7		
Protective, Communications and Special Installations				**16,200**	**52.26**
Lightning protection		item	1,300		
Fire alarm installation; smoke detectors; call points		item	4,850		
Telephone and data wireways; internal telephone system		item	2,700		
Security installation; intruder detection, CCTV to existing control unit etc.		item	7,300		
Builders Work				**6,500**	**20.97**
Forming holes, chases etc.		item	6,500		
Preliminaries and Contingency				**73,300**	**236.45**
Management costs; site establishment; site supervision @		12.0%			
Contractors contingency @		5.0%			
Construction cost (rate based on GIFA)				**490,000**	**1,580.67**

SECONDARY SCHOOL BLOCK

A small secondary school extension to provide a new science block with two teaching rooms attached to an existing building.

Gross internal floor area including tiers	200 m²
Model location is South East England	TPI = 561; LF = 0.98

This cost model is copyright of AECOM

Science Block and Classrooms	Quantity	Unit	Rate	Total (£)	Cost (£/m²)
Substructure				16,200	81.00
Excavation and disposal	45	m³	34		
Concrete strip foundations, masonry work below DPC; blockwork and facing brickwork	45	m	160		
Reinforced in situ concrete ground slab, vapour barrier; granular fill, insulation (0.25 W/m².K)	100	m²	75		
Upper Floors and Stairs				13,200	66.00
Pre-stressed precast concrete plank floor 200 mm thick	100	m²	60		
Precast concrete staircase with half landing	1	nr	5,500		
Nylon coated steel handrail	20	m	85		
Roof				17,700	88.50
Steel profiled sheet roof decking with single layer waterproof membrane, insulation and vapour control layer	100	m²	170		
Aluminium rainwater pipework and hopper heads	14	m	49		
External Walls				50,700	253.50
Facing brickwork, insulated cavity, 100 mm concrete blockwork inner leaf	85	m²	130		
Timber cladding on battens with breather membrane on block cavity wall	170	m²	210		
Aluminium flashings to timber boarding	115	m	34		
Windows and External Doors				30,500	152.50
Polyester powder coated double glazed aluminium windows, steel lintels, painted window boards	44	m²	470		
Polyester powder coated double glazed aluminium doors, steel lintels					
Single leaf external doors	3	nr	2,100		
Leaf and a half entrance doors	1	nr	3,550		
Internal Walls and Partitions				5,600	28.00
Lightweight aerated concrete blockwork 140 mm thick	130	m²	43		
Internal Doors				4,700	23.50
Painted solid core timber doors with vision panels, linings and ironmongery	6	nr	775		

SECONDARY SCHOOL BLOCK

Science Block and Classrooms	Quantity	Unit	Rate	Total (£)	Cost (£/m²)
Wall Finishes				**11,000**	**55.00**
Plaster and eggshell paint	600	m²	18		
Floor Finishes				**15,700**	**78.50**
Screed, latex and slip-resistant vinyl sheet flooring and skirtings to toilets	9	m²	65		
Screed, latex and vinyl safety flooring and skirtings to classrooms	170	m²	65		
Screed, latex and heavy contract polypropylene carpet to lobbies and landings	50	m²	49		
Aluminium stair edging	30	m	37		
Painted timber skirting	50	m	10		
Ceiling Finishes				**6,200**	**31.00**
Demountable suspended ceiling, sound-absorbing tiles	200	m²	31		
Furniture and Fittings				**17,200**	**86.00**
Black out blinds	9	nr	140		
Science laboratory furniture	2	nr	7,900		
Mirrors	3	nr	60		
Sanitary Fittings				**5,100**	**25.50**
Plastic laminate wall linings	3	nr	400		
WC suite	4	nr	430		
Wash hand basin	3	nr	310		
Warm air hand dryers	4	nr	310		
Disposal Installations				**4,600**	**23.00**
Waste, soil and vent installation; uPVC pipework and fittings	200	m²	23		
Water Installations				**5,400**	**27.00**
Hot and cold water supply	200	m²	27		
Space Heating and Air Treatment				**32,000**	**160.00**
Space heating, all costs associated with the supply and installation of the heating system, temperature control and distribution pipework	200	m²	160		
Ventilation Installations				**8,400**	**42.00**
Ventilation extraction to toilet areas	200	m²	42		
Electrical and Gas Installations				**42,100**	**210.50**
Distribution boards and sub-main	200	m²	18		
General power installation	200	m²	55		
General lighting installation	200	m²	110		
Gas installation	200	m²	27		

SECONDARY SCHOOL BLOCK

Science Block and Classrooms	Quantity	Unit	Rate	Total (£)	Cost (£/m²)
Protective, Communications and Special Installations				**24,800**	**124.00**
Lightning protection	200	m²	10		
Data installation	200	m²	21		
Fire alarm installation	200	m²	21		
Intruder alarm installation	200	m²	20		
CCTV installation	200	m²	22		
Door access installation	200	m²	31		
Builders Work				**15,900**	**79.50**
Forming holes, chases etc.	200	m²	15		
Alterations in connecting to existing building		item	13,000		
Preliminaries and Contingency				**67,000**	**335.00**
Management costs; site establishment; site supervision @		12.0%			
Contingency @		7.5%			
Construction cost (rate based on GIFA)				**394,000**	**1,970.00**

LABORATORY

This cost model is based on a university-funded research laboratory. the laboratories are primarily containment level 2 and are constructed to VC-A criteria. All floors have a 4.6 m storey height and the building achieves an overall BREEAM 'Excellent' rating. The building has a gross internal floor area of 11,000 m², of which 3,800 m² is net laboratory and associated technical space, and 3,000 m² is office, write-up and teaching space. Services are distributed vertically to the laboratory areas from a rooftop plant room. The offices are naturally ventilated along with under-floor displacement ventilation. The costs of site preparation, external works and external services are not included. Loose fittings, equipment and specialist laboratory equipment are also not included

Gross internal floor area including tiers	11,000 m²
Model location is South East	TPI = 561; LF = 0.98

This cost model is copyright of AECOM

Laboratory	Quantity	Unit	Rate	Total (£)	Cost (£/m²)
Substructure				**2,939,000**	**267.18**
CFA piles, 450 mm and 600 mm, including piling mat	3,400	m²	300		
Excavation, including lift pits, disposal off site	2,300	m³	80		
Formation of ground-floor slab, 350 mm thick, including pile caps and beams	3,400	m²	360		
Below slab drainage	3,400	m²	65		
Incoming service trenches; water, gas, electric and data		item	80,000		
Allowance for sundry groundwork items		item	210,000		
Upper Floors and Stairs				**2,914,500**	**264.95**
Reinforced concrete core/shear walls, 250–450 mm thick	1,350	m²	290		
Reinforced concrete columns and beams, various sizes	10,000	m²	70		
Reinforced concrete floor slabs; 275–400 mm thick	6,600	m²	180		
Structural steel columns and beams to roof structures, fire protection	175	tonnes	3,000		
Allowance for sundry items, fire stopping		item	110,000		
Roof				**1,585,000**	**144.09**
Reinforced concrete roof slab; 300 mm thick	3,400	m²	170		
Single-ply membrane roof including insulation, upstands, parapets and openings for services	2,400	m²	190		
Proprietary metal standing seam roof covering, including insulation	1,000	m²	270		
Feature glazed roof light		item	160,000		
Allowance for rainwater goods, gutters and downpipes		item	110,000		
Fall arrest		item	11,000		

Building Cost Models

LABORATORY

Laboratory	Quantity	Unit	Rate	Total (£)	Cost (£/m²)
Stairs and Ramps				554,000	50.36
Reinforced in situ concrete stair, general circulation, including handrails	11	nr	14,000		
Feature atrium stairs and balustrades	2,400	item	210,000		
Allowance for access ladders, steps and plant room walkways		item	110,000		
Balustrades and handrails to external balcony and courtyard		item	80,000		
External Walls and Doors				5,232,200	475.65
Metal rainscreen cladding system, two finishes, including metsec, concealed fixing's, flashings trims	2,780	m²	750		
Proprietary metal cassette cladding panels, inc metsec, concealed fixings, flashings, trims	1,320	m²	600		
Aluminium mesh screens to plant areas	720	m	440		
Allowance for sundry brickwork		item	80,000		
Curtain walling, powder coated, capped system, including opening windows to offices	2,840	m²	650		
Allowance for brise-soleil	3	nr	800		
External doors; powder coated aluminium glazed door sets including ironmongery		item	110,000		
Internal Walls and Partitions				1,560,400	141.85
Metal stud partitions; dB ratings 40 dB–65 dB	9,600	m²	90		
Glazed internal partitions	680	m²	480		
Allowance for forming holes for laboratory services through walls		item	160,000		
Allowance for fire protection through penetrations		item	210,000		
Internal Doors				795,000	72.27
Solid core doors, door frames, ironmongery, vision panels, PVC coated, to laboratories	160	nr	3,000		
Solid core doors, door frames, ironmongery, vision panels, veneered, to offices	180	nr	1,750		
Wall Finishes				595,900	54.17
Painting to walls, emulsion	22,700	m²	6		
Timber acoustic wall finishes to offices and atrium	1,560	m²	240		
Allowance for enhanced finishes to specialist laboratories		item	55,000		
Ceramic tiles to WCs and shower cores	250	m²	85		
Floor Finishes				1,090,300	99.12
Levelling screed and DPM	4,900	m²	37		
Medium duty raised access floor	3,650	m²	60		
Non-slip vinyl floor finish, including coved skirting	4,900	m²	80		

LABORATORY

Laboratory	Quantity	Unit	Rate	Total (£)	Cost (£/m²)
Carpet, heavy duty carpet tiles, including timber skirting	3,650	m²	60		
Ceramic floor tiles	300	m	80		
Allowance for sundry floor finishes		item	55,000		
Ceiling Finishes				**1,127,500**	**102.50**
Suspended metal plank system	4,900	m²	90		
Suspended perforated plasterboard systems, including acoustic treatment and painting	3,650	m²	130		
Suspended plasterboard system, including painting	300	m²	65		
Allowance for painting concrete soffits		item	27,500		
Allowance for fire protection above ceilings		item	110,000		
Allowance for sundry ceiling finishes		item	55,000		
Furniture and Fittings				**5,728,500**	**520.77**
Cold room fit-out	2	nr	42,500		
Autoclaves; single sided and pass through, various sizes	4	nr	130,000		
Allowance for fixed laboratory benching, shelving, service spines, under bench and over bench cupboards, lab sinks, wash hand basins and eye wash stations	3,800	m²	700		
Fume cupboards and safety cabinets	90	nr	11,000		
Allowance for fit-out to specialist laboratories		item	530,000		
WC fittings, IPS panelling, cubicles, mirrors, hand dryers etc.	30	nr	3,200		
Allowance for miscellaneous fixtures and fittings, reception desk, kitchenettes, copy stations, whiteboards etc.		item	160,000		
Window blinds	1,500	m²	120		
Lecture theatre seating, raked	250	nr	510		
Audio visual installation, to lecture theatre, meeting rooms, communal spaces		item	270,000		
Allowance for signage, statutory, directional and external building signage		item	110,000		
Sanitary Fittings and Disposal Installations				**381,200**	**34.65**
Allowances for sanitary appliances including WCs, urinals, wash hand basins, disabled access fittings, showers	11,000	m²	11		
Rainwater disposal	11,000	m²	4		
Above ground drainage and condensate	11,000	m²	9		
Laboratory drainage	3,800	m²	32		

LABORATORY

Laboratory	Quantity	Unit	Rate	Total (£)	Cost (£/m²)
Water Installations				**1,063,400**	**96.67**
Incoming mains and domestic cold water system	11,000	m²	26		
Domestic hot water system	11,000	m²	13		
Laboratory cold water installation	3,800	m²	27		
Purified water system to laboratory spaces	3,800	m²	55		
Allowance for irrigation and WC flushing system		item	21,000		
Allowance for centralized steam installation, serving autoclaves and laboratory appliances	3,800	m²	80		
Heat Source				**852,900**	**77.54**
Ground source heat pump installation, including boreholes, pipework and all associated plant		item	700,000		
Gas fired boilers including flues	11,000	m²	14		
Space Heating and Air Treatment				**3,907,000**	**355.18**
Low temperature hot water heating installation	11,000	m²	80		
Supply and extract ventilation to laboratory spaces including air handling units, heat recovery and ductwork	3,800	m²	420		
Supply and extract ventilation to office/write-up spaces including air handling units, heat recovery and ductwork	3,000	m²	37		
Extra for supply and extract ventilation to lecture theatre		item	55,000		
Chilled water system including air cooled chillers, plantroom installation, distribution pipework, fan coil units and chilled beams	11,000	m²	110		
DX cooling to IT rooms		item	55,000		
Ventilation Installations				**939,500**	**85.41**
Ventilation extraction to toilet areas		item	27,500		
Fume cupboard and safety equipment extract systems	3,800	m²	240		
Electrical and Gas Installations				**3,418,500**	**310.77**
HV installation including incoming main		item	47,500		
LV panel, sub-mains installation and distribution boards	11,000	m²	55		
Containment generally	11,000		48		
Lighting generally, including feature lighting and external lighting	11,000		130		
Small power generally including power provision to laboratory benches	11,000		48		
Standby generator and fuel supply		item	120,000		
Photovoltaic installation including inverters		item	160,000		

LABORATORY

Laboratory	Quantity	Unit	Rate	Total (£)	Cost (£/m²)
Protective, Communications and Special Installations				24,800	124.00
Lightning protection	200	m²	10		
Data installation	200	m²	21		
Fire alarm installation	200	m²	21		
Intruder alarm installation	200	m²	20		
CCTV installation	200	m²	22		
Door access installation	200	m²	31		
Fuel Installation				90,800	8.25
Gas installation generally including external supply		item	11,000		
Laboratory gas installation	3,800	m²	21		
Lift Installation				450,000	40.91
Goods lift	1	nr	190,000		
Passenger lift	2	nr	130,000		
Fire and Lightning Protection				44,100	4.01
Dry riser installation		item	21,000		
Lightning protection	11,000	m²	2		
Communications, Security and Control Systems				2,105,200	191.38
Fire alarm and detection system	11,000	m²	21		
Data installation, category 6 wiring	11,000	m²	32		
Allowance for disabled induction loops, disabled refuge and alarm system		item	16,000		
Security system; access control		item	130,000		
Security system; intruder alarm system	11,000	m²	3		
Security system; CCTV system		item	21,000		
Building management system	11,000	m²	120		
Specialist Installations				1,162,800	105.71
Compressed air system including packaged plantroom	3,800	m²	90		
Allowance for laboratory gases, including four reticulated, four directly fed to various rooms and gas detection system	3,800	m²	200		
Allowance for process chilled water system, complete with distribution pipework	3,800	m²	16		
Builders Work				432,500	39.32
Forming holes, chases etc.		3%			
Preliminaries and Contingency				13,145,800	1,195.07
Mechanical and electrical services contractor preliminaries and commissioning management @		15.0%			
Main contractor's preliminaries, overheads and profit @		20.0%			

LABORATORY

Laboratory	Quantity	Unit	Rate	Total (£)	Cost (£/m²)
Main contractor's pre-construction stage services	25	weeks			
Design reserve @		5.0%			
Construction cost (rate based on GIFA)				**52,116,000**	**4,737.78**

AFFORDABLE HOUSING

This cost model is based on a new build scheme of a mix of two and three bedroom houses. The buildings comply with Code for Sustainable Homes Level 3, are of a timber frame construction with a reinforced in situ concrete ground slab, facing brick external wall and concrete tiles on timber roof trusses. The softwood doors and windows are painted and internal subdivision is by load-bearing timber stud partition. Features for the Lifetimes Homes standard, National Housing Federation Standards and Housing Quality Indicators are also included.

Gross internal floor area   820 m²

Model location is South East England   TPI = 561; LF = 0.98

This updated cost model is copyright of AECOM

Affordable Housing	Quantity	Unit	Rate	Total (£)	Cost (£/m²)
Substructure				**90,600**	**110.49**
Excavation and disposal	144	m²	37		
Strip foundation with cavity masonry wall to dpc level	244	m	180		
Ground floor slab with thickenings, including insulation (0.13 W/m².K)	414	m²	100		
Frame, Upper Floors and Stairs				**142,700**	**174.02**
Structural timber frame	820	m²	130		
Timber cassette upper floor, including any lintels required	402	m²	60		
Wooden, straight flight stairs	10	nr	1,200		
Roof				**86,500**	**105.49**
Timber roof trusses and associated timbers	432	m²	90		
Insulation to roof at rafter level (0.13 W/m².K)	402	m²	40		
Concrete interlocking roof tiles	432	m²	46		
Entrance canopy, approx 3m²	10	nr	430		
uPVC rainwater down pipes	108	m	19		
uPVC gutters	220	m	24		
External Walls, Windows and Doors				**208,900**	**254.76**
Facing brick outer skin to structural timber frame	950	m²	85		
Forming cavity and fit insulation (0.22 W/m².K)	950	m²	21		
Application of enhanced construction details to reduce air tightness to 8.0 m³/(h.m²)	950	m²	13		
Party wall, 240 mm thick timber stud frame with Isowool quilt, enhanced insulation and plasterboard plank (0.2 W/m².K)	300	m²	90		
Softwood double glazed windows to meet secured by design requirements (0.15 W/m².K)	88	m²	540		
Softwood window boards	55	nr	57		
Single external doors (2.0 W/m².K)	20	nr	925		

AFFORDABLE HOUSING

Affordable Housing	Quantity	Unit	Rate	Total (£)	Cost (£/m²)
Internal Walls and Partitions				38,900	47.44
Loadbearing timber stud partition faced with plasterboard, skim finish	250	m²	80		
Non-loadbearing timber stud partition, faced with plasterboard, skim finish	270	m²	70		
Internal Doors				42,800	52.20
Internal doors, wood veneer, ironmongery	70	nr	400		
Wrot softwood storage cupboard door, ironmongery	20	nr	290		
Softwood door linings, architrave, painted	800	m	11		
Wall Finishes				23,400	28.54
Plaster skim with emulsion paint finish	2,200	m²	10		
Ceramic wall tiles	50	m²	46		
Floor Finishes				38,000	46.34
Screed	432	m²	23		
Carpet	520	m²	29		
Vinyl tiling including skirting	150	m²	31		
Softwood skirtings with gloss paint finish	670	m	13		
Ceiling Finishes				27,400	33.41
Plasterboard ceiling, plaster skim and emulsion paint finish	656	m²	32		
Moisture resistance plasterboard ceiling, plaster skim and emulsion paint finish	164	m²	39		
Fittings and Furnishings				24,000	29.27
General fittings – kitchen units, worktops, sink	10	nr	2,400		
Sanitary Fittings				21,000	25.61
Sanitary fittings, full fitted bathroom, wash hand basin, bath with shower over	10	nr	2,100		
Disposal Installations				7,900	9.63
Waste, soil and vent installation; uPVC pipework and fittings	820	m²	10		
Water Installations				56,400	68.78
Hot water	60	nr	510		
Cold water	60	nr	430		
Space Heating and Ventilation				49,200	60.00
Electric High efficiency condensing boiler, distribution pipework and panels with temperature control zones	820	m²	60		
Electrical and Gas Installations				37,600	45.85
Mains and sub-mains connection, small power distribution, cooker point, lighting distribution	820	m²	40		
Gas Installations, all costs associated with the supply and installation of gas supply	10	nr	480		

AFFORDABLE HOUSING

Affordable Housing	Quantity	Unit	Rate	Total (£)	Cost (£/m²)
Protective, Communications and Special Installations				8,000	9.76
Earthing and bonding	860	m²	2		
Smoke detectors; telephone points and TV aerial, including additional data and power sockets and wiring	820	m²	8		
Special Installations				49,100	59.88
Solar thermal hot water installation	10	nr	3,600		
Mechanical ventilation with heat recovery	820	m²	16		
Builders Work				18,500	22.56
Builders work in connection with services installation	10	nr	1,125		
Additional works to comply with Lifetime Homes requirements	10	nr	725		
Preliminaries and Contingency				149,100	181.83
Management costs; site establishment; site supervision @		10.0%			
Contingency @		5.0%			
Construction cost (rates based on GIFA)				1,120,000	1,365.86

Building Cost Models

APARTMENTS

This cost model is based on a mixed-tenure apartment building in a south-east location, featuring 65 open-market apartments and 35 flats for the affordable sector, in a mix of one and two-bedroom configurations. The scheme also features a 50-place semi-basement car park, providing secure spaces for the open-market element of the scheme. Demolition and site preparation, and external works are excluded

Apartment block: Gross internal floor area	7,000 m²
Open Market Apartments: Net internal floor area	3,660 m²
Affordable Apartments: Net internal floor area	1,930 m²
Car Park: Gross internal floor area	1,750 m²
Model location is South East	TPI = 561; LF = 0.98

This cost model is copyright of AECOM

Apartment Shell and Core	Quantity	Unit	Rate	Total (£)	Cost (£/m²)
Substructure				731,400	104.49
Substructure, piled foundations, pile caps, ground slab	1,000	m²	675		
Allowance for drainage	1,000	m²	44		
Allowance for lift pits etc.	2	nr	6,200		
Frame and Upper Floors				1,666,500	238.07
In situ reinforced concrete frame and upper floors	6,650	m²	190		
Balconies, primary and secondary frame, decking, balustrade	65	nr	6,200		
Roof				152,200	21.74
Flat roof coverings, single ply membrane, insulation, ballast; allowance for details to upstands	1,000	m²	95		
Extra for roof terraces and paving to terraces	350	m²	56		
Allowance for roof drainage, roof sundries	1,000	m²	19		
Roof access equipment, latchways, cat ladder, access hatch, safely balustrade		item	19,000		
Stairs				178,600	25.51
RC concrete stairs, mild steel balustrades and handrails	14	m²	8,700		
Extra over for enhanced finishes to entrance level staircases	2	m²	6,200		
Balustrade and parapet to terraces; polyester powder coated	120	m	370		
External Walls, Windows and Doors				2,322,000	331.71
Unitized curtain walling; powder coated insulated aluminium spandrel panels; double-glazed tilt and turn windows	4,000	m²	560		
Extra for doors to balconies, ironmongery	65	nr	1,000		

APARTMENTS

Apartment Shell and Core	Quantity	Unit	Rate	Total (£)	Cost (£/m²)
Entrance doors, aluminium framed glazed door and screen	1	nr	12,000		
External fire escape doors, metal, polyester powder coated	2	nr	2,500		
Internal Walls and Partitions				**344,400**	**49.20**
Core walls, in situ concrete, 225 thick	420	m²	120		
Party walls to apartments and corridors; dense concrete block; head restraint, fire stopping	4,900	m²	60		
Internal Doors				**168,000**	**24.00**
Fire doors to cores and corridors; hardwood architraves/frames, including basic ironmongery	60	nr	925		
Fire doors to risers; hardwood architraves/frames, including basic ironmongery	20	nr	625		
Apartment entrance doors; solid core doors, hardwood architraves/frames, including basic quality ironmongery	100	nr	1,000		
Wall Finishes				**110,600**	**15.80**
Plasterboard to concrete and blockwork, with specialist painted finish; entrance hall	500	nr	60		
Plasterboard to concrete and blockwork, with emulsion paint finish; lift lobbies and corridors	2,600	nr	31		
Floor Finishes				**116,700**	**16.67**
Feature ceramic floor tiles, sand cement screed; entrance hall	50	m²	150		
Heavy duty carpet, sand cement screed; corridors	1,050	m²	75		
Ceramic tile, sand cement screed; lift lobbies	170	m²	120		
Skirtings, surface fixed skirting, painted MDF	810	m	12		
Ceiling Finishes				**51,100**	**7.30**
Painted plasterboard with feature bulkheads; reception	50	m²	120		
Painted plasterboard on battens; lift lobbies and corridors	1,220	m²	37		
Furniture and Fittings				**19,000**	**2.71**
Allowance for reception area fittings; mailboxes, signage		item	19,000		
Sanitary Fittings and Disposal Installations				**130,600**	**18.66**
Allowance for cleaners sinks	14	nr	625		
Rainwater disposal	7,000	m²	4		
Soil, waste and overflow installations; stacks and connections to below ground drainage	7,000	m²	14		

APARTMENTS

Apartment Shell and Core	Quantity	Unit	Rate	Total (£)	Cost (£/m²)
Water Installations				**155,900**	**22.27**
Cold water storage tanks, booster pumps, mains distribution pipework, trace heating, water softener/conditioner etc..	7,000	m²	19		
Hot and cold water services to landlord's areas, including local water storage heaters	1,220	m²	21		
Space Heating and Ventilation				**200,800**	**28.69**
Electric panel heaters; landlord's areas	1,220	m²	4		
Central extract system for bathrooms; ductwork, extract fans	7,000	m²	25		
Reception area air treatment		item	9,300		
Supply and extract; plantroom areas		item	12,000		
Electrical Installations				**210,100**	**30.01**
Mains switchgear, cabling, containment and landlord's distribution boards	7,000	m²	12		
Small power; landlord's areas	1,220	m²	7		
Power supply to mechanical services	7,000	m²	6		
Lighting and emergency lighting to landlord's areas	1,220	m²	35		
Feature lighting to entrances		item	18,000		
Earthing and bonding	7,000	m²	2		
Lift Installation				**221,000**	**31.57**
Lift installation; 13 person fire fighting lifts serving 7 storeys	2	nr	110,000		
Protective, Communications and Special Installations				**251,300**	**35.90**
Allowance for dry riser inlets	1,220	m²	37		
Lightning protection	7,000	m²	3		
Fire alarm system to landlord's areas	1,220	m²	44		
Telephone containment only	7,000	m²	5		
TV/Satellite system; central aerial and distribution	7,000	m²	5		
Localized controls for cold water system	7,000	m²	5		
CCTV and access control to perimeter		item	30,000		
Builders Work				**59,000**	**8.43**
Forming holes and chases; firestopping @		5.0%			
Preliminaries and Contingency				**1,198,000**	**171.14**
Testing and commissioning of building services @		2.50%			
Contractor's overheads and profit, site establishment and supervision @		15.0%			
Contingency @		5.0%			
Construction cost (Apartment shell and core only, rate based on GIFA)				**8,297,100**	**1,185.29**

APARTMENTS

Open Market Apartment Fit-Out	Quantity	Unit	Rate	Total (£)	Cost (£/m²)
Internal Walls and Partitions and Doors				**469,500**	**128.28**
Metal stud partitions; 1 layer wall board each side; insulation; skim coat	4,575	m²	60		
Flush doors; non-fire-rated; single leaf; solid core hardwood veneered; softwood frames; decorations; ironmongery	260	nr	750		
Wall Finishes				**329,500**	**90.03**
Plasterboard dry lining; MF framing; to external façade; emulsion paint finish	650	m²	56		
Plasterboard; to concrete and blockwork walls; emulsion paint	3,730	m²	31		
Ceramic tiles to kitchens	290	m²	105		
Ceramic tiles to bathrooms	1,400	m²	105		
Floor Finishes				**382,600**	**104.54**
Suspended floor construction; ply on timber battens	3,080	m²	31		
Edge fixed carpet; PC sum £20/m²; underlay	3,080	m²	44		
Screed; ceramic tiling; to kitchens and bathrooms	730	m²	130		
Skirtings; surface fixed skirting, painted MDF	3,700	m	12		
Skirting; ceramic to match tiling	580	m	19		
Ceiling Finishes				**136,000**	**37.16**
Plasterboard suspended ceiling on battens; painting	3,660	m²	33		
Feature bulkhead to junction with external wall	420	m	18		
Plasterboard bulkhead for bathroom extract ductwork	65	nr	120		
Furniture and Fittings				**671,800**	**183.55**
Fully fitted kitchen to developer's specification with quality laminate worktops; appliances	65	nr	6,200		
Additional fittings to kitchens to 2 bed apartments	30	nr	2,500		
Built-in furniture to bedrooms; MDF, softwood frame and doors	95	nr	800		
Allowance for built-in cloak, meter and airing cupboards	95	nr	370		
Bathroom furniture, cistern enclosure; shelving	95	nr	500		
Bathroom accessories, mirrors etc.	95	nr	370		
Sanitary Fittings and Disposal Installations				**380,400**	**103.93**
Fully fitted bathroom; WC, bidet, washhand basin, pressed steel bath with power shower and screen	65	nr	3,250		
Fully fitted en-suite shower room; WC, washhand basin, power shower, tray and screen; including all fixtures and fittings	30	nr	2,650		
Kitchen sink; including all fixtures and fittings	65	nr	400		

APARTMENTS

Open Market Apartment Fit-Out	Quantity	Unit	Rate	Total (£)	Cost (£/m²)
Soil waste and vent installation within apartments; connections to stacks	545	nr	100		
Allowance for overflow pipework	3,660	nr	3		
Water Installations				**214,500**	**58.61**
Cold water supply; connection, meter	65	nr	200		
Cold water distribution within apartments; final connections with sanitary fittings and appliances	545	nr	130		
Domestic electric water heaters	65	nr	625		
Hot water distribution within apartments; final connections with sanitary fittings and appliances	450	nr	200		
Space Heating, Air Treatment and Ventilation				**239,100**	**65.33**
Electrical panel heaters; local thermostatic control; power supply measured separately	320	nr	270		
Electric heated towel rails	95	nr	470		
Kitchen and bathroom extract, centralized bathroom system; localized kitchen extract with vent to façade, extract fans	160	nr	675		
Electrical Installation				**282,900**	**77.30**
Mains and sub-mains; connection; LV distribution boards to apartments; meters	65	nr	470		
Small power distribution; sockets and fused connection points; wiring	1,685	nr	40		
Cooker point; wiring	65	nr	120		
Lighting; pendants, ceiling roses and bulkhead connections, wiring; to general areas	520	nr	34		
Lighting; low energy fluorescent and low voltage fittings, wiring; to kitchens and bathrooms	580	nr	120		
Shaving outlet; wiring	95	nr	85		
Lighting; 5 amp lighting sockets; wiring	390	nr	40		
Lighting distribution; switches and wiring	640	nr	37		
Extra for; kitchen pelmet lighting	65	nr	220		
Extra for; bathroom mirror lighting	95	nr	160		
Allowance for earthing and bonding	65	nr	200		
Communication Installation				**141,100**	**38.55**
Fire alarm; combined detector/sounder; mains supply	130	nr	270		
Phone points and wiring; 2 nr points	65	nr	140		
TV sockets and wiring; 2 nr sockets	65	nr	140		
Video entry phone system	65	nr	1,350		
Builders Work				**62,900**	**17.19**
Forming holes and chases; firestopping @		5.0%			

APARTMENTS

Open Market Apartment Fit-Out	Quantity	Unit	Rate	Total (£)	Cost (£/m²)
Preliminaries and Contingency				**726,200**	**198.42**
Testing and commissioning of building services @		2.50%			
Contractor's overheads and profit, site establishment and supervision @		16.0%			
Contingency @		5.0%			
Construction cost (Open Market Apartment fit-out only; rate based on NIA area)				**4,044,200**	**1,104.99**

Affordable Market Apartment Fit-Out	Quantity	Unit	Rate	Total(£)	Cost (£/m²)
Internal Walls and Partitions and Doors				**213,500**	**110.62**
Metal stud partitions; 1 layer wall board each side; insulation; skim coat	2,100	m²	60		
Flush doors; non-fire-rated; single leaf; solid core hardwood veneered; softwood frames; decorations; ironmongery	140	nr	625		
Wall Finishes				**133,900**	**69.38**
Plasterboard dry lining; MF framing; to external façade; emulsion paint finish	380	m²	56		
Plasterboard; to concrete and blockwork walls; emulsion paint	2,510	m²	31		
Ceramic tiles to kitchens	70	m²	85		
Ceramic tiles to bathrooms	340	m²	85		
Floor Finishes				**142,700**	**73.94**
Suspended floor construction; ply on timber battens	1,450	m²	31		
Edge fixed carpet; PC sum £20/m²; underlay	1,450	m²	31		
Sand cement screed; ceramic tiling; to kitchens and bathrooms	455	m²	65		
Skirtings; surface fixed skirting, painted MDF	1,410	m	12		
Skirting; ceramic to match tiling	460	m	12		
Ceiling Finishes				**66,700**	**34.56**
Plasterboard suspended ceiling on battens; painting	1,905	m²	35		
Furniture and Fittings				**173,300**	**89.79**
Kitchen fittings to housing association specifications	35	nr	3,100		
Additional fittings to kitchens to 2 bed apartments	15	nr	625		
Allowance for built-in furniture to bedrooms; MDF, softwood frame and doors	50	nr	625		
Allowance for built-in cloak, meter and airing cupboards	35	nr	310		
Allowance for bathroom furniture, cistern enclosure; shelving	35	nr	190		
Allowance for bathroom accessories, mirrors etc.	35	nr	190		

APARTMENTS

Affordable Market Apartment Fit-Out	Quantity	Unit	Rate	Total(£)	Cost (£/m²)
Sanitary Fittings and Disposal Installations				**93,700**	**48.55**
Fully fitted bathroom; WC, bidet, washhand basin, pressed steel bath with power shower and screen; including all fixtures and fittings	35	nr	1,600		
Kitchen sink; including all fixtures and fittings	35	nr	340		
Soil waste and vent installation within apartments; connections to stacks	210	nr	100		
Allowance for overflow pipework	1,930	nr	3		
Water Installations				**95,700**	**49.59**
Cold water supply; connection, meter	35	nr	200		
Cold water distribution within apartments; final connections with sanitary fittings and appliances	245	nr	130		
Domestic electric water heaters	35	nr	625		
Hot water distribution within apartments; final connections with sanitary fittings and appliances	175	nr	200		
Space Heating, Air Treatment and Ventilation				**105,100**	**54.46**
Electrical panel heaters; local thermostatic control; power supply measured separately	170	nr	270		
Electric heated towel rails; power supply measured separately	35	nr	340		
Kitchen and bathroom extract, centralized bathroom system; localized kitchen extract with vent to façade, extract fans	70	nr	675		
Electrical Installation				**85,400**	**44.25**
Mains and sub-mains; connection; LV distribution boards to apartments; meters	35	nr	470		
Small power distribution; sockets and fused connection points; wiring	905	nr	40		
Cooker point; wiring	35	nr	120		
Lighting; pendants, ceiling roses and bulkhead connections, wiring; to general areas	280	nr	34		
Shaving outlet; wiring	35	nr	85		
Lighting; 5 amp lighting sockets; wiring	245	nr	37		
Allowance for earthing and bonding	35	nr	200		
Communication Installation				**21,400**	**11.09**
Fire alarm; combined detector/sounder; mains supply	35	nr	270		
Phone points and wiring; 2 nr points	35	nr	75		
TV sockets and wiring; 2 nr sockets	35	nr	75		
Audio entry phone system	35	nr	190		
Builders Work				**20,100**	**10.41**
Forming holes and chases; firestopping @		5.0%			

APARTMENTS

Affordable Market Apartment Fit-Out	Quantity	Unit	Rate	Total(£)	Cost (£/m²)
Preliminaries and Contingency				**250,900**	**130.00**
Testing and commissioning of building services @		2.50%			
Contractor's overheads and profit, site establishment and supervision @		16.0%			
Contingency @		5.0%			
Construction cost (Affordable Market Apartment fit-out only, rate based on NIA)				**1,336,600**	**692.53**

Semi-Basement Car Park	Quantity	Unit	Rate	Total (£)	Cost (£/m²)
Substructure				**490,100**	**280.06**
Concrete retaining wall; temporary propping	510	m²	280		
Excavation and disposal, including dewatering	5,250	m²	60		
Tie in slab edge to retaining wall	170	m	190		
Frame and Upper Floors				**494,500**	**282.57**
Reinforced in situ concrete columns and suspended slab to ground floor	1,750	m²	220		
Extra for vehicle ramp		item	50,000		
Allowance for louvres for natural ventilation	175	m²	340		
Stairs				**19,800**	**11.31**
In situ concrete stairs and half landings; mild steel, polyester coated handrails and balustrades; finishes	2	nr	9,900		
Internal Walls and Partitions				**15,600**	**8.91**
Blockwork partitions; facework; 215 mm average thickness; emulsion paint finish	60	m²	60		
Reinforced concrete core walls; emulsion paint finish	100	m²	120		
Internal Doors				**10,600**	**6.06**
Flush doors; fire rated; double leaf; solid core ply faced; softwood frames; decorations; ironmongery; complete	2	nr	1,250		
Flush doors; fire rated; single leaf; solid core ply faced; softwood frames; decorations; ironmongery; complete	4	nr	1,000		
Fire shutters; 120 minutes fire resistance; frame and subframe; electric operation	2	nr	2,050		
Finishes				**60,400**	**34.51**
Emulsion paint to concrete and blockwork	320	m²	4		
Allowance for painted floor finish, with car parking demarcation	1,750	m²	9		
Allowance for insulation to underside of building footprint	1,000	m²	44		
Fittings				**25,200**	**14.40**
Car park barriers and operating system		item	19,000		
Protective bollards; kerbs; barriers; column guards etc.		item	6,200		

Building Cost Models

APARTMENTS

Semi-Basement Car Park	Quantity	Unit	Rate	Total (£)	Cost (£/m²)
Electrical Installations				**65,500**	**37.43**
Mains and sub-mains installation	1,750	m²	6		
Lighting and luminaires to car park areas	1,750	m²	25		
Emergency lighting and luminaires	1,750	m²	6		
Protective and Communications Installations				**98,700**	**56.40**
Sprinkler installation; ordinary hazard group 1	1,750	m²	44		
Fire, smoke detection and alarm system	1,750	m²	12		
Builders Work				**4,900**	**2.80**
Forming holes and chases; fire stopping @			3.0%		
Preliminaries and Contingency				**285,200**	**162.97**
Testing and commissioning of building services @			2.50%		
Overheads and profit, site establishment and supervision @			16.0%		
Contingency @			5.0%		
Construction cost (Semi-Basement car park only; rate based on GIFA)				**1,570,500**	**897.42**

COMMUNITY CENTRE

An 860 m² building built to achieve BREEAM Very Good level of performance. Main structure is a concrete ground bearing slab and steel frame on concrete pads. External walls are a mixture of facing bricks at ground level and render above. Roof construction is a single ply warm roof construction, with a small amount of patent glazing.

Gross internal floor area	860 m²
Model location is South East England	TPI = 561; LF = 0.98

This cost model is copyright of AECOM

Community Centre	Quantity	Unit	Rate	Total (£)	Cost (£/m²)
Substructure				**95,500**	**111.05**
Excavation & disposal off site	430	m³	22		
Reinforced concrete ground slab, hardcore, dpm, ground beams and column bases for steel frame (0.25 W/m².K)	860	m²	100		
Frame				**117,600**	**136.74**
Steel propped portal frame, cold rolled purlins, surface treatments (@ 60 kg/m²)	52	tonne	1,850		
Intumescent paint to give 30 minute fire protection to steelwork	52	tonne	370		
Allowance for miscellaneous works, protecting columns		item	3,050		
Roof				**147,700**	**171.74**
Ply roofing with rigid board insulation and single ply polymeric roof cladding (0.18 W/m².K)	887	m²	150		
Roof glazing, aluminium, double glazed (2.2 W/m².K)	16	m²	725		
Rainwater drainage, aluminium gutters and downpipes	50	m	60		
External Walls, Windows and Doors				**204,300**	**237.56**
Light gauge steel framing; board; natural insulation; clay facing brick externally (0.20 W/m².K)	47	m²	220		
Light gauge steel framing; board; natural insulation; rendered finish on cement profile substrate (0.20 W/m².K)	497	m²	180		
Curtain walling with powder coated aluminium frame	125	m²	725		
Aluminium triple glazed windows	25	m²	370		
Steel fire escape doors	2	nr	1,100		
Canopy over entrance	1	nr	2,450		
Internal Walls and Partitions				**62,100**	**72.21**
Sound reducing board and metal stud partitions; generally 2 layers of board each side; various levels of fire and sound insulations; part glazed as appropriate	700	m²	80		
Fireproofing between partitions and roof		item	6,100		

COMMUNITY CENTRE

Community Centre	Quantity	Unit	Rate	Total (£)	Cost (£/m²)
Internal Doors				**39,600**	**46.05**
Timber internal doorset with softwood frames and ironmongery; vision panels; including fire resisting where necessary	38	nr	850		
Timber acoustic sliding stacking panel partition; 6 m × 2.6 m high	1	nr	7,300		
Wall Finishes				**25,300**	**29.42**
Plaster blocks and board; emulsion paint generally	1,100	m²	16		
Ceramic wall tiles splashbacks to WC areas	130	m²	60		
Floor Finishes				**55,300**	**64.30**
Demountable stage	20	m²	180		
Vinyl flooring	300	m²	90		
Carpet tiling	275	m²	43		
Laminate timber	250	m²	49		
Entrance matwell	2	m²	310		
Ceiling Finishes				**32,900**	**38.26**
Concealed grid plasterboard suspended ceiling	500	m²	49		
Plasterboard fixed directly to roof purlins	350	m²	24		
Fittings and Furnishings				**30,000**	**34.88**
Kitchen fitting; worktops; shelving		item	30,000		
Sanitary Fittings				**22,900**	**26.63**
WC suite; urinals, including disabled facility including all sanitary and fittings	15	nr	1,200		
Equipment for shower rooms; including disabled person equipment	2	nr	2,450		
Disposal Installations				**4,700**	**4.52**
Waste, soil and vent installation; uPVC pipework and fittings	15	nr	310		
Water Installations				**37,000**	**43.02**
Hot and cold water supply to WC, kitchen	860	m²	43		
Space Heating and Air Treatment and Ventilation				**79,200**	**92.09**
Boiler		item	30,000		
Underfloor heating system	860	m²	55		
Toilet and kitchen extract		item	1,850		
Electrical and Gas Installations				**60,500**	**70.35**
Small power, basic and emergency lighting	860	m²	18		
General lighting	860	m²	24		
External building lighting generally		item	6,100		
Incoming services		item	18,000		

COMMUNITY CENTRE

Community Centre	Quantity	Unit	Rate	Total (£)	Cost (£/m²)
Protective, Communications and Special Installations				**51,300**	**59.65**
Lightning protection, earthing and bonding	860	m²	5		
Fire and intruder alarms, panic alarm buttons	860	m²	15		
Data installations and containment	860	m²	12		
Basic audio visual installation, projector, screen, satellite TV installation		item	24,000		
Builders Work				**9,200**	**10.70**
Forming holes, chases etc.		item	9,200		
Preliminaries and Contingency				**222,900**	**259.19**
Management costs; site establishment; site supervision @		15.0%			
Contingency @		5.0%			
Construction cost (rate based on GIFA)				**1,298,000**	**1,508.36**

CAR PARK

This cost model looks at a six-storey car park that forms part of a high-end multi-use development. it incorporates external cladding and an open roof, with external d-ramps and fully natural ventilation with BREEAM equal to very good. The multi-storey car park has a gross floor area of 40,000 m² has spaces for 1,300 cars.

Gross internal floor area	40,00 m²
Car parking spaces	1,300 nr
Model location is South East England	TPI = 561; LF = 0.98

This cost model is copyright of AECOM

Car Park	Quantity	Unit	Rate	Total (£)	Cost (£/m²)
Substructures				**1,880,300**	**47.01**
400 mm diameter CFA in situ concrete piles, average 18 m long, 335 nr	6,000	m	120,000		
Pile caps including excavation, disposal, concrete, reinforcement, formwork	4,800	m²	42		
Ground floor construction including ground beams/ post tensioned beams	4,250	m²	270		
tarmacadam surfacing to car park including type 1, base layer, binder and top coat	4,250	m²			
Frame, Upper Floors and Stairs				**9,900,000**	**247.50**
In situ reinforced concrete frame	40,000	m²	95		
160 mm thick post tensioned in situ reinforced concrete fat slab flooring	40,000	m²	150		
Pre-cast concrete stairs including balustrades and handrails	10	nr	8,300		
Allowance for upstands and movement joints		item	17,000		
Roof				**570,600**	**14.27**
Top finish including waterproofing to stair cores and plant rooms	280	m²	150		
Asphalt finish to top deck of car park	6,780	m²	75		
metal decking TATA Comflor CF80 including concrete, reinforcement, edge lipping to cores and plant rooms	280	m	60		
TATA D100 decking including insulation, waterproofing and top finish coat to cores	50	m	65		
External walls, windows and doors				**2,204,000**	**55.10**
Timber cladding on 140 mm blockwork substrate to stair cores	200	m²	350		
Single skin 100 mm Forticrete facing blockwork to party wall, stretcher bond	2,400	m²	90		
Timber louvres/blades on a metal carrier frame	5,000	m²	310		
140 mm dense fair faced blockwork to form staircase lobbies	830	m²	60		

CAR PARK

Car Park	Quantity	Unit	Rate	Total (£)	Cost (£/m²)
Capless stick system built external curtain walling system with glazed insulated clear glass units	100	m²	600		
Aluminium louvered screen	450	m²	330		
Double leaf external door sets including ironmongery	5	nr	2,500		
Flush steel faced external single door sets including ironmongery	37	nr	2,200		
Double leaf external door sets to risers and plant areas including ironmongery	4	nr	2,200		
Single leaf external door sets to risers including ironmongery	4	nr	1,750		
Internal Walls and Doors				**188,200**	**4.71**
Blockwork walls 100 mm thick	85	m²	55		
Blockwork walls 140 mm thick	450	m²	60		
Blockwork walls 200 mm thick	630	m²	95		
Single door sets including frame and ironmongery	23	nr	1,050		
Double door sets including ironmongery	50	nr	1,450		
Wall Finishes				**35,500**	**0.89**
Paint finish to exposed columns in car park	250	m²	6		
Decorations to walls/upstands	6,200	m²	6		
Floor Finishes				**235,800**	**5.90**
Asphalt finish to car park ramps and top level	360	m²	50		
Paint finish to ramps	1,440	m²	6		
screed to plant rooms	245	m²	39		
Paint finish to floor of plant rooms	245	m²	17		
Paint finish to pedestrian walkways and around disabled parking bays	3,164	m²	17		
Paint to floor stairs and landings	850	m²	8		
Parking lines including zebra crossing, arrows, parking spaces etc.	25,000	m²	6		
Ceiling Finishes				**216,700**	**5.42**
Paint to soffit of slabs and edges of exposed beams to car park	36,600	m²	6		
Paint to soffits of ramps	1,440	m²	6		
Paint to soffit of stairs and landings	850	m²	9		
Fittings and Furnishings				**166,000**	**4.15**
Statutory signage		item	11,000		
Directional signage		item	32,500		
Bollards		item	95,000		
Trolley bays tubular stainless steel guarding		item	27,500		

CAR PARK

Car Park	Quantity	Unit	Rate	Total (£)	Cost (£/m²)
Sanitary Fittings				7,800	0.20
Unisex disabled toilet suite including WC, wash hand basin, grab rails, and other fittings	2	nr	1,550		
Cleaners sink	6	nr	775		
Disposal Installation				290,000	7.25
Car park drainage/rainwater installation	40,000	m²	7		
Petrol interceptor		item	13,000		
Foul drainage installation to sanitary appliances		item	13,000		
Water Installation				20,700	0.52
Water installation for car washing	3	nr	2,550		
Hot and cold water supplies to sanitary appliances		item	13,000		
Space Heating and Air Treatment and Ventilation				65,200	1.63
Parking attendant heating		item	1,300		
Trace heating		item	6,300		
Staircase ventilation for firefighting purposes		item	45,000		
Plant room ventilation		item	6,300		
Toilet ventilation		item	6,300		
Electrical Installation				2,024,000	50.60
LV distribution	40,000	m²	12		
General lighting	40,000	m²	22		
Emergency lighting (self-contained)	40,000	m²	7		
Lighting control system	40,000	m²	7		
Small power installation	40,000	m²	2		
General earthing	40,000	m²	1		
Protective, Communications and Special Installations				1,611,100	40.28
Dry riser installations	12	landings	3,200		
Lightning protection	40,000	m²	1		
Fire alarm installation	40,000	m²	19		
Fire telephones and disabled refuge alarms		item	19,000		
CCTV installation	50	nr	4,200		
Car park barriers	4	nr	19,000		
VMS signage (one per floor)	6	nr	25,000		
Pay-on-foot machines	12	nr	9,400		
Electric charging points – active (20%)	260	nr	290		
Electric charging points – passive (20%)	26	nr	290		
Solar PV panels	200	m²	650		
Lift Installations				440,000	11.00
13 person passenger/goods lifts 1.0 mps	2	nr	90,000		
13 person passenger/fire fighting lifts 1.0 mps	2	nr	130,000		

CAR PARK

Car Park	Quantity	Unit	Rate	Total (£)	Cost (£/m²)
Builders Work				**110,000**	**2.75**
Forming holes, chases etc.		item	110,000		
Preliminaries and Contingency				**2,917,100**	**72.93**
Management costs; site establishment; site supervision @		14.5%			
Testing and commissioning		item	22,000		
Construction cost (rate based on GIFA)				**22,883,000**	**572.11**
Cost per parking space				**18,000**	

Structural Engineer's Pocket Book
Eurocodes, Third Edition

Fiona Cobb

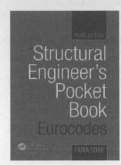

The hugely useful Structural Engineer's Pocket Book is now overhauled and revised in line with the Eurocodes. It forms a comprehensive pocket reference guide for professional and student structural engineers, especially those taking the IStructE Part 3 exam. With stripped-down basic material -- tables, data, facts, formulae and rules of thumb – it is directly usable for scheme design by structural engineers in the office, in transit or on site.

It brings together data from many different sources, and delivers a compact source of job-simplifying and time-saving information at an affordable price. It acts as a reliable first point of reference for information that is needed on a daily basis.

This third edition is referenced throughout to the structural Eurocodes. After giving general information, and details on actions on structures, it runs through reinforced concrete, steel, timber and masonry.

October 2014: 133 x 203: 408pp
Pb: 978-0-08-097121-6: £23.00

To Order: Tel: +44 (0) 1235 400524 Fax: +44 (0) 1235 400525
or Post: Taylor and Francis Customer Services,
Bookpoint Ltd, Unit T1, 200 Milton Park, Abingdon, Oxon, OX14 4TA UK
Email: book.orders@tandf.co.uk

For a complete listing of all our titles visit:
www.tandf.co.uk

Taylor & Francis
Taylor & Francis Group

Preliminaries Build-up Example

The number of items priced in the preliminaries section of Bills of Quantities and the manner in which they are priced vary considerably between contractors. Some contractors, by modifying their percentage factor for overheads and profit, attempt to cover the costs of preliminary items in their Prices for Measured Work. However, the cost of Preliminaries will vary widely according to job size and complexity, site location, accessibility, degree of mechanization practicable, position of the contractor's head office and relationships with local labour/domestic subcontractors. It is therefore usually far safer to price preliminary items separately on their merits according to the project.

It is not possible for the quantity surveyor/cost manager to quantify the main contractors' preliminaries and it is left for the contractor to interpret the tender documentation and ascertain his resources and method of working they require to complete the works.

This preliminaries bill is therefore a relatively simple pricing schedule of headings under which the contractor prices their preliminaries items.

An example in pricing preliminaries follows, and this assumes the form of contract used is the JCT Standard Building Contract With Quantities (SBC/Q) and the value, including preliminaries, is approximately £4,000,000. The contract is estimated to take 52 weeks to complete and the value is built up as follows:

	£
Contractors' Labour, Plant and Materials value	2,800,000
Subcontracts and Provisional Sums	730,000
Preliminaries, say	£470,000
Project value	**£4,000,000**

At the end of the section the data is summarized to give a total value of preliminaries for the project example.

NOTE the term 'Not priced', where used throughout this section means either that the cost implication is negligible or that it is usually included elsewhere in the tender.

1 Preliminaries

Part B Pricing Schedule

1.1   Employer's requirements
1.1.1 Site accommodation – not priced
1.1.2 Site records – not priced
1.1.3 Completion and post completion requirements

Defects after completion	Based on 0.2% of the contract sum	**£8,000.00**

1.2   Main Contractor's cost items
1.2.1 Management and staff – main contractor's project specific management and staff

Management and staff	Based on 6% of the tender (excluding preliminaries)	**£212,000**

1.2.2 Site establishment – main contractors and common user temporary site accommodation

Management and staff	Based on 0.35% of the tender (excluding preliminaries)	**£16,000**

1.2.3 Temporary services

Lighting and power	Based on 1% of the tender (excluding preliminaries)	**£36,000**
Water for the works (provided from mains)	Based on 0.1% of the tender (excluding preliminaries)	**£4,000**
Temporary telephones etc.	Based on 0.15% of the tender (excluding preliminaries)	**£6,000**

1.2.4 Security – allow for staff and security equipment

Security	Based on 0.15% of the tender (excluding preliminaries)	**£6,000**
Temporary hoarding, fencing etc.	2.4 m high fencing of OSB boarding one side; pair of gates	**£15,000**

1.2.5 Safety and environmental protection: – compliance with all welfare facilities, first aid etc.

Safety, health and welfare	Based on 0.1% of the tender (excluding preliminaries)	**£8,000**

1.2.6 Control and protection – allowance for setting out, protection of the works, sampling

Control and protection of the works	Based on 0.1% of the tender (excluding preliminaries)	**£4,000**
Drying the works	Based on 0.03% of the tender (excluding preliminaries)	**£2,000**

1.2.7 Mechanical plant – common user mechanical plant and equipment

Plant/Transport	Based on 2% of the tender (excluding preliminaries)	**£80,000**
Small plant and tools	Based on 0.2% of the tender (excluding preliminaries)	**£8,000**

1.2.8 Temporary works – common user access scaffolding

Temporary roads and walkways	Say 250 × 3.5 m hardcore/recycled concrete	**£9,000**
Access scaffolding	Based on 0.75% of the tender (excluding preliminaries)	**£27,000**

1.2.9 Site records – photographs, progress reporting etc. – not priced
1.2.10 Completion and post completion requirements – testing and commissioning, handover plan etc. – not priced

1.2.11 Cleaning – the cost would normally represent an allowance for final clearing of the works on completion with the residue for cleaning throughout the contract period. This amount being reduced with the introduction of waste management plans

Removing rubbish and cleaning	Based on 0.15% of the tender (excluding preliminaries)	**£6,000**

1.2.12 Fees and charges – any miscellaneous fees charges (rates etc.) – not priced
1.2.13 Site services – temporary works not specific to an element

Traffic regulations	Based on 0.05% of the tender (excluding preliminaries)	**£2,000**
Additional temporary items	Based on 0.25% of the tender (excluding preliminaries)	**£10,000**

1.2.14 Insurances, bonds, guarantees and warranties

Contract value (including preliminaries), say	£4,000,000
Estimated increased costs during contract period, say 2%	£80,000
	£4,080,000
Estimated increased costs incurred during period of reinstatement, say 5%	£204,000
	£4,202,400
Professional fees, say 15.0%	£642,600
	£4,926,600
Insurance of the works — Based on 0.15% of project value	**£8,000**

If at the Contractor's risk, the insurance cover must be sufficient to include the full cost of reinstatement, all increases in cost, professional fees and any consequential costs such as demolition. The average provision for fire risk is 0.15% of the value of the work after adding for increased costs and professional fees.

NOTE: Insurance premiums are liable to considerable variation, depending on the contractor, the nature of the work and the market in which the insurance is placed.

Summary of Preliminaries Costs Example (NB round to nearest £1000)	
Defects after completion	£8,000
Management and staff	£212,000
Site accommodation	£16,000
Lighting and power for the works	£36,000
Water for the works	£4,000
Temporary telephones	£6,000
Security	£6,000
Hoardings, fans, fencing, etc..	£15,000
Safety, health and welfare	£8,000
Protection of the works	£4,000
Drying the works	£2,000
Large plant/transport	£80,000
Small plant and tools	£38,000
Temporary roads and walkways	£9,000
Access scaffolding	£27,000
Removing rubbish, etc.., and cleaning	£6,000
Traffic regulations	£2,000
Additional temporary works	£10,000
Insurance	£8,000
TOTAL PRELIMINARIES	**£467,000**

It is emphasized that the above is an example only of the way in which Preliminaries may be priced and it is essential that for any particular contract or project the items set out in Preliminaries should be assessed on their respective values. The value of the Preliminaries items in recent tenders received by the editors varies from a 10% to 18% addition to all other costs. The above example represents approximately a 13.2% addition to the value of measured work.

Approximate Estimating Rates

Estimating by means of priced approximate quantities is always more accurate than by using overall building prices per square metre. Prices given in this section, which is arranged in elemental order, are derived from *Prices from Measured Works* section, but also include for all the incidental items and labours which are normally measured separately in Bills of Quantities. They have been established with a tender price level as indicated on the book cover. They include overheads and profit but *do not include for main contractors preliminaries*, for which an example build-up is given in an earlier section.

Whilst every effort is made to ensure the accuracy of these figures, they have been prepared for approximate estimating purposes and guidance only and on no account should they be used for the preparation of tenders.

Unless otherwise described units denoted as m² refer to appropriate unit areas (rather than gross internal floor areas).

As elsewhere in this edition prices do not include Value Added Tax or fees for professional services.

Approximate Estimating Rates are exclusive of any main contractor's preliminaries.

Cost plans can be developed from Elemental Cost Plans using both *Approximate Estimating Rates* and/or *Prices for Measured Works* depending upon the level of information available.

The cost targets within each formal cost plan approved by the employer will be used as the baseline for future cost comparisons. Each subsequent cost plan will require reconciliation with the preceding cost plan and explanations relating to changes made. In view of this, it is essential that records of any transfers made to or from the risk allowances and any adjustments made to cost targets are maintained, so that explanations concerning changes can be provided to both the employer and the project team.

PRELIMINARIES – TYPICAL PLANT HIRE RATES

Item	Unit	Range £		
PRELIMINARIES – TYPICAL PLANT HIRE RATES				
Minimum hire periods generally apply				
Typical rates for site accommodation				
Office cabins – 24 ft × 9 ft (20 m²)	week	25.00	to	32.00
Office cabins – 32 ft × 10 ft (30 m²)	week	32.00	to	40.50
Fire rated cabins – 24 ft × 9 ft (20 m²)	week	54.00	to	68.00
Fire rated cabins – 32 ft × 10 ft (30 m²)	week	73.00	to	92.00
Meeting room – 24 ft × 9 ft (20 m²)	week	25.00	to	32.00
Meeting room – 32 ft × 9 ft (30 m²)	week	32.00	to	40.50
Mess cabins (incl Furniture)	week	30.00	to	38.00
Drying rooms (incl Furniture)	week	34.00	to	43.00
Safestore-20ft × 8 ft (15 m²)	week	17.60	to	22.00
Container with padlock – 20ft × 8 ft (15 m²)	week	9.80	to	12.30
Toilets – 13 ft × 9 ft	week	39.00	to	49.00
Signs and notices	week	49.00	to	62.00
Fire extinguishers	week	7.80	to	9.85
Kitchen with cooker, fridge, sink, water heater; 32 ft × 10 ft	week	145.00	to	185.00
Mess room with wash basin, water heater, seating; 16 ft × 7 ft	week	88.00	to	110.00
Toilets – fitted to mains; three pan unit	week	155.00	to	200.00
Toilets – fitted to mains; four pan unit	week	185.00	to	235.00
Haulage to and from site; Site offices, Storage sheds, Toilets	item	4300.00	to	5400.00
Power and Lighting				
3 Kva transformer	week	9.70	to	12.30
5 Kva transformer	week	16.10	to	20.00
10 Kva transformer	week	32.50	to	41.00
4 Way distribution Box; 110 v	week	8.25	to	10.40
14 m extension cable	week	4.00	to	5.00
25 m extension cable	week	10.40	to	13.10
2 × 500 W floodlights	week	12.20	to	15.30
2 × 500 W floodlights on stands	week	15.20	to	19.20
Generators				
1 Kva 240 v portable (low noise)	week	38.00	to	48.00
2 Kva dual volt, portable petrol	week	26.00	to	33.00
5 Kva silenced	week	74.00	to	93.00
7.5 Kva silenced	week	110.00	to	140.00
15 Kva silenced	week	160.00	to	200.00
40 Kva silenced	week	235.00	to	300.00
60 Kva silenced	week	245.00	to	310.00
108 Kva silenced	week	305.00	to	385.00
165 Kva silenced	week	380.00	to	480.00
Mixers and Small Plant				
small mixer; 5–3 1/2 diesel	week	35.50	to	45.00
mixer; 7–5 diesel	week	44.00	to	56.00
compresser	week	93.00	to	120.00
Small dehumidifers	week	24.00	to	30.50
Large dehumidifers	week	64.00	to	81.00
Turbo dryer	week	40.00	to	51.00
Gas jet air heater; 260,000 Btu	week	48.00	to	61.00

1.1 SUBSTRUCTURE

Item	Unit	Range £		
1 SUBSTRUCTURE				
1.1 SUBSTRUCTURE				
Strip or trenchfill foundations with masonry up to 150 mm above floor level only; blinded hardcore bed; slab insulation; reinforced ground bearing slab 200 mm thick. To suit residential and small commercial developments with good ground bearing capacity				
shallow foundations up to 1.00 m deep	m^2	180.00	to	220.00
shallow foundations up to 1.50 m deep	m^2	210.00	to	250.00
Foundations in poor ground; mini piles; typically 300 mm diameter; 15 m long; 175 mm thick reinforced concrete slab; for single storey commercial type development				
minipiles to building columns only; 1 per column	m^2	140.00	to	170.00
minipiles to columns and perimeter ground floor beam; piles at 2.00 m centres to ground beams	m^2	265.00	to	320.00
minipiles to entire building at 2.00 m × 2.00 m grid	m^2	290.00	to	350.00
Raft foundations				
simple reinforced concrete raft on poorer ground for development up to two storey high	m^2	155.00	to	190.00
Basements				
Reinforced concrete basement floors; NOTE: excluding bulk excavation costs and hardcore fill				
200 mm thick waterproof concrete with 2 × layers of A252 mesh reinforcement	m^2	71.00	to	86.00
Reinforced concrete basement walls; NOTE: excluding bulk excavation costs and hardcore fill				
200 mm thick waterproof reinforced concrete	m^2	360.00	to	435.00
Trench fill foundations				
Machine excavation, disposal, plain in situ concrete 20.00 N/mm² – 20 mm aggregate (1:2:4) trench fill, 300 mm high cavity masonry in cement mortar (1:3), pitch polymer damp roof course				
With 3 courses of common brick outer skin up to dpc level (PC bricks @ £240/1000)				
600 mm × 1000 mm deep	m	110.00	to	140.00
600 mm × 1500 mm deep	m	155.00	to	185.00
With 3 courses of facing bricks brick outer skin up to dpc level (PC bricks @ £350/1000)				
600 mm × 1000 mm deep	m	115.00	to	140.00
600 mm × 1500 mm deep	m	160.00	to	190.00
With 3 courses of facing bricks brick outer skin up to dpc level (PC bricks @ £500/1000)				
600 mm × 1000 mm deep	m	120.00	to	140.00
600 mm × 1500 mm deep	m	160.00	to	190.00

1.1 SUBSTRUCTURE

Item	Unit	Range £		
1.1 SUBSTRUCTURE – cont				
Strip foundations				
Excavate trench 600 mm wide, partial backfill, partial disposal, earthwork support (risk item), compact base of trench, plain in situ concrete 20.00 N/mm² – 20 mm aggregate (1:2:4), 250 mm thick, 3 courses of common brick up to dpc level, cavity brickwork/blockwork in cement mortar (1:3), pitch polymer damp proof course machine excavation				
With 3 courses of common brick outer skin up to dpc level (PC bricks @ £240/1000)				
600 mm × 1000 mm deep	m	120.00	to	150.00
600 mm × 1500 mm deep	m	160.00	to	200.00
With 3 courses of common brick outer skin up to dpc level (PC bricks @ £350/1000)				
600 mm × 1000 mm deep	m	130.00	to	155.00
600 mm × 1500 mm deep	m	170.00	to	205.00
With 3 courses of common brick outer skin up to dpc level (PC bricks @ £500/1000)				
600 mm × 1000 mm deep	m	130.00	to	160.00
600 mm × 1500 mm deep	m	170.00	to	205.00
Column bases				
Excavate pit in firm ground by machine, partial backfill, partial disposal, support, compact base of pit; in situ concrete 25.00 N/mm²; reinforced at 50 kg/m³				
rates per base for base size				
600 mm × 600 mm × 300 mm; 1000 mm deep pit	nr	120.00	to	145.00
900 mm × 900 mm × 450 mm; 1250 mm deep pit	nr	190.00	to	230.00
1500 mm × 1500 mm × 600 mm; 1500 mm deep pit	nr	480.00	to	580.00
2700 mm × 2700 mm × 1000 mm; 1500 mm deep pit	nr	1875.00	to	2275.00
rates per m³ for base size				
600 mm × 600 mm × 300 mm; 1000 mm deep pit	m³	990.00	to	1175.00
900 mm × 900 mm × 450 mm; 1250 mm deep pit	m³	540.00	to	650.00
1500 mm × 1500 mm × 600 mm; 1500 mm deep pit	m³	370.00	to	450.00
2700 mm × 2700 mm × 1000 mm; 1500 mm deep pit	m³	270.00	to	330.00
extra for reinforcement at 75 kg/m³ concrete, base size	m³	83.00	to	100.00
extra for reinforcement at 100 kg/m³ concrete, base size	m³	105.00	to	125.00
Piling				
Enabling Works				
excavate to form piling mat; supply and lay imported hardcore – recycled brick and similar to form piling mat	m³	9.65	to	11.70
provision of plant (1nr rig); including bringing to and removing from site; maintenance, erection and dismantling at each pile position	item	15000.00	to	18000.00
Supply and install concrete Continuous Flight Auger (CFA) piles; cart away inactive spoil; measure total length of pile				
450 mm diameter reinforced concrete CFA piles	m	44.00	to	53.00
600 mm diameter reinforced concrete CFA piles	m	71.00	to	85.00
750 mm diameter reinforced concrete CFA piles	m	110.00	to	130.00
900 mm diameter reinforced concrete CFA piles	m	155.00	to	190.00

1.1 SUBSTRUCTURE

Item	Unit	Range £		
Pile testing				
using tension piles as reaction; typical loading test 1000 kN to 2000 kN	item	6700.00	to	8200.00
integrity testing; minimum 20 per visit	nr	11.40	to	13.80
Secant wall piling, 750 mm diameter piles' including site mobilization and demobilization; Measure total area of piling (total length × total depth)				
supply and install secant wall piling	m²	245.00	to	295.00
Steel sheet piling; Measure total area of piling (total length × total depth)				
interlocking steel sheet piling to excavation perimeter; Corus LX or similar; extraction on completion	m²	160.00	to	190.00
Pile caps				
Excavate pit in firm ground by machine, partial backfill, partial disposal, support, compaction, cut off pile and prepare reinforcement, formwork; reinforced in situ concrete cap 25 N/mm²; reinforcement at 50 kg/m³; rates per cap for cap size				
900 mm × 900 mm × 1000 mm; 1–2 piles	nr	360.00	to	440.00
2100 mm × 2100 mm × 1000 mm; 2–3 piles	nr	1175.00	to	1425.00
2700 mm × 2700 mm × 1500 mm; 3–5 piles	nr	2650.00	to	3200.00
Excavate pit in firm ground by machine, partial backfill, partial disposal, support, compaction, cut off pile and prepare reinforcement, formwork; reinforced in situ concrete cap 25 N/mm²; reinforcement at 50 kg/m³; rates per m3 for typical cap size				
up to 900 mm × 900 mm × 1000 mm; 1–2 piles	m³	415.00	to	500.00
up to 2100 mm × 2100 mm × 1000 mm; 2–3 piles	m³	270.00	to	330.00
up to 2700 mm × 2700 mm × 1500 mm; 3–5 piles	m³	245.00	to	300.00
extra for				
reinforcement at 75 kg/m³ concrete	m³	83.00	to	100.00
reinforcement at 100 kg/m³ concrete	m³	100.00	to	125.00
alternative strength concrete C30 (30.00 N/mm²)	m³	1.40	to	1.70
alternative strength concrete C40 (40.00 N/mm²)	m³	7.10	to	8.60
Concrete ground beams				
Reinforced in situ concrete ground beams; bar reinforcement; formwork				
300 mm × 300 mm, reinforcement at 180 kg/m³	m	46.00	to	56.00
450 mm × 450 mm, reinforcement at 200 kg/m³	m	90.00	to	110.00
450 mm × 600 mm, reinforcement at 270 kg/m³	m	130.00	to	160.00
Precast concrete ground beams; square ends and dowelled. Supplied and installed in lengths to span between stranchion bases; concrete strength C50; maximum UDL of 4.5 kN/m; in situ work at beam ends not included.				
350 mm × 425 mm	m	76.00	to	92.00
350 mm × 875 mm	m	180.00	to	220.00
350 mm × 1075 mm	m	250.00	to	300.00
450 mm × 400 mm	m	90.00	to	110.00
Concrete lift pits				
Excavate and disposal; reinforced concrete floor and walls; bitumen tanking as necessary				
1.65 m × 1.81 m × 1.60 m deep pit – 8 person lift/630 kg	nr	2125.00	to	2550.00
1.80 m × 2.50 m × 1.60 m deep pit – 13 person lift/1000 kg	nr	2650.00	to	3200.00
3.00 m × 2.50 m × 1.60 m deep pit – 21 person lift/1700 kg	nr	3100.00	to	3800.00

1.1 SUBSTRUCTURE

Item	Unit	Range £		
1.1 SUBSTRUCTURE – cont				
Ground floor				
Mechanical excavation to reduce levels, disposal, level and compact, hardcore bed blinded with sand, 1200 gauge polythene damp proof membrane, in situ concrete 20.00 N/mm² – 20 mm aggregate (1:2:4)				
150 mm thick concrete slab with 1 layer of A195 fabric reinforcement	m²	86.00	to	100.00
200 mm thick concrete slab with 1 layer of A252 fabric reinforcement	m²	93.00	to	110.00
250 mm thick concrete slab with 1 layer of A393 fabric reinforcement	m²	95.00	to	115.00
extra for				
every additional 50 mm thick concrete	m²	7.00	to	8.45
alternative strength concrete C30 (30.00 N/mm²)	m³	1.40	to	1.70
alternative strength concrete C40 (40.00 N/mm²)	m³	7.10	to	8.60
reinforcement at 25 kg/m³ concrete	m³	31.00	to	37.50
reinforcement at 50 kg/m³ concrete	m³	52.00	to	62.00
reinforcement at 75 kg/m³ concrete	m³	83.00	to	100.00
reinforcement at 100 kg/m³ concrete	m³	100.00	to	125.00
Warehouse ground floor				
Steel fibre reinforced floor slab placed using large pour construction techniques providing a finish floor flatness complying with FM2 special +/- 15 mm from datum. NOTE: Excavation, subbase and damp proof membrane not included				
nominal 200 mm thick in situ concrete floor slab, concrete grade C40, reinforced with steel fibres, surface power floated and cured with a spray application of curing and hardening agent	m²	29.00	to	35.50
Suspended ground floor				
Beam and block flooring				
suspended floor with 150 mm deep precast concrete beams and infill blocks	m²	52.00	to	63.00
suspended floor with 225 mm deep precast concrete beams and infill blocks	m²	56.00	to	68.00
Board or slab insulation				
Kingspan Thermafloor TF70 (Thermal conductivity 0.022 W/mK) rigid urethane floor insulation for solid concrete and suspended ground floors				
50 mm thick	m²	15.10	to	18.30
75 mm thick	m²	21.00	to	25.00
100 mm thick	m²	26.00	to	32.00
115 mm thick	m²	14.10	to	17.00
125 mm thick	m²	31.00	to	37.50
150 mm thick	m²	38.00	to	45.50
Styrofoam Floormate 500 (Thermal conductivity 0.033 W/mK) extruded polystyrene foam or other equal and approved				
50 mm thick	m²	18.70	to	22.50
80 mm thick	m²	24.00	to	29.00
120 mm thick	m²	30.00	to	36.50

2.1 FRAME

Item	Unit	Range £		

Underpinning
In stages not exceeding 1500 mm long from one side of existing wall and
foundation, excavate preliminary trench by machine and underpinning pit by
hand, partial backfill, partial disposal, earthwork support (open boarded), cutting
away projecting foundations, prepare underside of existing, compact base of pit,
plain in situ concrete 20.00 N/mm² – 20 mm aggregate (1:2:4), formwork,
brickwork in cement mortar (1:3), pitch polymer damp proof course, wedge and
pin to underside of existing with slates Commencing at 1.00 m below ground
level with common bricks, depth of underpinning

Item	Unit	Range £		
900 mm high, one brick wall	m²	360.00	to	435.00
1500 mm high, one brick wall	m²	520.00	to	630.00
extra for excavating commencing				
2.00 m below ground level	m²	71.00	to	86.00
3.00 m below ground level	m²	140.00	to	170.00
4.00 m below ground level	m²	190.00	to	235.00

Temporary works
Roadways
    formation of temporary roads to building perimeter comprising of geoxtile
    membrane and 300 mm MOT type 1; reduce level; spoil to heap on site within

Item	Unit	Range £		
25 m of excavation	m²	26.00	to	31.00
installation of wheel wash facility and maintenance	nr	2700.00	to	3300.00

2 SUPERSTRUCTURE

2.1 FRAME

Comparative Frame and Upper Floors; Upper floor area (unless otherwise
described)

Concrete frame; flat slab reinforced concrete floors up to 250 mm thick
Suspended slab; no coverings or finishes

Item	Unit	Range £		
up to six storeys	m²	125.00	to	150.00
six to twelve storeys	m²	135.00	to	165.00
thirteen to eighteen storeys	m²	160.00	to	190.00

Steel frame; composite beam and slab floors
Suspended slab; permanent steel shuttering with 130 mm thick concrete; no
coverings or finishes

Item	Unit	Range £		
up to six storeys	m²	135.00	to	160.00
seven to twelve storeys	m²	150.00	to	185.00
thirteen to eighteen storeys	m²	170.00	to	200.00

**Frame only; Reinforced in situ concrete columns, bar reinforcement,
formwork. Generally all formwork assumes four uses**
Reinforcement rate 180 kg/m³; column size

Item	Unit	Range £		
225 mm × 225 mm	m	75.00	to	90.00
300 mm × 600 mm	m	165.00	to	200.00
450 mm × 900 mm	m	305.00	to	370.00

2.1 FRAME

Item	Unit	Range £		
2.1 FRAME – cont				
Frame only – cont				
Reinforcement rate 240 kg/m³; column size				
225 mm × 225 mm	m	84.00	to	100.00
300 mm × 600 mm	m	180.00	to	220.00
450 mm × 900 mm	m	345.00	to	415.00
In situ concrete casing to steel column, formwork; column size				
225 mm × 225 mm	m	65.00	to	79.00
300 mm × 600 mm	m	155.00	to	190.00
450 mm × 900 mm	m	270.00	to	325.00
Frame only; Reinforced in situ concrete beams, bar reinforcement, formwork. Generally all formwork assumes four uses				
Reinforcement rate 200 kg/m³; beam size				
225 mm × 450 mm	m	96.00	to	115.00
300 mm × 600 mm	m	160.00	to	190.00
450 mm × 600 mm	m	205.00	to	250.00
600 mm × 600 mm	m	250.00	to	305.00
Reinforcement rate 240 kg/m³; beam size				
225 mm × 450 mm	m	105.00	to	130.00
300 mm × 600 mm	m	170.00	to	205.00
450 mm × 600 mm	m	215.00	to	260.00
600 mm × 600 mm	m	270.00	to	330.00
In situ concrete casing to steel beams, formwork; beam size				
225 mm × 450 mm	m	87.00	to	105.00
300 mm × 600 mm	m	130.00	to	160.00
450 mm × 600 mm	m	180.00	to	215.00
600 mm × 600 mm	m	215.00	to	260.00
Other floor and frame constructions				
reinforced concrete cantilevered balcony; up to 1.50 m wide × 1.20 m deep	nr	2650.00	to	3200.00
reinforced concrete cantilevered walkways; up to 1.00 m wide	m²	180.00	to	215.00
reinforced concrete walkways and supporting frame; up to 1.00 m wide × 2.50 m high	m²	230.00	to	280.00
Frame only; steel frame				
Fabricated steelwork erected on site with bolted connections, primed				
smaller sections ne 40 kg/m	tonne	1850.00	to	2250.00
universal beams; grade S275	tonne	1700.00	to	2075.00
universal beams; grade S355	tonne	1775.00	to	2175.00
universal columns; grade S275	tonne	1700.00	to	2075.00
universal columns; grade S355	tonne	1775.00	to	2175.00
composite columns	tonne	1675.00	to	2000.00
hollow section circular	tonne	1925.00	to	2325.00
hollow section square or rectangular	tonne	1925.00	to	2325.00
cellular beams (FABSEC)	tonne	1925.00	to	2325.00
lattice beams	tonne	2375.00	to	2900.00
roof trusses	tonne	2500.00	to	3000.00

2.2 UPPER FLOORS

Item	Unit	Range £		
Steel finishes				
grit blast and one coat zinc chromate primer	m²	7.50	to	9.10
touch up primer and one coat of two pack epoxy zinc phosphate primer	m²	5.00	to	6.05
blast cleaning	m²	2.50	to	3.00
galvanizing	m²	13.80	to	16.70
galvanizing	tonne	205.00	to	250.00
Other floor and frame constructions				
steel space deck on steel frame, unprotected	m²	250.00	to	300.00
18.00 m high bay warehouse; steel propped portal frame, cold rolled purlin sections, primed surface treatment only, excluding decorations and protection	m²	93.00	to	110.00
columns and beams to mansard roof, 60 min fire protection	m²	150.00	to	180.00
feature columns and beams to glazed atrium roof unprotected	m²	155.00	to	190.00
Fire protection to steelwork				
Sprayed mineral fibre; gross surface area				
60 minute protection	m²	15.80	to	19.10
90 minute protection	m²	28.00	to	34.00
Sprayed vermiculite cement; gross surface area				
60 minute protection	m²	18.10	to	22.00
90 minute protection	m²	26.50	to	32.00
Supply and fit fire resistant boarding to steel columns and beams; noggins, brackets and angles, intumescent paste. Beamclad or similar; measure board area				
30 minute protection for concealed applications	m²	24.00	to	29.00
60 minute protection for concealed applications	m²	51.00	to	62.00
30 minute protection left exposed for decoration	m²	26.50	to	32.00
60 minute protection left exposed for decoration	m²	61.00	to	73.00
Intumescent fire protection coating/decoration to exposed steelwork; Gross surface area (m²) or per tonne; On site application, spray applied				
30 minute protection per m²	m²	10.40	to	12.60
30 minute protection per tonne	tonne	265.00	to	320.00
60 minute protection per m²	m²	13.20	to	15.90
60 minute protection per tonne	tonne	335.00	to	405.00
Intumescent fire protection coating/decoration to exposed steelwork; Gross surface area (m²) or per tonne; Off site application, spray applied				
60 minute protection per m²	m²	21.00	to	25.50
60 minute protection per tonne	tonne	540.00	to	650.00
90 minute protection per m²	m²	41.00	to	49.00
90 minute protection per tonne	tonne	1025.00	to	1250.00
120 minute protection per m²	m²	63.00	to	76.00
120 minute protection per tonne	tonne	1600.00	to	1950.00
2.2 UPPER FLOORS				
Rates for area of flooring				
Composite steel and concrete upper floors (Note: all floor thicknesses are nominal); A142 mesh reinforcement				
TATA Slimdek SD225 steel decking 1.25 mm thick; 130 mm reinforced concrete topping	m²	87.00	to	105.00
Re-entrant type steel deck 0.90 mm thick; 150 mm reinforced concrete	m²	72.00	to	87.00
Re-entrant type steel deck 60 mm deep, 1.20 mm thick; 200 mm reinforced concrete	m²	78.00	to	94.00

2.3 ROOF

Item	Unit	Range £		
2.2 UPPER FLOORS – cont				
Trapezoidal steel decking 0.90 mm thick; 150 mm reinforced concrete topping	m²	74.00	to	89.00
Trapezoidal steel decking 1.20 mm thick; 150 mm reinforced concrete topping	m²	77.00	to	93.00
Composite precast concrete and in situ concrete upper floors				
Omnidec (Hanson Building Products) flooring system comprising of 50 mm thick precast concrete deck (as permanent shuttering); 210 mm polystyrene void formers; concrete topping and reinforcement	m²	105.00	to	130.00
Post-tensioned concrete upper floors				
reinforced post-tensioned suspended concrete slab 150–225 mm thick, 40 N/mm², reinforcement 60 kg/m³, formwork	m²	100.00	to	120.00
Precast concrete suspended floors				
1200 mm wide suspended slab; 75 mm thick screed; no coverings or finishes				
3.00 m span; 150 mm thick planks; 5.00 kN/m² loading	m²	67.00	to	77.00
6.00 m span; 150 mm thick planks; 5.00 kN/m² loading	m²	68.00	to	79.00
7.50 m span; 200 mm thick planks; 5.00 kN/m² loading	m²	71.00	to	83.00
9.00 m span; 250 mm thick planks; 5.00 kN/m² loading	m²	75.00	to	87.00
12.00 m span; 350 mm thick planks; 5.00 kN/m² loading	m²	79.00	to	91.00
3.00 m span; 150 mm thick planks; 8.50 kN/m² loading	m²	65.00	to	76.00
6.00 m span; 200 mm thick planks; 8.50 kN/m² loading	m²	67.00	to	78.00
7.50 m span; 250 mm thick planks; 8.50 kN/m² loading	m²	75.00	to	87.00
3.00 m span; 150 mm thick planks; 12.50 kN/m² loading	m²	68.00	to	79.00
6.00 m span; 250 mm thick planks; 12.50 kN/m² loading	m²	75.00	to	87.00
Softwood floors				
Joisted floor; plasterboard ceiling; skim; emulsion; t&g chipboard, sheet vinyl flooring and painted softwood skirtings	m²	61.00	to	79.00
2.3 ROOF				
Flat roof decking; structure only				
Softwood flat roofs; structure only				
comprising roof joists; 100 mm × 50 mm wall plates; herringbone strutting; 18 mm thick external quality plywood boarding	m²	46.00	to	56.00
Metal decking				
galvanized steel roof decking; insulation; three layer felt roofing and chippings; 0.70 mm thick steel decking (U-value = 0.25 W/m²K)	m²	70.00	to	84.00
aluminium roof decking; three layer felt roofing and chippings; 0.90 mm thick aluminium decking (U-value = 0.25 W/m²K)	m²	7.95	to	9.60
Concrete decking flat roofs; structure only				
precast concrete suspended slab with sand:cement screed over	m²	71.00	to	86.00
reinforced concrete slabs; on steel permanent steel shuttering; 150 mm reinforced concrete topping	m²	74.00	to	89.00
Screeds/Decks to receive roof coverings				
18 mm thick external quality plywood boarding	m²	24.50	to	29.50
50 mm thick cement and sand screed	m²	16.40	to	19.80
75 mm thick lightweight bituminous screed and vapour barrier	m²	57.00	to	69.00

2.3 ROOF

Item	Unit	Range £		
Softwood trussed pitched roofs; structure only				
Timber; roof plan area (unless otherwise described)				
comprising 75 mm × 50 mm Fink roof trusses at 600 mm centres (measured on plan)	m²	30.00	to	36.50
comprising 100 mm × 38 mm Fink roof trusses at 600 mm centres (measured on plan)	m²	34.00	to	41.00
Mansard type roof comprising 100 mm × 50 mm roof trusses at 600 mm centres; 70° pitch	m²	35.00	to	42.00
forming dormers	m²	580.00	to	700.00
Timber trusses with tile coverings				
Timber; roof plan area (unless otherwise described)				
Timber roof trusses; insulation; roof coverings; PVC rainwater goods; plasterboard; skim and emulsion to ceilings (U-value = 0.25 W/m²K)				
concrete interlocking tile coverings	m²	160.00	to	195.00
clay plain tile coverings	m²	170.00	to	210.00
clay pan tile coverings	m²	180.00	to	220.00
natural Welsh slate coverings	m²	210.00	to	255.00
reconstructed stone coverings	m²	200.00	to	240.00
Timber dormer roof trusses; insulation; roof coverings; PVC rainwater goods; plasterboard; skim and emulsion to ceilings (U-value = 0.25 W/m²K)				
concrete interlocking tile coverings	m²	175.00	to	210.00
clay pantile coverings	m²	180.00	to	220.00
plain clay tile coverings	m²	195.00	to	235.00
natural slate coverings	m²	205.00	to	250.00
composite slate coverings	m²	185.00	to	225.00
reconstructed stone coverings	m²	195.00	to	240.00
extra for end of terrace semi/detached configuration	m²	35.00	to	42.50
extra for hipped roof configuration	m²	38.00	to	46.00
Steel trussed with metal sheet cladding				
Steel roof trusses and beams; thermal and acoustic insulation (U-value = 0.25 W/m²K)				
aluminium profiled composite cladding	m²	240.00	to	290.00
copper roofing on boarding	m²	245.00	to	300.00
Flat roofing systems				
Includes insulation and vapour control barrier; excludes decking (U-value = 0.25 W/m²K)				
single layer polymer roofing membrane	m²	76.00	to	93.00
single layer polymer roofing membrane with tapered insulation	m²	135.00	to	165.00
20 mm thick polymer modified asphalt roofing including felt underlay	m²	79.00	to	96.00
high performance bitumen felt roofing system	m²	100.00	to	120.00
high performance polymer modified bitumen membrane	m²	105.00	to	130.00
Kingspan KS1000TD composite Single Ply roof panels for roof pitches greater than 0.7° (after deflection); 1.5 mm Single Ply External Covering, Internal Coating Bright White Polyester (steel)				
71 mm thick (U-value = 0.25 W/m²K)	m²	57.00	to	69.00
100 mm thick (U-value = 0.18 W/m²K)	m²	58.00	to	70.00
120 mm thick (U-value = 0.15 W/m²K)	m²	67.00	to	81.00

2.3 ROOF

Item	Unit	Range £		
2.3 ROOF – cont				
Flat roofing systems – cont				
Edges to felt flat roofs; softwood splayed fillet;				
280 mm × 25 mm painted softwood fascia; no gutter aluminium edge trim	m	36.00	to	44.00
Edges to flat roofs; code 4 lead drip dresses into gutter; 230 mm × 25 mm painted softwood fascia;				
100 mm uPVC gutter	m	67.00	to	81.00
100 mm cast iron gutter; decorated	m	56.00	to	68.00
Roof walkways				
600 mm × 600 mm × 50 mm precast concrete slabs on support system;				
pedestrian access	m²	45.00	to	54.00
extra for				
solar reflective paint	m²	2.55	to	3.10
limestone chipping finish	m²	4.50	to	5.45
grip tiles in hot bitumen	m²	31.00	to	38.00
Sheet roof claddings				
Fibre cement sheet profiled cladding				
Profile 6; single skin; natural grey finish	m²	24.00	to	29.50
P61 Insulated System; natural grey finish; metal inner lining panel (U-value = 0.25 W/m²K)	m²	48.00	to	58.00
extra for coloured fibre cement sheeting	m²	1.60	to	2.00
double skin GRP translucent roof sheets	m²	55.00	to	66.00
triple skin GRP translucent roof sheets	m²	61.00	to	74.00
Steel PVF2 coated galvanized trapezoidal profile cladding fixed to steel purlins (not included); for roof pitches greater than 4°				
single skin trapezoidal sheeting only	m²	15.00	to	18.20
composite insulated roofing system; 80 mm overall panel thickness (U-value = 0.25 W/m²K)	m²	43.50	to	53.00
composite insulated roofing system; 115 mm overall panel thickness (U-value = 0.18 W/m²K)	m²	46.00	to	55.00
composite insulated roofing system; 150 mm overall panel thickness (U-value = 0.14 W/m²K)	m²	62.00	to	74.00
For roof pitches greater than 1°				
standing seam joints composite insulated roofing system; 90 mm overall panel thickness (U-value = 0.25 W/m²K)	m²	70.00	to	85.00
standing seam joints composite insulated roofing system; 110 mm overall panel thickness (U-value = 0.20 W/m²K)	m²	76.00	to	92.00
standing seam joints composite insulated roofing system; 125 mm overall panel thickness (U-value = 0.18 W/m²K)	m²	79.00	to	96.00
Aluminium roofing; standing seam				
Kalzip standard natural aluminium; 0.9mm thick; 180 mm glassfibre insulation; vapour control layer; liner sheets; (U-value = 0.25 W/m²K)	m²	59.00	to	71.00
Copper roofing				
copper roofing with standing seam joints; insulation breather membrane or vapour barrier (U-value = 0.25 W/m²K)	m²	140.00	to	170.00
Stainless steel				
Terne-coated stainless steel roofing on and including Metmatt underlay (U-value = 0.25 W/m²K)	m²	115.00	to	140.00

2.3 ROOF

Item	Unit	Range £		
Lead				
Roof coverings in welded seam construction including Geotec underlay fixed to prevent lifting and distortion	m²	125.00	to	150.00
Comparative tiling and slating finishes				
Including underfelt, battening, eaves courses and ridges				
concrete troughed or bold roll interlocking tiles; sloping	m²	33.00	to	39.50
Tudor clay pantiles; sloping	m²	36.00	to	44.00
natural red pantiles; sloping	m²	42.00	to	51.00
blue composition (cement fibre) slates; sloping	m²	38.00	to	45.50
machine made clay plain tiles; sloping	m²	56.00	to	68.00
Welsh natural slates; sloping	m²	115.00	to	140.00
Spanish slates; sloping	m²	69.00	to	84.00
man made slates; sloping	m²	63.00	to	77.00
reconstructed stone slates; random slates; sloping	m²	54.00	to	66.00
handmade sandfaced plain tiles; sloping	m²	85.00	to	100.00
Landscaped roofs				
Polyester based elastomeric bitumen waterproofing and vapour equalization layer, copper lined bitumen membrane root barrier and waterproofing layer, separation and slip layers, protection layer, 50 mm thick drainage board, filter fleece, insulation board, Sedum vegetation blanket				
intensive (high maintenance – may include trees and shrubs require deeper substrate layers, are generally limited to flat roofs	m²	130.00	to	155.00
extensive (low maintenance – herbs, grasses, mosses and drought tolerant succulents such as Sedum)	m²	120.00	to	145.00
Ethylene tetrafluoroethylene (ETFE) systems				
Multiple layered ETFE inflated cushions supported by a lightweight aluminium or steel structure	m²	700.00	to	850.00
Insulation				
Glass fibre roll; Crown Loft Roll 44 (Thermal conductivity 0.044 W/mK) or other equal; laid loose				
100 mm thick	m²	2.45	to	2.95
150 mm thick	m²	3.10	to	3.80
200 mm thick	m²	3.80	to	4.60
Glass fibre quilt; Isover Modular roll (Thermal conductivity 0.043 W/mK) or other equal and approved; laid loose				
100 mm thick	m²	3.15	to	3.80
150 mm thick	m²	4.20	to	5.05
170 mm thick	m²	5.10	to	6.20
200 mm thick	m²	6.20	to	7.50
Crown Rafter Roll 32 (Thermal conductivity 0.032 W/mK) glass fibre flanged building roll; pinned vertically or to slope between timber framing				
50 mm thick	m²	5.35	to	6.50
75 mm thick	m²	7.00	to	8.45
100 mm thick	m²	8.55	to	10.30

2.4 STAIRS AND RAMPS

Item	Unit	Range £		
2.3 ROOF – cont				
Insulation – cont				
Thermafleec EcoRoll (0.039 W/mK)				
50 mm thick	m²	5.90	to	7.15
75 mm thick	m²	7.45	to	9.00
100 mm thick	m²	9.35	to	11.30
140 mm thick	m²	12.50	to	15.10
Rooflights/patent glazing and glazed roofs				
Rooflights				
individual polycarbonate rooflights; rectangular; fixed light	m²	440.00	to	530.00
individual polycarbonate rooflights; rectangular; manual opening	m²	610.00	to	730.00
individual polycarbonate rooflights; rectangular; electric opening	m²	910.00	to	1100.00
Velux style rooflights to traditional roof construction (tiles/slates)	m²	405.00	to	490.00
Patent glazing; including flashings, standard aluminium alloy bars; Georgian wired or laminated glazing; fixed lights				
single glazed	m²	170.00	to	210.00
low-e clear toughened and laminated double glazed units	m²	450.00	to	550.00
Rainwater disposal				
Gutters				
100 mm uPVC gutter	m	23.00	to	27.50
170 mm uPVC gutter	m	37.00	to	44.50
100 mm cast iron gutter; decorated	m	41.50	to	50.00
150 mm cast iron gutter; decorated	m	71.00	to	86.00
Rainwater downpipes pipes; fixed to backgrounds; including offsets and shoes				
68 mm diameter uPVC	m	19.40	to	23.50
110 mm diameter uPVC	m	35.50	to	43.00
75 mm diameter cast iron; decorated	m	62.00	to	75.00
100 mm diameter cats iron; decorated	m	81.00	to	98.00
2.4 STAIRS AND RAMPS				
Reinforced concrete construction				
Escape staircase; granolithic finish; mild steel balustrades and handrails				
3.00 m rise; dogleg	nr	5500.00	to	6700.00
plus or minus for each 300 mm variation in storey height	nr	540.00	to	650.00
Staircase; terrazzo finish; mild steel balustrades and handrails; plastered soffit; balustrades and staircase soffit decorated				
3.00 m rise; dogleg	nr	8400.00	to	10000.00
plus or minus for each 300 mm variation in storey height	nr	820.00	to	990.00
Staircase; terrazzo finish; stainless steel balustrades and handrails; plastered and decorated soffit				
3.00 m rise; dogleg	nr	10000.00	to	12000.00
plus or minus for each 300 mm variation in storey height	nr	990.00	to	1175.00
Staircase; high quality finishes; stainless steel and glass balustrades; plastered and decorated soffit				
3.00 m rise; dogleg	nr	15000.00	to	18500.00
plus or minus for each 300 mm variation in storey height	nr	1775.00	to	2175.00

2.4 STAIRS AND RAMPS

Item	Unit	Range £		
Metal construction				
Steel access/fire ladder				
3.00 m high	nr	630.00	to	770.00
4.00 m high; epoxide finished	nr	980.00	to	1175.00
Light duty metal staircase; galvanized finish; perforated treads; no risers; balustrades and handrails; decorated				
3.00 m rise; straight; 900 mm wide	nr	3600.00	to	4400.00
plus or minus for each 300 mm variation in storey height	nr	300.00	to	360.00
Light duty circular metal staircase; galvanized finish; perforated treads; no risers; balustrades and handrails; decorated				
3.00 m rise; straight; 1548 mm diameter	nr	4000.00	to	4900.00
plus or minus for each 300 mm variation in storey height	nr	355.00	to	430.00
Heavy duty cast iron staircase; perforated treads; no risers; balustrades and hand rails; decorated				
3.00 m rise; straight	nr	4600.00	to	5500.00
plus or minus for each 300 mm variation in storey height	nr	470.00	to	570.00
3.00 m rise; spiral; 1548 mm diameter	nr	5200.00	to	6300.00
plus or minus for each 300 mm variation in storey height	nr	520.00	to	630.00
Feature metal staircase; galvanized finish perforated treads; no risers; decorated				
3.00 m rise; spiral balustrades and handrails	nr	6000.00	to	7300.00
3.00 m rise; dogleg; hardwood balustrades and handrails	nr	7200.00	to	8700.00
3.00 m rise; dogleg; stainless steel balustrades and handrails	nr	9200.00	to	11000.00
plus or minus for each 300 mm variation in storey height	nr	610.00	to	740.00
galvanized steel catwalk; nylon coated balustrading 450 mm wide	m	340.00	to	410.00
Timber construction				
Softwood staircase; softwood balustrades and hardwood handrail; plasterboard; skim and emulsion to soffit				
2.60 m rise; standard; straight flight	nr	940.00	to	1150.00
2.60 m rise; standard; top three treads winding	nr	1075.00	to	1325.00
2.60 m rise; standard; dogleg	nr	1175.00	to	1450.00
Oak staircase; balustrades and handrails; plasterboard; skim and emulsion to soffit				
2.60 m rise; purpose made; dogleg	nr	8100.00	to	9700.00
plus or minus for each 300 mm variation in storey height	nr	1075.00	to	1275.00
Comparative finishes/balustrading				
Wall handrails				
Softwood handrail and brackets	m	71.00	to	86.00
Hardwood handrail and brackets	m	100.00	to	120.00
Mild steel handrail and brackets	m	130.00	to	160.00
Stainless steel handrail and brackets	m	170.00	to	200.00
Balustrading and handrails				
Mild steel balustrade and steel or timber handrail	m	260.00	to	315.00
Balustrade and handrail with metal infill panels	m	330.00	to	400.00
Balustrade and handrail with glass infill panels	m	360.00	to	440.00
Stainless steel balustrade and handrail	m	430.00	to	520.00
Stainless steel and structural glass balustrade	m	760.00	to	920.00

2.5 EXTERNAL WALLS

Item	Unit	Range £		
2.4 STAIRS AND RAMPS – cont				
Comparative finishes/balustrading – cont				
Finishes to treads and risers; including nosings etc.				
vinyl or rubber	m	51.00	to	61.00
carpet (PC sum £20/m²)	m	87.00	to	105.00
2.5 EXTERNAL WALLS				
Wall area (unless otherwise described)				
Brick/block walling; solid walls				
Common brick solid walls; bricks PC £450.00/1000				
half brick thick	m²	54.00	to	65.00
one brick thick	m²	100.00	to	120.00
one and a half brick thick	m²	145.00	to	175.00
two brick thick	m²	190.00	to	230.00
extra for fair face per side	m²	2.45	to	3.00
Engineering brick walls; class B; bricks PC £590.00/1000				
half brick thick	m²	59.00	to	72.00
one brick thick	m²	110.00	to	130.00
one and a half brick thick	m²	160.00	to	190.00
two brick thick	m²	205.00	to	250.00
Facing brick walls; sand faced facings; bricks PC £550.00/1000				
half brick thick; pointed one side	m²	69.00	to	84.00
one brick thick; pointed both sides	m²	135.00	to	165.00
Facing bricks solid walls; hand made facings; bricks PC £700.00/1000				
half brick thick; fair face one side	m²	75.00	to	91.00
one brick thick; fair face both sides	m²	165.00	to	200.00
add or deduct for each variation of £10.00/1000 in PC value				
half brick thick	m²	0.75	to	0.90
one brick thick	m²	1.50	to	1.80
one and a half brick thick	m²	2.20	to	2.70
two brick thick	m²	3.00	to	3.60
Aerated lightweight block walls				
100 mm thick	m²	31.00	to	37.50
140 mm thick	m²	40.50	to	49.00
215 mm thick	m²	76.00	to	91.00
Dense aggregate block walls				
100 mm thick	m²	33.00	to	40.00
140 mm thick	m²	43.00	to	52.00
Coloured dense aggregate masonry block walls; Lignacite or similar				
100 mm thick single skin, weathered face, self-coloured	m²	81.00	to	98.00
100 mm thick single skin, polished face	m²	125.00	to	150.00
Brick/block walling; cavity walls				
Cavity wall; facing brick outer skin; insulation; with plaster on standard weight block inner skin; emulsion (U-value = 0.30 W/m²K)				
machine made facings; PC £550.00/1000	m²	130.00	to	155.00
machine made facings; PC £700.00/1000	m²	135.00	to	160.00
self-finished masonry block, weathered finish outer skin and paint grade block inner skins; fair faced both sides	m²	130.00	to	160.00

2.5 EXTERNAL WALLS

Item	Unit	Range £		
self-finished masonry block, polished finish outer skin and paint grade block inner skins; fair faced both sides	m²	175.00	to	210.00
Cavity wall; rendered block outer skin; insulation; with plaster on standard weight block inner skin; emulsion (U-value = 0.30 W/m²K)				
block outer skin; insulation; lightweight block inner skin outer block rendered	m²	150.00	to	180.00
Cavity wall; facing brick outer skin; insulation; plasterboard on stud inner skin; emulsion (U-value = 0.30 W/m²K)				
machine made facings; PC £550.00/1000	m²	100.00	to	120.00
machine made facings; PC £700.00/1000	m²	105.00	to	130.00
Cavity wall; block; insulated to U-value = 0.30 W/m²K				
Reinforced concrete walling				
In situ reinforced concrete 25.00 N/mm²; 15 kg/m² reinforcement; formwork both sides				
150 mm thick	m²	110.00	to	130.00
225 mm thick	m²	125.00	to	150.00
300 mm thick	m²	140.00	to	170.00
Panelled walling				
Precast concrete panels; including insulation; lining and fixings generally 7.5 m × 0.15 m thick × storey height (U-value = 0.30 W/m²K)				
standard panels	m²	220.00	to	270.00
standard panels; exposed aggregate finish	m²	245.00	to	300.00
reconstructed stone faced panels	m²	280.00	to	340.00
brick clad panels (P.C £350.00/1000 for bricks)	m²	370.00	to	450.00
natural stone faced panels (Portland Stone or similar)	m²	480.00	to	580.00
marble or granite faced panels	m²	640.00	to	770.00
Kingspan TEK cladding panel; 142 mm thick (15 mm thick OSB board and 112 mm thick rigid eurathane core) fixed to frame; metal edge trim flashings	m²	290.00	to	355.00
Sheet claddings				
Cement fibre profiled cladding				
Profile 6; single skin; natural grey finish	m²	25.00	to	30.50
P61 Insulated System; natural grey finish; metal inner lining panel (U-value = 0.30 W/m²K)	m²	48.00	to	58.00
P61 Insulated System; natural grey finish; 100 mm lightweight concrete block (U-value = 0.30 W/m²K)	m²	87.00	to	105.00
P61 Insulated System; natural grey finish; 12mm self-finished plasterboard lining (U-value = 0.30 W/m²K)	m²	63.00	to	76.00
extra for coloured fibre cement sheeting	m²	1.45	to	1.75
Metal profiled cladding				
Standard trapazoidal profile				
composite insulated roofing system; 60 mm overall panel thickness (U-value = 0.35 W/m²K)	m²	43.50	to	53.00
composite insulated roofing system; 80 mm overall panel thickness (U-value = 0.26 W/m²K)	m²	46.00	to	56.00
composite insulated roofing system; 100 mm overall panel thickness (U-value = 0.20 W/m²K)	m²	49.00	to	59.00

2.5 EXTERNAL WALLS

Item	Unit	Range £		
Micro-rib profile				
composite insulated roofing system; micro-rib panel; 60 mm overall panel thickness (U-value = 0.35 W/m²K)	m²	120.00	to	145.00
composite insulated roofing system; micro-rib panel; 80 mm overall panel thickness (U-value = 0.26 W/m²K)	m²	130.00	to	160.00
composite insulated roofing system; micro-rib panel; 100 mm overall panel thickness (U-value = 0.20 W/m²K)	m²	140.00	to	170.00
Flat panel				
composite insulated roofing system; micro-rib panel; 60 mm overall panel thickness (U-value = 0.35 W/m²K)	m²	195.00	to	235.00
composite insulated roofing system; micro-rib panel; 80 mm overall panel thickness (U-value = 0.26 W/m²K)	m²	190.00	to	230.00
composite insulated roofing system; micro-rib panel; 100 mm overall panel thickness (U-value = 0.20 W/m²K)	m²	220.00	to	260.00
Glazed walling				
Curtain walling				
stick curtain walling with double glazed units, aluminium structural framing and spandrel rails. Standard colour powder coated capped	m²	470.00	to	570.00
unitized curtain walling system with double glazed units, aluminium structural framing and spandrel rails. Standard colour powder coated	m²	970.00	to	1175.00
unitized curtain walling bespoke project specific system with double glazed units, aluminium structural framing and spandrel rails. Standard colour powder coated.	m²	970.00	to	1175.00
visual mock-ups for project specific curtain walling solutions	item	31000.00	to	38000.00
Solar shading				
fixed aluminium Brise Soleil including uni-strut supports; 300 mm deep	m	155.00	to	190.00
Other systems				
lift surround of double glazed or laminated glass with aluminium or stainless steel framing	m²	850.00	to	1025.00
Other cladding systems				
Tiles (clay/slate/glass/ceramic)				
machine made clay tiles; including battens	m²	35.00	to	42.00
best hand made sand faced tiles; including battens	m²	40.00	to	48.50
concrete plain tiles; including battens	m²	49.00	to	60.00
natural slates; including battens	m²	75.00	to	91.00
20 mm × 20 mm thick mosaic glass or ceramic; in common colours; fixed on prepared surface	m²	105.00	to	130.00
Steel				
vitreous enamelled insulated steel sandwich panel system; with insulation board on inner face	m²	165.00	to	200.00
Formalux sandwich panel system; with coloured lining tray; on steel cladding rails	m²	190.00	to	230.00
Rainscreen				
25 mm thick tongued and grooved tanalized softwood boarding; including timber battens	m²	40.50	to	49.00
timber shingles, Western Red Cedar, preservative treated in random widths; not including any subframe or battens	m²	52.00	to	63.00
25 mm thick tongued and grooved Western Red Cedar boarding including timber battens	m²	46.00	to	56.00

2.5 EXTERNAL WALLS

Item	Unit	Range £		
25 mm thick Western Red Cedar tongued and grooved wall cladding on and including treated softwood battens on breather membrane, 10 mm Eternit Blueclad board and 50 mm insulation board; the whole fixed to Metsec frame system; including sealing all joints etc.	m²	110.00	to	140.00
Sucupira Preta timber boarding 19 mm thick × 75 mm wide slats with chamfered open joints fixed to 38 mm × 50 mm softwood battens with stainless steel screws on 100 mm Kingspan K2 insulations and 28 mm WBP plywoodd fxed to 50 mm × 1000 softwood studs; complete with breather membrane, aluminium cills and Sucupira Petra batten at external joints	m²	265.00	to	320.00
Trespa single skin rainscreen cladding, 8 mm thick panel, with secondary support/frame system; adhesive fixed panels; open joints; 100 mm Kingspan K15 insulation; aluminium subframe fixed to masonry/concrete	m²	230.00	to	280.00
Terracotta rainscreen cladding; aluminium support rails; anti-graffiti coating	m²	365.00	to	440.00
Corium brick tiles in metal tray system; standard colour; including all necessary angle trim on main support system; comprising of polythene vapour check; 75 mm × 50 mm timber studs; 50 mm thick rigid insulation with taped joints; 38 mm × 50 mm timber counter battens	m²	305.00	to	370.00

Comparative external finishes

Comparative concrete wall finishes

wrought formwork one side including rubbing down	m²	5.00	to	6.05
shotblasting to expose aggregate	m²	6.30	to	7.60
bush hammering to expose aggregate	m²	16.10	to	19.50

Comparative in situ finishes

two coats Sandtex Matt cement paint to render	m²	5.90	to	7.10
13 mm thick cement and sand plain face rendering	m²	22.50	to	27.00
ready-mixed self-coloured acrylic resin render on blockwork	m²	58.00	to	70.00
three coat Tyrolean rendering; including backing	m²	39.50	to	48.00
Stotherm Lamella; 6mm thick work in 2 coats	m²	78.00	to	95.00

Insulation

Crown Dritherm Cavity Slab 37 (Thermal conductivity 0.037 W/mK) glass fibre batt or other equal; as full or partial cavity fill; including cutting and fitting around wall ties and retaining discs

50 mm thick	m²	5.15	to	6.20
75 mm thick	m²	5.65	to	6.85
100 mm thick	m²	5.60	to	6.80

Crown Dritherm Cavity Slab 34 (Thermal conductivity 0.034 W/mK) glass fibre batt or other equal; as full or partial cavity fill; including cutting and fitting around wall ties and retaining discs

65 mm thick	m²	4.30	to	5.20
75 mm thick	m²	4.80	to	5.80
85 mm thick	m²	6.20	to	7.45
100 mm thick	m²	6.30	to	7.65

Crown Dritherm Cavity Slab 32 (Thermal conductivity 0.032 W/mK) glass fibre batt or other equal; as full or partial cavity fill; including cutting and fitting around wall ties and retaining discs

65 mm thick	m²	5.00	to	6.00
75 mm thick	m²	5.60	to	6.80
85 mm thick	m²	6.10	to	7.35
100 mm thick	m²	6.80	to	8.25

Approximate Estimating Rates

2.5 EXTERNAL WALLS

Item	Unit	Range £		
2.5 EXTERNAL WALLS – cont				
Insulation – cont				
Crown Frametherm Roll 40 (Thermal conductivity 0.040 W/mK) glass fibre semi-rigid or rigid batt or other equal; pinned vertically in timber frame construction				
90 mm thick	m²	5.70	to	6.95
140 mm thick	m²	7.70	to	9.35
Kay-Cel (Thermal conductivity 0.033 W/mK) expanded polystyrene board standard grade SD/N or other equal; fixed with adhesive				
20 mm thick	m²	4.10	to	4.95
25 mm thick	m²	4.20	to	5.10
30 mm thick	m²	4.30	to	5.20
40 mm thick	m²	4.80	to	5.80
50 mm thick	m²	5.20	to	6.30
60 mm thick	m²	5.65	to	6.85
75 mm thick	m²	6.20	to	7.50
100 mm thick	m²	7.05	to	8.50
Kingspan Kooltherm K8 (Thermal conductivity 0.022 W/mK) zero ODP rigid urethene insulation board or other equal; as partial cavity fill; including cutting and fitting around wall ties and retaining discs				
40 mm thick	m²	9.70	to	11.80
50 mm thick	m²	11.40	to	13.80
60 mm thick	m²	13.10	to	15.80
75 mm thick	m²	15.50	to	18.80
Kingspan Thermawall TW50 (Thermal conductivity 0.022 W/mK) zero ODP rigid urethene insulation board or other equal and approved; as partial cavity fill; including cutting and fitting around wall ties and retaining discs				
25 mm thick	m²	9.65	to	11.70
50 mm thick	m²	14.10	to	17.10
75 mm thick	m²	19.80	to	24.00
100 mm thick	m²	25.00	to	30.00
Thermafleece TF35 high density wool insulating batts (0.035 W/mK); 60% British wool, 30% recycled polyester and 10% polyester binder with a high recycled content				
50 mm thick	m²	14.90	to	18.00
75 mm thick	m²	18.40	to	22.00

2.6 WINDOWS AND EXTERNAL DOORS

Item	Unit	Range £		
2.6 WINDOWS AND EXTERNAL DOORS				
Window and external door area (unless otherwise described)				
Softwood windows (U-value = 1.6 W/m²K)				
Standard windows				
painted; double glazed; up to 1.50 m²	m²	310.00	to	370.00
painted; double glazed; over 1.50 m², up to 3.20 m²	m²	250.00	to	300.00
Purpose made windows				
painted; double glazed; up to 1.50 m²	m²	560.00	to	670.00
painted; double glazed; over 1.50 m²	m²	500.00	to	600.00
Hardwood windows (U-value = 1.6 W/m²K)				
Standard windows; stained				
double glazed	m²	530.00	to	640.00
Purpose made windows; stained				
double glazed	m²	600.00	to	720.00
Steel windows (U-value = 1.6 W/m²K)				
Standard windows				
double glazed; powder coated	m²	345.00	to	420.00
Purpose made windows				
double glazed; powder coated	m²	400.00	to	480.00
uPVC windows				
Windows; standard ironmongery; sills and factory glazed with low-e 24mm				
double glazing				
standard with low-e 24 mm double glazing	m²	200.00	to	240.00
WER A rating	m²	210.00	to	250.00
WER C rating	m²	200.00	to	240.00
Secured by Design accreditation	m²	200.00	to	245.00
extra for colour finish to uPVC	m²	30.50	to	37.00
Softwood external doors				
Standard external softwood doors and hardwood frames; doors painted;				
including ironmongery				
Matchboarded, framed, ledged and braced door, 838 mm × 1981 mm	nr	520.00	to	630.00
flush door; cellular core; plywood faced; 838 mm × 1981 mm	nr	540.00	to	650.00
Heavy duty solid flush door				
single leaf	nr	970.00	to	1175.00
double leaf	nr	1575.00	to	1925.00
single leaf; emergency fire exit	nr	1325.00	to	1600.00
double leaf; emergency fire exit	nr	1950.00	to	2350.00
Steel external doors				
Standard doors				
single external steel door, including frame, ironmongery, powder coated finish	nr	900.00	to	1075.00
single external steel security door, including frame, ironmongery, powder coated finish	nr	1775.00	to	2125.00
double external steel security door, including frame, ironmongery, powder coated finish	nr	2600.00	to	3200.00

2.6 WINDOWS AND EXTERNAL DOORS

Item	Unit	Range £		

2.6 WINDOWS AND EXTERNAL DOORS – cont

Steel external doors – cont
Overhead doors

single skin; manual	m^2	120.00	to	140.00
single skin; electric	m^2	205.00	to	250.00
insulated; electric; 1.4 W/m²K; 1 hour fire resistant	m^2	470.00	to	570.00
electric operation standard lift, 42 mm thick insulated sandwich panels	m^2	120.00	to	140.00
rapid lift fabric door, electric operation	m^2	700.00	to	850.00

uPVC external doors
Entrance doors; residential standard; PVCu frame; brass furniture (spyhole/
security chain/letter plate/draught excluder/multipoint locking)

overall 1480 × 2100 mm with glazed side panel	nr	600.00	to	720.00
overall 1800 × 2100 mm with glazed side panel	nr	570.00	to	690.00
overall 2430 × 2100 mm with glazed side panel each side	nr	680.00	to	830.00

Composite aluminium/timber windows, entrance screens and doors;
U value = 1.5 W/m²K
Purpose made windows; stainless steel ironmongery; Velfac System 200 or
similar

fixed windows up to 1.50 m^2	m^2	260.00	to	320.00
fixed windows over 1.50 m^2 up to 4.00 m^2	m^2	235.00	to	285.00
outward opening pivot windows up to 1.50 m^2	m^2	650.00	to	780.00
outward opening pivot windows over 1.50 m^2 up to 4.00 m^2	m^2	285.00	to	345.00
round porthole window 1.40 m diameter	nr	760.00	to	920.00
round porthole window 1.80 m diameter	nr	1150.00	to	1375.00
round porthole window 2.60 m diameter	nr	2175.00	to	2650.00
purpose made entrance screens and doors double glazed	m^2	970.00	to	1175.00

Purpose made doors

glazed single personnel door; stainless steel ironmongery; Velfac System100 or similar	nr	1725.00	to	2075.00
glazed double personnel door; stainless steel ironmongery; Velfac System 100 or similar	pair	3000.00	to	3600.00
revolving door; 2000 mm diameter; clear laminated glazing; 4nr wings; glazed curved walls	nr	4650.00	to	5600.00
automatic sliding door; bi-parting	m^2	2550.00	to	3100.00

Stainless steel entrance screens and doors
Purpose made screen; double glazed

with manual doors	m^2	1575.00	to	1925.00
with automatic doors	m^2	1925.00	to	2350.00
purpose made revolving door 2000 mm diameter; clear laminated glazing; 4nr wings; glazed curved walls	m^2	4300.00	to	5200.00
automatic sliding door; bi-parting	m^2	2350.00	to	2850.00

Shop fronts, shutters and grilles
Purpose made screen

temporary timber shop fronts	m^2	70.00	to	85.00
hardwood and glass; including high enclosed window beds	m^2	750.00	to	910.00
flat façade; glass in aluminium framing; manual centre doors only	m^2	650.00	to	790.00

2.7 INTERNAL WALLS AND PARTITIONS

Item	Unit	Range £		
grilles or shutters	m²	300.00	to	360.00
fire shutters; powers operated	m²	375.00	to	455.00
Automatic glazed slidiing doors				
Automatic glazed sliding doors in aluminium; polyester powder coated; linear sliding doors; inner pivoted pocket screens; glazed with safety units in accordance with BS; manifestation etched logo; automatic action with infra-red control complete with emergency stop and mechanical manual locking; fixing in accordance with the manufacturers instructions including connections to services; mastic pointing all round				
Opening 2.10 m × 2.10 m	nr	8600.00	to	11000.00
2.7 INTERNAL WALLS AND PARTITIONS				
Internal partition area (unless otherwise described)				
Frame and panel partitions				
Timber stud partitions				
structure only comprising 100 mm × 38 mm softwood studs at 600 mm centres; head and sole plates	m²	21.00	to	26.00
softwood stud comprising 100 mm × 38 mm softwood studs at 600 mm centres; head and sole plates; 12.5 mm thick plasterboard each side; tape and fill joints; emulsion finish	m²	52.00	to	63.00
Metal stud and board partitions; height range from 2.40 m to 3.30 m				
73 mm partition; 48 mm studs and channels; one layer of 12.5 mm Gyproc Wallboard each side; joints filled with joint filler and joint tape; emulsion paint finish; softwood skirtings with gloss finish	m²	40.00	to	48.00
102 mm partition; 70 mm studs and channels; one layer of 15 mm Gyproc Wallboard each side; joints filled with joint filler and joint tape; emulsion paint finish; softwood skirtings with gloss finish	m²	42.00	to	51.00
102 mm thick partition; 70 mm steel studs at 600 mm centres generally; 1 layer 15 mm Fireline board each side; joints filled with joint filler and joint tape; emulsion paint finish; softwood skirtings with gloss finish	m²	49.00	to	60.00
130 mm thick partition; 70 mm steel studs at 600 mm centres generally; 2 layers 15 mm Fireline board each side; joints filled with joint filler and joint tape; emulsion paint finish; softwood skirtings with gloss finish	m²	58.00	to	71.00
102 mm thick partition; 70 mm steel studs at 600 mm centres generally; 1 layer 15 mm Soundbloc board each side; joints filled with joint filler and joint tape; emulsion paint finish; softwood skirtings with gloss finish	m²	52.00	to	63.00
130 mm thick partition; 70 mm steel studs at 600 mm centres generally; 2 layers 15 mm Soundbloc board each side; joints filled with joint filler and joint tape; emulsion paint finish; softwood skirtings with gloss finish	m²	71.00	to	86.00
Alternative Board Finishes				
12 mm plywood boarding	m²	10.20	to	12.40
one coat Thistle board finish 3 mm thick work to walls; plasterboard base	m²	6.60	to	8.00
extra for curved work	%	15.80	to	21.00

2.7 INTERNAL WALLS AND PARTITIONS

Item	Unit	Range £		

2.7 INTERNAL WALLS AND PARTITIONS – cont

Acoustic insulation to partitions

Mineral fibre quilt; Isover Acoustic Partition Roll (APR 1200) or other equal;
pinned vertically to timber or plasterboard

Item	Unit		Range £	
25 mm thick	m²	2.40	to	2.90
50 mm thick	m²	3.10	to	3.80
Brick/block masonry partitions				
Brick				
common brick half brick thick wall; bricks PC £240.00/1000	m²	52.00	to	62.00
Block				
aerated/lightweight block partitions				
100 mm thick	m²	32.00	to	39.00
140 mm thick	m²	43.50	to	53.00
200/215 mm thick	m²	55.00	to	67.00
dense aggregate block walls				
100 mm thick	m²	33.00	to	40.00
140 mm thick	m²	43.00	to	52.00
extra for fair face (rate per side)	m²	1.00	to	1.20
extra for plaster and emulsion (rate per side)	m²	19.30	to	23.00
extra for curved work	%	15.80	to	21.00
Concrete partitions				
Reinforced concrete walls; C25 strength; standard finish; reinforced at 100 kg/m³				
150 mm thick	m²	140.00	to	165.00
225 mm thick	m²	160.00	to	190.00
300 mm thick	m²	180.00	to	215.00
extra for plaster and emulsion (rate per side)	m²	19.30	to	23.00
extra for curved work	%	21.00	to	26.00
Glass				
Hollow glass block walling; Pittsburgh Corning sealed Thinline or other equal and approved; in cement mortar joints; reinforced with 6 mm diameter stainless steel rods; pointed both sides with mastic or other equal and approved				
240 mm × 240 mm × 80 mm flemish; cross reeded or clear blocks	m²	390.00	to	470.00
190 mm × 190 mm × 100 mm glass blocks; 30 minute fire-rated	m²	590.00	to	720.00
190 mm × 190 mm × 100 mm glass blocks; 60 minute fire-rated	m²	1075.00	to	1275.00
Demountable/Folding partitions				
Demountable partitioning; aluminium framing; veneer finish doors				
medium quality; 46 mm thick panels factory finish vinyl faced	m²	305.00	to	370.00
high quality; 46 mm thick panels factory finish vinyl faced	m²	460.00	to	560.00
Demountable aluminium/steel partitioning and doors				
high quality; folding	m²	370.00	to	445.00
high quality; sliding	m²	780.00	to	950.00
Demountable fire partitions				
enamelled steel; half hour	m²	570.00	to	680.00
stainless steel; half hour	m²	890.00	to	1075.00
soundproof partitions; hardwood doors luxury veneered	m²	275.00	to	330.00

2.8 INTERNAL DOORS

Item	Unit	Range £		
Aluminium internal patent glazing				
single glazed laminated	m²	170.00	to	200.00
double glazed; 1 layer toughened and 1 layer laminated glass	m²	230.00	to	280.00
Stainless steel glazed manual doors and screens				
high quality; to inner lobby of malls	m²	750.00	to	910.00
Acoustic folding partition; headtrack suspension and bracing; aluminium framed with high density particle board panel with additional acoustic insulation; melamine laminate finish; acoustic seals. Nominal weight approximately 55 kg/m²				
sound reduction 48 db (Rw)	m²	510.00	to	620.00
sound reduction 55 db (Rw)	m²	610.00	to	740.00
Framed panel cubicles				
Changing and WC cubicles; high pressure laminate faced mdf; proprietary system				
standard quality WC cubicle partition sets; aluminium framing; melamine face chipboard dividing panels and doors; ironmongery; small range (up to 5 cubicles); standard cubicle set; (rate per cubicle)	nr	395.00	to	480.00
medium quality WC cubicle partition sets; stainless steel framing; real wood veneer face chipboard diiding panels and doors; ironmongery; small range (up to 6 cubicles); standard cubicle set; (rate per cubicle)	nr	800.00	to	970.00
high end specification flush fronted system floor to ceiling 44mm doors, no visible fixings; real wood veneer or HPL or high Gloss paint and lacquer; cubicle set; 800 mm × 1500 mm × 2400 mm high per cubicle, with satin finished stainless steel ironmongery; 30 mm high pressure laminated (HPL) chipboard divisions and 44 mm solid cored real wood veneered doors and pilasters; small range (up to 6 cubicles); standard cubicle set; (rate per cubicle)	nr	2200.00	to	2700.00
IPS duct panel systems; melamine finish chipboard; softwood timber subframe 2.70 m high IPS back panelling system; to accommodate wash hand basins or urinals; access hatch; frame support	m²	120.00	to	150.00
2.8 INTERNAL DOORS				
The following rates include for the supply and hang of doors, complete with all frames, architrave, typical medium standard ironmongery set and appropriate finish				
Standard doors				
Standard doors; cellular core; softwood; softwood architrave; aluminium ironmongery (latch only)				
single leaf; hardboard face; gloss paint finish	nr	280.00	to	340.00
single leaf; moulded panel; gloss paint finish	nr	290.00	to	350.00
single leaf; Sapele veneered finish	nr	300.00	to	360.00
Purpose-made doors				
Softwood panelled; softwood lining; softwood architrave; aluminium ironmongery (latch only); brass or stainless ironmongery (latch only); painting and polishing				
single leaf; four panels; mouldings	nr	415.00	to	500.00
double leaf; four panels; mouldings	nr	810.00	to	980.00

3.1 WALL FINISHES

Item	Unit	Range £		

2.8 INTERNAL DOORS – cont

Purpose-made doors – cont
Hardwood panelled; hardwood lining; hardwood architrave; aluminium
ironmongery (latch only); brass or stainless ironmongery (latch only); painting
and polishing

Item	Unit		Range £	
single leaf; four panelled doors; mouldings	nr	840.00	to	1000.00
double leaf; four panelled doors; mouldings	nr	1625.00	to	1950.00

Fire doors
Standard fire doors; cellular core; softwood lining; softwood architrave;
aluminuim ironmongery (lockable, self-closure); painting or polishing;

single leaf; flush hardboard faced; 30 min fire resistance; painted	nr	550.00	to	660.00
single leaf; Oak veneered; 30 min fire resistance; polished	nr	455.00	to	550.00
double leaf; Oak veneered; 30 min fire resistance; polished	nr	1100.00	to	1350.00
single leaf; flush hardboard faced; 60 min fire resistance; painted	nr	690.00	to	830.00
single leaf; Oak veneered; 60 min fire resistance; polished	nr	740.00	to	900.00
double leaf; Oak veneered; 60 min fire resistance; polished	nr	1375.00	to	1650.00

Ironmongery sets
Stainless steel ironmongery; euro locks; push plates; kick plates; signage;
closures; standard sets

office door; non-locking; fire rated	nr	290.00	to	350.00
office/store; lockable; fire rated	nr	330.00	to	400.00
classroom door; lockable; fire rated	nr	435.00	to	530.00
maintenance/plant room door; lockable; fire rated	nr	310.00	to	380.00
standard bathroom door (unisex)	nr	260.00	to	310.00
accessible toilet door	nr	150.00	to	180.00
fire escape door	nr	1575.00	to	1925.00

Overhead doors
High speed internal rapid action fabric door

High speed internal rapid action fabric door	m²	380.00	to	460.00

3 INTERNAL FINISHES

3.1 WALL FINISHES
Internal wall area (unless otherwise described)

In situ wall finishes
Comparative finishes

one mist and two coats emulsion paint	m²	3.80	to	4.60
two coats of lightweight plaster	m²	13.80	to	16.70
two coats of lightweight plaster with emulsion finish	m²	17.60	to	21.00
two coat sand cement render and emulsion finish	m²	24.00	to	29.50
plaster and vinyl wallpaper coverings	m²	22.50	to	27.00
polished plaster system; Armourcoat or similar; 11 mm thick first coat and 2 mm thick finishing coat with a polished finish	m²	110.00	to	130.00

3.2 FLOOR FINISHES

Item	Unit	Range £		
Rigid tile/panel/board finishes				
Timber boarding/panelling; on and including battens; plugged to wall				
12 mm thick softwood boarding	m²	37.00	to	45.00
hardwood panelling; t&g & v-jointed	m²	110.00	to	135.00
Ceramic wall tiles; including backing				
economical quality	m²	32.00	to	38.50
medium to high quality	m²	41.00	to	49.50
Porcelain; including backing				
Porcelain mosaic tiling; walls and floors	m²	76.00	to	92.00
Marble				
Roman Travertine marble wall linings 20 mm thick; polished	m²	270.00	to	330.00
Granite				
Dakota mahogany granite cladding 20 mm thick; polished finish; jointed and pointed in coloured mortar	m²	360.00	to	430.00
Dakota mahogany granite cladding 40 mm thick; polished finish; jointed and pointed in coloured mortar	m²	600.00	to	720.00
Comparative woodwork finishes				
Knot; one coat primer; two undercoats; gloss on wood surfaces; number of coats:				
two coats gloss	m²	9.55	to	11.60
three coats gloss	m²	10.50	to	12.70
three coats gloss; small girth n.e. 300 mm	m	6.90	to	8.30
Polyurethane lacquer				
two coats	m²	7.10	to	8.55
three coats	m²	7.80	to	9.40
Flame-retardant paint				
Unitherm or similar; two coats	m²	17.50	to	21.00
Polish				
wax polish; seal	m²	14.00	to	17.00
wax polish; stain and body in	m²	17.30	to	21.00
French polish; stain and body in	m²	25.50	to	31.00
Other wall finishes				
Wall coverings				
decorated paper backed vinyl wall paper	m²	8.60	to	10.40
PVC wall linings – Altro or similar; standard satins finish	m²	68.00	to	83.00
3.2 FLOOR FINISHES				
Internal floor area (unless otherwise described)				
In situ screed and floor finishes; laid level; over 300 mm wide				
Cement and sand (1:3) screeds; steel trowelled				
50 mm thick	m²	16.40	to	19.80
75 mm thick	m²	21.00	to	25.50
100 mm thick	m²	23.00	to	28.00
Flowing Screeds				
Latex screeds; self-colour; 3 mm to 5 mm thick	m²	7.30	to	8.85
Lafarge Gyvlon flowing screed 50 mm thick	m²	10.80	to	13.00
Treads, steps and the like (small areas)	m²	30.00	to	36.50

3.2 FLOOR FINISHES

Item	Unit	Range £		

3.2 FLOOR FINISHES – cont

In situ screed and floor finishes – cont

Item	Unit		Range £	
Granolithic; laid on green concrete				
20 mm thick	m²	23.50	to	28.00
38 mm thick	m²	27.00	to	33.00
Resin floor finish				
Altrotect; 2 coat application nominally 350–500 micron thick	m²	15.90	to	19.20
AltroFlow EP; 3 part solvent-system; up to 3mm thick	m²	34.00	to	41.00
Atro Screed (TB Screed; Quartz; Multiscreed), 3 mm to 4 mm thick	m²	45.00	to	54.00
Altrocrete heavy duty polyurethane screed, 6 mm to 8 mm screed	m²	50.00	to	60.00
Sheet/board flooring				
Chipboard				
18 mm to 22 mm thick chipboard flooring; t&g joints	m²	16.00	to	19.40
Softwood				
22 mm thick wrought softwood flooring; 150 mm wide; t&g joints;	m²	27.50	to	33.50
softwood skirting, gloss paint finish	m	15.50	to	18.80
mdf skirting, gloss paint finish	m	14.80	to	18.00
Hardwood				
Wrought hardwood t&g strip flooring; polished; including fillets	m²	96.00	to	115.00
Hardwood skirting, stained finish	m	19.80	to	24.00
Sprung floors				
Taraflex Combisport 85 System with Taraflex Sports M Plus; t&g plywood 22 mm thick on softwood battens and crumb rubber cradles	m²	155.00	to	190.00
Sprung composition block flooring (sports), court markings, sanding and sealing	m²	100.00	to	120.00
Rigid Tile/slab finishes (includes skirtings; excludes screeds)				
Quarry tile flooring	m²	55.00	to	66.00
Glazed ceramic tiled flooring				
standard plain tiles	m²	43.00	to	52.00
anti slip tiles	m²	47.00	to	57.00
designer tiles	m²	97.00	to	120.00
Terrazzo tile flooring 28 mm thick polished	m²	49.00	to	59.00
York stone 50 mm thick paving	m²	155.00	to	190.00
Slate tiles, smooth; straight cut	m²	58.00	to	70.00
Portland stone paving	m²	250.00	to	305.00
Roman Travertine marble paving; polished	m²	240.00	to	290.00
Granite paving 20 mm thick paving	m²	365.00	to	440.00
Parquet/wood block wrought hardwood block floorings; 25 mm thick; polished; t&g joints	m²	125.00	to	150.00
Flexible tiling; welded sheet or butt joint tiles; adhesive fixing				
vinyl floor tiling; 2.00 mm thick	m²	14.30	to	17.30
vinyl safety flooring; 2.00–2.50 mm thick	m²	42.50	to	52.00
vinyl safety flooring; 3.5 mm thick heavy duty	m²	49.00	to	59.00
linoleum tile flooring; 333 mm × 333 mm × 3.20 mm tiles	m²	38.00	to	46.00
linoleum sheet flooring; 2.00 mm thick	m²	31.00	to	37.50
rubber studded tile flooring; 500 mm × 500 mm × 2.50 mm thick	m²	38.00	to	46.00

3.3 CEILING FINISHES

Item	Unit	Range £		
Carpet tiles; including underlay, edge grippers				
heavy domestic duty; to floors; PC Sum £24/m²	m²	64.00	to	78.00
heavy domestic duty; to treads and risers; PC Sum £24/m²	m	49.50	to	60.00
heavy contract duty; Forbo Flotex HD	m²	44.00	to	53.00
Entrance matting				
door entrance and circulation matting; soil and water removal	m²	65.00	to	79.00
Entrance matting and matwell				
Gradus Topguard barrier matting with aluminium frame	m²	340.00	to	415.00
Nuway Tuftiguard barrier matting with aluminium frame	m²	360.00	to	435.00
Gradus Topguard barrier matting with stainless steel frame	m²	335.00	to	405.00
Nuway Tuftiguard barrier matting with stainless steel frame	m²	350.00	to	430.00
Gradus Topguard barrier matting with polished brass frame	m²	490.00	to	600.00
Nuway Tuftiguard barrier matting with polished brass frame	m²	510.00	to	620.00
Access floors and finishes				
Raised access floors: including 600 mm × 600 mm steel encased particle boards on height adjustable pedestals < 300 mm				
light grade duty	m²	40.00	to	48.00
medium grade duty	m²	40.00	to	48.00
heavy grade duty	m²	54.00	to	65.00
battened raft chipboard floor with sound insulation fixed to battens; medium quality carpeting	m²	90.00	to	110.00
Common floor coverings bonded to access floor panels				
anti static vinyl	m²	25.00	to	30.00
heavy duty fully flexible vinyl tiles	m²	23.00	to	28.00
needle punch carpet	m²	15.30	to	18.50
3.3 CEILING FINISHES				
Internal ceiling area (unless otherwise described)				
In situ finishes				
Decoration only to soffits; one mist and two coats emulsion paint				
to exposed steelwork (surface area)	m²	5.75	to	6.95
to concrete soffits (surface area)	m²	4.20	to	5.10
to plaster/plasterboard	m²	3.85	to	4.65
Plaster to soffits				
plaster skim coat to plasterboard ceilings	m²	6.80	to	8.25
lightweight plaster to concrete	m²	18.10	to	22.00
Plasterboard to soffits				
9 mm Gyproc board and skim coat	m²	27.00	to	33.00
12.50 mm Gyproc board and skim coat	m²	28.00	to	34.00
Other board finishes; with fire-resisting properties; excluding decoration				
12.50 mm thick Gyproc Fireline board	m²	21.00	to	25.00
15 mm thick Gyproc Fireline board	m²	21.50	to	26.00
12 mm thick Supalux	m²	52.00	to	63.00
Specialist plasters; to soffits				
sprayed acoustic plaster; self-finished	m²	75.00	to	91.00
rendering; Tyrolean finish	m²	75.00	to	91.00

3.3 CEILING FINISHES

Item	Unit	Range £		

3.3 CEILING FINISHES – cont

In situ finishes – cont
Other ceiling finishes

softwood timber t&g boarding	m	46.00	to	61.00

Suspended and integrated ceilings
Armstrong suspended ceiling; assume large rooms over 250 m²

mineral fibre; basic range; Cortega, Tatra, Academy; exposed grid	m²	19.00	to	25.00
mineral fibre; medium quality; Corline; exposed grid	m²	27.50	to	36.00
mineral fibre; medium quality; Corline; concealed grid	m²	41.00	to	54.00
mineral fibre; medium quality; Corline; silhouette grid	m²	51.00	to	68.00
mineral fibre; Specific; Bioguard; exposed grid	m²	23.00	to	30.00
mineral fibre; Specific; Bioguard; clean room grid	m²	36.00	to	47.50
mineral fibre; Specific; Hygiene, Cleanroom; exposed grid	m²	49.50	to	65.00
metal open cell; Cellio Global White; exposed grid	m²	46.50	to	61.00
metal open cell; Cellio Black; exposed grid	m²	56.00	to	74.00
wood veneer board; Microlook 8; plain; exposed grid	m²	61.00	to	80.00
wood veneer; plain; exposed grid	m²	84.00	to	110.00
wood veneers; perforated; exposed grid	m²	145.00	to	190.00
wood veneers; plain; concealed grid	m²	140.00	to	180.00
wood veneers; perforated; concealed grid	m²	190.00	to	250.00

Other suspended ceilings

perforated aluminium ceiling tiles 600 mm × 600 mm	m²	32.00	to	38.50
perforated aluminium ceiling tiles 1200 mm × 600 mm	m²	38.00	to	46.00
metal linear strip; Dampa/Luxalon	m²	48.50	to	59.00
metal linear strip micro perforated acoustic ceiling with Rockwool acoustic infill	m²	59.00	to	72.00
metal tray	m²	51.00	to	61.00
egg-crate	m²	78.00	to	94.00
open grid; Formalux/Dimension	m²	100.00	to	120.00

Integrated ceilings

coffered; with steel surfaces	m²	135.00	to	160.00
acoustic suspended ceilings on anti vibration mountings	m²	61.00	to	74.00

Comparative ceiling finishes
Emulsion paint to plaster

two coats	m²	4.30	to	5.20
one mist and two coats	m²	4.70	to	5.65

Gloss paint to timber or metal

primer and two coats	m²	6.30	to	7.65
primer and three coats	m²	9.20	to	11.10

Other coatings

Artex plastic compound one coat; textured to plasterboard	m²	7.15	to	8.65

4.1 FITTINGS, FURNISHINGS AND EQUIPMENT

Item	Unit	Range £		

4 FITTINGS, FURNISHINGS AND EQUIPMENT

4.1 FITTINGS, FURNISHINGS AND EQUIPMENT

Residential fittings (volume housing)
Kitchen fittings for residential units (not including white goods). NB quality and quantity of units can varies enormously. Always obtain costs from your preferred supplier. The following assume medium standard units from a large manufacturer and includes all units, worktops, stainless steel sink and taps, not including white goods

Item	Unit	Range £		
one person flat	nr	1900.00	to	2400.00
two person flat/house	nr	2125.00	to	2700.00
three person house	nr	3550.00	to	4500.00
four person house; includes utility area	nr	6700.00	to	8400.00
five person house; includes utility area	nr	9000.00	to	11000.00

Office furniture and equipment
There is a large quality variation for office furniture and we have assumed a medium level, even so prices will vary between suppliers
Reception desk

Item	Unit	Range £		
straight counter; 3500 mm long; 2 person	nr	1975.00	to	2500.00
curved counter; 3500 mm long; 2 person	nr	4750.00	to	6000.00
curved counter; 3500 mm long; 2 person; real wood veneer finish	nr	9500.00	to	12000.00

Furniture and equipment to general office area; standard off the shelf specification

Item	Unit	Range £		
workstation; 2000 mm long desk; drawer unit; task chair	nr	760.00	to	960.00

Hotel bathroom pods
Fully fitted out, finished and furnished bathroom pods; installed; suitable for business class hotels

Item	Unit	Range £		
standard pod (4.50 m plan area)	nr	4750.00	to	6000.00
accessible pod (4.50 m plan area)	nr	6400.00	to	8100.00

Window blinds
Louvre blind

Item	Unit	Range £		
89 mm louvres, manual chain operation, fixed to masonry	m²	55.00	to	70.00
127 mm louvres, manual chain operation, fixed to masonry	m²	48.00	to	61.00

Roller blind

Item	Unit	Range £		
fabric blinds, roller type, manual chain operation, fixed to masonry	m²	42.50	to	54.00
solar black out blinds, roller type, manual chain operation, fixed to masonry	m²	73.00	to	93.00

Fire curtains
Electrically operated automatic 2 hr fire curtains to form a virtually continuous barrier against both fire and smoke; size in excess of 4 m² (minimum size 800 mm)

Item	Unit	Range £		
	m²	1425.00	to	1800.00

5.1 SANITARY INSTALLATIONS

Item	Unit	Range £		
4.1 FITTINGS, FURNISHINGS AND EQUIPMENT – cont				
Dock Levellers and Shelters				
ASSA ABLOY Entrance Systems				
curtain mechanical shelter; 2 side curtains, one top curtain	nr	970.00	to	1150.00
teledock leveller; electro-hydraulic operation with two lifting cylinders; one lip ram cylinder; movable telescopic lip;tions; Control for door and leveller from one communal control panel, including interlocking of leveller/door	nr	2600.00	to	3150.00
inflatable mechanical shelter; top bag with polyester fabric panels, 1000 mm extension; side bags with polyester fabric panels, 650 mm extension, strong blower for straight movement of top and side bags; side section with steel guards; colour from standard range	nr	3650.00	to	4400.00
Stand-alone Load House; a complete loading unit attached externally to building; modular steel skin/insulated cladding; two wall elements, one roof element	nr	5500.00	to	6600.00
5 SERVICES				
5.1 SANITARY INSTALLATIONS				
Please refer to *Spon's Mechanical and Electrical Services Price Book* for a more comprehensive range of rates				
Rates are for gross internal floor area – GIFA unless otherwise described				
shopping malls and the like (not tenant fit-out works)	m²	0.95	to	1.30
office building – multi-storey	m²	7.70	to	10.80
office building – business park	m²	7.70	to	10.80
performing arts building (medium specification)	m²	14.20	to	19.80
sports hall	m²	11.30	to	15.60
hotel	m²	38.00	to	53.00
hospital; private	m²	22.50	to	31.00
school; secondary	m²	10.80	to	15.00
residential; multi-storey tower	m²	0.60	to	0.80
supermarket	m²	1.45	to	2.00
Comparative sanitary fittings/sundries				
Note: Material prices vary considerably, the following composite rates are based on average prices for mid priced fittings:				
Individual sanitary appliances (including fittings)				
low level WCs; vitreous china pan and cistern; black plastic seat; low pressure ball valve; plastic flush pipe; fixing brackets	nr	310.00	to	395.00
bowl type wall urinal; white glazed vitreous china flushing cistern; chromium plated flush pipes and spreaders; fixing brackets	nr	255.00	to	330.00
sink; glazed fireclay; chromium plated waste; plug and chain	nr	590.00	to	750.00
sink; stainless steel; chromium plated waste; plug and chain				
single drainer; double bowl (bowl and half)	nr	340.00	to	430.00
double drainer; double bowl (bowl and half)	nr	360.00	to	460.00
bath; reinforced acrylic; chromium plated taps; overflow; waste; chain and plug; P trap and overflow connections	nr	520.00	to	670.00
bath; enamelled steel; chromium plated taps; overflow; waste; chain and plug; P trap and overflow connections	nr	640.00	to	820.00
shower tray; glazed fireclay; chromium plated waste; riser pipe; rose and mixing valve	nr	480.00	to	620.00

5.7 VENTILATING SYSTEMS

Item	Unit	Range £		
Soil waste stacks; 3.15 m storey height; branch and connection to drain				
110 mm diameter PVC	nr	450.00	to	580.00
extra for additional floors	nr	590.00	to	750.00
100 mm diameter cast iron; decorated	nr	610.00	to	780.00
extra for additional floors	nr	450.00	to	580.00
5.4 WATER INSTALLATIONS				
Hot and cold water installations; mains supply; hot and cold water distribution.				
Gross internal floor area (unless described otherwise)				
shopping malls and the like (not tenant fit out works)	m²	11.40	to	15.90
office building – multi-storey	m²	16.70	to	23.00
office building – business park	m²	12.50	to	17.40
performing arts building (medium specification)	m²	19.70	to	27.00
sports hall	m²	14.90	to	20.50
hotel	m²	42.00	to	58.00
hospital; private	m²	50.00	to	70.00
school; secondary; potable and non-potable to labs, art rooms	m²	39.00	to	54.00
residential; multi-storey tower	m²	47.00	to	66.00
supermarket	m²	26.00	to	36.00
distribution centre	m²	1.80	to	2.50
5.6 SPACE HEATING AND AIR CONDITIONING				
Gross internal floor area (unless described otherwise)				
shopping malls and the like (not tenant fit-out works); LTHW, air conditioning, ventilation	m²	75.00	to	98.00
office building – multi-storey; LTHW, ductwork, chilled water, ductwork	m²	105.00	to	140.00
office building – business park; LTHW, ductwork, chilled water, ductwork	m²	120.00	to	160.00
performing arts building (medium specification); LTHW, ductwork, chilled water, ductwork	m²	210.00	to	275.00
sports hall; warm air to main hall, LTHW radiators to ancillary	m²	37.00	to	48.50
hotel; air conditioning	m²	140.00	to	180.00
hospital; private; LPHW	m²	210.00	to	275.00
school; secondary; LTHW, cooling to ICT server room	m²	115.00	to	150.00
residential; multi-storey tower; LTHW to each apartment	m²	63.00	to	83.00
supermarket	m²	16.00	to	21.00
distribution centre; LTHW to offices, displacement system to warehouse	m²	22.00	to	29.00
5.7 VENTILATING SYSTEMS				
Gross internal area (unless otherwise described)				
shopping malls and the like (not tenant fit-out works)	m²	36.00	to	47.50
office building – multi-storey; toilet and kitchen areas only	m²	6.20	to	8.10
office building – business park; toilet and kitchen areas only	m²	7.20	to	9.50
performing arts building (medium specification); toilet and kitchen areas, workshop	m²	14.30	to	18.90
sports hall	m²	13.40	to	17.60
hotel; general toilet extraction, bathrooms, kitchens	m²	42.00	to	55.00
hospital; private; toilet and kitchen areas only	m²	7.70	to	10.10
school; secondary; toilet and kitchen areas, science labs	m²	17.60	to	23.00
residential; multi-storey tower	m²	36.00	to	47.50
supermarket	m²	5.20	to	6.90
distribution centre; smoke extract system	m²	5.20	to	6.90

5.10 LIFT AND CONVEYOR INSTALLATIONS

Item	Unit	Range £		
5.8 ELECTRICAL INSTALLATIONS				
Including LV distribution, HV distribution, lighting, small power. Gross internal floor area (unless described otherwise)				
shopping malls and the like (not tenant fit-out works)	m²	115.00	to	160.00
office building – multi-storey	m²	99.00	to	140.00
office building – business park	m²	52.00	to	73.00
performing arts building (medium specification)	m²	190.00	to	260.00
sports hall	m²	58.00	to	80.00
hotel	m²	180.00	to	250.00
hospital; private	m²	210.00	to	290.00
school; secondary	m²	125.00	to	175.00
residential; multi-storey tower	m²	84.00	to	115.00
supermarket	m²	68.00	to	94.00
distribution centre	m²	63.00	to	88.00
Comparative fittings/rates per point				
Consumer control unit; 63–100 Amp 230 volt; switched and insulated; RCDB protection. Gross internal floor area (unless described otherwise)	nr	320.00	to	400.00
Fittings; excluding lamps or light fittings				
lighting point; PVC cables	nr	47.50	to	60.00
lighting point; PVC cables in screwed conduits	nr	54.00	to	68.00
lighting point; MICC cables	nr	71.00	to	90.00
Switch socket outlet; PVC cables				
single	nr	6.65	to	8.40
double	nr	78.00	to	98.00
Switch socket outlet; PVC cables in screwed conduit				
single	nr	88.00	to	110.00
double	nr	105.00	to	130.00
Switch socket outlet; MICC cables				
single	nr	86.00	to	110.00
double	nr	99.00	to	125.00
Other power outlets				
Immersion heater point (excluding heater)	nr	110.00	to	140.00
Cooker point; including control unit	nr	180.00	to	230.00
5.9 FUEL INSTALLATIONS				
Gas mains service to plantroom. Gross internal floor area (unless described otherwise)				
shopping mall/supermarket	m²	3.00	to	3.80
warehouse/distribution centre	m²	0.85	to	1.10
office/hotel	m²	1.40	to	1.80
5.10 LIFT AND CONVEYOR INSTALLATIONS				
Passenger lifts				
Passenger lifts (standard brushed stainless steel finish; 2 panel centre opening doors)				
8-person; 4 stops; 1.0 m/s speed	nr	74000.00	to	90000.00
8-person; 4 stops; 1.6 m/s speed	nr	78000.00	to	94000.00

5.11 FIRE AND LIGHTNING PROTECTION

Item	Unit	Range £		
13-person; 6 stops; 1.0 m/s speed	nr	93000.00	to	110000.00
13-person; 6 stops; 1.6 m/s speed	nr	100000.00	to	120000.00
13-person; 6 stops; 2.0 m/s speed	nr	100000.00	to	120000.00
21-person; 8 stops; 1.0 m/s speed	nr	130000.00	to	155000.00
21-person; 8 stops; 1.6 m/s speed	nr	130000.00	to	160000.00
21-person; 8 stops; 2.0 m/s speed	nr	130000.00	to	160000.00
extra for enhanced finishes; mirror; carpet	nr	3650.00	to	4400.00
extra for lift car LCD TV	nr	7900.00	to	9600.00
extra for intelligent group control; 5 cars; 11 stops	nr	38000.00	to	46000.00
Special installations				
wall climber lift; 10-person; 0.50 m/sec; 10 levels	nr	370000.00	to	450000.00
disabled platform lift single wheelchair; 400 kg; 4 stops; 0.16 m/s	nr	9100.00	to	11000.00
Non-passenger lifts				
Goods lifts; prime coated internal finish				
2000 kg load; 4 stops; 1.0 m/s speed	nr	94000.00	to	110000.00
2000 kg load; 4 stops; 1.6 m/s speed	nr	100000.00	to	120000.00
2500 kg load; 4 stops; 1.0 m/s speed	nr	100000.00	to	120000.00
2000 kg load; 8 stops; 1.0 m/s speed	nr	130000.00	to	155000.00
2000 kg load; 8 stops; 1.6 m/s speed	nr	130000.00	to	160000.00
Other goods lifts				
Hoist	nr	25500.00	to	31000.00
Kitchen service hoist 50 kg; 2 levels	nr	11000.00	to	13000.00
Escalators				
30° escalator; 0.50 m/sec; enamelled steel glass balustrades				
3.50 m rise; 800 mm step width	nr	81000.00	to	98000.00
4.60 m rise; 800 mm step width	nr	87000.00	to	105000.00
5.20 m rise; 800 mm step width	nr	90000.00	to	110000.00
6.00 m rise; 800 mm step width	nr	100000.00	to	120000.00
extra for enhanced finish; enamelled finish; glass balustrade	nr	10500.00	to	13000.00
5.11 FIRE AND LIGHTNING PROTECTION				
Gross internal floor area (unless described otherwise)				
Including lightning protection, sprinklers unless stated otherwise				
shopping malls and the like (not tenant fit-out works)	m²	14.70	to	20.00
office building – multi-storey	m²	31.50	to	44.00
office building – business park	m²	5.20	to	7.25
performing arts building (medium specification); lightning protection only	m²	1.60	to	2.20
sports hall; lightning protection only	m²	3.15	to	4.40
hospital; private; lightning protection only	m²	2.05	to	2.90
school; secondary; lightning protection only	m²	2.00	to	2.75
hotel	m²	36.00	to	50.00
residential; multi-storey tower	m²	52.00	to	73.00
supermarket; lightning protection only	m²	1.00	to	1.45
distribution centre	m²	40.50	to	56.00

5.14 BUILDER'S WORK IN CONNECTION WITH SERVICES

Item	Unit	Range £		
5.12 COMMUNICATION AND SECURITY INSTALLATIONS				
Gross internal floor area (unless described otherwise)				
Including fire alarm, public address system, security installation unless stated otherwise				
shopping malls and the like (not tenant fit-out works)	m²	41.00	to	54.00
office building – multi-storey	m²	29.50	to	39.00
office building – business park; fire alarm and wireways only	m²	14.30	to	18.80
performing arts building (medium specification)	m²	90.00	to	120.00
sports hall	m²	59.00	to	78.00
hotel	m²	100.00	to	130.00
hospital; private	m²	76.00	to	100.00
school; secondary	m²	57.00	to	75.00
residential; multi-storey tower	m²	57.00	to	75.00
supermarket	m²	17.60	to	23.00
distribution centre	m²	21.00	to	27.50
5.13 SPECIAL INSTALLATIONS				
Rates for area of material or per item				
Solar				
Residential solar water heating including collectors; dual coil cylinders; pump; controller (NB excludes any grant allowance)	m²	770.00	to	1025.00
Solar power including 2.2 kWp monocrystalline solar modules (12 × 185 Wp) on roof mounting kit; certified inverter; DC and AC isolation switches and connection to the grid and certification	nr	6800.00	to	9100.00
Window cleaning equipment				
twin track	m	160.00	to	210.00
manual trolley/cradle	nr	11000.00	to	14000.00
automatic trolley/cradle	nr	22000.00	to	29000.00
Sauna				
2.20 m × 2.20 m internal Finnish sauna; benching; heater; made of Sauna grade Aspen or similar	nr	8600.00	to	11000.00
Jacuzzi Installation				
2.20 m × 2.20 m × 0.95 m; 5 adults; 99 jets; lights; pump	nr	11000.00	to	15000.00
5.14 BUILDER'S WORK IN CONNECTION WITH SERVICES				
Gross internal floor area (unless described otherwise)				
Warehouses, sports halls and shopping malls. Gross internal floor area				
main supplies, lighting and power to landlord areas	m²	3.40	to	3.80
central heating and electrical installation	m²	10.30	to	11.60
central heating, electrical and lift installation	m²	11.60	to	13.20
air conditioning, electrical and ventilation installations	m²	26.00	to	29.50
Offices and hotels. Gross internal floor area				
main supplies, lighting and power to landlord areas	m²	9.90	to	13.20
central heating and electrical installation	m²	14.40	to	19.20
central heating, electrical and lift installation	m²	17.10	to	23.00
air conditioning, electrical and ventilation installations	m²	31.50	to	42.00

8.2 ROADS, PATHS, PAVINGS AND SURFACINGS

Item	Unit	Range £		
8 EXTERNAL WORKS8.2 ROADS, PATHS, PAVINGS AND SURFACINGS				
Paved areas				
Gravel paving rolled to falls and chambers paving on subbase; including excavation	m²	10.90	to	13.70
Resin bound paving				
16 mm–24 mm deep of natural gravel	m²	65.00	to	82.00
16 mm–24 mm deep of crushed rock	m²	66.00	to	83.00
16 mm–24 mm deep of marble chips	m²	71.00	to	89.00
Tarmacadam paving				
two layers; limestone or igneous chipping finish paving on subbase; including excavation and type 1 subbase	m²	85.00	to	110.00
Slab paving				
precast concrete paving slabs on subbase; including excavation	m²	57.00	to	72.00
precast concrete tactile paving slabs on subbase; including excavation	m²	78.00	to	98.00
York stone slab paving on subbase; including excavation	m²	130.00	to	165.00
imitation York stone slab paving on subbase; including excavation	m²	86.00	to	110.00
Brick/Block/Setts paving				
brick paviors on subbase; including excavation	m²	83.00	to	105.00
precast concrete block paviors to footways including excavation; 150 mm hardcore subbase with dry sand joints	m²	85.00	to	110.00
granite setts on 100 mm thick concrete subbase; including excavation; pointing mortar joints	m²	150.00	to	190.00
cobblestone paving cobblestones on 100 mm thick concrete subbase; including excavation; pointing mortar joints	m²	160.00	to	200.00
Reinforced grass construction				
Grasscrete or similar on 200 mm type 1 subbase; topsoil spread across units and grass seeded upon completion	m²	71.00	to	90.00
Car parking alternatives				
Surface parking; include drains, kerbs, lighting				
surface level parking	m²	87.00	to	110.00
surface car parking with landscape areas	m²	110.00	to	140.00
Rates per car				
surface level parking	car	1825.00	to	2275.00
surface car parking with landscape areas	car	2325.00	to	2950.00
Other surfacing options				
permeable concrete block paving; 80 mm thick Formpave Aquaflow or equal; maximum gradient 5%; laid on 50 mm sand/gravel; subbase 365 mm thick clean crushed angular non-plastic aggregate, geotextiles filter layer to top and bottom of subbases	m²	52.00	to	66.00
All purpose roads				
Tarmacadam or reinforced concrete roads, including all earthworks, drainage, pavements, lighting, signs, fencing and safety barriers				
single 7.30 m wide carriageway	m	1075.00	to	1375.00
wide single 10.00 m wide carriageway	m	4100.00	to	5200.00
dual two lane road 7.30 m wide carriageway	m	3100.00	to	3900.00
dual three lane road 11.00 m wide carriageway	m	4000.00	to	5000.00

Approximate Estimating Rates

8.3 SOFT LANDSCAPING, PLANTING AND IRRIGATION SYSTEMS

Item	Unit	Range £		
8.2 ROADS, PATHS, PAVINGS AND SURFACINGS – cont				
Road crossings				
NOTE: Costs include road markings, beacons, lights, signs, advance danger signs etc.				
Zebra crossing	nr	19000.00	to	24000.00
Pelican crossing	nr	33000.00	to	42000.00
Underpass				
Provision of underpasses to new roads, constructed as part of a road building programme				
Precast concrete pedestrian underpass				
3.00 m wide × 2.50 m high	m	5500.00	to	7000.00
Precast concrete vehicle underpass				
7.00 m wide × 5.00 m high	m	24000.00	to	30500.00
14.00 m wide × 5.00 m high	m	57000.00	to	72000.00
Sports Grounds				
Pitch plateau construction and pitch drainage not included				
Hockey and support soccer training (no studs)				
Sand filled polypropelene synthetic artificial pitch to FIH national standard. Notts Pad XC FIH shockpad; sport needlepunched polypropelene sand filled synthetic multi-sport carpet; white lining; 3 m high perimeter fence	m²	28.50	to	37.50
FIFA One Star Football and IRB reg 22 Rugby Union				
Lano Rugby Max 60 mm monofilament laid on shockpad on a dynamic base infill with 32 kg/m² sand and 11 kg/m² SBR rubber – IRB Clause 22 compliant; supply and spread to required infill rates, 2EW sand and 0.5–2 mm SBR Rubber. Sand (32 kg/m²)and rubber (11 kg/m²) spread and brushed into surface; white lining; 3.5 m high perimeter fence	m²	24.00	to	31.00
Tennis and Netball Court				
Lay final layer on existing subbase of 50 mm porous stone followed by 2 coats of porous bitumen macadam. 40 mm binder course; 14 mm aggregate and 20 mm surface course with 6mm aggregate. Laid to fine level tolerances and laser leveled, +/- 6 mm tolerance over 3 m; white lining; 3 m high perimeter fencing	m²	38.00	to	50.00
Natural Grass Winter Sports				
Strip topsoil and level existing surface (cut and fill); laser grade formation and proof roll. Load and spread topsoil, grade and consolidate; cultivate and grade topsoil. Supply and spread 25 mm approved sand and mix into topsoil by power harrow; apply pre-seeding fertilizer and sow winter rye sports grass	m²	8.55	to	11.30
8.3 SOFT LANDSCAPING, PLANTING AND IRRIGATION SYSTEMS				
Preparatory excavation and subbases				
Surface treatment				
spread and lightly consolidate top soil from spoil heap 150 mm thick; by machine	m²	1.80	to	2.20
spread and lightly consolidate top soil from spoil heap 150 mm thick; by hand	m²	7.50	to	9.05

8.4 FENCING, RAILINGS AND WALLS

Item	Unit	Range £		
Seeded and planted areas				
Plant supply, planting, maintenance and 12 months guarantee				
seeded areas	m²	4.60	to	5.55
turfed areas	m²	5.90	to	7.15
Planted areas (per m² of planted area)				
herbaceous plants	m²	5.30	to	6.45
climbing plants	m²	8.40	to	10.20
general planting	m²	20.00	to	24.50
woodland	m²	30.00	to	37.00
shrubbed planting	m²	52.00	to	62.00
dense planting	m²	52.00	to	63.00
shrubbed area including allowance for small trees	m²	67.00	to	81.00
Trees				
light standard bare root tree (PC £9.75)	nr	57.00	to	69.00
standard root balled tree (PC 23.50)	nr	67.00	to	82.00
heavy standard root ball tree (PC £39.25)	nr	125.00	to	150.00
semi-mature root balled tree (PC £125.00)	nr	400.00	to	480.00
Parklands				
NOTE: Work on parklands will involve different techniques of earth shifting and cultivation. The following rates include for normal surface excavation, they include for the provision of any land drainage.				
parklands, including cultivating ground, applying fertilizer, etc. and seeding with parks type grass	ha	20000.00	to	24000.00
Lakes including excavation average 1.0 m deep, laying 1000 micron sheet with welded joints; spreading top soil evenly on top 200 mm deep				
regular shaped lake	1000 m²	23000.00	to	27500.00
8.4 FENCING, RAILINGS AND WALLS				
Crib retaining walls				
Permacrib timber crib retaining walls on concrete foundation; granular infill material; measure face area m². Excludes backfill material to the rear of the wall				
wall heights up to 2.0 m	m²	175.00	to	210.00
wall heights up to 3.0 m	m²	190.00	to	225.00
wall heights up to 4.0 m	m²	200.00	to	240.00
wall heights up to 5.0 m	m²	210.00	to	250.00
wall heights up to 6.0 m	m²	220.00	to	265.00
Andacrib concrete crib retaining walls on concrete foundation; granular infill material; measure face area m². Excludes backfill material to the rear of the wall				
wall heights up to 2.0 m	m²	190.00	to	230.00
wall heights up to 3.0 m	m²	200.00	to	240.00
wall heights up to 4.0 m	m²	210.00	to	250.00
wall heights up to 5.0 m	m²	220.00	to	270.00
wall heights up to 6.0 m	m²	230.00	to	280.00
Textomur green faced reinforced soil system; geogrid/geotextile reinforcement and compaction of reinforced soil mass.				
wall heights up to 3.0 m	m²	125.00	to	150.00
wall heights up to 4.0 m	m²	130.00	to	160.00
wall heights up to 5.0 m	m²	140.00	to	165.00
wall heights up to 6.0 m	m²	140.00	to	170.00

8.5 EXTERNAL FIXTURES

Item	Unit	Range £		

8.4 FENCING, RAILINGS AND WALLS – cont

Crib retaining walls – cont
Titan/Geolock vertical modular block reinforced soil system; steel ladder.
Geogrid reinforcement and soil fill material

Item	Unit	Range £		
wall heights up to 3.0 m	m²	205.00	to	250.00
wall heights up to 4.0 m	m²	220.00	to	265.00
wall heights up to 5.0 m	m²	230.00	to	275.00
wall heights up to 6.0 m	m²	240.00	to	290.00
Guard rails and parking bollards etc.				
Post and rail fencing				
open metal post and rail fencing 1.00 m high	m	140.00	to	170.00
galvanized steel post and rail fencing 2.00 m high	m	165.00	to	200.00
Guard rails				
steel guard rails and vehicle barriers	m	57.00	to	69.00
Bollards and barriers				
parking bollards precast concrete or steel	nr	150.00	to	180.00
vehicle control barrier; manual pole	nr	860.00	to	1025.00
Chain link fencing; plastic coated				
1.20 m high	m	19.90	to	24.00
1.80 m high	m	26.50	to	32.00
Timber fencing				
1.20 m high chestnut pale facing	m	20.50	to	25.00
1.80 m high cross-boarded fencing	m	27.00	to	33.00
Screen walls; one brick thick; including foundations etc.				
1.80 m high facing brick screen wall	m	300.00	to	370.00
1.80 m high coloured masonry block boundary wall	m	330.00	to	400.00

8.5 EXTERNAL FIXTURES

Item	Unit	Range £		
Street Furniture				
Roadsigns				
reflected traffic signs 0.25 m² area on steel post	nr	120.00	to	150.00
internally illuminated traffic signs; dependent on area	nr	200.00	to	240.00
externally illuminated traffic signs; dependent on area	nr	770.00	to	930.00
Lighting				
lighting to pedestrian areas an estate roads on 4.00 m–6.00 m columns with				
up to 70 W lamps	nr	230.00	to	280.00
lighting to main roads				
10.00 m – 12.00 m columns with 250 W lamps	nr	485.00	to	590.00
12.00 m – 15.00 m columns with 400 W high pressure sodium lighting	nr	620.00	to	750.00
Benches; bolted to ground				
benches – hardwood and precast concrete	nr	1025.00	to	1225.00
Litter bins; bolted to ground				
precast concrete	nr	185.00	to	225.00
hardwood slatted	nr	190.00	to	230.00
cast iron	nr	380.00	to	460.00
large aluminium	nr	570.00	to	690.00
Bus stops including basic shelter	nr	2375.00	to	2900.00
Pillar box	nr	570.00	to	690.00
Galvanized steel cycle stand	nr	47.50	to	58.00

8.6 EXTERNAL DRAINAGE

Item	Unit	Range £		
Galvanized steel flag staff	nr	1125.00	to	1350.00
Playground equipment				
Modern swings with flat rubber safety seats: four seats; two bays	nr	1425.00	to	1725.00
Stainless steel slide, 3.40 m long	nr	1650.00	to	2000.00
Climbing frame – igloo type 3.20 m × 3.75 m on plan × 2.00 m high	nr	1075.00	to	1325.00
See-saw comprising timber plank on sealed ball bearings 3960 mm × 230 mm × 70 mm thick	nr	95.00	to	115.00
Wickstead Tumbleguard type safety surfacing around play equipment	m²	13.10	to	15.80
Bark particles type safety surfacing 150 mm thick on hardcore bed	nr	13.30	to	16.10
8.6 EXTERNAL DRAINAGE				
Overall £/m² of drained area allowances				
site drainage (per m² of paved area)	m²	18.80	to	24.00
building storm water drainage (per m² of gross internal floor area)	m²	13.10	to	16.50
Machine excavation, grade bottom, earthwork support, laying and jointing pipes and accessories, backfill and compact, disposal of surplus soil. **uPVC pipes and fittings, lip seal coupling joints**				
up to 1.50 m deep; nominal size				
100 mm diameter pipe	m	44.00	to	56.00
160 mm diameter pipe	m	53.00	to	67.00
over 1.50 m not exceeding 3.00 m deep; nominal size				
100 mm diameter pipe	m	79.00	to	99.00
160 mm diameter pipe	m	88.00	to	110.00
Machine excavation, grade bottom, earthwork support, laying and jointing pipes and accessories, backfill and compact, disposal of surplus soil. **uPVC Ultra-Rib ribbed pipes and fittings, sealed ring push fit joints**				
up to 1.50 m deep; nominal size				
150 mm diameter pipe	m	45.00	to	57.00
225 mm diameter pipe	m	57.00	to	72.00
300 mm diameter pipe	m	67.00	to	84.00
over 1.50 m not exceeding 3.00 m deep; nominal size				
150 mm diameter pipe	m	80.00	to	100.00
225 mm diameter pipe	m	96.00	to	120.00
300 mm diameter pipe	m	110.00	to	135.00
Machine excavation, grade bottom, earthwork support, laying and jointing pipes and accessories, backfill and compact, disposal of surplus soil. **Vitrified clay pipes and fittings, Hepseal socketted, with push fit flexible joints; shingle bed and surround**				
up to 1.00 m deep; nominal size				
150 mm diameter pipe	m	54.00	to	68.00
225 mm diameter pipe	m	75.00	to	95.00
300 mm diameter pipe	m	98.00	to	120.00

8.6 EXTERNAL DRAINAGE

Item	Unit	Range £		
8.6 EXTERNAL DRAINAGE – cont				
Machine excavation, grade bottom, earthwork support, laying and jointing pipes and accessories, backfill and compact, disposal of surplus soil. – cont				
over 1.50 m not exceeding 3.00 m deep; nominal size				
150 mm diameter pipe	m	88.00	to	110.00
225 mm diameter pipe	m	110.00	to	140.00
300 mm diameter pipe	m	130.00	to	165.00
Machine excavation, grade bottom, earthwork support, laying and jointing pipes and accessories, backfill and compact, disposal of surplus soil. Class M tested concrete centrifugally spun pipes and fittings, flexible joints; concrete bed and surround				
up to 3.00 m deep; nominal size				
300 mm diameter pipe	m	150.00	to	195.00
525 mm diameter pipe	m	240.00	to	300.00
900 mm diameter pipe	m	410.00	to	520.00
1200 mm diameter pipe	m	580.00	to	730.00
Machine excavation, grade bottom, earthwork support, laying and jointing pipes and accessories, backfill and compact, disposal of surplus soil. Cast iron Timesaver drain pipes and fittings, mechanical coupling joints				
up to 1.50 m deep; nominal size				
100 mm diameter pipe	m	91.00	to	115.00
150 mm diameter pipe	m	130.00	to	160.00
over 1.50 m not exceeding 3.00 m deep; nominal size				
100 mm diameter pipe	m	125.00	to	160.00
150 mm diameter pipe	m	165.00	to	210.00
Brick manholes				
Excavate pit in firm ground, partial backfill, partial disposal, earthwork support, compact base of pit, 150 mm plain in situ concrete 20.00 N/mm² – 20 mm aggregate (1:2:4) base, formwork, one brick wall of engineering bricks in cement mortar (1:3) finished fair face, vitrified clay channels, plain in situ concrete 25.00 N/mm² – 20 mm aggregate (1:2:4) cover and reducing slabs, fabric reinforcement, formwork step irons, medium duty cover and frame; Internal size of manhole				
600 mm × 450 mm; cover to invert				
not exceeding 1.00 m deep	nr	430.00	to	520.00
over 1.00 m not exceeding 1.50 m deep	nr	570.00	to	690.00
over 1.50 m not exceeding 2.00 m deep	nr	710.00	to	860.00
900 mm × 600 mm; cover to invert				
not exceeding 1.00 m deep	nr	560.00	to	680.00
over 1.00 m not exceeding 1.50 m	nr	770.00	to	940.00
over 1.50 m not exceeding 2.00 m	nr	980.00	to	1175.00
900 mm × 900 mm; cover to invert				
not exceeding 1.00 m deep	nr	670.00	to	810.00
over 1.00 m not exceeding 1.50 m	nr	910.00	to	1100.00
over 1.50 m not exceeding 2.00 m	nr	1150.00	to	1400.00

8.6 EXTERNAL DRAINAGE

Item	Unit	Range £		
1200 × 1800 mm; cover to invert				
not exceeding 1.00 m deep	nr	1350.00	to	1625.00
over 1.00 m not exceeding 1.50 m deep	nr	1775.00	to	2150.00
over 1.50 m not exceeding 2.00 m deep	nr	2175.00	to	2650.00
Concrete manholes				
Excavate pit in firm ground, disposal, earthwork support, compact base of pit, plain in situ concrete 20.00 N/mm² – 20 mm aggregate (1:2:4) base, formwork, reinforced precast concrete chamber and shaft rings, taper pieces and cover slabs bedded jointed and pointed in cement; mortar (1:3) weak mix concrete filling to working space, vitrified clay channels, plain in situ concrete 25.00 N/mm² – 20 mm aggregate (1:1:5:3) benchings, step irons, medium duty cover and frame; depth from cover to invert; Internal diameter of manhole 900 mm diameter; cover to invert				
over 1.00 m not exceeding 1.50 m deep	nr	620.00	to	750.00
1050 mm diameter; cover to invert				
over 1.00 m not exceeding 1.50 m deep	nr	690.00	to	830.00
over 1.50 m not exceeding 2.00 m deep	nr	780.00	to	950.00
over 2.00 m not exceeding 3.00 m deep	nr	960.00	to	1175.00
1500 mm diameter; cover to invert				
over 1.50 m not exceeding 2.00 m deep	nr	1275.00	to	1550.00
over 2.00 m not exceeding 3.00 m deep	nr	1625.00	to	1975.00
over 3.00 m not exceeding 4.00 m deep	nr	1975.00	to	2425.00
1800 mm diameter; cover to invert				
over 2.00 m not exceeding 3.00 m deep	nr	2225.00	to	2700.00
over 3.00 m not exceeding 4.00 m deep	nr	2700.00	to	3300.00
2100 mm diameter; cover to invert				
over 2.00 m not exceeding 3.00 m deep	nr	3300.00	to	4000.00
over 3.00 m not exceeding 4.00 m deep	nr	4100.00	to	4950.00
Polypropylene inspection chambers				
475 mm diameter PPIC inspection chamber including all excavations; earthwork support; cart away surplus spoil; concrete bed and surround; lightweight cover and frame				
600 mm deep	nr	250.00	to	305.00
900 mm deep	nr	300.00	to	365.00
Soakaways: stormwater management				
Soakaway crates; heavyweight (60 tonne) polypropylene high-void box units; including excavation and backfilling as required. Note final surfacing not included	m³	170.00	to	210.00
Soakaway crates; lightweight (20 tonne) polypropylene high-void box units; including excavation and backfilling as required. Not efinal surfacing not included	m³	170.00	to	200.00
Septic tanks				
Excavate for, supply and install Klargester glass fibre septic tank, complete with lockable cover				
2800 litre capacity	nr	3450.00	to	3450.00
3800 litre capacity	nr	4200.00	to	4200.00
3800 litre capacity	nr	4950.00	to	4950.00

8.7 EXTERNAL SERVICES

Item	Unit	Range £		
8.6 EXTERNAL DRAINAGE – cont				
Urban and landscape drainage				
Exacavte for and lay oil separators, complete with lockable cover				
Excavate for, supply and lay 1000 litre polyethelyne by pass oil separator	nr	3450.00	to	3450.00
Excavate for, supply and lay 1000 litre polyethelyne full retention oil separator	nr	3600.00	to	3600.00
Land drainage				
NOTE: If land drainage is required on a project, the propensity of the land to flood will decide the spacing of the land drains. Costs include for excavation and backfilling of trenches and laying agricultural clay drain pipes with 75 mm diameter lateral runs average 600 mm deep, and 100 mm diameter mains runs average 750 mm deep.				
land drainage to parkland with laterals at 30 m centres and main runs at 100 m centres	ha	7300.00	to	8500.00
8.7 EXTERNAL SERVICES				
Service runs				
Water main; all laid in trenches including excavation and backfill with excavated material				
up to 50 mm diameter MDPE pipe	m	110.00	to	140.00
Electric main; all laid in trenches including excavation and backfill with excavated material				
600/1000 volt cables. Two core 25 mm diameter cable including 100 mm diameter clayware duct	m	130.00	to	155.00
Gas main; all laid in trenches including excavation and backfill with excavated material				
150 mm diameter gas pipe	m	125.00	to	150.00
Telephone duct; all laid in trenches including excavation and backfill with excavated material				
100 mm diameter uPVC duct	m	70.00	to	85.00
Service connection charges				
The privatisation of telephone, water, gas and electricity has complicated the assessment of service connection charges. Typically, service connection charges will include the actual cost of the direct connection plus an assessment of distribution costs from the main. The latter cost is difficult to estimate as it depends on the type of scheme and the distance from the mains. In addition, service charges are complicated by discounts that maybe offered. For instance, the electricity boards will charge less for housing connections if the house is all electric. However, typical charges for a reasonably sized housing estate might be as follows:				
Water and Sewerage connections				
water connections; water main up to 2 m from property	house	570.00	to	690.00
water infrastructure charges for new properties	house	475.00	to	580.00
sewerage infrastructure charges for new properties	house	475.00	to	580.00

8.8 ANCILLARY BUILDINGS AND STRUCTURES

Item	Unit	Range £		
Electric				
all electric	house	1150.00	to	1375.00
pre-packaged substation housing	nr	28500.00	to	34500.00
Gas				
gas connection to house	house	285.00	to	345.00
governing station	nr	17000.00	to	21000.00
Telephone	house	110.00	to	140.00

8.8 ANCILLARY BUILDINGS AND STRUCTURES

Footbridges
Footbridge of either precast concrete or steel construction up to 6.00 m wide, 6.00 m high including deck, access stairs and ramp, parapets etc.

Item	Unit			
5 m span between piers or abutments	m²	1325.00	to	1675.00
20 m span between piers or abutments	m²	1900.00	to	2400.00
Footbridge of timber (stress graded with concrete piers)				
12 m span between piers or abutments	m²	950.00	to	1200.00

Roadbridges
Reinforced concrete bridge with precast beams; including all excavation, reinforcement, formwork, concrete, bearings, expansion joints, deck water proofing and finishings, parapets etc. deck area

Item	Unit			
10.00 m span	m²	1425.00	to	1800.00
15.00 m span	m²	1900.00	to	2400.00

Reinforced concrete bridge with prefabricated steel beams; including all excavation, reinforcement, formwork, concrete, bearings, expansion joints, deck water proofing and finishings, parapets etc. deck area

Item	Unit			
20.00 m span	m²	1225.00	to	1550.00
30.00 m span	m²	1175.00	to	1500.00

Multi-parking systems/stack parkers
Fully automatic systems

Item	Unit			
integrated robotic parking system using robotic car transporter to store vehicles	car	21000.00	to	26000.00

Semi-automatic systems

Item	Unit			
integrated parking system, transverse and vertical positioning; semi-automatic parking achieving 17 spaces in a 6 car width × 3 car height grid	car	11000.00	to	14000.00

Integrated stacker systems

Item	Unit			
integrated parking system, vertical positioning only; double width, double height pit stacker achieving 4 spaces with each car stacker	car	5700.00	to	7200.00
integrated parking system, vertical positioning only; triple stacker achieving 3 spaces with each car stacker, generally 1 below ground and 2 above ground	car	12000.00	to	15000.00
integrated parking system, vertical positioning only; triple height double width stacker, achieving 6 spaces with each car stacker, generally 2 below ground and 4 above ground	car	8600.00	to	11000.00

Designing Tall Buildings: Structure as Architecture, 2nd Edition

Mark Sarkisian

This second edition of Designing Tall Buildings, an accessible reference to guide you through the fundamental principles of designing high-rises, features two new chapters, additional sections, 400 images, project examples, and updated US and international codes. Each chapter focuses on a theme central to tall-building design, giving a comprehensive overview of the related architecture and structural engineering concepts. Author Mark Sarkisian, PE, SE, LEED® AP BD+C, provides clear definitions of technical terms and introduces important equations, gradually developing your knowledge. Projects drawn from SOM's vast portfolio of built high-rises, many of which Sarkisian engineered, demonstrate these concepts.

This book advises you to consider the influence of a particular site's geology, wind conditions, and seismicity. Using this contextual knowledge and analysis, you can determine what types of structural solutions are best suited for a tower on that site. You can then conceptualize and devise efficient structural systems that are not only safe, but also constructible and economical. Sarkisian also addresses the influence of nature in design, urging you to integrate structure and architecture for buildings of superior performance, sustainability, and aesthetic excellence

Jan 2016: 234X156 mm: 300 pp
Pb: 978-1-138-88671-1 £31.99

To Order: Tel: +44 (0) 1235 400524 Fax: +44 (0) 1235 400525
or Post: Taylor and Francis Customer Services,
Bookpoint Ltd, Unit T1, 200 Milton Park, Abingdon, Oxon, OX14 4TA UK
Email: book.orders@tandf.co.uk

For a complete listing of all our titles visit:
www.tandf.co.uk

Structural Competency for Architects

Hollee Hitchcock Becker

Structural Competency for Architects is a comprehensive volume covering topics from structural systems and typologies to statics, strength of materials, and component design. The book includes everything you need to know about structures for the design of components, as well as the logic for design of structural patterns, and selection of structural typologies.

Organized into six key modules, each chapter includes examples, problems, and labs, along with an answer key available on our website, so that you learn the fundamentals. Structural Competency for Architects will also help you pass your registration examinations.

June 2014: 229 x 286: 344pp
Pb: 978-0-415-81788-2: £53.99

To Order: Tel: +44 (0) 1235 400524 Fax: +44 (0) 1235 400525
or Post: Taylor and Francis Customer Services,
Bookpoint Ltd, Unit T1, 200 Milton Park, Abingdon, Oxon, OX14 4TA UK
Email: book.orders@tandf.co.uk

For a complete listing of all our titles visit:
www.tandf.co.uk

Taylor & Francis
Taylor & Francis Group

Materials for Architects and Builders

Arthur Lyons

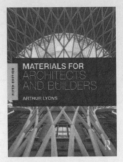

Materials for Architects and Builders provides a clear and concise introduction to the broad range of materials used within the construction industry and covers the essential details of their manufacture, key physical properties, specification and uses.

Understanding the basics of materials is a crucial part of undergraduate and diploma construction or architecture-related courses, and this established textbook helps the reader to do just that with the help of colour photographs and clear diagrams throughout.

This new edition has been completely revised and updated to include the latest developments in materials research, new images, appropriate technologies and relevant legislation. The ecological effects of building construction and lifetime use remain an important focus, and this new edition includes a wide range of energy saving building components.

August 2014: 246 x189: 448 pp
Pb: 978-0-415-70497-7: £42.99

To Order: Tel: +44 (0) 1235 400524 Fax: +44 (0) 1235 400525
or Post: Taylor and Francis Customer Services,
Bookpoint Ltd, Unit T1, 200 Milton Park, Abingdon, Oxon, OX14 4TA UK
Email: book.orders@tandf.co.uk

For a complete listing of all our titles visit:
www.tandf.co.uk

ESSENTIAL READING FROM TAYLOR AND FRANCIS

Concrete Design

Edited by Paul W. McMullin, Jonathan S. Price, Esra Hasanbas Persellin

Concrete Design covers concrete design fundamentals for architects and engineers, such as tension, flexural, shear, and compression elements, anchorage, lateral design, and footings. As part of the Architect's Guidebooks to Structures Series it provides a comprehensive overview using both imperial and metric units of measurement. Written by experienced professional structural engineers Concrete Design is beautifully illustrated, with more than 170 black and white images, contains clear examples that show all design steps, and provides rules of thumb and simple tables for initial sizing. A refreshing change in textbooks for architectural materials courses, it is an indispensable reference for practicing architects and students alike. As a compact summary of key ideas it is ideal for anyone needing a quick guide to concrete design.

May 2016: 186X123 mm: 366 pp
Pb: 9978-1-138-82997-8: £21.99

To Order: Tel: +44 (0) 1235 400524 Fax: +44 (0) 1235 400525
or Post: Taylor and Francis Customer Services,
Bookpoint Ltd, Unit T1, 200 Milton Park, Abingdon, Oxon, OX14 4TA UK
Email: book.orders@tandf.co.uk

For a complete listing of all our titles visit:
www.tandf.co.uk

Acoustics of Multi-Use Performing Arts Centers

MARK HOLDEN

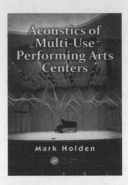

Employing the experiences of a world-renowned acoustician celebrated for the successful design of multi-use performing arts centers, Acoustics of Multi-Use Performing Arts Centers examines the complexities of this venue and discusses the challenges and solutions that arise in the concept, design, construction and commissioning phases.

This book addresses the various programming needs of a multi-use performing arts center (needs that can range from symphony, opera and ballet to highly-amplified concerts and Broadway productions) and provides instruction from the planning of the initial concept to the final tuning stages.

While assuming a basic understanding of the principals of sound, construction and performance, the author includes architectural drawings drawn to scale and presents case studies with in-depth discussion of undocumented halls. He also provides a full chapter on tuning multi-use halls and offers an inside look at design options for adjustable acoustics that include stage, pit and hall design.

Dec 2015: 246X174 mm: 400 pp
Hb: 978-0-415-51719-5: £77.00

To Order: Tel: +44 (0) 1235 400524 Fax: +44 (0) 1235 400525
or Post: Taylor and Francis Customer Services,
Bookpoint Ltd, Unit T1, 200 Milton Park, Abingdon, Oxon, OX14 4TA UK
Email: book.orders@tandf.co.uk

For a complete listing of all our titles visit:
www.tandf.co.uk

Spon's Asia Pacific Construction Costs Handbook, Fifth Edition

LANGDON & SEAH

In the last few years, the global economic outlook has continued to be shrouded in uncertainty and volatility following the financial crisis in the Euro zone. While the US and Europe are going through a difficult period, investors are focusing more keenly on Asia. This fifth edition provides overarching construction cost data for 16 countries: Brunei, Cambodia, China, Hong Kong, India, Indonesia, Japan, Malaysia, Myanmar, Philippines, Singapore, South Korea, Sri Lanka, Taiwan, Thailand and Vietnam.

May 2015: 234X156 mm: 452 pp
Hb: 978-1-4822-4358-1: £160.00

To Order: Tel: +44 (0) 1235 400524 Fax: +44 (0) 1235 400525
or Post: Taylor and Francis Customer Services,
Bookpoint Ltd, Unit T1, 200 Milton Park, Abingdon, Oxon, OX14 4TA UK
Email: book.orders@tandf.co.uk

For a complete listing of all our titles visit:
www.tandf.co.uk

Taylor & Francis
Taylor & Francis Group

It is one thing to
imagine a better world.
It's another to deliver it.

**Understanding change,
unlocking potential, creating
brilliant new communities.**

The Tate Modern extension
takes an iconic building and
adds to it. Cost management
provided by AECOM.

Built to deliver a better world

aecom.com

BIM and Quantity Surveying
PITTARD & SELL

The sudden arrival of Building Information Modelling (BIM) as a key part of the building industry is redefining the roles and working practices of its stakeholders. Many clients, designers, contractors, quantity surveyors, and building managers are still finding their feet in an industry where BIM compliance can bring great rewards.

This guide is designed to help quantity surveying practitioners and students understand what BIM means for them, and how they should prepare to work successfully on BIM compliant projects. The case studies show how firms at the forefront of this technology have integrated core quantity surveying responsibilities like cost estimating, tendering, and development appraisal into high profile BIM projects. In addition to this, the implications for project management, facilities management, contract administration and dispute resolution are also explored through case studies, making this a highly valuable guide for those in a range of construction project management roles.

Featuring a chapter describing how the role of the quantity surveyor is likely to permanently shift as a result of this development, as well as descriptions of tools used, this covers both the organisational and practical aspects of a crucial topic.

December 2015: 234 x 156 mm: 258 pp
Pb: 978-0-415-87043-6; £24.99

To Order: Tel: +44 (0) 1235 400524 Fax: +44 (0) 1235 400525
or Post: Taylor and Francis Customer Services,
Bookpoint Ltd, Unit T1, 200 Milton Park, Abingdon, Oxon, OX14 4TA UK
Email: book.orders@tandf.co.uk

For a complete listing of all our titles visit:
www.tandf.co.uk

Estimator's Pocket Book

Duncan Cartlidge

The Estimator's Pocket Book is a concise and practical reference covering the main pricing approaches, as well as useful information such as how to process sub-contractor quotations, tender settlement and adjudication. It is fully up-to-date with NRM2 throughout, features a look ahead to NRM3 and describes the implications of BIM for estimators.

It includes instructions on how to handle:

- the NRM order of cost estimate;
- unit-rate pricing for different trades;
- pro rata pricing and dayworks
- builders' quantities;
- approximate quantities.

Worked examples show how each of these techniques should be carried out in clear, easy-to-follow steps. This is the indispensible estimating reference for all quantity surveyors, cost managers, project managers and anybody else with estimating responsibilities. Particular attention is given to NRM2, but the overall focus is on the core estimating skills needed in practice.

May 2013 186x123: 310pp
Pb: 978-0-415-52711-8: £21.99

To Order: Tel: +44 (0) 1235 400524 Fax: +44 (0) 1235 400525
or Post: Taylor and Francis Customer Services,
Bookpoint Ltd, Unit T1, 200 Milton Park, Abingdon, Oxon, OX14 4TA UK
Email: book.orders@tandf.co.uk

For a complete listing of all our titles visit:
www.tandf.co.uk

PART 4

Prices for Measured Works

This part contains the following NRM2 work sections:

Building Construction Handbook, 11th Edition

CHUDLEY & GREENO

The Building Construction Handbook is THE authoritative reference for all construction students and professionals. Its detailed drawings clearly illustrate the construction of building elements, and have been an invaluable guide for builders since 1988. The principles and processes of construction are explained with the concepts of design included where appropriate. Extensive coverage of building construction practice, techniques, and regulations representing both traditional procedures and modern developments are included to provide the most comprehensive and easy to understand guide to building construction.

This new edition has been updated to reflect recent changes to the building regulations, as well as new material on the latest technologies used in domestic construction.

Building Construction Handbook is the essential, easy-to-use resource for undergraduate and vocational students on a wide range of courses including NVQ and BTEC National, through to Higher National Certificate and Diploma, to Foundation and three-year Degree level. It is also a useful practical reference for building designers, contractors and others engaged in the construction industry.

April 2016: 234 x 156 mm: 1012 pp
Pb: 978-1-138-90709-6: £28.99

INTRODUCTION

The rates contained in *Prices for Measured Works* are intended to apply to a project in the Outer London area costing about £4,000,000 including Preliminaries, and assume that reasonable quantities of all types of work are required. Similarly it has been necessary to assume that the size of the project warrants the subletting of all types of work normally sublet. Adjustments should be made to standard rates for time, location, local conditions, site constraints and all the other factors likely to affect costs of any individual project.

Adjustments may need to be made for different project vales. In *How to use this Book* you will find some guidance on adjustments accordance to contract cost.

The distinction between builders' work and work normally sublet is stressed because prices for work which can be sublet may well be inadequate for the contractor who is called upon to carry out relatively small quantities of such work themselves.

Measured Works prices are generally based upon wage rates and known material costs from May 2016. Built up prices and subcontractor rates include an allowance of 5% for overheads and profit.

As elsewhere in this edition, prices do not include Value Added Tax or professional services fees.

Measured Work rates are exclusive of any main contractor's preliminaries.

This year we have re-structured the *Prices for Measured Works* sections to reflect the recent introduction of NRM2.

To aid readers with the transition we have scheduled the SMM7 Work Sections and shown the new NRM2 Work Sections alongside:

SMM7 Work Section	NRM2 Work Section
	2 Off-site manufactured materials, components and buildings:
C Demolition/Alteration/Renovation	3 Demolitions
	4 Alterations, repairs and conservation
D Groundwork	5 Excavate and filling
D20 Excavation and Filling	7 Piling
D30 Cast In Place Piling	8 Underpinning
D32 Steel Piling	9 Diaphragm walls and embedded retaining walls
D40 Embedded Retaining Walling	10 Crib walls, gabions and reinforced earth
D41 Crib Walls/Gabions/Reinforced Earthworks	
D50 Underpinning	
E In Situ Concrete/Large Precast Concrete	11 In situ concrete works
E10 In Situ Concrete Construction	12 Precast/composite concrete
E20 Formwork For In Situ Concrete	13 Precast concrete
E30 Reinforcement For In Situ Concrete	
E40 Designed Joints In Situ Concrete	
E41 Worked Finishes/Cutting To In Situ Concrete	
E42 Accessories Cast Into In Situ Concrete	
E50 Precast Concrete Large Units	
E60 Precast/Composite Concrete Decking	

Prices for Measured Works

INTRODUCTION

SMM7 Work Section	NRM2 Work Section
F Masonry	14 Masonry
F10 Brick/Block Walling	
F11 Glass block Walling	
F20 Natural Stone Rubble Walling	
F22 Cast Stone Walling/Dressings	
F30 Accessories/Sundry Items For Brick/Block/Stone Walling	
F31 Precast Concrete Sills/Lintels/Coping features	
G Structural/Carcassing Metal/Timber	15 Structural metalwork
G10 Structural Steel Framing	16 Carpentry
G12 Isolated Structural Metal Members	
G20 Carpentry/Timber Framing/First Fixing	
H Cladding/Covering	17 Sheet roof coverings
H10 Patent Glazing	18 Tile and slate roof and wall coverings
H11 Curtain Walling	21 Cladding and covering
H20 Rigid Sheet Cladding	
H30 Fibre Cement Profile Sheet Cladding	
H31 Metal Profiled/Flat Sheet Cladding/Covering/Siding	
H32 Plastic Profiled Sheet Cladding/Covering/Siding	
H41 Glass Reinforced Plastic Panel Cladding Features	
H51 Natural Stone Cladding Features	
H53 Clay Slab/Cladding/Features	
H60 Plain Roof Tiling	
H61 Fibre Cement Slating	
H62 Natural Slating	
H63 Reconstructed Stone Slating/Tiling	
H64 Timber Shingling	
H71 Lead Sheet Coverings/Flashes	
H72 Aluminium Sheet Coverings/Flashings	
H73 Copper Strip Sheet Coverings/Flashings	
H74 Zinc Strip Sheet Coverings/Flashings	
H75 Stainless Steel Sheet Coverings/Flashings	
H76 Fibre Bitumen Thermoplastic Sheet Coverings/Flashings	
H92 Rainscreen Cladding	
J Waterproofing	19 Waterproofing
J10 Specialist Waterproof Rendering	
J20 Mastic Asphalt Tanking/Damp-Proof Membranes	
J21 Mastic Asphalt Roofing/Insulation/Finishes	
J30 Liquid Applied Tanking/Damp-Proof Membranes	

INTRODUCTION

SMM7 Work Section	NRM2 Work Section
J40 Flexible Sheet Tanking/Damp-Proof Membranes J41 Built Up Felt Roof Coverings J42 Single Layer Plastic Roof Coverings J43 Proprietary Roof Decking With Felt Finish	
K Linings/Sheathing/Dry partitioning K10 Plasterboard Dry Lining/Partitions/Ceilings K11 Rigid Sheet Flooring/Sheathing/Linings/Casings K13 Rigid Sheet Fine Linings/Panelling K14 Glass Reinforced Gypsum Linings/Panelling K20 Timber Board Flooring/Sheathing/Linings/Casings K30 Demountable Partitions K32 Framed Panel Cubicle Partitions K33 Concrete/Terrazzo Partitions K40 Demountable Suspended Ceilings K41 Raised Access Floors	20 Proprietary linings and partitions 30 Suspended ceilings
L Windows/Doors/Stairs L10 Windows/Rooflights/Screens/Louvres L20 Doors/Shutters/Hatches L30 Stairs/Walkways/Balustrades L40 General Glazing	23 Windows, screens and lights 24 Doors, shutters and hatches 25 Stairs, walkways and balustrades 27 Glazing
M Surface Finishes M10 Cement: Sand/Concrete Screeds/Granolithic Screeds/Topping M11 Mastic Asphalt Flooring/Floor Underlays M12 Trowelled Bitumen/Resin/Rubber Latex M20 Plastered/Rendered/Roughcast Coating M21 Insulation With Rendered Finish M22 Sprayed Mineral Fibre Coatings M30 Metal Mesh Lathing/Anchored Reinforcement For Plastered Ceilings M31 Fibrous Plaster M40 Stone/Concrete/Quarry/Ceramic Tiling M41 Terrazzo Tiling/In Situ Terrazzo M42 Wood Block/Composition Block/Parquet M50 Rubber/Plastic/Cork/Lino/Carpet Tiling/Sheeting M51 Edge Fixed Carpeting M52 Decorative Papers/Fabrics M60 Painting/Clear Finishing	28 Floor, wall, ceiling and roof finishings 29 Decoration

INTRODUCTION

SMM7 Work Section	NRM2 Work Section
N Furniture/Equipment	32 Furniture fittings and equipment
N10/11 General Fixtures/Kitchen Fittings	
N13 Sanitary Appliances/Fittings	
N15 Signs/Notices	
P Building Fabric Sundries	22 General joinery
P10 Sundry Insulation/Proofing Work/Fire Stops	31 Insulation, fire stopping and fire protection
P20 Unframed Isolated Trims/Skirtings/Sundry Items	
P21 Ironmongery	
P30 Trenches/Pipeways/Pits For Buried Engineering Services	
P31 Holes/Chases/Covers/Supports for Services	
Q Paving/Planting/Fencing/Site Furniture	35 Site work
Q10 Kerbs/Edgings/Channels/Paving Access	36 Fencing
Q20 Hardcore/Granular/Cement Bound Bases	37 Soft landscaping
Q21 In Situ Concrete Roads/Pavings	
Q22 Coated Macadam/Asphalt Roads/Pavings	
Q23 Gravel/Hoggin/Woodchip Roads/Pavings	
Q25 Slab/Brick/Block/Sett/Cobble Pavings	
Q26 Special Surfacings/Pavings For Sport	
Q30 Seeding/Turfing	
Q31 Planting	
Q40 Fencing	
R Disposal Systems	33 Drainage above ground
R10 Rainwater Pipework/Gutters	34 Drainage below ground
R11 Foul Drainage Above Ground	
R12 Drainage Below Ground	
R13 Land Drainage	
S Piped Supply Systems	
S10/S11 Hot And Cold Water	
S13 Pressurized Water	
T Mechanical Heating/Cooling/Refrigeration Systems	
T10 Gas/Oil Fired Boilers	
T31 Low Temperature Hot Water	
V Electrical Systems	39 Electrical services
V21/V22 General Lighting and Low Voltage Power	
W Security systems	
W20 Lightning Protection	
X Transport Systems	41 Builder's work in connection with services

03 DEMOLITIONS

Item	PC £	Labour hours	Labour £	Plant £	Material £	Unit	Total rate £
DEMOLITIONS							
NOTE: Demolition rates vary enormously from project to project, depending upon access, type of construction, method of demolition, redundant or recycleable materials etc. Always obtain specific quotations for each project under consideration. The following rates for simple demolition works may be of some assistance for comparative purposes. Generally scaffold and other access equipment is not included in the rates.							
Demolishing structures; by machine; note disposal not included							
Demolishing to ground level; single storey brick out-building; timber flat roof; grub up shallow foundations; volume							
single storey out building approximately 50 m^3	–	–	–	–	–	m^3	**66.15**
single storey out building approximately 200 m^3	–	–	–	–	–	m^3	**46.31**
single storey out building approximately 500 m^3	–	–	–	–	–	m^3	**22.05**
Demolishing to ground level; light steel framed and sheet roofed cycle shelter; grub up shallow foundations; volume cycle shelter approximately 120 m^3	–	–	–	–	–	m^3	**49.61**
Concrete framed multi-storey carpark	–	–	–	–	–	m^3	**33.07**
Demolishing to ground level; multi-storey masonry building; flat roof; 1–4 storeys; volume	–	–	–	–	–	m^3	**22.05**
Warehouse to slab level; 7 m eaves	–	–	–	–	–	m^2	**13.23**
Warehouse to below slab; 7 m eaves; grub up shallow foundations (pads, ground beams and ground slab)	–	–	–	–	–	m^2	**22.05**
Reinforced concrete frame building; easy access	–	–	–	–	–	m^3	**39.69**
Reinforced concrete frame building; restricted access	–	–	–	–	–	m^3	**73.87**
Concrete encased steel frame building; easy access	–	–	–	–	–	m^3	**27.56**
Concrete encased steel frame building; restricted access	–	–	–	–	–	m^3	**66.15**
Demolishing parts of structures; by hand; cost of skip included							
Breaking up plain concrete bed; load into skip							
100 mm thick	–	0.45	9.75	5.09	–	m^2	**14.84**
150 mm thick	–	0.67	14.43	9.96	–	m^2	**24.39**
200 mm thick	–	0.90	19.52	10.16	–	m^2	**29.68**
300 mm thick	–	1.33	28.88	14.92	–	m^2	**43.80**
Breaking up reinforced concrete bed; load into skip							
100 mm thick	–	0.50	10.93	5.68	–	m^2	**16.61**
150 mm thick	–	0.75	16.19	10.54	–	m^2	**26.73**
200 mm thick	–	1.00	21.65	11.36	–	m^2	**33.01**
300 mm thick	–	1.50	32.58	16.94	–	m^2	**49.52**
Demolishing reinforced concrete column or cutting away concrete casing to steel column; load into skip	–	9.99	216.55	75.97	–	m^3	**292.52**
Demolishing reinforced concrete beam or cutting away concrete casing to steel beam; load into skip	–	11.47	248.75	82.34	–	m^3	**331.09**

Prices for Measured Works

03 DEMOLITIONS

Item	PC £	Labour hours	Labour £	Plant £	Material £	Unit	Total rate £
DEMOLITIONS – cont							
Demolishing parts of structures – cont							
Demolishing reinforced concrete wall; load into skip							
100 mm thick	–	0.99	21.46	7.53	–	m²	**28.99**
150 mm thick	–	1.50	32.58	11.23	–	m²	**43.81**
225 mm thick	–	2.25	48.77	16.91	–	m²	**65.68**
300 mm thick	–	3.00	64.97	22.68	–	m²	**87.65**
Demolishing reinforced concrete suspended slabs; load into skip							
100 mm thick	–	0.84	18.14	7.05	–	m²	**25.19**
150 mm thick	–	1.25	27.12	10.29	–	m²	**37.41**
225 mm thick	–	1.87	40.58	15.42	–	m²	**56.00**
300 mm thick	–	2.50	54.23	20.85	–	m²	**75.08**
Breaking up small concrete plinth; make good structures; load into skip	–	3.83	83.11	37.74	5.69	m³	**126.54**
Breaking up small precast concrete kerb; make good structures; load into skip	–	0.41	8.98	3.08	5.69	m	**17.75**
Remove precast concrete window sill; set aside for reuse	–	1.33	28.88	–	–	m	**28.88**
Remove brick on-edge-coping; prepare walls for raising; load into skip							
one brick thick	–	0.38	10.34	0.59	–	m	**10.93**
one and a half brick thick	–	0.50	13.80	0.87	–	m	**14.67**
Demolishing brick chimney to 300 mm below roof level; sealing off flues with slates; piecing in treated sawn softwood rafters and making good roof coverings; load into skip (excluding scaffold access)							
680 mm × 680 mm × 900 mm high above roof	–	11.40	259.76	16.57	65.00	nr	**341.33**
add for each additional 300 mm height	–	2.08	43.68	4.42	–	nr	**48.10**
680 mm × 1030 mm × 900 mm high above roof	–	16.86	378.44	17.68	93.19	nr	**489.31**
add for each additional 300 mm height	–	3.11	65.41	6.63	–	nr	**72.04**
1030 mm × 1030 mm × 900 mm high above roof	–	25.52	566.58	27.63	125.84	nr	**720.05**
add for each additional 300 mm height	–	4.71	98.97	8.84	–	nr	**107.81**
Demolishing defective brick chimney to roof level; rebuild using 25% new facing bricks; new lead flashings; core flues; reset chimney pot; load into skip (excluding scaffold access)							
680 mm × 680 mm × 900 mm high above roof	–	10.73	227.51	16.57	103.94	nr	**348.02**
add for each additional 300 mm height	–	2.08	43.68	4.42	9.08	nr	**57.18**
680 mm × 1030 mm × 900 mm high above roof	–	15.66	329.06	17.68	125.65	nr	**472.39**
add for each additional 300 mm height	–	3.11	65.41	6.63	11.26	nr	**83.30**
1030 mm × 1030 mm × 900 mm high above roof	–	24.02	504.86	27.63	146.89	nr	**679.38**
add for each additional 300 mm height	–	4.71	98.97	8.84	14.57	nr	**122.38**
Demolishing external brick walls; load into skip							
half brick thick	–	0.87	17.95	4.34	–	m²	**22.29**
two half brick thick skins if cavity wall	–	1.46	29.99	8.40	–	m²	**38.39**
one brick thick	–	1.50	30.91	8.40	–	m²	**39.31**
one and a half brick thick	–	2.04	42.03	12.60	–	m²	**54.63**
two brick thick	–	2.62	53.88	16.79	–	m²	**70.67**
extra for plaster, render or pebbledash finish per side	–	0.08	1.67	0.87	–	m²	**2.54**

03 DEMOLITIONS

Item	PC £	Labour hours	Labour £	Plant £	Material £	Unit	Total rate £
Demolish external stone walls; load into skip							
300 mm thick	–	1.00	20.57	8.84	–	m²	**29.41**
400 mm thick	–	1.33	27.39	11.49	–	m²	**38.88**
600 mm thick	–	2.00	41.15	17.36	–	m²	**58.51**
Demolish external stone walls; clean off and set aside for reuse							
300 mm thick	–	1.50	30.86	2.21	–	m²	**33.07**
400 mm thick	–	2.00	41.15	3.32	–	m²	**44.47**
600 mm thick	–	3.00	61.72	4.42	–	m²	**66.14**
Remove fireplace surround and hearth							
breaking up 500 mm wide concrete hearth; load into skip	–	1.50	30.91	2.89	–	m	**33.80**
remove fire surround; tiled interior; load into skip	–	1.54	31.67	6.63	–	nr	**38.30**
cast iron surround; set aside for reuse	–	2.58	53.14	–	–	nr	**53.14**
stone surround; set aside for reuse	–	6.74	138.67	–	–	nr	**138.67**
Fill in opening with common bricks; air brick; 2 coat plaster; timber skirting; re-screed floor	–	0.95	26.01	–	64.22	m²	**90.23**
Removing roof timbers complete (NB not coverings); including rafters, purlins, ceiling joists, plates etc.; load into skip; measured flat on plan	–	0.28	6.29	3.39	–	m²	**9.68**
Remove softwood floor structure; load into skip							
joists at ground floor level	–	0.20	4.21	1.16	–	m²	**5.37**
joists at first floor level	–	0.41	8.52	1.16	–	m²	**9.68**
joists at second floor	–	0.58	12.04	1.16	–	m²	**13.20**
individual timber members	–	0.23	4.63	0.04	–	m	**4.67**
Remove boarding; withdraw nails; set aside for reuse							
softwood flooring at ground floor level	–	0.30	6.17	–	–	m²	**6.17**
softwood flooring at first floor level	–	0.51	10.53	–	–	m²	**10.53**
softwood flooring at second floor level	–	0.61	12.59	–	–	m²	**12.59**
fascia/barge board/gutter board second floor level; load into skip	–	0.52	10.59	0.04	–	m	**10.63**
Remove doors and windows; set aside for reuse							
solid single door only (frame left in place)	–	0.33	6.85	0.87	–	nr	**7.72**
solid single door; frame to skip	–	0.55	11.31	1.45	–	nr	**12.76**
solid double door only (frame left in place)	–	0.66	13.58	1.73	–	nr	**15.31**
solid double door; frame to skip	–	1.00	20.55	2.89	–	nr	**23.44**
glazed screen and doors	–	1.00	20.57	4.34	–	m²	**24.91**
casement window frame; set aside frame; glass into skip; up to 1.00 m²	–	0.75	15.42	0.22	–	m²	**15.64**
casement window frame; set aside frame; glass into skip; 1.00 to 2.00 m²	–	1.00	20.57	0.22	–	m²	**20.79**
casement window frame; set aside frame; glass into skip; 2.00 to 3.00 m²	–	1.10	22.63	0.22	–	m²	**22.85**
pair French windows and frame; upto 1200 mm × 1200 mm	–	3.33	68.50	4.34	–	nr	**72.84**
removing double hung sash window and frame; store for reuse	–	2.50	51.38	–	–	nr	**51.38**

03 DEMOLITIONS

Item	PC £	Labour hours	Labour £	Plant £	Material £	Unit	Total rate £
DEMOLITIONS – cont							
Demolishing parts of structures – cont							
Remove doors and windows; load into skip							
remove solid timber door	–	0.30	6.17	1.45	–	nr	**7.62**
remove door frame	–	0.10	2.06	1.45	–	nr	**3.51**
casement window frame; up to 1.00 m²	–	0.60	12.35	1.45	–	m²	**13.80**
casement window frame; 1.00 to 2.00 m²	–	0.80	16.46	1.45	–	m²	**17.91**
casement window frame; 2.00 to 3.00 m²	–	0.90	18.51	1.45	–	m²	**19.96**
Demolishing internal partitions; load into skip							
half brick thick brickwork	–	0.87	17.95	4.34	–	m²	**22.29**
one brick thick brickwork	–	1.50	30.91	8.40	–	m²	**39.31**
one and a half brick thick brickwork	–	2.04	42.03	12.60	–	m²	**54.63**
75 mm blockwork	–	0.58	12.04	3.17	–	m²	**15.21**
90 mm blockwork	–	0.62	12.78	3.76	–	m²	**16.54**
100 mm blockwork	–	0.67	13.70	4.34	–	m²	**18.04**
115 mm blockwork	–	0.71	14.63	4.34	–	m²	**18.97**
125 mm blockwork	–	0.75	15.37	4.63	–	m²	**20.00**
140 mm blockwork	–	0.79	16.31	4.91	–	m²	**21.22**
150 mm blockwork	–	0.84	17.22	5.49	–	m²	**22.71**
190 mm blockwork	–	0.98	20.18	6.93	–	m²	**27.11**
215 mm blockwork	–	1.08	22.22	7.52	–	m²	**29.74**
255 mm blockwork	–	1.25	25.71	8.97	–	m²	**34.68**
extra for plaster finish per side	–	0.08	1.67	0.87	–	m²	**2.54**
breaking up brick plinths	–	3.33	68.50	28.91	–	m³	**97.41**
stud walls (metal or timber); solid board each side	–	0.38	7.77	2.89	–	m²	**10.66**
stud walls (metal or timber); glazed panels	–	0.50	10.36	2.89	–	m²	**13.25**
lightweight steel mesh security screen	–	0.41	8.52	1.44	–	m²	**9.96**
solid steel demountable partition	–	0.62	12.78	2.03	–	m²	**14.81**
glazed demountable partition including removal of glass	–	0.84	17.22	2.89	–	m²	**20.11**
glazed screen including any doors incorporated	–	1.00	20.57	4.34	–	m²	**24.91**
remove large folding partition and track	–	9.90	203.67	43.97	–	m²	**247.64**
remove timber skirting, architrave, dado rail; load into skip	–	0.10	2.06	0.04	–	m	**2.10**
Removing timber staircase and balustrades							
single straight flight	–	2.92	59.99	28.91	–	nr	**88.90**
single dogleg flight	–	4.17	77.75	39.33	–	nr	**117.08**
Perimeter Hoarding							
security fencing; 2400 mm high painted plywood hoarding	–	–	–	–	–	m	**121.28**

03 DEMOLITIONS

Item	PC £	Labour hours	Labour £	Plant £	Material £	Unit	Total rate £
TEMPORARY SUPPORT OF STRUCTURES, ROADS AND THE LIKE							
Temporary support							
NOTE: The requirement for shoring and strutting for the formation of large openings are dependent upon a number of factors; for example the weight of the superimposed structure to be supported; the number of windows; number of floors; roof type; whether raking shores are required; depth to load bearing surface; duration support is to be left in place. Prices therefore are best built-up by assessing the use and waste of materials and the labours involved. The follow guide needs to be tested against each particular project.							
Support of structures not to be demolished							
strutting to window openings over proposed new openings	–	0.56	15.80	–	4.34	nr	**20.14**
plates, struts, braces and hardwood wedges in supports to floors and roof of openings	–	1.11	31.30	–	12.66	nr	**43.96**
dead shore and needle using die square timber with sole plates, braces, hardwood wedges and steel dogs	–	27.75	782.63	–	63.08	nr	**845.71**
set of two raking shores using die square timber with 50 mm thick wall pieces, hardwood wedges and steel dogs, including forming holes for needles and making good	–	33.30	939.15	–	64.70	nr	**1003.85**
cut holes through one brick wall for die square needle and make good on completion of works, including facings externally and plaster internally	–	5.36	139.72	–	20.80	nr	**160.52**

04 ALTERATIONS, REPAIRS AND CONSERVATION

Item	PC £	Labour hours	Labour £	Plant £	Material £	Unit	**Total rate £**
REMOVING							
Remove materials from existing buildings							
NOTE: It is highly unlikely for exactly the same composite items of alteration works will be encountered on different projects. The following spot items have been included to allow the reader to build up the composite rates for estimating purposes. Rates include for the removal of debris from site, but do not include for any temporary works, shoring, scaffolding etc. or for any redecorations.							
Removing coverings load into skip							
Roof coverings							
slates	–	0.45	8.33	0.53	–	m²	**8.86**
slates; set aside for reuse	–	0.54	10.14	–	–	m²	**10.14**
nibbed tiles	–	0.36	6.70	0.53	–	m²	**7.23**
nibbed tiles; set aside for reuse	–	0.45	8.33	–	–	m²	**8.33**
underfelt and nails	–	0.04	0.72	0.26	–	m²	**0.98**
felt or polymer membrane	–	0.22	4.16	0.53	–	m²	**4.69**
profiled metal sheeting; insulation; steel liner	–	0.75	14.03	1.00	–	m²	**15.03**
sheet metal coverings	–	0.45	8.33	0.53	–	m²	**8.86**
remove tiling battens	–	0.07	1.27	0.26	–	m²	**1.53**
Remove roof coverings; select and refix including providing 25% new tiles including nails							
natural slates; Welsh Blue	–	1.08	40.18	0.26	13.70	m²	**54.14**
clay plain tiles	–	0.99	36.92	0.26	7.27	m²	**44.45**
asbestos free artificial blue/black slates	–	0.99	36.92	0.26	5.28	m²	**42.46**
concrete interlocking tiles	–	0.63	23.53	0.26	2.43	m²	**26.22**
removing ridge or hip tile; provide and fit new	–	0.45	16.65	–	8.51	m	**25.16**
Fixtures and fittings; remove and load into skip							
Handrails and balustrades							
timber or metal handrail and brackets	–	0.27	5.59	0.28	–	m	**5.87**
timber balustrades	–	0.30	6.18	0.87	–	m	**7.05**
metal balustrades	–	0.45	9.19	0.87	–	m	**10.06**
Kitchen fittings							
wall units (up to 500 mm wide)	–	0.41	8.38	4.34	–	nr	**12.72**
floor units (up to 500 mm wide)	–	0.27	5.59	6.36	–	nr	**11.95**
larder units (up to 500 mm wide)	–	0.36	7.39	14.46	–	nr	**21.85**
Bathroom fittings							
bath panels and bearers	–	0.36	7.39	0.28	–	nr	**7.67**
toilet roll holders or soap dispensers; 2 or 3 screw fixings	–	0.27	6.21	0.04	–	nr	**6.25**
towel rails; 4 or 6 screw fixings	–	0.54	11.18	0.04	–	nr	**11.22**
mirror; up to 600 mm × 600 mm; 4 screw fixings	–	0.58	13.29	0.06	–	nr	**13.35**
pipe casings	–	0.27	5.59	1.16	–	m	**6.75**

04 ALTERATIONS, REPAIRS AND CONSERVATION

Item	PC £	Labour hours	Labour £	Plant £	Material £	Unit	Total rate £
Other fixtures and fittings							
shelves, window boards and the like; screw fixings	–	0.30	6.87	0.28	–	m	**7.15**
curtain track, plastic; screw fixing at 600 mm centres	–	0.10	2.28	–	–	m	**2.28**
small nameplates or individual numerals/letters; screw fixings	–	0.25	5.71	–	–	nr	**5.71**
small notice board and frame from walls; up to 600 mm × 900 mm; screw fixings	–	0.50	11.42	–	–	nr	**11.42**
Finishes; removing and load into skip unless stated otherwise							
Breaking up floor screeds							
asphalt paving	–	0.54	11.18	0.66	–	m²	**11.84**
floor screed up to 100 mm thick unreinforced	–	0.33	6.79	0.66	–	m²	**7.45**
granolithic floor screed up to 100 mm thick unreinforced	–	0.58	11.97	0.66	–	m²	**12.63**
Floor finishes							
carpet and underfelt	–	0.11	2.19	1.10	–	m²	**3.29**
carpetgrip edge fixing strip	–	0.01	0.23	–	–	m	**0.23**
vinyl or similar sheet flooring	–	0.09	1.80	0.80	–	m²	**2.60**
vinyl or similar tile flooring with tile remover	–	0.17	3.39	0.91	–	m²	**4.30**
ceramic floor tiles with tile remover	–	0.40	8.22	0.97	–	m²	**9.19**
woodblock flooring; set aside for reuse	–	0.67	13.77	–	–	m²	**13.77**
Hack off wall finishes with chipping hammer							
plaster to walls	–	0.33	6.79	0.88	–	m²	**7.67**
cement rendering, pebbledash or similar	–	0.25	5.13	0.61	–	m²	**5.74**
ceramic wall tiles	–	0.40	8.22	0.88	–	m²	**9.10**
Strip off wallpaper use steam paper stripper							
1 layer	–	0.10	2.06	0.12	–	m²	**2.18**
2 layers	–	0.14	2.88	0.12	–	m²	**3.00**
3 layers	–	0.17	3.60	0.12	–	m²	**3.72**
Remove wall linings including battening behind							
plain sheet boarding	–	0.27	5.59	1.16	–	m²	**6.75**
matchboarding	–	0.36	7.39	1.73	–	m²	**9.12**
Ceiling finishes							
plasterboard and skim; removes nails or screws	–	0.20	3.73	0.79	–	m²	**4.52**
wood lathe an plaster; removes nails or screws	–	0.22	4.11	1.31	–	m²	**5.42**
suspended ceilings; lay in grid; hangers and tee sections	–	0.45	8.33	1.31	–	m²	**9.64**
plain sheet boarding; including battening	–	0.41	7.60	1.05	–	m²	**8.65**
matchboarding; including battening	–	0.54	10.14	1.57	–	m²	**11.71**
Wall panelling							
remove timber wall panelling; clean off and set aside for reuse	–	0.58	11.97	–	–	m²	**11.97**

04 ALTERATIONS, REPAIRS AND CONSERVATION

Item	PC £	Labour hours	Labour £	Plant £	Material £	Unit	Total rate £
CUTTING OR FORMING OPENINGS							
Cutting openings or recesses; spoil into adjacent skip							
Through reinforced concrete walls; make good; not including any props or supports							
150 mm thick wall	–	5.02	112.12	5.62	6.01	m²	**123.75**
225 mm thick wall	–	6.95	159.00	8.13	8.38	m²	**175.51**
300 mm thick wall	–	8.75	192.49	7.55	10.68	m²	**210.72**
Through reinforced concrete suspended slab; make good; not including any props or supports							
150 mm thick wall	–	3.81	84.18	5.30	10.34	m²	**99.82**
225 mm thick wall	–	5.66	124.96	8.13	8.38	m²	**141.47**
300 mm thick wall	–	7.04	154.69	7.55	10.68	m²	**172.92**
Through brick or block walls or partitions							
half brick thick	–	2.38	46.56	4.68	–	m²	**51.24**
one brick thick	–	3.95	77.34	9.36	–	m²	**86.70**
one and a half brick thick	–	5.52	108.14	14.04	–	m²	**122.18**
two brick thick	–	7.09	138.93	18.72	–	m²	**157.65**
100 mm thick blockwork	–	1.75	34.21	4.13	–	m²	**38.34**
140 mm thick blockwork	–	2.10	41.24	5.79	–	m²	**47.03**
215 mm thick blockwork	–	2.60	50.94	8.81	–	m²	**59.75**
Cut opening through wall for new window (not included); 1200 mm × 1200 mm; quoining up jambs; cut and pin galvanized steel lintel; new soldier course in facing bricks							
one brick thick wall or two half brick thick skins	8.00	7.18	140.90	13.47	79.64	nr	**234.01**
one and a half brick thick wall	8.00	9.44	185.25	20.21	79.64	nr	**285.10**
two brick thick wall	8.00	11.70	229.58	26.96	79.64	nr	**336.18**
Cut opening through wall for new door (not included); 1200 mm × 2100 mm; quoining up jambs; cut and pin galvanized steel lintel; new soldier course in facing bricks							
one brick thick wall or two half brick thick skins	8.00	17.11	372.03	23.57	137.46	nr	**533.06**
one and a half brick thick wall	8.00	22.75	493.21	35.37	166.34	nr	**694.92**
two brick thick wall	8.00	28.39	615.47	47.18	195.23	nr	**857.88**
Diamond cutting							
Cutting to create openings in block wall							
up to 1.00 m girth; up to 150 mm thick	–	–	–	–	–	m	**20.67**
up to 1.00 m girth; 200 mm thick	–	–	–	–	–	m	**24.66**
up to 1.00 m girth; 250 mm thick	–	–	–	–	–	m	**28.66**
up to 1.00 m girth; 300 mm thick	–	–	–	–	–	m	**32.64**
3 m–3.5 m girth; up to 150 mm thick	–	–	–	–	–	m	**93.05**
3 m–3.5 m girth; 200 mm thick	–	–	–	–	–	m	**110.99**
3 m–3.5 m girth; 250 mm thick	–	–	–	–	–	m	**128.96**
3 m–3.5 m girth; 300 mm thick	–	–	–	–	–	m	**146.89**
9 m–10 m girth; up to 150 mm thick	–	–	–	–	–	m	**308.81**
9 m–10 m girth; 200 mm thick	–	–	–	–	–	m	**411.74**
9 m–10 m girth; 250 mm thick	–	–	–	–	–	m	**514.67**
9 m–10 m girth; 300 mm thick	–	–	–	–	–	m	**617.62**

04 ALTERATIONS, REPAIRS AND CONSERVATION

Item	PC £	Labour hours	Labour £	Plant £	Material £	Unit	Total rate £
Cutting to create openings in brick wall							
up to 1.00 m girth; up to 150 mm thick	–	–	–	–	–	m	25.85
up to 1.00 m girth; 200 mm thick	–	–	–	–	–	m	30.83
up to 1.00 m girth; 250 mm thick	–	–	–	–	–	m	35.82
up to 1.00 m girth; 300 mm thick	–	–	–	–	–	m	40.80
3 m–3.5 m girth; up to 150 mm thick	–	–	–	–	–	m	186.12
3 m–3.5 m girth; 200 mm thick	–	–	–	–	–	m	221.99
3 m–3.5 m girth; 250 mm thick	–	–	–	–	–	m	257.92
3 m–3.5 m girth; 300 mm thick	–	–	–	–	–	m	293.79
9 m–10 m girth; up to 150 mm thick	–	–	–	–	–	m	617.62
9 m–10 m girth; 200 mm thick	–	–	–	–	–	m	823.49
9 m–10 m girth; 250 mm thick	–	–	–	–	–	m	1029.37
9 m–10 m girth; 300 mm thick	–	–	–	–	–	m	1235.23
Cutting to create openings in concrete wall							
up to 1.00 m girth; up to 150 mm thick	–	–	–	–	–	m	103.40
up to 1.00 m girth; 200 mm thick	–	–	–	–	–	m	123.33
up to 1.00 m girth; 250 mm thick	–	–	–	–	–	m	143.28
up to 1.00 m girth; 300 mm thick	–	–	–	–	–	m	163.21
3 m–3.5 m girth; up to 150 mm thick	–	–	–	–	–	m	930.58
3 m–3.5 m girth; 200 mm thick	–	–	–	–	–	m	1109.93
3 m–3.5 m girth; 250 mm thick	–	–	–	–	–	m	1289.57
3 m–3.5 m girth; 300 mm thick	–	–	–	–	–	m	1468.94
9 m–10 m girth; up to 150 mm thick	–	–	–	–	–	m	3088.08
9 m–10 m girth; 200 mm thick	–	–	–	–	–	m	4117.43
9 m–10 m girth; 250 mm thick	–	–	–	–	–	m	5146.79
9 m–10 m girth; 300 mm thick	–	–	–	–	–	m	6176.14
Cutting to create openings in timber wall							
up to 1.00 m girth; up to 150 mm thick	–	–	–	–	–	m	31.02
up to 1.00 m girth; 200 mm thick	–	–	–	–	–	m	37.00
up to 1.00 m girth; 250 mm thick	–	–	–	–	–	m	42.98
up to 1.00 m girth; 300 mm thick	–	–	–	–	–	m	48.97
3 m–3.5 m girth; up to 150 mm thick	–	–	–	–	–	m	279.18
3 m–3.5 m girth; 200 mm thick	–	–	–	–	–	m	332.97
3 m–3.5 m girth; 250 mm thick	–	–	–	–	–	m	386.88
3 m–3.5 m girth; 300 mm thick	–	–	–	–	–	m	440.69
9 m–10 m girth; up to 150 mm thick	–	–	–	–	–	m	926.42
9 m–10 m girth; 200 mm thick	–	–	–	–	–	m	1235.23
9 m–10 m girth; 250 mm thick	–	–	–	–	–	m	1544.04
9 m–10 m girth; 300 mm thick	–	–	–	–	–	m	1852.85
Quoining up jambs to openings							
Common bricks							
half brick thick wall	–	0.90	23.52	–	6.88	m	30.40
one brick thick wall	–	1.35	35.14	–	13.77	m	48.91
one and a half brick thick wall	–	1.75	45.52	–	20.64	m	66.16
two brick thick wall	–	2.15	56.15	–	27.52	m	83.67
100 mm blockwork wall	–	0.63	16.43	–	7.74	m	24.17
140 mm blockwork wall	–	0.78	20.23	–	11.27	m	31.50
215 mm blockwork wall	–	0.97	25.28	–	19.92	m	45.20

04 ALTERATIONS, REPAIRS AND CONSERVATION

Item	PC £	Labour hours	Labour £	Plant £	Material £	Unit	Total rate £
CUTTING OR FORMING OPENINGS – cont							
Cutting back projections; spoil into skip							
Brick projection flush with adjacent wall							
225 mm × 112 mm	–	0.27	7.08	0.27	–	m	**7.35**
225 mm × 225 mm	–	0.45	11.63	0.56	–	m	**12.19**
337 mm × 112 mm	–	0.63	16.43	0.83	–	m	**17.26**
450 mm × 225 mm	–	0.81	20.99	1.10	–	m	**22.09**
chimney breast	–	1.57	40.96	8.81	–	m²	**49.77**
Temporary screens							
Providing and erecting; maintaining temporary dust proof screens; 50 mm × 75 mm softwood framing and 12 mm thick plywood covering to one side; single layer of polythene sheet to the other; clear away on completion	–	0.90	24.23	1.05	12.55	m²	**37.83**
FILLING IN OPENINGS							
Fill existing openings							
Fill in small holes							
make good hole where small (up to 12 mm dia.) pipe removed; cement mortar	–	0.19	4.96	–	0.13	nr	**5.09**
Larger holes fill in with common brickwork or blockwork							
half brick thick	–	1.66	43.24	–	27.98	m²	**71.22**
one brick thick	–	2.74	71.31	–	57.60	m²	**128.91**
one and a half brick thick	–	3.77	98.36	–	86.39	m²	**184.75**
two brick thick	–	4.71	122.89	–	115.20	m²	**238.09**
100 mm blockwork	–	0.94	24.53	–	18.38	m²	**42.91**
140 mm blockwork	–	1.13	29.33	–	25.43	m²	**54.76**
215 mm blockwork	–	1.44	37.42	–	38.55	m²	**75.97**
cavity wall; facing bricks; 100 mm cavity fill insulation; 140 mm block	–	4.13	107.72	–	68.72	m²	**176.44**
REMOVE EXISTING AND REPLACING							
Remove existing material for replacement							
NB Access equipment is not included							
Cutting out decayed, defective or cracked work and replacing with new common bricks; gauged mortar							
small areas; half brick thick walling	–	4.56	124.81	2.77	34.03	m²	**161.61**
small areas; one brick thick walling	–	8.88	243.05	4.42	69.36	m²	**316.83**
small areas; one and a half brick thick walling	–	12.58	344.33	7.74	104.06	m²	**456.13**
small areas; two brick thick walling	–	16.10	440.66	9.94	138.73	m²	**589.33**
individual bricks; half brick thick walling	–	0.28	7.66	–	0.66	nr	**8.32**

04 ALTERATIONS, REPAIRS AND CONSERVATION

Item	PC £	Labour hours	Labour £	Plant £	Material £	Unit	Total rate £
Cutting out decayed, defective or cracked work and replacing with new facing bricks; gauged mortar; facing to one side							
small areas; half brick thick walling (PC £ per 1000)	500.00	6.75	184.76	2.77	43.95	m²	**231.48**
small areas; half brick thick walling (PC £ per 1000)	700.00	6.75	184.76	2.77	60.49	m²	**248.02**
individual bricks; half brick thick walling (PC £ per 1000)	500.00	0.42	11.50	–	0.83	m²	**12.33**
Cutting out decayed, defective or cracked soldier course arch and replace with new.							
facing bricks (PC £ per 1000)	500.00	1.80	49.27	0.28	11.81	m²	**61.36**
facing bricks (PC £ per 1000)	700.00	1.80	49.27	0.28	16.22	m²	**65.77**
Cutting out raking cracks in brickwork; stitching in new common bricks							
half brick thick (PC £ per 1000)	380.00	2.96	81.02	2.32	15.45	m²	**98.79**
one brick thick (PC £ per 1000)	380.00	5.41	148.08	4.34	32.19	m²	**184.61**
one and a half brick thick (PC £ per 1000)	380.00	8.09	221.42	6.65	47.64	m²	**275.71**
Cutting out raking cracks in brickwork; stitching in new facing bricks; half brick thick; facing to one side							
facing bricks (PC £ per 1000)	500.00	4.40	120.43	2.32	21.98	m²	**144.73**
facing bricks (PC £ per 1000)	700.00	4.44	121.53	2.32	30.24	m²	**154.09**
Cutting out raking cracks in cavity brickwork; stitching in new common bricks one side; facing bricks the other side; both skins half brick thick; facing to one side							
facing bricks (PC £ per 1000)	500.00	7.59	207.74	4.63	35.86	m²	**248.23**
facing bricks (PC £ per 1000)	700.00	7.59	207.74	4.63	47.26	m²	**259.63**
Cutting away old angle fillets and replacing with new; cement mortar; 50 mm face width	–	0.23	6.30	–	3.92	m	**10.22**
Cutting out ends of joists and plates from walls; making good in common bricks in cement mortar							
175 mm joists; 400 mm centres	–	0.60	16.43	–	12.82	m	**29.25**
225 mm joists; 400 mm centres	–	0.74	20.25	–	14.92	m	**35.17**
Cutting and pinning to existing brickwork ends of joists	–	0.37	10.13	–	–	nr	**10.13**
Remove defective wall and rebuild with new materials							
defective parapet wall; 600 mm high; with two courses of tiles and brick-on-edge coping over; rebuilding with new facing bricks, tiles and coping	–	6.16	168.61	8.10	74.11	m	**250.82**
defective capping stones and haunching; replace stones and re-haunch in cement mortar	–	1.39	38.05	–	7.82	m	**45.87**

PREPARING EXISTING STRUCTURES FOR CONNECTION

Prepare existing structures for connection
Making good where intersecting wall has been removed

Item	PC £	Labour hours	Labour £	Plant £	Material £	Unit	Total rate £
half brick thick	–	0.28	7.30	–	1.24	m	**8.54**
one brick thick	–	0.37	9.65	–	2.49	m	**12.14**
100 mm blockwork	–	0.23	6.00	–	1.24	m	**7.24**

04 ALTERATIONS, REPAIRS AND CONSERVATION

Item	PC £	Labour hours	Labour £	Plant £	Material £	Unit	Total rate £
PREPARING EXISTING STRUCTURES FOR CONNECTION – cont							
Prepare existing structures for connection – cont							
Making good where intersecting wall has been removed – cont							
150 mm blockwork	–	0.27	7.04	–	1.24	m	**8.28**
215 mm blockwork	–	0.32	8.34	–	2.49	m	**10.83**
225 mm blockwork	–	0.36	9.39	–	2.49	m	**11.88**
REPAIRING							
Repairing concrete							
Reinstating plain concrete bed with site mixed in situ concrete; mix 20.00 N, where opening no longer required.							
100 mm thick	–	0.44	10.26	–	13.81	m²	**24.07**
150 mm thick	–	0.72	16.32	–	20.71	m²	**37.03**
Reinstating reinforced concrete bed with site mixed in situ concrete; mix 20.00 N; mesh reinforcement, where opening no longer required							
100 mm thick	–	0.66	15.01	–	17.02	m²	**32.03**
150 mm thick	–	0.91	20.44	–	23.92	m²	**44.36**
Reinstating reinforced concrete suspended floor with site mixed in situ concrete; mix 25.00 N; mesh reinforcement; formwork, where opening no longer required							
150 mm thick	–	2.96	68.51	–	27.79	m²	**96.30**
225 mm thick	–	3.12	71.98	–	38.47	m²	**110.45**
300 mm thick	–	3.60	82.38	–	49.14	m²	**131.52**
Reinstating small hole through concrete suspended slab with site mixed in situ concrete; mix 25.00 N, where opening no longer required							
50 mm dia.	–	0.39	8.45	–	0.06	nr	**8.51**
100 mm dia.	–	0.51	11.06	–	0.18	nr	**11.24**
150 mm dia.	–	0.65	14.09	–	0.40	nr	**14.49**
Clean out minor crack and fill with cement: mortar mixed with bonding agent	–	0.31	6.07	–	1.24	m	**7.31**
Clean out crack to form a 20 mm × 20 mm groove and fill with cement: mortar mixed with bonding agent	–	0.61	12.55	0.12	4.92	m	**17.59**
Repairing metal							
Overhauling and repairing metal casement windows adjust and oil ironmongery; prepare affected parts for redecoration	–	1.39	39.21	–	–	nr	**39.21**

04 ALTERATIONS, REPAIRS AND CONSERVATION

Item	PC £	Labour hours	Labour £	Plant £	Material £	Unit	Total rate £
Repairing timber							
Removing or punching projecting nails; refixing timber flooring							
loose boards	–	0.14	3.58	–	–	m²	**3.58**
refix floorboards previously set aside	–	0.74	18.93	–	0.83	m²	**19.76**
Remove damaged softwood flooring and fix new plain edge softwood flooring							
small areas	–	1.06	27.12	–	31.62	m²	**58.74**
individual boards 150 mm wide	–	0.28	7.16	–	3.21	m	**10.37**
Sanding down and resurfacing existing flooring; preparing; body in with wax polish							
softwood	–	0.70	15.26	–	1.42	m²	**16.68**
hardwood	–	0.85	18.53	–	1.42	m²	**19.95**
Fitting existing softwood skirting or architrave to new frames							
75 mm high	–	0.09	2.30	–	–	m	**2.30**
150 mm high	–	0.12	3.07	–	–	m	**3.07**
225 mm high	–	0.15	3.84	–	–	m	**3.84**
Piecing in new 25 mm × 150 mm moulded softwood skirtings to match existing where old has been removed; prepare for decoration	–	0.35	8.13	–	4.40	m	**12.53**
Repair doors							
Easing and adjusting softwood doors; oil ironmongery	–	0.59	15.09	–	–	nr	**15.09**
Remove softwood doors; easing and adjust; oil ironmongery; rehang	–	0.75	19.19	–	–	nr	**19.19**
Remove softwood door; plane 12 mm from bottom edge; rehang	–	1.11	28.39	–	–	nr	**28.39**
Take off existing softwood doorstops; supply and fit new 25 mm × 38 mm doorstop	–	0.10	2.56	–	1.39	m	**3.95**
Cutting out infected or decayed structural members; supply and fix new treated sawn softwood members							
Floors and flat roofs							
50 mm × 125 mm	–	0.37	9.46	–	4.94	m	**14.40**
50 mm × 150 mm	–	0.41	10.49	–	5.92	m	**16.41**
50 mm × 175 mm	–	0.44	11.26	–	5.52	m	**16.78**
Pitched roofs							
38 mm × 100 mm	–	0.33	8.44	–	3.15	m	**11.59**
50 mm × 100 mm	–	0.42	10.74	–	3.95	m	**14.69**
50 mm × 125 mm	–	0.46	11.77	–	4.94	m	**16.71**
50 mm × 150 mm	–	0.51	13.05	–	5.92	m	**18.97**
Kerb bearers							
50 mm × 75 mm	–	0.42	10.74	–	2.45	m	**13.19**
50 mm × 100 mm	–	0.52	13.30	–	3.95	m	**17.25**
75 mm × 100 mm	–	0.63	16.12	–	5.63	m	**21.75**

04 ALTERATIONS, REPAIRS AND CONSERVATION

Item	PC £	Labour hours	Labour £	Plant £	Material £	Unit	Total rate £
REPOINTING							
Repointing masonry							
Raking out decayed masonry joints and repointing in cement mortar							
brickwork walls generally	–	0.69	18.89	–	1.30	m²	**20.19**
cutting out staggered cracks and repointing to match existing along brick joints	–	0.37	10.13	–	–	m	**10.13**
brickwork and re-wedge horizontal flashing	–	0.23	6.30	–	0.66	m	**6.96**
brickwork and re-wedge stepped flashing	–	0.34	9.30	–	0.66	m	**9.96**
brickwork in chimney stacks	–	1.11	30.39	–	1.30	m²	**31.69**
uncoursed stonework	–	1.11	30.39	–	0.66	m	**31.05**
CLEANING SURFACES							
Cleaning surfaces by hand using hand held manual tools only (brushes, scrapers etc.)							
Masonry/Concrete							
Cleaning surface of moss and lichen from walls	–	0.28	5.75	–	–	m²	**5.75**
Cleaning bricks of mortar, sort and stack for reuse	–	9.25	190.29	–	–	1000	**190.29**
Clean surface of concrete to receive new damp-proof membrane	–	0.14	2.88	–	–	m²	**2.88**
DECONTAMINATION							
Insecticide or fungicidal treatments							
Treating individual timbers with two coats of proprietary insecticide and fungicide by brush or spray application as appropriate							
general boarding	–	–	–	–	–	m²	**10.34**
structural members	–	–	–	–	–	m²	**8.91**
skirtings, architraves etc.	–	–	–	–	–	m	**9.21**
remove cobwebs, dust and roof insulation; treat exposed timbers with proprietary insecticide and fungicide by spray application	–	–	–	–	–	m²	**17.19**
lifting necessary floorboards; treating flooring with two coats of proprietary insecticide and fungicide by spray application; refix boards	–	–	–	–	–	m²	**16.54**
treating surfaces of adjoining masonry or concrete with two coats of proprietary dry rot by spray application	–	–	–	–	–	m²	**10.11**

05 EXCAVATE AND FILLING

Item	PC £	Labour hours	Labour £	Plant £	Material £	Unit	Total rate £
SITE CLEARANCE AND PREPARATION							
Site clearance							
Prices are applicable to excavation in firm soil							
Fell and removing trees							
girth 500–1500 mm	–	18.50	244.98	14.70	–	nr	**259.68**
girth 1500–3000 mm	–	32.50	430.36	14.70	–	nr	**445.06**
girth exceeding 3000 mm	–	46.50	615.75	18.38	–	nr	**634.13**
Removing tree stumps							
girth 500 mm–1.50 m	–	2.00	26.48	17.49	–	nr	**43.97**
girth 1.50–3.00 m	–	3.00	39.72	25.00	–	nr	**64.72**
girth exceeding 3.00 m	–	6.00	79.45	35.07	–	nr	**114.52**
Clearing site vegetation							
bushes, scrub, undergrowth, hedges and trees and tree stumps not exceeding 500 mm girth	–	0.03	0.40	0.98	–	m²	**1.38**
Lifting turf for preservation, by hand							
stacking	–	0.32	4.24	–	–	m²	**4.24**
Topsoil for preservation; to spoil heap less than 50 m from excavations							
average depth 150 mm	–	0.02	0.26	0.69	–	m²	**0.95**
add or deduct for each 25 mm variation in average depth	–	0.01	0.14	0.16	–	m²	**0.30**
EXCAVATION							
Excavating by machine							
To reduce levels							
maximum depth not exceeding 0.25 m	–	0.03	0.44	0.54	–	m³	**0.98**
maximum depth not exceeding 1.00 m	–	0.03	0.44	0.54	–	m³	**0.98**
maximum depth not exceeding 2.00 m	–	0.04	0.48	0.59	–	m³	**1.07**
maximum depth not exceeding 4.00 m	–	0.04	0.53	0.64	–	m³	**1.17**
Basements and the like; commencing level exceeding 0.25 m below existing ground level							
maximum depth not exceeding 1.00 m	–	0.06	0.80	0.98	–	m³	**1.78**
maximum depth not exceeding 2.00 m	–	0.07	0.92	0.98	–	m³	**1.90**
maximum depth not exceeding 4.00 m	–	0.08	1.06	1.20	–	m³	**2.26**
maximum depth not exceeding 6.00 m	–	0.09	1.20	1.50	–	m³	**2.70**
maximum depth not exceeding 8.00 m	–	0.12	1.59	1.72	–	m³	**3.31**
Pits							
maximum depth not exceeding 0.25 m	–	0.31	4.11	3.53	–	m³	**7.64**
maximum depth not exceeding 1.00 m	–	0.33	4.37	3.53	–	m³	**7.90**
maximum depth not exceeding 2.00 m	–	0.39	5.17	3.98	–	m³	**9.15**
maximum depth not exceeding 4.00 m	–	0.47	6.23	4.50	–	m³	**10.73**
maximum depth not exceeding 6.00 m	–	0.49	6.49	4.74	–	m³	**11.23**
Extra over pit excavating for commencing level exceeding 0.25 m below existing ground level							
1.00 m below	–	0.03	0.40	0.53	–	m³	**0.93**
2.00 m below	–	0.05	0.66	0.76	–	m³	**1.42**
3.00 m below	–	0.06	0.80	0.98	–	m³	**1.78**
4.00 m below	–	0.09	1.20	1.28	–	m³	**2.48**

05 EXCAVATE AND FILLING

Item	PC £	Labour hours	Labour £	Plant £	Material £	Unit	Total rate £
EXCAVATION – cont							
Excavating by machine – cont							
Trenches; width not exceeding 0.30 m							
maximum depth not exceeding 0.25 m	–	0.26	3.44	2.78	–	m³	**6.22**
maximum depth not exceeding 1.00 m	–	0.28	3.71	2.78	–	m³	**6.49**
maximum depth not exceeding 2.00 m	–	0.33	4.37	3.23	–	m³	**7.60**
maximum depth not exceeding 4.00 m	–	0.40	5.29	3.98	–	m³	**9.27**
maximum depth not exceeding 6.00 m	–	0.46	6.09	4.74	–	m³	**10.83**
Trenches; width exceeding 0.30 m							
maximum depth not exceeding 0.25 m	–	0.23	3.04	2.48	–	m³	**5.52**
maximum depth not exceeding 1.00 m	–	0.25	3.31	2.48	–	m³	**5.79**
maximum depth not exceeding 2.00 m	–	0.30	3.97	3.00	–	m³	**6.97**
maximum depth not exceeding 4.00 m	–	0.35	4.63	3.53	–	m³	**8.16**
maximum depth not exceeding 6.00 m	–	0.43	5.69	4.50	–	m³	**10.19**
Extra over trench excavating for commencing level exceeding 0.25 m below existing ground level							
1.00 m below	–	0.03	0.40	0.53	–	m³	**0.93**
2.00 m below	–	0.05	0.66	0.76	–	m³	**1.42**
3.00 m below	–	0.06	0.80	0.98	–	m³	**1.78**
4.00 m below	–	0.09	1.20	1.28	–	m³	**2.48**
For pile caps and ground beams between piles							
maximum depth not exceeding 0.25 m	–	0.35	4.63	4.81	–	m³	**9.44**
maximum depth not exceeding 1.00 m	–	0.39	5.17	4.74	–	m³	**9.91**
maximum depth not exceeding 2.00 m	–	0.39	5.17	5.26	–	m³	**10.43**
To bench sloping ground to receive filling							
maximum depth not exceeding 0.25 m	–	0.07	0.92	1.28	–	m³	**2.20**
maximum depth not exceeding 1.00 m	–	0.09	1.20	1.28	–	m³	**2.48**
maximum depth not exceeding 2.00 m	–	0.09	1.20	1.50	–	m³	**2.70**
Extra over any types of excavating irrespective of depth							
excavating below ground water level	–	0.13	1.72	1.72	–	m³	**3.44**
next to existing services	–	0.35	4.63	0.98	–	m³	**5.61**
around existing services crossing excavation	–	0.60	7.95	2.78	–	m³	**10.73**
Extra over any types of excavating irrespective of depth for breaking out existing materials							
rock	–	2.95	39.06	15.59	–	m³	**54.65**
concrete	–	2.55	33.77	12.30	–	m³	**46.07**
reinforced concrete	–	3.60	47.67	17.92	–	m³	**65.59**
brickwork, blockwork or stonework	–	1.85	24.50	9.05	–	m³	**33.55**
Extra over any types of excavating irrespective of depth for breaking out existing hard pavings, 75 mm thick							
coated macadam or asphalt	–	0.19	2.52	0.78	–	m²	**3.30**
Extra over any types of excavating irrespective of depth for breaking out existing hard pavings, 150 mm thick							
concrete	–	0.39	5.17	1.87	–	m²	**7.04**
reinforced concrete	–	0.58	7.68	2.55	–	m²	**10.23**
coated macadam or asphalt and hardcore	–	0.26	3.44	0.87	–	m²	**4.31**

05 EXCAVATE AND FILLING

Item	PC £	Labour hours	Labour £	Plant £	Material £	Unit	Total rate £
Working space allowance to excavations 600 mm wide							
reduce levels, basements and the like	–	0.07	0.92	0.98	–	m²	**1.90**
pits	–	0.19	2.52	2.78	–	m²	**5.30**
trenches	–	0.18	2.38	2.48	–	m²	**4.86**
pile caps and ground beams between piles	–	0.20	2.65	2.78	–	m²	**5.43**
Extra over excavating for working space for backfilling in with special materials							
hardcore	–	0.13	1.72	1.23	17.28	m²	**20.23**
sand	–	0.13	1.72	1.23	21.12	m²	**24.07**
40 mm–20 mm gravel	–	0.13	1.72	1.23	27.29	m²	**30.24**
plain in situ ready mixed designated concrete C7.5–40 mm aggregate	–	0.93	14.44	1.28	57.26	m²	**72.98**
Excavating by hand							
Topsoil for preservation							
average depth 150 mm	–	0.23	3.04	–	–	m²	**3.04**
add or deduct for each 25 mm variation in average depth	–	0.03	0.40	–	–	m²	**0.40**
To reduce levels							
maximum depth not exceeding 0.25 m	–	1.44	19.07	–	–	m³	**19.07**
maximum depth not exceeding 1.00 m	–	1.63	21.59	–	–	m³	**21.59**
maximum depth not exceeding 2.00 m	–	1.80	23.84	–	–	m³	**23.84**
maximum depth not exceeding 4.00 m	–	1.99	26.36	–	–	m³	**26.36**
Basements and the like; commencing level exceeding 0.25 m below existing ground level							
maximum depth not exceeding 1.00 m	–	1.90	25.16	–	–	m³	**25.16**
maximum depth not exceeding 2.00 m	–	2.04	27.02	–	–	m³	**27.02**
maximum depth not exceeding 4.00 m	–	2.73	36.15	–	–	m³	**36.15**
maximum depth not exceeding 6.00 m	–	3.33	44.10	–	–	m³	**44.10**
maximum depth not exceeding 8.00 m	–	4.02	53.24	–	–	m³	**53.24**
Pits							
maximum depth not exceeding 0.25 m	–	2.13	28.20	–	–	m³	**28.20**
maximum depth not exceeding 1.00 m	–	2.75	36.41	–	–	m³	**36.41**
maximum depth not exceeding 2.00 m	–	3.30	43.70	–	–	m³	**43.70**
maximum depth not exceeding 4.00 m	–	4.18	55.36	–	–	m³	**55.36**
maximum depth not exceeding 6.00 m	–	5.17	68.46	–	–	m³	**68.46**
Extra over pit excavating for commencing level exceeding 0.25 m below existing ground level							
1.00 m below	–	0.42	5.57	–	–	m³	**5.57**
2.00 m below	–	0.88	11.65	–	–	m³	**11.65**
3.00 m below	–	1.30	17.21	–	–	m³	**17.21**
4.00 m below	–	1.71	22.65	–	–	m³	**22.65**
Trenches; width not exceeding 0.30 m							
maximum depth not exceeding 0.25 m	–	1.85	24.50	–	–	m³	**24.50**
maximum depth not exceeding 1.00 m	–	2.76	36.55	–	–	m³	**36.55**
Trenches; width exceeding 0.30 m							
maximum depth not exceeding 0.25 m	–	1.80	23.84	–	–	m³	**23.84**
maximum depth not exceeding 1.00 m	–	2.46	32.57	–	–	m³	**32.57**
maximum depth not exceeding 2.00 m	–	2.88	38.14	–	–	m³	**38.14**
maximum depth not exceeding 4.00 m	–	3.66	48.47	–	–	m³	**48.47**
maximum depth not exceeding 6.00 m	–	4.68	61.97	–	–	m³	**61.97**

05 EXCAVATE AND FILLING

Item	PC £	Labour hours	Labour £	Plant £	Material £	Unit	Total rate £
EXCAVATION – cont							
Excavating by hand – cont							
Extra over trench excavating for commencing level exceeding 0.25 m below existing ground level							
1.00 m below	–	0.42	5.57	–	–	m³	**5.57**
2.00 m below	–	0.88	11.65	–	–	m³	**11.65**
3.00 m below	–	1.30	17.21	–	–	m³	**17.21**
4.00 m below	–	1.71	22.65	–	–	m³	**22.65**
For pile caps and ground beams between piles							
maximum depth not exceeding 0.25 m	–	2.78	36.81	–	–	m³	**36.81**
maximum depth not exceeding 1.00 m	–	2.96	39.20	–	–	m³	**39.20**
maximum depth not exceeding 2.00 m	–	3.52	46.61	–	–	m³	**46.61**
To bench sloping ground to receive filling							
maximum depth not exceeding 0.25 m	–	1.30	17.21	–	–	m³	**17.21**
maximum depth not exceeding 1.00 m	–	1.48	19.59	–	–	m³	**19.59**
maximum depth not exceeding 2.00 m	–	1.67	22.11	–	–	m³	**22.11**
Extra over any types of excavating irrespective of depth							
excavating below ground water level	–	0.32	4.24	–	–	m³	**4.24**
next existing services	–	0.93	12.32	–	–	m³	**12.32**
around existing services crossing excavation	–	1.85	24.50	–	–	m³	**24.50**
Extra over any types of excavating irrespective of depth for breaking out existing materials							
rock	–	4.63	61.31	14.02	–	m³	**75.33**
concrete	–	4.16	55.08	11.68	–	m³	**66.76**
reinforced concrete	–	5.55	73.49	16.36	–	m³	**89.85**
brickwork, blockwork or stonework	–	2.78	36.81	7.00	–	m³	**43.81**
Extra over any types of excavating irrespective of depth for breaking out existing hard pavings, 60 mm thick							
precast concrete paving slabs	–	0.28	3.71	–	–	m²	**3.71**
Extra over any types of excavating irrespective of depth for breaking out existing hard pavings, 75 mm thick							
coated macadam or asphalt	–	0.37	4.90	0.93	–	m²	**5.83**
Extra over any types of excavating irrespective of depth for breaking out existing hard pavings, 150 mm thick							
concrete	–	0.65	8.61	1.64	–	m²	**10.25**
reinforced concrete	–	0.83	10.99	2.34	–	m²	**13.33**
coated macadam or asphalt and hardcore	–	0.46	6.09	1.18	–	m²	**7.27**
Working space allowance to excavations							
reduce levels, basements and the like	–	2.13	28.20	1.44	–	m²	**29.64**
pits	–	2.22	29.40	1.44	–	m²	**30.84**
trenches	–	1.94	25.69	1.44	–	m²	**27.13**
pile caps and ground beams between piles	–	2.31	30.59	1.44	–	m²	**32.03**
Extra over excavation for working space for backfilling with special materials							
hardcore	–	0.74	9.80	1.44	18.51	m²	**29.75**
sand	–	0.74	9.80	1.44	24.39	m²	**35.63**
40 mm–20 mm gravel	–	0.74	9.80	1.44	27.29	m²	**38.53**
plain in situ concrete ready mixed designated concrete; C7.5–40 mm aggregate	–	1.02	15.83	1.44	57.26	m²	**74.53**

05 EXCAVATE AND FILLING

Item	PC £	Labour hours	Labour £	Plant £	Material £	Unit	Total rate £
SUPPORT TO EXCAVATIONS							
Earthwork support (average risk prices)							
Maximum depth not exceeding 1.00 m							
distance between opposing faces not exceeding 2.00 m	–	0.10	1.26	–	0.19	m²	**1.45**
distance between opposing faces 2.00–4.00 m	–	0.10	1.39	–	0.22	m²	**1.61**
distance between opposing faces exceeding 4.00 m	–	0.11	1.51	–	0.27	m²	**1.78**
Maximum depth not exceeding 2.00 m							
distance between opposing faces not exceeding 2.00 m	–	0.11	1.51	–	0.22	m²	**1.73**
distance between opposing faces 2.00–4.00 m	–	0.12	1.64	–	0.27	m²	**1.91**
distance between opposing faces exceeding 4.00 m	–	0.13	1.76	–	0.35	m²	**2.11**
Maximum depth not exceeding 4.00 m							
distance between opposing faces not exceeding 2.00 m	–	0.15	2.02	–	0.27	m²	**2.29**
distance between opposing faces 2.00–4.00 m	–	0.15	2.02	–	0.35	m²	**2.37**
distance between opposing faces exceeding 4.00 m	–	0.17	2.27	–	0.43	m²	**2.70**
Maximum depth not exceeding 6.00 m							
distance between opposing faces not exceeding 2.00 m	–	0.17	2.27	–	0.33	m²	**2.60**
distance between opposing faces 2.00–4.00 m	–	0.18	2.39	–	0.43	m²	**2.82**
distance between opposing faces exceeding 4.00 m	–	0.21	2.77	–	0.55	m²	**3.32**
Maximum depth not exceeding 8.00 m							
distance between opposing faces not exceeding 2.00 m	–	0.22	2.90	–	0.43	m²	**3.33**
distance between opposing faces 2.00–4.00 m	–	0.27	3.52	–	0.55	m²	**4.07**
distance between opposing faces exceeding 4.00 m	–	0.31	4.15	–	0.65	m²	**4.80**
Earthwork support (open boarded)							
Maximum depth not exceeding 1.00 m							
distance between opposing faces not exceeding 2.00 m	–	0.27	3.52	–	0.39	m²	**3.91**
distance between opposing faces 2.00–4.00 m	–	0.29	3.90	–	0.43	m²	**4.33**
distance between opposing faces exceeding 4.00 m	–	0.33	4.40	–	0.55	m²	**4.95**
Maximum depth not exceeding 2.00 m							
distance between opposing faces not exceeding 2.00 m	–	0.33	4.40	–	0.43	m²	**4.83**
distance between opposing faces 2.00–4.00 m	–	0.37	4.90	–	0.53	m²	**5.43**
distance between opposing faces exceeding 4.00 m	–	0.42	5.53	–	0.65	m²	**6.18**
Maximum depth not exceeding 4.00 m							
distance between opposing faces not exceeding 2.00 m	–	0.42	5.53	–	0.49	m²	**6.02**
distance between opposing faces 2.00–4.00 m	–	0.47	6.29	–	0.61	m²	**6.90**
distance between opposing faces exceeding 4.00 m	–	0.53	7.05	–	0.77	m²	**7.82**

05 EXCAVATE AND FILLING

Item	PC £	Labour hours	Labour £	Plant £	Material £	Unit	Total rate £
SUPPORT TO EXCAVATIONS – cont							
Earthwork support (open boarded) – cont							
Maximum depth not exceeding 6.00 m							
distance between opposing faces not exceeding 2.00 m	–	0.53	7.05	–	0.55	m²	**7.60**
distance between opposing faces 2.00–4.00 m	–	0.58	7.68	–	0.68	m²	**8.36**
distance between opposing faces exceeding 4.00 m	–	0.67	8.81	–	0.87	m²	**9.68**
Maximum depth not exceeding 8.00 m							
distance between opposing faces not exceeding 2.00 m	–	0.70	9.31	–	0.71	m²	**10.02**
distance between opposing faces 2.00–4.00 m	–	0.79	10.44	–	0.82	m²	**11.26**
distance between opposing faces exceeding 4.00 m	–	0.92	12.20	–	1.09	m²	**13.29**
Earthwork support (close boarded)							
Maximum depth not exceeding 1.00 m							
distance between opposing faces not exceeding 2.00 m	–	0.70	9.31	–	0.77	m²	**10.08**
distance between opposing faces 2.00–4.00 m	–	0.77	10.18	–	0.87	m²	**11.05**
distance between opposing faces exceeding 4.00 m	–	0.85	11.32	–	1.09	m²	**12.41**
Maximum depth not exceeding 2.00 m							
distance between opposing faces not exceeding 2.00 m	–	0.88	11.70	–	0.87	m²	**12.57**
distance between opposing faces 2.00–4.00 m	–	0.97	12.83	–	1.04	m²	**13.87**
distance between opposing faces exceeding 4.00 m	–	1.05	13.97	–	1.30	m²	**15.27**
Maximum depth not exceeding 4.00 m							
distance between opposing faces not exceeding 2.00 m	–	1.10	14.60	–	0.98	m²	**15.58**
distance between opposing faces 2.00–4.00 m	–	1.24	16.35	–	1.22	m²	**17.57**
distance between opposing faces exceeding 4.00 m	–	1.36	17.99	–	1.52	m²	**19.51**
Maximum depth not exceeding 6.00 m							
distance between opposing faces not exceeding 2.00 m	–	1.37	18.11	–	1.09	m²	**19.20**
distance between opposing faces 2.00–4.00 m	–	1.49	19.75	–	1.38	m²	**21.13**
distance between opposing faces exceeding 4.00 m	–	1.67	22.14	–	1.74	m²	**23.88**
Maximum depth not exceeding 8.00 m							
distance between opposing faces not exceeding 2.00 m	–	1.67	22.14	–	1.42	m²	**23.56**
distance between opposing faces 2.00–4.00 m	–	1.84	24.40	–	1.64	m²	**26.04**
distance between opposing faces exceeding 4.00 m	–	2.11	27.93	–	1.96	m²	**29.89**
Extra over earthwork support for							
Curved	–	0.02	0.25	–	0.19	m²	**0.44**
Below ground water level	–	0.27	3.52	–	0.17	m²	**3.69**
Unstable ground	–	0.44	5.79	–	0.33	m²	**6.12**
Next to roadways	–	0.35	4.65	–	0.27	m²	**4.92**
Left in	–	0.57	7.55	–	7.62	m²	**15.17**

05 EXCAVATE AND FILLING

Item	PC £	Labour hours	Labour £	Plant £	Material £	Unit	Total rate £
Earthwork support (average risk prices – inside existing buildings)							
Maximum depth not exceeding 1.00 m							
distance between opposing faces not exceeding 2.00 m	–	0.18	2.38	–	0.27	m²	**2.65**
distance between opposing faces 2.00–4.00 m	–	0.19	2.52	–	0.32	m²	**2.84**
distance between opposing faces exceeding 4.00 m	–	0.22	2.91	–	0.39	m²	**3.30**
Maximum depth not exceeding 2.00 m							
distance between opposing faces not exceeding 2.00 m	–	0.22	2.91	–	0.32	m²	**3.23**
distance between opposing faces 2.00–4.00 m	–	0.24	3.18	–	0.42	m²	**3.60**
distance between opposing faces exceeding 4.00 m	–	0.32	4.24	–	0.46	m²	**4.70**
Maximum depth not exceeding 4.00 m							
distance between opposing faces not exceeding 2.00 m	–	0.28	3.71	–	0.42	m²	**4.13**
distance between opposing faces 2.00–4.00 m	–	0.31	4.11	–	0.49	m²	**4.60**
distance between opposing faces exceeding 4.00 m	–	0.34	4.50	–	0.58	m²	**5.08**
Maximum depth not exceeding 6.00 m							
distance between opposing faces not exceeding 2.00 m	–	0.34	4.50	–	0.47	m²	**4.97**
distance between opposing faces 2.00–4.00 m	–	0.38	5.03	–	0.58	m²	**5.61**
distance between opposing faces exceeding 4.00 m	–	0.43	5.69	–	0.68	m²	**6.37**
DISPOSAL							
Disposal of excavated material							
Basic Landfill tax rates (exclusive of site tipping charges)							
inactive waste	–	–	–	–	2.60	tonne	**2.60**
all other taxable wastes	–	–	–	–	84.40	tonne	**84.40**
Excavated material deposited off site							
inactive waste off site; to tip not exceeding 13 km (using lorries); including Landfill Tax	–	0.06	0.98	13.56	13.18	m³	**27.72**
active non-hazardous waste off site; to tip not exceeding 13 km (using lorries); including Landfill Tax	–	0.06	0.98	13.83	173.29	m³	**188.10**
RETAINING EXCAVATED MATERIAL ON SITE							
Excavated material deposited on site using dumpers to transport material							
on site; spreading; average 25 m distance	–	0.20	2.65	0.38	–	m³	**3.03**
on site; depositing in spoil heaps; average 50 m distance	–	–	–	0.90	–	m³	**0.90**
on site; spreading; average 50 m distance	–	0.20	2.65	0.69	–	m³	**3.34**

05 EXCAVATE AND FILLING

Item	PC £	Labour hours	Labour £	Plant £	Material £	Unit	Total rate £
RETAINING EXCAVATED MATERIAL ON SITE – cont							
on site; depositing in spoil heaps; average 100 m distance	–	–	–	1.61	–	m³	**1.61**
on site; spreading; average 100 m distance	–	0.20	2.65	1.07	–	m³	**3.72**
on site; depositing in spoil heaps; average 200 m distance	–	–	–	2.03	–	m³	**2.03**
on site; spreading; average 200 m distance	–	0.20	2.65	1.44	–	m³	**4.09**
FILLING OBTAINED FROM EXCAVATED MATERIAL							
Filling to make up levels							
By machine							
average thickness over 50 mm not exceeding 500 mm	–	0.22	2.86	1.72	–	m³	**4.58**
average thickness exceeding 500 mm	–	0.18	2.38	1.24	–	m³	**3.62**
By hand							
average thickness over 50 mm not exceeding 500 mm	–	1.25	16.55	3.38	–	m³	**19.93**
average thickness exceeding 500 mm	–	1.02	13.50	2.75	–	m³	**16.25**
Surface treatments							
Surface packing to filling							
To vertical or battered faces	–	0.17	2.25	0.17	–	m²	**2.42**
Compacting							
bottoms of excavations	–	0.04	0.53	0.04	–	m²	**0.57**
Trimming							
sloping surfaces; by hand	–	0.17	2.25	–	–	m²	**2.25**

05 EXCAVATE AND FILLING

Item	PC £	Labour hours	Labour £	Plant £	Material £	Unit	Total rate £
IMPORTED FILLING							
Basic material prices delivered to site							
Supply only in full loads							
D.O.T. type 1	–	–	–	–	24.34	tonne	**24.34**
D.O.T. type 2	–	–	–	–	18.81	tonne	**18.81**
10 mm single size aggregate	–	–	–	–	18.81	tonne	**18.81**
20 mm single size aggregate	–	–	–	–	18.81	tonne	**18.81**
20 mm all-in aggregate	–	–	–	–	18.81	tonne	**18.81**
40 mm all-in aggregate	–	–	–	–	18.81	tonne	**18.81**
40 mm scalpings	–	–	–	–	18.81	tonne	**18.81**
hardcore	–	–	–	–	15.31	tonne	**15.31**
soft/building sand	–	–	–	–	17.86	tonne	**17.86**
recycled type 1	–	–	–	–	13.70	tonne	**13.70**
E-blend (50% type 1 and 50% recycled type 1)	–	–	–	–	15.12	tonne	**15.12**
Filling to make up levels; by machine							
Average thickness over 50 mm not exceeding 500 mm							
obtained off site; imported topsoil	17.00	0.22	2.86	1.44	17.85	m³	**22.15**
obtained off site; hardcore	30.09	0.25	3.34	1.66	31.59	m³	**36.59**
obtained off site; granular fill type one	54.81	0.25	3.34	2.89	57.55	m³	**63.78**
obtained off site; granular fill type two	42.36	0.25	3.34	2.89	44.48	m³	**50.71**
obtained off site; sand	40.23	0.25	3.34	2.89	42.24	m³	**48.47**
Average thickness over 500 mm							
obtained off site; imported topsoil	17.00	0.18	2.38	1.62	17.85	m³	**21.85**
obtained off site; hardcore	29.78	0.22	2.86	0.32	31.27	m³	**34.45**
obtained off site; granular fill type one	54.81	0.22	2.86	2.26	57.55	m³	**62.67**
obtained off site; granular fill type two	42.36	0.22	2.86	2.26	44.48	m³	**49.60**
obtained off site; sand	40.23	0.22	2.86	2.26	42.24	m³	**47.36**
Filling to make up levels; by hand							
Average thickness over 50 mm not exceeding 500 mm							
obtained off site; imported topsoil	17.00	1.25	16.55	3.38	17.85	m³	**37.78**
obtained off site; hardcore	30.09	1.39	18.41	3.76	31.59	m³	**53.76**
obtained off site; granular fill type one	54.81	1.54	20.39	4.15	57.55	m³	**82.09**
obtained off site; granular fill type two	42.36	1.54	20.39	4.15	44.48	m³	**69.02**
obtained off site; sand	40.23	1.54	20.39	4.15	42.24	m³	**66.78**
Average thickness over 500 mm							
obtained off site; imported topsoil	17.00	1.02	13.50	2.75	17.85	m³	**34.10**
obtained off site; hardcore	30.09	1.34	17.74	3.62	31.59	m³	**52.95**
obtained off site; granular fill type one	54.81	1.43	18.93	3.89	57.55	m³	**80.37**
obtained off site; granular fill type two	42.36	1.43	18.93	3.89	44.48	m³	**67.30**
obtained off site; sand	40.23	1.43	18.93	3.89	42.24	m³	**65.06**
Surface treatments							
Surface packing to filling							
To vertical or battered faces	–	0.17	2.25	0.17	–	m²	**2.42**

05 EXCAVATE AND FILLING

Item	PC £	Labour hours	Labour £	Plant £	Material £	Unit	Total rate £
IMPORTED FILLING – cont							
Surface treatments – cont							
Compacting							
bottoms of excavations	–	0.04	0.53	0.04	–	m²	**0.57**
filling; blinding with sand 50 mm thick	–	0.04	0.53	0.04	1.72	m²	**2.29**
Trimming							
sloping surfaces; by hand	–	0.17	2.25	–	–	m²	**2.25**
sloping surfaces; in rock	–	0.93	12.22	3.27	–	m²	**15.49**
GEOTEXTILE FABRIC							
Erosion mats							
Filter membrane; one layer; laid on earth to receive granular material; over 500 mm wide							
Terram 500 filter membrane or other equal; one layer; laid on earth. NB mainly agricultural/ landscape use	0.21	0.04	0.53	–	0.24	m²	**0.77**
Terram 700 filter membrane or other equal and approved; one layer; laid on earth	0.40	0.04	0.53	–	0.45	m²	**0.98**
Terram 1000; filter membrane or other equal and approved; one layer; laid on earth	0.33	0.04	0.53	–	0.37	m²	**0.90**
Terram 2000; filter membrane or other equal and approved; one layer; laid on earth	0.87	0.04	0.53	–	1.05	m²	**1.58**
Terram High Visibility; orange top surface to warn of potential hazardous material underneath	0.52	0.04	0.53	–	0.62	m²	**1.15**
MEMBRANES							
Sheeting to prevent moisture loss/gas barrier							
Building paper; lapped joints							
subsoil grade 410; horizontal on foundations	–	0.02	0.32	–	0.77	m²	**1.09**
standard grade 420; horizontal on slabs	–	0.04	0.62	–	1.13	m²	**1.75**
Polythene sheeting; lapped joints; horizontal on slabs							
250 microns; 0.25 mm thick	–	0.04	0.62	–	0.50	m²	**1.12**
Visqueen sheeting or other equal and approved; lapped joints; horizontal on slabs							
250 microns; 0.25 mm thick	–	0.04	0.62	–	0.42	m²	**1.04**
300 microns; 0.30 mm thick	–	0.05	0.78	–	0.46	m²	**1.24**
gas-resistant damp-proof membrane taped joints	–	0.05	0.78	–	5.25	m²	**6.03**
sealing at perimeter of slab	–	0.10	1.55	–	1.35	m	**2.90**
sealing around penetrations; drain pipes etc.	–	1.00	15.52	–	4.13	nr	**19.65**

07 PILING

Item	PC £	Labour hours	Labour £	Plant £	Material £	Unit	Total rate £
INTERLOCKING SHEET PILES							
Arcelor 600 mm wide 'U' shaped steel sheeting piling; or other equal approved; pitched and driven							
Provision of all plant for sheet pile installation; including bringing to and removing from site; maintenance, erection and dismantling; assuming one rig for 1500 m² of piling							
leader rig with vibratory hammer	–	–	–	–	–	item	3730.00
conventional rig	–	–	–	–	–	item	4850.00
silent vibrationless rig	–	–	–	–	–	item	5400.00
Supply only of standard sheet pile sections							
PU12	–	–	–	–	–	m²	96.00
PU18–1	–	–	–	–	–	m²	105.00
PU22–1	–	–	–	–	–	m²	120.00
PU25	–	–	–	–	–	m²	136.00
PU32	–	–	–	–	–	m²	166.00
Pitching and driving of sheet piles; using the following plant							
leader rig with vibratory hammer	–	–	–	–	–	m²	21.00
conventional rig	–	–	–	–	–	m²	28.00
silent vibrationless rig	–	–	–	–	–	m²	44.00
Provision of all plant for sheet pile extraction; including bringing to and removing from site; maintenance, erection and dismantling; assuming one rig for 1500 m² of piling							
leader rig with vibratory hammer	–	–	–	–	–	item	3790.00
conventional rig	–	–	–	–	–	item	4870.00
silent vibrationless rig	–	–	–	–	–	item	4870.00
Extraction of sheet piles; using the following plant							
leader rig with vibratory hammer	–	–	–	–	–	m²	15.00
conventional rig	–	–	–	–	–	m²	18.00
silent vibrationless rig	–	–	–	–	–	m²	26.00
Credit on extracted piles; recovered in re-usable lengths; for sheet pile sections							
PU12	–	–	–	–	–	m²	68.00
PU18–1	–	–	–	–	–	m²	75.00
PU22–1	–	–	–	–	–	m²	85.00
PU25	–	–	–	–	–	m²	96.00
PU32	–	–	–	–	–	m²	118.00
BORED PILES							
NOTE: The following approximate prices, for the quantities of piling quoted, are for work on clear open sites with reasonable access. They are based on normal concrete mix 35.00 N/mm²; reinforced and include up to 0.15 m of projecting reinforcement at top of pile. The prices do not allow for removal of spoil.							

07 PILING

Item	PC £	Labour hours	Labour £	Plant £	Material £	Unit	Total rate £
BORED PILES – cont							
Minipile cast-in-place concrete piles							
300 mm nominal dia.							
Establish equipment on site							
Provision of all plant (1 nr rig) ; including bringing to and removing from site; maintenance, erection and dismantling at each pile position for 50 nr piles	–	–	–	–	–	item	4800.00
Bored piles							
300 mm dia. piles; nominally reinforced; 15 m long	–	–	–	–	–	nr	210.00
set up at each pile position	–	–	–	–	–	nr	22.00
add for additional piles length	–	–	–	–	–	m	14.00
deduct for reduction in pile length	–	–	–	–	–	m	5.00
Delays							
rig standing time	–	–	–	–	–	hour	190.00
Pile tests							
integrity tests (minimum 20 per visit)	–	–	–	–	–	nr	14.00
300 mm nominal dia.; small area with difficult access and limited headroom							
Establish equipment on site							
Provision of all plant (1 nr rig) ; including bringing to and removing from site; maintenance, erection and dismantling at each pile position for 50 nr piles	–	–	–	–	–	item	2600.00
Bored piles							
300 mm dia. piles; nominally reinforced; 15 m long	–	–	–	–	–	nr	210.00
set up at each pile position	–	–	–	–	–	nr	22.00
add for additional piles length	–	–	–	–	–	m	45.00
deduct for reduction in pile length	–	–	–	–	–	m	18.00
Delays							
rig standing time	–	–	–	–	–	hour	190.00
Pile tests							
integrity tests (minimum 20 per visit)	–	–	–	–	–	nr	14.00
450 mm nominal dia.							
Establish equipment on site							
Provision of all plant (2 nr rigs) ; including bringing to and removing from site; maintenance, erection and dismantling at each pile position for 100 nr piles	–	–	–	–	–	item	21500.00
Bored piles							
450 mm dia. piles; nominally reinforced; 20 m long	–	–	–	–	–	nr	3100.00
add for additional piles length	–	–	–	–	–	m	150.00
deduct for reduction in pile length	–	–	–	–	–	m	20.00
Blind bored piles							
450 mm dia.	–	–	–	–	–	m	130.00

07 PILING

Item	PC £	Labour hours	Labour £	Plant £	Material £	Unit	Total rate £
Delays							
rig standing time	–	–	–	–	–	hour	**190.00**
Extra over piling							
breaking through obstructions	–	–	–	–	–	hour	**190.00**
Pile tests							
working to 600 kN; using tension piles as reaction; first pile	–	–	–	–	–	nr	**10000.00**
working to 600 kN; using tension piles as reaction; subsequent piles	–	–	–	–	–	nr	**10000.00**
integrity tests (minimum 20 per visit)	–	–	–	–	–	nr	**14.00**
Rotary CFA bored cast-in-place concrete piles							
450 mm dia. piles							
Provision of plant							
Provision of all plant (1 nr rig); including bringing to and removing from site; maintenance, erection and dismantling at each pile position for 100 nr piles 20 m long	–	–	–	–	–	item	**11800.00**
Bored piles							
450 mm dia. piles; reinforced; 20 m long	–	–	–	–	–	nr	**810.00**
add for additional piles length	–	–	–	–	–	m	**40.00**
deduct for reduction in pile length	–	–	–	–	–	m	**16.00**
Delays							
rig standing time	–	–	–	–	–	hour	**390.00**
Pile tests							
working to 600 kN; using tension piles as reaction; first pile	–	–	–	–	–	nr	**5700.00**
working to 600 kN; using tension piles as reaction; subsequent piles	–	–	–	–	–	nr	**5700.00**
integrity tests (minimum 20 per visit)	–	–	–	–	–	nr	**12.00**
600 mm dia. piles							
Provision of all plant (1 nr rig); including bringing to and removing from site; maintenance, erection and dismantling at each pile position for 100 nr piles 20 m long	–	–	–	–	–	item	**11800.00**
Bored piles							
600 mm dia. piles; reinforced; 20 m long	–	–	–	–	–	nr	**1280.00**
add for additional piles length	–	–	–	–	–	m	**65.00**
deduct for reduction in pile length	–	–	–	–	–	m	**28.00**
Delays							
rig standing time	–	–	–	–	–	hour	**390.00**
Pile tests							
working to 1000 kN; using tension piles as reaction; first pile	–	–	–	–	–	nr	**7100.00**
working to 1000 kN; using tension piles as reaction; subsequent piles	–	–	–	–	–	nr	**7100.00**
integrity test 600 mm dia. piles; min 20 per visit	–	–	–	–	–	nr	**12.00**

Prices for Measured Works

07 PILING

Item	PC £	Labour hours	Labour £	Plant £	Material £	Unit	Total rate £
BORED PILES – cont							
750 mm dia. piles							
Provision of all plant (1 nr rig); including bringing to and removing from site; maintenance, erection and dismantling at each pile position for 100 nr piles 20 m long	–	–	–	–	–	item	**15500.00**
Bored piles							
750 mm dia. piles; reinforced; 20 m long	–	–	–	–	–	nr	**2000.00**
add for additional piles length	–	–	–	–	–	m	**100.00**
deduct for reduction in pile length	–	–	–	–	–	m	**40.00**
Delays							
rig standing time	–	–	–	–	–	hour	**470.00**
Pile tests							
working to 1500 kN; using tension piles as reaction; first pile	–	–	–	–	–	nr	**8600.00**
working to 1500 kN; using tension piles as reaction; subsequent piles	–	–	–	–	–	nr	**8600.00**
integrity test 600 mm dia. piles; min 20 per visit	–	–	–	–	–	nr	**12.00**
900 mm dia. piles							
Provision of all plant (1 nr rig); including bringing to and removing from site; maintenance, erection and dismantling at each pile position for 100 nr piles 20 m long	–	–	–	–	–	item	**15500.00**
Bored piles							
900 mm dia. piles; reinforced; 20 m long	–	–	–	–	–	nr	**2870.00**
add for additional piles length	–	–	–	–	–	m	**145.00**
deduct for reduction in pile length	–	–	–	–	–	m	**60.00**
Delays							
rig standing time	–	–	–	–	–	hour	**470.00**
Pile tests							
working to 2000 kN; using tension piles as reaction; first pile	–	–	–	–	–	nr	**10900.00**
working to 2000 kN; using tension piles as reaction; subsequent piles	–	–	–	–	–	nr	**10900.00**
integrity test 900 mm dia. piles; min 20 per visit	–	–	–	–	–	nr	**12.00**
CUTTING OFF TOPS OF PILES							
Cut off and prepare reinforcement to receive reinforced concrete pile cap; not including disposal							
450 mm dia. piles	–	1.00	22.89	–	–	m	**22.89**
600 mm dia. piles	–	1.20	27.47	–	–	m	**27.47**
750 mm dia. piles	–	1.50	34.34	–	–	m	**34.34**
900 mm dia. piles	–	1.80	41.20	–	–	m	**41.20**

08 UNDERPINNING

Item	PC £	Labour hours	Labour £	Plant £	Material £	Unit	Total rate £
UNDERPINNING							
Excavating; by machine							
Preliminary trenches							
maximum depth not exceeding 1.00 m	–	0.23	3.04	5.27	–	m³	**8.31**
maximum depth not exceeding 2.00 m	–	0.28	3.71	6.35	–	m³	**10.06**
maximum depth not exceeding 4.00 m	–	0.32	4.24	7.43	–	m³	**11.67**
Extra over preliminary trench excavating for breaking out existing hard pavings, 150 mm thick							
concrete	–	0.65	8.61	1.64	–	m²	**10.25**
Excavating; by hand							
Preliminary trenches							
maximum depth not exceeding 1.00 m	–	2.68	35.49	–	–	m³	**35.49**
maximum depth not exceeding 2.00 m	–	3.05	40.38	–	–	m³	**40.38**
maximum depth not exceeding 4.00 m	–	3.93	52.04	–	–	m³	**52.04**
Extra over preliminary trench excavating for breaking out existing hard pavings, 150 mm thick							
concrete	–	0.28	3.71	2.26	–	m²	**5.97**
Underpinning pits; commencing from 1.00 m below existing ground level							
maximum depth not exceeding 0.25 m	–	4.07	53.90	–	–	m³	**53.90**
maximum depth not exceeding 1.00 m	–	4.44	58.79	–	–	m³	**58.79**
maximum depth not exceeding 2.00 m	–	5.32	70.44	–	–	m³	**70.44**
Underpinning pits; commencing from 2.00 m below existing ground level							
maximum depth not exceeding 0.25 m	–	5.00	66.21	–	–	m³	**66.21**
maximum depth not exceeding 1.00 m	–	5.37	71.11	–	–	m³	**71.11**
maximum depth not exceeding 2.00 m	–	6.24	82.62	–	–	m³	**82.62**
Underpinning pits; commencing from 4.00 m below existing ground level							
maximum depth not exceeding 0.25 m	–	5.92	78.39	–	–	m³	**78.39**
maximum depth not exceeding 1.00 m	–	6.29	83.30	–	–	m³	**83.30**
maximum depth not exceeding 2.00 m	–	7.17	94.94	–	–	m³	**94.94**
Extra over any types of excavating irrespective of depth							
excavating below ground water level	–	0.32	4.24	–	–	m³	**4.24**
Earthwork support to preliminary trenches (open boarded – in 3.00 m lengths)							
Maximum depth not exceeding 1.00 m							
distance between opposing faces not exceeding 2.00 m	–	0.37	4.90	–	0.71	m²	**5.61**
Maximum depth not exceeding 2.00 m							
distance between opposing faces not exceeding 2.00 m	–	0.46	6.09	–	0.87	m²	**6.96**
Maximum depth not exceeding 4.00 m							
distance between opposing faces not exceeding 2.00 m	–	0.59	7.81	–	1.09	m²	**8.90**

08 UNDERPINNING

Item	PC £	Labour hours	Labour £	Plant £	Material £	Unit	Total rate £
UNDERPINNING – cont							
Earthwork support to underpinning pits (open boarded – in 3.00 m lengths)							
Maximum depth not exceeding 1.00 m							
distance between opposing faces not exceeding 2.00 m	–	0.41	5.43	–	0.77	m²	**6.20**
Maximum depth not exceeding 2.00 m							
distance between opposing faces not exceeding 2.00 m	–	0.51	6.75	–	0.98	m²	**7.73**
Maximum depth not exceeding 4.00 m							
distance between opposing faces not exceeding 2.00 m	–	0.65	8.61	–	1.20	m²	**9.81**
Earthwork support to preliminary trenches (closed boarded – in 3.00 m lengths)							
Maximum depth not exceeding 1.00 m							
1.00 m deep	–	0.93	12.32	–	1.20	m²	**13.52**
Maximum depth not exceeding 2.00 m							
distance between opposing faces not exceeding 2.00 m	–	1.16	15.36	–	1.52	m²	**16.88**
Maximum depth not exceeding 4.00 m							
distance between opposing faces not exceeding 2.00 m	–	1.43	18.93	–	1.85	m²	**20.78**
Earthwork support to underpinning pits (closed boarded – in 3.00 m lengths)							
Maximum depth not exceeding 1.00 m							
distance between opposing faces not exceeding 2.00 m	–	1.02	13.50	–	1.30	m²	**14.80**
Maximum depth not exceeding 2.00 m							
distance between opposing faces not exceeding 2.00 m	–	1.28	16.95	–	1.64	m²	**18.59**
Maximum depth not exceeding 4.00 m							
distance between opposing faces not exceeding 2.00 m	–	1.57	20.79	–	2.07	m²	**22.86**
Extra over earthwork support for							
Left in	–	0.69	9.13	–	7.62	m²	**16.75**
Cutting away existing projecting foundations							
Concrete							
maximum width 150 mm; maximum depth 150 mm	–	0.15	1.98	0.19	–	m	**2.17**
maximum width 150 mm; maximum depth 225 mm	–	0.22	2.91	0.28	–	m	**3.19**
maximum width 150 mm; maximum depth 300 mm	–	0.30	3.97	0.38	–	m	**4.35**
maximum width 300 mm; maximum depth 300 mm	–	0.58	7.68	0.74	–	m	**8.42**

08 UNDERPINNING

Item	PC £	Labour hours	Labour £	Plant £	Material £	Unit	Total rate £
Masonry							
maximum width one brick thick; maximum depth one course high	–	0.04	0.53	0.07	–	m	**0.60**
maximum width one brick thick; maximum depth two courses high	–	0.13	1.72	0.17	–	m	**1.89**
maximum width one brick thick; maximum depth three courses high	–	0.25	3.31	0.32	–	m	**3.63**
maximum width one brick thick; maximum depth four courses high	–	0.42	5.57	0.53	–	m	**6.10**
Preparing the underside of existing work to receive the pinning up of the new work							
Width of existing work							
380 mm wide	–	0.56	7.41	–	–	m	**7.41**
600 mm wide	–	0.74	9.80	–	–	m	**9.80**
900 mm wide	–	0.93	12.32	–	–	m	**12.32**
1200 mm wide	–	1.11	14.70	–	–	m	**14.70**
Disposal							
Excavated material							
off site; to tip not exceeding 13 km (using lorries); including Landfill Tax based on inactive waste	–	0.75	9.93	–	25.00	m³	**34.93**
Filling to excavations; by hand							
Average thickness exceeding 0.25 m							
arising from the excavations	–	0.93	12.32	–	–	m³	**12.32**
Surface treatments							
Compacting							
bottoms of excavations	–	0.04	0.53	0.04	–	m²	**0.57**
Plain in situ ready mixed designated concrete; C10–40 mm aggregate; poured against faces of excavation							
Underpinning							
thickness not exceeding 150 mm	–	3.42	53.09	–	106.15	m³	**159.24**
thickness 150–450 mm	–	2.87	44.55	–	106.15	m³	**150.70**
thickness exceeding 450 mm	–	2.50	38.81	–	106.15	m³	**144.96**
Plain in situ ready mixed designated concrete; C20–20 mm aggregate; poured against faces of excavation							
Underpinning							
thickness not exceeding 150 mm	–	3.42	53.09	–	110.29	m³	**163.38**
thickness 150–450 mm	–	2.87	44.55	–	110.29	m³	**154.84**
thickness exceeding 450 mm	–	2.50	38.81	–	110.29	m³	**149.10**
Extra for working around reinforcement	–	0.28	4.35	–	–	m³	**4.35**

08 UNDERPINNING

Item	PC £	Labour hours	Labour £	Plant £	Material £	Unit	Total rate £
UNDERPINNING – cont							
Sawn formwork; sides of foundations in underpinning							
Plain vertical							
height exceeding 1.00 m	–	1.48	27.16	–	2.72	m²	**29.88**
height not exceeding 250 mm	–	0.51	9.36	–	0.78	m²	**10.14**
height 250–500 mm	–	0.79	14.50	–	1.45	m²	**15.95**
height 500 mm–1.00 m	–	1.20	22.02	–	2.72	m²	**24.74**
Reinforcement bar; BS 4449 hot rolled deformed square high yield steel bars							
20 mm dia. nominal size							
straight	441.00	22.80	413.09	–	653.77	tonne	**1066.86**
bent	486.00	22.80	413.09	–	702.21	tonne	**1115.30**
16 mm dia. nominal size							
straight	441.00	24.70	447.96	–	676.17	tonne	**1124.13**
bent	486.00	24.70	447.96	–	724.59	tonne	**1172.55**
12 mm dia. nominal size							
straight	441.00	26.60	482.83	–	698.57	tonne	**1181.40**
bent	486.00	26.60	482.83	–	746.99	tonne	**1229.82**
10 mm dia. nominal size							
straight	441.00	28.50	517.70	–	737.19	tonne	**1254.89**
bent	486.00	28.50	517.70	–	785.62	tonne	**1303.32**
8 mm dia. nominal size							
straight	441.00	30.40	549.89	–	748.01	tonne	**1297.90**
bent	486.00	30.40	549.89	–	808.02	tonne	**1357.91**
Common bricks; in cement mortar (1:3)							
Walls in underpinning							
one brick thick (PC £ per 1000)	380.00	2.22	47.86	–	55.36	m²	**103.22**
one and a half brick thick	–	3.05	65.75	–	82.66	m²	**148.41**
two brick thick	–	3.79	81.71	–	113.03	m²	**194.74**
Class A engineering bricks; in cement mortar (1:3)							
Walls in underpinning							
one brick thick (PC £ per 1000)	750.00	2.22	47.86	–	71.44	m²	**119.30**
one and a half brick thick	–	3.05	65.75	–	106.80	m²	**172.55**
two brick thick	–	3.79	81.71	–	145.22	m²	**226.93**
Class B engineering bricks; in cement mortar (1:3)							
Walls in underpinning							
one brick thick (PC £ per 1000)	590.00	2.22	47.86	–	60.74	m²	**108.60**
one and a half brick thick	–	3.05	65.75	–	90.74	m²	**156.49**
two brick thick	–	3.79	81.71	–	123.82	m²	**205.53**
Add or deduct for variation of £10.00/1000 in PC of bricks							
one brick thick	–	–	–	–	1.26	m²	**1.26**
one and a half bricks thick	–	–	–	–	1.89	m²	**1.89**
two bricks thick	–	–	–	–	2.52	m²	**2.52**

08 UNDERPINNING

Item	PC £	Labour hours	Labour £	Plant £	Material £	Unit	Total rate £
Zedex CPT (Co-Polymer Thermoplastic) damp-proof course or other equal ; 200 mm laps; in gauged mortar (1:1:6)							
Horizontal							
width exceeding 225 mm	–	0.23	4.96	–	6.72	m²	**11.68**
width not exceeding 225 mm	–	0.46	9.91	–	6.72	m²	**16.63**
Hyload (pitch polymer) damp-proof course or similar; 150 mm laps; in cement mortar (1:3)							
Horizontal							
width exceeding 225 mm	–	0.23	4.96	–	5.09	m²	**10.05**
width not exceeding 225 mm	–	0.46	9.91	–	5.22	m²	**15.13**
Alumite aluminium cored bitumen gas retardant damp-proof course or other equal and approved; 200 mm laps; in gauged mortar (1:1;6)							
Horizontal							
width exceeding 225 mm	–	0.31	6.69	–	7.09	m²	**13.78**
width not exceeding 225 mm	–	0.60	12.94	–	7.09	m²	**20.03**
Two courses of slates in cement mortar (1:3)							
Horizontal							
width exceeding 225 mm	–	1.39	29.97	–	32.29	m²	**62.26**
width not exceeding 225 mm	–	2.31	49.80	–	33.02	m²	**82.82**
Wedging and pinning							
To underside of existing construction with slates in cement mortar (1:3)							
width of wall – half brick thick	–	1.02	21.99	–	7.36	m	**29.35**
width of wall – one brick thick	–	1.20	25.87	–	14.72	m	**40.59**
width of wall – one and a half brick thick	–	1.39	29.97	–	22.08	m	**52.05**

09 DIAPHRAGM WALLS AND EMBEDDED RETAINING WALLS

Item	PC £	Labour hours	Labour £	Plant £	Material £	Unit	Total rate £
EMBEDDED RETAINING WALLS							
Diaphragm walls; contiguous panel construction; panel lengths not exceeding 5 m							
Provision of all plant; including bringing to and removing from site; maintenance, erection and dismantling; assuming one rig for 1000 m² of walling	–	–	–	–	–	item	240000.00
Excavation for diaphragm wall; excavated material removed from site; Bentonite slurry supplied and disposed of							
600 mm thick walls	–	–	–	–	–	m³	650.00
1000 mm thick walls	–	–	–	–	–	m³	410.00
Ready mixed reinforced in situ concrete; normal portland cement; C30–10 mm aggregate in walls	–	–	–	–	–	m³	130.00
Reinforcement bar; BS 4449 cold rolled deformed square high yield steel bars; straight or bent							
25 mm–40 mm dia.	–	–	–	–	–	tonne	1200.00
20 mm dia.	–	–	–	–	–	tonne	1200.00
16 mm dia.	–	–	–	–	–	tonne	1200.00
Formwork 75 mm thick to form chases	–	–	–	–	–	m²	80.00
Construct twin guide walls in reinforced concrete; together with reinforcement and formwork along the axis of the diaphragm wall	–	–	–	–	–	m	500.00
Delays							
rig standing	–	–	–	–	–	hour	1600.00

10 CRIB WALLS, GABIONS AND REINFORCED EARTH

Item	PC £	Labour hours	Labour £	Plant £	Material £	Unit	Total rate £
GABION BASKET WALLS							
Gabion baskets							
Wire mesh gabion baskets; Maccaferri Ltd or other equal and approved; galvanized mesh 80 mm × 100 mm; filling with broken stones 125 mm–200 mm size							
2.00 × 1.00 × 0.50 m	18.45	1.00	22.89	10.27	81.38	nr	**114.54**
2.00 × 1.00 × 0.50 m PVC coated	23.33	1.00	22.89	10.27	87.24	nr	**120.40**
2.00 × 1.00 × 1.00 m	25.86	2.00	45.78	20.54	150.86	nr	**217.18**
2.00 × 1.00 × 1.00 m PVC coated	32.73	2.00	45.78	20.54	158.26	nr	**224.58**
Reno mattress gabion baskets or other equal and approved; Maccaferri Ltd; filling with broken stones 125 mm–200 mm size							
6.00 × 2.00 × 0.17 m	82.53	2.00	45.78	15.40	214.32	nr	**275.50**
6.00 × 2.00 × 0.23 m	89.33	2.50	57.24	20.54	265.94	nr	**343.72**
6.00 × 2.00 × 0.30 m	104.80	3.00	68.68	23.10	334.26	nr	**426.04**

11 IN SITU CONCRETE WORKS

Item	PC £	Labour hours	Labour £	Plant £	Material £	Unit	Total rate £
IN SITU CONCRETE							
Cement material prices							
Cements							
Ordinary Portland	–	–	–	–	119.31	tonne	**119.31**
high alumina	–	–	–	–	539.27	tonne	**539.27**
Sulfacrete sulphate-resisting	–	–	–	–	146.03	tonne	**146.03**
Ferrocrete rapid hardening	–	–	–	–	277.75	tonne	**277.75**
Snowcrete white cement	–	–	–	–	189.93	tonne	**189.93**
Cement admixtures							
Febtone colorant – red, marigold, yellow, brown, black	–	–	–	–	4.15	kg	**4.15**
Febproof waterproof	–	–	–	–	0.95	litre	**0.95**
Febond PVA bonding agent	–	–	–	–	2.27	litre	**2.27**
Febspeed frostproofer and hardener	–	–	–	–	1.25	litre	**1.25**
Basic mixed concrete prices							
NOTE: The following prices are for concrete ready for placing excluding any allowance for waste, discount or overheads and profit. Prices are based upon delivery to site within a 10 mile (16 km) radius of concrete mixing plant, using full loads							
GEN 0 (7.5N)	–	–	–	–	84.55	m^3	**84.55**
GEN 1 (10N)	–	–	–	–	85.50	m^3	**85.50**
GEN 2 (15N)	–	–	–	–	87.40	m^3	**87.40**
GEN 3 (20N)	–	–	–	–	88.83	m^3	**88.83**
RC20/25	–	–	–	–	93.58	m^3	**93.58**
RC25/30	–	–	–	–	95.00	m^3	**95.00**
RC30/37	–	–	–	–	96.42	m^3	**96.42**
RC35/45	–	–	–	–	101.17	m^3	**101.17**
RC40/50	–	–	–	–	102.13	m^3	**102.13**
RC50/60	–	–	–	–	104.03	m^3	**104.03**
FND3	–	–	–	–	101.17	m^3	**101.17**
FND4	–	–	–	–	103.08	m^3	**103.08**
Caltite waterprrof system to RC 35 concrete	–	–	–	–	189.00	m^3	**189.00**
Site Mixed Concrete (on site batching plant)							
mix 7.50 N/mm²; cement (1:8)	–	–	–	–	98.73	m^3	**98.73**
mix 10.00 N/mm²; cement (1:8)	–	–	–	–	100.87	m^3	**100.87**
mix 20.00 N/mm²; cement (1:2:4)	–	–	–	–	105.17	m^3	**105.17**
mix 25.00 N/mm²; cement to (1:1:5:3)	–	–	–	–	108.39	m^3	**108.39**
Add for							
rapid-hardening cement	–	–	–	–	10.64	m^3	**10.64**
polypropylene fibre additive	–	–	–	–	5.91	m^3	**5.91**
air entrained concrete	–	–	–	–	5.20	m^3	**5.20**
water repellent additive	–	–	–	–	5.55	m^3	**5.55**
foamed concrete	–	–	–	–	–	%	**26.25**
distance per mile in excess of 10 miles (16 km)	–	–	–	–	0.63	m^3	**0.63**

11 IN SITU CONCRETE WORKS

Item	PC £	Labour hours	Labour £	Plant £	Material £	Unit	Total rate £
SUPPLY AND LAY PRICES							
NOTE: The following concrete material prices include an allowance for shrinkage and waste. PC Sums are concrete supply only prices							
Plain in situ ready mixed designated concrete; C7.5–40 mm aggregate							
Foundations	84.55	1.02	15.91	–	93.22	m³	**109.13**
Isolated foundations	84.55	1.13	17.46	–	93.22	m³	**110.68**
Beds							
thickness not exceeding 300 mm	84.55	1.02	15.91	–	93.22	m³	**109.13**
thickness exceeding 300 mm	84.55	0.93	14.44	–	93.22	m³	**107.66**
Screeded beds; protection to compressible formwork exceeding 600 mm wide							
50 mm thick	84.55	0.10	1.55	–	4.66	m²	**6.21**
75 mm thick	84.55	0.15	2.33	–	6.99	m²	**9.32**
100 mm thick	84.55	0.20	3.11	–	9.32	m²	**12.43**
Filling hollow walls							
thickness not exceeding 150 mm	84.55	3.15	48.90	–	93.22	m³	**142.12**
Column casings							
stub columns beneath suspended ground slabs	84.55	4.50	69.86	–	93.22	m³	**163.08**
Plain in situ ready mixed designated concrete; C10–40 mm aggregate							
Foundations	85.50	1.02	15.91	–	89.78	m³	**105.69**
Isolated foundations	85.50	1.13	17.50	–	89.78	m³	**107.28**
Beds							
thickness not exceeding 300 mm	85.50	1.02	15.91	–	89.78	m³	**105.69**
thickness exceeding 300 mm	85.50	0.93	14.36	–	89.78	m³	**104.14**
Filling hollow walls							
thickness not exceeding 300 mm	85.50	3.15	48.90	–	94.27	m³	**143.17**
Plain in situ ready mixed designated concrete; C10–40 mm aggregate; poured on or against earth or unblinded hardcore							
Foundations	85.50	1.02	15.91	–	94.27	m³	**110.18**
Isolated foundations	85.50	1.13	17.50	–	94.27	m³	**111.77**
Beds							
thickness not exceeding 300 mm	85.50	1.02	15.91	–	94.27	m³	**110.18**
thickness exceeding 300 mm	85.50	0.93	14.36	–	94.27	m³	**108.63**
Plain in situ ready mixed designated concrete; C20–20 mm aggregate							
Foundations	88.83	0.98	15.21	–	93.27	m³	**108.48**
Isolated foundations	88.83	1.13	17.50	–	93.27	m³	**110.77**
Beds							
thickness not exceeding 300 mm	88.83	1.02	15.83	–	93.27	m³	**109.10**
thickness exceeding 300 mm	88.83	0.93	14.44	–	93.27	m³	**107.71**
Filling hollow walls							
thickness not exceeding 300 mm	88.83	3.00	46.57	–	97.93	m³	**144.50**

11 IN SITU CONCRETE WORKS

Item	PC £	Labour hours	Labour £	Plant £	Material £	Unit	Total rate £
IN SITU CONCRETE – cont							
Plain in situ ready mixed; 20N; poured on or against earth or unblinded hardcore							
Foundations	88.83	1.02	15.91	–	97.93	m³	**113.84**
Isolated foundations	88.83	1.13	17.50	–	97.93	m³	**115.43**
Beds							
thickness not exceeding 300 mm	88.83	1.02	15.91	–	99.89	m³	**115.80**
thickness exceeding 300 mm	88.83	0.93	14.44	–	93.27	m³	**107.71**
Reinforced in situ ready mixed designated concrete; 25N							
Foundations	95.00	1.02	15.91	–	104.74	m³	**120.65**
Ground beams	95.00	2.59	40.20	–	99.75	m³	**139.95**
Isolated foundations	95.00	1.13	17.46	–	104.74	m³	**122.20**
Beds							
thickness not exceeding 300 mm	95.00	1.02	15.83	–	104.74	m³	**120.57**
thickness exceeding 300 mm	95.00	0.93	14.44	–	99.75	m³	**114.19**
Slabs							
thickness not exceeding 300 mm	95.00	1.25	19.40	–	104.74	m³	**124.14**
thickness exceeding 300 mm	95.00	1.15	17.85	–	104.74	m³	**122.59**
Coffered and troughed slabs							
thickness not exceeding 300 mm	95.00	2.96	45.95	–	104.74	m³	**150.69**
thickness exceeding 300 mm	95.00	2.59	40.20	–	104.74	m³	**144.94**
Extra over for sloping work							
not exceeding 15°	–	0.23	3.57	–	–	m³	**3.57**
exceeding 15°	–	0.46	7.14	–	–	m³	**7.14**
Walls							
thickness not exceeding 300 mm	95.00	3.15	48.90	–	104.74	m³	**153.64**
thickness exceeding 300 mm	95.00	2.40	37.34	–	104.74	m³	**142.08**
Beams							
isolated	95.00	3.70	57.44	–	104.74	m³	**162.18**
isolated deep	95.00	4.07	63.18	–	104.74	m³	**167.92**
attached deep	95.00	3.70	57.44	–	104.74	m³	**162.18**
Beam casings							
isolated	95.00	4.07	63.18	–	107.24	m³	**170.42**
isolated deep	95.00	4.44	68.92	–	104.74	m³	**173.66**
attached deep	95.00	4.07	63.18	–	104.74	m³	**167.92**
Columns	95.00	4.20	65.19	–	104.74	m³	**169.93**
Column casings	95.00	4.90	76.06	–	99.75	m³	**175.81**
Staircases	95.00	5.25	81.50	–	104.74	m³	**186.24**
Upstands	95.00	3.30	51.23	–	104.74	m³	**155.97**
Reinforced in situ ready mixed designated concrete; 30N							
Foundations	96.42	1.02	15.91	–	106.31	m³	**122.22**
Ground beams	96.42	2.59	40.20	–	106.31	m³	**146.51**
Isolated foundations	96.42	1.13	17.55	–	106.31	m³	**123.86**

11 IN SITU CONCRETE WORKS

Item	PC £	Labour hours	Labour £	Plant £	Material £	Unit	Total rate £
Beds							
thickness not exceeding 300 mm	96.42	1.02	15.83	–	106.31	m^3	**122.14**
thickness exceeding 300 mm	96.42	0.95	14.74	–	106.31	m^3	**121.05**
Slabs							
thickness not exceeding 300 mm	96.42	1.25	19.40	–	106.31	m^3	**125.71**
thickness exceeding 300 mm	96.42	1.15	17.85	–	106.31	m^3	**124.16**
Coffered and troughed slabs							
thickness not exceeding 300 mm	96.42	2.96	45.95	–	106.31	m^3	**152.26**
thickness exceeding 300 mm	96.42	2.59	40.20	–	106.31	m^3	**146.51**
Extra over for sloping							
not exceeding 15°	–	0.23	3.57	–	–	m^3	**3.57**
exceeding 15°	–	0.46	7.14	–	–	m^3	**7.14**
Walls							
thickness not exceeding 300 mm	96.42	2.67	41.53	–	106.31	m^3	**147.84**
thickness exceeding 300 mm	96.42	2.41	37.37	–	106.31	m^3	**143.68**
Beams							
isolated	96.42	3.70	57.44	–	106.31	m^3	**163.75**
isolated deep	96.42	4.07	63.18	–	106.31	m^3	**169.49**
attached deep	96.42	3.70	57.44	–	106.31	m^3	**163.75**
Beam casings							
isolated	96.42	4.07	63.18	–	106.31	m^3	**169.49**
isolated deep	96.42	4.44	68.92	–	106.31	m^3	**175.23**
attached deep	96.42	4.07	63.18	–	106.31	m^3	**169.49**
Columns	96.42	4.44	68.92	–	106.31	m^3	**175.23**
Column casings	96.42	4.90	76.06	–	106.31	m^3	**182.37**
Staircases	96.42	5.55	86.15	–	106.31	m^3	**192.46**
Upstands	96.42	3.56	55.26	–	106.31	m^3	**161.57**
Reinforced in situ ready mixed designated concrete; 35N							
Foundations	101.17	1.02	15.91	–	111.54	m^3	**127.45**
Ground beams	101.17	2.59	40.20	–	111.54	m^3	**151.74**
Isolated foundations	101.17	1.13	17.55	–	111.54	m^3	**129.09**
Beds							
thickness not exceeding 300 mm	101.17	1.02	15.83	–	111.54	m^3	**127.37**
thickness exceeding 300 mm	101.17	0.95	14.74	–	111.54	m^3	**126.28**
Slabs							
thickness not exceeding 300 mm	101.17	1.25	19.40	–	111.54	m^3	**130.94**
thickness exceeding 300 mm	101.17	1.15	17.85	–	111.54	m^3	**129.39**
Coffered and troughed slabs							
thickness not exceeding 300 mm	101.17	2.96	45.95	–	111.54	m^3	**157.49**
thickness exceeding 300 mm	101.17	2.59	40.20	–	111.54	m^3	**151.74**
Extra over for sloping							
not exceeding 15°	–	0.23	3.57	–	–	m^3	**3.57**
exceeding 15°	–	0.46	7.14	–	–	m^3	**7.14**
Walls							
thickness not exceeding 300 mm	101.17	2.67	41.53	–	111.54	m^3	**153.07**
thickness exceeding 300 mm	101.17	2.41	37.37	–	111.54	m^3	**148.91**

11 IN SITU CONCRETE WORKS

Item	PC £	Labour hours	Labour £	Plant £	Material £	Unit	Total rate £
IN SITU CONCRETE – cont							
Reinforced in situ ready mixed designated concrete – cont							
Beams							
isolated	101.17	3.70	57.44	–	111.54	m³	**168.98**
isolated deep	101.17	4.07	63.18	–	111.54	m³	**174.72**
attached deep	101.17	3.70	57.44	–	111.54	m³	**168.98**
Beam casings							
isolated	101.17	4.07	63.18	–	111.54	m³	**174.72**
isolated deep	101.17	4.44	68.92	–	111.54	m³	**180.46**
attached deep	101.17	4.07	63.18	–	111.54	m³	**174.72**
Columns	101.17	4.44	68.92	–	111.54	m³	**180.46**
Column casings	101.17	4.90	76.06	–	111.54	m³	**187.60**
Staircases	101.17	5.55	86.15	–	111.54	m³	**197.69**
Upstands	101.17	3.56	55.26	–	111.54	m³	**166.80**
Reinforced in situ ready mixed designated concrete; 40N							
Foundations	102.13	1.02	15.91	–	112.59	m³	**128.50**
Isolated foundations	102.13	1.13	17.50	–	112.59	m³	**130.09**
Ground beams	102.13	2.59	40.20	–	112.59	m³	**152.79**
Beds							
thickness not exceeding 300 mm	102.13	0.95	14.74	–	112.59	m³	**127.33**
thickness exceeding 300 mm	102.13	0.93	14.44	–	112.59	m³	**127.03**
Slabs							
thickness not exceeding 300 mm	102.13	1.20	18.63	–	112.59	m³	**131.22**
thickness exceeding 300 mm	102.13	1.15	17.85	–	112.59	m³	**130.44**
Coffered and troughed slabs							
thickness not exceeding 300 mm	102.13	2.96	45.95	–	112.59	m³	**158.54**
thickness exceeding 300 mm	102.13	2.59	40.20	–	112.59	m³	**152.79**
Extra over for sloping							
not exceeding 15°	–	0.23	3.57	–	–	m³	**3.57**
exceeding 15°	–	0.46	7.14	–	–	m³	**7.14**
Walls							
thickness not exceeding 300 mm	102.13	2.73	42.38	–	112.59	m³	**154.97**
thickness exceeding 300 mm	102.13	2.41	37.41	–	112.59	m³	**150.00**
Beams							
isolated	102.13	3.70	57.44	–	112.59	m³	**170.03**
isolated deep	102.13	4.07	63.18	–	112.59	m³	**175.77**
attached deep	102.13	3.70	57.44	–	112.59	m³	**170.03**
Beam casings							
isolated	102.13	4.07	63.18	–	112.59	m³	**175.77**
isolated deep	102.13	4.44	68.92	–	112.59	m³	**181.51**
attached deep	102.13	4.07	63.18	–	112.59	m³	**175.77**
Columns	102.13	4.44	68.92	–	112.59	m³	**181.51**
Column casings	102.13	4.90	76.06	–	112.59	m³	**188.65**
Staircases	102.13	5.55	86.15	–	112.59	m³	**198.74**
Upstands	102.13	3.56	55.26	–	112.59	m³	**167.85**

11 IN SITU CONCRETE WORKS

Item	PC £	Labour hours	Labour £	Plant £	Material £	Unit	Total rate £
Sundry in situ concrete work							
Filling; plain in situ concrete; mixed on site							
mortices	–	0.09	1.40	0.19	0.55	nr	**2.14**
holes	–	0.23	3.57	0.49	113.81	m³	**117.87**
chases not exceeding 300 mm wide or thick	–	0.14	2.17	0.30	1.13	m	**3.60**
chases exceeding 300 mm wide or thick	–	0.19	2.95	0.41	113.81	m³	**117.17**
Proprietary voided Bubbledeck, Cobiax or other equal and approved slab; concrete mix RC35; to achieve design loadings of 5.0 kN/m² live and 3.0 kN/m² dead; with trowelled finish							
Beds							
360 mm overall thickness	–	–	–	–	–	m²	**116.84**
additional concrete up to 600 mm wide at edges where formers omitted at junctions with walls etc.	–	–	–	–	–	m	**47.80**
Extra over vibrated concrete for							
Reinforcement content over 5%	–	0.51	7.92	–	–	m³	**7.92**
Grouting with cement mortar (1:1)							
Stanchion bases							
10 mm thick	–	0.93	14.44	–	0.13	nr	**14.57**
25 mm thick	–	1.16	18.01	–	0.33	nr	**18.34**
Grouting with epoxy resin							
Stanchion bases							
10 mm thick	–	1.16	18.01	–	3.75	nr	**21.76**
25 mm thick	–	1.39	21.58	–	18.76	nr	**40.34**
Grouting with Conbextra GP cementitious grout							
Stanchion bases							
10 mm thick	–	1.16	18.01	–	1.27	nr	**19.28**
25 mm thick	–	1.39	21.58	–	3.24	nr	**24.82**
Grouting with Conbextra HF flowable cementitious grout							
Stanchion bases							
10 mm thick	–	1.16	18.01	–	1.56	nr	**19.57**
25 mm thick	–	1.39	21.58	–	4.00	nr	**25.58**
SURFACE FINISHES							
Worked finishes							
Tamping by mechanical means	–	0.02	0.32	0.10	–	m²	**0.42**
Power floating	–	0.16	2.49	0.33	–	m²	**2.82**
Trowelling	–	0.31	4.81	–	–	m²	**4.81**
Hacking							
by mechanical means	–	0.31	4.81	0.39	–	m²	**5.20**
by hand	–	0.65	10.09	–	–	m²	**10.09**
Lightly shot blasting surface of concrete	–	0.37	5.74	–	–	m²	**5.74**

11 IN SITU CONCRETE WORKS

Item	PC £	Labour hours	Labour £	Plant £	Material £	Unit	Total rate £
SURFACE FINISHES – cont							
Worked finishes – cont							
Blasting surface of concrete to produce textured finish	–	0.65	10.09	0.82	–	m²	10.91
Sand blasting (blast and vac method)	–	–	–	–	–	m²	38.37
Wood float finish	–	0.12	1.86	–	–	m²	1.86
Tamped finish							
level or to falls	–	0.06	0.93	–	–	m²	0.93
to falls	–	0.09	1.40	–	–	m²	1.40
Spade finish	–	0.14	2.17	–	–	m²	2.17
Diamond drilling/cutting							
Cutting holes in block wall							
up to 52 mm dia.; up to 150 mm thick	–	–	–	–	–	nr	15.38
up to 52 mm dia.; 200 mm thick	–	–	–	–	–	nr	19.16
up to 52 mm dia.; 250 mm thick	–	–	–	–	–	nr	22.96
up to 52 mm dia.; 300 mm thick	–	–	–	–	–	nr	26.75
79 mm–107 mm dia.; up to 150 mm thick	–	–	–	–	–	nr	24.13
79 mm–107 mm dia.; 200 mm thick	–	–	–	–	–	nr	28.78
79 mm–107 mm dia.; 250 mm thick	–	–	–	–	–	nr	33.43
79 mm–107 mm dia.; 300 mm thick	–	–	–	–	–	nr	38.09
251 mm–300 mm dia.; ne 150 mm thick	–	–	–	–	–	nr	59.54
251 mm–300 mm dia.; 200 mm thick	–	–	–	–	–	nr	75.41
251 mm–300 mm dia.; 250 mm thick	–	–	–	–	–	nr	91.30
251 mm–300 mm dia.; 300 mm thick	–	–	–	–	–	nr	107.16
Cutting holes in brick wall							
up to 52 mm dia.; up to 150 mm thick	–	–	–	–	–	nr	17.57
up to 52 mm dia.; 200 mm thick	–	–	–	–	–	nr	21.89
up to 52 mm dia.; 250 mm thick	–	–	–	–	–	nr	26.24
up to 52 mm dia.; 300 mm thick	–	–	–	–	–	nr	30.59
79 mm–107 mm dia.; up to 150 mm thick	–	–	–	–	–	nr	27.56
79 mm–107 mm dia.; 200 mm thick	–	–	–	–	–	nr	32.88
79 mm–107 mm dia.; 250 mm thick	–	–	–	–	–	nr	38.20
79 mm–107 mm dia.; 300 mm thick	–	–	–	–	–	nr	43.52
251 mm–300 mm dia.; ne 150 mm thick	–	–	–	–	–	nr	68.05
251 mm–300 mm dia.; 200 mm thick	–	–	–	–	–	nr	86.19
251 mm–300 mm dia.; 250 mm thick	–	–	–	–	–	nr	104.32
251 mm–300 mm dia.; 300 mm thick	–	–	–	–	–	nr	122.48
Cutting holes in concrete wall							
up to 52 mm dia.; up to 150 mm thick	–	–	–	–	–	nr	21.96
up to 52 mm dia.; 200 mm thick	–	–	–	–	–	nr	27.38
up to 52 mm dia.; 250 mm thick	–	–	–	–	–	nr	32.80
up to 52 mm dia.; 300 mm thick	–	–	–	–	–	nr	38.21
79 mm–107 mm dia.; up to150 mm thick	–	–	–	–	–	nr	34.46
79 mm–107 mm dia.; 200 mm thick	–	–	–	–	–	nr	41.10
79 mm–107 mm dia.; 250 mm thick	–	–	–	–	–	nr	47.75
79 mm–107 mm dia.; 300 mm thick	–	–	–	–	–	nr	54.41
251 mm–300 mm dia.; ne 150 mm thick	–	–	–	–	–	nr	85.05
251 mm–300 mm dia.; 200 mm thick	–	–	–	–	–	nr	107.73
251 mm–300 mm dia.; 250 mm thick	–	–	–	–	–	nr	130.41
251 mm–300 mm dia.; 300 mm thick	–	–	–	–	–	nr	153.08

11 IN SITU CONCRETE WORKS

Item	PC £	Labour hours	Labour £	Plant £	Material £	Unit	Total rate £
Cutting holes in timber wall							
up to 52 mm dia.; up to 150 mm thick	–	–	–	–	–	nr	**20.86**
up to 52 mm dia.; 200 mm thick	–	–	–	–	–	nr	**26.00**
up to 52 mm dia.; 250 mm thick	–	–	–	–	–	nr	**31.16**
up to 52 mm dia.; 300 mm thick	–	–	–	–	–	nr	**36.31**
79 mm–107 mm dia.; up to 150 mm thick	–	–	–	–	–	nr	**32.74**
79 mm–107 mm dia.; 200 mm thick	–	–	–	–	–	nr	**39.06**
79 mm–107 mm dia.; 250 mm thick	–	–	–	–	–	nr	**45.38**
79 mm–107 mm dia.; 300 mm thick	–	–	–	–	–	nr	**51.68**
251 mm–300 mm dia.; ne 150 mm thick	–	–	–	–	–	nr	**80.81**
251 mm–300 mm dia.; 200 mm thick	–	–	–	–	–	nr	**102.35**
251 mm–300 mm dia.; 250 mm thick	–	–	–	–	–	nr	**123.90**
251 mm–300 mm dia.; 300 mm thick	–	–	–	–	–	nr	**145.44**
Chasing for concealed conduits							
up to 35 mm deep, up to 50 mm wide in block wall	–	–	–	–	–	m	**7.14**
up to 35 mm deep, up to 50 mm wide in brick wall	–	–	–	–	–	m	**8.23**
up to 35 mm deep, up to 50 mm wide in concrete wall	–	–	–	–	–	m	**27.45**
up to 35 mm deep, up to 50 mm wide in concrete floor	–	–	–	–	–	m	**32.94**
up to 35 mm deep, 75 mm–100 mm wide in block wall	–	–	–	–	–	m	**11.15**
up to 35 mm deep, 75 mm–100 mm wide in brick wall	–	–	–	–	–	m	**12.86**
up to 35 mm deep, 75 mm–100 mm wide in concrete wall	–	–	–	–	–	m	**42.89**
up to 35 mm deep, 75 mm–100 mm wide in concrete floor	–	–	–	–	–	m	**51.48**
recess 100 mm × 100 mm × 35 mm deep in block wall	–	–	–	–	–	nr	**3.83**
recess 100 mm × 100 mm × 35 mm deep in brick wall	–	–	–	–	–	nr	**4.44**
recess 100 mm × 100 mm × 35 mm deep in concrete wall	–	–	–	–	–	nr	**17.74**
recess 100 mm × 100 mm × 35 mm deep in concrete floor	–	–	–	–	–	nr	**14.77**
Floor sawing							
diamond floor sawing; 25 mm depth	–	–	–	–	–	m	**5.49**
Ring sawing							
up to 25 mm deep in block wall	–	–	–	–	–	m	**7.14**
up to 25 mm deep in brick wall	–	–	–	–	–	m	**8.23**
up to 25 mm deep in concrete floor	–	–	–	–	–	m	**10.98**
up to 25 mm deep in concrete wall	–	–	–	–	–	m	**13.17**

11 IN SITU CONCRETE WORKS

Item	PC £	Labour hours	Labour £	Plant £	Material £	Unit	Total rate £
FORMWORK							
Sides of foundations; 18 mm thick external quality plywood; basic finish; four uses							
Plain vertical							
height not exceeding 250 mm	–	0.38	6.94	–	1.85	m	**8.79**
height not exceeding 250 mm; left in	–	0.38	6.94	–	3.28	m	**10.22**
height 250–500 mm	–	0.71	13.05	–	3.54	m	**16.59**
height 250–500 mm; left in	–	0.62	11.39	–	7.19	m	**18.58**
height 500 mm–1.00 m	–	1.00	18.33	–	4.52	m	**22.85**
height 500 mm–1.00 m; left in	–	0.95	17.50	–	11.19	m	**28.69**
height exceeding 1.00 m	–	1.33	24.44	–	4.52	m²	**28.96**
height exceeding 1.00 m; left in	–	1.17	21.47	–	11.19	m²	**32.66**
Sides of foundations; polystyrene sheet formwork; Cordek Claymaster or equal							
50 mm thick; plain vertical							
height not exceeding 250 mm; left in	–	0.09	1.65	–	1.81	m	**3.46**
height 250–500 mm; left in	–	0.14	2.65	–	3.61	m	**6.26**
height 500 mm–1.00 m; left in	–	0.22	3.96	–	7.22	m	**11.18**
height exceeding 1.00 m; left in	–	0.27	4.96	–	6.89	m²	**11.85**
75 mm thick; plain vertical							
height not exceeding 250 mm; left in	–	0.09	1.65	–	5.24	m	**6.89**
height 250–500 mm; left in	–	0.14	2.65	–	10.48	m	**13.13**
height 500 mm–1.00 m; left in	–	0.22	3.96	–	21.96	m	**25.92**
height exceeding 1.00 m; left in	–	0.27	4.96	–	20.96	m²	**25.92**
100 mm thick; plain vertical							
height not exceeding 250 mm; left in	–	0.09	1.65	–	7.37	m	**9.02**
height 250–500 mm; left in	–	0.14	2.65	–	14.74	m	**17.39**
height 500 mm–1.00 m; left in	–	0.22	3.96	–	29.47	m	**33.43**
height exceeding 1.00 m; left in	–	0.27	4.96	–	28.14	m²	**33.10**
150 mm thick; plain vertical							
height not exceeding 250 mm; left in	–	0.09	1.65	–	11.06	m	**12.71**
height 250–500 mm; left in	–	0.14	2.65	–	22.10	m	**24.75**
height 500 mm–1.00 m; left in	–	0.22	3.96	–	44.22	m	**48.18**
height exceeding 1.00 m; left in	–	0.28	5.05	–	42.21	m²	**47.26**
175 mm thick; plain vertical							
height not exceeding 250 mm; left in	–	0.10	1.84	–	12.86	m	**14.70**
height 250–500 mm; left in	–	0.15	2.75	–	25.71	m	**28.46**
height 500 mm–1.00 m; left in	–	0.22	4.03	–	51.43	m	**55.46**
height exceeding 1.00 m; left in	–	0.30	5.50	–	49.10	m²	**54.60**
200 mm thick; plain vertical							
height not exceeding 250 mm; left in	–	0.10	1.84	–	14.74	m	**16.58**
height 250–500 mm; left in	–	0.15	2.75	–	29.47	m	**32.22**
height 500 mm–1.00 m; left in	–	0.22	4.03	–	58.96	m	**62.99**
height exceeding 1.00 m; left in	–	0.30	5.50	–	56.27	m²	**61.77**
250 mm thick; plain vertical							
height not exceeding 250 mm; left in	–	0.15	2.75	–	18.43	m	**21.18**
height 250–500 mm; left in	–	0.17	3.21	–	36.84	m	**40.05**
height 500 mm–1.00 m; left in	–	0.25	4.59	–	73.69	m	**78.28**
height exceeding 1.00 m; left in	–	0.35	6.43	–	70.34	m²	**76.77**

11 IN SITU CONCRETE WORKS

Item	PC £	Labour hours	Labour £	Plant £	Material £	Unit	Total rate £
Combined heave pressure relief insulation and compressible board substructure formwork; Cordeck Cellcore CP or other equal and approved; butt joints; securely fixed in place							
Plain horizontal							
200 mm thick; beneath slabs; left in	–	0.54	9.91	–	19.65	m²	**29.56**
250 mm thick; beneath slabs; left in	–	0.58	10.73	–	21.86	m²	**32.59**
300 mm thick; beneath slabs; left in	–	0.63	11.56	–	23.56	m²	**35.12**
Dufaylite Clayboard void former; butt joints							
KN30; compressive strength of 30 kN/m²; board thickness; typically for residential							
60 mm	–	0.05	0.66	–	15.41	m²	**16.07**
90 mm	–	0.05	0.66	–	17.46	m²	**18.12**
110 mm	–	0.05	0.66	–	18.42	m²	**19.08**
160 mm	–	0.05	0.66	–	20.79	m²	**21.45**
KN90; compressive strength of 90 kN/m²; board thickness; typically for commercial							
60 mm	–	0.06	0.80	–	18.04	m²	**18.84**
90 mm	–	0.06	0.80	–	19.75	m²	**20.55**
110 mm	–	0.06	0.80	–	21.40	m²	**22.20**
160 mm	–	0.06	0.80	–	22.64	m²	**23.44**
600 mm voidpack pipe							
36 mm dia.	–	0.05	0.66	–	7.01	nr	**7.67**
Sides of ground beams and edges of beds; basic finish; 18 mm thick external quality plywood; basic finish; four uses							
Plain vertical							
height not exceeding 250 mm	–	0.41	7.60	–	1.82	m	**9.42**
height 250–500 mm	–	0.75	13.71	–	3.51	m	**17.22**
height 500 mm–1.00 m	–	1.04	19.16	–	4.48	m	**23.64**
height exceeding 1.00 m	–	1.38	25.27	–	4.48	m²	**29.75**
Edges of suspended slabs; basic finish; 18 mm thick external quality plywood; basic finish; four uses							
Plain vertical							
height not exceeding 250 mm	–	0.62	11.39	–	1.88	m	**13.27**
height 250–500 mm	–	0.92	16.84	–	2.97	m	**19.81**
height 500 mm–1.00 m	–	1.46	26.75	–	4.55	m	**31.30**
Sides of upstands; basic finish; 18 mm thick external quality plywood; basic finish; four uses							
Plain vertical							
height not exceeding 250 mm	–	0.52	9.58	–	1.94	m	**11.52**
height 250–500 mm	–	0.84	15.36	–	3.63	m	**18.99**
height 500 mm–1.00 m	–	1.46	26.75	–	5.42	m	**32.17**
height exceeding 1.00 m	–	1.67	30.56	–	5.42	m²	**35.98**

11 IN SITU CONCRETE WORKS

Item	PC £	Labour hours	Labour £	Plant £	Material £	Unit	Total rate £
FORMWORK – cont							
Steps in top surfaces; basic finish; 18 mm thick external quality plywood; basic finish; four uses							
Plain vertical							
height not exceeding 250 mm	–	0.41	7.60	–	1.97	m	**9.57**
height 250–500 mm	–	0.67	12.22	–	3.66	m	**15.88**
Steps in soffits; basic finish; 18 mm thick external quality plywood; basic finish; four uses							
Plain vertical							
height not exceeding 250 mm	–	0.46	8.42	–	1.53	m	**9.95**
height 250–500 mm	–	0.73	13.38	–	2.74	m	**16.12**
Machine bases and plinths; basic finish; 18 mm thick external quality plywood; basic finish; four uses							
Plain vertical							
height not exceeding 250 mm	–	0.41	7.60	–	1.82	m	**9.42**
height 250–500 mm	–	0.71	13.05	–	3.51	m	**16.56**
height 500 mm–1.00 m	–	1.04	19.16	–	4.48	m	**23.64**
height exceeding 1.00 m	–	1.33	24.44	–	4.48	m^2	**28.92**
Soffits of slabs; basic finish; 18 mm thick external quality plywood; basic finish; four uses							
Slab thickness not exceeding 200 mm							
horizontal; height to soffit not exceeding 1.50 m	–	1.50	27.58	–	4.14	m^2	**31.72**
horizontal; height to soffit 1.50–2.40 m	–	1.46	26.75	–	4.20	m^2	**30.95**
horizontal; height to soffit 2.40–2.70 m	–	1.38	25.27	–	3.49	m^2	**28.76**
horizontal; height to soffit 2.70–3.00 m	–	1.33	24.44	–	3.01	m^2	**27.45**
horizontal; height to soffit 3.00–4.50 m	–	1.41	25.92	–	4.36	m^2	**30.28**
horizontal; height to soffit 4.50–6.00 m	–	1.50	27.58	–	4.52	m^2	**32.10**
Slab thickness 200–300 mm							
horizontal; height to soffit 1.50–3.00 m	–	1.50	27.58	–	5.32	m^2	**32.90**
Slab thickness 300–400 mm							
horizontal; height to soffit 1.50–3.00 m	–	1.54	28.24	–	5.88	m^2	**34.12**
Slab thickness 400–500 mm							
horizontal; height to soffit 1.50–3.00 m	–	1.62	29.73	–	6.44	m^2	**36.17**
Slab thickness 500–600 mm							
horizontal; height to soffit 1.50–3.00 m	–	1.75	32.04	–	6.44	m^2	**38.48**
Extra over soffits of slabs for							
sloping not exceeding 15°	–	0.17	3.14	–	–	m^2	**3.14**
sloping exceeding 15°	–	0.33	6.11	–	–	m^2	**6.11**
Soffits of landings; basic finish; 18 mm thick external quality plywood; basic finish; four uses							
Slab thickness not exceeding 200 mm							
horizontal; height to soffit 1.50–3.00 m	–	1.50	27.58	–	4.42	m^2	**32.00**

11 IN SITU CONCRETE WORKS

Item	PC £	Labour hours	Labour £	Plant £	Material £	Unit	Total rate £
Slab thickness 200–300 mm							
horizontal; height to soffit 1.50–3.00 m	–	1.58	29.06	–	5.66	m²	**34.72**
Slab thickness 300–400 mm							
horizontal; height to soffit 1.50–3.00 m	–	1.62	29.73	–	6.27	m²	**36.00**
Slab thickness 400–500 mm							
horizontal; height to soffit 1.50–3.00 m	–	1.71	31.38	–	6.89	m²	**38.27**
Slab thickness 500–600 mm							
horizontal; height to soffit 1.50–3.00 m	–	1.84	33.69	–	6.89	m²	**40.58**
Extra over soffits of landings for							
sloping not exceeding 15°	–	0.17	3.14	–	–	m²	**3.14**
sloping exceeding 15°	–	0.33	6.11	–	–	m²	**6.11**
Top formwork; basic finish; 18 mm thick external quality plywood; basic finish; four uses							
Sloping exceeding 15°	–	1.25	22.95	–	3.23	m²	**26.18**
Walls; basic finish; 18 mm thick external quality plywood; basic finish; four uses							
Vertical	–	1.50	27.58	–	5.32	m²	**32.90**
Vertical; height exceeding 3.00 m above floor level	–	1.84	33.69	–	5.48	m²	**39.17**
Vertical; interrupted	–	1.75	32.04	–	5.48	m²	**37.52**
Vertical; to one side only	–	2.92	53.51	–	6.76	m²	**60.27**
Battered	–	2.33	42.78	–	5.76	m²	**48.54**
Beams; basic finish; 18 mm thick external quality plywood; basic finish; four uses							
Attached to slabs							
regular shaped; square or rectangular; height to soffit 1.50–3.00 m	–	1.84	33.69	–	5.12	m²	**38.81**
regular shaped; square or rectangular; height to soffit 3.00–4.50 m	–	1.92	35.18	–	5.32	m²	**40.50**
regular shaped; square or rectangular; height to soffit 4.50–6.00 m	–	2.00	36.67	–	5.48	m²	**42.15**
Attached to walls							
regular shaped; square or rectangular; height to soffit 1.50–3.00 m	–	1.92	35.18	–	5.12	m²	**40.30**
Isolated							
regular shaped; square or rectangular; height to soffit 1.50–3.00 m	–	2.00	36.67	–	5.12	m²	**41.79**
regular shaped; square or rectangular; height to soffit 3.00–4.50 m	–	2.08	38.15	–	5.32	m²	**43.47**
regular shaped; square or rectangular; height to soffit 4.50–6.00 m	–	2.17	39.81	–	5.48	m²	**45.29**
Extra over beams for							
regular shaped; sloping not exceeding 15°	–	0.25	4.62	–	0.53	m²	**5.15**
regular shaped; sloping exceeding 15°	–	0.50	9.25	–	1.06	m²	**10.31**
Beam casings; basic finish; 18 mm thick external quality plywood; basic finish; four uses							
Attached to slabs							
regular shaped; square or rectangular; height to soffit 1.50–3.00 m	–	1.92	35.18	–	5.12	m²	**40.30**
regular shaped; square or rectangular; height to soffit 3.00–4.50 m	–	2.00	36.67	–	5.32	m²	**41.99**

11 IN SITU CONCRETE WORKS

Item	PC £	Labour hours	Labour £	Plant £	Material £	Unit	Total rate £
FORMWORK – cont							
Beam casings – cont							
Attached to walls							
regular shaped; square or rectangular; height to soffit 1.50–3.00 m	–	2.00	36.67	–	5.12	m²	**41.79**
Isolated							
regular shaped; square or rectangular; height to soffit 1.50–3.00 m	–	2.08	38.15	–	5.12	m²	**43.27**
regular shaped; square or rectangular; height to soffit 3.00–4.50 m	–	2.17	39.81	–	5.32	m²	**45.13**
Extra over beam casings for							
regular shaped; sloping not exceeding 15°	–	0.25	4.62	–	0.53	m²	**5.15**
regular shaped; sloping exceeding 15°	–	0.50	9.25	–	1.06	m²	**10.31**
Columns; basic finish; 18 mm thick external quality plywood; basic finish; four uses							
Attached to walls							
regular shaped; square or rectangular; height to soffit 1.50–3.00 m	–	1.84	33.69	–	4.52	m²	**38.21**
Isolated							
regular shaped; square or rectangular; height to soffit 1.50–3.00 m	–	1.92	35.18	–	4.52	m²	**39.70**
regular shaped; circular; not exceeding 300 mm dia.; height to soffit 1.50–3.00 m	–	3.33	61.11	–	7.25	m²	**68.36**
regular shaped; circular; 300–600 mm dia.; height to soffit 1.50–3.00 m	–	3.12	57.31	–	6.44	m²	**63.75**
regular shaped; circular; 600–900 mm dia.; height to soffit 1.50–3.00 m	–	2.92	53.51	–	6.28	m²	**59.79**
Column casings; basic finish; 18 mm thick external quality plywood; basic finish; four uses							
Attached to walls							
regular shaped; square or rectangular; height to soffit 1.50–3.00 m	–	1.92	35.18	–	4.52	m²	**39.70**
Isolated							
regular shaped; square or rectangular; height to soffit 1.50–3.00 m	–	2.00	36.67	–	4.52	m²	**41.19**
Special shapes and finishes							
Recesses or rebates							
12 × 12 mm	–	0.05	0.99	–	0.33	m	**1.32**
25 × 25 mm	–	0.05	0.99	–	0.69	m	**1.68**
25 × 50 mm	–	0.05	0.99	–	0.90	m	**1.89**
50 × 50 mm	–	0.05	0.99	–	0.86	m	**1.85**
Nibs							
50 × 50 mm	–	0.46	8.42	–	0.59	m	**9.01**
100 × 100 mm	–	0.65	11.89	–	0.77	m	**12.66**
100 × 200 mm	–	0.86	15.86	–	11.72	m	**27.58**

11 IN SITU CONCRETE WORKS

Item	PC £	Labour hours	Labour £	Plant £	Material £	Unit	Total rate £
Extra over a basic finish for fine formed finishes							
Slabs	–	0.27	4.96	–	–	m²	**4.96**
Walls	–	0.27	4.96	–	–	m²	**4.96**
Beams	–	0.27	4.96	–	–	m²	**4.96**
Columns	–	0.27	4.96	–	–	m²	**4.96**
Add to prices for basic formwork for							
Curved radius 6.00 m	–	–	–	–	–	%	**52.50**
Curved radius 2.00 m	–	–	–	–	–	%	**105.00**
Coating with retardant agent	–	0.01	0.17	–	0.24	m²	**0.41**
Wall kickers; basic finish							
Height 150 mm	–	0.41	7.60	–	1.31	m	**8.91**
Height 225 mm	–	0.54	9.91	–	1.61	m	**11.52**
Suspended wall kickers; basic finish							
Height 150 mm	–	0.52	9.58	–	1.24	m	**10.82**
Wall ends, soffits and steps in walls; basic finish							
Plain							
width exceeding 1.00 m	–	1.58	29.06	–	5.32	m²	**34.38**
width not exceeding 250 mm	–	0.50	9.25	–	1.37	m	**10.62**
width 250–500 mm	–	0.79	14.53	–	2.98	m	**17.51**
width 500 mm–1.00 m	–	1.25	22.95	–	5.32	m	**28.27**
Openings in walls							
Plain							
width exceeding 1.00 m	–	1.75	32.04	–	5.32	m²	**37.36**
width not exceeding 250 mm	–	0.54	9.91	–	1.37	m	**11.28**
width 250–500 mm	–	0.92	16.84	–	2.98	m	**19.82**
width 500 mm–1.00 m	–	1.41	25.92	–	5.32	m	**31.24**
Stairflights							
Width 1.00 m; 150 mm waist; 150 mm undercut risers							
string, width 300 mm	–	4.17	76.47	–	11.87	m	**88.34**
Width 2.00 m; 200 mm waist; 150 mm undercut risers							
string, width 350 mm	–	7.50	137.57	–	62.09	m	**199.66**
Mortices							
Girth not exceeding 500 mm							
depth not exceeding 250 mm; circular	–	0.13	2.31	–	0.44	nr	**2.75**
Holes							
Girth not exceeding 500 mm							
depth not exceeding 250 mm; circular	–	0.17	3.14	–	0.59	nr	**3.73**
depth 250–500 mm; circular	–	0.25	4.62	–	1.39	nr	**6.01**
Girth 500 mm–1.00 m							
depth not exceeding 250 mm; circular	–	0.21	3.80	–	0.99	nr	**4.79**
depth 250–500 mm; circular	–	0.32	5.79	–	2.57	nr	**8.36**
Girth 1.00–2.00 m							
depth not exceeding 250 mm; circular	–	0.38	6.94	–	2.57	nr	**9.51**
depth 250–500 mm; circular	–	0.56	10.24	–	5.43	nr	**15.67**

Prices for Measured Works

11 IN SITU CONCRETE WORKS

Item	PC £	Labour hours	Labour £	Plant £	Material £	Unit	Total rate £
FORMWORK – cont							
Holes – cont							
Girth 2.00–3.00 m							
depth not exceeding 250 mm; circular	–	0.50	9.25	–	5.22	nr	**14.47**
depth 250–500 mm; circular	–	0.75	13.71	–	81.86	nr	**95.57**
REINFORCEMENT							
Bars; BS 4449; hot rolled deformed high steel bars; grade 500C							
40 mm dia. nominal size							
straight	–	14.44	265.03	–	666.63	tonne	**931.66**
bent	–	17.50	321.19	–	715.07	tonne	**1036.26**
32 mm dia. nominal size							
straight	–	15.34	281.55	–	672.23	tonne	**953.78**
bent	–	18.95	347.85	–	720.67	tonne	**1068.52**
25 mm dia. nominal size							
straight	–	16.25	298.16	–	681.07	tonne	**979.23**
bent	–	18.95	347.85	–	729.51	tonne	**1077.36**
20 mm dia. nominal size							
straight	–	16.25	298.16	–	699.66	tonne	**997.82**
bent	–	18.95	347.85	–	748.09	tonne	**1095.94**
16 mm dia. nominal size							
straight	–	19.86	364.41	–	721.50	tonne	**1085.91**
bent	–	22.56	414.11	–	769.92	tonne	**1184.03**
12 mm dia. nominal size							
straight	–	21.66	397.54	–	748.92	tonne	**1146.46**
bent	–	24.37	447.24	–	797.36	tonne	**1244.60**
10 mm dia. nominal size							
straight	–	23.46	430.67	–	792.59	tonne	**1223.26**
bent	–	26.17	480.36	–	841.03	tonne	**1321.39**
8 mm dia. nominal size							
straight	–	24.37	447.24	–	820.02	tonne	**1267.26**
links	–	27.07	496.93	–	890.84	tonne	**1387.77**
bent	–	27.07	496.93	–	868.46	tonne	**1365.39**
Bars; stainless steel; to EN 1.4301							
40 mm dia. nominal size							
straight	–	17.00	312.02	–	2991.74	tonne	**3303.76**
bent	–	21.00	379.76	–	3174.24	tonne	**3554.00**
32 mm dia. nominal size							
straight	–	17.00	312.02	–	2991.74	tonne	**3303.76**
bent	–	21.00	379.76	–	3174.24	tonne	**3554.00**
25 mm dia. nominal size							
straight	–	18.00	330.37	–	2991.91	tonne	**3322.28**
bent	–	18.00	330.37	–	3174.41	tonne	**3504.78**

11 IN SITU CONCRETE WORKS

Item	PC £	Labour hours	Labour £	Plant £	Material £	Unit	Total rate £
20 mm dia. nominal size							
straight	–	20.00	367.08	–	3007.98	tonne	**3375.06**
bent	–	20.00	367.08	–	3190.48	tonne	**3557.56**
16 mm dia. nominal size							
straight	–	22.00	403.79	–	3030.37	tonne	**3434.16**
bent	–	22.00	403.79	–	3212.87	tonne	**3616.66**
12 mm dia. nominal size							
straight	–	24.00	440.50	–	3052.76	tonne	**3493.26**
bent	–	24.00	440.50	–	3235.27	tonne	**3675.77**
10 mm dia. nominal size							
straight	–	26.00	477.19	–	3091.40	tonne	**3568.59**
bent	–	26.00	477.19	–	3273.90	tonne	**3751.09**
8 mm dia. nominal size							
straight	–	28.00	511.08	–	3113.79	tonne	**3624.87**
bent	–	28.00	511.08	–	3296.30	tonne	**3807.38**
Bars; stainless steel; to EN 1.4462							
40 mm dia. nominal size							
straight	–	17.00	312.02	–	4041.15	tonne	**4353.17**
bent	–	21.00	379.76	–	4269.28	tonne	**4649.04**
32 mm dia. nominal size							
straight	–	17.00	312.02	–	4041.15	tonne	**4353.17**
bent	–	21.00	379.76	–	4269.28	tonne	**4649.04**
25 mm dia. nominal size							
straight	–	18.00	330.37	–	3904.44	tonne	**4234.81**
bent	–	18.00	330.37	–	4132.57	tonne	**4462.94**
20 mm dia. nominal size							
straight	–	20.00	367.08	–	3920.50	tonne	**4287.58**
bent	–	20.00	367.08	–	4148.63	tonne	**4515.71**
16 mm dia. nominal size							
straight	–	22.00	403.79	–	3942.90	tonne	**4346.69**
bent	–	22.00	403.79	–	4171.03	tonne	**4574.82**
12 mm dia. nominal size							
straight	–	24.00	440.50	–	3965.29	tonne	**4405.79**
bent	–	24.00	440.50	–	4147.79	tonne	**4588.29**
10 mm dia. nominal size							
straight	–	26.00	477.19	–	4232.06	tonne	**4709.25**
bent	–	26.00	477.19	–	4368.93	tonne	**4846.12**
8 mm dia. nominal size							
straight	–	28.00	511.08	–	4254.45	tonne	**4765.53**
bent	–	28.00	511.08	–	4391.32	tonne	**4902.40**
Bars; stainless steel; to EN 1.4362 (low nickel alloys or Lean Duplexes)							
40 mm dia. nominal size							
straight	–	17.00	312.02	–	2854.87	tonne	**3166.89**
bent	–	21.00	379.76	–	3037.37	tonne	**3417.13**
32 mm dia. nominal size							
straight	–	17.00	312.02	–	2854.87	tonne	**3166.89**
bent	–	21.00	379.76	–	3037.37	tonne	**3417.13**

11 IN SITU CONCRETE WORKS

Item	PC £	Labour hours	Labour £	Plant £	Material £	Unit	Total rate £
REINFORCEMENT – cont							
Bars – cont							
25 mm dia. nominal size							
straight	–	18.00	330.37	–	2855.03	tonne	**3185.40**
bent	–	18.00	330.37	–	3037.55	tonne	**3367.92**
20 mm dia. nominal size							
straight	–	20.00	367.08	–	2871.10	tonne	**3238.18**
bent	–	20.00	367.08	–	3053.61	tonne	**3420.69**
16 mm dia. nominal size							
straight	–	22.00	403.79	–	2893.50	tonne	**3297.29**
bent	–	22.00	403.79	–	3076.01	tonne	**3479.80**
12 mm dia. nominal size							
straight	–	24.00	440.50	–	2915.89	tonne	**3356.39**
bent	–	24.00	440.50	–	3098.39	tonne	**3538.89**
10 mm dia. nominal size							
straight	–	26.00	477.19	–	2954.52	tonne	**3431.71**
bent	–	26.00	477.19	–	3137.02	tonne	**3614.21**
8 mm dia. nominal size							
straight	–	28.00	511.08	–	2976.92	tonne	**3488.00**
bent	–	28.00	511.08	–	3159.42	tonne	**3670.50**
Bars; stainless steel; to EN 1.4162 (low nickel alloys or Lean Duplexes)							
40 mm dia. nominal size							
straight	–	17.00	312.02	–	2717.98	tonne	**3030.00**
bent	–	21.00	379.76	–	2854.87	tonne	**3234.63**
32 mm dia. nominal size							
straight	–	17.00	312.02	–	2717.98	tonne	**3030.00**
bent	–	21.00	379.76	–	2854.87	tonne	**3234.63**
25 mm dia. nominal size							
straight	–	18.00	330.37	–	2718.16	tonne	**3048.53**
bent	–	18.00	330.37	–	2855.03	tonne	**3185.40**
20 mm dia. nominal size							
straight	–	20.00	367.08	–	2734.22	tonne	**3101.30**
bent	–	20.00	367.08	–	2871.10	tonne	**3238.18**
16 mm dia. nominal size							
straight	–	22.00	403.79	–	2756.62	tonne	**3160.41**
bent	–	22.00	403.79	–	2893.50	tonne	**3297.29**
12 mm dia. nominal size							
straight	–	24.00	440.50	–	2779.00	tonne	**3219.50**
bent	–	24.00	440.50	–	2915.89	tonne	**3356.39**
10 mm dia. nominal size							
straight	–	26.00	477.19	–	2817.63	tonne	**3294.82**
bent	–	26.00	477.19	–	2954.52	tonne	**3431.71**
8 mm dia. nominal size							
straight	–	28.00	511.08	–	2840.03	tonne	**3351.11**
bent	–	28.00	511.08	–	2976.92	tonne	**3488.00**

11 IN SITU CONCRETE WORKS

Item	PC £	Labour hours	Labour £	Plant £	Material £	Unit	Total rate £
Fabric; to BS4483 in standard 4.8 m × 2.4 m Sheets.							
Ref D49 (0.77 kg/m^2)							
100 mm minimum laps; bent	–	0.24	4.41	–	1.50	m^2	**5.91**
Ref D98 (1.54 kg/m^2)							
400 mm minimum laps	–	0.12	2.21	–	1.02	m^2	**3.23**
strips in one width; 600 mm width	–	0.15	2.75	–	1.02	m^2	**3.77**
strips in one width; 900 mm width	–	0.14	2.57	–	1.02	m^2	**3.59**
strips in one width; 1200 mm width	–	0.13	2.38	–	1.02	m^2	**3.40**
Ref A142 (2.22 kg/m^2)							
400 mm minimum laps	–	0.12	2.21	–	1.16	m^2	**3.37**
strips in one width; 600 mm width	–	0.15	2.75	–	1.16	m^2	**3.91**
strips in one width; 900 mm width	–	0.14	2.57	–	1.16	m^2	**3.73**
strips in one width; 1200 mm width	–	0.13	2.38	–	1.16	m^2	**3.54**
Ref A193 (3.02 kg/m^2)							
400 mm minimum laps	–	0.12	2.21	–	1.56	m^2	**3.77**
strips in one width; 600 mm width	–	0.15	2.75	–	1.56	m^2	**4.31**
strips in one width; 900 mm width	–	0.14	2.57	–	1.56	m^2	**4.13**
strips in one width; 1200 mm width	–	0.13	2.38	–	1.56	m^2	**3.94**
Ref A252 (3.95 kg/m^2)							
400 mm minimum laps	–	0.13	2.38	–	2.02	m^2	**4.40**
strips in one width; 600 mm width	–	0.16	2.94	–	2.02	m^2	**4.96**
strips in one width; 900 mm width	–	0.15	2.75	–	2.02	m^2	**4.77**
strips in one width; 1200 mm width	–	0.14	2.57	–	2.02	m^2	**4.59**
Ref A393 (6.16 kg/m^2)							
400 mm minimum laps	–	0.15	2.75	–	3.14	m^2	**5.89**
strips in one width; 600 mm width	–	0.18	3.31	–	3.14	m^2	**6.45**
strips in one width; 900 mm width	–	0.17	3.12	–	3.14	m^2	**6.26**
strips in one width; 1200 mm width	–	0.16	2.94	–	3.14	m^2	**6.08**
Ref B196 (3.05 kg/m^2)							
400 mm minimum laps	–	0.12	2.21	–	2.67	m^2	**4.88**
strips in one width; 600 mm width	–	0.15	2.75	–	2.67	m^2	**5.42**
strips in one width; 900 mm width	–	0.14	2.57	–	2.67	m^2	**5.24**
strips in one width; 1200 mm width	–	0.13	2.38	–	2.67	m^2	**5.05**
Ref B283 (3.73 kg/m^2)							
400 mm minimum laps	–	0.12	2.21	–	2.02	m^2	**4.23**
strips in one width; 600 mm width	–	0.15	2.75	–	2.02	m^2	**4.77**
strips in one width; 900 mm width	–	0.14	2.57	–	2.02	m^2	**4.59**
strips in one width; 1200 mm width	–	0.13	2.38	–	2.02	m^2	**4.40**
Ref B385 (4.53 kg/m^2)							
400 mm minimum laps	–	0.13	2.38	–	2.46	m^2	**4.84**
strips in one width; 600 mm width	–	0.16	2.94	–	2.46	m^2	**5.40**
strips in one width; 900 mm width	–	0.15	2.75	–	2.46	m^2	**5.21**
strips in one width; 1200 mm width	–	0.14	2.57	–	2.46	m^2	**5.03**
Ref B503 (5.93 kg/m^2)							
400 mm minimum laps	–	0.15	2.75	–	3.14	m^2	**5.89**
strips in one width; 600 mm width	–	0.18	3.31	–	3.14	m^2	**6.45**
strips in one width; 900 mm width	–	0.17	3.12	–	3.14	m^2	**6.26**
strips in one width; 1200 mm width	–	0.16	2.94	–	3.14	m^2	**6.08**

11 IN SITU CONCRETE WORKS

Item	PC £	Labour hours	Labour £	Plant £	Material £	Unit	Total rate £
REINFORCEMENT – cont							
Fabric – cont							
Ref B785 (8.14 kg/m^2)							
400 mm minimum laps	–	0.17	3.12	–	4.32	m^2	**7.44**
strips in one width; 600 mm width	–	0.20	3.67	–	4.32	m^2	**7.99**
strips in one width; 900 mm width	–	0.19	3.49	–	4.32	m^2	**7.81**
strips in one width; 1200 mm width	–	0.18	3.31	–	4.32	m^2	**7.63**
Ref B1131 (10.90 kg/m^2)							
400 mm minimum laps	–	0.18	3.31	–	5.78	m^2	**9.09**
strips in one width; 600 mm width	–	0.24	4.41	–	5.78	m^2	**10.19**
strips in one width; 900 mm width	–	0.22	4.04	–	5.78	m^2	**9.82**
strips in one width; 1200 mm width	–	0.20	3.67	–	5.78	m^2	**9.45**
DESIGNED JOINTS							
Formed; Fosroc impregnated fibreboard joint filler or other equal							
Width not exceeding 150 mm							
12.50 mm thick	–	0.14	2.57	–	1.55	m	**4.12**
20 mm thick	–	0.19	3.49	–	4.12	m	**7.61**
25 mm thick	–	0.23	4.22	–	2.69	m	**6.91**
Width 150–300 mm							
12.50 mm thick	–	0.23	4.22	–	2.32	m	**6.54**
20 mm thick	–	0.23	4.22	–	4.11	m	**8.33**
25 mm thick	–	0.23	4.22	–	4.88	m	**9.10**
Width 300–450 mm							
12.50 mm thick	–	0.28	5.13	–	3.23	m	**8.36**
20 mm thick	–	0.28	5.13	–	5.69	m	**10.82**
25 mm thick	–	0.28	5.13	–	6.76	m	**11.89**
Formed; waterproof bonded cork joint filler board							
Width not exceeding 150 mm							
10 mm thick	–	0.14	2.57	–	3.24	m	**5.81**
13 mm thick	–	0.14	2.57	–	3.29	m	**5.86**
19 mm thick	–	0.14	2.57	–	4.23	m	**6.80**
25 mm thick	–	0.14	2.57	–	4.83	m	**7.40**
Width 150–300 mm							
10 mm thick	–	0.19	3.49	–	5.76	m	**9.25**
13 mm thick	–	0.19	3.49	–	5.86	m	**9.35**
19 mm thick	–	0.19	3.49	–	7.75	m	**11.24**
25 mm thick	–	0.19	3.49	–	8.94	m	**12.43**
Width 300–450 mm							
10 mm thick	–	0.23	4.22	–	8.88	m	**13.10**
13 mm thick	–	0.23	4.22	–	9.02	m	**13.24**
19 mm thick	–	0.23	4.22	–	11.87	m	**16.09**
25 mm thick	–	0.23	4.22	–	13.63	m	**17.85**

11 IN SITU CONCRETE WORKS

Item	PC £	Labour hours	Labour £	Plant £	Material £	Unit	Total rate £
Sealants; Fosroc Pliastic 77 hot poured							
rubberized bituminous compound or other equal							
Width 10 mm							
25 mm depth	–	0.17	3.12	–	0.79	m	**3.91**
Width 12.50 mm							
25 mm depth	–	0.18	3.31	–	0.98	m	**4.29**
Width 20 mm							
25 mm depth	–	0.19	3.49	–	1.59	m	**5.08**
Width 25 mm							
25 mm depth	–	0.20	3.67	–	1.94	m	**5.61**
Sealants; Fosroc Thioflex 600 gun grade two part							
polysulphide or other equal							
Width 10 mm							
25 mm depth	–	0.05	0.91	–	3.78	m	**4.69**
Width 12.50 mm							
25 mm depth	–	0.06	1.10	–	4.74	m	**5.84**
Width 20 mm							
25 mm depth	–	0.07	1.28	–	7.57	m	**8.85**
Width 25 mm							
25 mm depth	–	0.08	1.47	–	9.46	m	**10.93**
Sealants; Grace Servicised Paraseal							
polysulphide compound or other equal; priming							
with Grace Servicised Primer P							
Width 10 mm							
25 mm depth	–	0.19	2.95	–	3.58	m	**6.53**
Width 13 mm							
25 mm depth	–	0.19	2.95	–	4.59	m	**7.54**
Width 19 mm							
25 mm depth	–	0.23	3.57	–	6.62	m	**10.19**
Width 25 mm							
25 mm depth	–	0.23	3.57	–	8.63	m	**12.20**
Waterstops; Grace Servicised or other equal							
Hydrophilic strip water stop; lapped joints; cast into							
concrete							
50 × 20 mm Adcor 500S	5.86	0.30	4.66	–	8.02	m	**12.68**
Servitite Internal 10 mm thick PVC water stop; flat							
dumbbell type; heat welded joints; cast into concrete							
Servitite 150; 150 mm wide	–	0.23	4.22	–	11.63	m	**15.85**
flat angle	–	0.28	5.13	–	27.69	nr	**32.82**
vertical angle	–	0.28	5.13	–	27.55	nr	**32.68**
flat three way intersection	–	0.37	6.79	–	36.79	nr	**43.58**
vertical three way intersection	–	0.37	6.79	–	45.50	nr	**52.29**
four way intersection	–	0.46	8.44	–	50.05	nr	**58.49**
Servitite 230; 230 mm wide	–	0.23	4.22	–	16.78	m	**21.00**
flat angle	–	0.28	5.13	–	34.01	nr	**39.14**
vertical angle	–	0.28	5.13	–	40.12	nr	**45.25**
flat three way intersection	–	0.37	6.79	–	49.70	nr	**56.49**
vertical three way intersection	–	0.37	6.79	–	85.58	nr	**92.37**

11 IN SITU CONCRETE WORKS

Item	PC £	Labour hours	Labour £	Plant £	Material £	Unit	Total rate £
DESIGNED JOINTS – cont							
Waterstops – cont							
Servitite Internal 10 mm thick PVC water stop – cont							
four way intersection	–	0.46	8.44	–	62.64	nr	**71.08**
Servitite AT200; 200 mm wide	–	0.23	4.22	–	18.51	m	**22.73**
flat angle	–	0.28	5.13	–	33.51	nr	**38.64**
vertical angle	–	0.28	5.13	–	36.23	nr	**41.36**
flat three way intersection	–	0.37	6.79	–	57.82	nr	**64.61**
vertical three way intersection	–	0.37	6.79	–	45.71	nr	**52.50**
four way intersection	–	0.46	8.44	–	68.83	nr	**77.27**
Servitite K305; 305 mm wide	–	0.28	5.13	–	27.29	m	**32.42**
flat angle	–	0.32	5.87	–	58.25	nr	**64.12**
vertical angle	–	0.32	5.87	–	62.70	nr	**68.57**
flat three way intersection	–	0.42	7.71	–	83.17	nr	**90.88**
vertical three way intersection	–	0.42	7.71	–	97.08	nr	**104.79**
four way intersection	–	0.51	9.36	–	112.62	nr	**121.98**
Serviseal External PVC water stop; PVC water stop; centre bulb type; heat welded joints; cast into concrete							
Serviseal 195; 195 mm wide	–	0.23	4.22	–	7.72	m	**11.94**
flat angle	–	0.28	5.13	–	17.39	nr	**22.52**
vertical angle	–	0.28	5.13	–	28.72	nr	**33.85**
flat three way intersection	–	0.37	6.79	–	29.17	nr	**35.96**
four way intersection	–	0.46	8.44	–	43.14	nr	**51.58**
Serviseal 240; 240 mm wide	–	0.23	4.22	–	9.65	m	**13.87**
flat angle	–	0.28	5.13	–	20.13	nr	**25.26**
vertical angle	–	0.28	5.13	–	30.67	nr	**35.80**
flat three way intersection	–	0.37	6.79	–	59.30	nr	**66.09**
four way intersection	–	0.46	8.44	–	48.63	nr	**57.07**
Serviseal AT240; 240 mm wide	–	0.23	4.22	–	21.73	m	**25.95**
flat angle	–	0.28	5.13	–	32.95	nr	**38.08**
vertical angle	–	0.28	5.13	–	31.62	nr	**36.75**
flat three way intersection	–	0.37	6.79	–	50.52	nr	**57.31**
four way intersection	–	0.46	8.44	–	73.76	nr	**82.20**
Serviseal K320; 320 mm wide	–	0.28	5.13	–	12.75	m	**17.88**
flat angle	–	0.32	5.87	–	42.85	nr	**48.72**
vertical angle	–	0.32	5.87	–	23.11	nr	**28.98**
flat three way intersection	–	0.42	7.71	–	62.98	nr	**70.69**
four way intersection	–	0.51	9.36	–	79.24	nr	**88.60**
ACCESSORIES CAST IN							
Foundation bolt boxes							
Temporary plywood; for group of 4 nr bolts							
75 × 75 × 150 mm	–	0.42	7.71	–	1.01	nr	**8.72**
75 × 75 × 250 mm	–	0.42	7.71	–	1.23	nr	**8.94**

11 IN SITU CONCRETE WORKS

Item	PC £	Labour hours	Labour £	Plant £	Material £	Unit	Total rate £
Expanded metal; Expamet Building Products Ltd or other equal and approved							
75 mm dia. × 150 mm long	–	0.28	5.13	–	1.11	nr	**6.24**
75 mm dia. × 300 mm long	–	0.28	5.13	–	1.39	nr	**6.52**
100 mm dia. × 450 mm long	–	0.28	5.13	–	2.40	nr	**7.53**
Foundation bolts and nuts							
Black hexagon							
10 mm dia. × 100 mm long	–	0.23	4.22	–	0.54	nr	**4.76**
12 mm dia. × 120 mm long	–	0.23	4.22	–	0.82	nr	**5.04**
16 mm dia. × 160 mm long	–	0.28	5.13	–	2.27	nr	**7.40**
20 mm dia. × 180 mm long	–	0.28	5.13	–	2.65	nr	**7.78**
Masonry slots							
Stainless steel; dovetail slots; 1.20 mm thick; 18G							
1000 mm long	–	0.25	4.59	–	6.67	m	**11.26**
100 mm long	–	0.07	1.28	–	0.30	nr	**1.58**
Stainless steel; metal insert slots; Halfen Ltd Ribslot or other equal and approved; 2.50 mm thick; end caps and foam filling							
41 × 41 mm; ref P3270	–	0.37	6.79	–	9.03	m	**15.82**
41 × 41 × 100 mm; ref P3250	–	0.09	1.65	–	1.35	nr	**3.00**
41 × 41 × 150 mm; ref P3251	–	0.09	1.65	–	0.99	nr	**2.64**
Cramps							
Stainless steel; once bent; one end shot fired into concrete; other end fanged and built into brickwork joint							
200 mm girth	–	0.14	2.82	–	1.23	nr	**4.05**
Column guards							
White nylon coated steel; Rigifix or other equal and approved; Huntley and Sparks Ltd; plugging; screwing to concrete; 1.50 mm thick							
75 × 75 × 1000 mm	–	0.74	13.58	–	16.10	nr	**29.68**
Galvanized steel; Rigifix or other equal and approved; Huntley and Sparks Ltd; 3 mm thick							
75 × 75 × 1000 mm	–	0.56	10.28	–	11.29	nr	**21.57**
Galvanized steel; Rigifix or other equal and approved; Huntley and Sparks Ltd; 4.50 mm thick							
75 × 75 × 1000 mm	–	0.56	10.28	–	15.17	nr	**25.45**
Stainless steel; HKW or other equal and approved; Halfen Ltd; 5 mm thick							
50 × 50 × 1200 mm	–	0.93	17.06	–	77.78	nr	**94.84**
50 × 50 × 2000 mm	–	1.11	20.37	–	128.54	nr	**148.91**
Channels							
Stainless steel; Halfen Ltd or other equal and approved							
ref 38/17/HTA	–	0.32	5.87	–	44.96	m	**50.83**
ref 41/22/HZA; 80 mm long; including T headed bolts and plate washers	–	0.09	1.65	–	29.07	nr	**30.72**

11 IN SITU CONCRETE WORKS

Item	PC £	Labour hours	Labour £	Plant £	Material £	Unit	Total rate £
ACCESSORIES CAST IN – cont							
Channel ties							
Stainless steel; Halfen Ltd or other equal and approved							
ref HTS – B12; 150 mm projection; including insulation retainer	–	0.03	0.65	–	0.57	nr	**1.22**
ref HTS – B12; 200 mm projection; including insulation retainer	–	0.03	0.65	–	0.67	nr	**1.32**

12 PRECAST/COMPOSITE CONCRETE

Item	PC £	Labour hours	Labour £	Plant £	Material £	Unit	Total rate £
PRECAST/COMPOSITE CONCRETE WORK							
Prestressed precast concrete structural suspended floors; Bison Hollowcore or other equal; supplied and fixed on hard level bearings, to areas of 500 m² per site visit; top surface screeding and ceiling finishes by others							
Floors to dwellings, offices, car parks, shop retail floors, hospitals, school teaching rooms, staff rooms and the like; superimposed load of 5.00 kN/m²							
floor spans up to 3.00 m; 1200 mm × 150 mm	–	–	–	–	–	m²	**52.40**
floor spans 3.00 m–6.00 m; 1200 mm × 150 mm	–	–	–	–	–	m²	**53.67**
floor spans 6.00 m–7.50 m; 1200 mm × 200 mm	–	–	–	–	–	m²	**57.36**
floor spans 7.50 m–9.50 m; 1200 mm × 250 mm	–	–	–	–	–	m²	**61.05**
floor spans 9.50 m–12.00 m; 1200 mm × 300 mm	–	–	–	–	–	m²	**63.09**
floor spans 12.00 m–12.50 m; 1200 mm × 350 mm	–	–	–	–	–	m²	**65.12**
floor spans 12.50 m–14.00 m; 1200 mm × 400 mm	–	–	–	–	–	m²	**72.24**
floor spans 14.00 m–15.00 m; 1200 mm × 450 mm	–	–	–	–	–	m²	**73.26**
Floors to shop stockrooms, light warehousing, schools, churches or similar places of assembly, light factory accommodation, laboratories and the like; superimposed load of 8.50 kN/m²							
floor spans up to 3.00 m; 1200 mm × 150 mm	–	–	–	–	–	m²	**50.88**
floor spans 3.00 m–6.00 m; 1200 mm × 200 mm	–	–	–	–	–	m²	**52.91**
floor spans 6.00 m–7.50 m; 1200 mm × 250 mm	–	–	–	–	–	m²	**61.05**
Floors to heavy warehousing, factories, stores and the like; superimposed load of 12.50 kN/m²							
floor spans up to 3.00 m; 1200 mm × 150 mm	–	–	–	–	–	m²	**53.93**
floor spans 3.00 m–6.00 m; 1200 mm × 250 mm	–	–	–	–	–	m²	**61.05**
Prestressed precast concrete staircase, supplied and fixed in conjunction with Bison Hollowcore flooring system or similar; comprising 2 nr 1100 mm wide flights with 7 nr 275 mm treads, 8 nr 185 mm risers and 150 mm waist; 1 nr 2200 mm × 1400 mm × 150 mm half landing and 1 nr top landing							
3.00 m storey height	–	–	–	–	–	nr	**2500.00**
Prestressed precast concrete beam and block floor; cement and sand grout brushed between beams and blocks; 440 × 215 × 100 mm concrete block infill							
Beam and block flooring at ground level							
beam and block flooring at ground level 150 mm thick beams; up to 3.30 m span	–	0.45	10.30	–	35.66	m²	**45.96**
beam and block flooring at ground level; 225 mm thick beams; up to 4.20 m span	–	0.45	10.30	–	38.68	m²	**48.98**

12 PRECAST/COMPOSITE CONCRETE

Item	PC £	Labour hours	Labour £	Plant £	Material £	Unit	Total rate £
PRECAST/COMPOSITE CONCRETE WORK – cont							
Composite floor comprising reinforced in situ ready-mixed concrete 30.00 N/mm^2; on and including steel deck permanent shutting; complete with A142 anti-crack mesh; NOTE temporary props may be required, but have not been included in the following rates							
0.9 mm thick re-entrant type deck; 60 mm deep							
150 mm thick suspended slab							
1.50 m–3.00 m high to soffit	–	0.91	14.84	–	50.72	m^2	**65.56**
3.00 m–4.50 m high to soffit	–	0.94	15.29	–	50.72	m^2	**66.01**
4.50 m–6.00 m high to soffit	–	0.98	16.00	–	50.72	m^2	**66.72**
200 mm thick suspended slab							
1.50 m–3.00 m high to soffit	–	0.95	15.46	–	56.15	m^2	**71.61**
3.00 m–4.50 m high to soffit	–	0.97	15.92	–	56.15	m^2	**72.07**
4.50 m–6.00 m high to soffit	–	1.00	16.32	–	56.15	m^2	**72.47**
1.2 mm thick re-entrant type deck							
150 mm thick suspended slab							
1.50 m–3.00 m high to soffit	–	0.91	14.84	–	54.35	m^2	**69.19**
3.00 m–4.50 m high to soffit	–	0.94	15.29	–	54.35	m^2	**69.64**
4.50 m–6.00 m high to soffit	–	0.98	16.00	–	54.35	m^2	**70.35**
200 mm thick suspended slab							
1.50 m–3.00 m high to soffit	–	0.95	15.46	–	59.79	m^2	**75.25**
3.00 m–4.50 m high to soffit	–	0.97	15.92	–	59.79	m^2	**75.71**
4.50 m–6.00 m high to soffit	–	1.00	16.32	–	59.79	m^2	**76.11**
0.9 mm thick trapezoidal deck 60 mm deep							
150 mm thick suspended slab							
1.50 m–3.00 m high to soffit	–	0.91	14.84	–	48.18	m^2	**63.02**
3.00 m–4.50 m high to soffit	–	0.94	15.29	–	48.18	m^2	**63.47**
4.50 m–6.00 m high to soffit	–	0.98	16.00	–	48.18	m^2	**64.18**
200 mm thick suspended slab							
1.50 m–3.00 m high to soffit	–	0.95	15.46	–	53.62	m^2	**69.08**
3.00 m–4.50 m high to soffit	–	0.97	15.92	–	53.62	m^2	**69.54**
4.50 m–6.00 m high to soffit	–	1.00	16.32	–	53.62	m^2	**69.94**
0.9 mm thick trapezoidal deck 80 mm deep							
150 mm thick suspended slab							
1.50 m–3.00 m high to soffit	–	0.91	14.84	–	49.39	m^2	**64.23**
3.00 m–4.50 m high to soffit	–	0.94	15.29	–	49.39	m^2	**64.68**
4.50 m–6.00 m high to soffit	–	0.98	16.00	–	49.39	m^2	**65.39**
200 mm thick suspended slab							
1.50 m–3.00 m high to soffit	–	0.95	15.46	–	54.83	m^2	**70.29**
3.00 m–4.50 m high to soffit	–	0.97	15.92	–	54.83	m^2	**70.75**
4.50 m–6.00 m high to soffit	–	1.00	16.32	–	54.83	m^2	**71.15**

12 PRECAST/COMPOSITE CONCRETE

Item	PC £	Labour hours	Labour £	Plant £	Material £	Unit	Total rate £
1.2 mm thick trapezoidal deck 80 mm deep							
150 mm thick suspended slab							
1.50 m–3.00 m high to soffit	–	0.91	14.84	–	52.70	m²	**67.54**
3.00 m–4.50 m high to soffit	–	0.94	15.29	–	52.70	m²	**67.99**
4.50 m–6.00 m high to soffit	–	0.98	16.00	–	52.70	m²	**68.70**
200 mm thick suspended slab							
1.50 m–3.00 m high to soffit	–	0.95	15.46	–	58.14	m²	**73.60**
3.00 m–4.50 m high to soffit	–	0.97	15.92	–	58.14	m²	**74.06**
4.50 m–6.00 m high to soffit	–	1.00	16.32	–	58.14	m²	**74.46**
Soffits of coffered or troughed slabs; basic finish							
Cordek Correx trough mould or other equal and							
approved; 300 mm deep; ribs of mould at 600 mm							
centres and cross ribs at centres of bay; slab							
thickness 300–400 mm							
horizontal; height to soffit 1.50–3.00 m	–	2.08	38.15	–	10.97	m²	**49.12**
horizontal; height to soffit 3.00–4.50 m	–	2.17	39.81	–	11.14	m²	**50.95**
horizontal; height to soffit 4.50–6.00 m	–	2.25	41.29	–	11.23	m²	**52.52**

13 PRECAST CONCRETE

Item	PC £	Labour hours	Labour £	Plant £	Material £	Unit	Total rate £
PRECAST CONCRETE GOODS							
Contractor designed precast concrete staircases and landings; including all associated steel supports and fixing in position							
Straight staircases; 280 mm treads; 170 mm undercut risers							
1200 mm wide; 2750 mm rise	–	–	–	–	–	nr	**1593.37**
1200 mm wide; 3750 mm rise	–	–	–	–	–	nr	**2124.49**
Dogleg staircases							
1200 mm wide; one full width half landing; 2750 mm rise	–	–	–	–	–	nr	**2443.15**
1200 mm wide; one full width half landing; 3750 mm rise	–	–	–	–	–	nr	**3186.73**
1200 mm wide 200 mm thick concrete landing support walls	–	–	–	–	–	nr	**743.56**
1800 mm wide; one full width half landing; 2750 mm rise	–	–	–	–	–	nr	**3452.29**
1800 mm wide; one full width half landing; 3750 mm rise	–	–	–	–	–	nr	**4514.53**
1800 mm wide 200 mm thick concrete landing support walls	–	–	–	–	–	nr	**1147.22**
Precast concrete sill, lintels, copings							
Lintels; plate; prestressed bedded							
100 mm × 70 mm × 600 mm long	5.92	0.37	7.98	–	6.24	nr	**14.22**
100 mm × 70 mm × 900 mm long	8.84	0.37	7.98	–	9.30	nr	**17.28**
100 mm × 70 mm × 1100 mm long	10.83	0.37	7.98	–	11.39	nr	**19.37**
100 mm × 70 mm × 1200 mm long	11.80	0.37	7.98	–	12.41	nr	**20.39**
100 mm × 70 mm × 1500 mm long	14.76	0.46	9.91	–	15.54	nr	**25.45**
100 mm × 70 mm × 1800 mm long	17.70	0.46	9.91	–	18.62	nr	**28.53**
100 mm × 70 mm × 2100 mm long	20.65	0.56	12.07	–	21.73	nr	**33.80**
140 mm × 70 mm × 1200 mm long	17.35	0.46	9.91	–	18.26	nr	**28.17**
140 mm × 70 mm × 1500 mm long	21.69	0.56	12.07	–	22.83	nr	**34.90**
Lintels; rectangular; reinforced with mild steel bars; bedded							
100 mm × 145 mm × 900 mm long	3.90	0.56	12.07	–	4.12	nr	**16.19**
100 mm × 145 mm × 1050 mm long	4.55	0.56	12.07	–	4.80	nr	**16.87**
100 mm × 145 mm × 1200 mm long	5.20	0.56	12.07	–	5.48	nr	**17.55**
225 mm × 145 mm × 1200 mm long	20.18	0.74	15.95	–	21.24	nr	**37.19**
225 mm × 225 mm × 1800 mm long	30.18	1.39	29.97	–	31.74	nr	**61.71**
Lintels; boot; reinforced with mild steel bars; bedded							
250 mm × 225 mm × 1200 mm long	22.35	1.11	23.93	–	23.52	nr	**47.45**
275 mm × 225 mm × 1800 mm long	36.89	1.67	36.00	–	38.79	nr	**74.79**
Padstones							
100 mm × 200 mm × 440 mm	20.00	0.28	6.04	–	21.02	nr	**27.06**
150 mm × 215 mm × 440 mm	19.99	0.37	7.98	–	21.03	nr	**29.01**
150 mm × 225 mm × 225 mm	15.70	0.56	12.07	–	16.64	nr	**28.71**

13 PRECAST CONCRETE

Item	PC £	Labour hours	Labour £	Plant £	Material £	Unit	**Total rate £**
Copings; once weathered; once throated; bedded and pointed							
152 mm × 76 mm	5.87	0.65	14.02	–	6.24	m	**20.26**
178 mm × 64 mm	6.48	0.65	14.02	–	6.89	m	**20.91**
305 mm × 76 mm	10.95	0.74	15.95	–	11.67	m	**27.62**
extra for fair ends	–	–	–	–	4.40	nr	**4.40**
extra for angles	–	–	–	–	5.00	nr	**5.00**
Copings; twice weathered; twice throated; bedded and pointed							
152 mm × 76 mm	5.87	0.65	14.02	–	6.24	m	**20.26**
178 mm × 64 mm	6.44	0.65	14.02	–	6.84	m	**20.86**
305 mm × 76 mm	10.95	0.74	15.95	–	11.67	m	**27.62**
extra for fair ends	–	–	–	–	4.40	nr	**4.40**
extra for angles	–	–	–	–	5.00	nr	**5.00**
Sills; splayed top edge, stooled ends; bedded and pointed							
200 mm × 90 mm	26.38	0.75	16.17	–	27.77	m	**43.94**
200 mm × 90 mm; slip sill	29.58	0.75	16.17	–	31.13	m	**47.30**

14 MASONRY

Item	PC £	Labour hours	Labour £	Plant £	Material £	Unit	Total rate £
BRICK WALLING							
Basic mortar prices							
Mortar materials only							
cement	–	–	–	–	119.31	tonne	**119.31**
sand	–	–	–	–	18.31	tonne	**18.31**
lime	–	–	–	–	159.60	tonne	**159.60**
white cement	–	–	–	–	199.50	tonne	**199.50**
Cemplas Super mortar plasticizer	–	–	–	–	6.73	litre	**6.73**
Coloured mortar materials; (excluding cement)							
light	–	–	–	–	52.87	tonne	**52.87**
medium	–	–	–	–	54.86	tonne	**54.86**
dark	–	–	–	–	65.84	tonne	**65.84**
extra dark	–	–	–	–	65.84	tonne	**65.84**
SUPPLY ONLY BRICK PRICES							
Alternative prices below are for supply only bricks							
NOTE These costs exclude labour, mortar wastage,							
overheads and profit recovery							
Forterra Brick							
Brown Rustic	–	–	–	–	408.50	1000	**408.50**
Rustic	–	–	–	–	408.50	1000	**408.50**
Brecken Grey	–	–	–	–	476.40	1000	**476.40**
Old English Brindled Red	–	–	–	–	369.00	1000	**369.00**
Chiltern	–	–	–	–	561.30	1000	**561.30**
Claydon Red Multi	–	–	–	–	491.90	1000	**491.90**
Cotswold	–	–	–	–	551.40	1000	**551.40**
Dapple Light	–	–	–	–	610.80	1000	**610.80**
Georgian	–	–	–	–	486.30	1000	**486.30**
Golden Buff	–	–	–	–	637.60	1000	**637.60**
Heather	–	–	–	–	561.30	1000	**561.30**
Hereward Light	–	–	–	–	506.10	1000	**506.10**
Honey Buff	–	–	–	–	467.90	1000	**467.90**
Ironstone	–	–	–	–	486.30	1000	**486.30**
Milton Buff	–	–	–	–	514.60	1000	**514.60**
Regency	–	–	–	–	528.70	1000	**528.70**
Sandfaced	–	–	–	–	541.50	1000	**541.50**
Saxon Gold	–	–	–	–	540.00	1000	**540.00**
Sunset Red	–	–	–	–	499.00	1000	**499.00**
Tudor	–	–	–	–	554.20	1000	**554.20**
Windsor	–	–	–	–	491.90	1000	**491.90**
Selected Regrades	–	–	–	–	248.70	1000	**248.70**
Sherbourne Red Pavers	–	–	–	–	24.83	m²	**24.83**
Coxmoor Rose Multi Pavers	–	–	–	–	24.83	m²	**24.83**
Ibstock Brick							
Aldridge Brown Blend	–	–	–	–	914.80	1000	**914.80**
Aldridge Leicester Anglican Red Rustic	–	–	–	–	914.80	1000	**914.80**
Chailey Stock	–	–	–	–	780.40	1000	**780.40**
Dorking Multi	–	–	–	–	731.30	1000	**731.30**
Funton Second Hard Stock	–	–	–	–	1168.80	1000	**1168.80**
Leicester Red Stock	–	–	–	–	701.90	1000	**701.90**
Roughdales Red Multi Rustic	–	–	–	–	914.80	1000	**914.80**

14 MASONRY

Item	PC £	Labour hours	Labour £	Plant £	Material £	Unit	Total rate £
Stourbridge Himley Mixed Russet	–	–	–	–	1386.80	1000	**1386.80**
Stourbridge Kenilworth Multi	–	–	–	–	885.90	1000	**885.90**
Strattford Red Rustic	–	–	–	–	914.80	1000	**914.80**
Swanage Handmade Restoration	–	–	–	–	1282.60	1000	**1282.60**
Tonbridge Handmade Multi	–	–	–	–	1946.70	1000	**1946.70**
SUPPLY AND LAY PRICES							
Common bricks							
In gauged mortar (1:1:6); prime cost for bricks	380.00	–	–	–	399.00	1000	**399.00**
Walls							
half brick thick	22.80	0.84	18.05	–	27.14	m²	**45.19**
half brick thick; building against other work; concrete	22.80	0.92	19.79	–	28.49	m²	**48.28**
half brick thick; building overhand	22.80	1.04	22.51	–	27.14	m²	**49.65**
half brick thick; curved; 6.00 m radii	22.80	1.08	23.28	–	27.14	m²	**50.42**
half brick thick; curved; 1.50 m radii	26.60	1.41	30.46	–	31.33	m²	**61.79**
one brick thick	45.60	1.41	30.46	–	54.28	m²	**84.74**
one brick thick; curved; 6.00 m radii	49.40	1.84	39.59	–	58.47	m²	**98.06**
one brick thick; curved; 1.50 m radii	49.40	2.29	49.29	–	59.15	m²	**108.44**
one and a half brick thick	68.40	1.92	41.33	–	81.44	m²	**122.77**
one and a half brick thick; battering	68.40	2.21	47.53	–	81.44	m²	**128.97**
two brick thick	91.20	2.33	50.25	–	108.58	m²	**158.83**
two brick thick; battering	91.20	2.75	59.18	–	108.58	m²	**167.76**
337 mm average thick; tapering, one side	68.40	2.41	52.00	–	81.44	m²	**133.44**
450 mm average thick; tapering, one side	91.20	3.12	67.33	–	108.58	m²	**175.91**
337 mm average thick; tapering, both sides	68.40	2.79	60.15	–	81.44	m²	**141.59**
450 mm average thick; tapering, both sides	91.20	3.50	75.47	–	109.24	m²	**184.71**
facework one side, half brick thick	22.80	0.92	19.79	–	27.14	m²	**46.93**
facework one side, one brick thick	45.60	1.50	32.40	–	54.28	m²	**86.68**
facework one side, one and a half brick thick	68.40	2.00	43.07	–	81.44	m²	**124.51**
facework one side, two brick thick	91.20	2.41	52.00	–	108.58	m²	**160.58**
facework both sides, half brick thick	22.80	1.00	21.54	–	27.14	m²	**48.68**
facework both sides, one brick thick	45.60	1.58	34.15	–	54.28	m²	**88.43**
facework both sides, one and a half brick thick	68.40	2.08	44.82	–	81.44	m²	**126.26**
facework both sides, two brick thick	91.20	2.50	53.94	–	108.58	m²	**162.52**
Isolated piers							
one brick thick	45.60	2.36	50.88	–	54.28	m²	**105.16**
two brick thick	91.20	3.70	79.77	–	109.24	m²	**189.01**
three brick thick	136.80	4.67	100.68	–	164.21	m²	**264.89**
Isolated casings							
half brick thick	22.80	1.20	25.87	–	27.14	m²	**53.01**
one brick thick	45.60	2.04	43.98	–	54.28	m²	**98.26**
Chimney stacks							
one brick thick	45.60	2.36	50.88	–	54.28	m²	**105.16**
two brick thick	91.20	3.70	79.77	–	109.24	m²	**189.01**
three brick thick	136.80	4.67	100.68	–	164.21	m²	**264.89**
Projections							
225 mm width; 112 mm depth; vertical	5.07	0.28	6.04	–	5.82	m	**11.86**
225 mm width; 225 mm depth; vertical	10.13	0.56	12.07	–	11.64	m	**23.71**
337 mm width; 225 mm depth; vertical	15.20	0.83	17.89	–	17.46	m	**35.35**
440 mm width; 225 mm depth; vertical	20.27	0.93	20.04	–	23.28	m	**43.32**

14 MASONRY

Item	PC £	Labour hours	Labour £	Plant £	Material £	Unit	Total rate £
BRICK WALLING – cont							
Common bricks – cont							
Closing cavities							
width of cavity 50 mm, closing with common brickwork half brick thick; vertical	–	0.28	6.04	–	1.38	m	**7.42**
width of cavity 50 mm, closing with common brickwork half brick thick; horizontal	–	0.28	6.04	–	4.33	m	**10.37**
width of cavity 50 mm, closing with common brickwork half brick thick; including damp-proof course; vertical	–	0.37	7.98	–	2.26	m	**10.24**
width of cavity 50 mm, closing with common brickwork half brick thick; including damp-proof course; horizontal	–	0.32	6.90	–	5.22	m	**12.12**
width of cavity 75 mm, closing with common brickwork half brick thick; vertical	–	0.28	6.04	–	2.03	m	**8.07**
width of cavity 75 mm, closing with common brickwork half brick thick; horizontal	–	0.28	6.04	–	6.43	m	**12.47**
width of cavity 75 mm, closing with common brickwork half brick thick; including damp-proof course; vertical	–	0.37	7.98	–	2.91	m	**10.89**
width of cavity 75 mm, closing with common brickwork half brick thick; including damp-proof course; horizontal	–	0.32	6.90	–	7.31	m	**14.21**
Bonding to existing							
half brick thick	–	0.28	6.04	–	1.53	m	**7.57**
one brick thick	–	0.42	9.05	–	3.06	m	**12.11**
one and a half brick thick	–	0.65	14.02	–	4.59	m	**18.61**
two brick thick	–	0.88	18.97	–	6.12	m	**25.09**
Arches							
height on face 102 mm, width of exposed soffit 102 mm, shape of arch – segmental, one ring	2.53	1.57	29.14	–	6.77	m	**35.91**
height on face 102 mm, width of exposed soffit 215 mm, shape of arch – segmental, one ring	5.07	2.04	39.27	–	9.66	m	**48.93**
height on face 102 mm, width of exposed soffit 102 mm, shape of arch – semi-circular, one ring	2.53	1.99	38.19	–	6.77	m	**44.96**
height on face 102 mm, width of exposed soffit 215 mm, shape of arch – semi-circular, one ring	5.07	2.50	49.18	–	9.66	m	**58.84**
height on face 215 mm, width of exposed soffit 102 mm, shape of arch – segmental, two ring	5.07	1.99	38.19	–	9.77	m	**47.96**
height on face 215 mm, width of exposed soffit 215 mm, shape of arch – segmental, two ring	10.13	2.45	48.11	–	15.65	m	**63.76**
height on face 215 mm, width of exposed soffit 102 mm, shape of arch – semi-circular, two ring	5.07	2.68	53.07	–	9.77	m	**62.84**
height on face 215 mm, width of exposed soffit 215 mm, shape of arch – semi-circular, two ring	10.13	3.05	61.05	–	15.65	m	**76.70**
ADD or DEDUCT to walls for variation of £10.00/ 1000 in prime cost of common bricks							
half brick thick	–	–	–	–	0.63	m²	**0.63**
one brick thick	–	–	–	–	1.26	m²	**1.26**
one and a half brick thick	–	–	–	–	1.89	m²	**1.89**
two brick thick	–	–	–	–	2.52	m²	**2.52**

14 MASONRY

Item	PC £	Labour hours	Labour £	Plant £	Material £	Unit	Total rate £
Class B engineering bricks							
In cement mortar (1:3); prime cost £ per 1000	590.00	–	–	–	619.50	1000	**619.50**
Walls							
half brick thick	25.19	0.92	19.79	–	30.01	m²	**49.80**
one brick thick	50.38	1.50	32.40	–	60.02	m²	**92.42**
one brick thick; building against other work	50.38	1.79	38.61	–	62.26	m²	**100.87**
one brick thick; curved; 6.00 m radii	50.38	2.00	43.07	–	60.02	m²	**103.09**
one and a half brick thick	75.56	2.00	43.07	–	90.04	m²	**133.11**
one and a half brick thick; building against other work	75.56	2.41	52.00	–	90.04	m²	**142.04**
two brick thick	100.75	2.50	53.94	–	120.05	m²	**173.99**
337 mm thick; tapering, one side	75.56	2.58	55.68	–	90.04	m²	**145.72**
450 mm thick; tapering, one side	100.75	3.33	71.79	–	120.05	m²	**191.84**
337 mm thick; tapering, both sides	75.56	3.00	64.62	–	90.04	m²	**154.66**
450 mm thick; tapering, both sides	100.75	3.79	81.69	–	120.79	m²	**202.48**
facework one side, half brick thick	25.19	1.00	21.54	–	30.01	m²	**51.55**
facework one side, one brick thick	50.38	1.58	34.15	–	60.02	m²	**94.17**
facework one side, one and a half brick thick	75.56	2.08	44.82	–	90.04	m²	**134.86**
facework one side, two brick thick	100.75	2.58	55.68	–	120.05	m²	**175.73**
facework both sides, half brick thick	25.19	1.08	23.28	–	30.01	m²	**53.29**
facework both sides, one brick thick	50.38	1.67	35.90	–	60.02	m²	**95.92**
facework both sides, one and a half brick thick	75.56	2.17	46.76	–	90.04	m²	**136.80**
facework both sides, two brick thick	100.75	2.66	57.44	–	120.05	m²	**177.49**
Isolated piers							
one brick thick	50.38	2.59	55.84	–	60.02	m²	**115.86**
two brick thick	100.75	4.07	87.75	–	120.79	m²	**208.54**
three brick thick	151.13	5.00	107.79	–	181.56	m²	**289.35**
Isolated casings							
half brick thick	25.19	1.30	28.02	–	30.01	m²	**58.03**
one brick thick	50.38	2.22	47.86	–	60.02	m²	**107.88**
Projections							
225 mm width; 112 mm depth; vertical	5.60	0.32	6.90	–	6.44	m	**13.34**
225 mm width; 225 mm depth; vertical	11.19	0.60	12.94	–	12.86	m	**25.80**
337 mm width; 225 mm depth; vertical	16.79	0.88	18.97	–	19.30	m	**38.27**
440 mm width; 225 mm depth; vertical	22.39	1.02	21.99	–	25.73	m	**47.72**
Bonding to existing							
half brick thick	–	0.32	6.90	–	1.66	m	**8.56**
one brick thick	–	0.46	9.91	–	3.31	m	**13.22**
one and a half brick thick	–	0.65	14.02	–	4.97	m	**18.99**
two brick thick	–	0.97	20.92	–	6.62	m	**27.54**
ADD or DEDUCT to walls for variation of £10.00/ 1000 in prime cost of bricks							
half brick thick	–	–	–	–	0.63	m²	**0.63**
one brick thick	–	–	–	–	1.26	m²	**1.26**
one and a half brick thick	–	–	–	–	1.89	m²	**1.89**
two brick thick	–	–	–	–	2.52	m²	**2.52**

14 MASONRY

Item	PC £	Labour hours	Labour £	Plant £	Material £	Unit	Total rate £
BRICK WALLING – cont							
Facing bricks; machine-made facings; in gauged mortar (1:1:6)							
In gauged mortar (1:1:6); prime cost £ per 1000	500.00	–	–	–	525.00	1000	**525.00**
Walls							
facework one side, half brick thick; stretcher bond	30.00	1.08	23.28	–	35.08	m²	**58.36**
facework one side, half brick thick, flemish bond with snapped headers	40.00	1.25	26.97	–	46.11	m²	**73.08**
facework one side, half brick thick, stretcher bond; building against other work; concrete	30.00	1.17	25.22	–	36.42	m²	**61.64**
facework one side, half brick thick; flemish bond with snapped headers; building against other work; concrete	40.00	1.33	28.72	–	47.45	m²	**76.17**
facework one side, half brick thick, stretcher bond; building overhand	30.00	1.33	28.72	–	35.08	m²	**63.80**
facework one side, half brick thick; flemish bond with snapped headers; building overhand	40.00	1.50	32.40	–	46.11	m²	**78.51**
facework one side, half brick thick; stretcher bond; curved; 6.00 m radii	30.00	1.58	34.15	–	35.08	m²	**69.23**
facework one side, half brick thick; flemish bond with snapped headers; curved; 6.00 m radii	40.00	1.79	38.61	–	46.11	m²	**84.72**
facework one side, half brick thick; stretcher bond; curved; 1.50 m radii	35.00	2.00	43.07	–	40.59	m²	**83.66**
facework one side, half brick thick; flemish bond with snapped headers; curved; 1.50 m radii	40.00	2.33	50.25	–	46.11	m²	**96.36**
facework both sides, one brick thick; two stretcher skins tied together	60.00	1.87	40.36	–	73.75	m²	**114.11**
facework both sides, one brick thick; flemish bond	60.00	1.92	41.33	–	70.16	m²	**111.49**
facework both sides, one brick thick; two stretcher skins tied together; curved; 6.00 m radii	60.00	2.58	55.68	–	73.75	m²	**129.43**
facework both sides, one brick thick; flemish bond; curved; 6.00 m radii	60.00	2.66	57.44	–	70.16	m²	**127.60**
facework both sides, one brick thick; two stretcher skins tied together; curved; 1.50 m radii	60.00	3.20	69.08	–	74.42	m²	**143.50**
facework both sides, one brick thick; flemish bond; curved; 1.50 m radii	60.00	3.33	71.79	–	70.83	m²	**142.62**
Isolated piers							
facework both sides, one brick thick; two stretcher skins tied together	60.00	2.45	52.81	–	75.80	m²	**128.61**
facework both sides, one brick thick; flemish bond	60.00	2.50	53.90	–	75.80	m²	**129.70**
Isolated casings							
facework one side, half brick thick; stretcher bond	30.00	1.85	39.88	–	35.08	m²	**74.96**
facework one side, half brick thick; flemish bond with snapped headers	40.00	2.04	43.98	–	46.11	m²	**90.09**
Projections							
225 mm width; 112 mm depth; stretcher bond; vertical	6.67	0.28	6.04	–	7.58	m	**13.62**
225 mm width; 112 mm depth; flemish bond with snapped headers; vertical	6.67	0.37	7.98	–	7.58	m	**15.56**
225 mm width; 225 mm depth; flemish bond; vertical	13.33	0.60	12.94	–	22.17	m	**35.11**

14 MASONRY

Item	PC £	Labour hours	Labour £	Plant £	Material £	Unit	Total rate £
328 mm width; 112 mm depth; stretcher bond; vertical	10.00	0.56	12.07	–	11.38	m	**23.45**
328 mm width; 112 mm depth; flemish bond with snapped headers; vertical	10.00	0.65	14.02	–	11.38	m	**25.40**
328 mm width; 225 mm depth; flemish bond; vertical	20.00	1.11	23.93	–	22.72	m	**46.65**
440 mm width; 112 mm depth; stretcher bond; vertical	13.33	0.83	17.89	–	15.17	m	**33.06**
440 mm width; 112 mm depth; flemish bond with snapped headers; vertical	13.33	0.88	18.97	–	15.17	m	**34.14**
440 mm width; 225 mm depth; flemish bond; vertical	26.67	1.62	34.92	–	30.33	m	**65.25**
Arches							
height on face 215 mm, width of exposed soffit 102 mm, shape of arch – flat	6.67	0.93	17.96	–	8.92	m	**26.88**
height on face 215 mm, width of exposed soffit 215 mm, shape of arch – flat	13.33	1.39	27.88	–	16.57	m	**44.45**
height on face 215 mm, width of exposed soffit 102 mm, shape of arch – segmental, one ring	6.67	1.76	32.71	–	11.39	m	**44.10**
height on face 215 mm, width of exposed soffit 215 mm, shape of arch segmental, one ring	13.33	2.13	40.69	–	18.64	m	**59.33**
height on face 215 mm, width of exposed soffit 102 mm, shape of arch – semi-circular, one ring	6.67	2.68	52.54	–	11.39	m	**63.93**
height on face 215 mm, width of exposed soffit 215 mm, shape of arch – semi-circular, one ring	13.33	3.61	72.60	–	18.64	m	**91.24**
height on face 215 mm, width of exposed soffit 102 mm, shape of arch – segmental, two ring	6.67	2.17	41.55	–	11.39	m	**52.94**
height on face 215 mm, width of exposed soffit 215 mm, shape of arch – segmental; two ring	13.33	2.82	55.56	–	18.64	m	**74.20**
height on face 215 mm, width of exposed soffit 102 mm, shape of arch – semi-circular, two ring	6.67	3.61	72.60	–	11.39	m	**83.99**
height on face 215 mm, width of exposed soffit 215 mm, shape of arch – semi-circular, two ring	13.33	5.00	102.55	–	18.64	m	**121.19**
Arches; cut voussoirs; prime cost £ per 1000	3450.00	–	–	–	3622.50	1000	**3622.50**
height on face 215 mm, width of exposed soffit 102 mm, shape of arch – segmental, one ring	46.00	1.80	33.57	–	54.76	m	**88.33**
height on face 215 mm, width of exposed soffit 215 mm, shape of arch – segmental, one ring	92.00	2.27	43.70	–	105.37	m	**149.07**
height on face 215 mm, width of exposed soffit 102 mm, shape of arch – semi-circular, one ring	46.00	2.04	38.74	–	54.76	m	**93.50**
height on face 215 mm, width of exposed soffit 215 mm, shape of arch – semi-circular, one ring	92.00	2.59	50.60	–	105.37	m	**155.97**
height on face 320 mm, width of exposed soffit 102 mm, shape of arch – segmental, one and a half ring	92.00	2.41	46.73	–	105.61	m	**152.34**
height on face 320 mm, width of exposed soffit 215 mm, shape of arch – segmental, one and a half ring	184.00	3.15	62.67	–	214.57	m	**277.24**
Arches; bullnosed specials; prime cost £ per 1000	2000.00	–	–	–	2100.00	1000	**2100.00**
height on face 215 mm, width of exposed soffit 102 mm, shape of arch – flat	26.67	0.97	18.82	–	30.98	m	**49.80**

14 MASONRY

Item	PC £	Labour hours	Labour £	Plant £	Material £	Unit	Total rate £
BRICK WALLING – cont							
Facing bricks – cont							
Arches – cont							
height on face 215 mm, width of exposed soffit 215 mm, shape of arch – flat	54.00	1.43	28.74	–	61.40	m	**90.14**
Bullseye windows; 600 mm dia.; facing bricks							
height on face 215 mm, width of exposed soffit 102 mm, two rings	14.00	4.63	94.58	–	17.52	nr	**112.10**
height on face 215 mm, width of exposed soffit 215 mm, two rings	28.00	6.48	134.46	–	33.59	nr	**168.05**
Bullseye windows; 600 mm; cut voussoirs; prime cost £ per 1000	3450.00	–	–	–	3622.50	1000	**3622.50**
height on face 215 mm, width of exposed soffit 102 mm, one ring	120.75	3.89	78.62	–	135.21	nr	**213.83**
height on face 215 mm, width of exposed soffit 215 mm, one ring	120.75	5.37	110.53	–	268.98	nr	**379.51**
Bullseye windows; 1200 mm dia.; facing bricks							
height on face 215 mm, width of exposed soffit 102 mm, two rings	28.00	7.22	150.42	–	36.79	nr	**187.21**
height on face 215 mm, width of exposed soffit 215 mm, two rings	56.00	10.36	218.12	–	69.57	nr	**287.69**
Bullseye windows; 1200 mm dia.; cut voussoirs	3450.00	–	–	–	3622.50	1000	**3622.50**
height on face 215 mm, width of exposed soffit 102 mm, one ring	207.00	6.11	126.49	–	234.81	nr	**361.30**
height on face 215 mm, width of exposed soffit 215 mm, one ring	414.00	8.70	182.32	–	464.27	nr	**646.59**
ADD or DEDUCT for variation of £10.00 per 1000 in PC of facing bricks in 102 mm high arches with 215 mm soffit	–	–	–	–	0.28	m	**0.28**
Facework sills							
150 mm × 102 mm; headers on edge; pointing top and one side; set weathering; horizontal	6.67	0.51	10.99	–	7.41	m	**18.40**
150 mm × 102 mm; cant headers on edge; pointing top and one side; set weathering; horizontal; prime cost £ per 1000	26.67	0.56	12.07	–	28.94	m	**41.01**
150 mm × 102 mm; bullnosed specials; headers on flat; pointing top and one side; horizontal; prime cost £ per 1000	26.67	0.46	9.91	–	28.94	m	**38.85**
Facework copings							
215 mm × 102 mm; headers on edge; pointing top and both sides; horizontal	6.67	0.42	9.05	–	7.69	m	**16.74**
260 mm × 102 mm; headers on edge; pointing top and both sides; horizontal	10.00	0.65	14.02	–	11.45	m	**25.47**
215 mm × 102 mm; double bullnose specials; headers on edge; pointing top and both sides; horizontal; prime cost £ per 1000	2000.00	0.46	9.91	–	28.34	m	**38.25**

14 MASONRY

Item	PC £	Labour hours	Labour £	Plant £	Material £	Unit	Total rate £
260 mm × 102 mm; single bullnose specials; headers on edge; pointing top and both sides; horizontal; prime cost £ per 1000	2000.00	0.65	14.02	–	56.42	m	**70.44**
ADD or DEDUCT for variation of £10.00 per 1000 in prime cost of facing bricks in copings 215 mm wide, 102 mm high	–	–	–	–	0.14	m	**0.14**
Extra over facing bricks for; facework ornamental bands and the like, plain bands							
flush; horizontal; 225 mm width; entirely of stretchers	–	0.19	4.10	–	0.74	m	**4.84**
Extra over facing brick for; facework quoins							
flush; mean girth 320 mm	–	0.28	6.04	–	0.74	m	**6.78**
Bonding to existing							
facework one side, half brick thick; stretcher bond	–	0.46	9.91	–	1.94	m	**11.85**
facework one side, half brick thick; flemish bond with snapped headers	–	0.46	9.91	–	1.94	m	**11.85**
facework both sides, one brick thick; two stretcher skins tied together	–	0.65	14.02	–	3.87	m	**17.89**
facework both sides, one brick thick; flemish bond	–	0.65	14.02	–	3.87	m	**17.89**
ADD or DEDUCT for variation of £10.00 per 1000 in prime cost of facing bricks; in walls built entirely of facings; in stretcher or flemish bond							
half brick thick	–	–	–	–	0.63	m²	**0.63**
one brick thick	–	–	–	–	1.26	m²	**1.26**
Facing bricks; handmade							
In gauged mortar (1:1:6); prime cost for bricks	700.00	–	–	–	735.00	1000	**735.00**
Walls							
facework one side, half brick thick; stretcher bond	42.00	1.08	23.28	–	39.71	m²	**62.99**
facework one side, half brick thick; flemish bond with snapped headers	56.00	1.25	26.97	–	63.75	m²	**90.72**
facework one side, half brick thick; stretcher bond; building against other work; concrete	42.00	1.17	25.22	–	49.65	m²	**74.87**
facework one side, half brick thick; flemish bond with snapped headers; building against other work; concrete	56.00	1.33	28.72	–	62.15	m²	**90.87**
facework one side, half brick thick; stretcher bond; building overhand	42.00	1.33	28.72	–	48.31	m²	**77.03**
facework one side, half brick thick; flemish bond with snapped headers; building overhand	56.00	1.50	32.40	–	63.75	m²	**96.15**
facework one side, half brick thick; stretcher bond; curved; 6.00 m radii	42.00	1.58	34.15	–	48.31	m²	**82.46**
facework one side, half brick thick; flemish bond with snapped headers; curved; 6.00 m radii	56.00	1.79	38.61	–	63.75	m²	**102.36**
facework one side, half brick thick; stretcher bond; curved 1.50 m radii	42.00	2.00	43.07	–	48.31	m²	**91.38**
facework one side, half brick thick; flemish bond with snapped headers; curved; 1.50 m radii	56.00	2.33	50.25	–	63.75	m²	**114.00**
facework both sides, one brick thick; two stretcher skins tied together	84.00	1.87	40.36	–	100.21	m²	**140.57**

14 MASONRY

Item	PC £	Labour hours	Labour £	Plant £	Material £	Unit	Total rate £
BRICK WALLING – cont							
Facing bricks – cont							
Walls – cont							
facework both sides, one brick thick; flemish bond	84.00	1.92	41.33	–	96.62	m²	**137.95**
facework both sides; one brick thick; two stretcher skins tied together; curved; 6.00 m radii	84.00	2.58	55.68	–	100.21	m²	**155.89**
facework both sides, one brick thick; flemish bond; curved; 6.00 m radii	84.00	2.66	57.44	–	96.62	m²	**154.06**
facework both sides, one brick thick; two stretcher skins tied together; curved; 1.50 m radii	84.00	3.20	69.08	–	100.88	m²	**169.96**
facework both sides, one brick thick; flemish bond; curved; 1.50 m radii	84.00	3.33	71.79	–	97.29	m²	**169.08**
Isolated piers							
facework both sides, one brick thick; two stretcher skins tied together	84.00	2.45	52.81	–	102.26	m²	**155.07**
facework both sides, one brick thick; flemish bond	84.00	2.50	53.90	–	102.26	m²	**156.16**
Isolated casings							
facework one side, half brick thick; stretcher bond	42.00	1.85	39.88	–	48.31	m²	**88.19**
facework one side, half brick thick; flemish bond with snapped headers	42.00	2.04	43.98	–	48.31	m²	**92.29**
Projections							
225 mm width; 112 mm depth; stretcher bond; vertical	9.33	0.28	6.04	–	10.52	m	**16.56**
225 mm width; 112 mm depth; flemish bond with snapped headers; vertical	9.33	0.37	7.98	–	10.52	m	**18.50**
225 mm width; 225 mm depth; flemish bond; vertical	18.67	0.60	12.94	–	21.05	m	**33.99**
328 mm width; 112 mm depth; stretcher bond; vertical	14.00	0.56	12.07	–	15.79	m	**27.86**
328 mm width; 112 mm depth; flemish bond with snapped headers; vertical	14.00	0.65	14.02	–	15.79	m	**29.81**
328 mm width; 225 mm depth; flemish bond; vertical	28.00	1.11	23.93	–	31.54	m	**55.47**
440 mm width; 112 mm depth; stretcher bond; vertical	18.67	0.83	17.89	–	21.05	m	**38.94**
440 mm width; 112 mm depth; flemish bond with snapped headers; vertical	18.67	0.88	18.97	–	21.05	m	**40.02**
440 mm width; 225 mm depth; flemish bond; vertical	37.33	1.62	34.92	–	42.09	m	**77.01**
Arches							
height on face 215 mm, width of exposed soffit 102 mm, shape of arch – flat	9.33	0.93	17.96	–	11.87	m	**29.83**
height on face 215 mm, width of exposed soffit 215 mm, shape of arch – flat	18.90	1.39	27.88	–	22.71	m	**50.59**
height on face 215 mm, width of exposed soffit 102 mm, shape of arch – segmental, one ring	9.33	1.76	32.71	–	14.33	m	**47.04**

14 MASONRY

Item	PC £	Labour hours	Labour £	Plant £	Material £	Unit	Total rate £
height on face 215 mm, width of exposed soffit 215 mm, shape of arch – segmental, one ring	18.67	2.13	40.69	–	24.52	m	**65.21**
height on face 215 mm, width of exposed soffit 102 mm, shape of arch – semi-circular, one ring	9.33	2.68	52.54	–	14.33	m	**66.87**
height on face 215 mm, width of exposed soffit 215 mm, shape of arch – semi-circular, one ring	18.67	3.61	72.60	–	24.52	m	**97.12**
height on face 215 mm, width of exposed soffit 102 mm, shape of arch – segmental, two ring	9.33	2.17	41.55	–	14.33	m	**55.88**
height on face 215 mm, width of exposed soffit 215 mm, shape of arch – segmental, two ring	18.67	2.82	55.56	–	24.52	m	**80.08**
height on face 215 mm, width of exposed soffit 102 mm, shape of arch – semi-circular, two ring	9.33	3.61	72.60	–	14.33	m	**86.93**
height on face 215 mm, width of exposed soffit 215 mm, shape of arch – semi-circular, two ring	18.67	5.00	102.55	–	24.52	m	**127.07**
Arches; cut voussoirs; prime cost £ per 1000	3450.00	–	–	–	3622.50	1000	**3622.50**
height on face 215 mm, width of exposed soffit 102 mm, shape of arch – segmental, one ring	46.00	1.80	33.57	–	54.76	m	**88.33**
height on face 215 mm, width of exposed soffit 215 mm, shape of arch – segmental, one ring	92.00	2.27	43.70	–	105.37	m	**149.07**
height on face 215 mm, width of exposed soffit 102 mm, shape of arch – semi-circular, one ring	46.00	2.04	38.74	–	54.76	m	**93.50**
height on face 215 mm, width of exposed soffit 215 mm, shape of arch – semi-circular, one ring	92.00	2.59	50.60	–	105.37	m	**155.97**
height on face 320 mm, width of exposed soffit 102 mm, shape of arch – segmental, one and a half ring	92.00	2.41	46.73	–	105.61	m	**152.34**
height on face 320 mm, width of exposed soffit 215 mm, shape of arch – segmental, one and a half ring	184.00	3.15	62.67	–	214.57	m	**277.24**
Arches; bullnosed specials; prime cost £ per 1000	2750.00	–	–	–	2887.50	1000	**2887.50**
height on face 215 mm, width of exposed soffit 102 mm, shape of arch – flat	36.67	0.97	18.82	–	42.00	m	**60.82**
height on face 215 mm, width of exposed soffit 215 mm, shape of arch – flat	74.25	1.43	28.74	–	83.74	m	**112.48**
Bullseye windows; 600 mm dia.							
height on face 215 mm, width of exposed soffit 102 mm, two ring	19.60	4.63	94.58	–	23.70	nr	**118.28**
height on face 215 mm, width of exposed soffit 215 mm, two ring	39.20	6.48	134.46	–	57.98	nr	**192.44**
Bullseye windows; 600 mm dia.; cut voussoirs; prime cost £ per 1000	4000.00	–	–	–	4200.00	1000	**4200.00**
height on face 215 mm, width of exposed soffit 102 mm, one ring	140.00	3.89	78.62	–	156.44	nr	**235.06**
height on face 215 mm, width of exposed soffit 215 mm, one ring	280.00	5.37	110.53	–	311.42	nr	**421.95**
Bullseye windows; 1200 mm dia.							
height on face 215 mm, width of exposed soffit 102 mm, two ring	39.20	7.22	150.42	–	49.14	nr	**199.56**
height on face 215 mm, width of exposed soffit 215 mm, two ring	160.72	10.36	218.12	–	94.27	nr	**312.39**

14 MASONRY

Item	PC £	Labour hours	Labour £	Plant £	Material £	Unit	Total rate £
BRICK WALLING – cont							
Facing bricks – cont							
Bullseye windows; 1200 mm dia.; cut voussoirs; prime cost £ per 1000	4000.00	–	–	–	4200.00	1000	**4200.00**
height on face 215 mm, width of exposed soffit 102 mm, one ring	240.00	6.11	126.49	–	271.19	nr	**397.68**
height on face 215 mm, width of exposed soffit 215 mm, one ring	480.00	8.70	182.32	–	537.03	nr	**719.35**
ADD or DEDUCT for variation of £10.00 per 1000 in prime cost of facing bricks in 102 mm high arches with 215 mm soffit	–	–	–	–	0.28	m	**0.28**
Facework sills							
150 mm × 102 mm; headers on edge; pointing top and one side; set weathering; horizontal; prime cost £ per m	9.33	0.51	10.99	–	10.52	m	**21.51**
150 mm × 102 mm; cant headers on edge; pointing top and one side; set weathering; prime cost £ per m	36.67	0.56	12.07	–	40.66	m	**52.73**
150 mm × 102 mm; bullnosed specials; headers on edge; pointing top and one side; horizontal; prime cost £ per m	36.67	0.46	9.91	–	40.66	m	**50.57**
Facework copings							
215 mm × 102 mm; headers on edge; pointing top and both sides; horizontal	9.33	0.42	9.05	–	10.63	m	**19.68**
260 mm × 102 mm; headers on edge; pointing top and both sides; horizontal	14.00	0.65	14.02	–	15.86	m	**29.88**
215 mm × 102 mm; double bullnose specials; headers on edge; pointing top and both sides; horizontal; prime cost £ per m	36.67	0.46	9.91	–	40.76	m	**50.67**
260 mm × 102 mm; single bullnose specials; headers on edge; pointing top and both sides; horizontal; prime cost £ per m	73.33	0.65	14.02	–	81.27	m	**95.29**
ADD or DEDUCT for variation of £10.00 per 1000 in prime cost of facing bricks in copings 215 mm wide, 102 mm high	–	–	–	–	0.14	m	**0.14**
Extra over facing bricks for; facework ornamental bands and the like, plain bands							
flush; horizontal; 225 mm width; entirely of stretchers; prime cost £ per m	–	0.19	4.10	–	0.74	m	**4.84**
Extra over facing bricks for; facework quoins							
flush mean girth 320 mm; prime cost £ per m	–	0.28	6.04	–	0.70	m	**6.74**
Bonding ends to existing							
facework one side, half brick thick; stretcher bond	–	0.46	9.91	–	1.94	m	**11.85**
facework one side, half brick thick; flemish bond with snapped headers	–	0.46	9.91	–	1.94	m	**11.85**
facework both sides, one brick thick; two stretcher skins tied together	–	0.65	14.02	–	3.87	m	**17.89**
facework both sides, one brick thick; flemish bond	–	0.65	14.02	–	3.87	m	**17.89**

14 MASONRY

Item	PC £	Labour hours	Labour £	Plant £	Material £	Unit	Total rate £
ADD or DEDUCT for variation of £10.00/1000 in prime cost of facing bricks; in walls built entirely of facings; in stretcher or flemish bond							
half brick thick	–	–	–	–	0.63	m²	0.63
one brick thick	–	–	–	–	1.26	m²	1.26
Facework steps							
Machine cut special in cement mortar (1:3); prime cost £ per 1000 with class A engineering bricks	2000.00	–	–	–	2100.00	1000	2100.00
Facework steps							
215 mm × 102 mm Class A engineering bricks; all headers-on-edge; edges set with bullnosed specials; pointing top and one side; set weathering; horizontal; specials	26.67	0.51	10.99	–	36.99	m	47.98
returned ends pointed	–	0.14	3.01	–	0.26	nr	3.27
430 mm × 102 mm; all headers-on-edge; edges set with bullnosed specials; pointing top and one side; set weathering; horizontal; engineering bricks	26.67	0.74	15.95	–	32.37	m	48.32
returned ends pointed	–	0.19	4.10	–	0.26	nr	4.36
Facing bricks; slips							
50 mm thick; in gauged mortar (1:1:6) built up against concrete including flushing up at back (ties not included)							
walls; prime cost £ per 1000	1350.00	1.85	39.88	–	91.31	m²	131.19
edges of suspended slabs; 200 mm wide	–	0.56	12.07	–	18.26	m	30.33
columns; 400 mm wide	–	1.11	23.93	–	36.52	m	60.45
Facing tile bricks; Ibstock Tilebrick or other equal and approved; in gauged mortar (1:1:6)							
Walls							
standard tile brick; facework one side; half brick thick; stretcher bond	–	0.87	18.75	–	88.94	m²	107.69
three quarter tile brick; facework one side; half brick thick; stretcher bond	–	0.87	18.75	–	103.87	m²	122.62
Extra over facing tile bricks for							
fair ends; 79 mm long	–	0.28	6.04	–	38.65	m	44.69
fair ends; 163 mm long	–	0.28	6.04	–	38.65	m	44.69
90° × ½ external return	–	0.28	6.04	–	81.47	m	87.51
90° internal return	–	0.28	6.04	–	96.11	m	102.15
30° or 45° or 60° external return	–	0.28	6.04	–	81.47	m	87.51
30° or 45° or 60° internal return	–	0.28	6.04	–	81.47	m	87.51
angled verge	–	0.28	6.04	–	45.38	m	51.42

14 MASONRY

Item	PC £	Labour hours	Labour £	Plant £	Material £	Unit	Total rate £
BLOCK WALLING							
SUPPLY ONLY BLOCK PRICES							
Alternative prices below are for supply only blocks.							
NOTE These costs exclude labour, mortar wastage,							
overheads and profit recovery							
Celcon Standard blocks; 440 mm × 215 mm							
100 mm Standard; 3.6N	–	–	–	–	10.10	m²	**10.10**
140 mm Standard; 3.6N	–	–	–	–	14.13	m²	**14.13**
150 mm Standard; 3.6N	–	–	–	–	15.14	m²	**15.14**
215 mm Standard; 3.6N	–	–	–	–	21.71	m²	**21.71**
100 mm Standard; 7N	–	–	–	–	14.71	m²	**14.71**
140 mm Standard; 7N	–	–	–	–	20.59	m²	**20.59**
150 mm Standard; 7N	–	–	–	–	22.07	m²	**22.07**
215 mm Standard; 7N	–	–	–	–	31.63	m²	**31.63**
215 mm × 65 mm × 100 mm coursing block	–	–	–	–	2.07	m	**2.07**
Celcon Solar blocks; 440 mm × 215 mm							
100 mm Solar; 3.6N	–	–	–	–	13.95	m²	**13.95**
140 mm Solar; 3.6N	–	–	–	–	19.04	m²	**19.04**
150 mm Solar; 3.6N	–	–	–	–	20.40	m²	**20.40**
215 mm Solar; 3.6N	–	–	–	–	28.49	m²	**28.49**
265 mm Solar; 3.6N	–	–	–	–	35.11	m²	**35.11**
Celcon Foundation Blocks							
300 mm × 440 mm × 215 mm; 3.6N	–	–	–	–	7.80	m	**7.80**
Durox Supabloc; Aerated concrete;							
620 mm × 215 mm (7 nr per m²).							
100 mm Supabloc; 3.6N	–	–	–	–	14.41	m²	**14.41**
140 mm Supabloc; 3.6N	–	–	–	–	20.18	m²	**20.18**
200 mm Supabloc; 3.6N	–	–	–	–	28.83	m²	**28.83**
215 mm Supabloc; 3.6N	–	–	–	–	30.98	m²	**30.98**
100 mm Supabloc4; 4.2N	–	–	–	–	17.11	m²	**17.11**
140 mm Supabloc4; 4.2N	–	–	–	–	23.95	m²	**23.95**
200 mm Supabloc4; 4.2N	–	–	–	–	34.22	m²	**34.22**
215 mm Supabloc4; 4.2N	–	–	–	–	36.78	m²	**36.78**
100 mm Supabloc7; 7.3N	–	–	–	–	18.46	m²	**18.46**
140 mm Supabloc7; 7.3N	–	–	–	–	25.84	m²	**25.84**
200 mm Supabloc7; 7.3N	–	–	–	–	36.91	m²	**36.91**
215 mm Supabloc7; 7.3N	–	–	–	–	39.68	m²	**39.68**
Forticrete painting quality blocks; 440 mm × 215 mm							
100 mm hollow	–	–	–	–	13.36	m²	**13.36**
100 mm solid	–	–	–	–	14.63	m²	**14.63**
140 mm hollow	–	–	–	–	18.47	m²	**18.47**
140 mm solid	–	–	–	–	20.48	m²	**20.48**
190 mm hollow	–	–	–	–	25.38	m²	**25.38**
190 mm solid	–	–	–	–	27.92	m²	**27.92**
215 mm hollow	–	–	–	–	28.72	m²	**28.72**
215 mm solid	–	–	–	–	31.45	m²	**31.45**
Forterra Fenlite Background dense block							
100 mm solid; 3.6N	–	–	–	–	9.70	m²	**9.70**
100 mm solid; 7.3N	–	–	–	–	11.80	m²	**11.80**

14 MASONRY

Item	PC £	Labour hours	Labour £	Plant £	Material £	Unit	Total rate £
Forterra Evalast Background Standard Dense Aggregate block							
100 mm solid	–	–	–	–	9.46	m²	**9.46**
140 mm hollow	–	–	–	–	13.25	m²	**13.25**
215 mm hollow	–	–	–	–	20.34	m²	**20.34**
Lignacite Lignacrete standard blocks; 440 mm × 215 mm; 7.3N							
100 mm	–	–	–	–	13.18	m²	**13.18**
140 mm	–	–	–	–	18.46	m²	**18.46**
150 mm	–	–	–	–	19.78	m²	**19.78**
190 mm	–	–	–	–	25.05	m²	**25.05**
215 mm	–	–	–	–	28.35	m²	**28.35**
Tarmac Hemelite; 440 mm × 215 mm							
100 mm solid; 3.50 N/mm²	–	–	–	–	9.39	m²	**9.39**
100 mm solid; 7.00 N/mm²	–	–	–	–	9.72	m²	**9.72**
140 mm solid; 7.00 N/mm²	–	–	–	–	13.82	m²	**13.82**
190 mm solid; 7.00 N/mm²	–	–	–	–	19.97	m²	**19.97**
215 mm solid; 7.00 N/mm²	–	–	–	–	24.60	m²	**24.60**
Tarmac Toplite standard blocks; 440 mm × 215 mm							
100 mm	–	–	–	–	9.18	m²	**9.18**
125 mm	–	–	–	–	11.48	m²	**11.48**
140 mm	–	–	–	–	12.85	m²	**12.85**
150 mm	–	–	–	–	13.77	m²	**13.77**
200 mm	–	–	–	–	18.36	m²	**18.36**
215 mm	–	–	–	–	19.74	m²	**19.74**
Tarmac Toplite GTI (thermal) blocks; 440 mm × 215 mm							
100 mm	–	–	–	–	13.19	m²	**13.19**
125 mm	–	–	–	–	16.49	m²	**16.49**
140 mm	–	–	–	–	18.47	m²	**18.47**
150 mm	–	–	–	–	19.79	m²	**19.79**
200 mm	–	–	–	–	26.38	m²	**26.38**
215 mm	–	–	–	–	28.36	m²	**28.36**
Plasmor concrete blocks; 440 mm × 215 mm							
100 mm Aglite, standard blocks 10.4N	–	–	–	–	13.91	m²	**13.91**
140 mm Aglite, standard blocks 10.4N	–	–	–	–	19.52	m²	**19.52**
190 mm Aglite, standard blocks 10.4N	–	–	–	–	31.89	m²	**31.89**
215 Aglite, standard blocks 10.4N	–	–	–	–	36.55	m²	**36.55**
100 mm Stranlite, standard blocks 10.4N	–	–	–	–	14.30	m²	**14.30**
140 mm Stranlite, standard blocks 10.4N	–	–	–	–	20.14	m²	**20.14**
190 mm Stranlite, standard blocks 10.4N	–	–	–	–	33.52	m²	**33.52**
215 mm Stranlite, standard blocks 10.4N	–	–	–	–	38.59	m²	**38.59**
100 mm Stranlite, paint grade blocks 7.3N	–	–	–	–	14.46	m²	**14.46**
140 mm Stranlite, paint grade blocks 7.3N	–	–	–	–	20.30	m²	**20.30**
190 mm Stranlite, paint grade blocks 7.3N	–	–	–	–	33.29	m²	**33.29**
215 mm Stranlite, paint grade blocks 7.3N	–	–	–	–	38.41	m²	**38.41**
100 mm Stranlite, paint grade blocks 10.4N	–	–	–	–	15.16	m²	**15.16**
140 mm Stranlite, paint grade blocks 10.4N	–	–	–	–	21.24	m²	**21.24**
190 mm Stranlite, paint grade blocks 10.4N	–	–	–	–	34.60	m²	**34.60**
215 mm Stranlite, paint grade blocks 10.4N	–	–	–	–	41.22	m²	**41.22**

14 MASONRY

Item	PC £	Labour hours	Labour £	Plant £	Material £	Unit	Total rate £
BLOCK WALLING – cont							
SUPPLY AND LAY PRICES							
Lightweight aerated concrete blocks; Thermalite Turbo blocks or other equal; in gauged mortar (1:2:9)							
Walls							
100 mm thick	14.90	0.41	8.92	–	17.25	m²	**26.17**
115 mm thick	17.14	0.41	8.92	–	19.83	m²	**28.75**
125 mm thick	18.63	0.41	8.92	–	21.57	m²	**30.49**
130 mm thick	19.37	0.41	8.92	–	22.42	m²	**31.34**
140 mm thick	20.86	0.46	9.89	–	24.15	m²	**34.04**
150 mm thick	22.35	0.46	9.89	–	25.87	m²	**35.76**
190 mm thick	28.31	0.50	10.87	–	32.77	m²	**43.64**
200 mm thick	29.80	0.50	10.87	–	34.50	m²	**45.37**
215 mm thick	32.04	0.50	10.87	–	37.10	m²	**47.97**
265 mm thick	39.49	0.55	11.85	–	45.69	m²	**57.54**
300 mm thick	44.70	0.55	11.85	–	51.75	m²	**63.60**
Isolated piers or chimney stacks							
190 mm thick	–	0.83	17.89	–	32.77	m²	**50.66**
215 mm thick	–	0.83	17.89	–	37.10	m²	**54.99**
Isolated casings							
100 mm thick	–	0.51	10.99	–	17.25	m²	**28.24**
115 mm thick	–	0.51	10.99	–	19.83	m²	**30.82**
125 mm thick	–	0.51	10.99	–	21.57	m²	**32.56**
140 mm thick	–	0.56	12.07	–	24.15	m²	**36.22**
Extra over for fair face; flush pointing							
walls; one side	–	0.04	0.86	–	–	m²	**0.86**
walls; both sides	–	0.09	1.94	–	–	m²	**1.94**
Closing cavities							
width of cavity 50 mm, closing with lightweight blockwork 100 mm thick; vertical	–	0.23	4.96	–	0.92	m	**5.88**
width of cavity 50 mm, closing with lightweight blockwork 100 mm thick; including damp-proof course; vertical	–	0.28	6.04	–	1.81	m	**7.85**
width of cavity 75 mm, closing with lightweight blockwork 100 mm thick; vertical	–	0.23	4.96	–	1.34	m	**6.30**
width of cavity 75 mm, closing with lightweight blockwork 100 mm thick; including damp-proof course; vertical	–	0.28	6.04	–	2.24	m	**8.28**
Bonding ends to common brickwork							
100 mm thick	–	0.14	3.01	–	1.95	m	**4.96**
115 mm thick	–	0.14	3.01	–	2.25	m	**5.26**
125 mm thick	–	0.23	4.96	–	2.45	m	**7.41**
130 mm thick	–	0.23	4.96	–	2.54	m	**7.50**
140 mm thick	–	0.23	4.96	–	2.74	m	**7.70**
150 mm thick	–	0.23	4.96	–	2.93	m	**7.89**
190 mm thick	–	0.28	6.04	–	3.71	m	**9.75**
200 mm thick	–	0.28	6.04	–	3.91	m	**9.95**
215 mm thick	–	0.32	6.90	–	4.21	m	**11.11**

14 MASONRY

Item	PC £	Labour hours	Labour £	Plant £	Material £	Unit	Total rate £
Lightweight aerated concrete blocks; Thermalite Aircrete Shield blocks or other equal; in thin joint mortar							
Walls							
90 mm thick	11.16	0.46	9.91	–	14.38	m²	**24.29**
100 mm thick	12.40	0.46	9.91	–	16.00	m²	**25.91**
115 mm thick	14.26	0.46	9.91	–	18.10	m²	**28.01**
125 mm thick	15.50	0.48	10.35	–	19.89	m²	**30.24**
140 mm thick	17.36	0.51	10.99	–	22.41	m²	**33.40**
150 mm thick	18.60	0.51	10.99	–	24.01	m²	**35.00**
190 mm thick	23.56	0.56	12.07	–	30.39	m²	**42.46**
200 mm thick	24.80	0.60	12.94	–	31.99	m²	**44.93**
Isolated piers or chimney stacks							
190 mm thick	–	0.60	12.94	–	30.39	m²	**43.33**
Isolated casings							
90 mm thick	–	0.35	7.55	–	14.60	m²	**22.15**
100 mm thick	–	0.35	7.55	–	16.00	m²	**23.55**
140 mm thick	–	0.38	8.19	–	22.41	m²	**30.60**
Lightweight aerated concrete blocks; Thermalite Aircrete Shield blocks or other equal; in gauged mortar (1:2:9)							
Walls							
90 mm thick	12.00	0.41	8.92	–	12.99	m²	**21.91**
100 mm thick	13.33	0.41	8.92	–	14.43	m²	**23.35**
115 mm thick	14.26	0.41	8.84	–	16.70	m²	**25.54**
125 mm thick	15.50	0.41	8.84	–	18.10	m²	**26.94**
140 mm thick	18.66	0.46	9.89	–	20.20	m²	**30.09**
150 mm thick	20.00	0.46	9.89	–	21.64	m²	**31.53**
190 mm thick	25.33	0.50	10.87	–	27.41	m²	**38.28**
200 mm thick	26.66	0.50	10.87	–	28.85	m²	**39.72**
Isolated piers or chimney stacks							
190 mm thick	–	0.83	17.89	–	27.41	m²	**45.30**
Isolated casings							
90 mm thick	–	0.51	10.99	–	12.99	m²	**23.98**
100 mm thick	–	0.51	10.99	–	14.43	m²	**25.42**
140 mm thick	–	0.56	12.07	–	20.20	m²	**32.27**
Extra over for fair face; flush pointing							
walls; one side	–	0.04	0.86	–	–	m²	**0.86**
walls; both sides	–	0.09	1.94	–	–	m²	**1.94**
Closing cavities							
width of cavity 50 mm, closing with lightweight blockwork 100 mm thick; vertical	–	0.23	4.96	–	0.78	m	**5.74**
width of cavity 50 mm, closing with lightweight blockwork 100 mm thick; including damp-proof course; vertical	–	0.28	6.04	–	1.67	m	**7.71**
width of cavity 75 mm, closing with lightweight blockwork 100 mm thick; vertical	–	0.23	4.96	–	1.13	m	**6.09**
width of cavity 75 mm, closing with lightweight blockwork 100 mm thick; including damp-proof course; vertical	–	0.28	6.04	–	2.02	m	**8.06**

14 MASONRY

Item	PC £	Labour hours	Labour £	Plant £	Material £	Unit	Total rate £
BLOCK WALLING – cont							
Lightweight aerated concrete blocks – cont							
Bonding ends to common brickwork							
90 mm thick	–	0.09	1.94	–	1.48	m	**3.42**
100 mm thick	–	0.14	3.01	–	1.64	m	**4.65**
140 mm thick	–	0.23	4.96	–	2.31	m	**7.27**
150 mm thick	–	0.23	4.96	–	2.46	m	**7.42**
190 mm thick	–	0.28	6.04	–	3.12	m	**9.16**
200 mm thick	–	0.28	6.04	–	3.29	m	**9.33**
Smooth face paint grade concrete block; in gauged mortar; flush pointing one side							
Walls							
100 mm thick	13.85	0.50	10.87	–	16.07	m²	**26.94**
140 mm thick	19.38	0.58	12.61	–	22.48	m²	**35.09**
150 mm thick	20.76	0.58	12.61	–	24.08	m²	**36.69**
190 mm thick	26.29	0.67	14.35	–	30.49	m²	**44.84**
200 mm thick	27.69	0.67	14.35	–	32.12	m²	**46.47**
215 mm thick	29.76	0.67	14.35	–	34.51	m²	**48.86**
100 mm thick; Paintgrade Smooth	16.54	0.55	11.88	–	19.10	m²	**30.98**
215 mm thick; Paintgrade Smooth	35.57	0.73	15.66	–	41.08	m²	**56.74**
Isolated piers or chimney stacks							
190 mm thick	–	0.93	20.04	–	30.49	m²	**50.53**
200 mm thick	–	0.93	20.04	–	32.12	m²	**52.16**
215 mm thick	–	0.93	20.04	–	34.51	m²	**54.55**
Isolated casings							
100 mm thick	–	0.69	14.88	–	16.07	m²	**30.95**
140 mm thick	–	0.74	15.95	–	22.48	m²	**38.43**
Extra over for fair face flush pointing							
walls; one side	–	0.04	0.86	–	–	m²	**0.86**
walls; both sides	–	0.09	1.94	–	–	m²	**1.94**
Bonding ends to common brickwork							
100 mm thick	–	0.23	4.96	–	1.86	m	**6.82**
140 mm thick	–	0.23	4.96	–	2.61	m	**7.57**
150 mm thick	–	0.28	6.04	–	2.79	m	**8.83**
190 mm thick	–	0.32	6.90	–	3.54	m	**10.44**
200 mm thick	–	0.32	6.90	–	3.73	m	**10.63**
215 mm thick	–	0.32	6.90	–	4.01	m	**10.91**
Smooth face paint grade aerated concrete blocks; Party Wall block; in gauged mortar; flush pointing one side							
Walls							
100 mm thick	10.19	0.56	12.07	–	11.67	m²	**23.74**
215 mm thick	21.90	0.74	15.95	–	25.07	m²	**41.02**
Isolated piers or chimney stacks							
215 mm thick	–	0.93	20.04	–	25.07	m²	**45.11**

14 MASONRY

Item	PC £	Labour hours	Labour £	Plant £	Material £	Unit	Total rate £
Isolated casings							
100 mm thick	–	0.69	14.88	–	11.67	m²	**26.55**
Extra over for fair face flush pointing							
walls; both sides	–	0.04	0.86	–	–	m²	**0.86**
Bonding ends to common brickwork							
100 mm thick	–	0.23	4.96	–	1.27	m	**6.23**
215 mm thick	–	0.32	6.90	–	2.75	m	**9.65**
Lightweight aerated high strength concrete blocks (7N); Thermalite High Strength 7 blocks or other equal and approved; in cement mortar (1:3)							
Walls							
100 mm thick	13.11	0.41	8.92	–	15.30	m²	**24.22**
140 mm thick	18.34	0.46	9.89	–	21.40	m²	**31.29**
150 mm thick	19.66	0.46	9.89	–	22.94	m²	**32.83**
190 mm thick	24.90	0.50	10.87	–	29.04	m²	**39.91**
200 mm thick	26.20	0.50	10.87	–	30.57	m²	**41.44**
215 mm thick	28.18	0.50	10.87	–	32.88	m²	**43.75**
Isolated piers or chimney stacks							
190 mm thick	–	0.83	17.89	–	29.04	m²	**46.93**
200 mm thick	–	0.83	17.89	–	30.57	m²	**48.46**
215 mm thick	–	0.83	17.89	–	32.88	m²	**50.77**
Isolated casings							
100 mm thick	–	0.51	10.99	–	15.30	m²	**26.29**
140 mm thick	–	0.56	12.07	–	21.40	m²	**33.47**
150 mm thick	–	0.56	12.07	–	22.94	m²	**35.01**
190 mm thick	–	0.69	14.88	–	29.04	m²	**43.92**
200 mm thick	–	0.69	14.88	–	30.57	m²	**45.45**
215 mm thick	–	0.69	14.88	–	32.88	m²	**47.76**
Extra over for flush pointing							
walls; one side	–	0.04	0.86	–	–	m²	**0.86**
walls; both sides	–	0.09	1.94	–	–	m²	**1.94**
Bonding ends to common brickwork							
100 mm thick	–	0.23	4.96	–	1.77	m	**6.73**
140 mm thick	–	0.23	4.96	–	2.50	m	**7.46**
150 mm thick	–	0.28	6.04	–	2.67	m	**8.71**
190 mm thick	–	0.32	6.90	–	3.38	m	**10.28**
200 mm thick	–	0.32	6.90	–	3.56	m	**10.46**
215 mm thick	–	0.32	6.90	–	3.84	m	**10.74**
Dense smooth faced concrete blocks; Lignacite standard and paint grade; in gauged mortar (1:2:9); flush pointing one side							
Walls							
100 mm thick; 3.6N	10.11	0.56	12.03	–	11.05	m²	**23.08**
140 mm thick; 3.6N	14.79	0.65	13.97	–	16.14	m²	**30.11**
100 mm thick; 7N	14.17	0.56	12.03	–	15.31	m²	**27.34**
140 mm thick; 7N	19.84	0.65	13.97	–	21.44	m²	**35.41**

14 MASONRY

Item	PC £	Labour hours	Labour £	Plant £	Material £	Unit	Total rate £
BLOCK WALLING – cont							
Dense smooth faced concrete blocks – cont							
Isolated piers or chimney stacks							
100 mm thick; 7N	–	1.14	24.58	–	15.70	m²	**40.28**
140 mm thick; 7N	–	1.26	27.16	–	21.77	m²	**48.93**
Isolated casings							
100 mm thick; 7N	–	0.78	16.82	–	15.31	m²	**32.13**
140 mm thick; 7N	–	0.90	19.40	–	21.44	m²	**40.84**
Dense aggregate concrete blocks; in cement mortar (1:2:9)							
Walls or partitions or skins of hollow walls							
100 mm thick; solid	12.40	0.62	13.39	–	14.43	m²	**27.82**
140 mm thick; solid	17.36	0.75	16.11	–	20.20	m²	**36.31**
140 mm thick; hollow	17.36	0.67	14.35	–	20.20	m²	**34.55**
190 mm thick; hollow	23.56	0.84	18.05	–	27.41	m²	**45.46**
215 mm thick; hollow	24.80	0.92	19.79	–	28.92	m²	**48.71**
Isolated piers or chimney stacks							
140 mm thick; hollow	–	1.02	21.99	–	20.20	m²	**42.19**
190 mm thick; hollow	–	1.34	28.89	–	27.41	m²	**56.30**
215 mm thick; hollow	–	1.53	32.98	–	28.92	m²	**61.90**
Isolated casings							
100 mm thick; solid	–	0.74	15.95	–	14.43	m²	**30.38**
140 mm thick; solid	–	0.93	20.04	–	20.20	m²	**40.24**
Extra over for fair face; flush pointing							
walls; one side	–	0.09	1.94	–	–	m²	**1.94**
walls; both sides	–	0.14	3.01	–	–	m²	**3.01**
Bonding ends to common brickwork							
100 mm thick solid	–	0.23	4.96	–	1.68	m	**6.64**
140 mm thick solid	–	0.28	6.04	–	2.35	m	**8.39**
140 mm thick hollow	–	0.28	6.04	–	2.35	m	**8.39**
190 mm thick hollow	–	0.32	6.90	–	3.18	m	**10.08**
215 mm thick hollow	–	0.37	7.98	–	3.38	m	**11.36**
Dense aggregate paint grade concrete blocks; (7 N) Forticrete or similar; in cement mortar (1:3)							
Walls							
75 mm thick; solid	10.02	0.50	10.87	–	11.69	m²	**22.56**
100 mm thick; hollow	8.18	0.62	13.39	–	9.73	m²	**23.12**
100 mm thick; solid	7.62	0.62	13.39	–	9.10	m²	**22.49**
140 mm thick; hollow	12.06	0.67	14.35	–	14.31	m²	**28.66**
140 mm thick; solid	12.06	0.75	16.11	–	14.31	m²	**30.42**
190 mm thick; hollow	16.32	0.84	18.05	–	19.36	m²	**37.41**
190 mm thick; solid	16.31	0.92	19.79	–	19.35	m²	**39.14**
215 mm thick; hollow	14.79	0.92	19.79	–	17.77	m²	**37.56**
215 mm thick; solid	17.86	0.94	20.25	–	21.23	m²	**41.48**
Dwarf support wall							
140 mm thick; solid	–	1.16	25.01	–	14.31	m²	**39.32**
190 mm thick; solid	–	1.34	28.89	–	19.35	m²	**48.24**
215 mm thick; solid	–	1.53	32.98	–	21.23	m²	**54.21**

14 MASONRY

Item	PC £	Labour hours	Labour £	Plant £	Material £	Unit	Total rate £
Isolated piers or chimney stacks							
140 mm thick; hollow	–	1.02	21.99	–	14.31	m²	**36.30**
190 mm thick; hollow	–	1.34	28.89	–	19.36	m²	**48.25**
215 mm thick; hollow	–	1.53	32.98	–	17.77	m²	**50.75**
Isolated casings							
75 mm thick; solid	–	0.69	14.88	–	11.69	m²	**26.57**
100 mm thick; solid	–	0.74	15.95	–	9.10	m²	**25.05**
140 mm thick; solid	–	0.93	20.04	–	14.31	m²	**34.35**
Extra over for fair face; flush pointing							
walls; one side	–	0.09	1.94	–	–	m²	**1.94**
walls; both sides	–	0.14	3.01	–	–	m²	**3.01**
Bonding ends to common brickwork							
75 mm thick solid	–	0.14	3.01	–	1.35	m	**4.36**
100 mm thick solid	–	0.23	4.96	–	1.06	m	**6.02**
140 mm thick solid	–	0.28	6.04	–	1.68	m	**7.72**
190 mm thick solid	–	0.32	6.90	–	2.26	m	**9.16**
215 mm thick solid	–	0.37	7.98	–	2.49	m	**10.47**
High strength concrete blocks (10N); Thermalite High Strength 10 blocks; in cement mortar (1:3)							
Walls							
100 mm thick (NB other thicknesses as a special order item)	–	0.50	10.87	–	19.67	m²	**30.54**
Thermalite Trenchblock 3.6N; with tongued and grooved joints; in cement mortar (1:4)							
Walls							
255 mm thick	–	0.60	12.94	–	29.53	m²	**42.47**
275 mm thick	–	0.65	14.02	–	31.84	m²	**45.86**
305 mm thick	–	0.70	15.09	–	34.70	m²	**49.79**
355 mm thick	–	0.75	16.17	–	40.94	m²	**57.11**
Concrete blocks; Thermalite Trenchblock, 7N; with tongued and grooved joints; in cement mortar (1:4)							
Walls							
255 mm thick	33.41	0.70	15.09	–	38.66	m²	**53.75**
275 mm thick	36.06	0.75	16.17	–	41.72	m²	**57.89**
305 mm thick	39.33	0.80	17.25	–	45.48	m²	**62.73**
355 mm thick	46.52	0.85	18.32	–	53.66	m²	**71.98**
Architectural blockwork							
Lignacite Ltd; 440 × 215 mm face size; solid blocks 17.5N; laid stretcher bond with class 3 Snowstorm mortar with white Portland cement and recessed joints							
Snowstorm Weathered							
100 mm thick weathered one face	–	1.00	21.56	–	48.11	m²	**69.67**
100 mm thick; weathered one face and one end	–	0.10	2.15	–	24.08	m	**26.23**
217 mm × 100 mm thick; cut half block; weathered one face	–	1.05	22.64	–	66.93	m²	**89.57**

14 MASONRY

Item	PC £	Labour hours	Labour £	Plant £	Material £	Unit	Total rate £
BLOCK WALLING – cont							
Architectural blockwork – cont							
Snowstorm Weathered – cont							
quoins; weathered two external faces;							
440 mm × 100 mm × 215 mm with 215 mm external return	–	0.11	2.37	–	97.53	m	**99.90**
100 mm × 440 mm × 65 mm Roman Bricks weathered one face	–	1.00	21.56	–	60.20	m²	**81.76**
Snowstorm Split							
100 mm thick; split one face	–	1.00	21.56	–	60.29	m²	**81.85**
100 mm thick; split one face and weathered one end	–	0.10	2.15	–	24.78	m	**26.93**
217 mm × 100 mm thick; cut half block; split one face	–	1.05	22.64	–	79.82	m²	**102.46**
quoins; split two external faces;							
440 mm × 215 mm × 100 mm thick with 215 mm external return	–	0.11	2.37	–	114.37	m	**116.74**
100 mm × 440 mm × 65 mm Roman Bricks; split one face	–	1.00	21.56	–	69.44	m²	**91.00**
Snowstorm Polished							
Note: polished dimensions reduced by 3 mm							
100 mm thick; polished one face	–	1.10	23.72	–	83.90	m²	**107.62**
100 mm thick; polished one face one end	–	0.11	2.37	–	68.16	m	**70.53**
217 mm × 100 mm thick; cut half block; polished one face	–	1.18	25.34	–	101.97	m²	**127.31**
quoins; polished two external faces;							
440 mm × 215 mm × 100 mm thick with 215 mm external return	–	0.12	2.58	–	166.77	m	**169.35**
100 mm × 440 mm × 65 mm Roman Bricks; polished one face	–	1.10	23.72	–	110.13	m²	**133.85**
Snowstorm Planished							
Note: planished dimensions reduced by 3 mm							
100 mm thick; planished one face	–	1.10	23.72	–	83.90	m²	**107.62**
100 mm thick; planished one face one end	–	0.11	2.37	–	68.16	m	**70.53**
100 mm thick; cut half block; planished one face	–	1.18	25.34	–	101.97	m²	**127.31**
quoins; planished two external faces;							
440 mm × 215 mm × 100 mm thick with 215 mm external return	–	0.12	2.58	–	166.77	m	**169.35**
Snowstorm and Crushed recycled Glass (Polished or Planished). Standard colours							
Note: planished dimensions reduced by 3 mm							
100 mm thick; planished one face	–	1.10	23.72	–	99.03	m²	**122.75**
100 mm thick; planished one face one end	–	0.11	2.37	–	73.88	m	**76.25**
100 mm thick; cut half block; planished one face	–	1.18	25.34	–	117.44	m²	**142.78**
quoins; planished two external faces;							
440 mm × 215 mm × 100 mm thick with 215 mm external return	–	0.12	2.58	–	173.49	m	**176.07**

14 MASONRY

Item	PC £	Labour hours	Labour £	Plant £	Material £	Unit	Total rate £
Midnight Polished (Note 7.3N strength) 10 mm Black Granite facing bonded to 90 mm thick backing block							
100 mm thick; polished one face	–	1.15	24.79	–	143.10	m²	**167.89**
100 mm thick; polished one face one end	–	0.12	2.58	–	227.88	m	**230.46**
100 mm thick; cut half block; polished one face	–	1.19	25.55	–	200.00	m²	**225.55**
quoins; polished two external faces;							
440 mm × 215 mm × 110 mm thick with 215 mm							
external return	–	0.14	3.01	–	256.55	m	**259.56**
Dense aggregate coloured concrete blocks; Forticrete Bathstone or equal; in coloured gauged mortar (1:1:6); flush pointing one side							
Walls							
100 mm thick hollow	30.87	0.74	15.95	–	34.59	m²	**50.54**
100 mm thick solid	30.87	0.74	15.95	–	34.59	m²	**50.54**
140 mm thick hollow	44.71	0.83	17.89	–	50.06	m²	**67.95**
140 mm thick solid	44.71	0.93	20.04	–	50.06	m²	**70.10**
215 mm thick hollow	51.11	1.16	25.01	–	57.54	m²	**82.55**
Isolated piers or chimney stacks							
140 mm thick solid	–	1.25	26.95	–	50.06	m²	**77.01**
215 mm thick hollow	–	1.57	33.85	–	57.54	m²	**91.39**
Extra over blocks for							
100 mm thick half lintel blocks; ref D14	–	0.23	4.96	–	25.69	m	**30.65**
140 mm thick half lintel blocks; ref H14	–	0.28	6.04	–	45.95	m	**51.99**
140 mm thick quoin blocks; ref H16	–	0.32	6.90	–	39.51	m	**46.41**
140 mm thick cavity closer blocks; ref H17	–	0.32	6.90	–	42.36	m	**49.26**
140 mm thick sill blocks; ref H21	–	0.28	6.04	–	31.09	m	**37.13**
Glazed finish blocks; Forticrete Astra-Glaze or equal; in gauged mortar (1:1:6); joints raked out; gun applied latex grout to joints							
Walls or partitions or skins of hollow walls							
100 mm thick; glazed one side	–	0.93	20.04	–	105.59	m²	**125.63**
extra for glazed square end return	–	0.37	7.98	–	30.43	m	**38.41**
100 mm thick; glazed both sides	–	1.11	23.93	–	132.34	m²	**156.27**
100 mm thick lintel 200 mm high; glazed one side	–	0.83	15.28	–	27.77	m	**43.05**
Fireborn terracotta blocks or other equal and approved; Ibstock Brick Ltd; in coloured gauged mortar (1:1:6); flush pointing one side							
Walls or partitions or skins of hollow walls							
102.50 mm thick; stretcher bond	–	0.33	7.12	–	49.45	m²	**56.57**
102.50 mm thick; stack bond	–	0.35	7.55	–	49.41	m²	**56.96**

GLASS BLOCK WALLING
NOTE: The following specialist prices for glass block walling; supplied by Roger Wilde Ltd; assume standard blocks in panels of 50 m²; work in straight walls at ground level; and all necessary ancillary fixing; strengthening; easy access; pointing and expansion materials etc.

14 MASONRY

Item	PC £	Labour hours	Labour £	Plant £	Material £	Unit	Total rate £
GLASS BLOCK WALLING – cont							
Hollow glass block walling; Pittsburgh Corning sealed Thinline or other equal and approved; in cement mortar joints; reinforced with 6 mm dia. stainless steel rods; pointed both sides with mastic or other equal and approved							
Walls; facework both sides							
115 mm × 115 mm × 80 mm flemish blocks	–	–	–	–	–	m²	630.00
190 mm × 190 mm × 80 mm flemish; cross reeded or clear blocks	–	–	–	–	–	m²	260.00
240 mm × 240 mm × 80 mm flemish; cross reeded or clear blocks	–	–	–	–	–	m²	410.00
240 mm × 115 mm × 80 mm flemish, or clear blocks	–	–	–	–	–	m²	2950.00
Fire-rated walls							
190 mm × 190 mm × 100 mm glass blocks; 30 minute fire-rated	–	–	–	–	–	m²	625.00
190 mm × 190 mm × 160 mm glass blocks; 60 minute fire-rated	–	–	–	–	–	m²	1130.00
NATURAL STONE RUBBLE WALLING							
Cotswold Guiting limestone or other equal and approved; laid dry							
Uncoursed random rubble walling							
275 mm thick	–	2.07	44.14	–	11.25	m²	55.39
350 mm thick	–	2.46	52.19	–	14.31	m²	66.50
425 mm thick	–	2.81	59.31	–	17.38	m²	76.69
500 mm thick	–	3.15	66.21	–	20.45	m²	86.66
Cotswold Guiting limestone or other equal; bedded; jointed and pointed in cement: lime mortar (1:2:9)							
Uncoursed random rubble walling; faced and pointed; both sides							
275 mm thick	–	1.98	42.08	–	15.16	m²	57.24
350 mm thick	–	2.18	45.78	–	19.30	m²	65.08
425 mm thick	–	2.39	49.70	–	23.44	m²	73.14
500 mm thick	–	2.59	53.39	–	27.57	m²	80.96
Coursed random rubble walling; rough dressed; faced and pointed one side							
114 mm thick	–	1.48	31.18	–	45.73	m²	76.91
150 mm thick	–	1.76	39.11	–	51.75	m²	90.86
Fair returns on walling							
114 mm wide	–	0.02	0.43	–	–	m	0.43
150 mm wide	–	0.03	0.65	–	–	m	0.65
275 mm wide	–	0.06	1.29	–	–	m	1.29
350 mm wide	–	0.08	1.72	–	–	m	1.72
425 mm wide	–	0.10	2.15	–	–	m	2.15
500 mm wide	–	0.12	2.58	–	–	m	2.58

14 MASONRY

Item	PC £	Labour hours	Labour £	Plant £	Material £	Unit	Total rate £
Fair raking cutting or circular cutting							
114 mm wide	–	0.20	4.34	–	6.62	m	**10.96**
150 mm wide	–	0.25	5.43	–	7.44	m	**12.87**
Level uncoursed rubble walling for damp-proof courses and the like							
275 mm wide	–	0.19	4.35	–	1.83	m	**6.18**
350 mm wide	–	0.20	4.58	–	2.27	m	**6.85**
425 mm wide	–	0.21	4.81	–	2.78	m	**7.59**
500 mm wide	–	0.22	5.04	–	3.29	m	**8.33**
Copings formed of rough stones; faced and pointed all round							
275 mm × 200 mm (average) high	–	0.56	12.16	–	6.56	m	**18.72**
350 mm × 250 mm (average) high	–	0.75	16.14	–	9.07	m	**25.21**
425 mm × 300 mm (average) high	–	0.97	20.73	–	12.37	m	**33.10**
500 mm × 300 mm (average) high	–	1.23	26.09	–	16.63	m	**42.72**
STONE WALLING/DRESSINGS							
Reconstructed walling; Bradstone masonry blocks; standard colours; or other equal and approved; laid to pattern or course recommended; bedded; jointed and pointed in approved coloured cement: lime mortar (1:2:9)							
Walls; facing and pointing one side							
Fyfestone Enviromasonry Rustic	–	1.00	21.56	–	23.47	m²	**45.03**
Fyfestone Enviromasonry Fairfaced	–	1.00	21.56	–	17.86	m²	**39.42**
Fyfestone Enviromasonry Split	–	1.10	23.72	–	28.54	m²	**52.26**
Fyfestone Enviromasonry Textured	–	1.10	23.72	–	30.67	m²	**54.39**
Fyfestone Enviromasonry Polished	–	2.00	43.11	–	75.52	m²	**118.63**
masonry blocks; random uncoursed	–	1.04	22.42	–	41.53	m²	**63.95**
extra for:							
returned ends	–	0.37	7.98	–	32.76	m	**40.74**
plain L shaped quoins	–	0.12	2.58	–	40.76	m	**43.34**
traditional walling; coursed squared	–	1.30	28.02	–	41.53	m²	**69.55**
squared coursed rubble	–	1.25	26.95	–	42.70	m²	**69.65**
squared random rubble	–	1.30	28.02	–	42.54	m²	**70.56**
squared and pitched rock faced walling; coursed	–	1.34	28.89	–	42.54	m²	**71.43**
ashlar; 440 × 215 × 100 mm thick	–	1.10	23.72	–	43.45	m²	**67.17**
rough hewn rockfaced walling; random	–	1.39	29.97	–	42.37	m²	**72.34**
extra for:							
returned ends	–	0.15	3.23	–	–	m	**3.23**
Isolated piers or chimney stacks; facing and pointing one side							
Enviromasonry Rustic	–	1.40	30.19	–	23.47	m²	**53.66**
masonry blocks; random uncoursed	–	1.43	30.83	–	41.53	m²	**72.36**
traditional walling; coursed squared	–	1.80	38.81	–	41.53	m²	**80.34**
squared coursed rubble	–	1.76	37.95	–	42.70	m²	**80.65**
squared random rubble	–	1.80	38.81	–	42.54	m²	**81.35**
squared and pitched rock faced walling; coursed	–	1.90	40.96	–	42.54	m²	**83.50**
ashlar; 440 × 215 × 100 mm thick	–	1.54	33.20	–	43.45	m²	**76.65**
rough hewn rockfaced walling; random	–	1.94	41.82	–	42.37	m²	**84.19**

14 MASONRY

Item	PC £	Labour hours	Labour £	Plant £	Material £	Unit	Total rate £
STONE WALLING/DRESSINGS – cont							
Reconstructed walling – cont							
Isolated casings; facing and pointing one side							
Enviromasonry Rustic	–	1.20	25.87	–	23.47	m²	**49.34**
masonry blocks; random uncoursed	–	1.25	26.95	–	41.53	m²	**68.48**
traditional walling; coursed squared	–	1.57	33.85	–	41.53	m²	**75.38**
squared coursed rubble	–	1.53	32.98	–	42.70	m²	**75.68**
squared random rubble	–	1.57	33.85	–	42.54	m²	**76.39**
squared and pitched rock faced walling; coursed	–	1.62	34.92	–	42.54	m²	**77.46**
ashlar; 440 × 215 × 100 mm thick	–	1.32	28.45	–	43.45	m²	**71.90**
rough hewn rockfaced walling; random	–	1.67	36.00	–	42.37	m²	**78.37**
Fair returns 100 mm wide							
Enviromasonry Rustic	–	0.10	2.15	–	–	m²	**2.15**
masonry blocks; random uncoursed	–	0.11	2.37	–	–	m²	**2.37**
traditional walling; coursed squared	–	0.14	3.01	–	–	m²	**3.01**
squared coursed rubble	–	0.13	2.80	–	–	m²	**2.80**
squared random rubble	–	0.14	3.01	–	–	m²	**3.01**
squared and pitched rock faced walling; coursed	–	0.14	3.01	–	–	m²	**3.01**
ashlar; 440 × 215 × 100 mm thick	–	0.14	3.01	–	–	m²	**3.01**
rough hewn rockfaced walling; random	–	0.15	3.23	–	–	m²	**3.23**
Fair raking cutting or circular cutting							
100 mm wide	–	0.17	3.66	–	–	m	**3.66**
Quoin							
ashlar; 440 × 215 × 100 mm thick	–	0.75	16.17	–	83.41	m	**99.58**
Reconstructed limestone dressings; Bradstone Architectural dressings in weathered Cotswold or North Cerney shades or other equal and approved; bedded, jointed and pointed in approved coloured cement: lime mortar (1:2:9)							
Copings; twice weathered and throated							
305 mm × 76 mm; type A	–	0.37	7.98	–	25.92	m	**33.90**
extra for;							
fair end	–	–	–	–	12.87	nr	**12.87**
returned mitred fair end	–	–	–	–	12.87	nr	**12.87**
Copings; once weathered and throated							
305 mm × 76 mm	–	0.37	7.98	–	25.55	m	**33.53**
356 mm × 76 mm	–	0.37	7.98	–	23.67	m	**31.65**
extra for;							
fair end	–	–	–	–	12.87	nr	**12.87**
returned mitred fair end	–	–	–	–	12.87	nr	**12.87**
Pier caps; four times weathered and throated							
305 mm × 305 mm	–	0.23	4.96	–	15.29	nr	**20.25**
381 mm × 381 mm	–	0.23	4.96	–	22.68	nr	**27.64**
457 mm × 457 mm	–	0.28	6.04	–	31.02	nr	**37.06**
533 mm × 533 mm	–	0.28	6.04	–	43.05	nr	**49.09**
Splayed corbels							
479 mm × 100 mm × 215 mm	–	0.14	3.01	–	25.36	nr	**28.37**
665 mm × 100 mm × 215 mm	–	0.19	4.10	–	35.06	nr	**39.16**

14 MASONRY

Item	PC £	Labour hours	Labour £	Plant £	Material £	Unit	Total rate £
100 mm × 140 mm lintels; rectangular; reinforced with mild steel bars							
all lengths to 2.07 m	–	0.26	5.61	–	40.50	m	**46.11**
100 mm × 215 mm lintels; rectangular; reinforced with mild steel bars							
all lengths to 2.85 m	–	0.30	6.47	–	43.24	m	**49.71**
Sills to suit standard windows; stooled 100 mm at ends							
150 mm ×140 mm; not exceeding 1.97 m long	–	0.28	6.04	–	54.66	m	**60.70**
197 mm ×140 mm; not exceeding 1.97 m long	–	0.28	6.04	–	61.38	m	**67.42**
Window surround; traditional with label moulding; for single light; sill 146 mm × 133 mm; jambs 146 mm × 146 mm; head 146 mm × 105 mm; including all dowels and anchors							
overall size 508 mm × 1479 mm	–	0.83	17.89	–	188.53	nr	**206.42**
Window surround; traditional with label moulding; three light; for windows 508 mm × 1219 mm; sill 146 mm × 133 mm; jambs 146 mm × 146 mm; head 146 mm × 103 mm; mullions 146 mm × 108 mm; including all dowels and anchors							
overall size 1975 mm × 1479 mm	–	2.17	46.79	–	443.70	nr	**490.49**
Door surround; moulded continuous jambs and head with label moulding; including all dowels and anchors							
door 839 mm × 1981 mm in 102 mm × 64 mm frame	–	1.53	32.98	–	401.87	nr	**434.85**
Portland Whitbed limestone bedded and jointed in cement–lime–mortar (1:2:9); slurrying with weak lime and stone dust mortar; flush pointing and cleaning on completion (cramps etc. not included)							
Facework; one face plain and rubbed; bedded against backing							
50 mm thick stones	–	–	–	–	–	m²	**347.53**
63 mm thick stones	–	–	–	–	–	m²	**395.85**
75 mm thick stones	–	–	–	–	–	m²	**447.68**
100 mm thick stones	–	–	–	–	–	m²	**491.44**
Fair returns on facework							
50 mm wide	–	–	–	–	–	m	**4.85**
63 mm wide	–	–	–	–	–	m	**6.06**
75 mm wide	–	–	–	–	–	m	**8.49**
100 mm wide	–	–	–	–	–	m	**10.92**
Fair raking cutting on facework							
50 mm thick	–	–	–	–	–	m	**21.85**
63 mm thick	–	–	–	–	–	m	**24.27**
75 mm thick	–	–	–	–	–	m	**29.12**
100 mm thick	–	–	–	–	–	m	**31.56**

14 MASONRY

Item	PC £	Labour hours	Labour £	Plant £	Material £	Unit	Total rate £
STONE WALLING/DRESSINGS – cont							
Portland Whitbed limestone bedded and jointed in cement–lime–mortar (1:2:9) – cont							
Copings; once weathered; and throated; rubbed; set horizontal or raking							
250 mm × 50 mm	–	–	–	–	–	m	169.87
extra for:							
external angle	–	–	–	–	–	nr	30.34
internal angle	–	–	–	–	–	nr	30.34
300 mm × 50 mm	–	–	–	–	–	m	179.59
extra for:							
external angle	–	–	–	–	–	nr	30.34
internal angle	–	–	–	–	–	nr	36.41
350 mm × 75 mm	–	–	–	–	–	m	200.22
extra for:							
external angle	–	–	–	–	–	nr	30.34
internal angle	–	–	–	–	–	nr	38.83
400 mm × 100 mm	–	–	–	–	–	m	240.26
extra for:							
external angle	–	–	–	–	–	nr	36.41
internal angle	–	–	–	–	–	nr	50.97
450 mm × 100 mm	–	–	–	–	–	m	291.23
extra for:							
external angle	–	–	–	–	–	nr	46.11
internal angle	–	–	–	–	–	nr	63.10
500 mm × 125 mm	–	–	–	–	–	m	442.90
extra for:							
external angle	–	–	–	–	–	nr	63.10
internal angle	–	–	–	–	–	nr	78.87
Band courses; plain; rubbed; horizontal							
225 mm × 112 mm	–	–	–	–	–	m	133.48
300 mm × 112 mm	–	–	–	–	–	m	178.38
extra for:							
stopped ends	–	–	–	–	–	nr	7.28
external angles	–	–	–	–	–	nr	7.28
Band courses; moulded 100 mm girth on face; rubbed; horizontal							
125 mm × 75 mm	–	–	–	–	–	m	151.68
extra for:							
stopped ends	–	–	–	–	–	nr	24.27
external angles	–	–	–	–	–	nr	30.34
internal angles	–	–	–	–	–	nr	60.68
150 mm × 75 mm	–	–	–	–	–	m	175.96
extra for:							
stopped ends	–	–	–	–	–	nr	24.27
external angles	–	–	–	–	–	nr	42.47
internal angles	–	–	–	–	–	nr	72.81
200 mm × 100 mm	–	–	–	–	–	m	200.22

14 MASONRY

Item	PC £	Labour hours	Labour £	Plant £	Material £	Unit	Total rate £
extra for:							
stopped ends	–	–	–	–	–	nr	24.27
external angles	–	–	–	–	–	nr	60.68
internal angles	–	–	–	–	–	nr	109.21
250 mm × 150 mm	–	–	–	–	–	m	291.23
extra for:							
stopped ends	–	–	–	–	–	nr	24.27
external angles	–	–	–	–	–	nr	60.68
internal angles	–	–	–	–	–	nr	121.34
300 mm × 250 mm	–	–	–	–	–	m	473.24
extra for:							
stopped ends	–	–	–	–	–	nr	24.27
external angles	–	–	–	–	–	nr	84.94
internal angles	–	–	–	–	–	nr	145.61
Coping apex block; two sunk faces; rubbed							
650 mm × 450 mm × 225 mm	–	–	–	–	–	nr	594.58
Coping kneeler block; three sunk faces; rubbed							
350 mm × 350 mm × 375 mm	–	–	–	–	–	nr	473.24
450 mm × 450 mm × 375 mm	–	–	–	–	–	nr	546.05
Corbel; turned and moulded; rubbed							
225 mm × 225 mm × 375 mm	–	–	–	–	–	nr	388.30
Slab surrounds to openings; one face splayed; rubbed							
75 mm × 100 mm	–	–	–	–	–	m	84.94
75 mm × 200 mm	–	–	–	–	–	m	115.28
100 mm × 100 mm	–	–	–	–	–	m	103.15
125 mm × 100 mm	–	–	–	–	–	m	115.28
125 mm × 150 mm	–	–	–	–	–	m	139.55
175 mm × 175 mm	–	–	–	–	–	m	169.87
225 mm × 175 mm	–	–	–	–	–	m	200.22
300 mm × 175 mm	–	–	–	–	–	m	242.69
300 mm × 225 mm	–	–	–	–	–	m	315.50
Slab surrounds to openings; one face sunk splayed; rubbed							
75 mm × 100 mm	–	–	–	–	–	m	109.21
75 mm × 200 mm	–	–	–	–	–	m	139.55
100 mm × 100 mm	–	–	–	–	–	m	127.42
125 mm × 100 mm	–	–	–	–	–	m	139.55
125 mm × 150 mm	–	–	–	–	–	m	163.82
175 mm × 175 mm	–	–	–	–	–	m	194.14
225 mm × 175 mm	–	–	–	–	–	m	224.49
300 mm × 175 mm	–	–	–	–	–	m	266.96
300 mm × 225 mm	–	–	–	–	–	m	339.77
extra for:							
throating	–	–	–	–	–	m	12.14
rebates and grooves	–	–	–	–	–	m	26.70
stooling	–	–	–	–	–	m	46.11

Prices for Measured Works

14 MASONRY

Item	PC £	Labour hours	Labour £	Plant £	Material £	Unit	Total rate £
STONE WALLING/DRESSINGS – cont							
Eurobrick insulated brick cladding systems or other equal and approved; extruded polystyrene foam insulation; brick slips bonded to insulation panels with Eurobrick gun applied adhesive or other equal and approved; pointing with formulated mortar grout							
25 mm insulation to walls							
over 300 mm wide; fixing with proprietary screws and plates to timber	–	1.39	29.28	–	46.48	m²	**75.76**
50 mm insulation to walls							
over 300 mm wide; fixing with proprietary screws and plates; to timber	–	1.39	29.28	–	50.78	m²	**80.06**
SUNDRY ITEMS AND ACCESSORIES							
Forming cavities							
In hollow walls							
width of cavity 50 mm; polypropylene ties; three wall ties per m²	–	0.05	1.08	–	0.32	m²	**1.40**
width of cavity 50 mm; galvanized steel twisted wall ties; three wall ties per m²	–	0.05	1.08	–	0.79	m²	**1.87**
width of cavity 50 mm; stainless steel butterfly wall ties; three wall ties per m²	–	0.05	1.08	–	0.58	m²	**1.66**
width of cavity 50 mm; stainless steel twisted wall ties; three wall ties per m²	–	0.05	1.08	–	0.72	m²	**1.80**
width of cavity 75 mm; polypropylene ties; three wall ties per m²	–	0.05	1.08	–	0.32	m²	**1.40**
width of cavity 75 mm; galvanized steel twisted wall ties; three wall ties per m²	–	0.05	1.08	–	0.84	m²	**1.92**
width of cavity 75 mm; stainless steel butterfly wall ties; three wall ties per m²	–	0.05	1.08	–	0.81	m²	**1.89**
width of cavity 75 mm; stainless steel twisted wall ties; three wall ties per m²	–	0.05	1.08	–	0.80	m²	**1.88**
Damp-proof courses							
Polythene damp-proof course or other equal and approved; 200 mm laps; in gauged mortar (1:1:6)							
100 mm width exceeding horizontal in brick or block course	–	0.01	0.22	–	0.11	m	**0.33**
150 mm width exceeding horizontal in brick or block course	–	0.01	0.22	–	0.14	m	**0.36**
300 mm width exceeding horizontal in brick or block course	–	0.01	0.22	–	0.27	m	**0.49**
width exceeding 225 mm; horizontal	–	0.23	4.96	–	0.38	m²	**5.34**
width exceeding 225 mm; forming cavity gutters in hollow walls; horizontal	–	0.37	7.98	–	0.38	m²	**8.36**
width not exceeding 225 mm; horizontal	–	0.46	9.91	–	0.38	m²	**10.29**
width not exceeding 225 mm; vertical	–	0.69	14.88	–	0.38	m²	**15.26**

14 MASONRY

Item	PC £	Labour hours	Labour £	Plant £	Material £	Unit	Total rate £
Icopal Polymeric damp-proof course; Xtra-Load Pro-Build or other equal and approved; 200 mm laps; in gauged morter (1:1:6)							
width exceeding 225 mm; horizontal	–	0.23	4.96	–	15.75	m²	**20.71**
width exceeding 225 mm; forming cavity gutters in hollow walls; horizontal	–	0.37	7.98	–	15.75	m²	**23.73**
width not exceeding 225 mm; horizontal	–	0.46	9.91	–	15.75	m²	**25.66**
width not exceeding 225 mm; vertical	–	0.69	14.88	–	15.75	m²	**30.63**
Zedex CPT (Co-Polymer Thermoplastic) damp-proof course or other equal and approved; 200 mm laps; in gauged mortar (1:1:6)							
width exceeding 225 mm; horizontal	–	0.23	4.96	–	6.72	m²	**11.68**
width exceeding 225 mm wide; forming cavity gutters in hollow walls; horizontal	–	0.37	7.98	–	6.72	m²	**14.70**
width not exceeding 225 mm; horizontal	–	0.46	9.91	–	6.72	m²	**16.63**
width not exceeding 225 mm; vertical	–	0.69	14.88	–	6.72	m²	**21.60**
Hyload (pitch polymer) damp-proof course or other equal and approved; 150 mm laps; in gauged mortar (1:1:6)							
width exceeding 225 mm; horizontal	–	0.23	4.96	–	4.74	m²	**9.70**
width exceeding 225 mm; forming cavity gutters in hollow walls; horizontal	–	0.37	7.98	–	4.74	m²	**12.72**
width not exceeding 225 mm; horizontal	–	0.46	9.91	–	4.74	m²	**14.65**
width not exceeding 225 mm; vertical	–	0.69	14.88	–	4.74	m²	**19.62**
Nubit bitumen and polyester-based damp-proof course or other equal and approved; 200 mm laps; in gauged mortar (1:1:6)							
width exceeding 225 mm; horizontal	–	0.23	4.96	–	9.46	m²	**14.42**
width exceeding 225 mm wide; forming cavity gutters in hollow walls; horizontal	–	0.37	7.98	–	9.46	m²	**17.44**
width not exceeding 225 mm; horizontal	–	0.46	9.91	–	9.46	m²	**19.37**
width not exceeding 225 mm; vertical	–	0.69	14.88	–	9.46	m²	**24.34**
Permabit bitumen polymer damp-proof course or other equal and approved; 150 mm laps; in gauged mortar (1:1:6)							
width exceeding 225 mm; horizontal	–	0.23	4.96	–	10.22	m²	**15.18**
width exceeding 225 mm; forming cavity gutters in hollow walls; horizontal	–	0.37	7.98	–	10.22	m²	**18.20**
width not exceeding 225 mm; horizontal	–	0.46	9.91	–	10.22	m²	**20.13**
width not exceeding 225 mm; vertical	–	0.69	14.88	–	10.22	m²	**25.10**
Alumite aluminium cored bitumen gas retardant damp-proof course or other equal and approved; 200 mm laps; in gauged mortar (1:1;6)							
width exceeding 225 mm; horizontal	–	0.31	6.69	–	7.09	m²	**13.78**
width exceeding 225 mm; forming cavity gutters in hollow walls; horizontal	–	0.49	10.56	–	7.09	m²	**17.65**
width not exceeding 225 mm; horizontal	–	0.60	12.94	–	7.09	m²	**20.03**
width not exceeding 225 mm; vertical	–	0.83	17.89	–	7.09	m²	**24.98**

14 MASONRY

Item	PC £	Labour hours	Labour £	Plant £	Material £	Unit	Total rate £
SUNDRY ITEMS AND ACCESSORIES – cont							
Damp-proof courses – cont							
Milled lead damp-proof course; BS 1178; 1.80 mm thick (code 4), 175 mm laps; in cement: lime mortar (1:2:9)							
width exceeding 225 mm; horizontal (PC £/kg)	–	1.85	39.88	–	37.69	m²	**77.57**
width not exceeding 225 mm; horizontal	–	2.78	59.93	–	37.69	m²	**97.62**
Two courses slates in cement: mortar (1:3)							
width exceeding 225 mm; horizontal	–	1.39	29.97	–	13.98	m²	**43.95**
width exceeding 225 mm; vertical	–	2.08	44.85	–	13.98	m²	**58.83**
Synthaprufe damp-proof membrane or other equal and approved; three coats brushed on							
width not exceeding 150 mm; vertical	–	0.31	4.07	–	7.19	m²	**11.26**
width 150 mm–225 mm; vertical	–	0.30	3.94	–	7.19	m²	**11.13**
width 225 mm–300 mm; vertical	–	0.28	3.67	–	7.19	m²	**10.86**
width exceeding 300 mm wide; vertical	–	0.26	3.41	–	7.19	m²	**10.60**
Joint reinforcement							
Brickforce galvanized steel joint reinforcement or other equal and approved							
width 60 mm; ref GBF40W60B25	–	0.05	1.08	–	1.46	m	**2.54**
width 100 mm; ref GBF40W100B25	–	0.07	1.51	–	1.71	m	**3.22**
width 175 mm; ref GBF40W175B25	–	0.10	2.15	–	3.43	m	**5.58**
Brickforce stainless steel joint reinforcement or other equal and approved							
width 60 mm; ref SBF35W60BSC	–	0.05	1.08	–	4.37	m	**5.45**
width 100 mm; ref SBF35W100BSC	–	0.07	1.51	–	6.11	m	**7.62**
width 175 mm; ref SBF35W175BSC	–	0.10	2.15	–	4.96	m	**7.11**
Wallforce stainless steel joint reinforcement or other equal and approved							
width 240 mm; ref SWF35W240	–	0.12	2.58	–	7.17	m	**9.75**
width 260 mm; ref SWF35W260	–	0.13	2.80	–	8.91	m	**11.71**
width 275 mm; ref SWF35W275	–	0.14	3.01	–	10.66	m	**13.67**
Weather fillets							
Weather fillets in cement: mortar (1:3)							
50 mm face width	–	0.11	2.37	–	0.05	m	**2.42**
100 mm face width	–	0.19	4.10	–	0.19	m	**4.29**
Angle fillets							
Angle fillets in cement: mortar (1:3)							
50 mm face width	–	0.11	2.37	–	0.05	m	**2.42**
100 mm face width	–	0.19	4.10	–	0.19	m	**4.29**
Pointing in							
Pointing with mastic							
wood frames or sills	–	0.09	1.47	–	0.95	m	**2.42**

14 MASONRY

Item	PC £	Labour hours	Labour £	Plant £	Material £	Unit	Total rate £
Pointing with polysulphide sealant							
wood frames or sills	–	0.09	1.47	–	2.40	m	**3.87**
Wedging and pinning							
To underside of existing construction with slates in cement: mortar (1:3)							
width of wall – one brick thick	–	0.74	15.95	–	3.09	m	**19.04**
width of wall – one and a half brick thick	–	0.93	20.04	–	6.16	m	**26.20**
width of wall – two brick thick	–	1.11	23.93	–	9.25	m	**33.18**
Joints							
Hacking joints and faces of brickwork or blockwork to form key for plaster	–	0.24	3.15	–	–	m²	**3.15**
Raking out joint in brickwork or blockwork for turned-in edge of flashing							
horizontal	–	0.14	3.01	–	–	m	**3.01**
stepped	–	0.19	4.10	–	–	m	**4.10**
Raking out and enlarging joint in brickwork or blockwork for nib of asphalt							
horizontal	–	0.19	4.10	–	–	m	**4.10**
Cutting grooves in brickwork or blockwork							
for water bars and the like	–	0.23	3.02	0.87	–	m	**3.89**
for nib of asphalt; horizontal	–	0.23	3.02	0.87	–	m	**3.89**
Preparing to receive new walls							
top existing 215 mm wall	–	0.19	4.10	–	–	m	**4.10**
Cleaning and priming both faces; filling with preformed closed cell joint filler and pointing one side with polysulphide sealant; 12 mm deep							
expansion joints; 12 mm wide	–	0.23	4.43	–	4.61	m	**9.04**
expansion joints; 20 mm wide	–	0.28	5.25	–	6.80	m	**12.05**
expansion joints; 25 mm wide	–	0.32	5.85	–	8.24	m	**14.09**
Fire-resisting horizontal expansion joints; filling with joint filler; fixed with high temperature slip adhesive; between top of wall and soffit							
wall not exceeding 215 mm wide; 10 mm wide joint with 30 mm deep filler (one hour fire seal)	–	0.23	4.96	–	5.44	m	**10.40**
wall not exceeding 215 mm wide; 10 mm wide joint with 30 mm deep filler (two hour fire seal)	–	0.23	4.96	–	5.44	m	**10.40**
wall not exceeding 215 mm wide; 20 mm wide joint with 45 mm deep filler (two hour fire seal)	–	0.28	6.04	–	8.41	m	**14.45**
wall not exceeding 215 mm wide; 30 mm wide joint with 75 mm deep filler (three hour fire seal)	–	0.32	6.90	–	21.63	m	**28.53**
Fire-resisting vertical expansiojn joints; filling with joint filler; fixed with high temperature slip adhesive; with polysulphide sealant one side; between end of wall and concrete							
wall not exceeding 215 mm wide; 20 mm wide joint with 45 mm deep filler (two hour fire seal)	–	0.37	7.45	–	13.21	m	**20.66**

14 MASONRY

Item	PC £	Labour hours	Labour £	Plant £	Material £	Unit	Total rate £
SUNDRY ITEMS AND ACCESSORIES – cont							
Slate and tile sills							
Sills; two courses of machine-made plain roofing tiles							
set weathering; bedded and pointed	–	0.56	12.07	–	6.12	m	**18.19**
Sundries							
Weep holes							
perpend units; plastic	–	0.02	0.43	–	119.37	nr	**119.80**
Chimney pots; red terracotta; plain roll top or cannon-head; setting and flaunching in cement mortar (1:3)							
185 mm dia. × 450 mm long	39.24	1.67	36.00	–	42.69	nr	**78.69**
185 mm dia. × 600 mm long	60.11	1.85	39.88	–	64.61	nr	**104.49**
185 mm dia. × 900 mm long	109.31	1.85	39.88	–	116.27	nr	**156.15**
Air bricks							
Air bricks; red terracotta; building into prepared openings							
215 mm × 65 mm	–	0.07	1.51	–	3.35	nr	**4.86**
215 mm × 140 mm	–	0.07	1.51	–	4.98	nr	**6.49**
215 mm × 215 mm	–	0.07	1.51	–	12.43	nr	**13.94**
Halfen channels and brick support (NB Supply only rates)							
cavity brick tie; Halfen HTS-C12; 225 mm stainless steel A2	–	–	–	–	0.42	nr	**0.42**
cavity brick tie; Halfen HTS-C12; 225 mm pre-galvanized	–	–	–	–	0.17	nr	**0.17**
frame cramp, with slot; Halfen HTS-FS12; 150 mm stainless steel A2	–	–	–	–	0.32	nr	**0.32**
frame cramp, with slot; Halfen HTS-FS12; 150 mm pre-galvanized	–	–	–	–	0.14	nr	**0.14**
head restraint; Halfen CHR-V telescopic concealed tube and HSC−9 tie, in A2/ 304 stainless steel	–	–	–	–	2.60	nr	**2.60**
brick tie channel; Halfen FRS 25/14 2700; supplied in 2700 mm lengths with fixing holes at 110 mm centres, grade 304 stainless steel	–	–	–	–	2.17	nr	**2.17**
channel tie; Halfen FRS−03; 150 mm long safety end to suit 25/14 FRS channel, stainless steel	–	–	–	–	0.42	nr	**0.42**
ribslot brick tie channel; Halfen; 3000 mm long, stainless steel A2, complete with polystyrene filler (vf) and nail holes, to suit HTS ties	–	–	–	–	12.65	nr	**12.65**
channel tie; Halfen HTS-B12; 200 mm long, grade 304 stainless steel safety end to suit 28/15, 25/17HH and ribslot channels	–	–	–	–	0.45	nr	**0.45**

14 MASONRY

Item	PC £	Labour hours	Labour £	Plant £	Material £	Unit	**Total rate £**
ribslot brick tie channel; Halfen; 3000 mm long, pre-galvanized mild steel, complete with polystyrene filler (vf) and nail holes, to suit HTS ties	–	–	–	–	6.76	nr	**6.76**
channel tie; Halfen HTS-B12; 150 mm long, safety end pre-galvanized to suit 28/15, 25/17 HH and ribslot channels	–	–	–	–	0.17	nr	**0.17**
brick support angle swystem; Halfen HK4-U-100 cavity/6 kN inc fixings stainless steel, A4	–	–	–	–	60.10	m	**60.10**
stone support anchor, grout-in; Halfen UMA-10-1; 150 mm stainless steel	–	–	–	–	1.54	nr	**1.54**
stone restraint anchor, grout-in; Halfen UHA-5-1; 180 mm stainless steel A4	–	–	–	–	0.85	nr	**0.85**
windpost, brickwork; Halfen BW1-1406 × 2500 mm complete with fixings & ties, stainless steel, A4	–	–	–	–	229.32	nr	**229.32**
windpost, cavity; Halfen CW2-3544 × 2500 mm stainless steel complete with fixings & ties	–	–	–	–	149.47	nr	**149.47**
cast-in CE marked channel, cold rolled; Halfen HTA-K-38/17; 200 mm anchor centres, stainless steel A2–70, complete with combi filler(kf) and nail holes supplied in 3000 mm lengths	–	–	–	–	21.45	m	**21.45**
'T' bolt and nut to suit; Halfen HS-38/17 channel; M12 × 50 mm, stainless steel A2–70	–	–	–	–	2.02	nr	**2.02**
cast-in CE marked channel, cold rolled; Halfen HTA-K-38/17; 200 mm anchor centres, hot-dip galvanized (fv), complete with combi filler (kf) and nail holes supplies in 3000 mm lengths	–	–	–	–	12.72	m	**12.72**
'T' bolt and nut to suit Halfen HS-38/17 channel; M12 × 50 mm, hot-dip galvanized (fv) 4.6 mild steel	–	–	–	–	0.79	nr	**0.79**
cast-in CE marked channel, hot rolled; Halfen HTA-K-40/22; hot-dip galvanized (fv) , complete with combi filler (kf) and nail holes supplied in 3000 mm lengths	–	–	–	–	42.89	m	**42.89**
'T' bolt and nut to suit Halfen HS-40/22 channel; M16 × 50 mm, hot-dip galvanized (fv) 4.6 mild steel	–	–	–	–	1.47	nr	**1.47**
cast-in channel, hot rolled; Halfen HTA-52/34; hot-dip galvanized (fv), complete with polystyrene filler (vf) and nail holes supplied in 6070 mm lengths	–	–	–	–	37.06	m	**37.06**

14 MASONRY

Item	PC £	Labour hours	Labour £	Plant £	Material £	Unit	Total rate £
SUNDRY ITEMS AND ACCESSORIES – cont							
Halfen channels and brick support – cont							
'T' bolt and nut, to suit Halfen HS – 52/34 channel; M16 × 60 mm hot-dip galvanized (fv) 4.6 mild steel	–	–	–	–	2.79	nr	**2.79**
cast-in CE marked channel, hot rolled; Halfen HTA−52/34; stainless steel A4, complete with polystyrene filler (vf) and nail holes supplied in 6070 mm lengths	–	–	–	–	114.36	m	**114.36**
'T' bolt and nut, to suit Halfen HS−52/34 channel; M16 × 60 mm stainless steel A4 (fv) 4.6 mild steel	–	–	–	–	7.41	nr	**7.41**
reinforcement, female coupler; Halfen HBM-F16/M20; 810 mm long	–	–	–	–	3.28	nr	**3.28**
reinforcement, male coupler; Halfen HBM-M16/M20; 770 mm long	–	–	–	–	4.40	nr	**4.40**
reinforcement continuity system; Halfen Kwikastrip M17Sb12C150; 1200 mm	–	–	–	–	36.36	nr	**36.36**
shear dowel set; Halfen CRET−124 stainless steel A4	–	–	–	–	56.10	nr	**56.10**
shear dowel set; Halfen CRET−10; 20 × 300 mm stainless steel A4	–	–	–	–	8.19	nr	**8.19**
shear sleeve; Halfen CRET-J−20; 160 stainless steel A4	–	–	–	–	6.75	nr	**6.75**
concrete balcony insulated connection Unit Halfen HIT HP; 80 mm unit for a 200 mm slab for a 1500 mm cantilever balcony (Passivhaus certified)	–	–	–	–	87.02	nr	**87.02**
concrete balcony insulated connection Unit Halfen HIT HP; 80 mm Unit for a 200 mm slab for a 2000 mm cantilever balcony (Passivhaus certified)	–	–	–	–	127.97	nr	**127.97**
steel balcony connection SBC-HBM; M20 × 4 BZP Cast-in Socket Anchor Assembly, 190 × 110 mm	–	–	–	–	38.90	nr	**38.90**
steel balcony connection; Halfen; SBC-TSS−10; 330 × 200 mm thermal separator pad, 4no. 22 mm holes at 190 × 110 mm	–	–	–	–	92.14	nr	**92.14**
precast lifting spread anchor CE marked; Halfen TPA-FS; 4.0 tonne × 180 mm carbon steel (wb)	–	–	–	–	2.10	nr	**2.10**
precast ring clutch CE marked; Halfen TPA R1; 5.0T (T)	–	–	–	–	116.71	nr	**116.71**
External door and window cavity closures							
Thermabate or equivalent; inclusive of flange clips; jointing strips; wall fixing ties and adhesive tape							
closing cavities; width of cavity 50 mm–60 mm	–	0.14	3.01	–	3.01	m	**6.02**
closing cavities; width of cavity 75 mm–84 mm	–	0.14	3.01	–	3.29	m	**6.30**
closing cavities; width of cavity 90 mm–99 mm	–	0.14	3.01	–	3.60	m	**6.61**
closing cavities; width of cavity 100 mm–110 mm	–	0.14	3.01	–	3.60	m	**6.61**

14 MASONRY

Item	PC £	Labour hours	Labour £	Plant £	Material £	Unit	Total rate £
Kooltherm or equivalent; inclusive of flange clips; jointing strips; wall fixing ties and adhesive tape; 1 hr fire-rating							
closing cavities; width of cavity 50 mm	–	0.14	3.01	–	3.83	m	**6.84**
closing cavities; width of cavity 75 mm	–	0.14	3.01	–	3.99	m	**7.00**
closing cavities; width of cavity 100 mm	–	0.14	3.01	–	4.27	m	**7.28**
closing cavities; width of cavity 125 mm	–	0.14	3.01	–	5.07	m	**8.08**
closing cavities; width of cavity 150 mm	–	0.14	3.01	–	5.31	m	**8.32**
Type H cavicloser or other equal and approved; uPVC universal cavity closer, insulator and damp-proof course by Cavity Trays Ltd; built into cavity wall as work proceeds, complete with face closer and ties							
closing cavities; width of cavity 50 mm–100 mm	–	0.07	1.51	–	1.60	m	**3.11**
Type L durropolyethelene lintel stop ends or other equal and approved; Cavity Trays Ltd; fixing with butyl anchoring strip; building in as the work proceeds							
adjusted to lintel as required	–	0.04	0.86	–	0.57	nr	**1.43**
Type W polypropylene weeps/vents or other equal and approved; Cavity Trays Ltd; built into cavity wall as work proceeds							
100/115 mm × 65 mm × 10 mm including lock fit wedges	–	0.04	0.86	–	0.42	nr	**1.28**
extra; extension duct							
200/225 mm × 65 mm × 10 mm	–	0.07	1.51	–	0.75	nr	**2.26**
Type X polypropylene abutment cavity tray or other equal and approved; Cavity Trays Ltd; built into facing brickwork as the work proceeds; complete with Code 4 flashing; intermediate/catchment tray with short leads (requiring soakers); to suit roof of							
17–20° pitch	–	0.05	1.08	–	5.28	nr	**6.36**
21–25° pitch	–	0.05	1.08	–	4.92	nr	**6.00**
26–45° pitch	–	0.05	1.08	–	4.70	nr	**5.78**
Type X polypropylene abutment cavity tray or other equal and approved; Cavity Trays Ltd; built into facing brickwork as the work proceeds; complete with Code 4 flashing; intermediate/catchment tray with long leads (suitable only for corrugated roof tiles); to suit roof of							
17–20° pitch	–	0.05	1.08	–	7.14	nr	**8.22**
21–25° pitch	–	0.05	1.08	–	6.57	nr	**7.65**
26–45° pitch	–	0.05	1.08	–	6.06	nr	**7.14**
Type X polypropylene abutment cavity tray or other equal and approved; Cavity Trays Ltd; built into facing brickwork as the work proceeds; complete with Code 4 flashing; ridge tray with short/long leads; to suit roof of							
17–20° pitch	–	0.05	1.08	–	12.03	nr	**13.11**
21–25° pitch	–	0.05	1.08	–	11.16	nr	**12.24**
26–45° pitch	–	0.05	1.08	–	9.92	nr	**11.00**

14 MASONRY

Item	PC £	Labour hours	Labour £	Plant £	Material £	Unit	Total rate £
SUNDRY ITEMS AND ACCESSORIES – cont							
External door and window cavity closures – cont							
Expamet stainless steel wall starters or other equal and approved; plugged and screwed							
to suit walls 60 mm–75 mm thick	–	0.23	3.02	0.12	2.12	m	**5.26**
to suit walls 100 mm–115 mm thick	–	0.23	3.02	0.12	21.41	m	**24.55**
to suit walls 125 mm–180 mm thick	–	0.37	4.86	0.18	2.89	m	**7.93**
to suit walls 190 mm–260 mm thick	–	0.46	6.05	0.23	3.58	m	**9.86**
Galvanized steel lintels; Catnic or other equal and approved; built into brickwork or blockwork							
70/125 Range CG open back lintel for cavity wall							
750 mm long	–	0.23	4.96	–	34.91	nr	**39.87**
900 mm long	–	0.28	6.04	–	41.69	nr	**47.73**
1200 mm long	–	0.32	6.90	–	54.78	nr	**61.68**
1500 mm long	–	0.37	7.98	–	68.93	nr	**76.91**
1800 mm long	–	0.42	9.05	–	94.58	nr	**103.63**
2100 mm long	–	0.46	9.91	–	111.36	nr	**121.27**
2400 mm long	–	0.56	12.07	–	153.75	nr	**165.82**
70/125 range CUB open back lintel for cavity wall							
2700 mm long	–	0.65	14.02	–	180.52	nr	**194.54**
3000 mm long	–	0.74	15.95	–	251.34	nr	**267.29**
70/125 range CU open back lintel for cavity wall							
3300 mm long	–	0.83	17.89	–	309.72	nr	**327.61**
3600 mm long	–	0.93	20.04	–	347.85	nr	**367.89**
3900 mm long	–	1.02	21.99	–	373.14	nr	**395.13**
4200 mm long	–	0.46	9.91	–	409.23	nr	**419.14**
90/125 range CG open back lintel for cavity wall							
750 mm long	–	0.23	4.96	–	38.83	nr	**43.79**
900 mm long	–	0.28	6.04	–	46.59	nr	**52.63**
1200 mm long	–	0.32	6.90	–	61.16	nr	**68.06**
1500 mm long	–	0.37	7.98	–	76.26	nr	**84.24**
1800 mm long	–	0.42	9.05	–	96.48	nr	**105.53**
2100 mm long	–	0.46	9.91	–	114.33	nr	**124.24**
2400 mm long	–	0.56	12.07	–	161.49	nr	**173.56**
90/125 range CUB open back lintel for cavity wall							
2700 mm long	–	0.65	14.02	–	186.86	nr	**200.88**
3000 mm long	–	0.74	15.95	–	268.74	nr	**284.69**
90/125 range CU open back lintel for cavity wall							
3300 mm long	–	0.83	17.89	–	334.48	nr	**352.37**
3600 mm long	–	0.93	20.04	–	372.53	nr	**392.57**
3900 mm long	–	1.02	21.99	–	397.44	nr	**419.43**
4200 mm long	–	0.46	9.91	–	426.02	nr	**435.93**
CN92 single lintel; for 75 mm internal walls							
1050 mm long	–	0.28	6.04	–	5.92	nr	**11.96**
1200 mm long	–	0.32	6.90	–	6.70	nr	**13.60**
CN102 single lintel; for 100 mm internal walls							
1050 mm long	–	0.28	6.04	–	7.48	nr	**13.52**
1200 mm long	–	0.32	6.90	–	8.25	nr	**15.15**

14 MASONRY

Item	PC £	Labour hours	Labour £	Plant £	Material £	Unit	Total rate £
CN100 single lintel; for 75 mm internal walls							
1050 mm long	–	0.28	6.04	–	18.24	nr	**24.28**
1200 mm long	–	0.32	6.90	–	22.66	nr	**29.56**
CN5XA single lintel; for 100 mm internal walls							
1050 mm long	–	0.28	6.04	–	22.32	nr	**28.36**
1200 mm long	–	0.32	6.90	–	23.34	nr	**30.24**
Precast concrete sill, lintels, copings							
Lintels; plate; prestressed bedded							
100 mm × 70 mm × 600 mm long	5.92	0.37	7.98	–	6.24	nr	**14.22**
100 mm × 70 mm × 900 mm long	8.84	0.37	7.98	–	9.30	nr	**17.28**
100 mm × 70 mm × 1100 mm long	10.83	0.37	7.98	–	11.39	nr	**19.37**
100 mm × 70 mm × 1200 mm long	11.80	0.37	7.98	–	12.41	nr	**20.39**
100 mm × 70 mm × 1500 mm long	14.76	0.46	9.91	–	15.54	nr	**25.45**
100 mm × 70 mm × 1800 mm long	17.70	0.46	9.91	–	18.62	nr	**28.53**
100 mm × 70 mm × 2100 mm long	20.65	0.56	12.07	–	21.73	nr	**33.80**
140 mm × 70 mm × 1200 mm long	17.35	0.46	9.91	–	18.26	nr	**28.17**
140 mm × 70 mm × 1500 mm long	21.69	0.56	12.07	–	22.83	nr	**34.90**
Lintels; rectangular; reinforced with mild steel bars; bedded							
100 mm × 145 mm × 900 mm long	3.90	0.56	12.07	–	4.12	nr	**16.19**
100 mm × 145 mm × 1050 mm long	4.55	0.56	12.07	–	4.80	nr	**16.87**
100 mm × 145 mm × 1200 mm long	5.20	0.56	12.07	–	5.48	nr	**17.55**
225 mm × 145 mm × 1200 mm long	20.18	0.74	15.95	–	21.24	nr	**37.19**
225 mm × 225 mm × 1800 mm long	30.18	1.39	29.97	–	31.74	nr	**61.71**
Lintels; boot; reinforced with mild steel bars; bedded							
250 mm × 225 mm × 1200 mm long	22.35	1.11	23.93	–	23.52	nr	**47.45**
275 mm × 225 mm × 1800 mm long	36.89	1.67	36.00	–	38.79	nr	**74.79**
Sundries – stone walling							
Coating backs of stones with brush applied cold bitumen solution; two coats							
limestone facework	–	0.19	2.95	–	2.39	m²	**5.34**
Cutting grooves in limestone masonry for							
water bars or the like	–	–	–	–	–	m	**12.14**
Mortices in limestone masonry for							
metal dowel	–	–	–	–	–	nr	**2.43**
metal cramp	–	–	–	–	–	nr	**4.85**

14 MASONRY

Item	PC £	Labour hours	Labour £	Plant £	Material £	Unit	Total rate £
FLUES AND FLUE LININGS							
Flues							
Scheidel Rite-Vent ICS Plus flue system; suitable for domestic multi-fuel appliances; stainless steel; twin wall; insulated; for use internally or externally							
80 mm pipes; including one locking band (fixing brackets measured separately)	–	0.90	13.15	–	89.11	m	**102.26**
Extra for							
Appliance Connecter	–	0.80	11.69	–	13.79	nr	**25.48**
30° Bend	–	1.80	26.29	–	67.08	nr	**93.37**
45° Bend	–	1.80	26.29	–	63.74	nr	**90.03**
135° Tee; fully welded	–	2.70	39.44	–	131.93	nr	**171.37**
Inspection Length	–	0.90	13.15	–	9.53	nr	**22.68**
Drain Plug and Support	–	1.00	14.61	–	61.79	nr	**76.40**
Damper	–	0.90	13.15	–	50.24	nr	**63.39**
Angled Flashing including Storm Collar	–	1.25	18.26	–	66.57	nr	**84.83**
Stub Terminal	–	1.00	14.61	–	22.04	nr	**36.65**
Tapered Terminal	–	1.00	14.61	–	45.92	nr	**60.53**
Floor Support (2 piece)	–	1.50	21.91	–	36.27	nr	**58.18**
Firestop Floor Support (2 piece)	–	1.50	21.91	–	20.33	nr	**42.24**
Wall Support (Stainless Steel)	–	1.00	14.61	–	77.01	nr	**91.62**
Wall Sleeve	–	1.20	17.52	–	30.57	nr	**48.09**
100 mm pipes; including one locking band (fixing brackets measured separately)	–	1.00	14.61	–	94.91	m	**109.52**
Extra for							
Appliance Connecter	–	0.90	13.15	–	15.22	nr	**28.37**
30° Bend	–	2.00	29.21	–	70.17	nr	**99.38**
45° Bend	–	2.00	29.21	–	66.62	nr	**95.83**
135° Tee; fully welded	–	3.00	43.83	–	128.09	nr	**171.92**
Inspection Length	–	1.00	14.61	–	218.44	nr	**233.05**
Drain Plug and Support	–	1.10	16.07	–	63.99	nr	**80.06**
Damper	–	1.00	14.61	–	52.90	nr	**67.51**
Angled Flashing including Storm Collar	–	1.40	20.45	–	74.17	nr	**94.62**
Stub Terminal	–	1.10	16.07	–	22.40	nr	**38.47**
Tapered Terminal	–	1.10	16.07	–	49.00	nr	**65.07**
Floor Support (2 piece)	–	1.65	24.11	–	41.24	nr	**65.35**
Firestop Floor Support (2 piece)	–	1.65	24.11	–	22.41	nr	**46.52**
Wall Support (Stainless Steel)	–	1.10	16.07	–	81.39	nr	**97.46**
Wall Sleeve	–	1.35	19.72	–	34.70	nr	**54.42**
150 mm pipes; including one locking band (fixing brackets measured separately)	–	1.10	16.07	–	110.81	m	**126.88**
Extra for							
Appliance Connecter	–	1.00	14.61	–	19.62	nr	**34.23**
30° Bend	–	2.20	32.14	–	84.22	nr	**116.36**
45° Bend	–	2.20	32.14	–	80.07	nr	**112.21**
135° Tee; fully welded	–	3.30	48.21	–	147.92	nr	**196.13**
Inspection Length	–	1.10	16.07	–	229.09	nr	**245.16**
Drain Plug and Support	–	1.20	17.52	–	81.71	nr	**99.23**
Damper	–	1.10	16.07	–	69.14	nr	**85.21**

14 MASONRY

Item	PC £	Labour hours	Labour £	Plant £	Material £	Unit	Total rate £
Angled Flashing including Storm Collar	–	1.55	22.64	–	75.26	nr	**97.90**
Stub Terminal	–	1.20	17.52	–	24.10	nr	**41.62**
Tapered Terminal	–	1.20	17.52	–	56.32	nr	**73.84**
Floor Support (2 piece)	–	1.80	26.29	–	41.24	nr	**67.53**
Firestop Floor Support (2 piece)	–	1.80	26.29	–	22.41	nr	**48.70**
Wall Support (Stainless Steel)	–	1.20	17.52	–	90.13	nr	**107.65**
Wall Sleeve	–	1.50	21.91	–	34.70	nr	**56.61**

15 STRUCTURAL METALWORK

Item	PC £	Labour hours	Labour £	Plant £	Material £	Unit	Total rate £
FRAMED MEMBERS, FRAMING, FABRICATION							
BASIC STEEL PRICES – MATERIAL SUPPLY ONLY							
TATA Steel have recently revised ther pricing structure and now publish pries for extras only. The extras cost needs to be added to their steel Base Price. The Base Price given here is a guide price only and subject to normal trading conditions with regards to discounts or plussages. The Base Price assumes a minimum order of 5 tonnes, S355JR steel and standard lengths.							
AUTHORS NOTE: AT THE TIME OF WRITING STEEL PRICES WERE INCREASING ON A MONTHLY BASIS. USERS ARE STRONGLY ADVISED TO CHECK FOR INCREASES ANNOUNCED BEYOND JUNE 2016							
Universal beams and columns							
Base price for steel	–	–	–	–	787.50	tonne	**787.50**
Extra to be added to Base Price for: (kg/m)							
1016 × 305 mm (222, 249, 272, 314, 349, 393, 438, 487) – Price on application	–	–	–	–	–	tonne	–
914 × 419 mm (343, 388)	–	–	–	–	210.00	tonne	**210.00**
914 × 305 mm (201, 224, 253, 289)	–	–	–	–	178.50	tonne	**178.50**
838 × 292 mm (226)	–	–	–	–	178.50	tonne	**178.50**
838 × 292 mm (176, 194)	–	–	–	–	168.00	tonne	**168.00**
762 × 267 mm (197)	–	–	–	–	168.00	tonne	**168.00**
762 × 267 mm (134, 147, 173)	–	–	–	–	147.00	tonne	**147.00**
686 × 254 mm (152, 170)	–	–	–	–	168.00	tonne	**168.00**
686 × 254 mm (125, 140)	–	–	–	–	147.00	tonne	**147.00**
610 × 305 mm (238)	–	–	–	–	178.50	tonne	**178.50**
610 × 305 mm (149, 179)	–	–	–	–	147.00	tonne	**147.00**
610 × 229 mm (125, 140)	–	–	–	–	10.50	tonne	**10.50**
610 × 229 mm (101, 113)	–	–	–	–	10.50	tonne	**10.50**
610 × 178 mm (82, 92, 100) – Price on application	–	–	–	–	–	tonne	–
533 × 312 mm (150, 182, 219, 272) – Price on application	–	–	–	–	–	tonne	–
533 × 210 mm (138)	–	–	–	–	189.00	tonne	**189.00**
533 × 210 mm (101, 109, 122)	–	–	–	–	10.50	tonne	**10.50**
533 × 210 mm (82, 92)	–	–	–	–	10.50	tonne	**10.50**
533 × 165 mm (66, 74, 85) – Price on application	–	–	–	–	–	tonne	–
457 × 191 mm (161)	–	–	–	–	210.00	tonne	**210.00**
457 × 191 mm (133)	–	–	–	–	189.00	tonne	**189.00**
457 × 191 mm (106)	–	–	–	–	52.50	tonne	**52.50**
457 × 191 mm (74, 82, 89, 98)	–	–	–	–	10.50	tonne	**10.50**
457 × 191 mm (67)	–	–	–	–	10.50	tonne	**10.50**
457 × 152 mm (74, 82)	–	–	–	–	52.50	tonne	**52.50**
457 × 152 mm (67)	–	–	–	–	52.50	tonne	**52.50**
457 × 152 mm (52, 60)	–	–	–	–	10.50	tonne	**10.50**

15 STRUCTURAL METALWORK

Item	PC £	Labour hours	Labour £	Plant £	Material £	Unit	Total rate £
406 × 178 mm (85)	–	–	–	–	52.50	tonne	**52.50**
406 × 178 mm (74)	–	–	–	–	10.50	tonne	**10.50**
406 × 178 mm (54, 60, 67)	–	–	–	–	10.50	tonne	**10.50**
406 × 140 mm (53)	–	–	–	–	52.50	tonne	**52.50**
406 × 140 mm (46)	–	–	–	–	10.50	tonne	**10.50**
406 × 140 mm (39)	–	–	–	–	10.50	tonne	**10.50**
356 × 171 mm (45, 51, 57, 67)	–	–	–	–	10.50	tonne	**10.50**
305 × 165 mm (40, 46, 54)	–	–	–	–	10.50	tonne	**10.50**
305 × 127 mm (42, 48)	–	–	–	–	10.50	tonne	**10.50**
305 × 127 mm (37)	–	–	–	–	10.50	tonne	**10.50**
305 × 102 mm (28, 33)	–	–	–	–	10.50	tonne	**10.50**
305 × 102 mm (25)	–	–	–	–	10.50	tonne	**10.50**
254 × 102 mm (28)	–	–	–	–	21.00	tonne	**21.00**
254 × 102 mm (22, 25)	–	–	–	–	10.50	tonne	**10.50**
254 × 146 (43)	–	–	–	–	10.50	tonne	**10.50**
254 × 146 (31, 37)	–	–	–	–	10.50	tonne	**10.50**
203 × 133 mm (25, 30)	–	–	–	–	10.50	tonne	**10.50**
203 × 102 mm (23)	–	–	–	–	10.50	tonne	**10.50**
178 × 102 mm (19)	–	–	–	–	21.00	tonne	**21.00**
152 × 89 mm (16)	–	–	–	–	21.00	tonne	**21.00**
127 × 76 mm (13)	–	–	–	–	21.00	tonne	**21.00**
Asymmetric Slimflor Beams (ASB) – Price on Application							
Universal columns (kg/m) – NB These are to be added to the base steel price. Base price for steel	–	–	–	–	787.50	tonne	**787.50**
356 × 406 mm (634)	–	–	–	–	315.00	tonne	**315.00**
356 × 406 mm (340, 393, 467, 551)	–	–	–	–	210.00	tonne	**210.00**
356 × 406 mm (235, 287)	–	–	–	–	178.50	tonne	**178.50**
356 × 368 mm (177, 202)	–	–	–	–	178.50	tonne	**178.50**
356 × 368 mm (153)	–	–	–	–	168.00	tonne	**168.00**
356 × 368 mm (129)	–	–	–	–	147.00	tonne	**147.00**
305 × 305 mm (240, 283)	–	–	–	–	21.00	tonne	**21.00**
305 × 305 mm (137, 158, 198)	–	–	–	–	10.50	tonne	**10.50**
305 × 305 mm (97, 118)	–	–	–	–	10.50	tonne	**10.50**
254 × 254 mm (132, 167)	–	–	–	–	10.50	tonne	**10.50**
254 × 254 mm (73, 89, 107)	–	–	–	–	10.50	tonne	**10.50**
203 × 203 mm (127)	–	–	–	–	189.00	tonne	**189.00**
203 × 203 mm (100, 113)	–	–	–	–	52.50	tonne	**52.50**
203 × 203 mm (71, 86)	–	–	–	–	10.50	tonne	**10.50**
203 × 203 mm (46, 52, 60)	–	–	–	–	10.50	tonne	**10.50**
152 × 152 mm (44, 51)	–	–	–	–	52.50	tonne	**52.50**
152 × 152 mm (23, 30, 37)	–	–	–	–	10.50	tonne	**10.50**
Channels (kg/m) – NB These are to be added to the base steel price. Base price for steel	–	–	–	–	787.50	tonne	**787.50**
430 × 100 mm (64.4)	–	–	–	–	309.75	tonne	**309.75**
380 × 100 mm (54.0)	–	–	–	–	288.75	tonne	**288.75**
300 × 100 mm (45.5)	–	–	–	–	120.75	tonne	**120.75**
300 × 90 mm (41.4)	–	–	–	–	131.25	tonne	**131.25**
260 × 90 mm (34.8)	–	–	–	–	120.75	tonne	**120.75**
260 × 75 mm (27.6)	–	–	–	–	73.50	tonne	**73.50**
230 × 90 mm (32.2)	–	–	–	–	131.25	tonne	**131.25**
230 × 75 mm (25.7)	–	–	–	–	73.50	tonne	**73.50**

15 STRUCTURAL METALWORK

Item	PC £	Labour hours	Labour £	Plant £	Material £	Unit	Total rate £
FRAMED MEMBERS, FRAMING, FABRICATION – cont							
Universal beams and columns – cont							
Channels (kg/m) – cont							
200 × 90 mm (29.7)	–	–	–	–	120.75	tonne	**120.75**
200 × 75 mm (23.4)	–	–	–	–	42.00	tonne	**42.00**
180 × 90 mm (26.1)	–	–	–	–	120.75	tonne	**120.75**
180 × 75 mm (20.3)	–	–	–	–	42.00	tonne	**42.00**
150 × 90 mm (23.9)	–	–	–	–	120.75	tonne	**120.75**
150 × 75 mm (17.9)	–	–	–	–	21.00	tonne	**21.00**
125 × 65 mm (14.8)	–	–	–	–	21.00	tonne	**21.00**
Equal angles (mm) – NB These are to be added to the base steel price. Base price for steel	–	–	–	–	787.50	tonne	**787.50**
200 × 200 mm (16,18,20,24)	–	–	–	–	68.25	tonne	**68.25**
150 × 150 mm (10,12,15,18)	–	–	–	–	68.25	tonne	**68.25**
120 × 120 mm (8, 10, 12, 15)	–	–	–	–	10.50	tonne	**10.50**
Unequal angles (mm) – NB These are to be added to the base steel price. Base price for steel	–	–	–	–	787.50	tonne	**787.50**
200 × 150 mm (12,15, 18)	–	–	–	–	120.75	tonne	**120.75**
200 × 100 mm (10, 12, 15)	–	–	–	–	68.25	tonne	**68.25**
Specification extras							
for Advance275JR steel	–	–	–	–	31.50	tonne	**31.50**
for Advance275JO steel	–	–	–	–	42.00	tonne	**42.00**
for Advance275J2 steel	–	–	–	–	84.00	tonne	**84.00**
for Advance355JO steel	–	–	–	–	10.50	tonne	**10.50**
for Advance355J2 steel	–	–	–	–	42.00	tonne	**42.00**
for Advance355 K2 steel	–	–	–	–	84.00	tonne	**84.00**
Hollow sections							
NOTE: The following prices are for basic quantities of 10 tonnes and over in one size, thickness, length, steelgrade and surface finish and include delivery (for delivery to outer London)							
circular hollow sections	–	–	–	–	890.40	tonne	**890.40**
rectangular hollow section	–	–	–	–	898.80	tonne	**898.80**
square hollow sections	–	–	–	–	868.35	tonne	**868.35**
SUPPLY AND FIX PRICES							
Framing, fabrication; weldable steel; BS EN 10025: 2004 Grade S275; hot rolled structural steel sections; welded fabrication							
Columns							
weight not exceeding 40 kg/m	–	–	–	–	–	tonne	**1521.00**
weight not exceeding 40 kg/m; cellular (Fabsec)	–	–	–	–	–	tonne	**1825.20**
weight not exceeding 40 kg/m; curved	–	–	–	–	–	tonne	**2112.30**
weight not exceeding 40 kg/m; square hollow section	–	–	–	–	–	tonne	**1980.00**
weight not exceeding 40 kg/m; circular hollow section	–	–	–	–	–	tonne	**2025.00**
weight 40–100 kg/m	–	–	–	–	–	tonne	**1361.70**
weight 40–100 kg/m; cellular (Fabsec)	–	–	–	–	–	tonne	**1588.50**

15 STRUCTURAL METALWORK

Item	PC £	Labour hours	Labour £	Plant £	Material £	Unit	Total rate £
weight 40–100 kg/m; curved	–	–	–	–	–	tonne	**1620.90**
weight 40–100 kg/m; square hollow section	–	–	–	–	–	tonne	**1593.00**
weight 40–100 kg/m; circular hollow section	–	–	–	–	–	tonne	**1593.00**
weight exceeding 100 kg/m	–	–	–	–	–	tonne	**1314.00**
weight exceeding 100 kg/m; cellular (Fabsec)	–	–	–	–	–	tonne	**1509.30**
weight exceeding 100 kg/m; curved	–	–	–	–	–	tonne	**1500.30**
weight exceeding 100 kg/m; square hollow section	–	–	–	–	–	tonne	**2574.00**
weight exceeding 100 kg/m; circular hollow section	–	–	–	–	–	tonne	**2574.00**
Beams							
weight not exceeding 40 kg/m	–	–	–	–	–	tonne	**1722.60**
weight not exceeding 40 kg/m; cellular (Fabsec)	–	–	–	–	–	tonne	**2065.50**
weight not exceeding 40 kg/m; curved	–	–	–	–	–	tonne	**2168.10**
weight not exceeding 40 kg/m; square hollow section	–	–	–	–	–	tonne	**2205.00**
weight not exceeding 40 kg/m; circular hollow section	–	–	–	–	–	tonne	**2205.00**
weight 40–100 kg/m	–	–	–	–	–	tonne	**1361.70**
weight 40–100 kg/m; cellular (Fabsec)	–	–	–	–	–	tonne	**1857.60**
weight 40–100 kg/m; curved	–	–	–	–	–	tonne	**1620.90**
weight 40–100 kg/m; square hollow section	–	–	–	–	–	tonne	**2056.50**
weight 40–100 kg/m; circular hollow section	–	–	–	–	–	tonne	**2056.50**
weight exceeding 100 kg/m	–	–	–	–	–	tonne	**1314.90**
weight exceeding 100 kg/m; cellular (Fabsec)	–	–	–	–	–	tonne	**1519.20**
weight exceeding 100 kg/m; curved	–	–	–	–	–	tonne	**1565.10**
weight exceeding 100 kg/m; square hollow section	–	–	–	–	–	tonne	**2575.80**
weight exceeding 100 kg/m; circular hollow section	–	–	–	–	–	tonne	**2575.80**
Bracings							
weight not exceeding 40 kg/m	–	–	–	–	–	tonne	**1722.60**
weight not exceeding 40 kg/m; square hollow section	–	–	–	–	–	tonne	**2205.00**
weight not exceeding 40 kg/m; circular hollow section	–	–	–	–	–	tonne	**2205.00**
weight 40–100 kg/m	–	–	–	–	–	tonne	**1361.70**
weight 40–100 kg/m; square hollow section	–	–	–	–	–	tonne	**2056.50**
weight 40–100 kg/m; circular hollow section	–	–	–	–	–	tonne	**2056.50**
weight exceeding 100 kg/m	–	–	–	–	–	tonne	**1314.90**
weight exceeding 100 kg/m; square hollow section	–	–	–	–	–	tonne	**2575.80**
weight exceeding 100 kg/m; circular hollow section	–	–	–	–	–	tonne	**2575.80**
Purlins and cladding rails							
weight not exceeding 40 kg/m	–	–	–	–	–	tonne	**1722.60**
weight not exceeding 40 kg/m; square hollow section	–	–	–	–	–	tonne	**2205.00**
weight not exceeding 40 kg/m; circular hollow section	–	–	–	–	–	tonne	**2205.00**
weight 40–100 kg/m	–	–	–	–	–	tonne	**1361.70**
weight 40–100 kg/m; square hollow section	–	–	–	–	–	tonne	**2056.50**
weight 40–100 kg/m; circular hollow section	–	–	–	–	–	tonne	**2056.50**
weight exceeding 100 kg/m	–	–	–	–	–	tonne	**1314.90**
weight exceeding 100 kg/m; square hollow section	–	–	–	–	–	tonne	**2575.80**
weight exceeding 100 kg/m; circular hollow section	–	–	–	–	–	tonne	**2575.80**
Grillages							
weight not exceeding 40 kg/m	–	–	–	–	–	tonne	**1722.60**
weight 40–100 kg/m	–	–	–	–	–	tonne	**1361.70**
weight exceeding 100 kg/m	–	–	–	–	–	tonne	**1314.90**

15 STRUCTURAL METALWORK

Item	PC £	Labour hours	Labour £	Plant £	Material £	Unit	Total rate £
FRAMED MEMBERS, FRAMING, FABRICATION – cont							
Framing, fabrication – cont							
Trestles, towers and built up columns							
straight	–	–	–	–	–	tonne	2205.00
Trusses and built up girders							
straight	–	–	–	–	–	tonne	2205.00
curved	–	–	–	–	–	tonne	2353.50
Fittings							
general steel fittings; nuts, bolts, plates etc.	–	–	–	–	–	tonne	2019.60
Add to the aforementioned prices for:							
grade 355 steelwork	–	–	–	–	–	%	3.00
Framing, erection							
Trial erection	–	–	–	–	–	tonne	275.00
Permanent erection on site	–	–	–	–	–	tonne	275.00
Surface preparation							
At works							
blast cleaning	–	–	–	–	–	m^2	2.63
Surface treatment							
At works							
galvanizing	–	–	–	–	–	tonne	215.00
galvanizing	–	–	–	–	–	m^2	14.50
shotblasting and priming to SA 2.5	–	–	–	–	–	m^2	7.20
touch up primer and one coat of two pack epoxy zinc phosphate primer	–	–	–	–	–	m^2	4.78
intumescent paint fire protection (30 minutes); spray applied	–	–	–	–	–	m^2	13.25
intumescent paint fire protection (60 minutes); spray applied	–	–	–	–	–	m^2	20.16
intumescent paint fire protection (90 minutes); spray applied	–	–	–	–	–	m^2	38.95
intumescent paint fire protection (120 minutes); spray applied	–	–	–	–	–	m^2	60.27
extra over for; separate decorative sealer top coat	–	–	–	–	–	m^2	3.78
On site							
intumescent paint fire protection (30 minutes); spray applied	–	–	–	–	–	m^2	9.94
intumescent paint fire protection (30 minutes) to circular columns etc.; spray applied	–	–	–	–	–	m^2	16.66
intumescent paint fire protection (60 minutes) to UBs etc.; spray applied	–	–	–	–	–	m^2	12.60
intumescent paint fire protection (60 minutes) to circular columns etc.; spray applied	–	–	–	–	–	m^2	21.14
extra for:							
separate decorative sealer top coat	–	–	–	–	–	m^2	3.31

15 STRUCTURAL METALWORK

Item	PC £	Labour hours	Labour £	Plant £	Material £	Unit	Total rate £
Metsec Lightweight Steel Framing System (SFS); or other equal and approved; as inner leaf to external wall; studs typically at 600 mm centres; including provision for all openings, abutments, junctions and head details etc.							
Inner leaf; with supports and perimeter sections; 12 mm plasterboard internally; 10 mm cement fibre substrate externally; (insulation and external cladding measured separately)							
100 mm thick steel walling	–	–	–	–	–	m²	**61.44**
150 mm thick steel walling	–	–	–	–	–	m²	**66.29**
200 mm thick steel walling	–	–	–	–	–	m²	**71.14**
Inner leaf; with 16 mm Pyroc sheething board							
100 mm thick steel walling	–	–	–	–	–	m²	**75.99**
150 mm thick steel walling	–	–	–	–	–	m²	**80.84**
200 mm thick steel walling	–	–	–	–	–	m²	**85.69**
16 mm Pyroc sheething board fixed to slab perimeter not exceeding 300 mm	–	–	–	–	–	m	**7.54**
Inner leaf; with 16 mm Pyroc sheething board and Thermawall TW50 insulation supported by halfen channels type 28/15 fixed to studs at 450 mm centres.							
100 mm thick steel walling with 50 mm insulation	–	–	–	–	–	m²	**87.40**
150 mm thick steel walling with 75 mm insulation	–	–	–	–	–	m²	**98.15**
200 mm thick steel walling with 100 mm insulatiom	–	–	–	–	–	m²	**108.20**
16 mm Pyroc sheething board and 40 mm Thermawall TW55 insulation fixed to slab perimeter not exceeding 300 mm	–	–	–	–	–	m	**8.62**
Cold formed galvanized steel; Kingspan Multibeam or other equal and approved							
Cold rolled purlins and cladding rails							
175 × 65 × 1.40 mm gauge purlins or rails; fixed to steelwork	–	0.04	0.74	–	6.53	m	**7.27**
175 × 65 × 1.60 mm gauge purlins or rails; fixed to steelwork	–	0.04	0.74	–	6.95	m	**7.69**
175 × 65 × 2.00 mm gauge purlins or rails; fixed to steelwork	–	0.04	0.74	–	8.40	m	**9.14**
205 × 65 × 1.40 mm gauge purlins or rails; fixed to steelwork	–	0.04	0.74	–	7.21	m	**7.95**
205 × 65 × 1.60 mm gauge purlins or rails; fixed to steelwork	–	0.04	0.74	–	7.85	m	**8.59**
205 × 65 × 2.00 mm gauge purlins or rails; fixed to steelwork	–	0.04	0.74	–	8.95	m	**9.69**
Cleats							
weld-on for 175 mm purlin or rail	–	0.10	1.84	–	1.89	nr	**3.73**
bolt-on for 175 mm purlin or rail; including fixing bolts	–	0.02	0.37	–	4.07	m	**4.44**
weld-on for 205 mm purlin or rail	–	0.10	1.84	–	2.15	nr	**3.99**
bolt-on for 205 mm purlin or rail; including fixing bolts	–	0.02	0.37	–	4.38	m	**4.75**

15 STRUCTURAL METALWORK

Item	PC £	Labour hours	Labour £	Plant £	Material £	Unit	Total rate £
FRAMED MEMBERS, FRAMING, FABRICATION – cont							
Cold formed galvanized steel – cont							
Tubular ties							
1800 mm long; bolted diagonally across purlins or cladding rails	–	0.02	0.37	–	4.79	m	**5.16**
Storage costs							
Costs for storing fabricated steelwork							
storage off site	–	–	–	–	–	t/week	**18.90**
storage on extending trailers	–	–	–	–	–	t/week	**31.50**
ISOLATED STRUCTURAL METAL MEMBERS							
Isolated structural member; weldable steel; BS EN 10025: 2004 Grade S275; hot rolled structural steel sections							
Plain member; beams							
weight not exceeding 40 kg/m	–	–	–	–	–	tonne	**1242.00**
weight 40–100 kg/m	–	–	–	–	–	tonne	**1242.00**
weight exceeding 100 kg/m	–	–	–	–	–	tonne	**1242.00**
Metsec open web steel lattice beams or other equal and approved; in single members; raised 3.50 m above ground; ends built in							
Beams; one coat zinc phosphate primer at works							
220 mm deep; to span 6.00 m (11.50 kg/m); ref B22	–	0.19	4.35	–	27.41	m	**31.76**
270 mm deep; to span 7.00 m (11.50 kg/m); ref B27	–	0.19	4.35	–	27.41	m	**31.76**
300 mm deep; to span 8.00 m (12.50 kg/m); ref B30	–	0.23	5.26	–	29.76	m	**35.02**
350 mm deep; to span 9.00 m (14.00 kg/m); ref B35	–	0.23	5.26	–	33.28	m	**38.54**
350 mm deep; to span 10.00 m (20.00 kg/m); ref D35	–	0.28	6.40	–	47.40	m	**53.80**
450 mm deep; to span 11.00 m (21.00 kg/m); ref D45	–	0.32	7.33	–	49.74	m	**57.07**
450 mm deep; to span 12.00 m (32.50 kg/m); ref G45	–	0.46	10.53	–	76.78	m	**87.31**
Beams; galvanized							
220 mm deep; to span 6.00 m (11.50 kg/m); ref B22	–	0.19	4.35	–	31.10	m	**35.45**
270 mm deep; to span 7.00 m (11.50 kg/m); ref B27	–	0.19	4.35	–	31.10	m	**35.45**
300 mm deep; to span 8.00 m (12.50 kg/m); ref B30	–	0.23	5.26	–	33.76	m	**39.02**

15 STRUCTURAL METALWORK

Item	PC £	Labour hours	Labour £	Plant £	Material £	Unit	Total rate £
350 mm deep; to span 9.00 m (14.00 kg/m); ref B35	–	0.23	5.26	–	37.79	m	**43.05**
350 mm deep; to span 10.00 m (20.00 kg/m); ref D35	–	0.28	6.40	–	53.79	m	**60.19**
450 mm deep; to span 11.00 m (21.00 kg/m); ref D45	–	0.32	7.33	–	56.48	m	**63.81**
450 mm deep; to span 12.00 m (32.50 kg/m); ref G45	–	0.46	10.53	–	87.18	m	**97.71**

PROFILED METAL DECKING

Soffits of slabs; galvanized steel permanent re-entrant type shuttering; including safety net

Item	PC £	Labour hours	Labour £	Plant £	Material £	Unit	Total rate £
Slab thickness not exceeding 200 mm							
0.9 mm decking; height to soffit 1.50–3.00 m	27.54	0.27	6.00	–	31.98	m²	**37.98**
0.9 mm decking; height to soffit 3.00–4.50 m	27.54	0.30	6.66	–	31.98	m²	**38.64**
1.2 mm decking; height to soffit 3.00–4.50 m	30.92	0.27	6.00	–	35.53	m²	**41.53**
Edge trim and restraints to decking							
Edge trim 1.2 mm × 300 mm girth	–	0.17	3.66	–	12.49	m	**16.15**
Edge trim 1.2 mm × 350 mm girth	–	0.17	3.66	–	12.49	m	**16.15**
Edge trim 1.2 mm × 400 mm girth	–	0.17	3.66	–	12.49	m	**16.15**
Bearings to decking; connection to steel work with thru-deck welded shear studs							
1995 × 95 mm high studs at 100 mm centres	–	–	–	–	13.95	m	**13.95**
1995 × 95 mm high studs at 200 mm centres	–	–	–	–	6.98	m	**6.98**
1995 × 95 mm high studs at 300 mm centres	–	–	–	–	4.65	m	**4.65**
19120 × 120 mm high studs at 100 mm centres	–	–	–	–	13.95	m	**13.95**
19120 × 120 mm high studs at 200 mm centres	–	–	–	–	6.98	m	**6.98**
19120 × 120 mm high studs at 300 mm centres	–	–	–	–	4.65	m	**4.65**

Soffits of slabs; galvanized steel permanent trapazoidal type shuttering; including safety net

Item	PC £	Labour hours	Labour £	Plant £	Material £	Unit	Total rate £
Slab thickness not exceeding 200 mm							
0.9 mm decking; height to soffit 1.50–3.00 m	25.18	0.27	6.00	–	29.52	m²	**35.52**
0.9 mm decking; height to soffit 3.00–4.50 m	25.18	0.30	6.66	–	29.52	m²	**36.18**
1.2 mm decking; height to soffit 3.00–4.50 m	29.38	0.27	6.00	–	33.93	m²	**39.93**

16 CARPENTRY

Item	PC £	Labour hours	Labour £	Plant £	Material £	Unit	Total rate £
TIMBER FRAMING							
SUPPLY ONLY TIMBER PRICES							
Hardwood; Joinery quality; 25 mm thicknes (£/m^3)							
Black Walnut	–	–	–	–	3517.50	m^3	**3517.50**
American White Ash	–	–	–	–	1470.00	m^3	**1470.00**
American White Oak	–	–	–	–	1837.50	m^3	**1837.50**
Pretreatment of timber by vacuum/pressure impregnation, excluding transport costs and any subsequent seasoning:							
interior work; minimum salt retention 4.00 kg/m^3	–	–	–	–	74.84	m^3	**74.84**
exterior work; minimum salt retention 5.30 kg/m^3	–	–	–	–	86.17	m^3	**86.17**
SUPPLY AND FIX PRICES							
Sawn softwood; untreated							
Floor members							
38 mm × 100 mm	–	0.11	2.23	–	2.07	m	**4.30**
38 mm × 150 mm	–	0.13	2.63	–	2.86	m	**5.49**
47 mm × 75 mm	–	0.11	2.23	–	1.32	m	**3.55**
47 mm × 100 mm	–	0.13	2.63	–	1.69	m	**4.32**
47 mm × 125 mm	–	0.13	2.63	–	2.16	m	**4.79**
47 mm × 150 mm	–	0.14	2.82	–	2.49	m	**5.31**
47 mm × 175 mm	–	0.14	2.82	–	2.94	m	**5.76**
47 mm × 200 mm	–	0.15	3.03	–	3.19	m	**6.22**
47 mm × 225 mm	–	0.15	3.03	–	3.67	m	**6.70**
47 mm × 250 mm	–	0.16	3.23	–	4.14	m	**7.37**
75 mm × 125 mm	–	0.15	3.03	–	4.98	m	**8.01**
75 mm × 150 mm	–	0.15	3.03	–	5.39	m	**8.42**
75 mm × 175 mm	–	0.15	3.03	–	6.48	m	**9.51**
75 mm × 200 mm	–	0.16	3.23	–	7.27	m	**10.50**
75 mm × 225 mm	–	0.16	3.23	–	7.92	m	**11.15**
75 mm × 250 mm	–	0.17	3.43	–	12.02	m	**15.45**
100 mm × 150 mm	–	0.20	4.04	–	6.93	m	**10.97**
100 mm × 200 mm	–	0.21	4.24	–	9.24	m	**13.48**
100 mm × 250 mm	–	0.23	4.64	–	11.58	m	**16.22**
100 mm × 300 mm	–	0.25	5.05	–	14.96	m	**20.01**
Wall or partition members							
25 mm × 25 mm	–	0.06	1.21	–	0.81	m	**2.02**
25 mm × 38 mm	–	0.06	1.21	–	0.91	m	**2.12**
25 mm × 75 mm	–	0.08	1.62	–	1.13	m	**2.75**
38 mm × 38 mm	–	0.08	1.62	–	1.07	m	**2.69**
38 mm × 50 mm	–	0.08	1.62	–	1.35	m	**2.97**
38 mm × 75 mm	–	0.11	2.23	–	1.66	m	**3.89**
38 mm × 100 mm	–	0.14	2.82	–	2.07	m	**4.89**
47 mm × 50 mm	–	0.11	2.23	–	1.04	m	**3.27**
47 mm × 75 mm	–	0.14	2.82	–	1.37	m	**4.19**
47 mm × 100 mm	–	0.17	3.43	–	1.73	m	**5.16**
47 mm × 125 mm	–	0.18	3.63	–	2.22	m	**5.85**
75 mm × 75 mm	–	0.17	3.43	–	3.07	m	**6.50**
75 mm × 100 mm	–	0.19	3.83	–	4.16	m	**7.99**
100 mm × 100 mm	–	0.19	3.83	–	5.06	m	**8.89**

16 CARPENTRY

Item	PC £	Labour hours	Labour £	Plant £	Material £	Unit	Total rate £
Joist strutting; herringbone							
47 mm × 50 mm; depth of joist 150 mm	–	0.46	9.29	–	2.55	m	**11.84**
47 mm × 50 mm; depth of joist 175 mm	–	0.46	9.29	–	2.59	m	**11.88**
47 mm × 50 mm; depth of joist 200 mm	–	0.46	9.29	–	2.64	m	**11.93**
47 mm × 50 mm; depth of joist 225 mm	–	0.46	9.29	–	2.69	m	**11.98**
47 mm × 50 mm; depth of joist 250 mm	–	0.46	9.29	–	2.73	m	**12.02**
Joist strutting; block							
47 mm × 150 mm; depth of joist 150 mm	–	0.28	5.66	–	3.10	m	**8.76**
47 mm × 175 mm; depth of joist 175 mm	–	0.28	5.66	–	3.55	m	**9.21**
47 mm × 200 mm; depth of joist 200 mm	–	0.28	5.66	–	3.80	m	**9.46**
47 mm × 225 mm; depth of joist 225 mm	–	0.28	5.66	–	4.27	m	**9.93**
47 mm × 250 mm; depth of joist 250 mm	–	0.28	5.66	–	4.74	m	**10.40**
Cleats							
225 mm × 100 mm × 75 mm	–	0.19	3.83	–	0.81	nr	**4.64**
Extra for stress grading to above timbers							
general structural (GS) grade	–	–	–	–	31.76	m³	**31.76**
special structural (SS) grade	–	–	–	–	63.52	m³	**63.52**
Extra for protecting and flameproofing timber with							
Celgard CF protection or other equal and approved							
small sections	–	–	–	–	125.90	m³	**125.90**
large sections	–	–	–	–	121.55	m³	**121.55**
Wrot surfaces							
plain; 50 mm wide	–	0.02	0.40	–	–	m	**0.40**
plain; 100 mm wide	–	0.03	0.61	–	–	m	**0.61**
plain; 150 mm wide	–	0.04	0.81	–	–	m	**0.81**
Sawn softwood; tanalized							
Floor members							
38 mm × 75 mm	–	0.11	2.23	–	1.82	m	**4.05**
38 mm × 100 mm	–	0.11	2.23	–	2.28	m	**4.51**
38 mm × 150 mm	–	0.13	2.63	–	3.17	m	**5.80**
47 mm × 75 mm	–	0.11	2.23	–	1.53	m	**3.76**
47 mm × 100 mm	–	0.13	2.63	–	1.96	m	**4.59**
47 mm × 125 mm	–	0.13	2.63	–	2.52	m	**5.15**
47 mm × 150 mm	–	0.14	2.82	–	2.91	m	**5.73**
47 mm × 175 mm	–	0.14	2.82	–	3.43	m	**6.25**
47 mm × 200 mm	–	0.15	3.03	–	3.75	m	**6.78**
47 mm × 225 mm	–	0.15	3.03	–	4.29	m	**7.32**
47 mm × 250 mm	–	0.16	3.23	–	4.83	m	**8.06**
75 mm × 125 mm	–	0.15	3.03	–	5.50	m	**8.53**
75 mm × 150 mm	–	0.15	3.03	–	6.02	m	**9.05**
75 mm × 175 mm	–	0.15	3.03	–	7.21	m	**10.24**
75 mm × 200 mm	–	0.16	3.23	–	8.11	m	**11.34**
75 mm × 225 mm	–	0.16	3.23	–	8.86	m	**12.09**
75 mm × 250 mm	–	0.17	3.43	–	13.07	m	**16.50**
100 mm × 150 mm	–	0.20	4.04	–	7.77	m	**11.81**
100 mm × 200 mm	–	0.21	4.24	–	10.35	m	**14.59**
100 mm × 250 mm	–	0.23	4.64	–	12.98	m	**17.62**
100 mm × 300 mm	–	0.25	5.05	–	16.63	m	**21.68**

16 CARPENTRY

Item	PC £	Labour hours	Labour £	Plant £	Material £	Unit	Total rate £
TIMBER FRAMING – cont							
Sawn softwood – cont							
Wall or partition members							
25 mm × 25 mm	–	0.06	1.21	–	0.84	m	**2.05**
25 mm × 38 mm	–	0.06	1.21	–	0.98	m	**2.19**
25 mm × 75 mm	–	0.08	1.62	–	1.24	m	**2.86**
38 mm × 38 mm	–	0.08	1.62	–	1.14	m	**2.76**
38 mm × 50 mm	–	0.08	1.62	–	1.45	m	**3.07**
38 mm × 75 mm	–	0.11	2.23	–	1.82	m	**4.05**
38 mm × 100 mm	–	0.14	2.82	–	2.28	m	**5.10**
47 mm × 50 mm	–	0.11	2.23	–	1.19	m	**3.42**
47 mm × 75 mm	–	0.14	2.82	–	1.58	m	**4.40**
47 mm × 100 mm	–	0.17	3.43	–	2.02	m	**5.45**
47 mm × 125 mm	–	0.18	3.63	–	2.56	m	**6.19**
75 mm × 75 mm	–	0.17	3.43	–	3.38	m	**6.81**
75 mm × 100 mm	–	0.19	3.83	–	4.58	m	**8.41**
100 mm × 100 mm	–	0.19	3.83	–	5.62	m	**9.45**
Roof members; flat							
38 mm × 75 mm	–	0.13	2.63	–	1.82	m	**4.45**
38 mm × 100 mm	–	0.13	2.63	–	2.28	m	**4.91**
38 mm × 125 mm	–	0.13	2.63	–	2.72	m	**5.35**
38 mm × 150 mm	–	0.13	2.63	–	3.17	m	**5.80**
47 mm × 100 mm	–	0.13	2.63	–	1.96	m	**4.59**
47 mm × 125 mm	–	0.13	2.63	–	2.52	m	**5.15**
47 mm × 150 mm	–	0.14	2.82	–	2.91	m	**5.73**
47 mm × 175 mm	–	0.14	2.82	–	3.43	m	**6.25**
47 mm × 200 mm	–	0.15	3.03	–	3.75	m	**6.78**
47 mm × 225 mm	–	0.15	3.03	–	4.29	m	**7.32**
47 mm × 250 mm	–	0.16	3.23	–	4.83	m	**8.06**
75 mm × 150 mm	–	0.15	3.03	–	6.02	m	**9.05**
75 mm × 175 mm	–	0.15	3.03	–	7.21	m	**10.24**
75 mm × 200 mm	–	0.16	3.23	–	8.11	m	**11.34**
75 mm × 225 mm	–	0.16	3.23	–	8.86	m	**12.09**
75 mm × 250 mm	–	0.17	3.43	–	13.07	m	**16.50**
Roof members; pitched							
25 mm × 100 mm	–	0.11	2.23	–	1.80	m	**4.03**
25 mm × 125 mm	–	0.11	2.23	–	2.45	m	**4.68**
25 mm × 150 mm	–	0.14	2.82	–	2.93	m	**5.75**
25 mm × 175 mm	–	0.16	3.23	–	3.43	m	**6.66**
25 mm × 200 mm	–	0.17	3.43	–	3.93	m	**7.36**
38 mm × 100 mm	–	0.14	2.82	–	2.28	m	**5.10**
38 mm × 125 mm	–	0.14	2.82	–	2.72	m	**5.54**
38 mm × 150 mm	–	0.14	2.82	–	3.17	m	**5.99**
38 mm × 175 mm	–	0.16	3.23	–	3.76	m	**6.99**
38 mm × 200 mm	–	0.17	3.43	–	4.32	m	**7.75**
47 mm × 50 mm	–	0.11	2.23	–	1.13	m	**3.36**
47 mm × 75 mm	–	0.14	2.82	–	1.53	m	**4.35**
47 mm × 100 mm	–	0.17	3.43	–	1.96	m	**5.39**
47 mm × 125 mm	–	0.17	3.43	–	2.52	m	**5.95**
47 mm × 150 mm	–	0.19	3.83	–	2.91	m	**6.74**

16 CARPENTRY

Item	PC £	Labour hours	Labour £	Plant £	Material £	Unit	Total rate £
47 mm × 175 mm	–	0.19	3.83	–	3.43	m	**7.26**
47 mm × 200 mm	–	0.19	3.83	–	3.75	m	**7.58**
47 mm × 225 mm	–	0.19	3.83	–	4.29	m	**8.12**
75 mm × 100 mm	–	0.23	4.64	–	4.48	m	**9.12**
75 mm × 125 mm	–	0.23	4.64	–	5.50	m	**10.14**
75 mm × 150 mm	–	0.23	4.64	–	6.02	m	**10.66**
100 mm × 150 mm	–	0.28	5.66	–	7.81	m	**13.47**
100 mm × 175 mm	–	0.28	5.66	–	9.10	m	**14.76**
100 mm × 200 mm	–	0.28	5.66	–	10.35	m	**16.01**
100 mm × 225 mm	–	0.31	6.26	–	11.63	m	**17.89**
100 mm × 250 mm	–	0.31	6.26	–	12.98	m	**19.24**
Plates							
38 mm × 75 mm	–	0.11	2.23	–	1.88	m	**4.11**
38 mm × 100 mm	–	0.14	2.82	–	2.28	m	**5.10**
47 mm × 75 mm	–	0.14	2.82	–	1.53	m	**4.35**
47 mm × 100 mm	–	0.17	3.43	–	1.96	m	**5.39**
75 mm × 100 mm	–	0.19	3.83	–	4.48	m	**8.31**
75 mm × 125 mm	–	0.22	4.44	–	5.45	m	**9.89**
75 mm × 150 mm	–	0.25	5.05	–	5.96	m	**11.01**
Plates; fixing by bolting							
38 mm × 75 mm	–	0.20	4.04	–	1.82	m	**5.86**
38 mm × 100 mm	–	0.23	4.64	–	2.28	m	**6.92**
47 mm × 75 mm	–	0.23	4.64	–	1.53	m	**6.17**
47 mm × 100 mm	–	0.26	5.25	–	1.96	m	**7.21**
75 mm × 100 mm	–	0.29	5.86	–	4.48	m	**10.34**
75 mm × 125 mm	–	0.31	6.26	–	5.45	m	**11.71**
75 mm × 150 mm	–	0.34	6.87	–	5.96	m	**12.83**
Joist strutting; herringbone							
47 mm × 50 mm; depth of joist 150 mm	–	0.46	9.29	–	2.84	m	**12.13**
47 mm × 50 mm; depth of joist 175 mm	–	0.46	9.29	–	2.89	m	**12.18**
47 mm × 50 mm; depth of joist 200 mm	–	0.46	9.29	–	2.94	m	**12.23**
47 mm × 50 mm; depth of joist 225 mm	–	0.46	9.29	–	2.99	m	**12.28**
47 mm × 50 mm; depth of joist 250 mm	–	0.46	9.29	–	3.03	m	**12.32**
Joist strutting; block							
47 mm × 150 mm; depth of joist 150 mm	–	0.28	5.66	–	3.52	m	**9.18**
47 mm × 175 mm; depth of joist 175 mm	–	0.28	5.66	–	4.04	m	**9.70**
47 mm × 200 mm; depth of joist 200 mm	–	0.28	5.66	–	4.36	m	**10.02**
47 mm × 225 mm; depth of joist 225 mm	–	0.28	5.66	–	4.90	m	**10.56**
47 mm × 250 mm; depth of joist 250 mm	–	0.28	5.66	–	5.43	m	**11.09**
Cleats							
225 mm × 100 mm × 75 mm	–	0.19	3.83	–	0.90	nr	**4.73**
Extra for stress grading to above timbers							
general structural (GS) grade	–	–	–	–	31.76	m³	**31.76**
special structural (SS) grade	–	–	–	–	63.52	m³	**63.52**
Extra for protecting and flameproofing timber with							
Celgard CF protection or other equal and approved							
small sections	–	–	–	–	125.90	m³	**125.90**
large sections	–	–	–	–	121.55	m³	**121.55**
Wrot surfaces							
plain; 50 mm wide	–	0.02	0.40	–	–	m	**0.40**
plain; 100 mm wide	–	0.03	0.61	–	–	m	**0.61**
plain; 150 mm wide	–	0.04	0.81	–	–	m	**0.81**

16 CARPENTRY

Item	PC £	Labour hours	Labour £	Plant £	Material £	Unit	Total rate £
TIMBER FRAMING – cont							
Trussed rafters, stress graded sawn softwood pressure impregnated; raised through two storeys and fixed in position. Roof trusses are always project specific and prices always need to be obtained from a manufacturer for any particular roof design							
W type truss (Fink); 22.5° pitch; 450 mm eaves overhang							
5.00 m span	–	1.48	29.89	–	48.65	nr	**78.54**
7.60 m span	–	1.62	32.72	–	64.40	nr	**97.12**
10.00 m span	–	1.85	37.36	–	116.90	nr	**154.26**
W type truss (Fink); 30° pitch; 450 mm eaves overhang							
5.00 m span	–	1.48	29.89	–	48.65	nr	**78.54**
7.60 m span	–	1.62	32.72	–	66.50	nr	**99.22**
10.00 m span	–	1.85	37.36	–	116.90	nr	**154.26**
W type truss (Fink); 45° pitch; 450 mm eaves overhang							
4.60 m span	–	1.48	29.89	–	106.40	nr	**136.29**
7.00 m span	–	1.62	32.72	–	206.15	nr	**238.87**
Mono type truss; 17.5° pitch; 450 mm eaves overhang							
3.30 m span	–	1.30	26.25	–	43.40	nr	**69.65**
5.60 m span	–	1.48	29.89	–	59.15	nr	**89.04**
7.00 m span	–	1.71	34.53	–	68.60	nr	**103.13**
Attic type truss; 45° pitch; 450 mm eaves overhang							
5.00 m span	–	2.91	58.77	–	53.90	nr	**112.67**
7.60 m span	–	3.05	61.59	–	80.15	nr	**141.74**
9.00 m span	–	3.24	65.44	–	232.40	nr	**297.84**
Moelven Toreboda glulam timber beams or other equal and approved; Moelven Laminated Timber Structures; LB grade whitewood; pressure impregnated; phenbol resorcinal adhesive; clean planed finish; fixed							
Laminated roof beams							
approximate rate for glulam beams	–	–	–	–	–	m³	**4510.00**
56 mm × 225 mm	–	–	–	–	–	m	**59.95**
66 mm × 315 mm	–	–	–	–	–	m	**98.91**
90 mm × 315 mm	–	–	–	–	–	m	**134.89**
90 mm × 405 mm	–	–	–	–	–	m	**173.45**
115 mm × 405 mm	–	–	–	–	–	m	**221.59**
115 mm × 495 mm	–	–	–	–	–	m	**270.52**
115 mm × 630 mm	–	–	–	–	–	m	**343.93**

16 CARPENTRY

Item	PC £	Labour hours	Labour £	Plant £	Material £	Unit	Total rate £
Masterboard or other equal and approved; 6 mm thick							
Eaves, verge soffit boards, fascia boards and the like							
over 300 mm wide	10.45	0.65	13.13	–	12.73	m²	**25.86**
75 mm wide	–	0.19	3.83	–	0.98	m	**4.81**
150 mm wide	–	0.22	4.44	–	1.91	m	**6.35**
225 mm wide	–	0.26	5.25	–	2.85	m	**8.10**
300 mm wide	–	0.28	5.66	–	3.77	m	**9.43**
Plywood; external quality; 12 mm thick							
Eaves, verge soffit boards, fascia boards and the like							
over 300 mm wide	8.07	0.76	15.35	–	10.04	m²	**25.39**
75 mm wide	–	0.23	4.64	–	0.78	m	**5.42**
150 mm wide	–	0.27	5.45	–	1.50	m	**6.95**
225 mm wide	–	0.31	6.26	–	2.24	m	**8.50**
300 mm wide	–	0.34	6.87	–	2.96	m	**9.83**
Plywood; external quality; 15 mm thick							
Eaves, verge soffit boards, fascia boards and the like							
over 300 mm wide	10.21	0.76	15.35	–	12.45	m²	**27.80**
75 mm wide	–	0.23	4.64	–	0.96	m	**5.60**
150 mm wide	–	0.27	5.45	–	1.87	m	**7.32**
225 mm wide	–	0.31	6.26	–	2.78	m	**9.04**
300 mm wide	–	0.34	6.87	–	3.69	m	**10.56**
Plywood; external quality; 18 mm thick							
Eaves, verge soffit boards, fascia boards and the like							
over 300 mm wide	11.96	0.76	15.35	–	14.43	m²	**29.78**
75 mm wide	–	0.23	4.64	–	1.10	m	**5.74**
150 mm wide	–	0.27	5.45	–	2.16	m	**7.61**
225 mm wide	–	0.31	6.26	–	3.22	m	**9.48**
300 mm wide	–	0.34	6.87	–	4.28	m	**11.15**
Plywood; marine quality; 18 mm thick							
Gutter boards; butt joints							
over 300 mm wide	11.78	0.86	17.37	–	14.23	m²	**31.60**
150 mm wide	–	0.31	6.26	–	2.13	m	**8.39**
225 mm wide	–	0.34	6.87	–	3.22	m	**10.09**
300 mm wide	–	0.38	7.68	–	4.26	m	**11.94**
Eaves, verge soffit boards, fascias boards and the like							
over 300 mm wide	11.78	0.76	15.35	–	14.23	m²	**29.58**
75 mm wide	–	0.23	4.64	–	1.09	m	**5.73**
150 mm wide	–	0.27	5.45	–	2.13	m	**7.58**
225 mm wide	–	0.31	6.26	–	3.18	m	**9.44**
300 mm wide	–	0.34	6.87	–	4.22	m	**11.09**

16 CARPENTRY

Item	PC £	Labour hours	Labour £	Plant £	Material £	Unit	Total rate £
TIMBER FRAMING – cont							
Plywood; marine quality; 25 mm thick							
Gutter boards; butt joints							
over 300 mm wide	16.36	0.93	18.78	–	19.40	m²	**38.18**
150 mm wide	–	0.32	6.46	–	2.91	m	**9.37**
225 mm wide	–	0.37	7.48	–	4.39	m	**11.87**
300 mm wide	–	0.42	8.48	–	5.82	m	**14.30**
Eaves, verge soffit boards, fascia baords and the like							
over 300 mm wide	16.36	0.81	16.36	–	19.40	m²	**35.76**
75 mm wide	–	0.24	4.85	–	1.48	m	**6.33**
150 mm wide	–	0.29	5.86	–	2.91	m	**8.77**
225 mm wide	–	0.29	5.86	–	4.35	m	**10.21**
300 mm wide	–	0.37	7.48	–	5.78	m	**13.26**
Sawn softwood; untreated							
Gutter boards; butt joints							
19 mm thick; 150 mm wide; sloping	–	1.16	23.43	–	12.05	m²	**35.48**
19 mm thick; 75 mm wide	–	0.32	6.46	–	0.93	m	**7.39**
19 mm thick; 150 mm wide	–	0.37	7.48	–	1.75	m	**9.23**
19 mm thick; 225 mm wide	–	0.42	8.48	–	3.17	m	**11.65**
25 mm thick; sloping	–	1.16	23.43	–	19.16	m²	**42.59**
25 mm thick; 75 mm wide	–	0.32	6.46	–	1.19	m	**7.65**
25 mm thick; 150 mm wide	–	0.37	7.48	–	2.81	m	**10.29**
25 mm thick; 225 mm wide	–	0.42	8.48	–	4.43	m	**12.91**
Cesspools with 25 mm thick sides and bottom							
225 mm × 225 mm × 150 mm	–	1.11	22.42	–	3.72	nr	**26.14**
300 mm × 300 mm × 150 mm	–	1.30	26.25	–	4.87	nr	**31.12**
Individual supports; firrings							
50 mm wide × 36 mm average depth	–	0.14	2.82	–	2.50	m	**5.32**
50 mm wide × 50 mm average depth	–	0.14	2.82	–	3.80	m	**6.62**
50 mm wide × 75 mm average depth	–	0.14	2.82	–	4.94	m	**7.76**
Individual supports; bearers							
25 mm × 50 mm	–	0.09	1.82	–	1.12	m	**2.94**
38 mm × 50 mm	–	0.09	1.82	–	1.44	m	**3.26**
50 mm × 50 mm	–	0.09	1.82	–	1.09	m	**2.91**
50 mm × 75 mm	–	0.09	1.82	–	1.42	m	**3.24**
Individual supports; angle fillets							
38 mm × 38 mm	–	0.09	1.82	–	1.00	m	**2.82**
50 mm × 50 mm	–	0.09	1.82	–	1.28	m	**3.10**
75 mm × 75 mm	–	0.11	2.23	–	2.61	m	**4.84**
Individual supports; tilting fillets							
19 mm × 38 mm	–	0.09	1.82	–	0.61	m	**2.43**
25 mm × 50 mm	–	0.09	1.82	–	0.99	m	**2.81**
38 mm × 75 mm	–	0.09	1.82	–	1.50	m	**3.32**
50 mm × 75 mm	–	0.09	1.82	–	1.92	m	**3.74**
75 mm × 100 mm	–	0.14	2.82	–	3.56	m	**6.38**

16 CARPENTRY

Item	PC £	Labour hours	Labour £	Plant £	Material £	Unit	Total rate £
Individual supports; grounds or battens							
13 mm × 19 mm	–	0.04	0.81	–	0.47	m	**1.28**
13 mm × 32 mm	–	0.04	0.81	–	0.47	m	**1.28**
25 mm × 50 mm	–	0.04	0.81	–	1.03	m	**1.84**
Individual supports; grounds or battens; plugged and screwed							
13 mm × 19 mm	–	0.14	2.82	–	0.46	m	**3.28**
13 mm × 32 mm	–	0.14	2.82	–	0.46	m	**3.28**
25 mm × 50 mm	–	0.14	2.82	–	1.02	m	**3.84**
Framed supports; open-spaced grounds or battens; at 300 mm centres one way							
25 mm × 50 mm	–	0.14	2.82	–	3.41	m²	**6.23**
25 mm × 50 mm; plugged and screwed	–	0.42	8.48	–	3.39	m²	**11.87**
Framed supports; at 300 mm centres one way and 600 mm centres the other way							
25 mm × 50 mm	–	0.69	13.93	–	5.12	m²	**19.05**
38 mm × 50 mm	–	0.69	13.93	–	6.70	m²	**20.63**
50 mm × 50 mm	–	0.69	13.93	–	4.92	m²	**18.85**
50 mm × 75 mm	–	0.69	13.93	–	6.56	m²	**20.49**
75 mm × 75 mm	–	0.69	13.93	–	15.05	m²	**28.98**
Framed supports; at 300 mm centres one way and 600 mm centres the other way; plugged and screwed							
25 mm × 50 mm	–	1.16	23.43	–	5.25	m²	**28.68**
38 mm × 50 mm	–	1.16	23.43	–	6.84	m²	**30.27**
50 mm × 50 mm	–	1.16	23.43	–	5.06	m²	**28.49**
50 mm × 75 mm	–	1.16	23.43	–	6.69	m²	**30.12**
75 mm × 75 mm	–	1.16	23.43	–	15.18	m²	**38.61**
Framed supports; at 500 mm centres both ways							
25 mm × 50 mm; to bath panels	–	0.83	16.76	–	6.67	m²	**23.43**
Framed supports; as bracketing and cradling around steelwork							
25 mm × 50 mm	–	1.30	26.25	–	7.23	m²	**33.48**
50 mm × 50 mm	–	1.39	28.07	–	6.96	m²	**35.03**
50 mm × 75 mm	–	1.48	29.89	–	9.25	m²	**39.14**
Sawn softwood; tanalized							
Gutter boards; butt joints							
19 mm thick; 150 mm; sloping	–	1.16	23.43	–	13.11	m²	**36.54**
19 mm thick; 75 mm wide	–	0.32	6.46	–	1.02	m	**7.48**
19 mm thick; 150 mm wide	–	0.37	7.48	–	1.91	m	**9.39**
19 mm thick; 225 mm wide	–	0.42	8.48	–	3.40	m	**11.88**
25 mm thick; sloping	–	1.16	23.43	–	20.55	m²	**43.98**
25 mm thick; 75 mm wide	–	0.32	6.46	–	1.29	m	**7.75**
25 mm thick; 150 mm wide	–	0.37	7.48	–	3.02	m	**10.50**
25 mm thick; 225 mm wide	–	0.42	8.48	–	4.75	m	**13.23**
Cesspools with 25 mm thick sides and bottom							
225 mm × 225 mm × 150 mm	–	1.11	22.42	–	3.99	nr	**26.41**
300 mm × 300 mm × 150 mm	–	1.30	26.25	–	5.25	nr	**31.50**

16 CARPENTRY

Item	PC £	Labour hours	Labour £	Plant £	Material £	Unit	Total rate £
TIMBER FRAMING – cont							
Sawn softwood – cont							
Individual supports; firrings							
50 mm wide × 36 mm average depth	–	0.14	2.82	–	2.60	m	**5.42**
50 mm wide × 50 mm average depth	–	0.14	2.82	–	3.95	m	**6.77**
50 mm wide × 75 mm average depth	–	0.14	2.82	–	5.15	m	**7.97**
Individual supports; bearers							
25 mm × 50 mm	–	0.09	1.82	–	1.20	m	**3.02**
38 mm × 50 mm	–	0.09	1.82	–	1.54	m	**3.36**
50 mm × 50 mm	–	0.09	1.82	–	1.23	m	**3.05**
50 mm × 75 mm	–	0.09	1.82	–	1.63	m	**3.45**
Individual supports; angle fillets							
38 mm × 38 mm	–	0.09	1.82	–	1.04	m	**2.86**
50 mm × 50 mm	–	0.09	1.82	–	1.35	m	**3.17**
75 mm × 75 mm	–	0.11	2.23	–	2.77	m	**5.00**
Individual supports; tilting fillets							
19 mm × 38 mm	–	0.09	1.82	–	0.63	m	**2.45**
25 mm × 50 mm	–	0.09	1.82	–	1.02	m	**2.84**
38 mm × 75 mm	–	0.09	1.82	–	1.58	m	**3.40**
50 mm × 75 mm	–	0.09	1.82	–	2.03	m	**3.85**
75 mm × 100 mm	–	0.14	2.82	–	3.78	m	**6.60**
Individual supports; grounds or battens							
13 mm × 19 mm	–	0.04	0.81	–	0.48	m	**1.29**
13 mm × 32 mm	–	0.04	0.81	–	0.49	m	**1.30**
25 mm × 50 mm	–	0.04	0.81	–	1.10	m	**1.91**
Individual supports; grounds or battens; plugged and screwed							
13 mm × 19 mm	–	0.14	2.82	–	0.47	m	**3.29**
13 mm × 32 mm	–	0.14	2.82	–	0.48	m	**3.30**
25 mm × 50 mm	–	0.14	2.82	–	1.09	m	**3.91**
Framed supports; open-spaced grounds or battens; at 300 mm centres one way							
25 mm × 50 mm	–	0.14	2.82	–	3.64	m²	**6.46**
25 mm × 50 mm; plugged and screwed	–	0.42	8.48	–	3.62	m²	**12.10**
Framed supports; at 300 mm centres one way and 600 mm centres the other way							
25 mm × 50 mm	–	0.69	13.93	–	5.47	m²	**19.40**
38 mm × 50 mm	–	0.69	13.93	–	7.23	m²	**21.16**
50 mm × 50 mm	–	0.69	13.93	–	5.62	m²	**19.55**
50 mm × 75 mm	–	0.69	13.93	–	7.60	m²	**21.53**
75 mm × 75 mm	–	0.69	13.93	–	16.61	m²	**30.54**
Framed supports; at 300 mm centres one way and 600 mm centres the other way; plugged and screwed							
25 mm × 50 mm	–	1.16	23.43	–	5.61	m²	**29.04**
38 mm × 50 mm	–	1.16	23.43	–	7.36	m²	**30.79**
50 mm × 50 mm	–	1.16	23.43	–	5.75	m²	**29.18**
50 mm × 75 mm	–	1.16	23.43	–	7.74	m²	**31.17**
75 mm × 75 mm	–	1.16	23.43	–	16.75	m²	**40.18**

16 CARPENTRY

Item	PC £	Labour hours	Labour £	Plant £	Material £	Unit	Total rate £
Framed supports; at 500 mm centres both ways							
25 mm × 50 mm; to bath panels	–	0.83	16.76	–	7.12	m²	**23.88**
Framed supports; as bracketing and cradling around steelwork							
25 mm × 50 mm	–	1.30	26.25	–	7.73	m²	**33.98**
50 mm × 50 mm	–	1.39	28.07	–	7.94	m²	**36.01**
50 mm × 75 mm	–	1.48	29.89	–	10.71	m²	**40.60**
Wrought softwood							
Gutter boards; tongued and grooved joints							
19 mm thick; 150 mm; sloping	–	1.39	28.07	–	18.19	m²	**46.26**
19 mm thick; 75 mm wide	–	0.37	7.48	–	1.32	m	**8.80**
19 mm thick; 150 mm wide	–	0.42	8.48	–	2.67	m	**11.15**
19 mm thick; 225 mm wide	–	0.46	9.29	–	3.90	m	**13.19**
25 mm thick; sloping	–	1.39	28.07	–	18.77	m²	**46.84**
25 mm thick; 75 mm wide	–	0.37	7.48	–	1.47	m	**8.95**
25 mm thick; 150 mm wide	–	0.42	8.48	–	2.65	m	**11.13**
25 mm thick; 225 mm wide	–	0.46	9.29	–	3.97	m	**13.26**
Eaves, verge soffit boards, fascia boards and the like							
19 mm thick; over 300 mm wide	–	1.15	23.23	–	18.28	m²	**41.51**
19 mm thick; 150 mm wide; once grooved	–	0.19	3.83	–	3.08	m	**6.91**
25 mm thick; 150 mm wide; once grooved	–	0.19	3.83	–	3.77	m	**7.60**
25 mm thick; 175 mm wide; once grooved	–	0.19	3.83	–	3.71	m	**7.54**
32 mm thick; 225 mm wide; once grooved	–	0.23	4.64	–	6.06	m	**10.70**
Wrought softwood; tanalized							
Gutter boards; tongued and grooved joints							
19 mm thick; 150 mm; sloping	–	1.39	28.07	–	19.25	m²	**47.32**
19 mm thick; 75 mm wide	–	0.37	7.48	–	1.40	m	**8.88**
19 mm thick; 150 mm wide	–	0.42	8.48	–	2.84	m	**11.32**
19 mm thick; 225 mm wide	–	0.46	9.29	–	4.14	m	**13.43**
25 mm thick; sloping	–	1.39	28.07	–	20.16	m²	**48.23**
25 mm thick; 75 mm wide	–	0.37	7.48	–	1.58	m	**9.06**
25 mm thick; 150 mm wide	–	0.42	8.48	–	2.86	m	**11.34**
25 mm thick; 225 mm wide	–	0.46	9.29	–	4.28	m	**13.57**
Eaves, verge soffit boards, fascia boards and the like							
19 mm thick; over 300 mm wide	–	1.15	23.23	–	19.33	m²	**42.56**
19 mm thick; 150 mm wide; once grooved	–	0.19	3.83	–	3.23	m	**7.06**
25 mm thick; 150 mm wide; once grooved	–	0.19	3.83	–	3.98	m	**7.81**
25 mm thick; 175 mm wide; once grooved	–	0.20	4.04	–	3.95	m	**7.99**
32 mm thick; 225 mm wide; once grooved	–	0.23	4.64	–	6.46	m	**11.10**
Straps; mild steel; galvanized							
Standard twisted vertical restraint; fixing to softwood and brick or blockwork							
27.5 mm × 2.5 mm × 400 mm girth	–	0.23	4.64	–	1.65	nr	**6.29**
27.5 mm × 2.5 mm × 600 mm girth	–	0.24	4.85	–	2.28	nr	**7.13**
27.5 mm × 2.5 mm × 800 mm girth	–	0.25	5.05	–	3.30	nr	**8.35**
27.5 mm × 2.5 mm × 1000 mm girth	–	0.28	5.66	–	4.27	nr	**9.93**
27.5 mm × 2.5 mm × 1200 mm girth	–	0.29	5.86	–	5.17	nr	**11.03**

16 CARPENTRY

Item	PC £	Labour hours	Labour £	Plant £	Material £	Unit	Total rate £
TIMBER FRAMING – cont							
Hangers; mild steel; galvanized							
Joist hangers 0.90 mm thick; The Expanded Metal Company Ltd Speedy or other equal and approved; for fixing to softwood; joist sizes							
50 mm wide; all sizes to 225 mm deep	2.13	0.11	2.23	–	2.45	nr	**4.68**
75 mm wide; all sizes to 225 mm deep	2.23	0.14	2.82	–	2.65	nr	**5.47**
100 mm wide; all sizes to 225 mm deep	2.40	0.17	3.43	–	2.93	nr	**6.36**
Joist hangers 2.50 mm thick; for building in; joist sizes							
50 mm × 100 mm	3.94	0.07	1.44	–	4.47	nr	**5.91**
50 mm × 125 mm	3.95	0.07	1.44	–	4.49	nr	**5.93**
50 mm × 150 mm	3.71	0.09	1.85	–	4.29	nr	**6.14**
50 mm × 175 mm	3.88	0.09	1.85	–	4.48	nr	**6.33**
50 mm × 200 mm	4.51	0.11	2.25	–	5.01	nr	**7.26**
50 mm × 225 mm	4.57	0.11	2.25	–	5.31	nr	**7.56**
75 mm × 150 mm	5.71	0.09	1.85	–	6.50	nr	**8.35**
75 mm × 175 mm	5.36	0.09	1.85	–	6.12	nr	**7.97**
75 mm × 200 mm	5.71	0.11	2.25	–	6.57	nr	**8.82**
75 mm × 225 mm	6.13	0.11	2.25	–	7.04	nr	**9.29**
75 mm × 250 mm	6.49	0.13	2.66	–	7.50	nr	**10.16**
100 mm × 200 mm	7.11	0.11	2.25	–	8.12	nr	**10.37**
Metal connectors; mild steel; galvanized							
Round toothed plate; for 10 mm or 12 mm dia. bolts							
38 mm dia.; single sided	–	0.01	0.20	–	0.60	nr	**0.80**
38 mm dia.; double sided	–	0.01	0.20	–	0.66	nr	**0.86**
50 mm dia.; single sided	–	0.01	0.20	–	0.64	nr	**0.84**
50 mm dia.; double sided	–	0.01	0.20	–	0.71	nr	**0.91**
63 mm dia.; single sided	–	0.01	0.20	–	0.93	nr	**1.13**
63 mm dia.; double sided	–	0.01	0.20	–	1.04	nr	**1.24**
75 mm dia.; single sided	–	0.01	0.20	–	1.39	nr	**1.59**
75 mm dia.; double sided	–	0.01	0.20	–	1.44	nr	**1.64**
framing anchor	–	0.14	2.82	–	1.13	nr	**3.95**
Bolts; mild steel; galvanized							
Fixing only bolts; 50 mm–200 mm long							
6 mm dia.	–	0.03	0.61	–	–	nr	**0.61**
8 mm dia.	–	0.03	0.61	–	–	nr	**0.61**
10 mm dia.	–	0.04	0.81	–	–	nr	**0.81**
12 mm dia.	–	0.04	0.81	–	–	nr	**0.81**
16 mm dia.	–	0.05	1.01	–	–	nr	**1.01**
20 mm dia.	–	0.05	1.01	–	–	nr	**1.01**
Bolts							
Expanding bolts; Rawlbolt projecting type or other equal and approved; Rawl Fixings; plated; one nut; one washer							
6 mm dia.; ref M6 10P	–	0.09	1.82	–	0.62	nr	**2.44**
6 mm dia.; ref M6 25P	–	0.09	1.82	–	0.72	nr	**2.54**
6 mm dia.; ref M6 60P	–	0.09	1.82	–	0.72	nr	**2.54**

16 CARPENTRY

Item	PC £	Labour hours	Labour £	Plant £	Material £	Unit	Total rate £
8 mm dia.; ref M8 25P	–	0.09	1.82	–	0.71	nr	**2.53**
8 mm dia.; ref M8 60P	–	0.09	1.82	–	0.72	nr	**2.54**
10 mm dia.; ref M10 15P	–	0.09	1.82	–	0.96	nr	**2.78**
10 mm dia.; ref M10 30P	–	0.09	1.82	–	1.00	nr	**2.82**
10 mm dia.; ref M10 60P	–	0.09	1.82	–	0.98	nr	**2.80**
12 mm dia.; ref M12 15P	–	0.09	1.82	–	1.76	nr	**3.58**
12 mm dia.; ref M12 30P	–	0.10	2.02	–	0.15	nr	**2.17**
12 mm dia.; ref M12 75P	–	0.09	1.82	–	0.20	nr	**2.02**
16 mm dia.; ref M16 35P	–	0.09	1.82	–	3.75	nr	**5.57**
16 mm dia.; ref M16 75P	–	0.09	1.82	–	3.86	nr	**5.68**
Expanding bolts; Rawlbolt loose bolt type or other equal; Rawl Fixings; plated; one bolt; one washer							
6 mm dia.; ref M6 10L	–	0.09	1.82	–	0.66	nr	**2.48**
6 mm dia.; ref M6 25L	–	0.09	1.82	–	0.66	nr	**2.48**
6 mm dia.; ref M6 40L	–	0.09	1.82	–	0.78	nr	**2.60**
8 mm dia.; ref M8 25L	–	0.09	1.82	–	0.86	nr	**2.68**
8 mm dia.; ref M8 40L	–	0.09	1.82	–	0.90	nr	**2.72**
10 mm dia.; ref M10 10L	–	0.09	1.82	–	0.91	nr	**2.73**
10 mm dia.; ref M10 25L	–	0.09	1.82	–	1.01	nr	**2.83**
10 mm dia.; ref M10 50L	–	0.09	1.82	–	1.01	nr	**2.83**
10 mm dia.; ref M10 75L	–	0.09	1.82	–	1.11	nr	**2.93**
12 mm dia.; ref M12 10L	–	0.09	1.82	–	1.75	nr	**3.57**
12 mm dia.; ref M12 25L	–	0.09	1.82	–	1.82	nr	**3.64**
12 mm dia.; ref M12 40L	–	0.09	1.82	–	1.80	nr	**3.62**
12 mm dia.; ref M12 60L	–	0.09	1.82	–	2.28	nr	**4.10**
16 mm dia.; ref M16 30L	–	0.09	1.82	–	3.83	nr	**5.65**
16 mm dia.; ref M16 60L	–	0.09	1.82	–	4.48	nr	**6.30**
Truss clips							
Truss clips; fixing to softwood; joist size							
38 mm wide	0.84	0.14	2.82	–	1.29	nr	**4.11**
50 mm wide	0.79	0.14	2.82	–	1.25	nr	**4.07**
Sole plate angles; mild steel galvanized							
Sole plate angle; fixing to softwood and concrete							
112 mm × 40 mm × 76 mm	0.96	0.19	3.83	–	2.43	nr	**6.26**
Chemical anchors							
R-CAS Spin-in epoxy acrylate capsules and standard studs or other equal and approved; Rawl Fixings; with nuts and washers; drilling masonry							
capsule ref 60–408; stud ref 60–448	–	0.25	5.05	–	0.97	nr	**6.02**
capsule ref 60–410; stud ref 60–454	–	0.28	5.66	–	1.11	nr	**6.77**
capsule ref 60–412; stud ref 60–460	–	0.31	6.26	–	1.51	nr	**7.77**
capsule ref 60–416; stud ref 60–472	–	0.34	6.87	–	1.88	nr	**8.75**
capsule ref 60–420; stud ref 60–478	–	0.36	7.27	–	2.44	nr	**9.71**
capsule ref 60–424; stud ref 60–484	–	0.40	8.07	–	7.78	nr	**15.85**

Prices for Measured Works

16 CARPENTRY

Item	PC £	Labour hours	Labour £	Plant £	Material £	Unit	Total rate £
TIMBER FRAMING – cont							
Chemical anchors – cont							
R-CAS Spin-in epoxy acrylate capsules and stainless steel studs or other equal and approved; Rawl Fixings; with nuts and washers; drilling masonry							
capsule ref 60–408; stud ref 60–905	–	0.25	5.05	–	1.94	nr	**6.99**
capsule ref 60–410; stud ref 60–910	–	0.28	5.66	–	3.42	nr	**9.08**
capsule ref 60–412; stud ref 60–915	–	0.31	6.26	–	4.35	nr	**10.61**
capsule ref 60–416; stud ref 60–920	–	0.34	6.87	–	9.27	nr	**16.14**
capsule ref 60–420; stud ref 60–925	–	0.36	7.27	–	18.04	nr	**25.31**
capsule ref 60–424; stud ref 60–930	–	0.40	8.07	–	32.11	nr	**40.18**
R-CAS Spin-in epoxy acrylate capsules and standard internal threaded sockets or other equal and approved; Rawl Fixings; drilling masonry							
capsule ref 60–408; socket ref 60–650	–	0.25	5.05	–	1.32	nr	**6.37**
capsule ref 60–410; socket ref 60–656	–	0.28	5.66	–	1.38	nr	**7.04**
capsule ref 60–412; socket ref 60–662	–	0.31	6.26	–	1.47	nr	**7.73**
capsule ref 60–416; socket ref 60–668	–	0.34	6.87	–	2.08	nr	**8.95**
capsule ref 60–420; socket ref 60–674	–	0.36	7.27	–	1.52	nr	**8.79**
capsule ref 60–424; socket ref 60–676	–	0.40	8.07	–	4.78	nr	**12.85**
R-CAS Spin-in epoxy acrylate capsules and stainless steel internal threaded sockets or other equal and approved; Rawl Fixings; drilling masonry							
capsule ref 60–408; socket ref 60–943	–	0.25	5.05	–	2.73	nr	**7.78**
capsule ref 60–410; socket ref 60–945	–	0.28	5.66	–	2.77	nr	**8.43**
capsule ref 60–412; socket ref 60–947	–	0.31	6.26	–	3.45	nr	**9.71**
capsule ref 60–416; socket ref 60–949	–	0.34	6.87	–	5.04	nr	**11.91**
capsule ref 60–420; socket ref 60–951	–	0.36	7.27	–	5.34	nr	**12.61**
capsule ref 60–424; socket ref 60–955	–	0.40	8.07	–	12.84	nr	**20.91**
R-CAS Spin-in epoxy acrylate capsules, perforated sleeves and standard studs or other equal and approved; Rawl Fixings; in low density material; with nuts and washers; drilling masonry							
capsule ref 60–408; sleeve ref 60–538; stud ref 60–448	–	0.25	5.05	–	2.39	nr	**7.44**
capsule ref 60–410; sleeve ref 60–544; stud ref 60–454	–	0.28	5.66	–	2.73	nr	**8.39**
capsule ref 60–412; sleeve ref 60–550; stud ref 60–460	–	0.31	6.26	–	3.18	nr	**9.44**
capsule ref 60–416; sleeve ref 60–562; stud ref 60–472	–	0.34	6.87	–	3.56	nr	**10.43**

16 CARPENTRY

Item	PC £	Labour hours	Labour £	Plant £	Material £	Unit	Total rate £
R-CAS Spin-in epoxy acrylate capsules, perforated sleeves and stainless steel studs or other equal and approved; Rawl Fixings; in low density material; with nuts and washers; drilling masonry							
capsule ref 60–408; sleeve ref 60–538; stud ref 60–905	–	0.25	5.05	–	3.38	nr	**8.43**
capsule ref 60–410; sleeve ref 60–544; stud ref 60–910	–	0.28	5.66	–	5.03	nr	**10.69**
capsule ref 60–412; sleeve ref 60–550; stud ref 60–915	–	0.31	6.26	–	6.02	nr	**12.28**
capsule ref 60–416; sleeve ref 60–562; stud ref 60–920	–	0.34	6.87	–	10.95	nr	**17.82**
R-CAS Spin-in epoxy acrylate capsules, perforated sleeves and standard internal threaded sockets or other equal and approved; The Rawlplug Company; in low density material; with nuts and washers; drilling masonry							
capsule ref 60–408; sleeve ref 60–538; socket ref 60–650	–	0.25	5.05	–	2.76	nr	**7.81**
capsule ref 60–410; sleeve ref 60–544; socket ref 60–656	–	0.28	5.66	–	2.98	nr	**8.64**
capsule ref 60–412; sleeve ref 60–550; socket ref 60–662	–	0.31	6.26	–	3.13	nr	**9.39**
R-CAS Spin-in epoxy acrylate capsules, perforated sleeves and stainless steel internal threaded sockets or other equal and approved; The Rawlplug Company; in low density material; drilling masonry							
capsule ref 60–416; sleeve ref 60–562; socket ref 60–668	–	0.34	6.87	–	3.76	nr	**10.63**
capsule ref 60–408; sleeve ref 60–538; socket ref 60–943	–	0.25	5.05	–	4.17	nr	**9.22**
capsule ref 60–410; sleeve ref 60–544; socket ref 60–945	–	0.28	5.66	–	4.39	nr	**10.05**
capsule ref 60–412; sleeve ref 60–550; socket ref 60–947	–	0.31	6.26	–	5.11	nr	**11.37**
capsule ref 60–416; sleeve ref 60–562; socket ref 60–949	–	0.34	6.87	–	6.72	nr	**13.59**
BOARDING TO FLOORS							
Rigid sheet boarding							
Chipboard boarding and flooring							
Boarding to floors; butt joints							
18 mm thick	4.30	0.28	5.66	–	5.32	m²	**10.98**
Boarding to floors; tongued and grooved joints							
18 mm thick	5.42	0.30	6.06	–	6.58	m²	**12.64**
22 mm thick	6.30	0.32	6.46	–	7.58	m²	**14.04**

16 CARPENTRY

Item	PC £	Labour hours	Labour £	Plant £	Material £	Unit	Total rate £
BOARDING TO FLOORS – cont							
Acoustic chipboard flooring							
Boarding to floors; tongued and grooved joints							
chipboard on blue bat bearers	–	–	–	–	–	m²	**21.85**
chipboard on New Era levelling system	–	–	–	–	–	m²	**29.52**
Laminated engineered board flooring; 180 or 240 mm face widths; with 6 mm wear surface down to tongue; pre-finished laquered, oiled or untreated.							
Boarding to floors; micro bevel or square edge							
Country laquered; on 10 mm Pro Foam	–	–	–	–	–	m²	**64.25**
Rustic laquered; on 10 mm Pro Foam	–	–	–	–	–	m²	**61.04**
Plywood flooring							
Boarding to floors; tongued and grooved joints							
18 mm thick	8.93	0.41	8.28	–	10.55	m²	**18.83**
22 mm thick	10.83	0.45	9.09	–	12.68	m²	**21.77**
Strip boarding							
Wrought softwood							
Boarding to floors; butt joints							
19 mm × 75 mm boards	–	0.56	11.31	–	17.21	m²	**28.52**
19 mm × 125 mm boards	–	0.51	10.30	–	13.21	m²	**23.51**
22 mm × 150 mm boards	–	0.46	9.29	–	14.68	m²	**23.97**
25 mm × 100 mm boards	–	0.51	10.30	–	16.04	m²	**26.34**
25 mm × 150 mm boards	–	0.46	9.29	–	16.32	m²	**25.61**
Boarding to floors; tongued and grooved joints							
19 mm × 75 mm boards	–	0.65	13.13	–	18.53	m²	**31.66**
19 mm × 125 mm boards	–	0.60	12.12	–	14.86	m²	**26.98**
22 mm × 150 mm boards	–	0.56	11.31	–	15.10	m²	**26.41**
25 mm × 100 mm boards	–	0.60	12.12	–	19.06	m²	**31.18**
25 mm × 150 mm boards	–	0.56	11.31	–	17.86	m²	**29.17**
BOARDING TO CEILINGS							
Rigid sheet boarding							
Plywood (Eastern European); internal quality							
Lining to ceilings 4 mm thick							
over 300 mm wide	2.29	0.46	9.29	–	2.96	m²	**12.25**
not exceeding 300 mm wide	–	0.30	6.06	–	0.91	m	**6.97**
holes for pipes and the like	–	0.02	0.40	–	–	nr	**0.40**
Lining to ceilings 6 mm thick							
over 300 mm wide	3.31	0.49	9.89	–	4.12	m²	**14.01**
not exceeding 300 mm wide	–	0.32	6.46	–	1.26	m	**7.72**
holes for pipes and the like	–	0.02	0.40	–	–	nr	**0.40**

16 CARPENTRY

Item	PC £	Labour hours	Labour £	Plant £	Material £	Unit	Total rate £
Lining to ceilings 12 mm thick							
over 300 mm wide	6.16	0.56	11.31	–	7.32	m²	**18.63**
not exceeding 300 mm wide	–	0.37	7.48	–	2.23	m	**9.71**
holes for pipes and the like	–	0.03	0.61	–	–	nr	**0.61**
Lining to ceilings 18 mm thick							
over 300 mm wide	8.99	0.60	12.12	–	10.52	m²	**22.64**
not exceeding 300 mm wide	–	0.40	8.07	–	3.18	m	**11.25**
holes for pipes and the like	–	0.03	0.61	–	–	nr	**0.61**
Plywood (Eastern European); external quality							
Lining to ceilings 4 mm thick							
over 300 mm wide	5.68	0.46	9.29	–	6.78	m²	**16.07**
not exceeding 300 mm wide	–	0.30	6.06	–	2.06	m	**8.12**
holes for pipes and the like	–	0.02	0.40	–	–	nr	**0.40**
Lining to ceilings 6.5 mm thick							
over 300 mm wide	6.35	0.49	9.89	–	7.54	m²	**17.43**
not exceeding 300 mm wide	–	0.32	6.46	–	2.29	m	**8.75**
holes for pipes and the like	–	0.02	0.40	–	–	nr	**0.40**
Lining to ceilings 9 mm thick							
over 300 mm wide	8.17	0.53	10.70	–	9.60	m²	**20.30**
not exceeding 300 mm wide	–	0.34	6.87	–	2.91	m	**9.78**
holes for pipes and the like	–	0.03	0.61	–	–	nr	**0.61**
Lining to ceilings 12 mm thick							
over 300 mm wide	10.21	0.56	11.31	–	11.90	m²	**23.21**
not exceeding 300 mm wide	–	0.37	7.48	–	3.60	m	**11.08**
holes for pipes and the like	–	0.03	0.61	–	–	nr	**0.61**
Extra over Linings fixed with screws	–	0.10	2.02	–	0.07	m²	**2.09**
Non-asbestos board; Masterboard or other equal and approved; sanded finish							
Lining to ceilings 6 mm thick							
over 300 mm wide	10.45	0.41	8.28	–	12.07	m²	**20.35**
not exceeding 300 mm wide	–	0.25	5.05	–	3.63	m	**8.68**
holes for pipes and the like	–	0.02	0.40	–	–	nr	**0.40**
Lining to ceilings 9 mm thick							
over 300 mm wide	24.15	0.42	8.48	–	27.53	m²	**36.01**
not exceeding 300 mm wide	–	0.27	5.45	–	8.27	m	**13.72**
holes for pipes and the like	–	0.03	0.61	–	–	nr	**0.61**
Supalux or other equal and approved; sanded finish							
Lining to ceilings 6 mm thick							
over 300 mm wide	20.16	0.41	8.28	–	23.04	m²	**31.32**
not exceeding 300 mm wide	–	0.25	5.05	–	6.92	m	**11.97**
holes for pipes and the like	–	0.03	0.61	–	–	nr	**0.61**
Lining to ceilings 9 mm thick							
over 300 mm wide	25.67	0.42	8.48	–	29.25	m²	**37.73**
not exceeding 300 mm wide	–	0.27	5.45	–	8.79	m	**14.24**
holes for pipes and the like	–	0.03	0.61	–	–	nr	**0.61**

16 CARPENTRY

Item	PC £	Labour hours	Labour £	Plant £	Material £	Unit	Total rate £
BOARDING TO CEILINGS – cont							
Supalux or other equal and approved – cont							
Lining to ceilings 12 mm thick							
over 300 mm wide	34.85	0.49	9.89	–	39.62	m²	**49.51**
not exceeding 300 mm wide	–	0.30	6.06	–	11.90	m	**17.96**
holes for pipes and the like	–	0.04	0.81	–	–	nr	**0.81**
Extra over Linings fixed with screws	–	0.10	2.02	–	0.07	m²	**2.09**
Strip boarding							
Wrought softwood							
Boarding to internal ceilings							
12 mm × 100 mm boards	–	0.93	18.78	–	16.94	m²	**35.72**
16 mm × 100 mm boards	–	0.93	18.78	–	18.39	m²	**37.17**
19 mm × 100 mm boards	–	0.93	18.78	–	20.92	m²	**39.70**
19 mm × 125 mm boards	–	0.88	17.78	–	19.11	m²	**36.89**
19 mm × 125 mm boards; chevron pattern	–	1.30	26.25	–	19.11	m²	**45.36**
25 mm × 125 mm boards	–	0.88	17.78	–	16.43	m²	**34.21**
12 mm × 100 mm boards; knotty pine	–	0.93	18.78	–	10.42	m²	**29.20**
BOARDING TO ROOFS							
Rigid sheet boarding							
Plywood; external quality; 18 mm thick							
Boarding to roofs; butt joints							
flat to falls	11.96	0.37	7.48	–	13.97	m²	**21.45**
sloping	11.96	0.40	8.07	–	13.97	m²	**22.04**
vertical	11.96	0.53	10.70	–	13.97	m²	**24.67**
Plywood; external quality; 12 mm thick							
Boarding to roofs; butt joints							
flat to falls	8.07	0.37	7.48	–	9.58	m²	**17.06**
sloping	8.07	0.40	8.07	–	9.58	m²	**17.65**
vertical	8.07	0.53	10.70	–	9.58	m²	**20.28**
Strip boarding							
Sawn softwood; untreated							
Boarding to roofs; 150 mm wide boards; butt joints							
19 mm thick; flat; over 600 mm wide	–	0.42	8.48	–	11.36	m²	**19.84**
19 mm thick; flat; not exceeding 600 mm wide	–	0.55	11.11	–	6.91	m	**18.02**
19 mm thick; sloping; over 600 mm wide	–	0.46	9.29	–	11.36	m²	**20.65**
19 mm thick; sloping; not exceeding 600 mm wide	–	0.62	12.52	–	6.91	m	**19.43**
19 mm thick; sloping; laid diagonally; over 600 mm wide	–	0.58	11.72	–	11.36	m²	**23.08**
19 mm thick; sloping; laid diagonally; not exceeding 600 mm wide	–	0.75	15.15	–	6.91	m	**22.06**
25 mm thick; flat; over 600 mm wide	–	0.42	8.48	–	18.46	m²	**26.94**

16 CARPENTRY

Item	PC £	Labour hours	Labour £	Plant £	Material £	Unit	Total rate £
25 mm thick; flat; not exceeding 600 mm wide	–	0.56	11.31	–	11.17	m	**22.48**
25 mm thick; sloping; over 600 mm wide	–	0.46	9.29	–	18.46	m²	**27.75**
25 mm thick; sloping; not exceeding 600 mm wide	–	0.62	12.52	–	11.17	m	**23.69**
25 mm thick; sloping; laid diagonally; over 600 mm wide	–	0.58	11.72	–	18.46	m²	**30.18**
25 mm thick; sloping; laid diagonally; not exceeding 600 mm wide	–	0.74	14.94	–	11.17	m	**26.11**
Boarding to tops or cheeks of dormers; 150 mm wide boards; butt joints							
19 mm thick; laid diagonally; over 600 mm wide	–	0.74	14.94	–	11.36	m²	**26.30**
19 mm thick; laid diagonally; not exceeding 600 mm wide	–	0.92	18.59	–	6.91	m	**25.50**
19 mm thick; laid diagonally; area not exceeding 1.00 m² irrespective of width	–	0.93	18.78	–	10.75	nr	**29.53**
Sawn softwood; tanalized							
Boarding to roofs; 150 wide boards; butt joints							
19 mm thick; flat; over 600 mm wide	–	0.42	8.48	–	12.41	m²	**20.89**
19 mm thick; flat; not exceeding 600 mm wide	–	0.56	11.31	–	7.55	m	**18.86**
19 mm thick; sloping; over 600 mm wide	–	0.46	9.29	–	12.41	m²	**21.70**
19 mm thick; sloping; not exceeding 600 mm wide	–	0.62	12.52	–	7.55	m	**20.07**
19 mm thick; sloping; laid diagonally; over 600 mm wide	–	0.58	11.72	–	12.41	m²	**24.13**
19 mm thick; sloping; laid diagonally; not exceeding 600 mm wlde	–	0.74	14.94	–	7.55	m	**22.49**
25 mm thick; flat; over 600 mm wide	–	0.42	8.48	–	19.84	m²	**28.32**
25 mm thick; flat; not exceeding 600 mm wide	–	0.56	11.31	–	11.80	m	**23.11**
25 mm thick; sloping; over 600 mm wide	–	0.46	9.29	–	19.84	m²	**29.13**
25 mm thick; sloping; not exceeding 600 mm wide	–	0.62	12.52	–	11.80	m	**24.32**
25 mm thick; sloping; laid diagonally; over 600 mm wide	–	0.58	11.72	–	19.84	m²	**31.56**
25 mm thick; sloping; laid diagonally; not exceeding 600 mm wide	–	0.74	14.94	–	11.80	m	**26.74**
Boarding to tops or cheeks of dormers; 150 mm wide boards; butt joints							
19 mm thick; laid diagonally; over 600 mm wide	–	0.74	14.94	–	12.41	m²	**27.35**
19 mm thick; laid diagonally; not exceeding 600 mm wide	–	0.92	18.59	–	7.55	m	**26.14**
19 mm thick; laid diagonally; area not exceeding 1.00 m² irrespective of width	–	0.93	18.78	–	11.81	nr	**30.59**
Wrought softwood							
Boarding to roofs; tongued and grooved joints							
19 mm thick; flat to falls	–	0.51	10.30	–	17.48	m²	**27.78**
19 mm thick; sloping	–	0.56	11.31	–	17.48	m²	**28.79**
19 mm thick; sloping; laid diagonally	–	0.72	14.54	–	17.48	m²	**32.02**
25 mm thick; flat to falls	–	0.51	10.30	–	17.61	m²	**27.91**
25 mm thick; sloping	–	0.56	11.31	–	17.61	m²	**28.92**
Boarding to tops or cheeks of dormers; tongued and grooved joints							
19 mm thick; laid diagonally	–	0.93	18.78	–	17.48	m²	**36.26**

16 CARPENTRY

Item	PC £	Labour hours	Labour £	Plant £	Material £	Unit	Total rate £
BOARDING TO ROOFS – cont							
Wrought softwood; tanalized							
Boarding to roofs; tongued and grooved joints							
19 mm thick; flat to falls	–	0.51	10.30	–	18.54	m²	**28.84**
19 mm thick; sloping	–	0.56	11.31	–	18.54	m²	**29.85**
19 mm thick; sloping; laid diagonally	–	0.72	14.54	–	18.54	m²	**33.08**
25 mm thick; flat to falls	–	0.51	10.30	–	18.99	m²	**29.29**
25 mm thick; sloping	–	0.56	11.31	–	18.99	m²	**30.30**
Boarding to tops or cheeks of dormers; tongued and grooved joints							
19 mm thick; laid diagonally	–	0.93	18.78	–	18.54	m²	**37.32**
CASINGS							
Chipboard (plain)							
Two-sided 15 mm thick pipe casing; to softwood framing (not included)							
300 mm girth	–	0.56	11.31	–	1.22	m	**12.53**
600 mm girth	–	0.65	13.13	–	2.15	m	**15.28**
Three-sided 15 mm thick pipe casing; to softwood framing (not included)							
450 mm girth	–	1.16	23.43	–	1.83	m	**25.26**
900 mm girth	–	1.39	28.07	–	3.28	m	**31.35**
extra for 400 mm × 400 mm removable access panel; brass cups and screws; additional framing	–	0.93	18.78	–	1.38	nr	**20.16**
Plywood (Eastern European); internal quality							
Two-sided 6 mm thick pipe casings; to softwood framing (not included)							
300 mm girth	–	0.74	14.94	–	1.40	m	**16.34**
600 mm girth	–	0.93	18.78	–	2.52	m	**21.30**
Three-sided 6 mm thick pipe casing; to softwood framing (not included)							
450 mm girth	–	1.06	21.41	–	2.10	m	**23.51**
900 mm girth	–	1.25	25.24	–	3.83	m	**29.07**
Plywood (Eastern European); external quality							
Two-sided 6.5 mm thick pipe casings; to softwood framing (not included)							
300 mm girth	–	0.74	14.94	–	2.43	m	**17.37**
600 mm girth	–	0.93	18.78	–	4.58	m	**23.36**
Three-sided 6.5 mm thick pipe casing; to softwood framing (not included)							
450 mm girth	–	1.06	21.41	–	3.64	m	**25.05**
900 mm girth	–	1.25	25.24	–	6.92	m	**32.16**

16 CARPENTRY

Item	PC £	Labour hours	Labour £	Plant £	Material £	Unit	Total rate £
Two-sided 12 mm thick pipe casing; to softwood framing (not included)							
300 mm girth	–	0.69	13.93	–	3.74	m	**17.67**
600 mm girth	–	0.83	16.76	–	7.19	m	**23.95**
Three-sided 12 mm thick pipe casing; to softwood framing (not included)							
450 mm girth	–	0.93	18.78	–	5.61	m	**24.39**
900 mm girth	–	1.11	22.42	–	10.84	m	**33.26**
extra for 400 mm × 400 mm removable access panel; brass cups and screws; additional framing	–	1.00	20.19	–	1.38	nr	**21.57**
Preformed white melamine faced plywood casings; Pendock Profiles Ltd or other equal and approved; to softwood battens (not included)							
Skirting trunking profile; plain butt joints in the running length							
45 mm × 150 mm; ref TK150	–	0.11	2.23	–	27.75	m	**29.98**
extra for stop end	–	0.04	0.81	–	17.28	nr	**18.09**
extra for external corner	–	0.09	1.82	–	23.87	nr	**25.69**
extra for internal corner	–	0.09	1.82	–	14.50	nr	**16.32**
Casing profiles							
150 mm × 150 mm; ref MX150/150; 5 mm thick	–	0.11	2.23	–	23.14	m	**25.37**
extra for stop end	–	0.04	0.81	–	5.97	nr	**6.78**
extra for external corner	–	0.09	1.82	–	36.54	nr	**38.36**
extra for internal corner	–	0.09	1.82	–	14.50	nr	**16.32**

17 SHEET ROOF COVERINGS

Item	PC £	Labour hours	Labour £	Plant £	Material £	Unit	Total rate £
BUILT UP FELT ROOF COVERINGS							
NOTE: The following items of felt roofing, unless otherwise described, include for conventional lapping, laying and bonding between layers and to base; and laying flat or to falls, crossfalls or to slopes not exceeding 10° – but exclude any insulation etc.							
Reinforced bitumen membranes							
Three layer coverings							
type S1P1 bitumen glass fibre based felt	–	–	–	–	–	m²	**16.34**
cover with and bed in hot bitumen 13 mm thick stone chippings	–	–	–	–	–	m²	**4.74**
two base layers type S2P3 bitumen polyester-based felt; top layer type S4P4 polyester-based mineral surfaced felt; 10 mm stone chipping covering; bitumen bonded	–	–	–	–	–	m²	**27.83**
cover with and bed in hot bitumen 300 mm × 300 mm × 8 mm g.r.p. tiles	–	–	–	–	–	m²	**50.62**
Skirtings; three layer; top layer mineral surfaced; dressed over tilting fillet; turned into groove							
not exceeding 200 mm girth	–	–	–	–	–	m	**12.17**
200 mm–400 mm girth	–	–	–	–	–	m	**15.04**
Coverings to kerbs; three layer							
400 mm–600 mm girth	–	–	–	–	–	m	**19.47**
Linings to gutters; three layer							
400 mm–600 mm girth	–	–	–	–	–	m	**23.64**
Collars around pipes and the like; three layer mineral surface; 150 mm high							
not exceeding 55 mm nominal size	–	–	–	–	–	nr	**12.92**
55 mm–110 mm nominal size	–	–	–	–	–	nr	**12.92**
Outlets and dishing to gullies							
300 mm dia.	–	–	–	–	–	nr	**14.01**
Polyster based roofing systems							
Andersons high performance polyester-based roofing system or other equal							
two layer coverings; first layer HT 125 underlay; second layer HT 350; fully bonded to wood; fibre or cork base	–	–	–	–	–	m²	**23.07**
top layer mineral surfaced	–	–	–	–	–	m²	**1.96**
13 mm thick stone chippings	–	–	–	–	–	m²	**4.74**
third layer of type 3B as underlay for concrete or screeded base	–	–	–	–	–	m²	**6.03**
working into outlet pipes and the like	–	–	–	–	–	nr	**14.00**
Skirtings; two layer; top layer mineral surfaced; dressed over tilting fillet; turned into groove							
not exceeding 200 mm girth	–	–	–	–	–	m	**118.83**
200 mm–400 mm girth	–	–	–	–	–	m	**15.38**

17 SHEET ROOF COVERINGS

Item	PC £	Labour hours	Labour £	Plant £	Material £	Unit	Total rate £
Coverings to kerbs; two layer							
400 mm–600 mm girth	–	–	–	–	–	m	**19.93**
Linings to gutters; three layer							
400 mm–600 mm girth	–	–	–	–	–	m	**21.42**
Collars around pipes and the like; two layer; 150 mm high							
not exceeding 55 mm nominal size	–	–	–	–	–	nr	**14.00**
55 mm–110 mm nominal size	–	–	–	–	–	nr	**14.00**
Ruberoid Challenger SBS high performance roofing or other equal							
two layer coverings; first and second layers Ruberglas 120 GP; fully bonded to wood, fibre or cork base	–	–	–	–	–	m²	**15.18**
top layer mineral surfaced	–	–	–	–	–	m²	**5.37**
13 mm thick stone chippings	–	–	–	–	–	m²	**4.74**
third layer of Rubervent 3G as underlay for concrete or screeded base	–	–	–	–	–	m²	**6.01**
working into outlet pipes and the like	–	–	–	–	–	nr	**13.89**
Skirtings; two layer; top layer mineral surfaced; dressed over tilting fillet; turned into groove							
not exceeding 200 mm girth	–	–	–	–	–	m	**11.59**
200 mm–400 mm girth	–	–	–	–	–	m	**15.16**
Coverings to kerbs; two layer							
400 mm–600 mm girth	–	–	–	–	–	m	**19.66**
Linings to gutters; three layer							
400 mm–600 mm girth	–	–	–	–	–	m	**21.04**
Collars around pipes and the like; two layer, 150 mm high							
not exceeding 55 mm nominal size	–	–	–	–	–	nr	**13.89**
55 mm–110 mm nominal size	–	–	–	–	–	nr	**13.89**
Ruberfort HP 350 high performance roofing or other equal							
two layer coverings; first layer Ruberfort HP 180; second layer Ruberfort HP 350; fully bonded; to wood; fibre or cork base	–	–	–	–	–	m²	**17.89**
top layer mineral surfaced	–	–	–	–	–	m²	**7.40**
13 mm thick stone chippings	–	–	–	–	–	m²	**4.74**
third layer of 'Rubervent 3G'; as underlay for concrete or screeded base	–	–	–	–	–	m²	**6.01**
working into outlet pipes and the like	–	–	–	–	–	nr	**14.06**
Skirtings; two layer; top layer mineral surface; dressed over tilting fillet; turned into groove							
not exceeding 200 mm girth	–	–	–	–	–	m	**11.82**
200 mm–400 mm girth	–	–	–	–	–	m	**15.46**
Coverings to kerbs; two layer							
400 mm–600 mm girth	–	–	–	–	–	m	**20.05**
Linings to gutters; three layer							
400 mm–600 mm girth	–	–	–	–	–	m	**25.90**

17 SHEET ROOF COVERINGS

Item	PC £	Labour hours	Labour £	Plant £	Material £	Unit	Total rate £
BUILT UP FELT ROOF COVERINGS – cont							
Polyster based roofing systems – cont							
Collars around pipes and the like; two layer; 150 mm high							
not exceeding 55 mm nominal size	–	–	–	–	–	nr	**14.06**
55 mm–110 mm nominal size	–	–	–	–	–	nr	**14.06**
Ruberoid Superflex Firebloc high performance roofing or other equal (15 year guarantee specification)							
two layer coverings; first layer Superflex 180; second layer Superflex 250; fully bonded to wood; fibre or cork base	–	–	–	–	–	m²	**22.12**
top layer mineral surfaced	–	–	–	–	–	m²	**5.26**
13 mm thick stone chippings	–	–	–	–	–	m²	**4.74**
third layer of Rubervent 3G as underlay for concrete or screeded base	–	–	–	–	–	m²	**6.01**
working into outlet pipes and the like	–	–	–	–	–	nr	**15.96**
Skirtings; two layer; top layer mineral surfaced; dressed over tilting fillet; turned into groove							
not exceeding 200 mm girth	–	–	–	–	–	m	**13.82**
200 mm–400 mm girth	–	–	–	–	–	m	**18.23**
Coverings to kerbs; two layer							
400 mm–600 mm girth	–	–	–	–	–	m	**24.34**
Linings to gutters; three layer							
400 mm–600 mm girth	–	–	–	–	–	m	**26.35**
Collars around pipes and the like; two layer; 150 mm high							
not exceeding 55 mm nominal size	–	–	–	–	–	nr	**15.96**
55 mm–110 mm nominal size	–	–	–	–	–	nr	**15.96**
Ruberoid Ultra Prevent high performance roofing or other equal							
two layer coverings; first layer Ultra prevENt underlay; second layer Ultra prevENt mineral surface cap sheet.	–	–	–	–	–	m²	**40.71**
extra over for;							
third layer of Rubervent 3G as underlay for concrete or screeded base	–	–	–	–	–	m²	**6.01**
working into outlet pipes and the like	–	–	–	–	–	nr	**19.30**
Skirtings; two layer; dressed over tilting fillet; turned into groove							
not exceeding 200 mm girth	–	–	–	–	–	m	**17.30**
200 mm–400 mm girth	–	–	–	–	–	m	**23.04**
Coverings to kerbs; two layer							
400 mm–600 mm girth	–	–	–	–	–	m	**31.80**
Linings to gutters; three layer							
400 mm–600 mm girth	–	–	–	–	–	m	**33.24**
Collars around pipes and the like; two layer; 150 mm high							
not exceeding 55 mm nominal size	–	–	–	–	–	nr	**19.28**
55 mm–110 mm nominal size	–	–	–	–	–	nr	**19.28**

17 SHEET ROOF COVERINGS

Item	PC £	Labour hours	Labour £	Plant £	Material £	Unit	Total rate £
Accessories							
Eaves trim; extruded aluminium alloy; working felt into trim							
Rubertrim; type FL/G; 65 mm face	–	–	–	–	–	m	**14.16**
extra over for:							
external angle	–	–	–	–	–	nr	**14.25**
Roof screed ventilator – aluminium alloy							
Extr-aqua-vent or other equal and approved – set on screed over and including dished sinking and collar	–	–	–	–	–	nr	**44.78**
Insulation board underlays							
Vapour barrier							
reinforced; metal lined	–	–	–	–	–	m²	**13.54**
Rockwool; Duorock flat insulation board							
140 mm thick (0.25 U-value)	–	–	–	–	–	m²	**37.10**
Kingspan Thermaroof TR21 zero OPD urethene insulation board							
50 mm thick	–	–	–	–	–	m²	**22.80**
90 mm thick	–	–	–	–	–	m²	**39.28**
100 mm thick (0.25 U-value)	–	–	–	–	–	m²	**43.63**
Wood fibre boards; impregnated; density 220–350 kg/m³							
12.70 mm thick	–	–	–	–	–	m²	**6.15**
Tapered insulation board underlays							
Tapered insulation £/m² prices can vary dramatically depending upon the factors which determine the scheme layout; these primarily being gutter/outlet locations and the length of fall involved. The following guide assumes a U-value of 0.18 W/m² K as a benchmark.							
As the required insulation value will vary from project to project the required U-value should be determined by calculating the buildings energy consumption at the design stage. Due to tapered insulation scheme prices varying by project, the following prices are indicative. Please contact a specialist for a project specific quotation. U-Value must be calculated in accordance with BSENISO 6946:2007 Annex C							
Tapered PIR (Polyisocyanurate) boards; bedded in hot bitumen							
effective thickness achieving 0.18 W/m² K	27.55	–	–	–	–	m²	**58.23**
minimum thickness achieving 0.18 W/m² K	32.77	–	–	–	–	m²	**64.65**
Tapered PIR (Polyisocyanurate) board; mechanically fastened							
effective thickness achieving 0.18 W/m² K	27.55	–	–	–	–	m²	**60.78**
Tapered Aspire; Hybrid EPS/PIR boards; bedded in hot bitumen							
effective thickness achieving 0.18 W/m² K	25.65	–	–	–	–	m²	**58.23**
minimum thickness achieving 0.18 W/m² K	30.88	–	–	–	–	m²	**64.65**

17 SHEET ROOF COVERINGS

Item	PC £	Labour hours	Labour £	Plant £	Material £	Unit	Total rate £
BUILT UP FELT ROOF COVERINGS – cont							
Tapered insulation board underlays – cont							
Tapered PIR (Polyisocyanurate) board; mechanically fastened							
effective thickness achieving 0.18 W/m² K	25.65	–	–	–	–	m²	**60.78**
minimum thickness achieving 0.18 W/m² K	30.88	–	–	–	–	m²	**67.20**
Tapered Rockwool boards; bedded in hot bitumen							
effective thickness achieving 0.18 W/m² K	49.21	–	–	–	–	m²	**85.66**
minimum thickness achieving 0.18 W/m² K	60.80	–	–	–	–	m²	**92.47**
Tapered Rockwool boards; mechanically fastened							
effective thickness achieving 0.18 W/m² K	49.21	–	–	–	–	m²	**88.22**
minimum thickness achieving 0.18 W/m² K	60.80	–	–	–	–	m²	**95.71**
Tapered EPS (Expanded polystyrene) boards; bedded in hot bitumen							
effective thickness achieving 0.18 W/m² K	22.52	–	–	–	–	m²	**51.88**
minimum thickness achieving 0.18 W/m² K	27.55	–	–	–	–	m²	**58.23**
Tapered EPS (Expanded polystyrene) boards; mechanicaly fastened							
effective thickness achieving 0.18 W/m² K	22.52	–	–	–	–	m²	**54.36**
minimum thickness achieving 0.18 W/m² K	27.55	–	–	–	–	m²	**60.77**
Insulation board overlays							
Dow Roofmate SL extruded polystyrene foam boards or other equal and approved; Thermal conductivity – 0.028 W/mK							
50 mm thick	–	–	–	–	–	m²	**15.09**
140 mm thick	–	–	–	–	–	m²	**26.34**
160 mm thick	–	–	–	–	–	m²	**28.53**
Dow Roofmate LG extruded polystyrene foam boards or other equal and approved; Thermal conductivity – 0.028 W/mK							
80 mm thick	–	–	–	–	–	m²	**53.74**
100 mm thick	–	–	–	–	–	m²	**57.60**
120 mm thick	–	–	–	–	–	m²	**61.51**
SINGLE LAYER PLASTIC ROOF COVERINGS							
Kingspan KS1000TD composite single ply roof panels for roof pitches greater than 0.7° (after deflection)							
1.5 mm single ply external covering, internal coating bright white polyester (steel)							
71 mm thick panel; U-value 0.25 W/m² K	–	–	–	–	–	m²	**50.33**
100 mm thick panel; U-value 0.18 W/m² K	–	–	–	–	–	m²	**55.67**
120 mm thick panel; U-value 0.15 W/m² K	–	–	–	–	–	m²	**58.88**
Trocal S PVC roofing or other equal and approved							
Coverings	–	–	–	–	–	m²	**17.25**

17 SHEET ROOF COVERINGS

Item	PC £	Labour hours	Labour £	Plant £	Material £	Unit	Total rate £
Skirtings; dressed over metal upstands							
not exceeding 200 mm girth	–	–	–	–	–	m	**13.41**
200 mm–400 mm girth	–	–	–	–	–	m	**16.47**
Coverings to kerbs							
400 mm–600 mm girth	–	–	–	–	–	m	**30.16**
Collars around pipes and the like; 150 mm high							
not exceeding 55 mm nominal size	–	–	–	–	–	nr	**9.22**
55 mm–110 mm nominal size	–	–	–	–	–	nr	**9.22**
Trocal metal upstands or other equal and approved							
Sarnafil polymeric waterproofing membrane; cold roof							
Roof coverings							
pitch not exceeding 5°; to metal decking or the like	–	–	–	–	–	m²	**30.66**
pitch not exceeding 5°; to concrete base or the like; prime concrete with spririt priming solution	–	–	–	–	–	m²	**30.66**
Sarnafil polymeric waterproofing membrane; 1.2 mm thick fleece backed membrane; cold roof							
Roof coverings							
pitch not exceeding 5°; to metal decking or the like	–	–	–	–	–	m²	**30.66**
pitch not exceeding 5°; to concrete base or the like; prime concrete with spririt priming solution	–	–	–	–	–	m²	**30.66**
Sarnafil polymeric waterproofing membrane; 120 mm thick Sarnaform G CFC & HCFC free insulation board; vapour control layer; prime concrete with spirit priming solution							
Mechanically fastened system							
Roof coverings							
pitch not exceeding 5°; to metal decking or the like	–	–	–	–	–	m²	**47.92**
pitch not exceeding 5°; to concrete base or the like	–	–	–	–	–	m²	**55.97**
Coverings to kerbs; parapet flashing; Sarnatrim 50 mm deep on face 100 mm fixing arm; standard Sarnafil detail 1.1							
not exceeding 200 mm girth	–	–	–	–	–	m	**26.87**
200 mm–400 mm girth	–	–	–	–	–	m	**30.62**
400 mm–600 mm girth	–	–	–	–	–	m	**34.34**
Eaves detail; Sarnatrim drip edge to gutter; standard Sarnafil detail 1.3							
not exceeding 200 mm girth	–	–	–	–	–	m	**25.84**
Skirtings/Upstands; skirting to brickwork with galvanized steel counter flashing to top edge; standard Sarnafil detail 2.3							
not exceeding 200 mm girth	–	–	–	–	–	m	**31.43**
200 mm–400 mm girth	–	–	–	–	–	m	**37.80**
400 mm–600 mm girth	–	–	–	–	–	m	**44.24**

17 SHEET ROOF COVERINGS

Item	PC £	Labour hours	Labour £	Plant £	Material £	Unit	Total rate £
SINGLE LAYER PLASTIC ROOF COVERINGS – cont							
Sarnafil polymeric waterproofing membrane – cont							
Skirtings/Upstands; skirting to brickwork with Sarnametal Raglet to chase; standard Sarnafil detail 2.8							
not exceeding 200 mm girth	–	–	–	–	–	m	**31.43**
200 mm–400 mm girth	–	–	–	–	–	m	**37.80**
400 mm–600 mm girth	–	–	–	–	–	m	**44.24**
Collars around pipe standards, and the like							
50 mm dia. × 150 mm high	–	–	–	–	–	nr	**40.73**
100 mm dia. × 150 mm high	–	–	–	–	–	nr	**40.73**
Outlets and dishing to gullies							
fix Sarnadrain PVC rainwater outlet; 110 mm dia.; weld membrane to same; fit plastic leafguard	–	–	–	–	–	nr	**96.27**
Fully adhered system							
Roof coverings							
pitch not exceeding 5°; to metal decking or the like	–	–	–	–	–	m²	**51.94**
pitch not exceeding 5°; to concrete base or the like	–	–	–	–	–	m²	**57.94**
Coverings to kerbs; parapet flashing; Sarnatrim 50 mm deep on face 100 mm fixing arm; standard Sarnafil detail 1.1							
not exceeding 200 mm girth	–	–	–	–	–	m	**24.85**
200 mm–400 mm girth	–	–	–	–	–	m	**28.59**
400 mm–600 mm girth	–	–	–	–	–	m	**32.34**
Eaves detail; Sarnametal drip edge to gutter; standard Sarnafil detail 1.3							
not exceeding 200 mm girth	–	–	–	–	–	m	**23.81**
Skirtings/Upstands; skirting to brickwork with galvanized steel counter flashing to top edge; standard Sarnafil detail 2.3							
not exceeding 200 mm girth	–	–	–	–	–	m	**29.40**
200 mm–400 mm girth	–	–	–	–	–	m	**35.82**
400 mm–600 mm girth	–	–	–	–	–	m	**40.10**
Skirtings/Upstands; skirting to brickwork with Sarnametal Raglet to chase; standard Sarnafil detail 2.8							
not exceeding 200 mm girth	–	–	–	–	–	m	**30.48**
200 mm–400 mm girth	–	–	–	–	–	m	**42.28**
400 mm–600 mm girth	–	–	–	–	–	m	**42.28**
Collars around pipe standards, and the like							
50 mm dia. × 150 mm high	–	–	–	–	–	nr	**40.92**
100 mm dia. × 150 mm high	–	–	–	–	–	nr	**40.92**
Outlets and dishing to gullies							
Fix Sarnadrain PVC rainwater outlet; 110 mm dia.; weld membrane to same; fit plastic leafguard	–	–	–	–	–	nr	**96.74**

17 SHEET ROOF COVERINGS

Item	PC £	Labour hours	Labour £	Plant £	Material £	Unit	Total rate £
Options							
Extra over for 1.2 mm fleece backed membrane	–	–	–	–	–	m²	**5.27**
Landscape roofing							
SarnaVert extensive biodiverse roof; sedum blanket; 100 mm growing medium; aquafrain; 1.5 mm thick membrane; 120 mm thick insulation board; vapour control layer							
Pitch not exceeding 5°; to metal decking or the like	–	–	–	–	–	m²	**127.90**
Pitch not exceeding 5°; to concrete base or the like; prime concrete with spririt priming solution	–	–	–	–	–	m²	**133.09**
Kerb and eaves; standard Sarnafil details							
Sarnafil kerb; 150 mm above roof level	–	–	–	–	–	m	**31.67**
Sarnafil eaves detail with gravel stop ne 200 mm girth	–	–	–	–	–	m	**52.49**
Collars around pipes and the like							
50 mm dia. × 150 mm high	–	–	–	–	–	nr	**40.72**
100 mm dia. × 150 mm high	–	–	–	–	–	nr	**40.72**
Sarnafil rainwater outlet	–	–	–	–	–	nr	**163.80**
GLASS REINFORCED PLASTIC PANEL							
Glass fibre translucent sheeting grade AB class 3							
Roof cladding; sloping not exceeding 50°; fixing to timber purlins with drive screws; to suit							
Profile 3 or other equal and approved	12.54	0.18	4.12	–	18.44	m²	**22.56**
Profile 6 or other equal and approved	12.54	0.23	5.26	–	18.44	m²	**23.70**
Roof cladding; sloping not exceeding 50°; fixing to timber purlins with hook bolts; to suit							
Profile 3 or other equal and approved	12.54	0.23	5.26	–	19.31	m²	**24.57**
Profile 6 or other equal and approved	12.54	0.28	6.40	–	19.31	m²	**25.71**
Longrib 1000 or other equal and approved	12.54	0.28	6.40	–	19.31	m²	**25.71**
LEAD SHEET COVERINGS/FLASHINGS							
Milled Lead; BS EN 12588; on and including Geotec underlay							
Roof and dormer coverings							
1.80 mm thick (code 4) roof coverings							
flat (in wood roll construction (Prime Cost £ per kg)	3.16	0.90	26.51	–	45.68	m²	**72.19**
pitched (in wood roll construction)	–	1.00	29.46	–	45.91	m²	**75.37**
pitched (in welded seam construction)	–	0.90	26.51	–	45.68	m²	**72.19**
vertical (in welded seam construction)	–	1.00	29.46	–	43.60	m²	**73.06**
1.80 mm thick (code 4) dormer coverings							
flat (in wood roll construction) (Prime Cost £ per kg)	3.16	0.68	19.89	–	45.16	m²	**65.05**
pitched (in wood roll construction)	–	0.75	22.09	–	45.33	m²	**67.42**
pitched (in welded seam construction)	–	0.68	19.89	–	45.16	m²	**65.05**
vertical (in welded seam construction)	–	1.50	44.19	–	43.60	m²	**87.79**

17 SHEET ROOF COVERINGS

Item	PC £	Labour hours	Labour £	Plant £	Material £	Unit	Total rate £
LEAD SHEET COVERINGS/FLASHINGS – cont							
Roof and dormer coverings – cont							
2.24 mm thick (code 5) roof coverings							
flat (in wood roll construction) (Prime Cost £ per kg)	3.16	0.94	27.85	–	55.06	m²	**82.91**
pitched (in wood roll construction)	–	1.05	30.93	–	55.31	m²	**86.24**
pitched (in welded seam construction)	–	0.94	27.85	–	55.06	m²	**82.91**
vertical (in welded seam construction)	–	1.05	30.93	–	52.88	m²	**83.81**
2.24 mm thick (code 5) dormer coverings							
flat (in wood roll construction) (Prime Cost £ per kg)	3.16	0.71	20.88	–	54.52	m²	**75.40**
pitched (in wood roll construction)	–	0.79	23.22	–	54.70	m²	**77.92**
pitched (in welded seam construction)	–	0.71	20.88	–	54.52	m²	**75.40**
vertical (in welded seam construction)	–	1.57	46.40	–	52.88	m²	**99.28**
2.65 mm thick (code 6) roof coverings							
flat (in wood roll construction) (Prime Cost £ per kg)	3.16	0.99	29.17	–	63.82	m²	**92.99**
pitched (in wood roll construction)	–	1.10	32.40	–	64.07	m²	**96.47**
pitched (in welded seam construction)	–	0.99	29.17	–	63.82	m²	**92.99**
vertical (in welded seam construction)	–	1.10	32.40	–	61.53	m²	**93.93**
2.65 mm thick (code 6) dormer coverings							
flat (in wood roll construction) (Prime Cost £ per kg)	3.16	0.74	21.89	–	63.25	m²	**85.14**
pitched (in wood roll construction)	–	0.82	24.31	–	63.43	m²	**87.74**
pitched (in welded seam construction)	–	0.74	21.89	–	63.25	m²	**85.14**
vertical (in welded seam construction)	–	1.65	48.61	–	61.53	m²	**110.14**
3.15 mm thick (code 7) roof coverings							
flat (in wood roll construction) (Prime Cost £ per kg)	3.16	1.06	31.17	–	74.52	m²	**105.69**
pitched (in wood roll construction)	–	1.18	34.62	–	74.79	m²	**109.41**
pitched (in welded seam construction)	–	1.06	31.17	–	74.52	m²	**105.69**
vertical (in welded seam construction)	–	1.18	34.62	–	72.07	m²	**106.69**
3.15 mm thick (code 7) dormer coverings							
flat (in wood roll construction) (Prime Cost £ per kg)	3.16	0.79	23.36	–	73.91	m²	**97.27**
pitched (in wood roll construction)	–	0.88	25.96	–	74.11	m²	**100.07**
pitched (in welded seam construction)	–	0.79	23.36	–	73.91	m²	**97.27**
vertical (in welded seam construction)	–	1.76	51.94	–	72.07	m²	**124.01**
3.55 mm thick (code 8) roof coverings							
flat (in wood roll construction) (Prime Cost £ per kg)	3.16	1.15	33.82	–	83.17	m²	**116.99**
pitched (in wood roll construction)	–	1.27	37.56	–	83.46	m²	**121.02**
pitched (in welded seam construction)	–	1.15	33.82	–	83.17	m²	**116.99**
vertical (in welded seam construction)	–	1.27	37.56	–	80.52	m²	**118.08**
3.55 mm thick (code 8) dormer coverings							
flat (in wood roll construction) (Prime Cost £ per kg)	3.16	0.86	25.37	–	82.51	m²	**107.88**
pitched (in wood roll construction)	–	0.96	28.16	–	82.73	m²	**110.89**
pitched (in welded seam construction)	–	0.86	25.37	–	82.51	m²	**107.88**
vertical (in welded seam construction)	–	1.91	56.36	–	80.52	m²	**136.88**
Sundries							
patination oil to finished work surfaces	–	0.03	0.74	–	0.23	m²	**0.97**
chalk slurry to underside of panels	–	0.33	9.81	–	2.02	m²	**11.83**
provision of 45 × 45 mm wood rolls at 600 mm centres (per m)	–	0.10	2.95	–	0.96	m	**3.91**
dressing over glazing bars and glass	–	0.25	7.36	–	0.62	m	**7.98**
soldered nail head	–	0.01	0.23	–	0.05	nr	**0.28**

17 SHEET ROOF COVERINGS

Item	PC £	Labour hours	Labour £	Plant £	Material £	Unit	Total rate £
1.32 mm thick (code 3) lead flashings, etc.							
Soakers							
200 × 200 mm	–	0.02	0.44	–	1.03	nr	**1.47**
300 × 300 mm	–	0.02	0.44	–	2.33	nr	**2.77**
1.80 mm thick (code 4) lead flashings, etc.							
Flashings; wedging into grooves							
150 mm girth	–	0.25	7.36	–	5.92	m	**13.28**
200 mm girth	–	0.25	7.36	–	7.89	m	**15.25**
240 mm girth	–	0.25	7.36	–	9.47	m	**16.83**
300 mm girth	–	0.25	7.36	–	11.84	m	**19.20**
Stepped flashings; wedging into grooves							
180 mm girth	–	0.50	14.73	–	7.10	m	**21.83**
270 mm girth	–	0.50	14.73	–	10.65	m	**25.38**
Linings to sloping gutters							
390 mm girth	–	0.40	11.78	–	15.39	m	**27.17**
450 mm girth	–	0.45	13.26	–	17.76	m	**31.02**
600 mm girth	–	0.55	16.20	–	23.67	m	**39.87**
Cappings to hips or ridges							
450 mm girth	–	0.50	14.73	–	17.76	m	**32.49**
600 mm girth	–	0.60	17.68	–	23.67	m	**41.35**
Saddle flashings; at intersections of hips and ridges; dressing and bossing							
450 × 450 mm	–	0.50	14.73	–	10.30	nr	**25.03**
600 × 200 mm	–	0.50	14.73	–	16.51	nr	**31.24**
Slates; with 150 mm high collar							
450 × 450 mm; to suit 50 mm dia. pipe	–	0.75	22.09	–	12.39	nr	**34.48**
450 × 450 mm; to suit 100 mm dia. pipe	–	0.75	22.09	–	13.31	nr	**35.40**
450 × 450 mm; to suit 150 mm dia. pipe	–	0.75	22.09	–	14.24	nr	**36.33**
2.24 mm thick (code 5) lead flashings, etc.							
Flashings; wedging into grooves							
150 mm girth	–	0.25	7.36	–	7.27	m	**14.63**
200 mm girth	–	0.25	7.36	–	9.69	m	**17.05**
240 mm girth	–	0.25	7.36	–	11.63	m	**18.99**
300 mm girth	–	0.25	7.36	–	14.54	m	**21.90**
Stepped flashings; wedging into grooves							
180 mm girth	–	0.50	14.73	–	8.72	m	**23.45**
270 mm girth	–	0.50	14.73	–	13.09	m	**27.82**
Linings to sloping gutters							
390 mm girth	–	0.40	11.78	–	18.91	m	**30.69**
450 mm girth	–	0.45	13.26	–	21.81	m	**35.07**
600 mm girth	–	0.55	16.20	–	29.08	m	**45.28**
Cappings to hips or ridges							
450 mm girth	–	0.50	14.73	–	21.81	m	**36.54**
600 mm girth	–	0.60	17.68	–	29.08	m	**46.76**
Saddle flashings; at intersections of hips and ridges; dressing and bossing							
450 × 450 mm	–	0.50	14.73	–	10.97	nr	**25.70**
600 × 200 mm	–	0.50	14.73	–	18.60	nr	**33.33**

17 SHEET ROOF COVERINGS

Item	PC £	Labour hours	Labour £	Plant £	Material £	Unit	Total rate £
LEAD SHEET COVERINGS/FLASHINGS – cont							
2.24 mm thick (code 5) lead flashings – cont							
Slates; with 150 mm high collar							
450 × 450 mm; to suit 50 mm dia. pipe	–	0.75	22.09	–	12.69	nr	**34.78**
450 × 450 mm; to suit 100 mm dia. pipe	–	0.75	22.09	–	13.83	nr	**35.92**
450 × 450 mm; to suit 150 mm dia. pipe	–	0.75	22.09	–	14.98	nr	**37.07**
ALUMINIUM SHEET COVERINGS/FLASHINGS							
Aluminium roofing; commercial grade; on and including Geotec underlay							
The following rates are based upon nett 'deck' or 'wall' areas							
Roof, dormer and wall coverings							
0.7 mm thick roof coverings; mill finish							
flat (in wood roll construction) (Prime Cost rate £ per kg)	4.86	1.00	29.46	–	20.02	m²	**49.48**
eaves detail ED1	–	0.20	5.89	–	2.48	m	**8.37**
abutment upstands at perimeters	–	0.33	9.72	–	0.99	m	**10.71**
pitched over 3° (in standing seam construction)	–	0.75	22.09	–	16.75	m²	**38.84**
vertical (in angled or flat seam construction)	–	0.80	23.57	–	16.75	m²	**40.32**
0.7 mm thick dormer coverings; mill finish							
flat (in wood roll construction)	–	1.50	44.19	–	19.72	m²	**63.91**
eaves detail ED1	–	0.20	5.89	–	2.48	m	**8.37**
pitched over 3° (in standing seam construction)	–	1.25	36.82	–	16.75	m²	**53.57**
vertical (in angled or flat seam construction)	–	1.35	39.77	–	16.75	m²	**56.52**
0.7 mm thick roof coverings; Pvf2 finish							
flat (in wood roll construction) (Prime Cost rate £ per kg)	6.06	1.00	29.46	–	23.49	m²	**52.95**
eaves detail ED1	–	0.20	5.89	–	3.09	m	**8.98**
abutment upstands at perimeters	–	0.33	9.72	–	1.23	m	**10.95**
pitched over 3° (in standing seam construction)	–	0.75	22.09	–	20.40	m²	**42.49**
vertical (in angled or flat seam construction)	–	0.80	23.57	–	20.40	m²	**43.97**
0.7 mm thick dormer coverings; Pvf2 finish							
flat (in wood roll construction)	–	1.50	44.19	–	23.49	m²	**67.68**
eaves detail ED1	–	0.20	5.89	–	3.09	m	**8.98**
pitched over 3° (in standing seam construction)	–	1.25	36.82	–	20.40	m²	**57.22**
vertical (in angled or flat seam construction)	–	1.35	39.77	–	20.40	m²	**60.17**
0.7 mm thick aluminium flashings, etc.							
Flashings; wedging into grooves; mill finish							
150 mm girth (PC per kg)	4.86	0.25	7.36	–	1.49	m	**8.85**
240 mm girth	–	0.25	7.36	–	2.38	m	**9.74**
300 mm girth	–	0.25	7.36	–	2.98	m	**10.34**
Stepped flashings; wedging into grooves; mill finish							
180 mm girth	–	0.50	14.73	–	1.79	m	**16.52**
270 mm girth	–	0.50	14.73	–	2.68	m	**17.41**

17 SHEET ROOF COVERINGS

Item	PC £	Labour hours	Labour £	Plant £	Material £	Unit	Total rate £
Flashings; wedging into grooves; Pvf2 finish							
150 mm girth (PC per kg)	6.06	0.25	7.36	–	1.85	m	**9.21**
240 mm girth	–	0.25	7.36	–	2.96	m	**10.32**
300 mm girth	–	0.25	7.36	–	3.70	m	**11.06**
Stepped flashings; wedging into grooves; Pvf2 finish							
180 mm girth	–	0.50	14.73	–	2.22	m	**16.95**
270 mm girth	–	0.50	14.73	–	3.33	m	**18.06**
Sundries							
provision of square batten roll at 500 mm centres (per m)	–	0.10	2.95	–	1.22	m	**4.17**
Standing seam aluminium roofing							
Kalzip; 65 mm seam, 400 cover width, Ref BS AW 3004 standard natural aluminium, stucco embossed finish, 0.9 mm thick; ST Clips fixed with stainless steel fasteners; 37 Plus 180 mm Glassfibre Insulation compressed to 165 mm (0.25 U Value); vapour control layer, clear reinforced polyethelyne 530 MNs/g all laps sealed; Liner Sheets, profiled steel, 1000 mm cover width, bright white polyester paint finish Ref TR35/200S, 0.7 mm thick, fixed with stainless steel fasteners.							
roof coverings (twin skin construction); pitch not less than 1.5°; fixed to cold rolled purlins (not included	–	–	–	–	–	m²	**59.59**
Eaves details							
40 × 20 mm extruded aluminium drip angle fixed to Kalzip sheet using aluminium blind sealed rivets; black solid rubber eaves filler blocks; ST clips fixed with stainless steel fasteners	–	–	–	–	–	m	**19.32**
0.90 mm thick stucco embossed natural aluminium external eaves closure; 375 mm girth twice bent	–	–	–	–	–	m	**8.11**
0.70 mm thick bright white polyester liner sheet closure internal flashing; 200 mm girth once bent with stainless steel fasteners, black solid rubber profiled liner small flute filler sealed top and bottom with sealant tape	–	–	–	–	–	m	**9.34**

17 SHEET ROOF COVERINGS

Item	PC £	Labour hours	Labour £	Plant £	Material £	Unit	Total rate £
ALUMINIUM SHEET COVERINGS/FLASHINGS – cont							
Standing seam aluminium roofing – cont							
Verge details							
extruded aluminium gable end channel fixed to kalzip seam using aluminium blind rivets; extruded aluminium gable end clips fixed to St Clips with stainless steel fasteners; extruded aluminium gable tolerence clip hooked over gable end channel	–	–	–	–	–	m	**15.76**
0.90 mm thick stucco embossed natural aluminium external verge closure 600 mm girth four times bent, fixed to extruded aluminium gable tolerence clip and vertical cladding with stainless steel fasteners, black profiled filler blocks to vertical clddding	–	–	–	–	–	m	**15.82**
0.70 mm thick bright white polyester liner sheet closure internal flashing 200 mm girth once bent fixed with stainless steel fasteners, black solid rubber profiled filler blocks sealed top and bottom with sealant tape	–	–	–	–	–	m	**9.50**
Duo-Ridge details							
2 Nr Extruded aluminium zed sections fixed to kalzip seams using aluminium blind sealed rivets; 2 Nr natural aluminium stucco embossed U Type ridge closures fixed to kalzip seams using aluminium blind sealed rivets; 2 Nr black solid rubber ridge filler blocks, 2 Nr ST clips fixed with stainless steel fasteners; fix seam of kaizip sheet to ST clips using aluminium blind sealed rivets (for fixed point); turn up kalzip 400 sheets both sides	–	–	–	–	–	m	**30.51**
0.90 mm thick stucco embossed natural aluminium external ridge closure; 600 mm girth three times bent, fixed to U Type Ridge closure with stainless steel fasteners	–	–	–	–	–	m	**12.12**
0.70 mm thick bright white polyester liner sheet closure flashing 600 mm girth once bent fixed with stainless steel fasteners, black solid rubber profiled filler blocks sealed top and bottom with sealant tape	–	–	–	–	–	m	**10.57**
Accessories							
extra over for;							
smooth curving Kalzip sheets	–	–	–	–	–	m²	**7.01**
crimp curving liner (below 52.5 m convex radius)	–	–	–	–	–	sheet	**11.83**
polyster coating Kalzip sheets	–	–	–	–	–	m²	**3.65**
PvDF coating Kalzip sheets	–	–	–	–	–	m²	**4.21**
vapour control layer, foil encapsulated polythene 4300 MNs/g	–	–	–	–	–	m²	**0.58**
200 mm thick thermal insulation quilt	–	–	–	–	–	m²	**0.19**
30 mm thick semi-rigid acoustic insulation slab	–	–	–	–	–	m²	**4.37**

17 SHEET ROOF COVERINGS

Item	PC £	Labour hours	Labour £	Plant £	Material £	Unit	Total rate £
1.0 mm Flashings etc.; fixing/wedging into grooves							
flashing; 500 mm girth	–	–	–	–	–	m	**12.92**
flashing; 750 mm girth	–	–	–	–	–	m	**17.06**
flashing; 1000 mm girth	–	–	–	–	–	m	**24.00**
1.2 mm Flashings etc.; fixing/wedging into grooves							
flashing; 500 mm girth	–	–	–	–	–	m	**14.21**
flashing; 750 mm girth	–	–	–	–	–	m	**18.77**
flashing; 1000 mm girth	–	–	–	–	–	m	**23.72**
1.4 mm Flashings etc.; fixing/wedging into grooves							
flashing; 500 mm girth	–	–	–	–	–	m	**16.34**
flashing; 750 mm girth	–	–	–	–	–	m	**21.58**
flashing; 1000 mm girth	–	–	–	–	–	m	**32.43**
Aluminium Alumasc Skyline coping system; polyester powder coated							
Coping; fixing straps plugged and screwed to brickwork							
362 mm wide; for parapet wall 241–300 mm wide	–	0.50	8.65	–	35.67	m	**44.32**
extra over for;							
90° angle	–	0.25	4.33	–	87.08	nr	**91.41**
90° tee junction	–	0.35	6.06	–	95.86	nr	**101.92**
stop end	–	0.15	2.59	–	44.39	nr	**46.98**
stop end upstand	–	0.20	3.46	–	48.72	nr	**52.18**
COPPER STRIP SHEET COVERINGS/FLASHINGS							
Copper roofing; BS EN 504; on and including Geotec underlay							
The following rates are based upon nett deck or wall areas							
Roof and dormer coverings							
0.6 mm thick roof coverings; mill finish							
flat (in wood roll construction) (Prime Cost £ per kg)	5.67	1.10	32.40	–	54.89	m²	**87.29**
eaves detail ED1	–	0.20	5.89	–	6.66	m	**12.55**
abutment upstands at perimeters	–	0.33	9.72	–	3.33	m	**13.05**
pitched over 3° (in standing seam construction)	–	0.85	25.04	–	44.89	m²	**69.93**
vertical (in angled or flat seam construction)	–	0.90	26.51	–	44.89	m²	**71.40**
0.6 mm thick dormer coverings; mill finish							
flat (in wood roll construction)	5.67	1.60	47.13	–	54.89	m²	**102.02**
eaves detail ED1	–	0.20	5.89	–	6.66	m	**12.55**
pitched over 3° (in standing seam construction)	–	1.25	36.82	–	44.89	m²	**81.71**
vertical (in angled or flat seam construction)	–	1.35	39.77	–	44.89	m²	**84.66**
0.6 mm thick roof coverings; oxide finish							
flat (in wood roll construction) (Prime Cost £ per kg)	6.95	1.10	32.40	–	66.56	m²	**98.96**
eaves detail ED1	–	0.20	5.89	–	8.18	m	**14.07**
abutment upstands at perimeters	–	0.33	9.72	–	4.09	m	**13.81**
pitched over 3° (in standing seam construction)	–	0.85	25.04	–	54.66	m²	**79.70**
vertical (in angled or flat seam construction)	–	0.80	23.57	–	54.66	m²	**78.23**

17 SHEET ROOF COVERINGS

Item	PC £	Labour hours	Labour £	Plant £	Material £	Unit	Total rate £
COPPER STRIP SHEET COVERINGS/ FLASHINGS – cont							
Roof and dormer coverings – cont							
0.6 mm thick dormer coverings; oxide finish							
flat (in wood roll construction)	6.95	1.50	44.19	–	66.56	m²	**110.75**
eaves detail ED1	–	0.20	5.89	–	8.18	m	**14.07**
pitched over 3° (in standing seam construction)	–	1.25	36.82	–	54.66	m²	**91.48**
vertical (in angled or flat seam construction)	–	1.35	39.77	–	54.66	m²	**94.43**
0.6 mm thick roof coverings; KME pre-patinated finish							
flat (in wood roll construction)	58.20	1.10	32.40	–	98.31	m²	**130.71**
eaves detail ED1	–	0.20	5.89	–	12.44	m	**18.33**
abutment upstands at perimeters	–	0.33	9.72	–	6.22	m	**15.94**
pitched over 3° (in standing seam construction)	–	0.85	25.04	–	79.65	m²	**104.69**
vertical (in angled or flat seam construction)	–	0.90	26.51	–	79.65	m²	**106.16**
0.6 mm thick dormer coverings; KME pre-patinated finish							
flat (in wood roll construction)	58.20	1.50	44.19	–	98.31	m²	**142.50**
eaves detail ED1	–	0.20	5.89	–	12.44	m	**18.33**
pitched over 3° (in standing seam construction)	–	1.25	36.82	–	79.65	m²	**116.47**
vertical (in angled or flat seam construction)	–	1.35	39.77	–	79.65	m²	**119.42**
0.7 mm thick roof coverings; mill finish							
flat (in wood roll construction) (Prime Cost £ per kg)	5.67	1.00	29.46	–	62.16	m²	**91.62**
eaves detail ED1	–	0.20	5.89	–	7.63	m	**13.52**
abutment upstands at perimeters	–	0.33	9.72	–	3.82	m	**13.54**
pitched over 3° (in standing seam construction)	–	0.75	22.09	–	50.71	m²	**72.80**
vertical (in angled or flat seam construction)	–	0.80	23.57	–	50.71	m²	**74.28**
0.7 mm thick dormer coverings; mill finish							
flat (in wood roll construction)	5.67	1.50	44.19	–	62.16	m²	**106.35**
eaves detail ED1	–	0.20	5.89	–	7.63	m	**13.52**
pitched over 3° (in standing seam construction)	–	1.25	36.82	–	50.68	m²	**87.50**
vertical (in angled or flat seam construction)	–	1.35	39.77	–	50.65	m²	**90.42**
0.7 mm thick roof coverings; oxide finish							
flat (in wood roll construction) (Prime Cost £ per kg)	6.95	1.00	29.46	–	84.40	m²	**113.86**
eaves detail ED1	–	0.20	5.89	–	9.37	m	**15.26**
abutment upstands at perimeters	–	0.33	9.72	–	4.68	m	**14.40**
pitched over 3° (in standing seam construction)	–	0.75	22.09	–	68.79	m²	**90.88**
vertical (in angled or flat seam construction)	–	0.80	23.57	–	61.35	m²	**84.92**
0.7 mm thick dormer coverings; oxide finish							
flat (in wood roll construction)	6.95	1.50	44.19	–	75.48	m²	**119.67**
eaves detail ED1	–	0.20	5.89	–	9.37	m	**15.26**
pitched over 3° (in standing seam construction)	–	1.25	36.82	–	61.35	m²	**98.17**
vertical (in angled or flat seam construction)	–	1.35	39.77	–	61.35	m²	**101.12**
0.7 mm thick roof coverings; KME pre-patinated finish							
flat (in wood roll construction)	66.95	1.10	32.40	–	112.81	m²	**145.21**
eaves detail ED1	–	0.20	5.89	–	14.31	m	**20.20**
abutment upstands at perimeters	–	0.33	9.72	–	7.16	m	**16.88**
pitched over 3° (in standing seam construction)	–	0.85	25.04	–	91.34	m²	**116.38**
vertical (in angled or flat seam construction)	–	0.90	26.51	–	91.34	m²	**117.85**

17 SHEET ROOF COVERINGS

Item	PC £	Labour hours	Labour £	Plant £	Material £	Unit	Total rate £
0.7 mm thick dormer coverings; KME pre-patinated finish							
flat (in wood roll construction)	66.95	1.50	44.19	–	112.81	m²	**157.00**
eaves detail ED1	–	0.20	5.89	–	14.31	m	**20.20**
pitched over 3° (in standing seam construction)	–	1.25	36.82	–	91.34	m²	**128.16**
vertical (in angled or flat seam construction)	–	1.35	39.77	–	91.34	m²	**131.11**
0.6 mm thick copper flashings, etc.							
Flashings; wedging into grooves; mill finish							
150 mm girth (Prime Cost £ per kg)	5.95	0.25	7.36	–	4.00	m	**11.36**
240 mm girth	–	0.25	7.36	–	7.99	m	**15.35**
300 mm girth	–	0.25	7.36	–	9.99	m	**17.35**
Stepped flashings; wedging into grooves; mill finish							
180 mm girth	–	0.50	14.73	–	5.99	m	**20.72**
270 mm girth	–	0.50	14.73	–	8.99	m	**23.72**
Flashings; wedging into grooves; oxide finish							
150 mm girth (Prime Cost £ per kg)	7.30	0.25	7.36	–	6.13	m	**13.49**
240 mm girth	–	0.25	7.36	–	9.81	m	**17.17**
300 mm girth	–	0.25	7.36	–	12.26	m	**19.62**
Stepped flashings; wedging into grooves; oxide finish							
180 mm girth	–	0.50	14.73	–	7.36	m	**22.09**
270 mm girth	–	0.50	14.73	–	11.04	m	**25.77**
Flashings; wedging into grooves; KME pre-patinated finish							
150 mm girth (Prime Cost £ per kg)	61.11	0.25	7.36	–	9.33	m	**16.69**
240 mm girth	–	0.25	7.36	–	14.93	m	**22.29**
300 mm girth	–	0.25	7.36	–	18.66	m	**26.02**
Stepped flashings; wedging into grooves; KME pre-patinated finish							
180 mm girth	–	0.50	14.73	–	11.20	m	**25.93**
270 mm girth	–	0.50	14.73	–	16.80	m	**31.53**
0.7 mm thick copper flashings, etc.							
Flashings; wedging into grooves; mill finish							
150 mm girth (Prime Cost £ per kg)	5.67	0.25	7.36	–	5.72	m	**13.08**
240 mm girth	–	0.25	7.36	–	9.16	m	**16.52**
300 mm girth	–	0.25	7.36	–	11.45	m	**18.81**
Stepped flashings; wedging into grooves; mill finish							
180 mm girth	–	0.50	14.73	–	6.87	m	**21.60**
270 mm girth	–	0.50	14.73	–	10.30	m	**25.03**
Flashings; wedging into grooves; oxide finish							
150 mm girth (Prime Cost £ per kg)	6.95	0.25	7.36	–	7.02	m	**14.38**
240 mm girth	–	0.25	7.36	–	11.24	m	**18.60**
300 mm girth	–	0.25	7.36	–	14.05	m	**21.41**
Stepped flashings; wedging into grooves; oxide finish							
180 mm girth	–	0.50	14.73	–	8.43	m	**23.16**
270 mm girth	–	0.50	14.73	–	12.64	m	**27.37**

17 SHEET ROOF COVERINGS

Item	PC £	Labour hours	Labour £	Plant £	Material £	Unit	Total rate £
COPPER STRIP SHEET COVERINGS/ FLASHINGS – cont							
0.7 mm thick copper flashings – cont							
Flashings; wedging into grooves; KME pre-patinated finish							
150 mm girth (Prime Cost £ per kg)	66.95	0.25	7.36	–	10.74	m	**18.10**
240 mm girth	–	0.25	7.36	–	17.18	m	**24.54**
300 mm girth	–	0.25	7.36	–	21.47	m	**28.83**
Stepped flashings; wedging into grooves; KME pre-patinated finish							
180 mm girth	–	0.50	14.73	–	12.88	m	**27.61**
270 mm girth	–	0.50	14.73	–	19.32	m	**34.05**
Sundries							
provision of square batten roll at 500 mm centres (per m)	–	0.10	2.95	–	1.22	m	**4.17**
ZINC STRIP SHEET COVERINGS/FLASHINGS							
Zinc roofing; BS 849; on and including Delta Trella underlay							
The following rates are based upon nett deck or wall areas							
Natural Bright Rheinzink							
Roof, dormer and wall coverings							
0.7 mm thick roof coverings							
flat (in wood roll construction) (Prime Cost £ per kg)	2.99	1.00	29.46	–	28.56	m²	**58.02**
eaves detail ED1	–	0.20	5.89	–	4.82	m	**10.71**
abutment upstands at perimeters	–	0.33	9.72	–	2.41	m	**12.13**
pitched over 3° (in standing seam construction)	–	0.75	22.09	–	23.73	m²	**45.82**
0.7 mm thick dormer coverings							
flat (in wood roll construction) (Prime Cost £ per kg)	2.99	1.50	44.19	–	28.56	m²	**72.75**
eaves detail ED1	–	0.20	5.89	–	4.82	m	**10.71**
pitched over 3° (in standing seam construction)	–	1.25	36.82	–	23.73	m²	**60.55**
0.8 mm thick wall coverings							
vertical (in angled or flat seam construction)	–	0.80	23.57	–	26.57	m²	**50.14**
0.8 mm thick dormer coverings							
vertical (in angled or flat seam construction)	–	1.35	39.77	–	26.66	m²	**66.43**
0.8 mm thick zinc flashings, etc.; Natural Bright Rheinzink							
Flashings; wedging into grooves							
150 mm girth	–	0.25	7.36	–	2.77	m	**10.13**
240 mm girth	–	0.25	7.36	–	4.43	m	**11.79**
300 mm girth	–	0.25	7.36	–	5.53	m	**12.89**

17 SHEET ROOF COVERINGS

Item	PC £	Labour hours	Labour £	Plant £	Material £	Unit	Total rate £
Stepped flashings; wedging into grooves							
180 mm girth	–	0.50	14.73	–	3.32	m	**18.05**
270 mm girth	–	0.50	14.73	–	4.98	m	**19.71**
Integral box gutter							
900 mm girth; 2 × bent; 2 × welted	–	1.00	29.46	–	22.88	m	**52.34**
Valley gutter							
600 mm girth; 2 × bent; 2 × welted	–	0.75	22.09	–	13.73	m	**35.82**
Hips and ridges							
450 mm girth; 2 × bent; 2 × welted	–	1.00	29.46	–	8.30	m	**37.76**
Natural Bright Rheinzink PRO							
Roof, dormer and wall coverings							
0.7 mm thick roof coverings							
flat (in wood roll construction) (Prime Cost £ per kg)	4.07	1.00	29.46	–	38.89	m²	**68.35**
eaves detail ED1	–	0.20	5.89	–	6.56	m	**12.45**
abutment upstands at perimeters	–	0.33	9.72	–	3.28	m	**13.00**
pitched over 3° (in standing seam construction)	–	0.75	22.09	–	32.31	m²	**54.40**
0.7 mm thick dormer coverings							
flat (in wood roll construction) (Prime Cost £ per kg)	4.07	1.50	44.19	–	38.89	m²	**83.08**
eaves detail ED1	–	0.20	5.89	–	6.56	m	**12.45**
pitched over 3° (in standing seam construction)	–	1.25	36.82	–	32.30	m²	**69.12**
0.8 mm thick wall coverings							
vertical (in angled or flat seam construction)	–	0.80	23.57	–	36.19	m²	**59.76**
0.8 mm thick dormer coverings							
vertical (in angled or flat seam construction)	–	1.35	39.77	–	36.19	m²	**75.96**
0.8 mm thick zinc flashings, etc.; Natural Bright Rheinzink PRO							
Flashings; wedging into grooves							
150 mm girth	–	0.25	7.36	–	3.77	m	**11.13**
240 mm girth	–	0.25	7.36	–	6.03	m	**13.39**
300 mm girth	–	0.25	7.36	–	7.54	m	**14.90**
Stepped flashings; wedging into grooves							
180 mm girth	–	0.50	14.73	–	4.52	m	**19.25**
270 mm girth	–	0.50	14.73	–	6.79	m	**21.52**
Integral box gutter							
900 mm girth; 2 × bent; 2 × welted	–	1.00	29.46	–	31.16	m	**60.62**
Valley gutter							
600 mm girth; 2 × bent; 2 × welted	–	0.75	22.09	–	18.69	m	**40.78**
Hips and ridges							
450 mm girth; 2 × bent; 2 × welted	–	1.00	29.46	–	11.30	m	**40.76**
Pre-weathered Rheinzink							
Roof, dormer and wall coverings							
0.7 mm thick roof coverings; pre-weathered Rheinzink							
flat (in wood roll construction) (Prime Cost £ per kg)	3.45	1.00	29.46	–	32.97	m²	**62.43**
eaves detail ED1	–	0.20	5.89	–	5.57	m	**11.46**
abutment upstands at perimeters	–	0.33	9.72	–	2.78	m	**12.50**
pitched over 3° (in standing seam construction)	–	0.75	22.09	–	27.39	m²	**49.48**

17 SHEET ROOF COVERINGS

Item	PC £	Labour hours	Labour £	Plant £	Material £	Unit	Total rate £
ZINC STRIP SHEET COVERINGS/FLASHINGS – **cont**							
Roof, dormer and wall coverings – cont							
0.7 mm thick dormer coverings; pre-weathered Rheinzink							
flat (in wood roll construction) (Prime Cost £ per kg)	3.45	1.50	44.19	–	32.97	m²	**77.16**
eaves detail ED1	–	0.20	5.89	–	5.57	m	**11.46**
pitched over 3° (in standing seam construction)	–	1.25	36.82	–	27.39	m²	**64.21**
0.8 mm thick wall coverings; pre-weathered Rheinzink							
vertical (in angled or flat seam construction)	–	0.80	23.57	–	30.68	m²	**54.25**
0.8 mm thick dormer coverings; pre-weathered Rheinzink							
vertical (in angled or flat seam construction)	–	1.35	39.77	–	30.68	m²	**70.45**
0.8 mm thick zinc flashings, etc.; pre-weathered **Rheinzink**							
Flashings; wedging into grooves							
150 mm girth	–	0.25	7.36	–	3.20	m	**10.56**
240 mm girth	–	0.25	7.36	–	5.12	m	**12.48**
300 mm girth	–	0.25	7.36	–	6.39	m	**13.75**
Stepped flashings; wedging into grooves							
180 mm girth	–	0.50	14.73	–	3.83	m	**18.56**
270 mm girth	–	0.50	14.73	–	5.75	m	**20.48**
Integral box gutter							
900 mm girth; 2 × bent; 2 × welted	–	1.00	29.46	–	26.42	m	**55.88**
Valley gutter							
600 mm girth; 2 × bent; 2 × welted	–	0.75	22.09	–	15.85	m	**37.94**
Hips and ridges							
450 mm girth; 2 × bent; 2 × welted	–	1.00	29.46	–	9.58	m	**39.04**
Pre-weathered Rheinzink PRO							
Roof, dormer and wall coverings							
0.7 mm thick roof coverings							
flat (in wood roll construction) (Prime Cost £ per kg)	4.53	1.00	29.46	–	43.31	m²	**72.77**
eaves detail ED1	–	0.20	5.89	–	7.31	m	**13.20**
abutment upstands at perimeters	–	0.33	9.72	–	3.65	m	**13.37**
pitched over 3° (in standing seam construction)	–	0.75	22.09	–	35.98	m²	**58.07**
0.7 mm thick dormer coverings							
flat (in wood roll construction) (Prime Cost £ per kg)	4.53	1.50	44.19	–	43.31	m²	**87.50**
eaves detail ED1	–	0.20	5.89	–	7.31	m	**13.20**
pitched over 3° (in standing seam construction)	–	1.25	36.82	–	35.98	m²	**72.80**
0.8 mm thick wall coverings							
vertical (in angled or flat seam construction)	–	0.80	23.57	–	40.30	m²	**63.87**
0.8 mm thick dormer coverings							
vertical (in angled or flat seam construction)	–	1.35	39.77	–	40.30	m²	**80.07**

17 SHEET ROOF COVERINGS

Item	PC £	Labour hours	Labour £	Plant £	Material £	Unit	Total rate £
0.8 mm thick zinc flashings, etc.; Pre-weathered Rheinzink PRO							
Flashings; wedging into grooves							
150 mm girth	–	0.25	7.36	–	4.20	m	**11.56**
240 mm girth	–	0.25	7.36	–	6.72	m	**14.08**
300 mm girth	–	0.25	7.36	–	8.39	m	**15.75**
Stepped flashings; wedging into grooves							
180 mm girth	–	0.50	14.73	–	5.03	m	**19.76**
270 mm girth	–	0.50	14.73	–	7.56	m	**22.29**
Integral box gutter							
900 mm girth; 2 × bent; 2 × welted	–	–	–	–	–	m	**34.70**
Valley gutter							
600 mm girth; 2 × bent; 2 × welted	–	0.75	22.09	–	20.81	m	**42.90**
Hips and ridges							
450 mm girth; 2 × bent; 2 × welted	–	1.00	29.46	–	12.59	m	**42.05**
VM Natural Bright							
Roof, dormer and wall coverings							
0.7 mm thick roof coverings							
flat (in wood roll construction) (Prime Cost £ per kg)	2.99	1.00	29.46	–	28.56	m²	**58.02**
eaves detail ED1	–	0.20	5.89	–	4.82	m	**10.71**
abutment upstands at perimeters	–	0.33	9.72	–	2.41	m	**12.13**
pitched over 3° (in standing seam construction)	–	0.75	22.09	–	23.73	m²	**45.82**
0.7 mm thick dormer coverings							
flat (in wood roll construction) (Prime Cost £ per kg)	2.99	1.50	44.19	–	28.56	m²	**72.75**
eaves detail ED1	–	0.20	5.89	–	4.82	m	**10.71**
pitched over 3° (in standing seam construction)	–	1.25	36.82	–	23.73	m²	**60.55**
0.8 mm thick wall coverings							
vertical (in angled or flat seam construction)	–	0.80	23.57	–	26.57	m²	**50.14**
0.8 mm thick dormer coverings							
vertical (in angled or flat seam construction)	–	1.35	39.77	–	26.57	m²	**66.34**
0.8 mm thick zinc flashings, etc.; VM Natural Bright							
Flashings; wedging into grooves							
150 mm girth	–	0.25	7.36	–	2.77	m	**10.13**
240 mm girth	–	0.25	7.36	–	4.43	m	**11.79**
300 mm girth	–	0.25	7.36	–	5.53	m	**12.89**
Stepped flashings; wedging into grooves							
180 mm girth	–	0.50	14.73	–	3.32	m	**18.05**
270 mm girth	–	0.50	14.73	–	4.98	m	**19.71**
Integral box gutter							
900 mm girth; 2 × bent; 2 × welted	–	1.00	29.46	–	22.88	m	**52.34**
Valley gutter							
600 mm girth; 2 × bent; 2 × welted	–	0.75	22.09	–	13.73	m	**35.82**

17 SHEET ROOF COVERINGS

Item	PC £	Labour hours	Labour £	Plant £	Material £	Unit	Total rate £
ZINC STRIP SHEET COVERINGS/FLASHINGS – cont							
0.8 mm thick zinc flashings – cont							
Hips and ridges							
450 mm girth; 2 × bent; 2 × welted	–	1.00	29.46	–	7.53	m	**36.99**
VM Natural Bright PLUS							
Roof, dormer and wall coverings							
0.7 mm thick roof coverings							
flat (in wood roll construction) (Prime Cost £ per kg)	4.07	1.00	29.46	–	38.89	m²	**68.35**
eaves detail ED1	–	0.20	5.89	–	6.56	m	**12.45**
abutment upstands at perimeters	–	0.33	9.72	–	3.28	m	**13.00**
pitched over 3° (in standing seam construction)	–	0.75	22.09	–	32.31	m²	**54.40**
0.7 mm thick dormer coverings							
flat (in wood roll construction) (Prime Cost £ per kg)	4.07	1.50	44.19	–	38.89	m²	**83.08**
eaves detail ED1	–	0.20	5.89	–	6.56	m	**12.45**
pitched over 3° (in standing seam construction)	–	1.25	36.82	–	32.31	m²	**69.13**
0.8 mm thick wall coverings							
vertical (in angled or flat seam construction)	–	0.80	23.57	–	36.19	m²	**59.76**
0.8 mm thick dormer coverings							
vertical (in angled or flat seam construction)	–	1.35	39.77	–	36.19	m²	**75.96**
0.8 mm thick zinc flashings, etc.; VM Natural Bright PLUS							
Flashings; wedging into grooves							
150 mm girth	–	0.25	7.36	–	3.77	m	**11.13**
240 mm girth	–	0.25	7.36	–	6.03	m	**13.39**
300 mm girth	–	0.25	7.36	–	7.54	m	**14.90**
Stepped flashings; wedging into grooves							
180 mm girth	–	0.50	14.73	–	4.52	m	**19.25**
270 mm girth	–	0.50	14.73	–	6.79	m	**21.52**
Integral box gutter							
900 mm girth; 2 × bent; 2 × welted	–	1.00	29.46	–	31.16	m	**60.62**
Valley gutter							
600 mm girth; 2 × bent; 2 × welted	–	0.75	22.09	–	18.69	m	**40.78**
Hips and ridges							
450 mm girth; 2 × bent; 2 × welted	–	1.00	29.46	–	11.30	m	**40.76**

17 SHEET ROOF COVERINGS

Item	PC £	Labour hours	Labour £	Plant £	Material £	Unit	Total rate £
VM Quartz (pre-weathered)							
Roof, dormer and wall coverings							
0.7 mm thick roof coverings							
flat (in wood roll construction) (Prime Cost £ per kg)	3.71	1.00	29.46	–	35.51	m²	**64.97**
eaves detail ED1	–	0.20	5.89	–	5.99	m	**11.88**
abutment upstands at perimeters	–	0.33	9.72	–	3.00	m	**12.72**
pitched over 3° (in standing seam construction)	–	0.75	22.09	–	29.50	m²	**51.59**
0.7 mm thick dormer coverings							
flat (in wood roll construction) (Prime Cost £ per kg)	3.71	1.50	44.19	–	35.51	m²	**79.70**
eaves detail ED1	–	0.20	5.89	–	5.99	m	**11.88**
pitched over 3° (in standing seam construction)	–	1.25	36.82	–	29.50	m²	**66.32**
0.8 mm thick wall coverings							
vertical (in angled or flat seam construction)	–	0.80	23.57	–	33.04	m²	**56.61**
0.8 mm thick dormer coverings							
vertical (in angled or flat seam construction)	–	1.35	39.77	–	33.04	m²	**72.81**
0.8 mm thick zinc flashings, etc.; VM Quartz (pre-weathered)							
Flashings; wedging into grooves							
150 mm girth	–	0.25	7.36	–	3.44	m	**10.80**
240 mm girth	–	0.25	7.36	–	5.51	m	**12.87**
300 mm girth	–	0.25	7.36	–	6.88	m	**14.24**
Stepped flashings; wedging into grooves							
180 mm girth	–	0.50	14.73	–	4.13	m	**18.86**
270 mm girth	–	0.50	14.73	–	6.20	m	**20.93**
Integral box gutter							
900 mm girth; 2 × bent; 2 × welted	–	1.00	29.46	–	28.45	m	**57.91**
Valley gutter							
600 mm girth; 2 × bent; 2 × welted	–	0.75	22.09	–	17.07	m	**39.16**
Hips and ridges							
450 mm girth; 2 × bent; 2 × welted	–	1.00	29.46	–	10.32	m	**39.78**
VM Quartz (pre-weathered) PLUS							
Roof, dormer and wall coverings							
0.7 mm thick roof coverings							
flat (in wood roll construction) (Prime Cost £ per kg)	4.79	1.00	29.46	–	45.84	m²	**75.30**
eaves detail ED1	–	0.20	5.89	–	7.74	m	**13.63**
abutment upstands at perimeters	–	0.33	9.72	–	3.87	m	**13.59**
pitched over 3° (in standing seam construction)	–	0.75	22.09	–	38.09	m²	**60.18**
0.7 mm thick dormer coverings							
flat (in wood roll construction) (Prime Cost £ per kg)	4.79	1.50	44.19	–	45.84	m²	**90.03**
eaves detail ED1	–	0.20	5.89	–	7.74	m	**13.63**
pitched over 3° (in standing seam construction)	–	1.25	36.82	–	38.09	m²	**74.91**
0.8 mm thick wall coverings							
vertical (in angled or flat seam construction)	–	0.80	23.57	–	42.66	m²	**66.23**
0.8 mm thick dormer coverings							
vertical (in angled or flat seam construction)	–	1.35	39.77	–	42.66	m²	**82.43**

17 SHEET ROOF COVERINGS

Item	PC £	Labour hours	Labour £	Plant £	Material £	Unit	Total rate £
ZINC STRIP SHEET COVERINGS/FLASHINGS – cont							
0.8 mm thick zinc flashings, etc.; VM Quartz (pre-weathered) PLUS							
Flashings; wedging into grooves							
150 mm girth	–	0.25	7.36	–	4.44	m	**11.80**
240 mm girth	–	0.25	7.36	–	7.11	m	**14.47**
300 mm girth	–	0.25	7.36	–	8.88	m	**16.24**
Stepped flashings; wedging into grooves							
180 mm girth	–	0.50	14.73	–	5.33	m	**20.06**
270 mm girth	–	0.50	14.73	–	8.00	m	**22.73**
Integral box gutter							
900 mm girth; 2 × bent; 2 × welted	–	1.00	29.46	–	36.73	m	**66.19**
Valley gutter							
600 mm girth; 2 × bent; 2 × welted	–	0.75	22.09	–	22.03	m	**44.12**
Hips and ridges							
450 mm girth; 2 × bent; 2 × welted	–	1.00	29.46	–	13.32	m	**42.78**
Sundries and accessories							
Klober breather membrane/underlay	–	0.10	2.95	–	3.48	m²	**6.43**
Delta Trela Chestwig underlay	–	0.10	2.95	–	5.94	m²	**8.89**
Delta Trela Football Studs underlay	–	0.10	2.95	–	1.10	m²	**4.05**
Trapezoidal batten roll at 500 mm centres (per m)	–	0.10	2.95	–	1.10	m	**4.05**
Zinflash; 0.6 mm thick lead look flashing (no patination oil required)							
Flashings; wedging into grooves							
150 mm girth	–	0.25	7.36	–	4.91	m	**12.27**
250 mm girth	–	0.25	7.36	–	8.18	m	**15.54**
300 mm girth	–	0.25	7.36	–	9.82	m	**17.18**
380 mm girth	–	0.25	7.36	–	12.43	m	**19.79**
450 mm girth	–	0.25	7.36	–	14.73	m	**22.09**
Stepped flashings; wedging into grooves							
150 mm girth	–	0.50	14.73	–	4.91	m	**19.64**
250 mm girth	–	0.50	14.73	–	8.18	m	**22.91**
300 mm girth	–	0.50	14.73	–	9.82	m	**24.55**
380 mm girth	–	0.50	14.73	–	12.43	m	**27.16**
450 mm girth	–	0.50	14.73	–	14.73	m	**29.46**
STAINLESS STEEL SHEET COVERINGS/ FLASHINGS							
Terne-coated stainless steel roofing; Associated Lead Mills Ltd; or other equal and approved: on and including Metmatt underlay							
The following rates are based upon nett deck or wall areas							

17 SHEET ROOF COVERINGS

Item	PC £	Labour hours	Labour £	Plant £	Material £	Unit	Total rate £
Roof, dormer and wall coverings in Uginox grade 316; marine							
0.4 mm thick roof coverings							
flat (in wood roll construction) (Prime Cost £ per kg)	6.95	1.00	29.46	–	37.80	m²	**67.26**
eaves detail ED1	–	0.20	5.89	–	4.46	m	**10.35**
abutment upstands at perimeters	–	0.33	9.72	–	2.23	m	**11.95**
pitched over 3° (in standing seam construction)	–	0.75	22.09	–	31.11	m²	**53.20**
0.5 mm thick dormer coverings							
flat (in wood roll construction) (Prime Cost £ per kg)	6.49	1.50	44.19	–	43.85	m²	**88.04**
eaves detail ED1	–	0.20	5.89	–	4.16	m	**10.05**
pitched over 3° (in standing seam construction)	–	1.25	36.82	–	35.95	m²	**72.77**
0.5 mm thick wall coverings							
vertical (in angled or flat seam construction)	6.49	0.80	23.57	–	35.95	m²	**59.52**
vertical (with Coulisseau joint construction)	–	1.25	36.82	–	37.11	m²	**73.93**
0.5 mm thick Uginox grade 316 flashings, etc.							
Flashings; wedging into grooves							
150 mm girth (Prime Cost £ per kg)	6.49	0.25	7.36	–	3.95	m	**11.31**
240 mm girth	–	0.25	7.36	–	6.33	m	**13.69**
300 mm girth	–	0.25	7.36	–	7.91	m	**15.27**
Stepped flashings; wedging into grooves							
180 mm girth	–	0.50	14.73	–	4.74	m	**19.47**
270 mm girth	–	0.50	14.73	–	7.12	m	**21.85**
Fan apron							
250 mm girth	–	0.25	7.36	–	6.59	m	**13.95**
Integral box gutter							
900 mm girth; 2 × bent; 2 × welted	–	1.00	29.46	–	27.15	m	**56.61**
Valley gutter							
600 mm girth; 2 × bent; 2 × welted	–	0.75	22.09	–	19.29	m	**41.38**
Hips and ridges							
450 mm girth; 2 × bent; 2 × welted	–	1.00	29.46	–	11.86	m	**41.32**
Roof, dormer and wall coverings in Ugitop grade 304							
0.4 mm thick roof coverings							
flat (in wood roll construction) (Prime Cost £ per kg)	5.25	1.00	29.46	–	32.43	m²	**61.89**
eaves detail ED1	–	0.20	5.89	–	3.37	m	**9.26**
abutment upstands at perimeters	–	0.33	9.72	–	1.69	m	**11.41**
pitched over 3° (in standing seam construction)	–	0.75	22.09	–	24.57	m²	**46.66**
0.5 mm thick dormer coverings							
flat (in wood roll construction) (Prime Cost £ per kg)	5.00	1.50	44.19	–	34.75	m²	**78.94**
eaves detail ED1	–	0.20	5.89	–	3.20	m	**9.09**
pitched over 3° (in standing seam construction)	–	1.25	36.82	–	28.67	m²	**65.49**
0.5 mm thick wall coverings							
vertical (in angled or flat seam construction)	–	0.80	23.57	–	28.67	m²	**52.24**
vertical (with Coulisseau joint construction)	–	1.25	36.82	–	29.56	m²	**66.38**

17 SHEET ROOF COVERINGS

Item	PC £	Labour hours	Labour £	Plant £	Material £	Unit	Total rate £
STAINLESS STEEL SHEET COVERINGS/ FLASHINGS – cont							
0.5 mm thick Ugitop grade 304 flashings, etc.							
Flashings; wedging into grooves							
150 mm girth (Prime Cost £ per kg)	5.00	0.25	7.36	–	2.94	m	**10.30**
240 mm girth	–	0.25	7.36	–	5.12	m	**12.48**
300 mm girth	–	0.25	7.36	–	6.40	m	**13.76**
Stepped flashings; wedging into grooves							
180 mm girth	–	0.50	14.73	–	3.84	m	**18.57**
270 mm girth	–	0.50	14.73	–	5.76	m	**20.49**
Fan apron							
250 mm girth	–	0.25	7.36	–	5.33	m	**12.69**
Integral box gutter							
900 mm girth; 2 × bent; 2 × welted	–	1.00	29.46	–	21.98	m	**51.44**
Valley gutter							
600 mm girth; 2 × bent; 2 × welted	–	0.75	22.09	–	15.62	m	**37.71**
Hips and ridges							
450 mm girth; 2 × bent; 2 × welted	–	1.00	29.46	–	9.60	m	**39.06**
Roof, dormer and wall coverings in Ugitop grade 316							
0.4 mm thick roof coverings							
flat (in wood roll construction) (Prime Cost £ per kg)	6.49	1.00	29.46	–	39.00	m²	**68.46**
eaves detail ED1	–	0.20	5.89	–	4.16	m	**10.05**
abutment upstands at perimeters	–	0.33	9.72	–	2.08	m	**11.80**
pitched over 3° (in standing seam construction)	–	0.75	22.09	–	29.30	m²	**51.39**
0.5 mm thick dormer coverings							
flat (in wood roll construction) (Prime Cost £ per kg)	6.18	1.50	44.19	–	41.99	m²	**86.18**
eaves detail ED1	–	0.20	5.89	–	3.96	m	**9.85**
pitched over 3° (in standing seam construction)	–	1.25	36.82	–	34.46	m²	**71.28**
0.5 mm thick wall coverings							
vertical (in angled or flat seam construction)	–	0.80	23.57	–	34.46	m²	**58.03**
vertical (with Coulisseau joint construction)	–	1.25	36.82	–	29.56	m²	**66.38**
0.5 mm thick Ugitop grade 316 flashings, etc.							
Flashings; wedging into grooves							
150 mm girth	–	0.25	7.36	–	3.95	m	**11.31**
240 mm girth	–	0.25	7.36	–	6.33	m	**13.69**
300 mm girth	–	0.25	7.36	–	7.91	m	**15.27**
Stepped flashings; wedging into grooves							
180 mm girth	–	0.50	14.73	–	4.74	m	**19.47**
270 mm girth	–	0.50	14.73	–	7.12	m	**21.85**
Fan apron							
250 mm girth	–	0.25	7.36	–	6.59	m	**13.95**
Integral box gutter							
900 mm girth; 2 × bent; 2 × welted	–	1.00	29.46	–	27.15	m	**56.61**

17 SHEET ROOF COVERINGS

Item	PC £	Labour hours	Labour £	Plant £	Material £	Unit	Total rate £
Valley gutter							
600 mm girth; 2 × bent; 2 × welted	–	0.75	22.09	–	19.29	m	**41.38**
Hips and ridges							
450 mm girth; 2 × bent; 2 × welted	–	1.00	29.46	–	11.86	m	**41.32**
Sundries							
provision of square batten roll at 500 mm centres (per m)	–	0.10	2.95	–	1.22	m	**4.17**
FIBRE BITUMEN THERMOPLASTIC SHEET COVERINGS/FLASHINGS							
Glass fibre reinforced bitumen strip slates; Ruberglas 105 or other equal and approved; 1000 mm × 336 mm mineral finish; to external quality plywood boarding (boarding not included)							
Roof coverings	11.18	0.23	5.26	–	13.09	m²	**18.35**
Wall coverings	11.18	0.37	8.47	–	13.09	m²	**21.56**
Extra over coverings for;							
double course at eaves; felt soaker	–	0.19	4.35	–	8.75	m	**13.10**
verges; felt soaker	–	0.14	3.20	–	7.27	m	**10.47**
valley slate; cut to shape; felt soaker and cutting both sides	–	0.42	9.62	–	11.40	m	**21.02**
ridge slate; cut to shape	–	0.28	6.40	–	7.27	m	**13.67**
hip slate; cut to shape; felt soaker and cutting both sides	–	0.42	9.62	–	11.33	m	**20.95**
holes for pipes and the like	–	0.48	10.99	–	–	nr	**10.99**
Bostik Findley Flashband Plus sealing strips and flashings or other equal and approved; special grey finish							
Flashings; wedging at top if required; pressure bonded; to walls							
100 mm girth	–	0.23	3.98	–	0.75	m	**4.73**
150 mm girth	–	0.31	5.37	–	0.97	m	**6.34**
225 mm girth	–	0.37	6.40	–	1.50	m	**7.90**
300 mm girth	–	0.42	7.27	–	2.23	m	**9.50**
450 mm girth	–	0.45	7.78	–	3.03	m	**10.81**
600 mm girth	–	0.47	8.22	–	4.55	m	**12.77**

18 TILE AND SLATE ROOF AND WALL COVERINGS

Item	PC £	Labour hours	Labour £	Plant £	Material £	Unit	Total rate £
ALTERNATIVE SUPPLY ONLY TILE PRICES							
Clay tiles; plain, interlocking and pantiles							
Dreadnought							
Handformed Classic Staffordshire Blue	–	–	–	–	708.75	1000	**708.75**
Handformed Classic Purple Brown	–	–	–	–	661.50	1000	**661.50**
Handformed Classic Bronze	–	–	–	–	661.50	1000	**661.50**
Handformed Classic Deep Red	–	–	–	–	637.88	1000	**637.88**
Red smooth/sandfaced	–	–	–	–	330.75	1000	**330.75**
Country brown smooth/sandfaced	–	–	–	–	368.55	1000	**368.55**
Brown Antique smooth/sandfaced	–	–	–	–	382.73	1000	**382.73**
Blue/Dark Heather	–	–	–	–	401.63	1000	**401.63**
Sandtoft pantiles							
Bridgewater Double Roman	–	–	–	–	5194.35	1000	**5194.35**
Gaelic	–	–	–	–	2043.83	1000	**2043.83**
Arcadia	–	–	–	–	1245.09	1000	**1245.09**
William Blyth pantiles							
Barco Bold Roll	–	–	–	–	947.63	1000	**947.63**
Celtic (French)	–	–	–	–	1073.31	1000	**1073.31**
Concrete tiles; plain and interlocking							
Marley Eternit roof tiles							
Anglia	–	–	–	–	608.48	1000	**608.48**
Ashmore	–	–	–	–	733.21	1000	**733.21**
Duo Modern	–	–	–	–	917.70	1000	**917.70**
Pewter Mendip	–	–	–	–	1047.38	1000	**1047.38**
Malvern	–	–	–	–	977.55	1000	**977.55**
Plain	–	–	–	–	349.13	1000	**349.13**
Redland roof tiles							
Redland 49	–	–	–	–	707.80	1000	**707.80**
Mini Stoneworld	–	–	–	–	748.13	1000	**748.13**
Grovebury	–	–	–	–	1006.42	1000	**1006.42**
PLAIN TILING							
SUPPLY AND FIX PRICES							
NOTE: The following items of tile roofing unless otherwise described, include for conventional fixing assuming normal exposure with appropriate nails and/or rivets or clips to pressure impregnated softwood battens fixed with galvanized nails; prices also include for all bedding and pointing at verges, beneath ridge tiles, etc.							
Clay interlocking plain tiles; Sandtoft 20/20 natural red faced or other equal; 75 mm lap; on 25 mm × 38 mm battens and type 1F reinforced underlay							
Tiles 370 mm × 223 mm (Prime Cost £ per 1000)	867.00	–	–	–	910.35	1000	**910.35**
roof coverings	13.01	0.50	11.45	–	19.12	m²	**30.57**

18 TILE AND SLATE ROOF AND WALL COVERINGS

Item	PC £	Labour hours	Labour £	Plant £	Material £	Unit	Total rate £
extra over coverings for;							
fixing every tile	–	0.02	0.46	–	1.10	m²	1.56
double course at eaves	–	0.28	6.40	–	13.28	m	19.68
verges; extra single undercloak course of plain tiles	–	0.28	6.40	–	5.51	m	11.91
open valleys; cutting both sides	–	0.17	3.90	–	3.82	m	7.72
dry ridge tiles	–	0.56	12.82	–	15.75	m	28.57
dry hips; cutting both sides	–	0.69	15.79	–	13.24	m	29.03
holes for pipes and the like	–	0.19	4.35	–	–	nr	4.35
Clay pantiles; Sandtoft Old English; red sand faced or other equal; 75 mm lap; on 25 mm × 38 mm battens and type 1F reinforced underlay							
Pantiles 342 mm × 241 mm (Prime Cost £ per 1000)	1045.50	–	–	–	1097.77	1000	1097.77
roof coverings	16.73	0.50	11.45	–	23.77	m²	35.22
extra over coverings for;							
fixing every tile	–	0.02	0.46	–	2.75	m²	3.21
other colours	–	–	–	–	1.95	m²	1.95
double course at eaves	–	0.31	7.10	–	5.52	m	12.62
verges; extra single undercloak course of plain tiles	–	0.28	6.40	–	14.68	m	21.08
open valleys; cutting both sides	–	0.17	3.90	–	4.61	m	8.51
ridge tiles; tile slips	–	0.56	12.82	–	44.37	m	57.19
hips; cutting both sides	–	0.69	15.79	–	48.98	m	64.77
holes for pipes and the like	–	0.19	4.35	–	–	nr	4.35
Clay pantiles; William Blyth's Lincoln natural or other equal; 75 mm lap; on 19 mm × 38 mm battens and type 1F reinforced underlay							
Pantiles 343 mm × 280 mm (Prime Cost £ per 1000)	1220.80	–	–	–	1281.84	1000	1281.84
roof coverings	18.92	0.50	11.45	–	26.20	m²	37.65
extra over coverings for;							
fixing every tile	–	0.02	0.46	–	2.75	m²	3.21
other colours	–	–	–	–	1.70	m²	1.70
double course at eaves	–	0.31	7.10	–	6.30	m	13.40
verges; extra single undercloak course of plain tiles	–	0.28	6.40	–	13.89	m	20.29
open valleys; cutting both sides	–	0.17	3.90	–	5.39	m	9.29
ridge tiles; tile slips	–	0.56	12.82	–	29.21	m	42.03
hips; cutting both sides	–	0.69	15.79	–	34.59	m	50.38
holes for pipes and the like	–	0.19	4.35	–	–	nr	4.35
Clay plain tiles; Hinton, Perry and Davenhill Dreadnought smooth red machine-made or other equal; on 19 mm × 38 mm battens and type 1F reinforced underlay							
Tiles 265 mm × 165 mm (Prime Cost £ per 1000)	315.00	–	–	–	330.75	1000	330.75
roof coverings; to 64 mm lap	18.90	0.50	11.45	–	31.94	m²	43.39
wall coverings; to 38 mm lap	16.70	0.75	17.17	–	27.55	m²	44.72

Prices for Measured Works

18 TILE AND SLATE ROOF AND WALL COVERINGS

Item	PC £	Labour hours	Labour £	Plant £	Material £	Unit	Total rate £
PLAIN TILING – cont							
Clay plain tiles – cont							
extra over coverings for;							
ornamental tiles	–	–	–	–	21.88	m²	**21.88**
double course at eaves	–	0.23	5.26	–	3.91	m	**9.17**
verges	–	0.28	6.40	–	1.10	m	**7.50**
swept valleys; cutting both sides	–	0.60	13.73	–	5.55	m	**19.28**
bonnet hips; cutting both sides	–	0.74	16.94	–	55.65	m	**72.59**
external vertical angle tiles; supplementary nail fixings	–	0.37	8.47	–	70.60	m	**79.07**
half round ridge tiles	–	0.56	12.82	–	12.78	m	**25.60**
holes for pipes and the like	–	0.19	4.35	–	–	nr	**4.35**
Concrete plain tiles; BS EN 490 group A; on 25 mm × 38 mm battens and type 1F reinforced underlay							
Tiles 267 mm × 165 mm (Prime Cost £ per 1000)	400.50	–	–	–	420.52	1000	**420.52**
roof coverings; to 64 mm lap	24.03	0.25	5.72	–	37.60	m²	**43.32**
wall coverings; to 38 mm lap	21.23	0.30	6.87	–	32.55	m²	**39.42**
extra over coverings for;							
ornamental tiles	–	–	–	–	24.59	m²	**24.59**
double course at eaves	–	0.23	5.26	–	4.47	m	**9.73**
verges	–	0.31	7.10	–	1.51	m	**8.61**
swept valleys; cutting both sides	–	0.60	13.73	–	42.93	m	**56.66**
bonnet hips; cutting both sides	–	0.74	16.94	–	43.01	m	**59.95**
external vertical angle tiles; supplementary nail fixings	–	0.37	8.47	–	31.61	m	**40.08**
half round ridge tiles	–	0.46	10.53	–	9.32	m	**19.85**
third round hip tiles; cutting both sides	–	0.46	10.53	–	11.98	m	**22.51**
holes for pipes and the like	–	0.19	4.35	–	–	nr	**4.35**
INTERLOCKING TILING							
SUPPLY AND FIX PRICES							
NOTE: The following items of tile roofing unless otherwise described, include for conventional fixing assuming normal exposure with appropriate nails and/or rivets or clips to pressure impregnated softwood battens fixed with galvanized nails; prices also include for all bedding and pointing at verges, beneath ridge tiles, etc.							
Clay pantiles; Sandtoft Old English; red sand faced or other equal; 75 mm lap; on 25 mm × 38 mm battens and type 1F reinforced underlay							
Tiles 342 mm × 241 mm (Prime Cost £ per 1000)	1045.50	–	–	–	1097.77	1000	**1097.77**
roof coverings	16.73	0.42	9.62	–	23.77	m²	**33.39**

18 TILE AND SLATE ROOF AND WALL COVERINGS

Item	PC £	Labour hours	Labour £	Plant £	Material £	Unit	Total rate £
extra over coverings for;							
fixing every tile	–	0.02	0.46	–	2.75	m²	**3.21**
other colours	–	–	–	–	1.95	m²	**1.95**
double course at eaves	–	0.31	7.10	–	5.52	m	**12.62**
verges; extra single undercloak course of plain tiles	–	0.28	6.40	–	14.68	m	**21.08**
open valleys; cutting both sides	–	0.17	3.90	–	4.61	m	**8.51**
ridge tiles; tile slips	–	0.56	12.82	–	44.37	m	**57.19**
hips; cutting both sides	–	0.69	15.79	–	48.98	m	**64.77**
holes for pipes and the like	–	0.19	4.35	–	–	nr	**4.35**
Clay pantiles; William Blyth's Lincoln natural or other equal; 75 mm lap; on 19 mm × 38 mm battens and type 1F reinforced underlay							
Tiles 343 mm × 280 mm (Prime Cost £ per 1000)	1220.80	–	–	–	1281.84	1000	**1281.84**
roof coverings	19.87	0.42	9.62	–	26.20	m²	**35.82**
extra over coverings for;							
fixing every tile	–	0.02	0.46	–	2.75	m²	**3.21**
other colours	–	–	–	–	1.70	m²	**1.70**
double course at eaves	–	0.31	7.10	–	6.30	m	**13.40**
verges; extra single undercloak course of plain tiles	–	0.28	6.40	–	13.89	m	**20.29**
open valleys; cutting both sides	–	0.17	3.90	–	5.39	m	**9.29**
ridge tiles; tile slips	–	0.56	12.82	–	29.21	m	**42.03**
hips; cutting both sides	–	0.69	15.79	–	34.59	m	**50.38**
holes for pipes and the like	–	0.19	4.35	–	–	nr	**4.35**
Concrete interlocking tiles; Marley Eternit Anglia granule finish tiles or other equal; 75 mm lap; on 25 mm × 38 mm battens and type 1F reinforced underlay							
Tiles 387 mm × 230 mm (Prime Cost £ per 1000)	505.40	–	–	–	530.67	1000	**530.67**
roof coverings	7.93	0.42	9.62	–	13.41	m²	**23.03**
extra over coverings for;							
fixing every tile	–	0.02	0.46	–	0.46	m²	**0.92**
eaves; eaves filler	–	0.04	0.91	–	11.34	m	**12.25**
verges; 150 mm wide asbestos free strip undercloak	–	0.21	4.81	–	1.91	m	**6.72**
valley trough tiles; cutting both sides	–	0.51	11.68	–	26.17	m	**37.85**
segmental ridge tiles; tile slips	–	0.51	11.68	–	13.06	m	**24.74**
segmental hip tiles; tile slips; cutting both sides	–	0.65	14.88	–	14.74	m	**29.62**
dry ridge tiles; segmental including batten sections; unions and filler pieces	–	0.28	6.40	–	19.28	m	**25.68**
segmental mono-ridge tiles	–	0.51	11.68	–	20.70	m	**32.38**
gas ridge terminal	–	0.46	10.53	–	69.90	nr	**80.43**
holes for pipes and the like	–	0.19	4.35	–	–	nr	**4.35**

18 TILE AND SLATE ROOF AND WALL COVERINGS

Item	PC £	Labour hours	Labour £	Plant £	Material £	Unit	Total rate £
INTERLOCKING TILING – cont							
Concrete interlocking tiles – cont							
Concrete interlocking tiles; Marley Eternit Ludlow Major granule finish tiles or other equal; 75 mm lap; on 25 mm × 38 mm battens and type 1F reinforced underlay							
Tiles 420 mm × 330 mm (Prime Cost £ per 1000)	866.40	–	–	–	909.72	1000	**909.72**
roof coverings	8.92	0.32	7.33	–	13.17	m²	**20.50**
extra over coverings for;							
fixing every tile	–	0.02	0.46	–	0.46	m²	**0.92**
eaves; eaves filler	–	0.04	0.91	–	0.40	m	**1.31**
verges; 150 mm wide asbestos free strip							
undercloak	–	0.21	4.81	–	1.91	m	**6.72**
dry verge system; extruded white PVC	–	0.14	3.20	–	12.05	m	**15.25**
segmental ridge cap to dry verge	–	0.02	0.46	–	3.98	m	**4.44**
valley trough tiles; cutting both sides	–	0.51	11.68	–	26.96	m	**38.64**
segmental ridge tiles	–	0.46	10.53	–	8.30	m	**18.83**
segmental hip tiles; cutting both sides	–	0.60	13.73	–	11.15	m	**24.88**
dry ridge tiles; segmental including batten							
sections; unions and filler pieces	–	0.28	6.40	–	19.31	m	**25.71**
segmental mono-ridge tiles	–	0.46	10.53	–	17.91	m	**28.44**
gas ridge terminal	–	0.46	10.53	–	69.90	nr	**80.43**
holes for pipes and the like	–	0.19	4.35	–	–	nr	**4.35**
Concrete interlocking tiles; Marley Eternit Mendip granule finish double pantiles or other equal; 75 mm lap; on 22 mm × 38 mm battens and type 1F reinforced underlay							
Tiles 420 mm × 330 mm (Prime Cost £ per 1000)	812.30	–	–	–	852.91	1000	**852.91**
roof coverings (PC £ per 1000)	8.27	0.32	7.33	–	12.49	m²	**19.82**
extra over coverings for;							
fixing every tile	–	0.02	0.46	–	0.46	m²	**0.92**
eaves; eaves filler	–	0.02	0.46	–	11.07	m	**11.53**
verges; 150 mm wide asbestos free strip							
undercloak	–	0.21	4.81	–	1.91	m	**6.72**
dry verge system; extruded white PVC	–	0.14	3.20	–	12.05	m	**15.25**
segmental ridge cap to dry verge	–	0.02	0.46	–	3.98	m	**4.44**
valley trough tiles; cutting both sides	–	0.51	11.68	–	26.85	m	**38.53**
segmental ridge tiles	–	0.51	11.68	–	13.06	m	**24.74**
segmental hip tiles; cutting both sides	–	0.65	14.88	–	15.75	m	**30.63**
dry ridge tiles; segmental including batten							
sections; unions and filler pieces	–	0.28	6.40	–	19.31	m	**25.71**
segmental mono-ridge tiles	–	0.46	10.53	–	20.30	m	**30.83**
gas ridge terminal	–	0.46	10.53	–	69.90	nr	**80.43**
holes for pipes and the like	–	0.19	4.35	–	–	nr	**4.35**

18 TILE AND SLATE ROOF AND WALL COVERINGS

Item	PC £	Labour hours	Labour £	Plant £	Material £	Unit	Total rate £
Concrete interlocking tiles; Marley Eternit Modern smooth finish tiles or other equal; 75 mm lap; on 25 mm × 38 mm battens and type 1F reinforced underlay							
Tiles 420 mm × 220 mm (Prime Cost £ per 1000)	722.00	–	–	–	758.10	1000	**758.10**
roof coverings	7.51	0.32	7.33	–	12.02	m²	**19.35**
extra over coverings for;							
fixing every tile	–	0.02	0.46	–	0.46	m²	**0.92**
verges; 150 wide asbestos free strip undercloak	–	0.21	4.81	–	1.91	m	**6.72**
dry verge system; extruded white PVC	–	0.19	4.35	–	12.05	m	**16.40**
Modern ridge cap to dry verge	–	0.02	0.46	–	3.98	m	**4.44**
valley trough tiles; cutting both sides	–	0.51	11.68	–	26.65	m	**38.33**
Modern ridge tiles	–	0.46	10.53	–	10.50	m	**21.03**
Modern hip tiles; cutting both sides	–	0.60	13.73	–	12.89	m	**26.62**
dry ridge tiles; Modern; including batten sections; unions and filler pieces	–	0.28	6.40	–	21.52	m	**27.92**
Modern mono-ridge tiles	–	0.46	10.53	–	17.91	m	**28.44**
gas ridge terminal	–	0.46	10.53	–	69.90	nr	**80.43**
holes for pipes and the like	–	0.19	4.35	–	–	nr	**4.35**
Concrete interlocking tiles; Marley Eternit Ecologic Ludlow Major granule finish tiles or other equal; 75 mm lap; on 25 mm × 38 mm battens and type 1F reinforced underlay							
Tiles 420 mm × 330 mm (Prime Cost £ per 1000)	857.40	–	–	–	900.27	1000	**900.27**
roof coverings	8.40	0.32	7.33	–	13.07	m²	**20.40**
extra over coverings for:							
fixing every tile	–	0.02	0.46	–	0.46	m²	**0.92**
eaves; eaves filler	–	0.04	0.91	–	0.40	m	**1.31**
verges; 150 mm wide asbestos free strip undercloak	–	0.21	4.81	–	1.91	m	**6.72**
dry verge system; extruded white PVC	–	0.14	3.20	–	12.05	m	**15.25**
segmental ridge cap to dry verge	–	0.02	0.46	–	3.98	m	**4.44**
valley trough tiles; cutting both sides	–	0.51	11.68	–	26.94	m	**38.62**
segmental ridge tiles	–	0.46	10.53	–	8.30	m	**18.83**
segmental hip tiles; cutting both sides	–	0.60	13.73	–	11.13	m	**24.86**
dry ridge tiles; segmental including batten sections; unions and filler pieces	–	0.28	6.40	–	19.31	m	**25.71**
segmental mono-ridge tiles	–	0.46	10.53	–	17.91	m	**28.44**
gas ridge terminal	–	0.46	10.53	–	69.90	nr	**80.43**
holes for pipes and the like	–	0.19	4.35	–	–	nr	**4.35**
Concrete interlocking tiles; Marley Eternit Wessex smooth finish tiles or other equal; 75 mm lap; on 25 mm × 38 mm battens and type 1F reinforced underlay							
Tiles 413 mm × 330 mm (Prime Cost £ per 1000)	1263.50	–	–	–	1326.67	1000	**1326.67**
roof coverings	12.51	0.32	7.33	–	17.92	m²	**25.25**

18 TILE AND SLATE ROOF AND WALL COVERINGS

Item	PC £	Labour hours	Labour £	Plant £	Material £	Unit	Total rate £
INTERLOCKING TILING – cont							
Concrete interlocking tiles – cont							
extra over coverings for;							
fixing every tile	–	0.02	0.46	–	0.46	m²	**0.92**
verges; 150 mm wide asbestos free strip							
undercloak	–	0.21	4.81	–	1.91	m	**6.72**
dry verge system; extruded white PVC	–	0.19	4.35	–	12.05	m	**16.40**
Modern ridge cap to dry verge	–	0.02	0.46	–	3.98	m	**4.44**
valley trough tiles; cutting both sides	–	0.51	11.68	–	27.85	m	**39.53**
Modern ridge tiles	–	0.46	10.53	–	10.50	m	**21.03**
Modern hip tiles; cutting both sides	–	0.60	13.73	–	14.68	m	**28.41**
dry ridge tiles; Modern; including batten sections;							
unions and filler pieces	–	0.28	6.40	–	21.49	m	**27.89**
Modern mono-ridge tiles	–	0.46	10.53	–	17.91	m	**28.44**
gas ridge terminal	–	0.46	10.53	–	69.90	nr	**80.43**
holes for pipes and the like	–	0.19	4.35	–	–	nr	**4.35**
Concrete interlocking tiles; Fenland Pantile							
smooth finish pantiles or other equal; 75 mm lap;							
on 25 mm × 38 mm battens and type 1F reinforced							
underlay							
Tiles 381 mm × 229 mm (Prime Cost £ per 1000)	693.90	–	–	–	728.60	1000	**728.60**
roof coverings	12.97	0.42	9.62	–	18.32	m²	**27.94**
extra over coverings for;							
fixing every tile	–	0.04	0.91	–	0.19	m²	**1.10**
eaves; eaves filler	–	0.04	0.91	–	1.46	m	**2.37**
verges; extra single undercloak course of plain							
tiles	–	0.28	6.40	–	8.91	m	**15.31**
valley trough tiles; cutting both sides	–	0.56	12.82	–	40.28	m	**53.10**
universal ridge tiles	–	0.46	10.53	–	16.71	m	**27.24**
universal hip tiles; cutting both sides	–	0.60	13.73	–	20.53	m	**34.26**
universal gas flue ridge tile	–	0.46	10.53	–	88.81	nr	**99.34**
universal ridge vent tile with 110 mm dia. adaptor	–	0.50	11.45	–	114.61	nr	**126.06**
holes for pipes and the like	–	0.19	4.35	–	–	nr	**4.35**
Concrete interlocking tiles; Redland Renown							
granule finish tiles or other equal and approved;							
418 mm × 330 mm; to 75 mm lap; on							
25 mm × 38 mm battens and type 1F reinforced							
underlay							
Tiles 418 mm × 330 mm (Prime Cost £ per 1000)	940.50	–	–	–	987.52	1000	**987.52**
roof coverings	9.12	0.32	7.33	–	14.29	m²	**21.62**
extra over coverings for;							
fixing every tile	–	0.02	0.46	–	0.23	m²	**0.69**
verges; extra single undercloak course of plain							
tiles	–	0.23	5.26	–	5.02	m	**10.28**
cloaked verge system	–	0.14	3.20	–	10.57	m	**13.77**
valley trough tiles; cutting both sides	–	0.51	11.68	–	39.56	m	**51.24**
universal ridge tiles	–	0.46	10.53	–	16.71	m	**27.24**
universal hip tiles; cutting both sides	–	0.60	13.73	–	19.81	m	**33.54**
dry ridge system; universal ridge tiles	–	0.23	5.26	–	45.67	m	**50.93**

18 TILE AND SLATE ROOF AND WALL COVERINGS

Item	PC £	Labour hours	Labour £	Plant £	Material £	Unit	Total rate £
universal half round mono-pitch ridge tiles	–	0.51	11.68	–	36.64	m	48.32
universal gas flue ridge tile	–	0.46	10.53	–	88.81	nr	99.34
universal ridge vent tile with 110 mm dia. adaptor	–	0.46	10.53	–	114.61	nr	125.14
holes for pipes and the like	–	0.19	4.35	–	–	nr	4.35
Concrete interlocking tiles; Redland Regent granule finish bold roll tiles or other equal; 75 mm lap; on 25 mm × 38 mm battens and type 1F reinforced underlay							
Tiles 418 mm × 332 mm (Prime Cost £ per 1000)	975.60	–	–	–	1024.38	1000	1024.38
roof coverings	9.46	0.32	7.33	–	14.66	m²	21.99
extra over coverings for;							
fixing every tile	–	0.03	0.68	–	0.78	m²	1.46
eaves; eaves filler	–	0.04	0.91	–	1.19	m	2.10
verges; extra single undercloak course of plain tiles	–	0.23	5.26	–	4.13	m	9.39
cloaked verge system	–	0.14	3.20	–	10.47	m	13.67
valley trough tiles; cutting both sides	–	0.51	11.68	–	39.68	m	51.36
universal ridge tiles	–	0.46	10.53	–	16.71	m	27.24
universal hip tiles; cutting both sides	–	0.60	13.73	–	19.93	m	33.66
dry ridge system; universal ridge tiles	–	0.23	5.26	–	57.29	m	62.55
universal half round mono-pitch ridge tiles	–	0.51	11.68	–	36.64	m	48.32
universal gas flue ridge tile	–	0.46	10.53	–	88.81	nr	99.34
universal ridge vent tile with 110 mm dia. adaptor	–	0.46	10.53	–	114.61	nr	125.14
holes for pipes and the like	–	0.19	4.35	–	–	nr	4.35
Concrete interlocking slates; Redland Mini Stonewold concrete slates or other equal; 75 mm lap; on 25 mm × 38 mm battens and type 1F reinforced underlay							
Slates 418 mm × 334 mm (Prime Cost £ per 1000)	712.50	–	–	–	748.13	1000	748.13
roof coverings	7.62	0.32	7.33	–	12.68	m²	20.01
extra over coverings for;							
fixing every tile	–	0.02	0.46	–	2.70	m²	3.16
eaves; eaves filler	–	0.02	0.46	–	3.72	m	4.18
verges; extra single undercloak course of plain tiles	–	0.28	6.40	–	4.88	m	11.28
ambi-dry verge system	–	0.19	4.35	–	15.87	m	20.22
ambi-dry verge eave/ridge end piece	–	0.02	0.46	–	5.69	m	6.15
valley trough tiles; cutting both sides	–	0.51	11.68	–	38.69	m	50.37
universal angle ridge tiles	–	0.46	10.53	–	13.00	m	23.53
universal hip tiles; cutting both sides	–	0.60	13.73	–	27.24	m	40.97
dry ridge system; universal angle ridge tiles	–	0.23	5.26	–	21.18	m	26.44
universal mono-pitch angle ridge tiles	–	0.51	11.68	–	26.08	m	37.76
universal gas flue angle ridge tile	–	0.46	10.53	–	93.68	nr	104.21
universal angle ridge vent tile with 110 mm dia. adaptor	–	0.46	10.53	–	119.76	nr	130.29
holes for pipes and the like	–	0.19	4.35	–	–	nr	4.35

18 TILE AND SLATE ROOF AND WALL COVERINGS

Item	PC £	Labour hours	Labour £	Plant £	Material £	Unit	Total rate £
INTERLOCKING TILING – cont							
Concrete interlocking slates; Redland Richmond 10 Slates laid broken bonded, normally from right to left; smooth finish tiles or other equal; 75 mm lap; on 25 mm × 38 mm battens and type 1F reinforced underlay							
Slates 418 mm × 330 mm (Prime Cost £ per 1000)	1169.50	–	–	–	1227.98	1000	**1227.98**
roof coverings	12.53	0.32	7.33	–	18.30	m²	**25.63**
extra over coverings for;							
fixing every tile	–	0.02	0.46	–	0.84	m²	**1.30**
eaves; eaves filler	–	0.02	0.46	–	3.48	m	**3.94**
verges; extra single undercloak course of plain tiles	–	0.23	5.26	–	5.02	m	**10.28**
ambi-dry verge system	–	0.19	4.35	–	15.87	m	**20.22**
ambi-dry verge eave/ridge end piece	–	0.02	0.46	–	5.69	m	**6.15**
universal valley trough tiles; cutting both sides	–	0.56	12.82	–	43.51	m	**56.33**
universal hip tiles; cutting both sides	–	0.60	13.73	–	16.87	m	**30.60**
universal angle ridge tiles	–	0.46	10.53	–	13.00	m	**23.53**
dry ridge system; universal angle ridge tiles	–	0.23	5.26	–	21.18	m	**26.44**
universal mono-pitch angle ridge tiles	–	0.51	11.68	–	26.08	m	**37.76**
gas ridge terminal	–	0.46	10.53	–	92.95	nr	**103.48**
ridge vent with 110 mm dia. flexible adaptor	–	0.46	10.53	–	119.02	nr	**129.55**
holes for pipes and the like	–	0.19	4.35	–	–	nr	**4.35**
Concrete interlocking slates; Redland Stonewold II smooth finish tiles or other equal; 75 mm lap; on 25 mm × 38 mm battens and type 1F reinforced underlay							
Slates 430 mm × 380 mm (Prime Cost £ per 1000)	4303.50	–	–	–	4518.68	1000	**4518.68**
roof coverings	41.74	0.32	7.33	–	51.20	m²	**58.53**
extra over coverings for;							
fixing every tile	–	0.02	0.46	–	2.70	m²	**3.16**
eaves; eaves filler	–	0.02	0.46	–	3.72	m	**4.18**
verges; extra single undercloak course of plain tiles	–	0.28	6.40	–	5.18	m	**11.58**
ambi-dry verge system	–	0.19	4.35	–	15.87	m	**20.22**
ambi-dry verge eave/ridge end piece	–	0.02	0.46	–	5.69	m	**6.15**
valley trough tiles; cutting both sides	–	0.51	11.68	–	50.68	m	**62.36**
universal angle ridge tiles	–	0.46	10.53	–	13.00	m	**23.53**
universal hip tiles; cutting both sides	–	0.60	13.73	–	27.24	m	**40.97**
dry ridge system; universal angle ridge tiles	–	0.23	5.26	–	21.18	m	**26.44**
universal mono-pitch angle ridge tiles	–	0.51	11.68	–	26.08	m	**37.76**
universal gas flue angle ridge tile	–	0.46	10.53	–	93.68	nr	**104.21**
universal angle ridge vent tile with 110 mm dia. adaptor	–	0.46	10.53	–	119.76	nr	**130.29**
holes for pipes and the like	–	0.19	4.35	–	–	nr	**4.35**

18 TILE AND SLATE ROOF AND WALL COVERINGS

Item	PC £	Labour hours	Labour £	Plant £	Material £	Unit	Total rate £
Concrete interlocking slates; Redland Cambrian Slates made from 60% recycled Welsh slate or other equal; 50 mm lap; on 25 mm × 38 mm battens and type 1F reinforced underlay							
Slates 300 mm × 336 mm (Prime Cost £ per 1000)	1843.00	–	–	–	1935.15	1000	**1935.15**
roof coverings	29.30	0.32	7.33	–	39.75	m²	**47.08**
extra over coverings for;							
fixing every tile	–	0.02	0.46	–	1.19	m²	**1.65**
eaves; eaves filler	–	0.02	0.46	–	4.13	m	**4.59**
verges; extra single undercloak course of plain tiles	–	0.23	5.26	–	11.87	m	**17.13**
ambi-dry verge system	–	0.19	4.35	–	15.87	m	**20.22**
ambi-dry verge eave/ridge end piece	–	0.02	0.46	–	5.69	m	**6.15**
universal valley trough tiles; cutting both sides	–	0.56	12.82	–	43.51	m	**56.33**
universal hip tiles; cutting both sides	–	0.60	13.73	–	18.81	m	**32.54**
universal angle ridge tiles	–	0.46	10.53	–	13.00	m	**23.53**
dry ridge system; universal angle ridge tiles	–	0.23	5.26	–	21.18	m	**26.44**
universal mono-pitch angle ridge tiles	–	0.51	11.68	–	26.08	m	**37.76**
gas ridge terminal	–	0.46	10.53	–	92.95	nr	**103.48**
ridge vent with 110 mm dia. flexible adaptor	–	0.46	10.53	–	119.02	nr	**129.55**
holes for pipes and the like	–	0.19	4.35	–	–	nr	**4.35**
Sundries and accessories							
Hip irons							
galvanized mild steel; fixing with screws	–	0.09	2.06	–	3.30	nr	**5.36**
Rytons Clip strip or other equal and approved; continuous soffit ventilator							
51 mm wide; plastic; code CS351	–	0.28	6.40	–	1.00	m	**7.40**
Rytons over fascia ventilator or other equal and approved; continuous eaves ventilator							
40 mm wide; plastic; code OFV890	–	0.09	2.06	–	1.41	m	**3.47**
Rytons roof ventilator or other equal and approved; to suit rafters at 600 mm centres							
250 mm deep × 43 mm high; plastic; code TV600	–	0.09	2.06	–	1.61	m	**3.67**
Rytons push and lock ventilators or other equal and approved; circular							
83 mm dia.; plastic; code PL235	–	0.04	0.81	–	0.36	nr	**1.17**
Fixing only							
lead soakers (supply cost not included)	–	0.07	1.21	–	–	nr	**1.21**
Pressure impregnated softwood counter battens; 25 mm × 50 mm							
450 mm centres	–	0.06	1.38	–	2.58	m²	**3.96**
600 mm centres	–	0.04	0.91	–	1.95	m²	**2.86**
Underlay; BS EN 13707 type 1B; bitumen felt weighing 14 kg/10 m²; 75 mm laps							
To sloping or vertical surfaces	0.60	0.02	0.46	–	1.01	m²	**1.47**
Underlay; BS EN 13707 type 1F; reinforced bitumen felt weighing 22.50 kg/10 m²; 75 mm laps							
To sloping or vertical surfaces	0.76	0.02	0.46	–	1.18	m²	**1.64**

18 TILE AND SLATE ROOF AND WALL COVERINGS

Item	PC £	Labour hours	Labour £	Plant £	Material £	Unit	Total rate £
INTERLOCKING TILING – cont							
Underlay; Visqueen HP vapour barrier or other equal; multi-layer reinforce LDPE membrane with an aluminium core							
To sloping or vertical surfaces	3.38	0.02	0.46	–	4.29	m²	**4.75**
Underlay; reinforced vapour permeable membrane; 75 mm laps							
To sloping or vertical surfaces	0.81	0.02	0.46	–	1.23	m²	**1.69**
Underlay; Klober Permo air breathable reinforced vapour permeable membrane; 75 mm laps							
To sloping or vertical surfaces	1.75	0.03	0.58	–	2.22	m²	**2.80**
UV Breather membranes; Web Dynamics Ltd							
Web UV 10 standard grade breather membrane draped between open rafters; 100 mm vertical laps, 150 mm horizontal laps, secured with galvanized nails	0.76	0.03	0.68	–	0.98	m²	**1.66**
Web UV 15 professional grade breather membrane draped between open rafters or on sarking board; 100 mm vertical laps, 150 mm horizontal laps secured with galvanized nails	0.75	0.03	0.68	–	0.97	m²	**1.65**
Web UV 25 heavy dutyl grade breather membrane draped between open rafters or on sarking board; 100 mm vertical laps, 150 mm horizontal laps secured with galvanized nails	1.54	0.04	0.80	–	1.88	m²	**2.68**
FIBRE CEMENT SLATING							
Fibre cement artificial slates; Eternit Garsdale/ E2000T or other equal to 75 mm lap; on 19 mm × 50 mm battens and type 1F reinforced underlay							
Coverings; 600 mm × 600 mm slates							
roof coverings	–	0.60	13.73	–	16.14	m²	**29.87**
wall coverings	–	0.74	16.94	–	17.24	m²	**34.18**
Coverings; 600 mm × 300 mm slates							
roof coverings	–	0.46	10.53	–	17.35	m²	**27.88**
wall coverings	–	0.60	13.73	–	17.35	m²	**31.08**
Extra over slate coverings for							
double course at eaves	–	0.23	5.26	–	4.15	m	**9.41**
verges; extra single undercloak course	–	0.31	7.10	–	0.77	m	**7.87**
open valleys; cutting both sides	–	0.19	4.35	–	3.09	m	**7.44**
stop end	–	0.09	2.06	–	10.28	nr	**12.34**
roll top ridge tiles	–	0.56	12.82	–	28.80	m	**41.62**
stop end	–	0.09	2.06	–	15.95	nr	**18.01**
mono-pitch ridge tiles	–	0.46	10.53	–	35.66	m	**46.19**

18 TILE AND SLATE ROOF AND WALL COVERINGS

Item	PC £	Labour hours	Labour £	Plant £	Material £	Unit	Total rate £
stop end	–	0.09	2.06	–	51.92	nr	**53.98**
duo-pitch ridge tiles	–	0.46	10.53	–	31.10	m	**41.63**
stop end	–	0.09	2.06	–	26.82	nr	**28.88**
half round hip tiles; cutting both sides	–	0.19	4.35	–	67.21	m	**71.56**
holes for pipes and the like	–	0.19	4.35	–	–	nr	**4.35**

NATURAL SLATING
NOTE: The following items of slate roofing unless otherwise described, include for conventional fixing assuming normal exposure with appropriate nails and/or rivets or clips to pressure impregnated softwood battens fixed with galvanized nails; prices also include for all bedding and pointing at verges; beneath verge tiles etc.

Natural slates; BS EN 12326 Part 2; Spanish blue grey; uniform size; to 75 mm lap; on 25 mm × 50 mm battens and type 1F reinforced underlay

Item	PC £	Labour hours	Labour £	Plant £	Material £	Unit	Total rate £
Coverings; 400 mm × 250 mm slates (Prime Cost £ per 1000)	549.00	–	–	–	576.45	1000	**576.45**
roof coverings	13.45	0.73	16.72	–	23.50	m²	**40.22**
wall coverings	13.45	1.06	24.27	–	23.50	m²	**47.77**
Coverings; 500 mm × 250 mm slates (Prime Cost £ per 1000)	909.00	–	–	–	954.45	1000	**954.45**
roof coverings	17.00	0.60	13.73	–	25.54	m²	**39.27**
wall coverings	17.00	0.88	20.15	–	25.54	m²	**45.69**
Coverings; 600 mm × 300 mm slates (Prime Cost £ per 1000)	1791.00	–	–	–	1880.55	1000	**1880.55**
roof coverings	22.57	0.50	11.45	–	30.48	m²	**41.93**
wall coverings	22.57	0.69	15.79	–	30.48	m²	**46.27**
Extra over coverings for;							
double course at eaves	–	0.28	6.40	–	7.79	m	**14.19**
verges; extra single undercloak course	–	0.39	8.92	–	4.04	m	**12.96**
open valleys; cutting both sides	–	0.20	4.58	–	15.79	m	**20.37**
blue/black glass reinforced concrete 152 mm half round ridge tiles	–	0.46	10.53	–	15.28	m	**25.81**
blue/black glass reinforced concrete 125 mm × 125 mm plain angle ridge tiles	–	0.46	10.53	–	15.28	m	**25.81**
mitred hips; cutting both sides	–	0.20	4.58	–	15.79	m	**20.37**
blue/black glass reinforced concrete 152 mm half round hip tiles; cutting both sides	–	0.65	14.88	–	31.07	m	**45.95**
blue/black glass reinforced concrete 125 mm × 125 mm plain angle hip tiles; cutting both sides	–	0.65	14.88	–	31.06	m	**45.94**
holes for pipes and the like	–	0.19	4.35	–	–	nr	**4.35**

18 TILE AND SLATE ROOF AND WALL COVERINGS

Item	PC £	Labour hours	Labour £	Plant £	Material £	Unit	Total rate £
NATURAL SLATING – cont							
Natural slates; BS EN 12326 Part 2; Welsh blue grey; 7 mm nominal thickness; uniform size; to 75 mm lap; on 25 mm × 50 mm battens and type 1F reinforced underlay							
Coverings; 400 mm × 250 mm slates (Prime Cost £ per 1000)	1400.00	–	–	–	1470.00	1000	**1470.00**
roof coverings	34.30	0.70	16.02	–	46.48	m²	**62.50**
wall coverings	34.30	1.00	22.89	–	46.48	m²	**69.37**
Coverings; 500 mm × 250 mm slate (Prime Cost £ per 1000)	2373.00	–	–	–	2491.65	1000	**2491.65**
roof coverings	44.38	0.60	13.73	–	55.72	m²	**69.45**
wall coverings	44.38	0.80	18.31	–	55.72	m²	**74.03**
Coverings; 500 mm × 300 mm slates (Prime Cost £ per 1000)	2261.00	–	–	–	2374.05	1000	**2374.05**
roof coverings	35.27	0.60	13.73	–	45.69	m²	**59.42**
wall coverings	35.27	0.75	17.17	–	45.69	m²	**62.86**
Coverings; 600 mm × 300 mm slates (Prime Cost £ per 1000)	2695.00	–	–	–	2829.75	1000	**2829.75**
roof coverings	33.96	0.50	11.45	–	43.04	m²	**54.49**
wall coverings	33.96	0.65	14.88	–	43.04	m²	**57.92**
Extra over coverings for;							
double course at eaves	–	0.25	5.72	–	11.08	m	**16.80**
verges; extra single undercloak course	–	0.35	8.01	–	6.04	m	**14.05**
open valleys; cutting both sides	–	0.20	4.58	–	23.77	m	**28.35**
blue/black glazed ware 152 mm half round ridge tiles	–	0.46	10.53	–	9.65	m	**20.18**
blue/black glazed ware 125 mm × 125 mm plain angle ridge tiles	–	0.46	10.53	–	27.63	m	**38.16**
mitred hips; cutting both sides	–	0.20	4.58	–	23.77	m	**28.35**
blue/black glazed ware 152 mm half round hip tiles; cutting both sides	–	0.65	14.88	–	33.42	m	**48.30**
blue/black glazed ware 125 mm × 125 mm plain angle hip tiles; cutting both sides	–	0.65	14.88	–	51.40	m	**66.28**
holes for pipes and the like	–	0.19	4.35	–	–	nr	**4.35**
Natural slates; Westmoreland green; random lengths; 457 mm–229 mm proportionate widths to 75 mm lap; in diminishing courses; on 25 mm × 50 mm battens and type 1F underlay							
Coverings (Prime Cost £ per tonne; approximately 17.5 m²)	2605.85	–	–	–	2736.14	tonne	**2736.14**
roof coverings	148.53	1.00	22.89	–	171.36	m²	**194.25**
wall coverings	148.53	1.30	29.76	–	171.36	m²	**201.12**
Extra over coverings for;							
double course at eaves	–	0.60	13.73	–	30.14	m	**43.87**
verges; extra single undercloak course slates 152 mm wide	–	0.67	15.34	–	26.09	m	**41.43**
holes for pipes and the like	–	0.25	5.72	–	–	nr	**5.72**

18 TILE AND SLATE ROOF AND WALL COVERINGS

Item	PC £	Labour hours	Labour £	Plant £	Material £	Unit	Total rate £
NATURAL OR ARTIFICIAL STONE SLATING							
Reconstructed stone slates; Hardrow Slates or other equal and approved; standard colours; or similar; 75 mm lap; on 25 mm × 50 mm battens and type 1F reinforced underlay							
Coverings; 457 mm × 305 mm slates (Prime Cost £ per 1000)	1422.00	–	–	–	1493.10	1000	**1493.10**
roof coverings	24.46	0.74	16.94	–	34.76	m²	**51.70**
wall coverings	24.46	0.93	21.29	–	34.76	m²	**56.05**
Coverings; 457 mm × 457 mm slates (Prime Cost £ per 1000)	1854.00	–	–	–	1946.70	1000	**1946.70**
roof coverings	21.32	0.60	13.73	–	31.03	m²	**44.76**
wall coverings	21.32	0.79	18.08	–	31.03	m²	**49.11**
Extra over 457 mm × 305 mm coverings for;							
double course at eaves	–	0.28	6.40	–	6.48	m	**12.88**
verges; pointed	–	0.39	8.92	–	−0.74	m	**8.18**
open valleys; cutting both sides	–	0.20	4.58	–	15.68	m	**20.26**
ridge tiles	–	0.46	10.53	–	44.71	m	**55.24**
hip tiles; cutting both sides	–	0.65	14.88	–	35.82	m	**50.70**
holes for pipes and the like	–	0.19	4.35	–	–	nr	**4.35**
Reconstructed stone slates; Bradstone Cotswold style or other equal and approved; random lengths 550 mm–300 mm; proportional widths; to 80 mm lap; in diminishing courses; on 25 mm × 50 mm battens and type 1F reinforced underlay							
Roof coverings (all-in rate inclusive of eaves and verges), (Prime Cost £ per m²)	31.89	0.97	22.21	–	42.96	m²	**65.17**
Extra over coverings for;							
open valleys/mitred hips; cutting both sides	–	0.42	9.62	–	15.82	m²	**25.44**
ridge tiles	–	0.61	13.97	–	20.96	m	**34.93**
hip tiles; cutting both sides	–	0.97	22.21	–	35.73	m	**57.94**
holes for pipes and the like	–	0.28	6.40	–	–	nr	**6.40**
Reconstructed stone slates; Bradstone Moordale style or other equal and approved; random lengths 550 mm–450 mm; proportional widths; to 80 mm lap; in diminishing course; on 25 mm × 50 mm battens and type 1F reinforced underlay							
Roof coverings (all-in rate inclusive of eaves and verges), (Prime Cost £ per m²)	30.00	0.97	22.21	–	40.87	m²	**63.08**
Extra over coverings for;							
open valleys/mitred hips; cutting both sides	–	0.42	9.62	–	14.89	m²	**24.51**
ridge tiles	–	0.61	13.97	–	20.96	m	**34.93**
holes for pipes and the like	–	0.28	6.40	–	–	nr	**6.40**

18 TILE AND SLATE ROOF AND WALL COVERINGS

Item	PC £	Labour hours	Labour £	Plant £	Material £	Unit	Total rate £
TIMBER SHINGLES							
Red Cedar sawn shingles preservative treated; uniform length 400 mm; to 125 mm gauge; on 25 mm × 38 mm battens and type 1F reinforced underlay							
Shingles; uniform length 400 mm (Prime Cost £ per bundle)	29.16	–	–	–	30.62	bundle	**30.62**
roof coverings; 125 mm gauge (2.28 m^2 per bundle)	12.80	0.97	22.21	–	24.27	m^2	**46.48**
wall coverings; 190 mm gauge (3.47 m^2/bundle)	8.40	0.74	16.94	–	16.37	m^2	**33.31**
extra over for;							
double course at eaves	–	0.19	4.35	–	1.91	m	**6.26**
open valleys; cutting both sides	–	0.19	4.35	–	3.45	m	**7.80**
preformed ridge capping	–	0.28	6.40	–	4.74	m	**11.14**
preformed hip capping; cutting both sides	–	0.46	10.53	–	8.19	m	**18.72**
double starter course to cappings	–	0.09	2.06	–	4.50	m	**6.56**
holes for pipes and the like	–	0.14	3.20	–	–	nr	**3.20**

19 WATERPROOFING

Item	PC £	Labour hours	Labour £	Plant £	Material £	Unit	Total rate £
MASTIC ASPHALT ROOFING							
Mastic asphalt to BS 6925 Type R 988							
20 mm thick two coat coverings; felt isolating membrane; to concrete (or timber) base; flat or to falls or slopes not exceeding 10° from horizontal							
over 300 mm wide	–	–	–	–	–	m²	19.95
225 mm–300 mm wide	–	–	–	–	–	m²	30.81
150 mm–225 mm wide	–	–	–	–	–	m²	36.00
not exceeding 150 mm wide	–	–	–	–	–	m²	46.33
Add to the above for covering with:							
10 mm thick limestone chippings in hot bitumen	–	–	–	–	–	m²	3.25
coverings with solar reflective paint	–	–	–	–	–	m²	3.66
300 mm × 300 mm × 8 mm g.r.p. tiles in hot bitumen	–	–	–	–	–	m²	55.18
Cutting to line; jointing to old asphalt	–	–	–	–	–	m	6.32
13 mm thick two coat skirtings to brickwork base							
not exceeding 150 mm girth	–	–	–	–	–	m	13.60
150 mm–225 mm girth	–	–	–	–	–	m	15.61
225 mm–300 mm girth	–	–	–	–	–	m	19.10
13 mm thick three coat skirtings; expanded metal lathing reinforcement nailed to timber base							
not exceeding 150 mm girth	–	–	–	–	–	m	22.86
150 mm–225 mm girth	–	–	–	–	–	m	27.23
225 mm–300 mm girth	–	–	–	–	–	m	31.86
13 mm thick two coat fascias to concrete base							
not exceeding 150 mm girth	–	–	–	–	–	m	13.60
150 mm–225 mm girth	–	–	–	–	–	m	15.61
20 mm thick two coat linings to channels to concrete base							
not exceeding 150 mm girth	–	–	–	–	–	m	29.90
150 mm–225 mm girth	–	–	–	–	–	m	34.00
225 mm–300 mm girth	–	–	–	–	–	m	34.99
20 mm thick two coat lining to cesspools							
250 mm × 150 mm × 150 mm deep	–	–	–	–	–	nr	29.30
Collars around pipes, standards and like members	–	–	–	–	–	nr	20.96
Accessories							
Eaves trim; extruded aluminium alloy; working asphalt into trim							
Alutrim; type A roof edging or other equal and approved	–	–	–	–	–	m	12.49
extra; angle	–	–	–	–	–	nr	6.99
Roof screed ventilator – aluminium alloy							
Extr-aqua-vent or other equal and approved; set on screed over and including dished sinking; working collar around ventilator	–	–	–	–	–	nr	24.21

19 WATERPROOFING

Item	PC £	Labour hours	Labour £	Plant £	Material £	Unit	Total rate £
MASTIC ASPHALT ROOFING – cont							
Bituminous lightweight insulating roof screeds							
Bit-Ag or similar roof screed or other equal and approved; to falls or cross-falls; bitumen felt vapour barrier; over 300 mm wide							
75 mm (average) thick	–	–	–	–	–	m²	**49.75**
100 mm (average) thick	–	–	–	–	–	m²	**63.05**
APPLIED LIQUID APPLIED TANKING							
Tanking and damp-proofing							
Synthaprufe or other equal and approved; blinding with sand; horizontal on slabs							
two coats	–	0.19	2.95	–	3.36	m²	**6.31**
three coats	–	0.26	4.03	–	4.96	m²	**8.99**
One coat Vandex Super 0.75 kg/m² slurry or other equal and approved; one consolidating coat of Vandex BB75 1 kg/m² slurry or other equal and approved; horizontal on beds							
over 225 mm wide	–	0.32	4.97	–	8.62	m²	**13.59**
Intergritank; Methacrylate resin based structural waterproffing membrane; in two separate colour coded coats; minimumm 2 mm overalll dry film finish; on a primed substrate. Typical project size between 250 m² and 10,000 m²							
over 250 mm wide							
100 m²–499 m²	–	–	–	–	–	m²	**47.25**
500 m²–1999 m²	–	–	–	–	–	m²	**39.00**
over 2000 m²	–	–	–	–	–	m²	**31.50**
SPECIALIST WATERPROOF RENDERING							
Sika waterproof rendering or other equal; steel trowelled							
20 mm work to walls; three coat; to concrete base							
over 300 mm wide	–	–	–	–	–	m²	**34.56**
not exceeding 300 mm wide	–	–	–	–	–	m²	**52.36**
25 mm work to walls; three coat; to concrete base							
over 300 mm wide	–	–	–	–	–	m²	**40.84**
not exceeding 300 mm wide	–	–	–	–	–	m²	**62.83**
40 mm work to walls; four coat; to concrete base							
over 300 mm wide	–	–	–	–	–	m²	**60.22**
not exceeding 300 mm wide	–	–	–	–	–	m²	**94.26**

19 WATERPROOFING

Item	PC £	Labour hours	Labour £	Plant £	Material £	Unit	Total rate £
Sto External render only system; comprising glassfibre mesh reinforcement embedded in 10 mm Sto Levell Cote with Sto Armat Classic Basecoat Render and Stolit K 1.5 Decorative Topcoat Render (white)							
15 mm thick work to walls; two coats; to brickwork or blockwork base							
over 300 mm wide	–	–	–	–	–	m²	52.52
extra over for;							
bellcast bead	–	–	–	–	–	m	5.00
external angle with PVC mesh angle bead	–	–	–	–	–	m	4.61
internal angle with Sto Armor angle	–	–	–	–	–	m	4.61
render stop bead	–	–	–	–	–	m	4.61
K-Rend render or similar through-colour render system							
18 mm thick work to walls; two coats; to brickwork or blockwork base; first coat 8 mm standard base coat; second coat 10 mm K-rend silicone WP/FT							
over 300 mm wide	–	–	–	–	–	m²	61.76
MASTIC ASPHALT TANKING AND DAMP-PROOF MEMBRANES							
Mastic asphalt to BS 6925 Type T 1097							
13 mm thick one coat coverings to concrete base; flat; subsequently covered							
over 300 mm wide	–	–	–	–	–	m²	14.99
225 mm–300 mm wide	–	–	–	–	–	m²	43.06
150 mm–225 mm wide	–	–	–	–	–	m²	47.22
not exceeding 150 mm wide	–	–	–	–	–	m²	58.99
20 mm thick two coat coverings to concrete base; flat; subsequently covered							
over 300 mm wide	–	–	–	–	–	m²	18.87
225 mm–300 mm wide	–	–	–	–	–	m²	38.87
150 mm–225 mm wide	–	–	–	–	–	m²	54.38
not exceeding 150 mm wide	–	–	–	–	–	m²	63.54
30 mm thick three coat coverings to concrete base; flat; subsequently covered							
over 300 mm wide	–	–	–	–	–	m²	30.27
225 mm–300 mm wide	–	–	–	–	–	m²	62.39
150 mm–225 mm wide	–	–	–	–	–	m²	67.70
not exceeding 150 mm wide	–	–	–	–	–	m²	82.49
13 mm thick two coat coverings to brickwork base; vertical; subsequently covered							
over 300 mm wide	–	–	–	–	–	m²	41.65
225 mm–300 mm wide	–	–	–	–	–	m²	59.88
150 mm–225 mm wide	–	–	–	–	–	m²	64.66
not exceeding 150 mm wide	–	–	–	–	–	m²	84.50

19 WATERPROOFING

Item	PC £	Labour hours	Labour £	Plant £	Material £	Unit	Total rate £
MASTIC ASPHALT TANKING AND DAMP-PROOF MEMBRANES – cont							
Mastic asphalt to BS 6925 Type T 1097 – cont							
20 mm thick three coat coverings to brickwork base; vertical; subsequently covered							
over 300 mm wide	–	–	–	–	–	m²	**67.38**
225 mm–300 mm wide	–	–	–	–	–	m²	**80.67**
150 mm–225 mm wide	–	–	–	–	–	m²	**88.52**
not exceeding 150 mm wide	–	–	–	–	–	m²	**114.77**
Turning into groove 20 mm deep	–	–	–	–	–	m	**0.80**
Internal angle fillets; subsequently covered	–	–	–	–	–	m	**4.68**
FLEXIBLE SHEET TANKING AND DAMP-PROOF MEMBRANES							
Sheet tanking							
Preprufe pre-applied self-adhesive waterproofing membranes for use below concrete slabs or behind concrete walls. HPE film with a pressure sensitive adhesive and weather resistant protective coating							
Preprufe 300R heavy duty grade for use below slabs and on rafts							
over 300 mm wide; horizontal	–	0.10	1.55	–	14.41	m²	**15.96**
not exceeding 300 mm wide; horizontal	–	0.11	1.71	–	6.04	m	**7.75**
Preprufe 160R thinner grade for use blindside, zero property line applications against soil retention							
over 300 mm wide; horizontal	–	0.10	1.55	–	12.65	m²	**14.20**
not exceeding 300 mm wide; horizontal	–	0.11	1.71	–	5.30	m	**7.01**
Preprufe 800PA reinforced cross laminated HDPE film for use below ground car parks, basements, underground reservoirs and tanks							
over 300 mm wide; horizontal	–	0.15	2.33	–	8.37	m²	**10.70**
not exceeding 300 mm wide; horizontal	–	0.17	2.56	–	3.51	m	**6.07**
Visqueen self-adhesive damp-proof membrane							
over 300 mm wide; horizontal	–	–	–	–	–	m²	**6.70**
not exceeding 300 mm wide; horizontal	–	–	–	–	–	m	**2.58**
Tanking primer for self-adhesive dpm							
over 300 mm wide; horizontal	–	–	–	–	–	m²	**4.63**
not exceeding 300 mm wide; horizontal	–	–	–	–	–	m	**2.11**
Bituthene sheeting or other equal and approved; lapped joints; horizontal on slabs							
8000 grade	–	0.10	1.55	–	8.12	m²	**9.67**
5000HD heavy duty grade	–	0.12	1.86	–	7.51	m²	**9.37**
Bituthene sheeting or other equal and approved; lapped joints; dressed up vertical face of concrete							
8000 grade	–	0.17	2.64	–	8.12	m²	**10.76**

19 WATERPROOFING

Item	PC £	Labour hours	Labour £	Plant £	Material £	Unit	Total rate £
RIW Structureseal tanking and damp-proof membrane; or other equal and approved							
over 300 mm wide; horizontal	–	–	–	–	–	m²	**7.26**
Structureseal Fillet							
40 mm × 40 mm	–	–	–	–	–	m	**5.48**
Ruberoid Plasfrufe 2000SA self-adhesive damp-proof membrane							
over 300 mm wide; horizontal	–	–	–	–	–	m²	**12.74**
not exceeding 300 mm wide; horizontal	–	–	–	–	–	m	**4.89**
Extra for 50 mm thick sand blinding	–	–	–	–	–	m²	**2.24**
Servipak protection board or other equal; butt jointed; taped joints; to horizontal surfaces;							
3 mm thick	–	0.14	2.17	–	9.43	m²	**11.60**
6 mm thick	–	0.14	2.17	–	10.77	m²	**12.94**
12 mm thick	–	0.19	2.95	–	18.44	m²	**21.39**
Servipak protection board or other equal; butt jointed; taped joints; to vertical surfaces							
3 mm thick	–	0.19	2.95	–	9.43	m²	**12.38**
6 mm thick	–	0.19	2.95	–	10.77	m²	**13.72**
12 mm thick	–	0.23	3.57	–	18.44	m²	**22.01**
Expandite Famflex hot bitumen bonded waterproof tanking or other equal and approved; 150 mm laps							
horizontal; over 300 mm wide	–	0.37	5.74	–	12.62	m²	**18.36**
vertical; over 300 mm wide	–	0.60	9.31	–	12.62	m²	**21.93**

20 PROPRIETARY LININGS AND PARTITIONS

Item	PC £	Labour hours	Labour £	Plant £	Material £	Unit	Total rate £
PLASTERBOARD DRY LINING WALLS AND PARTITIONS							
SUPPLY ONLY SHEET LINING – ALTERNATIVE MATERIAL PRICES							
Fibreboard; 19 mm Decorative faced							
Ash	–	–	–	–	11.05	m²	**11.05**
Beech	–	–	–	–	10.56	m²	**10.56**
Oak	–	–	–	–	10.89	m²	**10.89**
Edgings; self-adhesive							
22 mm Ash	–	–	–	–	0.32	m	**0.32**
22 mm Beech	–	–	–	–	0.32	m	**0.32**
22 mm Oak	–	–	–	–	0.32	m	**0.32**
Chipboard Standard Grade							
12 mm	–	–	–	–	2.08	m²	**2.08**
18 mm	–	–	–	–	2.94	m²	**2.94**
22 mm	–	–	–	–	3.60	m²	**3.60**
25 mm	–	–	–	–	4.12	m²	**4.12**
Chipboard; melamine faced							
15 mm	–	–	–	–	3.07	m²	**3.07**
18 mm	–	–	–	–	3.34	m²	**3.34**
Medium density fibreboard; external quality							
6 mm	–	–	–	–	4.34	m²	**4.34**
9 mm	–	–	–	–	5.76	m²	**5.76**
19 mm	–	–	–	–	9.33	m²	**9.33**
25 mm	–	–	–	–	13.03	m²	**13.03**
Wallboard plank							
9.5 mm	–	–	–	–	2.65	m²	**2.65**
12.5 mm	–	–	–	–	2.65	m²	**2.65**
15 mm	–	–	–	–	3.18	m²	**3.18**
Moisture-resistant board							
12.5 mm	–	–	–	–	4.33	m²	**4.33**
15 mm	–	–	–	–	5.21	m²	**5.21**
Fireline board							
12.5 mm	–	–	–	–	3.38	m²	**3.38**
15 mm	–	–	–	–	4.06	m²	**4.06**
SUPPLY AND FIX PRICES							
Dry wall linings							
Linings; Gyproc GypLyner IWL independent walling system or other equal; comprising 48 mm wide metal I stud frame; 50 mm wide metal C stud floor and head channels; plugged and screwed to concrete							
62.5 mm partition; outer skin of 12.50 mm thick tapered edge wallboard one side; joints filled with joint filler and joint tape to receive direct decoration							
average height 2.00 m	–	0.53	10.70	–	7.74	m²	**18.44**
average height 3.00 m	–	0.53	10.70	–	7.61	m²	**18.31**
average height 4.00 m	–	0.53	10.70	–	7.58	m²	**18.28**

20 PROPRIETARY LININGS AND PARTITIONS

Item	PC £	Labour hours	Labour £	Plant £	Material £	Unit	**Total rate £**
Rate per m (SMM7 measurement rule)							
height 2.10 m–2.40 m	–	1.30	26.25	–	17.57	m	**43.82**
height 2.40 m–2.70 m	–	1.45	29.28	–	20.88	m	**50.16**
height 2.70 m–3.00 m	–	1.60	32.31	–	22.84	m	**55.15**
height 3.00 m–3.30 m	–	1.75	35.34	–	24.84	m	**60.18**
height 3.30 m–3.60 m	–	1.90	38.37	–	26.90	m	**65.27**
height 3.60 m–3.90 m	–	2.05	41.40	–	28.96	m	**70.36**
height 3.90 m–4.20 m	–	2.20	44.43	–	31.02	m	**75.45**
62.5 mm partition; outer skin of 15.00 mm thick tapered edge wallboard one side; joints filled with joint filler and joint tape to receive direct decoration							
average height 2.00 m	–	0.53	10.70	–	8.58	m²	**19.28**
average height 3.00 m	–	0.53	10.70	–	8.33	m²	**19.03**
average height 4.00 m	–	0.53	10.70	–	8.02	m²	**18.72**
Rate per m (SMM7 measurement rule)							
height 2.10 m–2.40 m	–	1.30	26.25	–	18.92	m	**45.17**
height 2.40 m–2.70 m	–	1.45	29.28	–	22.42	m	**51.70**
height 2.70 m–3.00 m	–	1.60	32.31	–	24.53	m	**56.84**
height 3.00 m–3.30 m	–	1.75	35.34	–	26.71	m	**62.05**
height 3.30 m–3.60 m	–	1.90	38.37	–	28.94	m	**67.31**
height 3.60 m–3.90 m	–	2.05	41.40	–	31.16	m	**72.56**
height 3.90 m–4.20 m	–	2.20	44.43	–	33.39	m	**77.82**
62.5 mm partition; outer skin of 12.50 mm thick tapered edge Fireline board one side; joints filled with joint filler and joint tape to receive direct decoration							
average height 2.00 m	–	0.53	10.70	–	8.63	m²	**19.33**
average height 3.00 m	–	0.53	10.70	–	8.26	m²	**18.96**
average height 4.00 m	–	0.53	10.70	–	8.10	m²	**18.80**
Rate per m (SMM7 measurement rule)							
height 2.10 m–2.40 m	–	1.30	26.52	–	19.37	m	**45.89**
height 2.40 m–2.70 m	–	1.45	29.56	–	22.92	m	**52.48**
height 2.70 m–3.00 m	–	1.60	32.58	–	25.09	m	**57.67**
height 3.00 m–3.30 m	–	1.75	35.62	–	27.33	m	**62.95**
height 3.30 m–3.60 m	–	1.90	38.64	–	29.61	m	**68.25**
height 3.60 m–3.90 m	–	2.05	41.67	–	31.90	m	**73.57**
height 3.90 m–4.20 m	–	2.20	44.70	–	34.18	m	**78.88**
62.5 mm partition; outer skin of 15.00 mm thick tapered edge Fireline board one side; joints filled with joint filler and joint tape to receive direct decoration							
average height 2.00 m	–	0.53	10.70	–	9.33	m²	**20.03**
average height 3.00 m	–	0.53	10.70	–	8.97	m²	**19.67**
average height 4.00 m	–	0.53	10.70	–	8.80	m²	**19.50**
Rate per m (SMM7 measurement rule)							
height 2.10 m–2.40 m	–	1.30	26.52	–	21.06	m	**47.58**
height 2.40 m–2.70 m	–	1.45	29.56	–	24.82	m	**54.38**
height 2.70 m–3.00 m	–	1.60	32.58	–	27.21	m	**59.79**
height 3.00 m–3.30 m	–	1.75	35.62	–	29.65	m	**65.27**
height 3.30 m–3.60 m	–	1.90	38.64	–	32.14	m	**70.78**
height 3.60 m–3.90 m	–	2.05	41.67	–	34.63	m	**76.30**
height 3.90 m–4.20 m	–	2.20	44.70	–	37.13	m	**81.83**

20 PROPRIETARY LININGS AND PARTITIONS

Item	PC £	Labour hours	Labour £	Plant £	Material £	Unit	Total rate £
PLASTERBOARD DRY LINING WALLS AND PARTITIONS – cont							
Linings – cont							
62.5 mm partition; outer skin of 12.50 mm thick tapered edge wallboard one side; filling cavity with 50 mm Isover insulation; wallboard joints filled with joint filler and joint tape to receive direct decoration							
average height 2.00 m	–	0.65	13.13	–	9.54	m²	**22.67**
average height 3.00 m	–	0.65	13.13	–	9.16	m²	**22.29**
average height 4.00 m	–	0.65	13.13	–	9.07	m²	**22.20**
Rate per m (SMM7 measurement rule)							
height 2.10 m–2.40 m	–	1.42	28.95	–	21.63	m	**50.58**
height 2.40 m–2.70 m	–	1.60	32.58	–	25.45	m	**58.03**
height 2.70 m–3.00 m	–	1.65	33.33	–	27.90	m	**61.23**
height 3.00 m–3.30 m	–	1.92	38.94	–	30.42	m	**69.36**
height 3.30 m–3.60 m	–	2.08	42.27	–	32.98	m	**75.25**
height 3.60 m–3.90 m	–	2.25	45.61	–	35.54	m	**81.15**
height 3.90 m–4.20 m	–	2.41	48.94	–	38.10	m	**87.04**
62.5 mm partition; outer skin of 15.00 mm thick tapered edge wallboard one side; filling cavity with 50 mm Isover slabs; wallboard joints filled with joint filler and joint tape to receive direct decoration							
average height 2.00 m	–	0.65	13.13	–	10.11	m²	**23.24**
average height 3.00 m	–	0.65	13.13	–	9.72	m²	**22.85**
average height 4.00 m	–	0.65	13.13	–	9.64	m²	**22.77**
Rate per m (SMM7 measurement rule)							
height 2.10 m–2.40 m	–	1.42	28.95	–	22.98	m	**51.93**
height 2.40 m–2.70 m	–	1.60	32.58	–	26.97	m	**59.55**
height 2.70 m–3.00 m	–	1.65	33.33	–	29.60	m	**62.93**
height 3.00 m–3.30 m	–	1.92	38.94	–	32.28	m	**71.22**
height 3.30 m–3.60 m	–	2.08	42.27	–	35.02	m	**77.29**
height 3.60 m–3.90 m	–	2.25	45.61	–	37.75	m	**83.36**
height 3.90 m–4.20 m	–	2.41	48.94	–	40.48	m	**89.42**
Gypsum plasterboard to BS EN 520; fixing on dabs or with nails; joints filled with joint filler and joint tape to receive direct decoration; to softwood base (measured elsewhere)							
Plain grade tapered edge wallboard							
9.50 mm board to walls							
average height 2.00 m	–	0.16	3.23	–	3.69	m²	**6.92**
average height 3.00 m	–	0.16	3.13	–	3.70	m²	**6.83**
average height 4.00 m	–	0.15	3.03	–	3.70	m²	**6.73**
Rate per m (SMM7 measurement rule)							
wall height 2.40 m–2.70 m	–	0.43	8.96	–	9.95	m	**18.91**
wall height 2.70 m–3.00 m	–	0.46	9.55	–	11.08	m	**20.63**
wall height 3.00 m–3.30 m	–	0.50	10.36	–	12.18	m	**22.54**
wall height 3.30 m–3.60 m	–	0.59	12.18	–	13.37	m	**25.55**

20 PROPRIETARY LININGS AND PARTITIONS

Item	PC £	Labour hours	Labour £	Plant £	Material £	Unit	Total rate £
9.50 mm board to reveals and soffits of openings and recesses							
not exceeding 300 mm wide	–	0.14	2.96	–	2.00	m	**4.96**
300 mm–600 mm wide	–	0.22	4.58	–	2.88	m	**7.46**
9.50 mm board to faces of columns – 4 nr faces							
not exceeding 300 mm total girth	–	0.20	4.04	–	3.16	m	**7.20**
300 mm–600 mm total girth	–	0.20	4.04	–	4.05	m	**8.09**
600 mm–900 mm total girth	–	0.20	4.04	–	4.96	m	**9.00**
900 mm–1200 mm total girth	–	0.20	4.04	–	5.86	m	**9.90**
12.50 mm board to walls							
average height 2.00 m	–	0.16	3.23	–	5.59	m²	**8.82**
average height 3.00 m	–	0.16	3.13	–	5.59	m²	**8.72**
average height 4.00 m	–	0.15	3.03	–	5.59	m²	**8.62**
Rate per m (SMM7 measurement rule)							
wall height 2.40 m–2.70 m	–	0.45	9.09	–	15.22	m	**24.31**
wall height 2.70 m–3.00 m	–	0.50	10.10	–	16.92	m	**27.02**
wall height 3.00 m–3.30 m	–	0.55	11.11	–	18.71	m	**29.82**
wall height 3.30 m–3.60 m	–	0.60	12.12	–	20.38	m	**32.50**
12.50 mm board to reveals and soffits of openings and recesses							
not exceeding 300 mm wide	–	0.19	4.11	–	2.60	m	**6.71**
300 mm–600 mm wide	–	0.37	8.01	–	4.05	m	**12.06**
12.50 mm board to faces of columns – 4 nr faces							
not exceeding 300 mm total girth	–	0.30	6.06	–	3.71	m	**9.77**
300 mm–600 mm total girth	–	0.30	6.06	–	5.26	m	**11.32**
600 mm–900 mm total girth	–	0.30	6.06	–	6.75	m	**12.81**
900 mm–1200 mm total girth	–	0.30	6.06	–	8.24	m	**14.30**
Tapered edge wallboard TEN							
12.50 mm board to walls							
average height 2.00 m	–	0.16	3.23	–	4.17	m²	**7.40**
average height 3.00 m	–	0.16	3.13	–	4.17	m²	**7.30**
average height 4.00 m	–	0.15	3.03	–	4.17	m²	**7.20**
Rate per m (SMM7 measurement rule)							
wall height 2.40 m–2.70 m	–	0.47	9.77	–	11.40	m	**21.17**
wall height 2.70 m–3.00 m	–	0.56	11.72	–	12.67	m	**24.39**
wall height 3.00 m–3.30 m	–	0.65	13.67	–	13.95	m	**27.62**
wall height 3.30 m–3.60 m	–	0.78	16.42	–	15.29	m	**31.71**
12.50 mm board to reveals and soffits of openings and recesses							
not exceeding 300 mm wide	–	0.19	4.11	–	2.18	m	**6.29**
300 mm–600 mm wide	–	0.37	8.01	–	3.21	m	**11.22**
12.50 mm board to faces of columns – 4 nr faces							
not exceeding 300 mm total girth	–	0.25	5.72	–	4.30	m	**10.02**
300 mm–600 mm total girth	–	0.20	4.04	–	5.31	m	**9.35**
600 mm–900 mm total girth	–	0.20	4.04	–	6.42	m	**10.46**
900 mm–1200 mm total girth	–	0.20	4.04	–	7.62	m	**11.66**

20 PROPRIETARY LININGS AND PARTITIONS

Item	PC £	Labour hours	Labour £	Plant £	Material £	Unit	Total rate £
PLASTERBOARD DRY LINING WALLS AND PARTITIONS – cont							
Tapered edge plank							
19 mm plank to walls							
average height 2.00 m	–	0.16	3.23	–	6.75	m²	**9.98**
average height 3.00 m	–	0.16	3.13	–	6.75	m²	**9.88**
average height 4.00 m	–	0.15	3.03	–	6.75	m²	**9.78**
Rate per m (SMM7 measurement rule)							
wall height 2.40 m–2.70 m	–	1.02	22.22	–	18.36	m	**40.58**
wall height 2.70 m–3.00 m	–	1.20	26.12	–	20.40	m	**46.52**
wall height 3.00 m–3.30 m	–	1.30	28.41	–	22.45	m	**50.86**
wall height 3.30 m–3.60 m	–	1.53	33.33	–	24.56	m	**57.89**
19 mm plank to reveals and soffits of openings and recesses							
not exceeding 300 mm wide	–	0.20	4.30	–	2.96	m	**7.26**
300 mm–600 mm wide	–	0.42	9.02	–	4.76	m	**13.78**
19 mm plank to faces of columns – 4 nr faces							
not exceeding 300 mm total girth	–	0.80	18.31	–	4.98	m	**23.29**
300 mm–600 mm total girth	–	0.80	16.16	–	6.96	m	**23.12**
600 mm–900 mm total girth	–	0.80	16.16	–	8.95	m	**25.11**
900 mm–1200 mm total girth	–	0.80	16.16	–	10.83	m	**26.99**
ThermaLine plus board (0.19 W/mK)							
27 mm board to walls							
average height 2.00 m	–	0.16	3.23	–	10.77	m²	**14.00**
average height 3.00 m	–	0.16	3.13	–	10.77	m²	**13.90**
average height 4.00 m	–	0.15	3.03	–	10.77	m²	**13.80**
Rate per m (SMM7 measurement rule)							
wall height 2.40 m–2.70 m	–	1.06	21.41	–	29.22	m	**50.63**
wall height 2.70 m–3.00 m	–	1.23	24.84	–	32.48	m	**57.32**
wall height 3.00 m–3.30 m	–	1.34	27.06	–	35.73	m	**62.79**
wall height 3.30 m–3.60 m	–	1.62	32.72	–	39.04	m	**71.76**
27 mm board to reveals and soffits of openings and recesses							
not exceeding 300 mm wide	–	0.21	4.24	–	4.17	m	**8.41**
300 mm–600 mm wide	–	0.43	8.68	–	7.17	m	**15.85**
27 mm board to faces of columns – 4 nr faces							
not exceeding 300 mm total girth	–	0.52	10.50	–	8.37	m	**18.87**
300 mm–600 mm total girth	–	0.65	13.13	–	8.37	m	**21.50**
600 mm–900 mm total girth	–	0.75	15.15	–	14.40	m	**29.55**
900 mm–1200 mm total girth	–	1.00	20.19	–	14.46	m	**34.65**
48 mm board to walls							
average height 2.00 m	–	0.16	3.23	–	14.55	m²	**17.78**
average height 3.00 m	–	0.16	3.13	–	14.55	m²	**17.68**
average height 4.00 m	–	0.15	3.03	–	14.55	m²	**17.58**
Rate per m (SMM7 measurement rule)							
wall height 2.40 m–2.70 m	–	1.06	21.41	–	39.55	m	**60.96**
wall height 2.70 m–3.00 m	–	1.30	26.25	–	43.97	m	**70.22**
wall height 3.00 m–3.30 m	–	1.43	28.88	–	48.37	m	**77.25**
wall height 3.30 m–3.60 m	–	1.71	34.53	–	52.84	m	**87.37**

20 PROPRIETARY LININGS AND PARTITIONS

Item	PC £	Labour hours	Labour £	Plant £	Material £	Unit	Total rate £
48 mm board to reveals and soffits of openings and recesses							
not exceeding 300 mm wide	–	0.23	4.64	–	5.32	m	**9.96**
300 mm–600 mm wide	–	0.46	9.29	–	9.48	m	**18.77**
48 mm board to faces of columns – 4 nr faces							
not exceeding 300 mm total girth	–	0.52	10.50	–	10.64	m	**21.14**
300 mm–600 mm total girth	–	0.65	13.13	–	10.64	m	**23.77**
600 mm–900 mm total girth	–	0.75	15.15	–	18.93	m	**34.08**
900 mm–1200 mm total girth	–	1.00	20.19	–	18.99	m	**39.18**
ThermaLine super boards							
50 mm board to walls							
average height 2.00 m	–	0.40	8.07	–	19.82	m²	**27.89**
average height 3.00 m	–	0.43	8.68	–	19.82	m²	**28.50**
average height 4.00 m	–	0.47	9.60	–	19.82	m²	**29.42**
Rate per m (SMM7 measurement rule)							
wall height 2.40 m–2.70 m	–	1.06	21.41	–	53.50	m	**74.91**
wall height 2.70 m–3.00 m	–	1.30	26.25	–	59.46	m	**85.71**
wall height 3.00 m–3.30 m	–	1.43	28.88	–	65.41	m	**94.29**
wall height 3.30 m–3.60 m	–	1.71	34.53	–	71.43	m	**105.96**
50 mm board to reveals and soffits of openings and recesses							
not exceeding 300 mm wide	–	0.52	10.50	–	6.87	m	**17.37**
300 mm–600 mm wide	–	0.46	9.29	–	12.58	m	**21.87**
50 mm board to faces of columns – 4 nr faces							
not exceeding 300 mm total girth	–	0.52	10.50	–	8.16	m	**18.66**
300–600 mm total girth	–	0.65	13.13	–	13.82	m	**26.95**
600 mm–900 mm total girth	–	0.75	15.15	–	19.57	m	**34.72**
900 mm–1200 mm total girth	–	1.00	20.19	–	25.32	m	**45.51**
60 mm board to walls							
average height 2.00 m	–	0.40	8.07	–	23.38	m²	**31.45**
average height 3.00 m	–	0.43	8.68	–	23.38	m²	**32.06**
average height 4.00 m	–	0.47	9.60	–	23.38	m²	**32.98**
Rate per m (SMM7 measurement rule)							
wall height 2.40 m–2.70 m	–	1.06	21.41	–	63.12	m	**84.53**
wall height 2.70 m–3.00 m	–	1.30	26.25	–	70.14	m	**96.39**
wall height 3.00 m–3.30 m	–	1.43	28.88	–	77.16	m	**106.04**
wall height 3.30 m–3.60 m	–	1.71	34.53	–	84.25	m	**118.78**
60 mm board to reveals and soffits of openings and recesses							
not exceeding 300 mm wide	–	0.23	4.64	–	7.94	m	**12.58**
300 mm–600 mm wide	–	0.46	9.29	–	14.71	m	**24.00**
60 mm board to faces of columns – 4 nr faces							
not exceeding 300 mm total girth	–	0.52	10.50	–	9.22	m	**19.72**
300–600 mm total girth	–	0.65	13.13	–	15.96	m	**29.09**
600 mm–900 mm total girth	–	0.75	15.15	–	22.77	m	**37.92**
900 mm–1200 mm total girth	–	1.00	20.19	–	29.59	m	**49.78**
80 mm board to walls							
average height 2.00 m	–	0.40	8.07	–	28.43	m²	**36.50**
average height 3.00 m	–	0.43	8.68	–	28.43	m²	**37.11**
average height 4.00 m	–	0.47	9.60	–	28.43	m²	**38.03**

20 PROPRIETARY LININGS AND PARTITIONS

Item	PC £	Labour hours	Labour £	Plant £	Material £	Unit	Total rate £
PLASTERBOARD DRY LINING WALLS AND PARTITIONS – cont							
ThermaLine super boards – cont							
Rate per m (SMM7 measurement rule)							
wall height 2.40 m–2.70 m	–	1.06	21.41	–	76.74	m	**98.15**
wall height 2.70 m–3.00 m	–	1.30	26.25	–	85.29	m	**111.54**
wall height 3.00 m–3.30 m	–	1.43	28.88	–	93.82	m	**122.70**
wall height 3.30 m–3.60 m	–	1.71	34.53	–	102.43	m	**136.96**
80 mm board to reveals and soffits of openings and recesses							
not exceeding 300 mm wide	–	0.23	4.64	–	9.45	m	**14.09**
300 mm–600 mm wide	–	0.46	9.29	–	17.74	m	**27.03**
80 mm board to faces of columns – 4 nr faces							
not exceeding 300 mm total girth	–	0.52	10.50	–	10.74	m	**21.24**
300–600 mm total girth	–	0.65	13.13	–	18.98	m	**32.11**
600 mm–900 mm total girth	–	0.75	15.15	–	27.32	m	**42.47**
900 mm–1200 mm total girth	–	1.00	20.19	–	35.65	m	**55.84**
90 mm board to walls							
average height 2.00 m	–	0.40	8.07	–	32.55	m²	**40.62**
average height 3.00 m	–	0.43	8.68	–	32.55	m²	**41.23**
average height 4.00 m	–	0.47	9.60	–	32.55	m²	**42.15**
Rate per m (SMM7 measurement rule)							
wall height 2.40 m–2.70 m	–	1.06	21.41	–	87.85	m	**109.26**
wall height 2.70 m–3.00 m	–	1.30	26.25	–	97.64	m	**123.89**
wall height 3.00 m–3.30 m	–	1.43	28.88	–	107.40	m	**136.28**
wall height 3.30 m–3.60 m	–	1.71	34.53	–	117.24	m	**151.77**
90 mm board to reveals and soffits of openings and recesses							
not exceeding 300 mm wide	–	0.23	4.64	–	10.69	m	**15.33**
300 mm–600 mm wide	–	0.46	9.29	–	20.21	m	**29.50**
90 mm board to faces of columns – 4 nr faces							
not exceeding 300 mm total girth	–	0.52	10.50	–	11.97	m	**22.47**
300–600 mm total girth	–	0.65	13.13	–	21.45	m	**34.58**
600 mm–900 mm total girth	–	0.75	15.15	–	31.02	m	**46.17**
900 mm–1200 mm total girth	–	1.00	20.19	–	40.59	m	**60.78**
Kingspan Kooltherm K18 insulated plasterboard; fixing with nails; joints filled with joint filler and joint tape to receive direct decoration; to softwood base (measured elsewhere)							
12.5 mm plasterboard bonded to CFC/HCFC free rigid phenolic insulation; overall thickness (0.21 W/m.K)							
32.5 mm thick panel	–	0.23	4.55	–	9.75	m²	**14.30**
37.5 mm thick panel	–	0.23	4.55	–	10.01	m²	**14.56**
42.5 mm thick panel	–	0.23	4.55	–	11.58	m²	**16.13**
52.5 mm thick panel	–	0.25	5.05	–	10.82	m²	**15.87**
62.5 mm thick panel	–	0.25	5.05	–	12.19	m²	**17.24**
72.5 mm thick panel	–	0.25	5.05	–	12.75	m²	**17.80**
82.5 mm thick panel	–	0.25	5.05	–	16.84	m²	**21.89**
92.5 mm thick panel	–	0.28	5.55	–	19.27	m²	**24.82**
112.5 mm thick panel	–	0.28	5.55	–	24.14	m²	**29.69**

20 PROPRIETARY LININGS AND PARTITIONS

Item	PC £	Labour hours	Labour £	Plant £	Material £	Unit	Total rate £
White plastic faced gypsum plasterboard to BS EN 520; industrial grade square edge wallboard; fixing on dabs or with screws; butt joints; to softwood base							
12.50 mm board to walls							
average height 2.00 m	–	0.21	4.24	–	7.55	m²	**11.79**
average height 3.00 m	–	0.25	5.05	–	7.55	m²	**12.60**
average height 4.00 m	–	0.31	6.22	–	7.55	m²	**13.77**
Rate per m (SMM7 measurement rule)							
wall height 2.40 m–2.70 m	–	0.69	13.93	–	20.39	m	**34.32**
wall height 2.70 m–3.00 m	–	0.83	16.76	–	22.65	m	**39.41**
wall height 3.00 m–3.30 m	–	0.97	19.59	–	24.91	m	**44.50**
wall height 3.30 m–3.60 m	–	1.11	22.42	–	27.16	m	**49.58**
12.50 mm board to reveals and soffits of openings and recesses							
not exceeding 300 mm wide	–	0.15	3.03	–	2.29	m	**5.32**
300 mm–600 mm wide	–	0.30	6.06	–	4.54	m	**10.60**
12.50 mm board to faces of columns – 4 nr							
not exceeding 300 mm total girth	–	0.30	6.06	–	2.39	m	**8.45**
300–600 mm total girth	–	0.39	7.88	–	4.04	m	**11.92**
600 mm–900 mm total girth	–	0.60	12.12	–	6.89	m	**19.01**
900 mm–1200 mm total girth	–	0.75	15.15	–	9.13	m	**24.28**
Plasterboard jointing system; filling joint with jointing compounds							
To walls and ceilings							
to suit 9.50 mm or 12.50 mm thick boards	–	0.09	1.82	–	1.87	m	**3.69**
Angle trim; plasterboard edge support system							
To walls and ceilings							
to suit 9.50 mm or 12.50 mm thick boards	–	0.09	1.82	–	1.80	m	**3.62**
Gyproc SoundBloc tapered edge plasterboard with higher density core; fixing on dabs or with nails; joints filled with joint filler and joint tape to receive direct decoration; to softwood base							
12.50 mm board to walls							
average height 2.00 m	–	0.36	7.27	–	6.29	m²	**13.56**
average height 3.00 m	–	0.38	7.68	–	6.29	m²	**13.97**
average height 4.00 m	–	0.40	8.07	–	6.37	m²	**14.44**
Rate per m (SMM7 measurement rule)							
wall height 2.40 m–2.70 m	–	0.97	19.59	–	16.97	m	**36.56**
wall height 2.70 m–3.00 m	–	1.11	22.42	–	18.87	m	**41.29**
wall height 3.00 m–3.30 m	–	1.32	26.66	–	20.76	m	**47.42**
wall height 3.30 m–3.60 m	–	1.43	28.88	–	22.71	m	**51.59**
12.50 mm board to ceilings							
over 300 mm wide	–	0.41	8.28	–	6.29	m²	**14.57**
15.00 mm board to walls							
average height 2.00 m	–	0.36	7.27	–	7.40	m²	**14.67**
average height 3.00 m	–	0.38	7.68	–	7.40	m²	**15.08**
average height 4.00 m	–	0.40	8.07	–	7.49	m²	**15.56**

20 PROPRIETARY LININGS AND PARTITIONS

Item	PC £	Labour hours	Labour £	Plant £	Material £	Unit	Total rate £
PLASTERBOARD DRY LINING WALLS AND PARTITIONS – cont							
Gyproc SoundBloc tapered edge plasterboard with higher density core – cont							
Rate per m (SMM7 measurement rule)							
wall height 2.40 m–2.70 m	–	1.00	20.19	–	20.12	m	**40.31**
wall height 2.70 m–3.00 m	–	1.14	23.03	–	22.35	m	**45.38**
wall height 3.00 m–3.30 m	–	1.27	25.65	–	24.60	m	**50.25**
wall height 3.30 m–3.60 m	–	1.46	29.48	–	26.90	m	**56.38**
15.00 mm board to reveals and soffits of openings and recesses							
not exceeding 300 mm wide	–	0.20	4.04	–	3.15	m	**7.19**
300 mm–600 mm wide	–	0.38	7.68	–	5.15	m	**12.83**
15.00 mm board to ceilings							
over 300 mm wide	–	0.43	8.68	–	7.45	m²	**16.13**
Two layers of gypsum plasterboard to BS 1230; plain grade square and tapered edge wallboard; fixing on dabs or with nails; joints filled with joint filler and joint tape; top layer to receive direct decoration; to softwood base							
19 mm two layer board to walls							
average height 2.00 m	–	0.48	9.69	–	6.63	m²	**16.32**
average height 3.00 m	–	0.51	10.30	–	6.63	m²	**16.93**
average height 4.00 m	–	0.54	10.91	–	6.71	m²	**17.62**
Rate per m (SMM7 measurement rule)							
wall height 2.40 m–2.70 m	–	1.30	26.25	–	17.87	m	**44.12**
wall height 2.70 m–3.00 m	–	1.48	29.89	–	19.87	m	**49.76**
wall height 3.00 m–3.30 m	–	1.67	33.73	–	21.86	m	**55.59**
wall height 3.30 m–3.60 m	–	1.94	39.18	–	23.92	m	**63.10**
19 mm two layer board to reveals and soffits of openings and recesses							
not exceeding 300 mm wide	–	0.28	5.66	–	2.94	m	**8.60**
300 mm–600 mm wide	–	0.56	11.31	–	4.67	m	**15.98**
19 mm two layer board to faces of columns – 4 nr							
not exceeding 300 mm total girth	–	0.50	10.10	–	6.10	m	**16.20**
300–600 mm total girth	–	0.69	13.93	–	5.95	m	**19.88**
600 mm–900 mm total girth	–	0.75	15.15	–	7.64	m	**22.79**
900 mm–1200 mm total girth	–	1.00	20.19	–	10.49	m	**30.68**
25 mm two layer board to walls							
average height 2.00 m	–	0.48	9.69	–	8.52	m²	**18.21**
average height 3.00 m	–	0.51	10.30	–	8.52	m²	**18.82**
average height 4.00 m	–	0.54	10.91	–	8.61	m²	**19.52**
Rate per m (SMM7 measurement rule)							
wall height 2.40 m–2.70 m	–	1.39	28.07	–	23.13	m	**51.20**
wall height 2.70 m–3.00 m	–	1.57	31.71	–	25.71	m	**57.42**
wall height 3.00 m–3.30 m	–	1.76	35.54	–	28.30	m	**63.84**
wall height 3.30 m–3.60 m	–	2.04	41.20	–	30.93	m	**72.13**

20 PROPRIETARY LININGS AND PARTITIONS

Item	PC £	Labour hours	Labour £	Plant £	Material £	Unit	Total rate £
25 mm two layer board to reveals and soffits of openings and recesses							
not exceeding 300 mm wide	–	0.28	5.66	–	3.55	m	**9.21**
300 mm–600 mm wide	–	0.56	11.31	–	5.85	m	**17.16**
25 mm two layer board to faces of columns – 4 nr							
not exceeding 300 mm total girth	–	0.35	7.07	–	4.66	m	**11.73**
300–600 mm total girth	–	0.69	13.93	–	7.17	m	**21.10**
600 mm–900 mm total girth	–	1.00	20.19	–	9.41	m	**29.60**
900 mm–1200 mm total girth	–	1.25	25.28	–	11.45	m	**36.73**
Gyproc Dri-Wall dry lining system or other equal or approved; plain grade tapered edge wallboard; fixed to walls with adhesive; joints filled with joint filler and joint tape; to receive direct decoration							
9.50 mm board to walls							
average height 2.00 m	–	0.41	8.31	–	4.97	m²	**13.28**
average height 3.00 m	–	0.43	8.75	–	4.97	m²	**13.72**
average height 4.00 m	–	0.51	10.22	–	4.97	m²	**15.19**
Rate per m (SMM7 measurement rule)							
wall height 2.40 m–2.70 m	–	1.11	22.42	–	13.42	m	**35.84**
wall height 2.70 m–3.00 m	–	1.28	25.85	–	14.89	m	**40.74**
wall height 3.00 m–3.30 m	–	1.43	28.88	–	16.36	m	**45.24**
wall height 3.30 m–3.60 m	–	1.67	33.73	–	17.90	m	**51.63**
9.50 mm board to reveals and soffits of openings and recesses							
not exceeding 300 mm wide	–	0.23	4.91	–	2.33	m	**7.24**
300 mm–600 mm wide	–	0.46	9.83	–	3.58	m	**13.41**
9.50 mm board to faces of columns – 4 nr faces							
300 mm total girth	–	0.45	9.09	–	2.38	m	**11.47**
300–600 mm total girth	–	0.58	11.72	–	4.65	m	**16.37**
600 mm–900 mm total girth	–	0.75	15.15	–	5.57	m	**20.72**
900 mm–1200 mm total girth	–	0.95	19.18	–	6.43	m	**25.61**
Angle; with joint tape bedded and covered with Jointex or other equal							
internal	–	0.05	1.14	–	0.54	m	**1.68**
external	–	0.11	2.52	–	0.54	m	**3.06**
Gyproc Dri-Wall M/F dry lining system or other equal; mild steel furrings fixed to walls with adhesive; tapered edge wallboard screwed to furrings; joints filled with joint filler and joint tape							
9.50 mm board to walls							
average height 2.00 m	–	0.55	11.11	–	7.90	m²	**19.01**
average height 3.00 m	–	0.57	11.61	–	7.89	m²	**19.50**
average height 4.00 m	–	0.62	12.45	–	7.91	m²	**20.36**
Rate per m (SMM7 measurement rule)							
wall height 2.40 m–2.70 m	–	1.48	29.89	–	21.30	m	**51.19**
wall height 2.70 m–3.00 m	–	1.69	36.01	–	23.66	m	**59.67**
wall height 3.00 m–3.30 m	–	1.90	38.37	–	26.03	m	**64.40**
wall height 3.30 m–3.60 m	–	2.22	44.84	–	28.47	m	**73.31**

20 PROPRIETARY LININGS AND PARTITIONS

Item	PC £	Labour hours	Labour £	Plant £	Material £	Unit	Total rate £
PLASTERBOARD DRY LINING WALLS AND PARTITIONS – cont							
Gyproc Dri-Wall M/F dry lining system or other equal – cont							
9.50 mm board to reveals and soffits of openings and recesses							
not exceeding 300 mm wide	–	0.23	4.91	–	2.04	m	**6.95**
300 mm–600 mm wide	–	0.46	9.83	–	2.99	m	**12.82**
12.50 mm board to walls							
average height 2.00 m	–	0.55	11.11	–	9.79	m²	**20.90**
average height 3.00 m	–	0.57	11.61	–	9.79	m²	**21.40**
average height 4.00 m	–	0.62	12.45	–	9.81	m²	**22.26**
Rate per m (SMM7 measurement rule)							
wall height 2.40 m–2.70 m	–	1.48	29.89	–	26.43	m	**56.32**
wall height 2.70 m–3.00 m	–	1.69	36.01	–	29.35	m	**65.36**
wall height 3.00 m–3.30 m	–	1.90	38.37	–	32.29	m	**70.66**
wall height 3.30 m–3.60 m	–	2.22	44.84	–	35.29	m	**80.13**
12.50 mm board to reveals and soffits of openings and recesses							
not exceeding 300 mm wide	–	0.23	4.91	–	2.60	m	**7.51**
300 mm–600 mm wide	–	0.46	9.83	–	4.13	m	**13.96**
Lafarge plasterboard to BS 1230; fixing on dabs or with screws; joints filled with joint filler and joint tape to receive direct decoration; to softwood							
Megadeco wallboard							
12.50 mm board to walls							
average height 2.00 m	–	0.36	7.27	–	3.98	m²	**11.25**
average height 3.00 m	–	0.38	7.65	–	3.98	m²	**11.63**
average height 4.00 m	–	0.40	8.07	–	3.98	m²	**12.05**
Rate per m (SMM7 measurement rule)							
wall height 2.40 m–2.70 m	–	0.97	19.59	–	12.44	m	**32.03**
wall height 2.70 m–3.00 m	–	1.11	24.31	–	13.84	m	**38.15**
wall height 3.00 m–3.30 m	–	1.25	25.24	–	15.22	m	**40.46**
wall height 3.30 m–3.60 m	–	1.43	28.88	–	16.60	m	**45.48**
12.50 mm board to ceilings							
over 300 mm wide	–	0.41	8.82	–	4.61	m²	**13.43**
15 mm board to walls							
average height 2.00 m	–	0.38	7.63	–	4.61	m²	**12.24**
average height 3.00 m	–	0.40	8.03	–	4.61	m²	**12.64**
average height 4.00 m	–	0.42	8.48	–	4.61	m²	**13.09**
Rate per m (SMM7 measurement rule)							
wall height 2.40 m–2.70 m	–	1.02	20.57	–	12.44	m	**33.01**
wall height 2.70 m–3.00 m	–	1.17	25.52	–	13.84	m	**39.36**
wall height 3.00 m–3.30 m	–	1.31	26.50	–	15.22	m	**41.72**
wall height 3.30 m–3.60 m	–	1.50	30.32	–	16.60	m	**46.92**
15 mm board to ceilings							
over 300 mm wide	–	0.43	9.26	–	4.61	m²	**13.87**

20 PROPRIETARY LININGS AND PARTITIONS

Item	PC £	Labour hours	Labour £	Plant £	Material £	Unit	Total rate £
Gypsum cladding; Glasroc Firecase S board or other equal; fixed with adhesive; joints pointed in adhesive							
25 mm thick column linings, faces = 4; 2 hour fire protection rating							
not exceeding 300 mm girth	–	0.30	6.06	–	18.35	m	**24.41**
300 mm–600 mm girth	–	0.30	6.06	–	24.98	m	**31.04**
600 mm–900 mm girth	–	0.45	9.09	–	31.59	m	**40.68**
900 mm–1200 mm girth	–	0.55	11.11	–	38.21	m	**49.32**
1200 mm–1500 mm girth	–	0.60	12.12	–	44.82	m	**56.94**
30 mm thick beam linings, faces = 3; 2 hour fire protection rating							
not exceeding 300 mm girth	–	0.60	12.12	–	16.64	m	**28.76**
300 mm–600 mm girth	–	0.60	12.12	–	24.12	m	**36.24**
600 mm–900 mm girth	–	0.75	15.15	–	31.59	m	**46.74**
900 mm–1200 mm girth	–	0.90	18.18	–	39.07	m	**57.25**
1200 mm–1500 mm girth	–	1.20	24.23	–	46.55	m	**70.78**
Vermiculite gypsum cladding; Vermiculux board or other equal approved; fixed with adhesive; joints pointed in adhesive							
25 mm thick column linings, faces = 4; 2 hour fire protection rating							
not exceeding 300 mm girth	–	0.20	4.04	–	10.43	m	**14.47**
300 mm–600 mm girth	–	0.30	6.06	–	20.85	m	**26.91**
600 mm–900 mm girth	–	0.45	9.09	–	30.93	m	**40.02**
900 mm–1200 mm girth	–	0.60	12.12	–	41.24	m	**53.36**
1200 mm–1500 mm girth	–	0.70	14.13	–	51.44	m	**65.57**
30 mm thick beam linings, faces = 3; 2 hour fire protection rating							
not exceeding 300 mm girth	–	0.50	10.10	–	13.70	m	**23.80**
300 mm–600 mm girth	–	0.60	12.12	–	27.42	m	**39.54**
600 mm–900 mm girth	–	0.90	18.18	–	43.92	m	**62.10**
900 mm–1200 mm girth	–	1.00	20.19	–	54.27	m	**74.46**
1200 mm–1500 mm girth	–	1.20	24.23	–	67.75	m	**91.98**
55 mm thick column linings, faces = 4 ; 4 hour fire protection rating							
not exceeding 300 mm girth	–	0.60	12.12	–	29.18	m	**41.30**
300 mm–600 mm girth	–	0.75	15.15	–	58.37	m	**73.52**
600 mm–900 mm girth	–	0.80	16.16	–	87.20	m	**103.36**
900 mm–1200 mm girth	–	1.00	20.19	–	116.27	m	**136.46**
1200 mm–1500 mm girth	–	1.02	20.60	–	145.23	m	**165.83**
60 mm thick beam linings, faces = 3; 4 hour fire protection rating							
not exceeding 300 mm girth	–	0.60	12.12	–	31.54	m	**43.66**
300 mm–600 mm girth	–	0.75	15.15	–	63.08	m	**78.23**
600 mm–900 mm girth	–	0.80	16.16	–	94.27	m	**110.43**
900 mm–1200 mm girth	–	1.00	20.19	–	125.70	m	**145.89**
1200 mm–1500 mm girth	–	1.02	20.60	–	157.01	m	**177.61**

20 PROPRIETARY LININGS AND PARTITIONS

Item	PC £	Labour hours	Labour £	Plant £	Material £	Unit	Total rate £
PLASTERBOARD DRY LINING WALLS AND PARTITIONS – cont							
Vermiculite gypsum cladding – cont							
Add to the above rates for working at height							
for work 3.50 m–5.00 m high	–	–	–	–	–	%	**2.63**
for work 5.00 m–6.50 m high	–	–	–	–	–	%	**5.25**
for work 6.50 m–8.00 m high	–	–	–	–	–	%	**13.13**
for work over 8.00 m high	–	–	–	–	–	%	**18.38**
Cutting and fitting around steel joints, angles, trunking, ducting, ventilators, pipes, tubes, etc.							
not exceeding 0.30 m girth	–	0.28	5.66	–	–	nr	**5.66**
0.30 m–1 m girth	–	0.37	7.48	–	–	nr	**7.48**
1 m–2 m girth	–	0.51	10.30	–	–	nr	**10.30**
over 2 m girth	–	0.42	8.48	–	–	m	**8.48**
PARTITIONING							
Gyproc metal stud proprietary partitions or other equal; metal Gypframe C studs at 600 mm centres; floor and ceiling channels							
73 mm partition; 48 mm studs and channels; one layer of 12.5 mm Gyproc Wallboard each side; joints filled with joint filler and joint tape to receive direct decoration							
average height 2.00 m	–	0.63	14.31	–	14.02	m²	**28.33**
average height 3.00 m	–	0.58	13.36	–	12.98	m²	**26.34**
Rate per m (SMM7 measurement rule)							
height 2.10 m–2.40 m	–	1.50	34.34	–	33.64	m	**67.98**
height 2.40 m–2.70 m	–	1.69	38.69	–	36.49	m	**75.18**
height 2.70 m–3.00 m	–	1.75	40.06	–	38.92	m	**78.98**
78 mm partition; 48 mm studs and channels; one layer of 15 mm Gyproc Wallboard each side; joints filled with joint filler and joint tape to receive direct decoration							
average height 2.00 m	–	0.65	14.88	–	15.20	m²	**30.08**
average height 3.00 m	–	0.61	13.97	–	14.15	m²	**28.12**
Rate per m (SMM7 measurement rule)							
height 2.10 m–2.40 m	–	1.57	36.06	–	36.49	m	**72.55**
height 2.40 m–2.70 m	–	1.77	40.52	–	39.69	m	**80.21**
height 2.70 m–3.00 m	–	1.99	45.56	–	46.53	m	**92.09**
95 mm partition; 70 mm studs and channels; one layer of 12.5 mm Gyproc Wallboard each side; joints filled with joint filler and joint tape to receive direct decoration							
average height 2.00 m	–	0.63	14.31	–	14.31	m²	**28.62**
average height 3.00 m	–	0.58	13.36	–	13.43	m²	**26.79**

20 PROPRIETARY LININGS AND PARTITIONS

Item	PC £	Labour hours	Labour £	Plant £	Material £	Unit	Total rate £
Rate per m (SMM7 measurement rule)							
height 2.10 m–2.40 m	–	1.50	34.34	–	34.36	m	**68.70**
height 2.40 m–2.70 m	–	1.69	38.69	–	37.37	m	**76.06**
height 2.70 m–3.00 m	–	1.75	40.06	–	40.30	m	**80.36**
height 3.00 m–3.30 m	–	2.00	45.78	–	43.21	m	**88.99**
100 mm partition; 70 mm studs and channels; one layer of 15 mm Gyproc Wallboard each side; joints filled with joint filler and joint tape to receive direct decoration							
average height 2.00 m	–	0.65	14.88	–	15.48	m²	**30.36**
average height 3.00 m	–	0.61	13.97	–	15.88	m²	**29.85**
Rate per m (SMM7 measurement rule)							
height 2.10 m–2.40 m	–	1.57	36.06	–	37.20	m	**73.26**
height 2.40 m–2.70 m	–	1.77	40.52	–	40.56	m	**81.08**
height 2.70 m–3.00 m	–	1.99	45.56	–	47.62	m	**93.18**
height 3.00 m–3.30 m	–	2.10	48.07	–	50.60	m	**98.67**
120 mm partition; 70 mm studs and channels; two layers of 12.5 mm Gyproc Wallboard each side; joints filled with joint filler and joint tape to receive direct decoration							
average height 2.00 m	–	0.73	16.72	–	20.25	m²	**36.97**
average height 3.00 m	–	0.68	15.65	–	17.21	m²	**32.86**
Rate per m (SMM7 measurement rule)							
height 2.10 m–2.40 m	–	1.75	40.06	–	48.61	m	**88.67**
height 2.40 m–2.70 m	–	1.99	45.56	–	53.41	m	**98.97**
height 2.70 m–3.00 m	–	2.05	46.92	–	58.13	m	**105.05**
height 3.00 m–3.30 m	–	2.50	57.24	–	62.82	m	**120.06**
130 mm partition; 70 mm studs and channels; two layers of 15 mm Gyproc Wallboard each side; joints filled with joint filler and joint tape to receive direct decoration							
average height 2.00 m	–	0.80	18.31	–	22.63	m²	**40.94**
average height 3.00 m	–	0.75	17.17	–	21.82	m²	**38.99**
Rate per m (SMM7 measurement rule)							
height 2.10 m–2.40 m	–	1.82	41.66	–	54.31	m	**95.97**
height 2.40 m–2.70 m	–	2.07	47.39	–	40.91	m	**88.30**
height 2.70 m–3.00 m	–	2.29	52.43	–	69.01	m	**121.44**
height 3.00 m–3.30 m	–	2.40	54.95	–	74.13	m	**129.08**
Gypwall metal stud proprietary partitions or other equal; metal GypWall Acoustic C studs at 600 mm centres; deep flange floor and ceiling channels							
95 mm partition; 70 mm studs and channels; one layer of 12.5 mm Gyproc Soundbloc each side; joints filled with joint filler and joint tape to receive direct decoration							
average height 2.00 m	–	0.63	14.31	–	24.66	m²	**38.97**
average height 3.00 m	–	0.58	13.36	–	23.86	m²	**37.22**

20 PROPRIETARY LININGS AND PARTITIONS

Item	PC £	Labour hours	Labour £	Plant £	Material £	Unit	Total rate £
PARTITIONING – cont							
Gypwall metal stud proprietary partitions or other equal – cont							
Rate per m (SMM7 measurement rule)							
height 2.10 m – 2.40 m	–	1.50	34.34	–	59.20	m	**93.54**
height 2.40 m – 2.70 m	–	1.69	38.69	–	64.88	m	**103.57**
height 2.70 m – 3.00 m	–	1.75	40.06	–	71.58	m	**111.64**
100 mm partition; 70 mm studs and channels; one layer of 15 mm Gyproc Soundbloc each side; joints filled with joint filler and joint tape to receive direct decoration							
average height 2.00 m	–	0.63	14.31	–	27.00	m²	**41.31**
average height 3.00 m	–	0.58	13.36	–	26.19	m²	**39.55**
Rate per m (SMM7 measurement rule)							
height 2.10 m–2.40 m	–	1.57	36.06	–	64.78	m	**100.84**
height 2.40 m–2.70 m	–	1.77	40.52	–	71.16	m	**111.68**
height 2.70 m–3.00 m	–	1.99	45.56	–	77.12	m	**122.68**
120 mm partition; 70 mm studs and channels; two layers of 12.5 mm Gyproc Soundbloc each side; joints filled with joint filler and joint tape to receive direct decoration							
average height 2.00 m	–	0.80	18.31	–	36.10	m²	**54.41**
average height 3.00 m	–	0.75	17.17	–	35.29	m²	**52.46**
Rate per m (SMM7 measurement rule)							
height 2.10 m–2.40 m	–	2.00	45.78	–	87.08	m	**132.86**
height 2.40 m–2.70 m	–	2.20	50.36	–	96.10	m	**146.46**
height 2.70 m– 3.00 m	–	2.30	52.66	–	106.27	m	**158.93**
130 mm partition; 70 mm studs and channels; two layers of 15 mm Gyproc Soundbloc each side; joints filled with joint filler and joint tape to receive direct decoration							
average height 2.00 m	–	0.80	18.31	–	40.75	m²	**59.06**
average height 3.00 m	–	0.75	17.17	–	39.94	m²	**57.11**
Rate per m (SMM7 measurement rule)							
height 2.10 m–2.40 m	–	2.00	45.78	–	98.24	m	**144.02**
height 2.40 m–2.70 m	–	2.20	50.36	–	108.65	m	**159.01**
height 2.70 m–3.00 m	–	2.30	52.66	–	120.22	m	**172.88**
Gyproc metal stud proprietary partitions or other equal; metal Gypframe C studs at 600 mm centres; floor and ceiling channels							
95 mm partition; 70 mm studs and channels; one layer of 12.5 mm Gyproc Fireline each side; joints filled with joint filler and joint tape to receive direct decoration							
average height 2.00 m	–	0.80	18.31	–	28.02	m²	**46.33**
average height 3.00 m	–	0.75	17.17	–	27.22	m²	**44.39**
Rate per m (SMM7 measurement rule)							
height 2.10 m–2.40 m	–	1.50	34.34	–	49.51	m	**83.85**
height 2.40 m–2.70 m	–	1.70	38.91	–	53.98	m	**92.89**
height 2.70 m–3.00 m	–	1.75	40.06	–	58.03	m	**98.09**

20 PROPRIETARY LININGS AND PARTITIONS

Item	PC £	Labour hours	Labour £	Plant £	Material £	Unit	Total rate £
100 mm partition; 70 mm studs and channels; one layer of 15 mm Gyproc Fireline each side; joints filled with joint filler and joint tape to receive direct decoration							
average height 2.00 m	–	0.80	18.31	–	30.96	m²	**49.27**
average height 3.00 m	–	0.75	17.17	–	30.16	m²	**47.33**
Rate per m (SMM7 measurement rule)							
height 2.10 m–2.40 m	–	1.50	34.34	–	53.04	m	**87.38**
height 2.40 m–2.70 m	–	1.70	38.91	–	57.95	m	**96.86**
height 2.70 m–3.00 m	–	1.75	40.06	–	62.44	m	**102.50**
120 mm partition; 70 mm studs and channels; two layers of 12.5 mm Gyproc Fireline each side; joints filled with joint filler and joint tape to receive direct decoration							
average height 2.00 m	–	1.10	25.18	–	28.15	m²	**53.33**
average height 3.00 m	–	1.15	26.32	–	26.27	m²	**52.59**
Rate per m (SMM7 measurement rule)							
height 2.10 m–2.40 m	–	2.00	45.78	–	67.57	m	**113.35**
height 2.40 m–2.70 m	–	2.20	50.36	–	74.29	m	**124.65**
height 2.70 m–3.00 m	–	2.30	52.66	–	78.80	m	**131.46**
130 mm partition; 70 mm studs and channels; two layers of 15 mm Gyproc Fireline each side; joints filled with joint filler and joint tape to receive direct decoration							
average height 2.00 m	–	1.10	25.18	–	31.09	m²	**56.27**
average height 3.00 m	–	1.15	26.32	–	29.21	m²	**55.53**
Rate per m (SMM7 measurement rule)							
height 2.10 m–2.40 m	–	2.00	45.78	–	74.62	m	**120.40**
height 2.40 m–2.70 m	–	2.20	50.36	–	82.24	m	**132.60**
height 2.70 m–3.00 m	–	2.30	52.66	–	89.00	m	**141.66**
Gypwall Rapid/db Plus metal stud housing partitioning system; or other equal; floor and ceiling channels plugged and screwed to concrete or nailed to timber							
75 mm partition; 43 mm studs at 600 mm centres; one layer of 15 mm SoundBloc Rapid each side; joints filled with joint filler and joint tape to receive direct decoration							
average height 2.00 m	–	0.63	12.62	–	21.17	m²	**33.79**
average height 3.00 m	–	0.83	16.83	–	27.32	m²	**44.15**
Rate per m (SMM7 measurement rule)							
height 2.10 m–2.40 m	–	1.50	30.29	–	50.80	m	**81.09**
height 2.10 m–2.40 m	–	1.75	35.34	–	58.56	m	**93.90**
height 2.40 m–2.70 m	–	2.00	40.39	–	64.80	m	**105.19**
Angles, corners and similar							
T-junctions	–	0.17	3.54	–	8.21	m	**11.75**
splayed corners	–	0.13	2.52	–	5.50	m	**8.02**
corners	–	0.10	2.02	–	2.79	m	**4.81**
fair ends	–	0.11	2.23	–	0.82	m	**3.05**

20 PROPRIETARY LININGS AND PARTITIONS

Item	PC £	Labour hours	Labour £	Plant £	Material £	Unit	Total rate £
PARTITIONING – cont							
Cavity insulation to stud partitions pinned to board							
25 mm Isover APR							
average height 2.00 m	–	0.06	1.26	–	1.13	m	**2.39**
average height 3.00	–	0.07	1.42	–	1.13	m	**2.55**
Rate per m (SMM7 measurement rule)							
height 2.10 m–2.40 m	–	0.15	3.03	–	2.72	m	**5.75**
height 2.40 m–2.70 m	–	0.16	3.23	–	3.06	m	**6.29**
height 3.00 m–3.30 m	–	0.17	3.43	–	3.74	m	**7.17**
50 mm Isover APR							
average height 2.00 m	–	0.06	1.26	–	1.86	m	**3.12**
average height 3.00 m	–	0.07	1.42	–	1.86	m	**3.28**
Rate per m (SMM7 measurement rule)							
height 2.10 m–2.40 m	–	0.15	3.03	–	4.46	m	**7.49**
height 2.40 m–2.70 m	–	0.16	3.23	–	5.02	m	**8.25**
height 3.00 m–3.30 m	–	0.17	3.43	–	6.13	m	**9.56**
GLASS REINFORCED GYPSUM LININGS							
Glass reinforced gypsum Glasroc Multi-board or other equal and approved; fixing with nails; joints filled with joint filler and joint tape; finishing with Jointex or other equal and approved to receive decoration; to softwood base							
10 mm board to walls							
average wall height 2.00 m	–	0.37	7.40	–	26.41	m²	**33.81**
average wall height 3.00 m	–	0.36	7.34	–	26.41	m²	**33.75**
average wall height 4.00 m	–	0.36	7.27	–	26.41	m²	**33.68**
Rate per m (SMM7 measurement rule)							
wall height 2.40 m–2.70 m	–	0.93	18.78	–	71.30	m	**90.08**
wall height 2.70 m–3.00 m	–	1.06	21.41	–	79.22	m	**100.63**
wall height 3.00 m–3.30 m	–	1.07	21.55	–	87.15	m	**108.70**
wall height 3.30 m–3.60 m	–	1.39	28.07	–	95.15	m	**123.22**
12.50 mm board to walls							
average wall height 2.00 m	–	0.39	7.78	–	34.49	m²	**42.27**
average wall height 3.00 m	–	0.38	7.72	–	34.49	m²	**42.21**
average wall height 4.00 m	–	0.38	7.63	–	34.49	m²	**42.12**
Rate per m (SMM7 measurement rule)							
wall height 2.40 m–2.70 m	–	0.97	19.59	–	93.27	m	**112.86**
wall height 2.70 m–3.00 m	–	1.11	22.42	–	103.65	m	**126.07**
wall height 3.00 m–3.30 m	–	1.25	25.24	–	114.01	m	**139.25**
wall height 3.30 m–3.60 m	–	1.14	23.08	–	124.45	m	**147.53**

20 PROPRIETARY LININGS AND PARTITIONS

Item	PC £	Labour hours	Labour £	Plant £	Material £	Unit	Total rate £
RIGID BOARD LININGS							
Blockboard (Birch faced)							
Lining to walls 18 mm thick							
over 600 wide	6.86	0.46	9.29	–	7.58	m²	**16.87**
not exceeding 600 wide	–	0.60	12.12	–	4.60	m	**16.72**
holes for pipes and the like	–	0.04	0.81	–	–	nr	**0.81**
Chipboard (plain)							
Lining to walls 12 mm thick							
over 600 mm wide	2.53	0.35	7.07	–	3.02	m²	**10.09**
not exceeding 600 mm wide	–	0.40	8.07	–	1.73	m	**9.80**
holes for pipes and the like	–	0.02	0.40	–	–	nr	**0.40**
Lining to walls 15 mm thick							
over 600 mm wide	2.98	0.37	7.48	–	3.50	m²	**10.98**
not exceeding 600 mm wide	–	0.44	8.88	–	2.15	m	**11.03**
holes for pipes and the like	–	0.03	0.61	–	–	nr	**0.61**
Lining to walls 18 mm thick							
over 600 mm wide	3.56	0.39	7.88	–	4.21	m²	**12.09**
not exceeding 600 mm wide	–	0.50	10.10	–	2.53	m	**12.63**
holes for pipes and the like	–	0.04	0.81	–	–	nr	**0.81**
Fire-retardant chipboard/mdf; Class 1 spread of flame							
Lining to walls 12 mm thick							
over 600 mm wide	–	0.35	7.07	–	6.88	m²	**13.95**
not exceeding 600 mm wide	–	0.40	8.07	–	4.18	m	**12.25**
holes for pipes and the like	–	0.02	0.40	–	–	nr	**0.40**
Lining to walls 18 mm thick							
over 600 mm wide	–	0.39	7.88	–	10.20	m²	**18.08**
not exceeding 600 mm wide	–	0.50	10.10	–	6.17	m	**16.27**
holes for pipes and the like	–	0.04	0.81	–	–	nr	**0.81**
Lining to walls 25 mm thick							
over 600 mm wide	–	0.41	8.28	–	14.29	m²	**22.57**
not exceeding 600 mm wide	–	0.56	11.31	–	8.63	m	**19.94**
holes for pipes and the like	–	0.05	1.01	–	–	nr	**1.01**
Chipboard Melamine faced; white matt finish; laminated masking strips							
Lining to walls 15 mm thick							
over 600 mm wide	3.83	0.97	19.59	–	4.70	m²	**24.29**
not exceeding 600 mm wide	–	1.26	25.44	–	2.90	m	**28.34**
holes for pipes and the like	–	0.06	1.21	–	–	nr	**1.21**
Insulation board to BS EN 622							
Lining to walls 12 mm thick							
over 600 mm wide	2.04	0.22	4.44	–	2.51	m²	**6.95**
not exceeding 600 mm wide	–	0.26	5.25	–	1.56	m	**6.81**
holes for pipes and the like	–	0.01	0.20	–	–	nr	**0.20**

20 PROPRIETARY LININGS AND PARTITIONS

Item	PC £	Labour hours	Labour £	Plant £	Material £	Unit	Total rate £
RIGID BOARD LININGS – cont							
Masterboard or other equal; sanded finish							
Lining to walls 6 mm thick							
over 600 mm wide	11.23	0.31	6.26	–	12.07	m²	**18.33**
not exceeding 600 mm wide	–	0.38	7.68	–	7.27	m	**14.95**
Lining to walls 9 mm thick							
over 600 mm wide	25.96	0.33	6.67	–	27.53	m²	**34.20**
not exceeding 600 mm wide	–	0.38	7.68	–	16.54	m	**24.22**
Supalux or other equal; sanded finish							
Lining to walls 6 mm thick							
over 600 mm wide	21.67	0.31	6.26	–	23.04	m²	**29.30**
not exceeding 600 mm wide	–	0.38	7.68	–	13.84	m	**21.52**
Lining to walls 9 mm thick							
over 600 mm wide	27.59	0.33	6.67	–	29.25	m²	**35.92**
not exceeding 600 mm wide	–	0.38	7.68	–	17.57	m	**25.25**
Lining to walls 12 mm thick							
over 600 mm wide	37.47	0.37	7.48	–	39.62	m²	**47.10**
not exceeding 600 mm wide	–	0.44	8.88	–	23.79	m	**32.67**
Monolux 40 or other equal; 6 mm × 50 mm Supalux cover fillets or other equal and approved one side							
Lining to walls 19 mm thick							
over 600 mm wide	75.62	0.65	13.13	–	82.95	m²	**96.08**
not exceeding 600 mm wide	–	0.92	18.59	–	51.95	m	**70.54**
Lining to walls 25 mm thick							
over 600 mm wide	90.67	0.69	13.93	–	98.76	m²	**112.69**
not exceeding 600 mm wide	54.40	0.98	19.79	–	61.44	m	**81.23**
Cement particle consisting of a mixture of wood fibre, cement and additive, Class 0, Cemboard or similar rigid high performance board as a sheathing							
Lining to walls 8 mm thick							
over 600 mm wide	5.38	0.65	13.13	–	9.20	m²	**22.33**
not exceeding 600 mm wide	–	0.92	18.59	–	7.70	m	**26.29**
Lining to walls 10 mm thick							
over 600 mm wide	6.19	0.65	13.13	–	10.06	m²	**23.19**
not exceeding 600 mm wide	–	0.92	18.59	–	8.21	m	**26.80**
Lining to walls 12 mm thick							
over 600 mm wide	7.34	0.65	13.13	–	11.26	m²	**24.39**
not exceeding 600 mm wide	–	0.92	18.59	–	8.94	m	**27.53**
Lining to walls 16 mm thick							
over 600 mm wide	10.21	0.65	13.13	–	14.27	m²	**27.40**
not exceeding 600 mm wide	–	0.92	18.59	–	10.74	m	**29.33**
Lining to walls 18 mm thick							
over 600 mm wide	11.37	0.65	13.13	–	15.49	m²	**28.62**
not exceeding 600 mm wide	–	0.92	18.59	–	11.48	m	**30.07**

20 PROPRIETARY LININGS AND PARTITIONS

Item	PC £	Labour hours	Labour £	Plant £	Material £	Unit	Total rate £
Lining to walls 24 mm thick							
over 600 mm wide	16.10	0.65	13.13	–	20.45	m²	**33.58**
not exceeding 600 mm wide	–	0.92	18.59	–	14.45	m	**33.04**
Fibre cement board consisting of fibre cement sheet; smooth finish for paint; A1 Non-combustible; high performance board							
Lining to walls 6 mm thick							
over 600 mm wide	5.40	0.65	13.13	–	9.22	m²	**22.35**
not exceeding 600 mm wide	–	0.92	18.59	–	7.71	m	**26.30**
Lining to walls 9 mm thick							
over 600 mm wide	8.43	0.65	13.13	–	12.40	m²	**25.53**
not exceeding 600 mm wide	–	0.92	18.59	–	9.62	m	**28.21**
Lining to walls 12 mm thick							
over 600 mm wide	11.17	0.65	13.13	–	15.28	m²	**28.41**
not exceeding 600 mm wide	–	0.92	18.59	–	11.35	m	**29.94**
Plywood (Eastern European); internal quality							
Lining to walls 4 mm thick							
over 600 mm wide	2.47	0.34	6.87	–	2.96	m²	**9.83**
not exceeding 600 mm wide	–	0.44	8.88	–	1.84	m	**10.72**
Lining to walls 6 mm thick							
over 600 mm wide	3.56	0.37	7.48	–	4.12	m²	**11.60**
not exceeding 600 mm wide	–	0.48	9.69	–	2.52	m	**12.21**
Lining to walls 12 mm thick							
over 600 mm wide	6.62	0.43	8.68	–	7.32	m²	**16.00**
not exceeding 600 mm wide	–	0.56	11.31	–	4.45	m	**15.76**
Lining to walls 18 mm thick							
over 600 mm wide	9.66	0.46	9.29	–	10.52	m²	**19.81**
not exceeding 600 mm wide	–	0.60	12.12	–	6.36	m	**18.48**
Plywood (Eastern European); external quality							
Lining to walls 4 mm thick							
over 600 mm wide	6.11	0.34	6.87	–	6.78	m²	**13.65**
not exceeding 600 mm wide	–	0.44	8.88	–	4.13	m	**13.01**
Lining to walls 6.5 mm thick							
over 600 mm wide	6.83	0.37	7.48	–	7.54	m²	**15.02**
not exceeding 600 mm wide	–	0.48	9.69	–	4.58	m	**14.27**
Lining to walls 9 mm thick							
over 600 mm wide	8.79	0.40	8.07	–	9.60	m²	**17.67**
not exceeding 600 mm wide	–	0.52	10.50	–	5.82	m	**16.32**
Lining to walls 12 mm thick							
over 600 mm wide	10.97	0.43	8.68	–	11.90	m²	**20.58**
not exceeding 600 mm wide	–	0.56	11.31	–	7.19	m	**18.50**
holes for pipes and the like	–	0.03	0.61	–	–	nr	**0.61**
Extra over Linings fixed with screws	–	0.10	2.02	–	0.07	m²	**2.09**

20 PROPRIETARY LININGS AND PARTITIONS

Item	PC £	Labour hours	Labour £	Plant £	Material £	Unit	Total rate £
RIGID BOARD LININGS – cont							
Internal quality American Cherry veneered plywood; 6 mm thick							
Lining to walls							
over 600 mm wide	8.00	0.41	8.28	–	8.68	m²	**16.96**
not exceeding 600 mm wide	–	0.54	10.91	–	5.32	m	**16.23**
Glazed hardboard to BS EN 622; on and including 38 mm × 38 mm sawn softwood framing							
3.20 mm thick panel							
to side of bath	–	1.67	33.73	–	8.62	nr	**42.35**
to end of bath	–	0.65	13.13	–	2.71	nr	**15.84**
RIGID SHEET ACOUSTIC PANEL LININGS							
Perforated steel acoustic wall panels; Eckel type HD EFP or other equal and approved; polyurethene enamel finish; fibrous glass acoustic insulation							
Walls							
average height 3.00 m; fixed to timber or masonry	–	–	–	–	–	m²	**178.79**
DEMOUNTABLE PARTITIONS							
Fire-resistant concertina partition							
Insulated panel and two-hour fire wall system for warehouses etc., comprising white polyester coated galvanized steel frame and 0.55 mm galvanized steel panels either side of rockwool infill							
100 mm thick wall: 31 Rw dB acoustic rating	–	–	–	–	–	m²	**42.49**
150 mm thick wall: 31 Rw dB acoustic rating	–	–	–	–	–	m²	**45.68**
intumescent mastic sealant; bedding frames at							
perimeter of metal fire walls	–	–	–	–	–	m	**3.72**
Room divider moveable wall							
Laminated both sides top hung movable acoustic panel wall with concealed uPVC vertical edge profiles, 1106 m × 3000 mm panels and two point panel support system							
105 mm thick wall: 47 Rw dB acoustic rating	–	–	–	–	–	m²	**393.03**
105 mm thick wall: 50 Rw dB acoustic rating	–	–	–	–	–	m²	**424.90**
105 mm thick wall: 53 Rw dB acoustic rating	–	–	–	–	–	m²	**456.76**
TERRAZZO FACED PARTITIONS							
Terrazzo faced partitions; polished on two faces							
Precast reinforced terrazzo faced WC partitions							
38 mm thick; over 300 mm wide	–	–	–	–	–	m²	**300.00**
50 mm thick; over 300 mm wide	–	–	–	–	–	m²	**310.00**

20 PROPRIETARY LININGS AND PARTITIONS

Item	PC £	Labour hours	Labour £	Plant £	Material £	Unit	Total rate £
Wall post; once rebated							
64 mm × 102 mm	–	–	–	–	–	m	138.00
64 mm × 152 mm	–	–	–	–	–	m	150.00
Centre post; twice rebated							
64 mm × 102 mm	–	–	–	–	–	m	145.00
64 mm × 152 mm	–	–	–	–	–	m	160.00
Lintel; once rebated							
64 mm × 102 mm	–	–	–	–	–	m	150.00
Pair of brass topped plates or sockets cast into posts for fixings (not included)	–	–	–	–	–	nr	35.00
Brass indicator bolt lugs cast into posts for fixings (not included)	–	–	–	–	–	nr	15.00
TIMBER FLOORING AND WALL LININGS							
Wrought softwood							
Boarding to internal walls; tongued and grooved and V-jointed							
12 mm × 100 mm boards	–	0.74	14.94	–	16.94	m²	31.88
16 mm × 100 mm boards	–	0.74	14.94	–	18.39	m²	33.33
19 mm × 100 mm boards	–	0.74	14.94	–	20.92	m²	35.86
19 mm × 125 mm boards	–	0.69	13.93	–	19.11	m²	33.04
19 mm × 125 mm boards; chevron pattern	–	1.11	22.42	–	19.11	m²	41.53
25 mm × 125 mm boards	–	0.69	13.93	–	16.43	m²	30.36
12 mm × 100 mm boards; knotty pine	–	0.74	14.94	–	10.42	m²	25.36
Wood strip; 22 mm thick; Junckers All in Beech Sylva Sport Premium pre-treated or other equal and approved; tongued and grooved joints; on bearers etc.; level fixing to cement and sand base							
Strip flooring; over 300 mm wide							
on 45 × 45 mm blue bat bearers	–	–	–	–	–	m²	63.32
on 10 mm Pro Foam	–	–	–	–	–	m²	72.32
on Uno bat 50 mm bearers	–	–	–	–	–	m²	66.24
on New Era levelling system	–	–	–	–	–	m²	68.38
on Uno bat 62 mm bearers	–	–	–	–	–	m²	67.31
on Duo bat 110 mm bearers	–	–	–	–	–	m²	112.96
Wood strip; 22 mm thick; Junckers pre-treated or other equal and approved flooring systems; tongued and grooved joints; on bearers etc.; level fixing to cement and sand base							
Strip flooring; over 300 mm wide							
Sylva Squash Beech untreated on blue bat bearers	–	–	–	–	–	m²	66.39
Classic Beech clip system	–	–	–	–	–	m²	87.95
Harmoni Oak clip system	–	–	–	–	–	m²	92.09
Classic Beech on blue bat bearers	–	–	–	–	–	m²	88.88
Harmoni Oak on blue bat bearers	–	–	–	–	–	m²	91.02

20 PROPRIETARY LININGS AND PARTITIONS

Item	PC £	Labour hours	Labour £	Plant £	Material £	Unit	Total rate £
TIMBER FLOORING AND WALL LININGS – cont							
Unfinished wood strip; 22 mm thick; Havwoods or other equal and approved; tongued and grooved joints; secret fixed; laid on semi-sprung bearers; fixing to cement and sand base; sanded and sealed							
Strip flooring; over 300 mm wide							
Prime Iroko	–	–	–	–	–	m²	**85.66**
Prime Maple	–	–	–	–	–	m²	**85.66**
American Oak	–	–	–	–	–	m²	**93.90**
FRAMED PANEL CUBICLE PARTITIONS							
Toilet cubicle partitions and IPS systems; Amwells or other equal and approved; standard colours and ironmongery; assembling and screwing to floor and wall							
Axis MFC standard cubic system, for use in small offices, retail and community halls etc.; standard cubicle set; 800 mm × 1500 mm × 1980 mm high per cubicle, with polished aluminium framing; 19 mm melamine-faced chipboard divisions and doors							
One cubicle set; 2 nr panels; 1 nr door	–	3.50	121.11	–	355.35	nr	**476.46**
range of 3 cubicle sets; 4 nr panels; 3 nr doors	–	10.75	371.97	–	854.33	nr	**1226.30**
range of 6 cubicle sets; 7 nr panels; 6 nr doors	–	21.25	735.29	–	1602.78	nr	**2338.07**
reduction of 1 nr panel for end unit adjoining side wall	–	–	–	–	–81.20	nr	**–81.20**
Splash SGL cubic system for heavy use environments like schools, swimming pools and prisons; standard cubicle set; 800 mm × 1500 mm × 1980 mm high per cubicle, with polished aluminium framing; 19 mm melamine-faced chipboard divisions and doors							
One cubicle set; 2 nr panels; 1 nr door	–	3.50	121.11	–	730.75	nr	**851.86**
range of 3 cubicle sets; 4 nr panels; 3 nr doors	–	10.75	371.97	–	1686.09	nr	**2058.06**
range of 6 cubicle sets; 7 nr panels; 6 nr door	–	21.25	735.29	–	3119.10	nr	**3854.39**
reduction of 1 nr panel for end unit adjoining side wall	–	–	–	–	–224.61	nr	**–224.61**
Minima Stylish stainless steel framed cubicle system in MFC, HPL, SGL, Real Wood veneer or Glass; 800 mm × 1500 mm × 2100 mm high per cubicle, with satin polished stainless steel framing; 18 mm high pressure laminated (HPL) chipboard divisions and doors							
One cubicle set; 2 nr panels; 1 nr door	–	3.50	121.11	–	854.11	nr	**975.22**
range of 3 cubicle sets; 4 nr panels; 3 nr doors	–	10.75	371.97	–	1903.72	nr	**2275.69**
range of 6 cubicle set; 7 nr panels; 6 nr doors	–	21.25	735.29	–	3478.14	nr	**4213.43**
reduction of 1 nr panel for end unit adjoining side wall	–	–	–	–	–189.80	nr	**–189.80**

20 PROPRIETARY LININGS AND PARTITIONS

Item	PC £	Labour hours	Labour £	Plant £	Material £	Unit	Total rate £
Urban Flush fronted, floor to ceiling option cubicle system in either 13 mm or 20 mm panels, material options HPL, SGL or Real Wood Veneer: cubicle set; 800 mm × 1500 mm × 2400 mm high per cubicle, with sating finished stainless steel ironmongery; 20 mm divisions, doors and pilasters							
One cubicle set; 2 nr panels; 1 nr door	–	4.00	138.41	–	1300.15	nr	**1438.56**
range of 3 cubicle set; 4 nr panels; 3 nr doors	–	11.75	406.57	–	3198.18	nr	**3604.75**
range of 6 cubicle sets; 7 nr panels; 6 nr doors	–	23.40	809.69	–	6045.24	nr	**6854.93**
reduction of 1 nr panel for end unit adjoining side wall	–	–	–	–	–252.02	nr	**−252.02**
Sylan High end specification flush fronted system floor to ceiling 44 mm doors, no visible fixings. Finishes in Real wood Veneer, HPL, high Gloss paint and lacquer; cubicle set; 800 mm × 1500 mm × 2400 mm high per cubicle, with satin finished stainless steel ironmongery; 30 mm high pressure laminated (HPL) chipboard divisions and 44 mm solid cored real wood veneered doors and pilasters							
One cubicle set; 2 nr panels; 1 nr door	–	5.00	173.01	–	2416.83	nr	**2589.84**
range of 3 cubicle set; 4 nr panels; 3 nr doors	–	15.00	519.04	–	5693.05	nr	**6212.09**
range of 6 cubicle sets; 7 nr panels; 6 nr doors	–	30.00	1038.06	–	10606.84	nr	**11644.90**
reduction of 1 nr panel for end unit adjoining side wall	–	–	–	–	–419.67	nr	**−419.67**
IPS Panel Systems; 12.50 mm melamine faced chipboard; fixed to existing timber subframe							
IPS back panel system to accomodate urinals; to conceal pipework; approx 2.70 m high × 2.10 m wide	–	–	–	–	–	nr	**865.92**
IPS back panel system to accomodate wash hand basins; to conceal pipework; approx 2.70 m high × 5.70 m wide	–	–	–	–	–	nr	**1707.18**
IPS back panel system to accomodate wash hand basins; to conceal pipework; approx 2.70 m high × 2.10 m wide	–	–	–	–	–	nr	**964.62**

21 CLADDING AND COVERING

Item	PC £	Labour hours	Labour £	Plant £	Material £	Unit	Total rate £
ROOF CLADDING – METAL							
Galvanized steel strip troughed sheets; Corus Products or other equal							
Roof cladding or decking; sloping not exceeding 50°; fixing to steel purlins with plastic headed self-tapping screws							
0.7 mm thick; 46 profile	–	–	–	–	–	m²	11.68
0.7 mm thick; 60 profile	–	–	–	–	–	m²	12.79
0.7 mm thick; 100 profile	–	–	–	–	–	m²	13.90
Galvanized steel strip troughed sheets; PMF Strip Mill Products or other equal							
Roof cladding; sloping not exceeding 50°; fixing to steel purlins with plastic headed self-tapping screws							
0.7 mm thick type HPS200 13.5/3 corrugated	–	–	–	–	–	m²	13.17
0.7 mm thick type HPS200 R32/1000	–	–	–	–	–	m²	11.95
0.7 mm thick type Arcline 40; plasticol finished	–	–	–	–	–	m²	17.57
Extra over last for aluminium roof cladding or decking	–	–	–	–	–	m²	7.22
Accessories for roof cladding							
HPS200 Drip flashing; 250 mm girth	–	–	–	–	–	m	4.22
HPS200 Ridge flashing; 375 mm girth	–	–	–	–	–	m	5.34
HPS200 Gable flashing; 500 mm girth	–	–	–	–	–	m	6.79
HPS200 Internal angle; 625 mm girth	–	–	–	–	–	m	7.84
GRP Transluscent rooflights; factory assembled							
Rooflight; vertical fixing to steel purlins (measured elsewhere)							
double skin; class 3 over 1	–	–	–	–	–	m²	48.18
triple skin; class 3 over 1	–	–	–	–	–	m²	53.53
Lightweight galvanized steel roof tiles; Decra Roof Systems; or other equal; coated finish							
Zinc coated tile panels, dry fixed							
Classic profile; clay or concrete tile appearance	–	0.23	5.26	–	20.44	m²	25.70
Stratos profile; slate or concrete tiles appearance	–	0.23	5.26	–	22.91	m²	28.17
Accessories for roof cladding							
pitched D ridge	–	0.09	2.06	–	6.43	m	8.49
barge cover (handed)	–	0.09	2.06	–	7.63	m	9.69
in line air vent	–	0.09	2.06	–	55.41	nr	57.47
in line soil vent	–	0.09	2.06	–	55.41	nr	57.47
gas flue terminal	–	0.19	4.35	–	108.02	nr	112.37
Kingspan KS100RW composite roof panels for roof pitches greater than 4° (after deflection)							
External coating XL Forte (steel), internal coating bright white polyester (steel)							
80 mm thick panel; U-value 0.25	–	–	–	–	–	m²	41.76
115 mm thick panel; U-value 0.18	–	–	–	–	–	m²	43.85
150 mm thick panel; U-value 0.14	–	–	–	–	–	m²	58.88
KS1000PC polycarbonate rooflight	–	–	–	–	–	m²	12.85

21 CLADDING AND COVERING

Item	PC £	Labour hours	Labour £	Plant £	Material £	Unit	Total rate £
Associated flashings							
Eaves – 200 mm girth	–	–	–	–	–	m	3.21
Ridge – 620 mm girth	–	–	–	–	–	m	19.27
Hip – 620 mm girth	–	–	–	–	–	m	19.27
Kingspan KS1000 KZ composite Standing Seam roof panels for roof pitches greater than 1.5° (after deflection)							
External Coating XL Forte (Steel) Internal Coating Bright White Polyester (steel)							
90 mm thick panel; U-value 0.25	–	–	–	–	–	m²	67.45
110 mm thick panel; U-value 0.20	–	–	–	–	–	m²	72.81
125 mm thick panel; U-value 0.18	–	–	–	–	–	m²	76.02
Associated flashings							
Eaves – 200 mm girth	–	–	–	–	–	m	3.21
Ridge – 620 mm girth	–	–	–	–	–	m	19.27
Hip – 620 mm girth	–	–	–	–	–	m	19.27
ROOF – PATENT GLAZING							
Patent glazing; aluminium alloy bars 2.55 m long at 622 mm centres; fixed to supports							
Roof cladding							
single glazed with 6.4 mm laminated glass	–	–	–	–	–	m²	145.00
single glazed with 7 mm thick Georgian wired cast glass	–	–	–	–	–	m²	150.00
thermally broken and double glazed with low-e clear toughened and laminated double glazed units; aluminium finished RAL matt colour	–	–	–	–	–	m²	395.00
Extra for opening roof vents							
600 mm × 900 mm top hung opening roof vent; manually operated	–	–	–	–	–	nr	445.00
600 mm × 900 mm top hung opening roof vent; electrically operated	–	–	–	–	–	nr	570.00
Skylight							
Self-supporting hipped or gable ended lantern/ skylight thermally broken and double glazed with low-e clear toughened and laminated double glazed units; aluminium finished RAL matt colour	–	–	–	–	–	m²	790.00
Associated code 4 lead flashings							
top flashing; 210 mm girth	–	–	–	–	–	m	62.00
bottom flashing; 240 mm girth	–	–	–	–	–	m	71.00
end flashing; 300 mm girth	–	–	–	–	–	m	78.00
Wall cladding							
single glazed with 6.4 mm laminated glass	–	–	–	–	–	m²	144.00
single glazed with 7 mm thick Georgian wired cast glass	–	–	–	–	–	m²	160.00
thermally broken and double glazed with low-e clear toughened and laminated double glazed units; aluminium finished RAL matt colour	–	–	–	–	–	m²	412.00

21 CLADDING AND COVERING

Item	PC £	Labour hours	Labour £	Plant £	Material £	Unit	Total rate £
ROOF – PATENT GLAZING – cont							
Patent glazing – cont							
Extra for aluminium alloy perimeter members							
38 mm × 38 mm × 3 mm angle jamb	–	–	–	–	–	m	**22.00**
pressed sill member	–	–	–	–	–	m	**44.00**
pressed channel head and PVC case	–	–	–	–	–	m	**44.00**
ROOF CLADDING – FIBRE CEMENT SHEET							
Fibre cement corrugated sheets; Eternit 2000 or other equal							
Roof cladding; sloping not exceeding 50°; fixing to steel purlins with hook bolts							
Profile 3; natural grey	–	0.23	5.26	–	19.97	m²	**25.23**
Profile 3; coloured	–	0.23	5.26	–	22.75	m²	**28.01**
Profile 6; natural grey	–	0.28	6.40	–	14.93	m²	**21.33**
Profile 6; coloured	–	0.28	6.40	–	16.66	m²	**23.06**
Profile 6; natural grey; insulated 100 mm glass fibre infill; white finish steel lining panel	–	0.46	10.53	–	30.33	m²	**40.86**
Profile 6; coloured; insulated 100 mm glass fibre infill; white finish steel lining panel	–	0.46	10.53	–	31.61	m²	**42.14**
Accessories; to Profile 3 cladding; natural grey							
eaves filler	–	0.09	2.06	–	11.96	m	**14.02**
external corner piece	–	0.11	2.52	–	8.84	m	**11.36**
apron flashing	–	0.11	2.52	–	11.96	m	**14.48**
plain wing or close fitting two piece adjustable capping to ridge	–	0.16	3.66	–	11.22	m	**14.88**
ventilating two piece adjustable capping to ridge	–	0.16	3.66	–	17.25	m	**20.91**
Accessories; to Profile 6 cladding; natural grey							
eaves filler	–	0.09	2.06	–	7.21	m	**9.27**
external corner piece	–	0.11	2.52	–	8.21	m	**10.73**
apron flashing	–	0.11	2.52	–	8.02	m	**10.54**
underglazing flashing	–	0.11	2.52	–	10.56	m	**13.08**
plain cranked crown to ridge	–	0.16	3.66	–	21.47	m	**25.13**
plain wing or close fitting two piece adjustable capping to ridge	–	0.16	3.66	–	14.42	m	**18.08**
ventilating two piece adjustable capping to ridge	–	0.16	3.66	–	18.44	m	**22.10**
WALL CLADDING – METAL							
Extended, hard skinned, foamed PVC-UE profiled sections; Swish Celuka or other equal and approved; Class 1 fire-rated to BS 476; Part 7; in white finish							
Wall cladding; vertical; fixing to timber							
100 mm shiplap profiles; Code 001	–	0.35	7.07	–	62.67	m²	**69.74**
150 mm shiplap profiles; Code 002	–	0.32	6.46	–	55.53	m²	**61.99**
125 mm feather-edged profiles; Code C208	–	0.34	6.87	–	60.52	m²	**67.39**
vertical angles	–	0.19	3.83	–	6.43	m	**10.26**
raking cutting	–	0.14	2.82	–	–	m	**2.82**

21 CLADDING AND COVERING

Item	PC £	Labour hours	Labour £	Plant £	Material £	Unit	Total rate £
Kingspan KS1000RW Composite Wall Panels							
External coating XL Forte (steel) internal coating							
bright white polyester (steel)							
60 mm thick; U-value 0.35	–	–	–	–	–	m²	**41.72**
80 mm thick; U-value 0.26	–	–	–	–	–	m²	**43.91**
100 mm thick; U-value 0.20	–	–	–	–	–	m²	**46.65**
Kingspan KS1000MR Composite Wall Panels							
External coating Spectum (steel) internal coating							
bright white polyester (steel)							
60 mm thick; U-value 0.35	–	–	–	–	–	m²	**74.64**
80 mm thick; U-value 0.26	–	–	–	–	–	m²	**80.14**
100 mm thick; U-value 0.20	–	–	–	–	–	m²	**85.63**
Kingspan KS900MR Composite Wall Panels							
External coating Spectum (steel) internal coating							
bright white polyester (steel)							
60 mm thick; U-value 0.35	–	–	–	–	–	m²	**85.63**
80 mm thick; U-value 0.26	–	–	–	–	–	m²	**92.21**
100 mm thick; U-value 0.20	–	–	–	–	–	m²	**98.79**
Kingspan KS600MR Composite Wall Panels							
External coating Sprectum (steel) internal coating							
bright white polyester (steel)							
60 mm thick; U-value 0.35	–	–	–	–	–	m²	**115.26**
80 mm thick; U-value 0.26	–	–	–	–	–	m²	**125.14**
100 mm thick; U-value 0.20	–	–	–	–	–	m²	**135.02**
Kingspan KS1000 Optimo FLAT Composite Wall Panels							
External coating Spectum (steel) internal coating							
bright white polyester (steel)							
60 mm thick; U-value 0.35	–	–	–	–	–	m²	**125.14**
80 mm thick; U-value 0.26	–	–	–	–	–	m²	**131.72**
100 mm thick; U-value 0.20	–	–	–	–	–	m²	**138.32**
Kingspan KS900 Optimo FLAT Composite Wall Panels							
External coating Spectrum (steel) internal coating							
bright white polyester (steel)							
60 mm thick; U-value 0.35	–	–	–	–	–	m²	**142.71**
80 mm thick; U-value 0.26	–	–	–	–	–	m²	**149.29**
100 mm thick; U-value 0.20	–	–	–	–	–	m²	**155.87**
Kingspan KS600 Optimo FLAT Composite Wall Panels							
External coating Spectrum (steel) internal coating							
bright white polyester (steel)							
60 mm thick; U-value 0.35	–	–	–	–	–	m²	**186.62**
80 mm thick; U-value 0.26	–	–	–	–	–	m²	**197.59**
100 mm thick; U-value 0.20	–	–	–	–	–	m²	**208.57**

21 CLADDING AND COVERING

Item	PC £	Labour hours	Labour £	Plant £	Material £	Unit	Total rate £
WALL CLADDING – METAL – cont							
Associated Flashings							
0.7 mm pre-coated steel							
cill; 300 mm girth	–	–	–	–	–	m	**19.76**
composite top hat	–	–	–	–	–	m	**23.05**
Miscellaneous items							
panel bearers at 1500 mm centres	–	–	–	–	–	nr	**8.78**
preformed corners to horizontally laid panels	–	–	–	–	–	nr	**98.79**
WALL CLADDING – RIGID SHEET CLADDING							
Resoplan sheet or other equal and approved; Eternit UK Ltd; flexible neoprene gasket joints; fixing with stainless steel screws and coloured caps							
6 mm thick cladding to walls							
over 300 mm wide	–	1.94	39.18	–	57.09	m²	**96.27**
not exceeding 300 mm wide	–	0.65	13.13	–	20.89	m	**34.02**
Eternit 2000 Glasal sheet or other equal and approved; Eternit UK Ltd; flexible neoprene gasket joints; fixing with stainless steel screws and coloured caps							
7.50 mm thick cladding to walls							
over 300 mm wide	–	1.94	39.18	–	50.30	m²	**89.48**
not exceeding 300 mm wide	–	0.65	13.13	–	18.85	m	**31.98**
external angle trim	–	0.09	1.82	–	9.72	m	**11.54**
7.50 mm thick cladding to eaves, verge soffit boards, fascia boards or the like							
100 mm wide	–	0.46	9.29	–	9.63	m	**18.92**
200 mm wide	–	0.56	11.31	–	14.24	m	**25.55**
300 mm wide	–	0.65	13.13	–	18.85	m	**31.98**
WALL CLADDING – WEATHERBOARDING							
Prodema ProdEX high density resin-bonded cellulose fibre weatherboarding panels; including secondary supports and fixing							
Walls							
8 mm Panels face fixed on to timber battens	–	–	–	–	–	m²	**164.80**
8 mm Panels face fixed on to aluminium rails	–	–	–	–	–	m²	**190.55**
8 mm Panels adhesive fixed on to timber battens or aluminium rails	–	–	–	–	–	m²	**195.70**
10 mm Panels secret fixed on to helping hand aluminium system	–	–	–	–	–	m²	**231.75**

21 CLADDING AND COVERING

Item	PC £	Labour hours	Labour £	Plant £	Material £	Unit	Total rate £
WALL CLADDING – RAINSCREEN							
Timber Western Red Cedar tongued and grooved wall cladding on and including treated softwrood battens on breather mambrane, 10 mm Eternit Blueclad board and 50 mm insulation board; the whole fixed to Metsec frame system; including sealing all joints etc.							
26 mm thick cladding to walls; boards laid horizontally	–	–	–	–	–	m²	**95.60**
Aluminium Reynobond rainscreen cladding; aluminium composite material cassettes with thermoplastic cores, back ventilated, including insulation, vapour control membrane and aluminium support system							
4 mm thick cladding; fixed to walls	–	–	–	–	–	m²	**169.96**
Clay Terracotta clay rainscreen cladding; including insulation, vapour control membrane and aluminium support system							
400 × 200 × 30 mm tile cladding; fixed to walls	–	–	–	–	–	m²	**308.05**
CURTAIN WALLING							
NOTE: Toughened and heat soak tested glass costs are comparable with heat strengthened glass. We have removed all reference to toughened glass in this issue due to the ever present risk of nickel sulphide inclusions. Consultation with mechanical services engineers is encouraged to ensure Part L compliance and user comfort criteria are understood, i.e. when to use high performance coated glass and when to add in external shading							
Stick curtain walling system; proprietary solution from system supplier; e.g. Schuco or equivalent Polyester powder coated solid colour matt finish or natural anodized finish mullions spaced 1.5 m apart and spanning typical storey height of 3.8 m. Floor to ceiling glass sealed units with 8.8 mm low-e coated laminated inner pane, air filled cavity and 8 mm clear monolithic heat strengthened outer pane, retained by external pressure plates and caps. Rates to include glass fronted solid spandrel panels, all brackets, membranes, fire stopping between floors, trade contractor preliminaries, including external access equipment							
Flat system	–	–	–	–	–	m²	**450.00**

21 CLADDING AND COVERING

Item	PC £	Labour hours	Labour £	Plant £	Material £	Unit	Total rate £
CURTAIN WALLING – cont							
Stick curtain walling system – cont							
Polyester powder coated solid colour matt finish or natural anodized finish mullions spaced 1.5 m apart and spanning typical storey height of 3.8 m – cont							
extra over for							
neutral selective high performance coating on surface #2 in lieu of low-e coating on surface #3 (G-values circa 0.25–0.30); for assisting in solar control	–	–	–	–	–	m²	37.00
inner laminated glass to be heat strengthened laminated to mitigate thermal fracture risk	–	–	–	–	–	m²	37.00
outer glass to be heat strengthened laminated in lieu of monolithic heat strengthened	–	–	–	–	–	m²	37.00
ceramic fritting glass on surface #2 for visual and/or performance requirements (minimal G-value improvement)	–	–	–	–	–	m²	50.00
flush glass finish without external face caps, achieved by concealed toggle fixings locating within perimeter channels within sealed units including silicone sealing between glass panes	–	–	–	–	–	m²	50.00
typical coping detail, including pressed aluminium profiles, membranes, seals, etc.	–	–	–	–	–	m	270.00
typical sill detail, including pressed aluminium profiles, membranes, seals, etc.	–	–	–	–	–	m	220.00
intermediate transoms (per transom)	–	–	–	–	–	m	48.00
Unitized curtain walling system; proprietary solution from system supplier e.g. Schuco or equivalent alternatively pre-designed system solution from specialist facade contractor							
Polyester powder coated solid colour matt finish or natural anodized curtain walling Element widths of 1.5 m spanning typical storey height of 3.8 m. Floor to ceiling glass sealed units with 8.8 mm low-e coated laminated inner pane, air filled cavity and 8 mm monolithic heat strengthed outer pane, retained by external beading system. Rates include 1.1 m solid spandrel panels, all brackets, membranes, fire stopping between floors, trade contractor preliminaries, including external access equipment							
flat system	–	–	–	–	–	m²	930.00

21 CLADDING AND COVERING

Item	PC £	Labour hours	Labour £	Plant £	Material £	Unit	Total rate £
extra over for							
neutral selective high performance coating on surface #2 in lieu of low-e coating on surface #3 (G-values circa 0.25–0.30); for assisting in solar control	–	–	–	–	–	m²	37.00
inner laminated glass to be heat strengthened laminated to mitigate thermal fracture risk	–	–	–	–	–	m²	37.00
flush glass finish without external face caps, achieved by carrier frames with glass sealed units factory silicone bonded; often referred to as SSG (Structural Silicone Glazing)	–	–	–	–	–	m²	60.00
typical coping detail, including pressed aluminium profiles, membranes, seals, etc.	–	–	–	–	–	m	270.00
typical sill detail, including pressed aluminium profiles, membranes, seals, etc.	–	–	–	–	–	m	220.00
Other curtain walling systems/costs							
Unitized curtain walling system; bespoke project specific solution via specialist facade contractor mostly based in mainland Europe.							
generally as described for system supplier solution but profiles catered to specific performance and visual requirements, thus additional design development. Note: These rates are subject to currency fluctuations between £ and €.	–	–	–	–	–	m²	930.00
flush glass finish finish (SEG) assuming direct bonding in the factory but with base chassis design to accommodate carrier frame for glass replacement, avoiding site bending	–	–	–	–	–	m²	30.00
project specific unitized curtain walling generally requires project specific performance testing. This rate is for a single wall type	–	–	–	–	–	nr	90000.00
visual mock-ups are often required for project specific unitized curtain walling solutions and in cases for proprietary unitized and stick curtain walling projects. This rate is for a single wall type	–	–	–	–	–	nr	30000.00
all curtain walling should be site hose tested. The rate depends upon the quantum of joints to be tested, generally 5%. Assume 5 days @ £1000	–	–	–	–	–	nr	6000.00
extra over for							
neutral selective high performance coating on surface #2 in lieu of low-e coating on surface #3 (G-values circa 0.25–0.30); for assisting in solar control	–	–	–	–	–	m²	37.00
inner laminated glass to be heat strengthened laminated to mitigate thermal fracture risk	–	–	–	–	–	m²	37.00
outer glass to be heat strengthened laminated in lieu of monolithic heat strengthened	–	–	–	–	–	m²	37.00
alternative solutions for achieving Part L and comfort criteria, include increasing the amount of solidity	–	–	–	–	–	m²	90.00

22 GENERAL JOINERY

Item	PC £	Labour hours	Labour £	Plant £	Material £	Unit	Total rate £
SOFTWOOD							
Wrought softwood							
Skirtings, picture rails, dado rails and the like;							
splayed or moulded							
19 mm × 44 mm; splayed	–	0.09	1.82	–	2.90	m	**4.72**
19 mm × 44 mm; moulded	–	0.09	1.82	–	3.08	m	**4.90**
19 mm × 69 mm; splayed	–	0.09	1.82	–	3.12	m	**4.94**
19 mm × 69 mm; moulded	–	0.09	1.82	–	3.12	m	**4.94**
19 mm × 94 mm; splayed	–	0.09	1.82	–	3.49	m	**5.31**
19 mm × 94 mm; moulded	–	0.09	1.82	–	3.49	m	**5.31**
19 mm × 144 mm; moulded	–	0.11	2.23	–	4.08	m	**6.31**
19 mm × 169 mm; moulded	–	0.11	2.23	–	4.33	m	**6.56**
25 mm × 50 mm; moulded	–	0.09	1.82	–	3.02	m	**4.84**
25 mm × 69 mm; splayed	–	0.09	1.82	–	3.34	m	**5.16**
25 mm × 94 mm; splayed	–	0.09	1.82	–	3.73	m	**5.55**
25 mm × 144 mm; splayed	–	0.11	2.23	–	4.49	m	**6.72**
25 mm × 144 mm; moulded	–	0.11	2.23	–	4.49	m	**6.72**
25 mm × 169 mm; moulded	–	0.11	2.23	–	4.99	m	**7.22**
25 mm × 219 mm; moulded	–	0.13	2.63	–	6.48	m	**9.11**
returned ends	–	0.14	2.82	–	–	nr	**2.82**
mitres	–	0.09	1.82	–	–	nr	**1.82**
Architraves, cover fillets and the like; half round;							
splayed or moulded							
13 mm × 25 mm; half round	–	0.11	2.23	–	2.72	m	**4.95**
13 mm × 50 mm; moulded	–	0.11	2.23	–	2.90	m	**5.13**
16 mm × 32 mm; half round	–	0.11	2.23	–	3.02	m	**5.25**
16 mm × 38 mm; moulded	–	0.11	2.23	–	3.02	m	**5.25**
16 mm × 50 mm; moulded	–	0.11	2.23	–	3.02	m	**5.25**
19 mm × 50 mm; splayed	–	0.11	2.23	–	3.02	m	**5.25**
19 mm × 63 mm; splayed	–	0.11	2.23	–	3.12	m	**5.35**
19 mm × 69 mm; splayed	–	0.11	2.23	–	3.32	m	**5.55**
25 mm × 44 mm; splayed	–	0.11	2.23	–	2.98	m	**5.21**
25 mm × 50 mm; moulded	–	0.11	2.23	–	3.13	m	**5.36**
25 mm × 63 mm; splayed	–	0.11	2.23	–	3.27	m	**5.50**
25 mm × 69 mm; splayed	–	0.11	2.23	–	3.71	m	**5.94**
32 mm × 88 mm; moulded	–	0.11	2.23	–	3.73	m	**5.96**
38 mm × 38 mm; moulded	–	0.11	2.23	–	3.31	m	**5.54**
50 mm × 50 mm; moulded	–	0.11	2.23	–	3.83	m	**6.06**
returned ends	–	0.14	2.82	–	–	nr	**2.82**
mitres	–	0.09	1.82	–	–	nr	**1.82**
Stops; screwed on							
16 mm × 38 mm	–	0.09	1.82	–	1.31	m	**3.13**
16 mm × 50 mm	–	0.09	1.82	–	1.42	m	**3.24**
19 mm × 38 mm	–	0.09	1.82	–	1.31	m	**3.13**
25 mm × 38 mm	–	0.09	1.82	–	1.43	m	**3.25**
25 mm × 50 mm	–	0.09	1.82	–	1.47	m	**3.29**

22 GENERAL JOINERY

Item	PC £	Labour hours	Labour £	Plant £	Material £	Unit	Total rate £
Glazing beads and the like							
13 mm × 16 mm	–	0.04	0.81	–	1.61	m	**2.42**
13 mm × 19 mm	–	0.04	0.81	–	1.61	m	**2.42**
13 mm × 25 mm	–	0.04	0.81	–	1.65	m	**2.46**
13 mm × 25 mm; screwed	–	0.04	0.81	–	2.67	m	**3.48**
13 mm × 25 mm; fixing with brass cups and screws	–	0.04	0.81	–	3.44	m	**4.25**
16 mm × 25 mm; screwed	–	0.04	0.81	–	2.67	m	**3.48**
16 mm quadrant	–	0.04	0.81	–	2.45	m	**3.26**
19 mm quadrant or scotia	–	0.04	0.81	–	2.45	m	**3.26**
19 mm × 36 mm; screwed	–	0.04	0.81	–	2.71	m	**3.52**
25 mm × 38 mm; screwed	–	0.04	0.81	–	2.84	m	**3.65**
25 mm quadrant or scotia	–	0.04	0.81	–	2.58	m	**3.39**
38 mm scotia	–	0.04	0.81	–	3.12	m	**3.93**
50 mm scotia	–	0.04	0.81	–	3.64	m	**4.45**
Isolated shelves, worktops, seats and the like							
19 mm × 150 mm	–	0.15	3.03	–	3.40	m	**6.43**
19 mm × 200 mm	–	0.20	4.04	–	4.68	m	**8.72**
25 mm × 150 mm	–	0.15	3.03	–	3.87	m	**6.90**
25 mm × 200 mm	–	0.20	4.04	–	5.50	m	**9.54**
32 mm × 150 mm	–	0.15	3.03	–	4.53	m	**7.56**
32 mm × 200 mm	–	0.20	4.04	–	6.17	m	**10.21**
Isolated shelves, worktops, seats and the like; cross-tongued joints							
19 mm × 300 mm	–	0.26	5.25	–	13.97	m	**19.22**
19 mm × 450 mm	–	0.31	6.26	–	21.05	m	**27.31**
19 mm × 600 mm	–	0.37	7.48	–	27.24	m	**34.72**
25 mm × 300 mm	–	0.26	5.25	–	14.97	m	**20.22**
25 mm × 450 mm	–	0.31	6.26	–	22.69	m	**28.95**
25 mm × 600 mm	–	0.37	7.48	–	29.52	m	**37.00**
32 mm × 300 mm	–	0.26	5.25	–	15.87	m	**21.12**
32 mm × 450 mm	–	0.31	6.26	–	24.12	m	**30.38**
32 mm × 600 mm	–	0.37	7.48	–	31.48	m	**38.96**
Isolated shelves, worktops, seats and the like; slatted with 50 wide slats at 75 mm centres							
19 mm thick	–	0.60	12.12	–	34.35	m	**46.47**
25 mm thick	–	0.60	12.12	–	35.10	m	**47.22**
32 mm thick	–	0.60	12.12	–	35.77	m	**47.89**
Window boards, nosings, bed moulds and the like; rebated and rounded							
19 mm × 75 mm	–	0.17	3.43	–	4.43	m	**7.86**
19 mm × 150 mm	–	0.19	3.83	–	5.45	m	**9.28**
19 mm × 225 mm; in one width	–	0.24	4.85	–	6.71	m	**11.56**
19 mm × 300 mm; cross-tongued joints	–	0.28	5.66	–	15.34	m	**21.00**
25 mm × 75 mm	–	0.17	3.43	–	4.68	m	**8.11**
25 mm × 150 mm	–	0.19	3.83	–	5.97	m	**9.80**
25 mm × 225 mm; in one width	–	0.24	4.85	–	7.52	m	**12.37**
25 mm × 300 mm; cross-tongued joints	–	0.28	5.66	–	16.58	m	**22.24**
32 mm × 75 mm	–	0.17	3.43	–	4.90	m	**8.33**
32 mm × 150 mm	–	0.19	3.83	–	6.43	m	**10.26**
32 mm × 225 mm; in one width	–	0.24	4.85	–	8.20	m	**13.05**

22 GENERAL JOINERY

Item	PC £	Labour hours	Labour £	Plant £	Material £	Unit	Total rate £
SOFTWOOD – cont							
32 mm × 300 mm; cross-tongued joints	–	0.28	5.66	–	17.62	m	**23.28**
38 mm × 75 mm	–	0.17	3.43	–	5.36	m	**8.79**
38 mm × 150 mm	–	0.19	3.83	–	7.39	m	**11.22**
38 mm × 225 mm; in one width	–	0.24	4.85	–	9.57	m	**14.42**
38 mm × 300 mm; cross-tongued joints	–	0.28	5.66	–	19.72	m	**25.38**
returned and fitted ends	–	0.14	2.82	–	–	nr	**2.82**
Handrails; mopstick							
50 mm dia.	–	0.23	4.64	–	9.74	m	**14.38**
Handrails; rounded							
44 mm × 50 mm	–	0.23	4.64	–	9.44	m	**14.08**
50 mm × 75 mm	–	0.25	5.05	–	10.31	m	**15.36**
63 mm × 87 mm	–	0.28	5.66	–	11.43	m	**17.09**
75 mm × 100 mm	–	0.32	6.46	–	14.01	m	**20.47**
Handrails; moulded							
44 mm × 50 mm	–	0.23	4.64	–	9.44	m	**14.08**
50 mm × 75 mm	–	0.25	5.05	–	10.31	m	**15.36**
63 mm × 87 mm	–	0.28	5.66	–	11.43	m	**17.09**
75 mm × 100 mm	–	0.32	6.46	–	14.01	m	**20.47**
Sundries on softwood							
Extra over fixing with nails for							
gluing and pinning	–	0.02	0.35	–	0.03	m	**0.38**
masonry nails	–	0.02	0.36	–	0.11	m	**0.47**
steel screws	–	0.02	0.34	–	0.09	m	**0.43**
self-tapping screws	–	0.02	0.35	–	0.09	m	**0.44**
steel screws; gluing	–	0.03	0.59	–	0.09	m	**0.68**
steel screws; sinking; filling heads	–	0.04	0.75	–	0.09	m	**0.84**
steel screws; sinking; pellating over	–	0.08	1.63	–	0.09	m	**1.72**
brass cups and screws	–	0.10	2.02	–	0.23	m	**2.25**
Extra over for							
countersinking	–	0.01	0.30	–	–	m	**0.30**
pellating	–	0.07	1.42	–	–	m	**1.42**
Head or nut in softwood							
let in flush	–	0.04	0.75	–	–	nr	**0.75**
Head or nut; in hardwood							
let in flush	–	0.06	1.11	–	–	nr	**1.11**
let in over; pellated	–	0.13	2.60	–	–	nr	**2.60**
HARDWOOD							
Selected Sapele							
Skirtings, picture rails, dado rails and the like; splayed or moulded							
19 mm × 44 mm; splayed	3.99	0.13	2.63	–	4.38	m	**7.01**
19 mm × 44 mm; moulded	3.99	0.13	2.63	–	4.38	m	**7.01**
19 mm × 69 mm; splayed	4.65	0.13	2.63	–	5.07	m	**7.70**
19 mm × 69 mm; moulded	4.65	0.13	2.63	–	5.07	m	**7.70**
19 mm × 94 mm; splayed	5.43	0.13	2.63	–	5.88	m	**8.51**
19 mm × 94 mm; moulded	5.43	0.13	2.63	–	5.88	m	**8.51**
19 mm × 144 mm; moulded	7.21	0.15	3.03	–	7.75	m	**10.78**

22 GENERAL JOINERY

Item	PC £	Labour hours	Labour £	Plant £	Material £	Unit	Total rate £
19 mm × 169 mm; moulded	7.98	0.15	3.03	–	8.56	m	**11.59**
25 mm × 44 mm; moulded	4.49	0.13	2.63	–	4.90	m	**7.53**
25 mm × 69 mm; splayed	5.35	0.13	2.63	–	5.81	m	**8.44**
25 mm × 94 mm; splayed	6.56	0.13	2.63	–	7.08	m	**9.71**
25 mm × 144 mm; splayed	8.60	0.15	3.03	–	9.22	m	**12.25**
25 mm × 144 mm; moulded	8.60	0.15	3.03	–	9.22	m	**12.25**
25 mm × 169 mm; moulded	9.64	0.15	3.03	–	10.31	m	**13.34**
25 mm × 219 mm; moulded	11.02	0.17	3.43	–	11.75	m	**15.18**
returned ends	–	0.20	4.04	–	–	nr	**4.04**
mitres	–	0.14	2.82	–	–	nr	**2.82**
Architraves, cover fillets and the like; half round; splayed or moulded							
13 mm × 25 mm; half round	2.43	0.15	3.03	–	2.73	m	**5.76**
13 mm × 50 mm; moulded	3.88	0.15	3.03	–	4.26	m	**7.29**
16 mm × 32 mm; half round	2.46	0.15	3.03	–	2.76	m	**5.79**
16 mm × 38 mm; moulded	3.74	0.15	3.03	–	4.11	m	**7.14**
16 mm × 50 mm; moulded	3.99	0.15	3.03	–	4.38	m	**7.41**
19 mm × 50 mm; splayed	3.99	0.15	3.03	–	4.38	m	**7.41**
19 mm × 63 mm; splayed	4.33	0.15	3.03	–	4.74	m	**7.77**
19 mm × 69 mm; splayed	4.65	0.15	3.03	–	5.07	m	**8.10**
25 mm × 44 mm; splayed	4.32	0.15	3.03	–	4.73	m	**7.76**
25 mm × 50 mm; moulded	4.49	0.15	3.03	–	4.90	m	**7.93**
25 mm × 63 mm; splayed	5.02	0.15	3.03	–	5.45	m	**8.48**
25 mm × 69 mm; splayed	5.35	0.15	3.03	–	5.81	m	**8.84**
32 mm × 88 mm; moulded	6.52	0.15	3.03	–	7.04	m	**10.07**
38 mm × 38 mm; moulded	5.12	0.15	3.03	–	5.57	m	**8.60**
50 mm × 50 mm; moulded	6.70	0.15	3.03	–	7.21	m	**10.24**
returned ends	–	0.20	4.04	–	–	nr	**4.04**
mitres	–	0.14	2.82	–	–	nr	**2.82**
Stops; screwed on							
16 mm × 38 mm	1.65	0.14	2.82	–	1.73	m	**4.55**
16 mm × 50 mm	1.77	0.14	2.82	–	1.86	m	**4.68**
19 mm × 38 mm	1.65	0.14	2.82	–	1.73	m	**4.55**
25 mm × 38 mm	2.08	0.14	2.82	–	2.18	m	**5.00**
25 mm × 50 mm	2.42	0.14	2.82	–	2.54	m	**5.36**
Glazing beads and the like							
13 mm × 16 mm	2.15	0.06	1.21	–	2.26	m	**3.47**
13 mm × 19 mm	2.15	0.06	1.21	–	2.26	m	**3.47**
13 mm × 25 mm	2.32	0.06	1.21	–	2.44	m	**3.65**
13 mm × 25 mm; screwed	3.13	0.06	1.21	–	3.29	m	**4.50**
13 mm × 25 mm; fixing with brass cups and screws	3.87	0.06	1.21	–	4.06	m	**5.27**
16 mm × 25 mm; screwed	3.13	0.06	1.21	–	3.29	m	**4.50**
16 mm quadrant	3.02	0.06	1.21	–	3.17	m	**4.38**
19 mm quadrant or scotia	3.02	0.06	1.21	–	3.17	m	**4.38**
19 mm × 36 mm; screwed	3.87	0.06	1.21	–	4.06	m	**5.27**
25 mm × 38 mm; screwed	4.22	0.06	1.21	–	4.43	m	**5.64**
25 mm quadrant or scotia	3.45	0.06	1.21	–	3.62	m	**4.83**
38 mm scotia	5.12	0.06	1.21	–	5.38	m	**6.59**
50 mm scotia	6.70	0.06	1.21	–	7.04	m	**8.25**

22 GENERAL JOINERY

Item	PC £	Labour hours	Labour £	Plant £	Material £	Unit	Total rate £
HARDWOOD – cont							
Selected Sapele – cont							
Isolated shelves; worktops, seats and the like							
19 mm × 150 mm	7.34	0.20	4.04	–	7.71	m	**11.75**
19 mm × 200 mm	8.68	0.28	5.66	–	9.11	m	**14.77**
25 mm × 150 mm	8.60	0.20	4.04	–	9.03	m	**13.07**
25 mm × 200 mm	10.33	0.28	5.66	–	10.85	m	**16.51**
32 mm × 150 mm	9.71	0.20	4.04	–	10.20	m	**14.24**
32 mm × 200 mm	11.74	0.28	5.66	–	12.33	m	**17.99**
Isolated shelves, worktops, seats and the like; cross-tongued joints							
19 mm × 300 mm	19.86	0.35	7.07	–	20.85	m	**27.92**
19 mm × 450 mm	31.09	0.42	8.48	–	32.64	m	**41.12**
19 mm × 600 mm	41.29	0.51	10.30	–	43.35	m	**53.65**
25 mm × 300 mm	22.30	0.35	7.07	–	23.41	m	**30.48**
25 mm × 450 mm	35.03	0.42	8.48	–	36.78	m	**45.26**
25 mm × 600 mm	46.56	0.51	10.30	–	48.89	m	**59.19**
32 mm × 300 mm	24.39	0.35	7.07	–	25.61	m	**32.68**
32 mm × 450 mm	38.40	0.42	8.48	–	40.32	m	**48.80**
32 mm × 600 mm	51.05	0.51	10.30	–	53.60	m	**63.90**
Isolated shelves, worktops, seats and the like; slatted with 50 wide slats at 75 mm centres							
19 mm thick	57.22	0.80	16.16	–	60.77	m²	**76.93**
25 mm thick	61.25	0.80	16.16	–	65.02	m²	**81.18**
32 mm thick	64.71	0.80	16.16	–	68.65	m²	**84.81**
Window boards, nosings, bed moulds and the like; rebated and rounded							
19 mm × 75 mm	5.66	0.22	4.44	–	6.36	m	**10.80**
19 mm × 150 mm	8.21	0.25	5.05	–	9.04	m	**14.09**
19 mm × 225 mm; in one width	10.08	0.33	6.67	–	11.00	m	**17.67**
19 mm × 300 mm; cross-tongued joints	20.54	0.37	7.48	–	21.99	m	**29.47**
25 mm × 75 mm	6.23	0.22	4.44	–	6.96	m	**11.40**
25 mm × 150 mm	9.14	0.25	5.05	–	10.02	m	**15.07**
25 mm × 225 mm; in one width	12.01	0.33	6.67	–	13.03	m	**19.70**
25 mm × 300 mm; cross-tongued joints	23.86	0.37	7.48	–	25.47	m	**32.95**
32 mm × 75 mm	6.77	0.22	4.44	–	7.53	m	**11.97**
32 mm × 150 mm	10.18	0.25	5.05	–	11.11	m	**16.16**
32 mm × 225 mm; in one width	13.56	0.33	6.67	–	14.66	m	**21.33**
32 mm × 300 mm; cross-tongued joints	26.25	0.37	7.48	–	27.98	m	**35.46**
returned and fitted ends	–	0.21	4.24	–	–	nr	**4.24**
Handrails; rounded							
44 mm × 50 mm	12.46	0.31	6.26	–	13.08	m	**19.34**
50 mm × 75 mm	15.03	0.33	6.67	–	15.78	m	**22.45**
63 mm × 87 mm	17.63	0.37	7.48	–	18.51	m	**25.99**
75 mm × 100 mm	21.91	0.42	8.48	–	23.01	m	**31.49**
Handrails; moulded							
44 mm × 50 mm	13.87	0.31	6.26	–	14.56	m	**20.82**
50 mm × 75 mm	16.42	0.33	6.67	–	17.24	m	**23.91**
63 mm × 87 mm	19.04	0.37	7.48	–	19.99	m	**27.47**
75 mm × 100 mm	23.31	0.42	8.48	–	24.48	m	**32.96**

22 GENERAL JOINERY

Item	PC £	Labour hours	Labour £	Plant £	Material £	Unit	Total rate £
Sundries on hardwood							
Extra over fixing with nails for							
gluing and pinning	–	0.02	0.35	–	0.03	m	**0.38**
masonry nails	–	0.02	0.36	–	0.11	m	**0.47**
steel screws	–	0.02	0.34	–	0.09	m	**0.43**
self-tapping screws	–	0.02	0.35	–	0.09	m	**0.44**
steel screws; gluing	–	0.03	0.59	–	0.09	m	**0.68**
steel screws; sinking; filling heads	–	0.04	0.75	–	0.09	m	**0.84**
steel screws; sinking; pellating over	–	0.08	1.63	–	0.09	m	**1.72**
brass cups and screws	–	0.10	2.02	–	0.23	m	**2.25**
Extra over for							
countersinking	–	0.01	0.30	–	–	m	**0.30**
pellating	–	0.07	1.42	–	–	m	**1.42**
Head or nut in softwood							
let in flush	–	0.04	0.75	–	–	nr	**0.75**
Head or nut in hardwood							
let in flush	–	0.06	1.11	–	–	nr	**1.11**
let in over; pellated	–	0.13	2.60	–	–	nr	**2.60**
MEDIUM DENSITY FIBREBOARD							
Medium density fibreboard; Sapele veneered one side; 18 mm thick							
Window boards and the like; rebated; hardwood lipped on one edge							
18 mm × 200 mm	–	0.25	5.05	–	15.03	m	**20.08**
18 mm × 250 mm	–	0.28	5.66	–	15.80	m	**21.46**
18 mm × 300 mm	–	0.31	6.26	–	16.19	m	**22.45**
18 mm × 350 mm	–	0.33	6.67	–	17.37	m	**24.04**
returned and fitted ends	–	0.20	4.04	–	2.91	nr	**6.95**
Medium density fibreboard; American White Ash veneered one side; 18 mm thick							
Window boards and the like; rebated; hardwood lipped on one edge							
18 mm × 200 mm	–	0.25	5.05	–	15.61	m	**20.66**
18 mm × 250 mm	–	0.28	5.66	–	16.59	m	**22.25**
18 mm × 300 mm	–	0.31	6.26	–	17.07	m	**23.33**
18 mm × 350 mm	–	0.33	6.67	–	18.52	m	**25.19**
returned and fitted ends	–	0.20	4.04	–	2.91	nr	**6.95**
Medium density fibreboard							
Skirtings, picture rails, dado rails and the like; splayed or moulded							
18 mm × 50 mm; splayed	–	0.09	1.82	–	2.85	m	**4.67**
18 mm × 50 mm; moulded	–	0.09	1.82	–	2.85	m	**4.67**
18 mm × 75 mm; splayed	–	0.09	1.82	–	2.96	m	**4.78**
18 mm × 75 mm; moulded	–	0.09	1.82	–	2.96	m	**4.78**
18 mm × 100 mm; splayed	–	0.09	1.82	–	3.08	m	**4.90**
18 mm × 100 mm; moulded	–	0.09	1.82	–	3.08	m	**4.90**
18 mm × 150 mm; moulded	–	0.11	2.23	–	3.35	m	**5.58**

22 GENERAL JOINERY

Item	PC £	Labour hours	Labour £	Plant £	Material £	Unit	Total rate £
MEDIUM DENSITY FIBREBOARD – cont							
Medium density fibreboard – cont							
Skirtings, picture rails, dado rails and the like – cont							
18 mm × 175 mm; moulded	–	0.11	2.23	–	3.46	m	**5.69**
22 mm × 100 mm; splayed	–	0.09	1.82	–	4.90	m	**6.72**
25 mm × 50 mm; moulded	–	0.09	1.82	–	2.98	m	**4.80**
25 mm × 75 mm; splayed	–	0.09	1.82	–	3.14	m	**4.96**
25 mm × 100 mm; splayed	–	0.09	1.82	–	3.34	m	**5.16**
25 mm × 150 mm; splayed	–	0.11	2.23	–	3.75	m	**5.98**
25 mm × 150 mm; moulded	–	0.11	2.23	–	3.75	m	**5.98**
25 mm × 175 mm; moulded	–	0.11	2.23	–	3.95	m	**6.18**
25 mm × 225 mm; moulded	–	0.13	2.63	–	4.22	m	**6.85**
returned ends	–	0.14	2.82	–	–	nr	**2.82**
mitres	–	0.09	1.82	–	–	nr	**1.82**
Architraves, cover fillets and the like; half round; splayed or moulded							
12 mm × 25 mm; half round	–	0.11	2.23	–	2.72	m	**4.95**
12 mm × 50 mm; moulded	–	0.11	2.23	–	2.80	m	**5.03**
15 mm × 32 mm; half round	–	0.11	2.23	–	2.73	m	**4.96**
15 mm × 38 mm; moulded	–	0.11	2.23	–	2.75	m	**4.98**
15 mm × 50 mm; moulded	–	0.11	2.23	–	2.80	m	**5.03**
18 mm × 50 mm; splayed	–	0.11	2.23	–	2.80	m	**5.03**
18 mm × 63 mm; splayed	–	0.11	2.23	–	2.93	m	**5.16**
18 mm × 75 mm; splayed	–	0.11	2.23	–	2.98	m	**5.21**
25 mm × 44 mm; splayed	–	0.11	2.23	–	2.98	m	**5.21**
25 mm × 50 mm; moulded	–	0.11	2.23	–	2.98	m	**5.21**
25 mm × 63 mm; splayed	–	0.11	2.23	–	3.08	m	**5.31**
25 mm × 75 mm; splayed	–	0.11	2.23	–	3.17	m	**5.40**
30 mm × 88 mm; moulded	–	0.11	2.23	–	4.08	m	**6.31**
38 mm × 38 mm; moulded	–	0.11	2.23	–	3.52	m	**5.75**
50 mm × 50 mm; moulded	–	0.11	2.23	–	3.70	m	**5.93**
returned ends	–	0.14	2.82	–	–	nr	**2.82**
mitres	–	0.09	1.82	–	–	nr	**1.82**
Stops; screwed on							
15 mm × 38 mm	–	0.09	1.82	–	1.51	m	**3.33**
15 mm × 50 mm	–	0.09	1.82	–	1.55	m	**3.37**
18 mm × 38 mm	–	0.09	1.82	–	1.54	m	**3.36**
25 mm × 38 mm	–	0.09	1.82	–	1.62	m	**3.44**
25 mm × 50 mm	–	0.09	1.82	–	1.70	m	**3.52**
Glazing beads and the like							
12 mm × 16 mm	–	0.04	0.81	–	1.73	m	**2.54**
12 mm × 19 mm	–	0.04	0.81	–	1.74	m	**2.55**
12 mm × 25 mm	–	0.04	0.81	–	1.76	m	**2.57**
12 mm × 25 mm; screwed	–	0.04	0.81	–	2.52	m	**3.33**
12 mm × 25 mm; fixing with brass cups and screws	–	0.04	0.81	–	2.89	m	**3.70**
15 mm × 25 mm; screwed	–	0.04	0.81	–	2.61	m	**3.42**
15 mm quadrant	–	0.04	0.81	–	2.50	m	**3.31**

22 GENERAL JOINERY

Item	PC £	Labour hours	Labour £	Plant £	Material £	Unit	Total rate £
18 mm quadrant or scotia	–	0.04	0.81	–	2.51	m	**3.32**
18 mm × 36 mm; screwed	–	0.04	0.81	–	2.66	m	**3.47**
25 mm × 38 mm; screwed	–	0.04	0.81	–	2.78	m	**3.59**
25 mm quadrant or scotia	–	0.04	0.81	–	2.61	m	**3.42**
38 mm scotia	–	0.04	0.81	–	2.48	m	**3.29**
50 mm scotia	–	0.04	0.81	–	2.88	m	**3.69**
Isolated shelves, worktops, seats and the like							
18 mm × 150 mm	–	0.15	3.03	–	3.27	m	**6.30**
18 mm × 200 mm	–	0.20	4.04	–	3.45	m	**7.49**
25 mm × 150 mm	–	0.15	3.03	–	3.76	m	**6.79**
25 mm × 200 mm	–	0.20	4.04	–	4.02	m	**8.06**
30 mm × 150 mm	–	0.15	3.03	–	5.27	m	**8.30**
30 mm × 200 mm	–	0.20	4.04	–	5.84	m	**9.88**
Isolated shelves, worktops, seats and the like; cross-tongued joints							
18 mm × 300 mm	–	0.26	5.25	–	10.68	m	**15.93**
18 mm × 450 mm	–	0.31	6.26	–	12.18	m	**18.44**
18 mm × 600 mm	–	0.37	7.48	–	20.37	m	**27.85**
25 mm × 300 mm	–	0.26	5.25	–	11.25	m	**16.50**
25 mm × 450 mm	–	0.31	6.26	–	13.74	m	**20.00**
25 mm × 600 mm	–	0.37	7.48	–	19.84	m	**27.32**
30 mm × 300 mm	–	0.26	5.25	–	12.98	m	**18.23**
30 mm × 450 mm	–	0.31	6.26	–	15.45	m	**21.71**
30 mm × 600 mm	–	0.37	7.48	–	22.48	m	**29.96**
Isolated shelves, worktops, seats and the like; slatted with 50 wide slats at 75 mm centres							
18 mm thick	–	0.60	12.12	–	33.72	m	**45.84**
25 mm thick	–	0.60	12.12	–	35.76	m	**47.88**
30 mm thick	–	0.60	12.12	–	37.61	m	**49.73**
Window boards, nosings, bed moulds and the like; rebated and rounded							
18 mm × 75 mm	–	0.17	3.43	–	3.34	m	**6.77**
18 mm × 150 mm	–	0.19	3.83	–	3.75	m	**7.58**
18 mm × 225 mm	–	0.24	4.85	–	4.06	m	**8.91**
18 mm × 300 mm	–	0.28	5.66	–	4.47	m	**10.13**
25 mm × 75 mm	–	0.17	3.43	–	3.48	m	**6.91**
25 mm × 150 mm	–	0.19	3.83	–	4.06	m	**7.89**
25 mm × 225 mm	–	0.24	4.85	–	4.54	m	**9.39**
25 mm × 300 mm	–	0.28	5.66	–	5.07	m	**10.73**
30 mm × 75 mm	–	0.17	3.43	–	4.69	m	**8.12**
30 mm × 150 mm	–	0.19	3.83	–	5.75	m	**9.58**
30 mm × 225 mm	–	0.24	4.85	–	6.58	m	**11.43**
30 mm × 300 mm	–	0.28	5.66	–	7.55	m	**13.21**
38 mm × 75 mm	–	0.17	3.43	–	5.29	m	**8.72**
38 mm × 150 mm	–	0.19	3.83	–	6.57	m	**10.40**
38 mm × 225 mm	–	0.24	4.85	–	7.55	m	**12.40**
38 mm × 300 mm	–	0.28	5.66	–	8.73	m	**14.39**
returned and fitted ends	–	–	–	–	1.02	nr	**1.02**

22 GENERAL JOINERY

Item	PC £	Labour hours	Labour £	Plant £	Material £	Unit	Total rate £
MEDIUM DENSITY FIBREBOARD – cont							
Pin-boards; medium board							
Sundeala A pin-board or other equal and approved; fixed with adhesive to backing (not included); over 300 mm wide							
6 mm thick	–	0.56	11.31	–	6.53	m²	**17.84**
9 mm thick Colourboard	–	0.56	11.31	–	11.19	m²	**22.50**
ASSOCIATED METALWORK							
Metalwork; mild steel							
Angle section bearers; for building in							
90 mm × 90 mm × 6 mm	–	0.31	7.10	–	–	m	**7.10**
120 mm × 120 mm × 8 mm	–	0.32	7.33	–	0.16	m	**7.49**
200 mm × 150 mm × 12 mm	–	0.37	8.47	–	4.10	m	**12.57**
Metalwork; mild steel; galvanized							
Waterbars; groove in timber							
6 mm × 30 mm	–	0.46	9.29	–	4.60	m	**13.89**
6 mm × 40 mm	–	0.46	9.29	–	5.78	m	**15.07**
6 mm × 50 mm	–	0.46	9.29	–	4.26	m	**13.55**
Angle section bearers; for building in							
90 mm × 90 mm × 6 mm	–	0.31	7.10	–	13.42	m	**20.52**
120 mm × 120 mm × 8 mm	–	0.32	7.33	–	23.68	m	**31.01**
200 mm × 150 mm × 12 mm	–	0.37	8.47	–	52.77	m	**61.24**
Dowels; mortice in timber							
8 mm dia. × 100 mm long	–	0.04	0.81	–	0.67	nr	**1.48**
10 mm dia. × 50 mm long	–	0.04	0.81	–	1.06	nr	**1.87**
Cramps							
25 mm × 3 mm × 230 mm girth; one end bent, holed and screwed to softwood; other end fishtailed for building in	–	0.06	1.21	–	1.60	nr	**2.81**
Metalwork; stainless steel							
Angle section bearers; for building in							
90 mm × 90 mm × 6 mm	–	0.31	7.10	–	31.24	m	**38.34**
120 mm × 120 mm × 8 mm	–	0.32	7.33	–	55.14	m	**62.47**
200 mm × 150 mm × 12 mm	–	0.37	8.47	–	120.92	m	**129.39**

23 WINDOWS, SCREENS AND LIGHTS

Item	PC £	Labour hours	Labour £	Plant £	Material £	Unit	Total rate £
WINDOWS							
SUPPLY ONLY PRICES							
NOTE: The following supply only prices are for purpose-made components, to which fixings, sealants etc. labour and overheads and profit need to be added, before they may be used to arrive at a guide price for a complete window. The reader is then referred to the subsequent SUPPLY AND FIX pages for fixing costs based on the overall window size.							
Purpose-made window casements; treated wrought softwood							
Casements; rebated; moulded							
44 mm thick	–	–	–	–	47.75	m²	**47.75**
57 mm thick	–	–	–	–	50.13	m²	**50.13**
Casements; rebated; moulded; in medium panes							
44 mm thick	–	–	–	–	76.31	m²	**76.31**
57 mm thick	–	–	–	–	79.52	m²	**79.52**
Casements; rebated; moulded; with semi-circular head							
44 mm thick	–	–	–	–	99.79	m²	**99.79**
57 mm thick	–	–	–	–	102.91	m²	**102.91**
Casements; rebated; moulded; to bullseye window							
44 mm thick; 600 mm dia.	–	–	–	–	158.18	nr	**158.18**
44 mm thick; 900 mm dia.	–	–	–	–	188.44	nr	**188.44**
57 mm thick; 600 mm dia.	–	–	–	–	165.63	nr	**165.63**
57 mm thick; 900 mm dia.	–	–	–	–	22.80	nr	**22.80**
Fitting and hanging casements (in factory)							
square or rectangular	–	–	–	–	11.16	nr	**11.16**
semi-circular	–	–	–	–	18.13	nr	**18.13**
bullseye	–	–	–	–	22.80	nr	**22.80**
Purpose-made window casements; selected Sapele							
Casements; rebated; moulded							
44 mm thick	–	–	–	–	54.26	m²	**54.26**
57 mm thick	–	–	–	–	59.45	m²	**59.45**
Casements; rebated; moulded; in medium panes							
44 mm thick	–	–	–	–	87.87	m²	**87.87**
57 mm thick	–	–	–	–	94.81	m²	**94.81**
Casements; rebated; moulded with semi-circular head							
44 mm thick	–	–	–	–	110.70	m²	**110.70**
57 mm thick	–	–	–	–	117.46	m²	**117.46**
Casements; rebated; moulded; to bullseye window							
44 mm thick; 600 mm dia.	–	–	–	–	195.64	nr	**195.64**
44 mm thick; 900 mm dia.	–	–	–	–	235.43	nr	**235.43**
57 mm thick; 600 mm dia.	–	–	–	–	211.78	nr	**211.78**
57 mm thick; 900 mm dia.	–	–	–	–	255.99	nr	**255.99**

23 WINDOWS, SCREENS AND LIGHTS

Item	PC £	Labour hours	Labour £	Plant £	Material £	Unit	Total rate £
WINDOWS – cont							
Purpose-made window casements – cont							
Fitting and hanging casements (in factory)							
square or rectangular	–	–	–	–	12.10	nr	**12.10**
semi-circular	–	–	–	–	19.99	nr	**19.99**
bullseye	–	–	–	–	25.57	nr	**25.57**
Purpose-made window frames; treated wrought softwood							
Frames; rounded; rebated check grooved							
44 mm × 69 mm	–	–	–	–	12.99	m	**12.99**
44 mm × 94 mm	–	–	–	–	13.61	m	**13.61**
44 mm × 119 mm	–	–	–	–	14.23	m	**14.23**
57 mm × 94 mm	–	–	–	–	14.25	m	**14.25**
69 mm × 144 mm	–	–	–	–	18.78	m	**18.78**
90 mm × 140 mm	–	–	–	–	26.67	m	**26.67**
Mullions and transoms; twice rounded, rebated and check grooved							
57 mm × 69 mm	–	–	–	–	15.12	m	**15.12**
57 mm × 94 mm	–	–	–	–	15.89	m	**15.89**
69 mm × 94 mm	–	–	–	–	17.96	m	**17.96**
69 mm × 144 mm	–	–	–	–	26.39	m	**26.39**
Sill; sunk weathered, rebated and grooved							
69 mm × 94 mm	–	–	–	–	31.52	m	**31.52**
69 mm × 144 mm	–	–	–	–	33.38	m	**33.38**
Purpose-made window frames; selected Sapele							
Frames; rounded; rebated check grooved							
44 mm × 69 mm	–	–	–	–	16.57	m	**16.57**
44 mm × 94 mm	–	–	–	–	17.83	m	**17.83**
44 mm × 119 mm	–	–	–	–	19.07	m	**19.07**
57 mm × 94 mm	–	–	–	–	20.86	m	**20.86**
69 mm × 144 mm	–	–	–	–	29.88	m	**29.88**
90 mm × 140 mm	–	–	–	–	42.45	m	**42.45**
Mullions and transoms; twice rounded, rebated and check grooved							
57 mm × 69 mm	–	–	–	–	18.83	m	**18.83**
57 mm × 94 mm	–	–	–	–	21.91	m	**21.91**
69 mm × 94 mm	–	–	–	–	26.27	m	**26.27**
69 mm × 144 mm	–	–	–	–	39.93	m	**39.93**
Sill; sunk weathered, rebated and grooved							
69 mm × 94 mm	–	–	–	–	37.21	m	**37.21**
69 mm × 144 mm	–	–	–	–	41.25	m	**41.25**

23 WINDOWS, SCREENS AND LIGHTS

Item	PC £	Labour hours	Labour £	Plant £	Material £	Unit	Total rate £
SUPPLY AND FIX PRICES							
Standard windows; treated wrought softwood; Jeld-Wen or other equal and approved							
Side-hung casement windows; factory glazed with low-e 24 mm double glazing (U-value = 1.6 W/m² K); with 140 mm wide softwood sills; opening casements and ventilators hung on rustproof hinges; fitted with aluminized lacquered finish casement stays and fasteners							
488 mm × 750 mm; ref LEWN07V	172.83	0.65	13.13	–	181.61	nr	**194.74**
488 mm × 900 mm; ref LEWN09V	175.82	0.74	14.94	–	184.75	nr	**199.69**
630 mm × 750 mm; ref LEW107C	156.25	0.74	14.94	–	164.19	nr	**179.13**
630 mm × 750 mm; ref LEW107V	191.65	0.74	14.94	–	201.36	nr	**216.30**
630 mm × 900 mm; ref LEW109V	198.56	0.83	16.76	–	208.62	nr	**225.38**
630 mm × 900 mm; ref LEW109CH	168.51	0.74	14.94	–	177.07	nr	**192.01**
630 mm × 1050 mm; ref LEW110C	174.69	0.93	18.78	–	183.60	nr	**202.38**
630 mm × 1050 mm; ref LEW110V	207.75	0.74	14.94	–	218.32	nr	**233.26**
915 mm × 900 mm; ref LEW2NO9W	238.88	1.02	20.60	–	250.96	nr	**271.56**
915 mm × 1050 mm; ref LEW2N1OW	250.95	1.06	21.41	–	263.68	nr	**285.09**
915 mm × 1200 mm; ref LEW2N12W	271.55	1.11	22.42	–	285.31	nr	**307.73**
915 mm × 1350 mm; ref LEW2N13W	284.69	1.25	25.24	–	299.10	nr	**324.34**
915 mm × 1500 mm; ref LEW2N15W	323.02	1.30	26.25	–	339.40	nr	**365.65**
1200 mm × 750 mm; ref LEW2O7C	245.52	1.06	21.41	–	257.97	nr	**279.38**
1200 mm × 750 mm; ref LEW2O7CV	310.54	1.06	21.41	–	326.25	nr	**347.66**
1200 mm × 900 mm; ref LEW2O9C	261.18	1.11	22.42	–	274.42	nr	**296.84**
1200 mm × 900 mm; ref LEW2O9W	274.84	1.11	22.42	–	288.76	nr	**311.18**
1200 mm × 900 mm; ref LEW2O9CV	324.50	1.11	22.42	–	340.89	nr	**363.31**
1200 mm × 1050 mm; ref LEW210C	276.50	1.25	25.24	–	290.56	nr	**315.80**
1200 mm × 1050 mm; ref LEW210W	291.33	1.25	25.24	–	306.12	nr	**331.36**
1200 mm × 1050 mm; ref LEW210T	343.74	1.25	25.24	–	361.15	nr	**386.39**
1200 mm × 1050 mm; ref LEW210CV	337.12	1.25	25.24	–	354.20	nr	**379.44**
1200 mm × 1200 mm; ref LEW212C	297.28	1.34	27.06	–	312.41	nr	**339.47**
1200 mm × 1200 mm; ref LEW212W	308.15	1.34	27.06	–	323.83	nr	**350.89**
1200 mm × 1200 mm; ref LEW212TX	400.45	1.34	27.06	–	420.74	nr	**447.80**
1200 mm × 1200 mm; ref LEW212CV	355.13	1.34	27.06	–	373.15	nr	**400.21**
1200 mm × 1350 mm; ref LEW213W	330.80	1.43	28.88	–	347.60	nr	**376.48**
1200 mm × 1350 mm; ref LEW213CV	392.75	1.43	28.88	–	412.66	nr	**441.54**
1200 mm × 1500 mm; ref LEW215W	383.62	1.57	31.71	–	403.12	nr	**434.83**
1770 mm × 750 mm; ref LEW307CC	360.56	1.30	26.25	–	378.85	nr	**405.10**
1770 mm × 900 mm; ref LEW309CC	383.63	1.57	31.71	–	403.08	nr	**434.79**
1770 mm × 1050 mm; ref LEW310C	363.60	1.62	32.72	–	382.04	nr	**414.76**
1770 mm × 1050 mm; ref LEW310T	429.24	1.57	31.71	–	450.96	nr	**482.67**
1770 mm × 1050 mm; ref LEW310CC	402.60	1.30	26.25	–	422.99	nr	**449.24**
1770 mm × 1050 mm; ref LEW310CW	423.94	1.30	26.25	–	445.40	nr	**471.65**
1770 mm × 1200 mm; ref LEW312C	388.31	1.67	33.73	–	408.04	nr	**441.77**
1770 mm × 1200 mm; ref LEW312T	454.70	1.67	33.73	–	477.75	nr	**511.48**
1770 mm × 1200 mm; ref LEW312CC	433.89	1.67	33.73	–	455.89	nr	**489.62**
1770 mm × 1200 mm; ref LEW312CW	449.41	1.67	33.73	–	472.18	nr	**505.91**

23 WINDOWS, SCREENS AND LIGHTS

Item	PC £	Labour hours	Labour £	Plant £	Material £	Unit	Total rate £
WINDOWS – cont							
Standard windows – cont							
1770 mm × 1200 mm; ref LEW312CVC	506.26	1.67	33.73	–	531.88	nr	**565.61**
1770 mm × 1350 mm; ref LEW313CC	491.32	1.76	35.54	–	516.20	nr	**551.74**
1770 mm × 1350 mm; ref LEW313CW	489.36	1.76	35.54	–	514.14	nr	**549.68**
1770 mm × 1350 mm; ref LEW313CVC	551.22	1.76	35.54	–	579.09	nr	**614.63**
1770 mm × 1500 mm; ref LEW315T	584.61	1.85	37.36	–	614.14	nr	**651.50**
2340 mm × 1050 mm; ref LEW410CWC	559.23	1.80	36.35	–	587.51	nr	**623.86**
2340 mm × 1200 mm; ref LEW412CWC	594.07	1.90	38.37	–	624.13	nr	**662.50**
2340 mm × 1350 mm; ref LEW413CWC	653.24	2.04	41.20	–	686.30	nr	**727.50**
Top-hung casement windows; factory glazed with low-e 24 mm double glazing (U-value = 1.6 W/m^{2}K); with 140 mm wide softwood sills; opening casements and ventilators hung on rustproof hinges; fitted with aluminized lacquered finish casement stays							
630 mm × 750 mm; ref LEW107 A	168.45	0.74	14.94	–	177.00	nr	**191.94**
630 mm × 900 mm; ref LEW109 A	177.81	0.83	16.76	–	186.83	nr	**203.59**
630 mm × 1050 mm; ref LEW110 A	188.36	0.93	18.78	–	197.96	nr	**216.74**
915 mm × 750 mm; ref LEW2N07 A	204.58	0.97	19.59	–	214.93	nr	**234.52**
915 mm × 900 mm; ref LEW2N09 A	229.54	1.02	20.60	–	241.14	nr	**261.74**
915 mm × 1050 mm; ref LEW2N10 A	244.21	1.06	21.41	–	256.60	nr	**278.01**
915 mm × 1350 mm; ref LEW2N13 AS	308.09	1.25	25.24	–	323.72	nr	**348.96**
1200 mm × 750 mm; ref LEW207 A	240.71	1.06	21.41	–	252.92	nr	**274.33**
1200 mm × 900 mm; ref LEW209 A	262.71	1.11	22.42	–	276.02	nr	**298.44**
1200 mm × 1050 mm; ref LEW210 A	280.86	1.25	25.24	–	295.12	nr	**320.36**
1200 mm × 1200 mm; ref LEW212 A	304.41	1.34	27.06	–	319.85	nr	**346.91**
1200 mm × 1350 mm; ref LEW213 AS	355.43	1.43	28.88	–	373.47	nr	**402.35**
1200 mm × 1500 mm; ref LEW215 AS	387.88	1.57	31.71	–	407.54	nr	**439.25**
1770 mm × 1050 mm; ref LEW310 AE	388.86	1.57	31.71	–	408.58	nr	**440.29**
1770 mm × 1200 mm; ref LEW312 AE	414.42	1.67	33.73	–	435.41	nr	**469.14**
High performance Hi-Profile top-hung reversible windows; factory glazed with low-e 24 mm double glazing (U-value = 1.4 W/m^{2}K); weather stripping; opening panes hung on rustproof hinges; fitted with aluminized lacquered espagnolette bolts							
600 mm × 900 mm; ref LEXC0609 AR	317.47	0.83	16.76	–	333.48	nr	**350.24**
600 mm × 1050 mm; ref LEXC0610 AR	333.46	0.93	18.78	–	350.30	nr	**369.08**
600 mm × 1200 mm; ref LEXC0612 AR	350.76	1.03	20.80	–	368.52	nr	**389.32**
600 mm × 1350 mm; ref LEXC0613 AR	367.53	1.11	22.42	–	386.13	nr	**408.55**
1200 mm × 900 mm; ref LEXC1209 AGR	496.80	1.11	22.42	–	521.82	nr	**544.24**
1200 mm × 1050 mm; ref LEXC1210 AGR	524.80	1.25	25.24	–	551.26	nr	**576.50**
1200 mm × 1200 mm; ref LEXC1212 AGR	553.29	1.34	27.06	–	581.22	nr	**608.28**
1200 mm × 1350 mm; ref LEXC1213 AGR	581.66	1.43	28.88	–	611.01	nr	**639.89**
1800 mm × 900 mm; ref LEXC1809 AGAR	762.15	1.57	31.71	–	800.53	nr	**832.24**
1800 mm × 1050 mm; ref LEXC1810 AGAR	805.97	1.62	32.72	–	846.53	nr	**879.25**
1800 mm × 1200 mm; ref LEXC1812 AGAR	850.77	1.67	33.73	–	893.61	nr	**927.34**
1800 mm × 1350 mm; ref LEXC1813 AGAR	895.34	1.76	35.54	–	940.41	nr	**975.95**

23 WINDOWS, SCREENS AND LIGHTS

Item	PC £	Labour hours	Labour £	Plant £	Material £	Unit	Total rate £
High performance double-hung sash windows with glazing bars; factory glazed with low-e 24 mm double glazing (U-value = 1.6 W/m²K; solid frames; 63 mm × 175 mm softwood sills; standard flush external linings; spiral spring balances and sash catch							
635 mm × 1050 mm; ref LESV0610B	536.76	1.85	37.36	–	563.78	nr	**601.14**
635 mm × 1350 mm; ref LESV0613B	591.58	2.04	41.20	–	621.38	nr	**662.58**
635 mm × 1650 mm; ref LESV0616B	656.98	2.27	45.84	–	690.10	nr	**735.94**
860 mm × 1050 mm; ref LESV0810B	602.70	2.13	43.02	–	633.00	nr	**676.02**
860 mm × 1350 mm; ref LESV0813B	663.90	2.41	48.67	–	697.33	nr	**746.00**
860 mm × 1650 mm; ref LESV0816B	754.38	2.78	56.14	–	792.36	nr	**848.50**
1085 mm × 1050 mm; ref LESV1010B	703.48	2.41	48.67	–	738.83	nr	**787.50**
1085 mm × 1350 mm; ref LESV1013B	747.08	2.78	56.14	–	784.65	nr	**840.79**
1085 mm × 1650 mm; ref LESV1016B	876.51	3.42	69.07	–	920.61	nr	**989.68**
1725 mm × 1050 mm; ref LESV1710B	1206.57	3.42	69.07	–	1267.12	nr	**1336.19**
1725 mm × 1350 mm; ref LESV1713B	1346.82	4.26	86.04	–	1414.43	nr	**1500.47**
1725 mm × 1650 mm; ref LESV1716B	1545.58	4.35	87.85	–	1623.17	nr	**1711.02**
Standard windows; Jeld-Wen Hardwood or other equal and approved; factory applied preservative stain base coat Side-hung casement windows; factory glazed with low-e 24 mm double glazing (U-Value 1.6 W/m²K); 45 mm × 140 mm hardwood sills; weather stripping; opening sashes on canopy hinges; fitted with fasteners; brown finish ironmongery							
630 mm × 750 mm; ref LEW107CH	188.11	0.88	17.78	–	197.65	nr	**215.43**
630 mm × 900 mm; ref LEW109CH	198.15	1.11	22.42	–	208.19	nr	**230.61**
630 mm × 900 mm; ref LEW109VH	217.35	0.88	17.78	–	228.35	nr	**246.13**
630 mm × 1050 mm; ref LEW2110VH	222.33	1.20	24.23	–	233.57	nr	**257.80**
915 mm × 900 mm; ref LEWN09WH	326.63	1.39	28.07	–	343.14	nr	**371.21**
915 mm × 1050 mm; ref LEWN10WH	339.10	1.48	29.89	–	356.28	nr	**386.17**
915 mm × 1200 mm; ref LEWN12WH	352.26	1.57	31.71	–	370.10	nr	**401.81**
1200 mm × 900 mm; ref LEW209CH	285.49	1.57	31.71	–	299.89	nr	**331.60**
1200 mm × 900 mm; ref LEW209WH	361.46	1.57	31.71	–	379.67	nr	**411.38**
1200 mm × 1050 mm; ref LEW210CH	297.95	1.67	33.73	–	312.98	nr	**346.71**
1200 mm × 1050 mm; ref LEW210WH	364.22	1.67	33.73	–	382.56	nr	**416.29**
1200 mm × 1200 mm; ref LEW212CH	311.11	1.80	36.35	–	326.84	nr	**363.19**
1200 mm × 1200 mm; ref LEW212WH	305.57	1.80	36.35	–	321.03	nr	**357.38**
1200 mm × 1350 mm; ref LEW213WH	310.54	1.94	39.18	–	326.25	nr	**365.43**
1200 mm × 1550 mm; ref LEW215WH	315.50	2.04	41.20	–	331.45	nr	**372.65**
1770 mm × 1050 mm; ref LEW310CCH	530.55	2.08	42.01	–	557.30	nr	**599.31**
1770 mm × 1200 mm; ref LEW312CCH	551.90	2.22	44.84	–	579.72	nr	**624.56**
2339 mm × 1200 mm; ref LEW412CMCH	1165.76	2.41	48.67	–	1224.31	nr	**1272.98**

23 WINDOWS, SCREENS AND LIGHTS

Item	PC £	Labour hours	Labour £	Plant £	Material £	Unit	Total rate £
WINDOWS – cont							
Standard windows – cont							
Top-hung casement windows; factory glazed with low-e 24 mm double glazing (U-value = 1.6 W/m² K); 45 mm × 140 mm hardwood sills; weather stripping; opening sashes on canopy hinges; fitted with fasteners; brown finish ironmongery							
630 mm × 900 mm; ref LEW109 AH	192.86	0.88	17.78	–	202.63	nr	**220.41**
630 mm × 1050 mm; ref LEW110 AH	203.37	1.20	24.23	–	213.72	nr	**237.95**
915 mm × 900 mm; ref LEW2N09 AH	237.23	1.39	28.07	–	249.23	nr	**277.30**
915 mm × 1050 mm; ref LEW2N10 AH	247.76	1.48	29.89	–	260.33	nr	**290.22**
915 mm × 1350 mm; ref LEW2N13 ASH	304.44	1.67	33.73	–	319.88	nr	**353.61**
1200 mm × 1050 mm; ref LEW210 AH	279.81	1.57	31.71	–	293.98	nr	**325.69**
1200 mm × 1350 mm; ref LEW213 ASH	342.03	1.67	33.73	–	359.31	nr	**393.04**
1770 mm × 1050 mm; ref LEW310 AEH	357.64	1.80	36.35	–	375.74	nr	**412.09**
Purpose-made double-hung sash windows; treated wrought softwood							
Cased frames of 100 mm × 25 mm grooved inner linings; 114 mm × 25 mm grooved outer linings; 125 mm × 38 mm twice rebated head linings; 125 mm × 32 mm twice rebated grooved pulley stiles; 150 mm × 13 mm linings; 50 mm × 19 mm parting slips; 25 mm × 19 mm inside beads; 150 mm × 75 mm Oak twice sunk weathered throated sill; 50 mm thick rebated and moulded sashes; moulded horns							
over 1.25 m² each; both sashes in medium panes; including spiral spring balances	342.30	2.08	42.01	–	424.62	m²	**466.63**
up to 1.25 m² each; both sashes in medium panes; including spiral spring balances with cased mullions	389.07	2.31	46.65	–	473.73	m²	**520.38**
Purpose-made double-hung sash windows; selected Sapele							
Cased frames of 100 mm × 25 mm grooved inner linings; 114 mm × 25 mm grooved outer linings; 125 mm × 38 mm twice rebated head linings; 125 mm × 32 mm twice rebated grooved pulley stiles; 150 mm × 13 mm linings; 50 mm × 19 mm parting slips; 25 mm × 19 mm inside beads; 150 mm × 75 mm Oak twice sunk weathered throated sill; 50 mm thick rebated and moulded sashes; moulded horns							
over 1.25 m² each; both sashes in medium panes; including spiral sash balances	380.94	2.78	56.14	–	465.19	m²	**521.33**
over 1.25 m² each; both sashes in medium panes; including spiral sash balances with cased mullions	408.76	3.08	62.20	–	494.40	m²	**556.60**

23 WINDOWS, SCREENS AND LIGHTS

Item	PC £	Labour hours	Labour £	Plant £	Material £	Unit	Total rate £
Clements EB24 range of factory finished steel fixed light; casement and fanlight windows and doors; with a U-value of 2.0 W/m² K (part L compliant); to EN ISO 9001 2000 ; polyester powder coated; factory glazed with low-e double glazing; fixed in position; including lugs plugged and screwed to brickwork or blockwork							
Basic fixed light including easy-glaze snap-on beads							
508 mm × 292 mm	143.00	2.00	69.21	–	150.23	nr	**219.44**
508 mm × 457 mm	156.00	2.00	69.21	–	163.88	nr	**233.09**
508 mm × 628 mm	169.00	2.00	69.21	–	177.59	nr	**246.80**
508 mm × 923 mm	195.00	2.00	69.21	–	204.93	nr	**274.14**
508 mm × 1218 mm	221.00	2.50	86.51	–	232.27	nr	**318.78**
Basic 'Tilt and Turn' window; including easy-glaze snap-on beads							
508 mm × 292 mm	338.00	2.00	69.21	–	354.98	nr	**424.19**
508 mm × 457 mm	351.00	2.00	69.21	–	368.63	nr	**437.84**
508 mm × 628 mm	364.00	2.00	69.21	–	382.34	nr	**451.55**
508 mm × 923 mm; including fixed light	429.00	2.00	69.21	–	450.63	nr	**519.84**
508 mm × 1218 mm; including fixed light	455.00	2.50	86.51	–	477.97	nr	**564.48**
Basic casement; including easy-glaze snap-on beads							
508 mm × 628 mm	403.00	2.00	69.21	–	423.29	nr	**492.50**
508 mm × 923 mm	429.00	2.00	69.21	–	450.63	nr	**519.84**
508 mm × 1218 mm	455.00	2.50	86.51	–	477.97	nr	**564.48**
Double door							
1143 mm × 2057 mm	2548.00	3.50	121.11	–	2675.72	nr	**2796.83**
Extra for							
pressed steel sills; to suit above windows	39.00	0.50	10.10	–	40.99	m	**51.09**
G + bar	78.00	–	–	–	81.90	m	**81.90**
simulated leaded light	78.00	–	–	–	81.90	m	**81.90**
Thermally broken composite double glazed aluminium/ timber windows; Velfac 200 or other approved; with a maximum glazing U value of 1.5 W/m² K; argon filled cavity; low-e glazing with laminated glass unless otherwise specified; including multipoint espagnolette locking mechanisms and other ironmongery							
Standard fixed casement windows							
900 mm × 900 mm single fixed pane; low-e glass 6/14/4	203.62	2.70	54.53	–	218.50	nr	**273.03**
900 mm × 2000 mm single fixed pane; low-e glass 6/14/4	434.44	5.94	119.96	–	463.25	nr	**583.21**
1200 mm × 1200 mm single fixed pane; low-e glass 6/14/4	278.46	4.75	95.93	–	298.38	nr	**394.31**
1200 mm × 2200 mm three fixed panes; low-e glass 6/14/4	556.92	7.90	159.55	–	593.24	nr	**752.79**
2200 mm × 2200 mm single fixed pane; low-e glass 6/12/6	706.86	12.00	242.34	–	754.34	nr	**996.68**

23 WINDOWS, SCREENS AND LIGHTS

Item	PC £	Labour hours	Labour £	Plant £	Material £	Unit	Total rate £
WINDOWS – cont							
Thermally broken composite double glazed aluminium/ timber windows – cont							
Outward opening standard sash casement windows							
900 mm × 900 mm top-hung sash; low-e glass 6/14/4	483.52	2.67	53.92	–	512.40	nr	**566.32**
900 mm × 2200 mm with small top-hung sash; fixed lower pane; low-e glass 6/14/4	483.52	5.90	119.15	–	511.99	nr	**631.14**
900 mm × 3000 mm with small top-hung sash; fixed lower pane; low-e glass 6/14/4	613.03	8.00	161.56	–	653.17	nr	**814.73**
1600 mm × 1600 mm with two side-hung sashes; low-e glass 4/16/4	567.63	2.60	52.51	–	603.57	nr	**656.08**
1600 mm × 1600 mm with two side-hung projecting sashes; low-e glass 6/14/4	631.89	2.60	52.51	–	667.34	nr	**719.85**
1800 mm × 900 mm with two side-hung projecting sashes; low-e glass 6/14 E	418.63	5.50	111.08	–	442.28	nr	**553.36**
1800 mm × 3000 mm with two side-hung projecting sashes; two top-hung sashes; low-e glass 6/14/4	1300.00	15.00	302.94	–	1377.18	nr	**1680.12**
2000 mm × 1600 mm with one side-hung sash next to a top-hung projecting sash over a fixed sash; low-e glass 6/16/4	696.15	9.50	191.86	–	737.25	nr	**929.11**
1200 mm × 2200 mm with fixed lower sash and top-hung projecting upper sash; lower low-e glass 4 toughened/16/6.4; upper low-e glass 6/14/4	551.57	7.90	159.55	–	588.28	nr	**747.83**
1200 mm × 2200 mm with fixed lower sash and fully reversible upper sash; lower 6 toughened/14/4; upper low-e glass 4/16/4	599.76	7.90	159.55	–	638.88	nr	**798.43**
1800 mm × 900 mm with two side-hung projecting sashes; low-e glass 6/14/4; 60 minute fire integrity	631.89	5.50	111.08	–	668.48	nr	**779.56**
Outward opening standard doors							
1800 mm × 2200 mm French casement patio door; low-e toughened glass 4/16/4; deadlock; handles; cylinder	2750.00	13.00	262.54	–	2899.59	nr	**3162.13**
Allternative cavity fill							
Extra for Krypton cavity fill (over Argon)	–	–	–	–	22.49	m²	**22.49**
uPVC windows; Profile 22 or other equal and approved; reinforced where appropriate with aluminium alloy; including standard ironmongery; sills and factory glazed with low-e 24 mm double glazing; fixed in position; including lugs plugged and screwed to brickwork or blockwork							
Casement/fixed light; including e.p.d.m. glazing gaskets and weather seals							
630 mm × 900 mm; ref P109C	72.29	1.25	43.25	–	76.08	nr	**119.33**
630 mm × 1200 mm; ref P112V	83.64	1.25	43.25	–	88.04	nr	**131.29**
1200 mm × 1200 mm; ref P212C	127.80	1.25	43.25	–	134.46	nr	**177.71**
1770 mm × 1200 mm; ref P312CC	255.12	1.50	51.90	–	268.14	nr	**320.04**

23 WINDOWS, SCREENS AND LIGHTS

Item	PC £	Labour hours	Labour £	Plant £	Material £	Unit	Total rate £
Casement/fixed light; including vents; e.p.d.m. glazing gaskets and weather seals							
630 mm × 900 mm; ref P109V	73.13	1.25	43.25	–	76.97	nr	**120.22**
630 mm × 1200 mm; ref P112C	50.71	1.25	43.25	–	53.47	nr	**96.72**
1200 mm × 1200 mm; ref P212W	77.31	1.25	43.25	–	81.44	nr	**124.69**
1200 mm × 1200 mm; ref P212CV	131.29	1.25	43.25	–	138.13	nr	**181.38**
1770 mm × 1200 mm; ref P312WW	166.92	1.50	51.90	–	175.53	nr	**227.43**
Secured by design accreditation							
Casement/fixed light; including e.p.d.m. glazing gaskets and weather seals							
630 mm × 900 mm; ref P109C	73.76	1.25	43.25	–	77.63	nr	**120.88**
630 mm × 1200 mm; ref P112V	85.34	1.25	43.25	–	89.83	nr	**133.08**
1200 mm × 1200 mm; ref P212C	130.41	1.25	43.25	–	137.19	nr	**180.44**
1770 mm × 1200 mm; ref P312CC	260.32	1.50	51.90	–	273.61	nr	**325.51**
Casement/fixed light; including vents; e.p.d.m. glazing gaskets and weather seals							
630 mm × 900 mm; ref P109V	74.63	1.25	43.25	–	78.54	nr	**121.79**
630 mm × 1200 mm; ref P112C	51.74	1.25	43.25	–	54.55	nr	**97.80**
1200 mm × 1200 mm; ref P212W	78.88	1.25	43.25	–	83.09	nr	**126.34**
1200 mm × 1200 mm; ref P212CV	133.97	1.25	43.25	–	140.93	nr	**184.18**
1770 mm × 1200 mm; ref P312WW	170.32	1.50	51.90	–	179.11	nr	**231.01**
WER A rating							
Casement/fixed light; including e.p.d.m. glazing gaskets and weather seals							
630 mm × 900 mm; ref P109C	76.78	1.25	43.25	–	80.80	nr	**124.05**
630 mm × 1200 mm; ref P112V	88.85	1.25	43.25	–	93.51	nr	**136.76**
1200 mm × 1200 mm; ref P212C	135.76	1.25	43.25	–	142.81	nr	**186.06**
1770 mm × 1200 mm; ref P312CC	270.99	1.50	51.90	–	284.80	nr	**336.70**
Casement/fixed light; including vents; e.p.d.m. glazing gaskets and weather seals							
630 mm × 900 mm; ref P109V	77.69	1.25	43.25	–	81.75	nr	**125.00**
630 mm × 1200 mm; ref P112C	53.87	1.25	43.25	–	56.78	nr	**100.03**
1200 mm × 1200 mm; ref P212W	82.12	1.25	43.25	–	86.50	nr	**129.75**
1200 mm × 1200 mm; ref P212CV	139.46	1.25	43.25	–	146.71	nr	**189.96**
1770 mm × 1200 mm; ref P312WW	177.31	1.50	51.90	–	186.45	nr	**238.35**
WER C rating							
Casement/fixed light; including e.p.d.m. glazing gaskets and weather seals							
630 mm × 900 mm; ref P109C	72.29	1.25	43.25	–	76.08	nr	**119.33**
630 mm × 1200 mm; ref P112V	83.64	1.25	43.25	–	88.04	nr	**131.29**
1200 mm × 1200 mm; ref P212C	127.80	1.25	43.25	–	134.46	nr	**177.71**
1770 mm × 1200 mm; ref P312CC	255.12	1.50	51.90	–	268.14	nr	**320.04**
Casement/fixed light; including vents; e.p.d.m. glazing gaskets and weather seals							
630 mm × 900 mm; ref P109V	73.13	1.25	43.25	–	76.97	nr	**120.22**
630 mm × 1200 mm; ref P112C	50.71	1.25	43.25	–	53.47	nr	**96.72**
1200 mm × 1200 mm; ref P212W	77.31	1.25	43.25	–	81.44	nr	**124.69**
1200 mm × 1200 mm; ref P212CV	131.29	1.25	43.25	–	138.13	nr	**181.38**
1770 mm × 1200 mm; ref P312WW	166.92	1.50	51.90	–	175.53	nr	**227.43**

23 WINDOWS, SCREENS AND LIGHTS

Item	PC £	Labour hours	Labour £	Plant £	Material £	Unit	Total rate £
WINDOWS – cont							
Colour finish uPVC windows							
Casement/fixed light; including e.p.d.m. glazing gaskets and weather seals							
630 mm × 900 mm; ref P109C	84.30	1.25	43.25	–	88.69	nr	**131.94**
630 mm × 1200 mm; ref P112V	97.54	1.25	43.25	–	102.64	nr	**145.89**
1200 mm × 1200 mm; ref P212C	149.04	1.25	43.25	–	156.75	nr	**200.00**
1770 mm × 1200 mm; ref P312CC	297.50	1.50	51.90	–	312.65	nr	**364.55**
Casement/fixed light; including vents; e.p.d.m. glazing gaskets and weather seals							
630 mm × 900 mm; ref P109V	85.29	1.25	43.25	–	89.73	nr	**132.98**
630 mm × 1200 mm; ref P112C	59.13	1.25	43.25	–	62.31	nr	**105.56**
1200 mm × 1200 mm; ref P212W	90.16	1.25	43.25	–	94.93	nr	**138.18**
1200 mm × 1200 mm; ref P212CV	153.10	1.25	43.25	–	161.03	nr	**204.28**
1770 mm × 1200 mm; ref P312WW	194.67	1.25	43.25	–	204.67	nr	**247.92**
uPVC windows; Profile 22 or other equal and approved; reinforced where appropriate with aluminium alloy; in refurbishment work, including standard ironmongery; sills and factory glazed with low-e 24 mm double glazing; removing existing windows and fixing new in position; including lugs plugged and screwed to brickwork or blockwork							
Casement/fixed light; including e.p.d.m. glazing gaskets and weather seals							
630 mm × 900 mm; ref P109C	72.29	2.50	86.51	–	76.08	nr	**162.59**
630 mm × 1200 mm; ref P112V	50.71	2.50	86.51	–	53.47	nr	**139.98**
1200 mm × 1200 mm; ref P212C	127.80	3.00	103.80	–	134.46	nr	**238.26**
1770 mm × 1200 mm; ref P312CC	255.12	3.25	112.45	–	268.14	nr	**380.59**
Casement/fixed light; including vents; e.p.d.m. glazing gaskets and weather seals							
630 mm × 900 mm; ref P109V	73.13	2.50	86.51	–	76.97	nr	**163.48**
630 mm × 1200 mm; ref P112C	83.64	2.75	95.15	–	88.04	nr	**183.19**
1200 mm × 1200 mm; ref P212W	77.31	3.00	103.80	–	81.44	nr	**185.24**
1200 mm × 1200 mm; ref P212CV	131.29	3.00	103.80	–	138.13	nr	**241.93**
1770 mm × 1200 mm; ref P312WW	166.92	3.25	112.45	–	175.53	nr	**287.98**
1770 mm × 1200 mm; ref P312CV	159.85	3.25	112.45	–	168.10	nr	**280.55**
Aluminium windows; Schuco AWS 50 (or similar) proprietary system or equal and approved							
Polyester powder coated solid colour matt finish or natural anodized window system of glass sealed units with 6.4 mm low-e coated laminated inner pane, air filled cavity and 6 mm clear annealed outer pane. Rates to include all brackets, membranes, cills, silicone seals, trade contractor preliminaries, including external access equipment							
Ribbon construction windows 1.5 m high	–	–	–	–	–	m²	**475.00**
Punched hole windows fixing into prepared apertures by others	–	–	–	–	–	m²	**510.00**

23 WINDOWS, SCREENS AND LIGHTS

Item	PC £	Labour hours	Labour £	Plant £	Material £	Unit	Total rate £
Extra for							
1.25 m wide × 1.5 m high opening vents, assuming tilt and turn operation	–	–	–	–	–	m²	**150.00**
neutral selective high performance coating in lieu of low-e, for assisting in solar control	–	–	–	–	–	m²	**40.00**
outer glass pane to be toughened and heat soak tested or heat strengthened in lieu of annealed	–	–	–	–	–	m²	**28.00**
inner laminated glass to be toughened and heat soak tested laminated, or heat strengthened laminated	–	–	–	–	–	m²	**55.00**
ROOFLIGHTS							
Rooflights, skylights, roof windows and frames; pre-glazed; treated Nordic Red Pine and aluminium trimmed Velux windows or other equal and approved; type U flashings and soakers (for tiles and pantiles up to 45 mm deep), and sealed double glazing unit (trimming opening not included)							
Roof windows; u-value = 1.4 W/m²k							
550 mm × 780 mm; ref GGL–3073-C02	239.52	1.85	37.36	–	251.62	nr	**288.98**
550 mm × 980 mm; ref GGL–3073-C04	249.82	2.08	42.01	–	262.45	nr	**304.46**
660 mm × 1180 mm; ref GGL–3073-F06	289.31	2.31	46.65	–	303.93	nr	**350.58**
780 mm × 980 mm; ref GGL–3073-M04	274.72	2.31	46.65	–	288.61	nr	**335.26**
780 mm × 1180 mm; ref GGL–3073-M06	315.93	2.78	56.14	–	331.96	nr	**388.10**
780 mm × 1400 mm; ref GGL–3073-M08	329.66	2.31	46.65	–	346.35	nr	**393.00**
940 mm × 1600 mm; ref GGL–3073-P10	406.93	2.78	56.14	–	427.51	nr	**483.65**
1140 mm × 1180 mm; ref GGL–3073-S06	389.76	2.78	56.14	–	409.48	nr	**465.62**
1340 mm × 980 mm; ref GGL–3073-U04	389.76	2.78	56.14	–	409.48	nr	**465.62**
extra for electric powered windows plugged in to adjacent power supply	–	–	–	–	247.89	nr	**247.89**
Rooflights, skylights, roof windows and frames; uPVC; plugged and screwed to concrete; or screwed to timber							
Rooflight; Cox Suntube range or other equal and approved; double skin polycarbonate dome							
230 mm dia.; for flat roof using felt or membrane	239.28	2.50	50.48	–	251.80	nr	**302.28**
230 mm dia.; for up to 30° pitch roof with standard tiles	264.56	3.00	60.59	–	278.24	nr	**338.83**
230 mm dia.; for up to 30° pitch roof with bold roll tiles	245.60	3.00	60.59	–	258.37	nr	**318.96**
300 mm dia.; for flat roof using felt or membrane	388.26	2.50	50.48	–	408.23	nr	**458.71**
300 mm dia.; for up to 30° pitch roof with standard tiles	388.26	3.00	60.59	–	408.12	nr	**468.71**
300 mm dia.; for up to 30° pitch roof with bold roll tiles	416.25	3.00	60.59	–	437.56	nr	**498.15**

23 WINDOWS, SCREENS AND LIGHTS

Item	PC £	Labour hours	Labour £	Plant £	Material £	Unit	Total rate £
ROOFLIGHTS – cont							
Rooflights, skylights, roof windows and frames – cont							
Rooflight; Cox Galaxy range or other equal and approved; double skin polycarbonate dome only; fitting to existing kerb							
600 mm × 600 mm	111.06	1.50	30.29	–	116.92	nr	**147.21**
900 mm × 900 mm	204.96	1.75	35.34	–	215.61	nr	**250.95**
1200 mm × 1800 mm	601.35	2.00	40.39	–	631.92	nr	**672.31**
Rooflight; Cox Galaxy range or other equal and approved; triple skin polycarbonate dome only; fitting to existing kerb							
600 mm × 600 mm rooflight	171.56	1.50	30.29	–	180.43	nr	**210.72**
900 mm × 900 mm rooflight	307.90	1.75	35.34	–	323.69	nr	**359.03**
1200 mm × 1800 mm rooflight	930.92	2.00	40.39	–	977.97	nr	**1018.36**
Rooflight; Cox Trade range or other equal and approved; double skin polycarbonate fixed light dome on 150 mm PVC upstand							
600 mm × 600 mm	190.52	2.00	40.39	–	200.39	nr	**240.78**
900 mm × 900 mm	307.90	2.25	45.44	–	323.74	nr	**369.18**
1200 mm × 1800 mm	609.48	2.50	50.48	–	640.50	nr	**690.98**
Rooflight; Cox Trade range or other equal and approved; double skin polycarbonate manual opening light dome on 150 mm PVC upstand							
600 mm × 600 mm	297.97	2.50	50.48	–	313.61	nr	**364.09**
900 mm × 900 mm	431.60	2.75	55.53	–	454.03	nr	**509.56**
1200 mm × 1800 mm	811.73	3.10	62.60	–	853.27	nr	**915.87**
Rooflight; Cox Trade range or other equal and approved; double skin polycarbonate electric opening light dome on 150 mm PVC upstand (not including adjacent power supply)							
600 mm × 600 mm	506.54	2.75	55.53	–	532.61	nr	**588.14**
900 mm × 900 mm	674.48	3.00	60.59	–	709.05	nr	**769.64**
1200 mm × 1800 mm	1131.37	3.50	70.69	–	1188.88	nr	**1259.57**
Rooflight; Cox Trade range or other equal and approved; triple skin polycarbonate fixed light dome on 150 mm PVC upstand							
600 mm × 600 mm	223.92	2.00	40.39	–	235.47	nr	**275.86**
900 mm × 900 mm	368.39	2.00	40.39	–	387.16	nr	**427.55**
1200 m × 1800 mm	746.73	2.50	50.48	–	784.61	nr	**835.09**
Rooflight; Cox Trade range or other equal and approved; triple skin polycarbonate manual opening light dome on 150 mm PVC upstand							
600 mm × 600 mm	331.37	2.50	50.48	–	348.69	nr	**399.17**
900 mm × 900 mm	492.09	2.50	50.48	–	517.44	nr	**567.92**
1200 m × 1800 mm	948.98	3.10	62.60	–	997.38	nr	**1059.98**

23 WINDOWS, SCREENS AND LIGHTS

Item	PC £	Labour hours	Labour £	Plant £	Material £	Unit	Total rate £
Rooflight; Cox Trade range or other equal and approved; triple skin polycarbonate electric opening light dome on 150 mm PVC upstand							
600 mm × 600 mm light	539.95	2.75	55.53	–	567.69	nr	**623.22**
900 mm × 900 mm light	734.98	2.75	55.53	–	772.47	nr	**828.00**
1200 mm × 1800 mm light	1268.62	3.50	70.69	–	1333.00	nr	**1403.69**
Rooflight; Cox 2000 range or other equal and approved double skin polycarbonate fixed light dome on 235 mm solid core PVC upstand							
600 mm × 600 mm	609.48	2.00	40.39	–	640.30	nr	**680.69**
900 mm × 900 mm	826.59	2.50	50.48	–	868.27	nr	**918.75**
1200 mm × 1800 mm	2204.73	3.00	60.59	–	2315.37	nr	**2375.96**
Rooflight; Cox 2000 range or other equal and approved double skin polycarbonate manual opening light dome on 235 mm solid core PVC upstand							
600 mm × 600 mm	837.02	2.50	50.48	–	879.62	nr	**930.10**
900 mm × 900 mm	1054.14	3.00	60.59	–	1107.59	nr	**1168.18**
1200 mm × 1800 mm	2641.75	3.60	72.70	–	2774.64	nr	**2847.34**
Rooflight; Cox 2000 range or other equal and approved double skin polycarbonate electric opening light dome on 235 mm solid core PVC upstand (not including adjacent power supply)							
600 mm × 600 mm	1101.57	2.75	55.53	–	1157.39	nr	**1212.92**
900 mm × 900 mm	1318.68	3.25	65.64	–	1385.37	nr	**1451.01**
1200 mm × 1800 mm	2974.03	4.00	80.78	–	3123.53	nr	**3204.31**
Rooflight; Cox 2000 range or other equal and approved triple skin polycarbonate fixed light dome on 235 mm solid core PVC upstand							
600 mm × 600 mm	820.76	2.00	40.39	–	862.14	nr	**902.53**
900 mm × 900 mm	1199.14	2.50	50.48	–	1259.44	nr	**1309.92**
1200 mm × 1800 mm	2945.25	3.00	60.59	–	3092.91	nr	**3153.50**
Rooflight; Cox 2000 range or other equal and approved triple skin polycarbonate manual opening light dome on 235 mm solid core PVC upstand							
600 mm × 600 mm	1048.31	2.50	50.48	–	1101.47	nr	**1151.95**
900 mm × 900 mm	1426.68	3.00	60.59	–	1498.77	nr	**1559.36**
1200 mm × 1800 mm	3382.27	3.60	72.70	–	3552.18	nr	**3624.88**
Rooflight; Cox 2000 range or other equal and approved triple skin polycarbonate electric opening light dome on 235 mm solid core PVC upstand (not incluing adjacent power supply)							
600 mm × 600 mm	1312.85	2.75	55.53	–	1379.24	nr	**1434.77**
900 mm × 900 mm	1691.23	3.25	65.64	–	1776.54	nr	**1842.18**
1200 mm × 1800 mm	3714.55	4.00	80.78	–	3901.08	nr	**3981.86**

23 WINDOWS, SCREENS AND LIGHTS

Item	PC £	Labour hours	Labour £	Plant £	Material £	Unit	Total rate £
LOUVRES							
Louvres, Brise Soleils and frames; polyester powder coated aluminium; fixing in position including brackets							
Brise soleil, to mitigate the effects of solar gain. This rate assumes a single natural anodized extruded aluminium fin, with brackets and orientated either horizontally or vertically. The quantity of fins per storey height should be calculated to achieve desired shading.							
300 mm deep	–	–	–	–	–	m	**150.00**

24 DOORS, SHUTTERS AND HATCHES

Item	PC £	Labour hours	Labour £	Plant £	Material £	Unit	Total rate £
EXTERNAL DOORS AND FRAMES							
EXTERNAL DOORS							
Note: door ironmogery is not included unless stated							
Doors; standard matchboarded; wrought softwood							
Matchboarded, framed, ledged and braced doors; 44 mm thick overall; 19 mm thick tongued, grooved and V-jointed boarding; one side vertical boarding							
762 mm × 1981 mm	77.77	1.67	33.73	–	81.66	nr	**115.39**
838 mm × 1981 mm	84.97	1.67	33.73	–	89.22	nr	**122.95**
Flush door; external quality; skeleton or cellular core; plywood faced both sides; lipped all round							
762 mm × 1981 mm × 54 mm	99.42	1.62	32.72	–	104.39	nr	**137.11**
838 mm × 1981 mm × 54 mm	101.58	1.62	32.72	–	106.66	nr	**139.38**
Fire doors							
Flush door; half-hour fire-resisting; external quality; skeleton or cellular core; plywood faced both sides; lipped on all four edges							
762 mm × 1981 mm × 44 mm	209.51	1.71	34.53	–	219.99	nr	**254.52**
838 mm × 1981 mm × 44 mm	211.65	1.71	34.53	–	222.23	nr	**256.76**
Flush door; half-hour fire-resisting; external quality wlth 6 mm Georgian wired polished plate glass opening; skeleton or cellular core; plywood faced both sides; lipped on all four edges; including glazing beads							
762 mm × 1981 mm × 44 mm	209.51	1.71	34.53	–	219.99	nr	**254.52**
838 mm × 1981 mm × 44 mm	211.65	1.71	34.53	–	222.23	nr	**256.76**
726 mm × 2040 mm × 44 mm	209.51	1.71	34.53	–	219.99	nr	**254.52**
826 mm × 2040 mm × 44 mm	211.74	1.71	34.53	–	222.33	nr	**256.86**
926 mm × 2040 mm × 44 mm	223.42	1.71	34.53	–	234.59	nr	**269.12**
External softwood door frame composite standard joinery sets							
External door frame composite set; 56 mm × 78 mm wide (finished); for external doors							
762 mm × 1981 mm × 44 mm	55.22	0.75	15.15	–	58.16	nr	**73.31**
813 mm × 1981 mm × 44 mm	56.70	0.75	15.15	–	59.70	nr	**74.85**
838 mm × 1981 mm × 44 mm	56.31	0.75	15.15	–	59.30	nr	**74.45**

24 DOORS, SHUTTERS AND HATCHES

Item	PC £	Labour hours	Labour £	Plant £	Material £	Unit	Total rate £
EXTERNAL DOORS AND FRAMES – cont							
Doorsets; anti-vandal security door and frame units; Bastion Security Ltd or other equal and approved; to BS 5051; factory primed; fixing with frame anchors to masonry; cutting mortices; external							
46 mm thick insulated door with birch grade plywood; sheet steel bonded into door core; 2 mm thick polyester coated laminate finish; hardwood lippings all edges; 95 mm × 65 mm hardwood frame; polyester coated standard ironmongery; weather stripping all round; low projecting aluminium threshold; plugging; screwing							
for 980 mm × 2100 mm structural opening; single doorsets; panic bolt	–	–	–	–	–	nr	**15.00**
for 1830 mm × 2100 mm structural opening; double doorsets; panic bolt	–	–	–	–	–	nr	**2400.00**
Doorsets; galvanized steel door and frame units; treated softwood frame, primed hardwood sill; fixing in position; plugged and screwed to brickwork or blockwork							
Door and frame							
838 mm × 1981 mm	–	2.78	56.14	–	470.07	nr	**526.21**
Doorsets; steel security door and frame; Hormann or other equal and approved; including ironmongery, weather seals and all necessary fixing accessories							
Horman ref E55–1 doorset.							
To suit structural opening 1100 × 2105 mm; fire-rating 30 minutes; acoustic rating 38 dB; including stainless steel ironmongery	–	–	–	–	–	nr	**1858.92**
To suit structural opening 2000 × 2105 mm; fire-rating 30 minutes; acoustic rating 38 dB; including stainless steel ironmongery	–	–	–	–	–	nr	**2761.83**
Doorsets; steel bullet-resistant door and frame units; Wormald Doors or other equal and approved; Medite laquered panels; ironmongery							
Door and frame							
1000 mm × 2060 mm overall; fixed to masonry	–	–	–	–	–	nr	**3877.19**

24 DOORS, SHUTTERS AND HATCHES

Item	PC £	Labour hours	Labour £	Plant £	Material £	Unit	Total rate £
uPVC doors; Profile 22 or other equal and approved; reinforced where appropriate with aluminium alloy; including standard ironmongery; thresholds and factory glazed with low-e 24 mm double glazing; fixed in position; including lugs plugged and screwed to brickwork or blockwork							
Fixed light; including e.p.d.m. glazing gaskets and weather seals							
2100 × 900 uPVC door with midrail; half glazed	206.44	1.50	51.90	–	216.95	nr	**268.85**
2100 × 900 uPVC door with midrail; glazed	248.57	1.50	51.90	–	261.22	nr	**313.12**
2100 × 900 solid composite door	405.51	1.50	51.90	–	426.01	nr	**477.91**
Secured by design accrediation							
Fixed light; including e.p.d.m. glazing gaskets and weather seals							
2100 × 900 uPVC door with midrail; half glazed	238.35	2.00	69.21	–	250.46	nr	**319.67**
2100 × 900 uPVC door with midrail; glazed	248.57	1.50	51.90	–	261.22	nr	**313.12**
WER A rating							
Fixed light; including e.p.d.m. glazing gaskets and weather seals							
2100 × 900 uPVC door with midrail; half glazed	219.29	2.00	69.21	–	230.44	nr	**299.65**
2100 × 900 uPVC door with midrail; glazed	264.03	1.50	51.90	–	277.45	nr	**329.35**
WER C rating							
Fixed light; including e.p.d.m. glazing gaskets and weather seals							
2100 × 900 uPVC door with midrail; half glazed	206.44	2.00	69.21	–	216.95	nr	**286.16**
2100 × 900 uPVC door with midrail; glazed	248.57	1.50	51.90	–	261.22	nr	**313.12**
Colour finish							
Fixed light; including e.p.d.m. glazing gaskets and weather seals							
2100 × 900 upvc door with midrail; half glazed	200.22	2.00	69.21	–	210.42	nr	**279.63**
2100 × 900 upvc door with midrail; glazed	288.17	1.50	51.90	–	302.80	nr	**354.70**
Sliding/folding; aluminium double glazed sliding patio doors; white acrylic finish; with and including 18 thick annealed double glazing; fixed in position; including lugs plugged and screwed to brickwork or blockwork							
Patio doors							
1800 mm × 2100 mm; ref PF1821	1040.70	8.00	276.82	–	1093.27	nr	**1370.09**
2400 m × 2100 mm; ref PF2421	1123.12	8.00	276.82	–	1179.81	nr	**1456.63**
2700 mm × 2100 mm; ref PF2721	1272.60	8.00	276.82	–	1336.77	nr	**1613.59**

24 DOORS, SHUTTERS AND HATCHES

Item	PC £	Labour hours	Labour £	Plant £	Material £	Unit	Total rate £
EXTERNAL DOORS AND FRAMES – cont							
External softwood door frame composite standard joinery sets							
External door frame composite set; 56 mm × 78 mm wide (finished); for external doors							
762 mm × 1981 mm × 44 mm	55.22	0.75	15.15	–	58.16	nr	**73.31**
813 mm × 1981 mm × 44 mm	56.70	0.75	15.15	–	59.70	nr	**74.85**
838 mm × 1981 mm × 44 mm	56.31	0.75	15.15	–	59.30	nr	**74.45**
External door frame composite set; 56 mm × 78 mm wide (finished); with 45 mm × 140 mm (finished) hardwood sill; for external doors							
686 mm × 1981 mm × 44 mm	68.43	1.00	20.19	–	72.03	nr	**92.22**
762 mm × 1981 mm × 44 mm	70.85	1.00	20.19	–	74.57	nr	**94.76**
838 mm × 1981 mm × 44 mm	83.53	1.00	20.19	–	87.89	nr	**108.08**
826 mm × 2040 mm × 44 mm	73.25	1.00	20.19	–	77.09	nr	**97.28**
INTERNAL DOORS AND FRAMES							
INTERNAL DOORS							
Note: door ironmogery is not included							
Moulded panel doors; white based coated facings suitable for paint finish only; two, four or six panel options							
526 mm × 2040 mm × 40 mm	37.48	0.75	15.15	–	39.35	nr	**54.50**
626 mm × 2040 mm × 40 mm	37.48	0.75	15.15	–	39.35	nr	**54.50**
726 mm × 2040 mm × 40 mm	37.48	0.75	15.15	–	39.35	nr	**54.50**
826 mm × 2040 mm × 40 mm	41.14	0.75	15.15	–	43.20	nr	**58.35**
926 mm × 2040 mm × 40 mm	46.63	0.75	15.15	–	48.96	nr	**64.11**
Doors; standard flush; softwood composition							
Flush door; internal quality; skeleton or cellular core; hardboard faced both sides; lipped on two long edges; Jeld-Wen or other equal and approved							
457 mm × 1981 mm × 35 mm	31.18	1.16	23.43	–	32.74	nr	**56.17**
533 mm × 1981 mm × 35 mm	31.18	1.16	23.43	–	32.74	nr	**56.17**
610 mm × 1981 mm × 35 mm	31.18	1.16	23.43	–	32.74	nr	**56.17**
686 mm × 1981 mm × 35 mm	31.18	1.16	23.43	–	32.74	nr	**56.17**
762 mm × 1981 mm × 35 mm	31.18	1.16	23.43	–	32.74	nr	**56.17**
838 mm × 1981 mm × 35 mm	35.33	1.16	23.43	–	37.10	nr	**60.53**
626 mm × 2040 mm × 40 mm	33.26	1.16	23.43	–	34.92	unr	**58.35**
726 mm × 2040 mm × 40 mm	33.26	1.16	23.43	–	34.92	nr	**58.35**
826 mm × 2040 mm × 40 mm	33.26	1.16	23.43	–	34.92	nr	**58.35**
926 mm × 2040 mm × 40 mm	35.33	1.16	23.43	–	37.10	nr	**60.53**
Flush door; internal quality; skeleton or cellular core; faced both sides; lipped on two long edges; Jeld-Wen paint grade veneer or other equal and approved							
457 mm × 1981 mm × 35 mm	35.33	1.16	23.43	–	37.10	nr	**60.53**
533 mm × 1981 mm × 35 mm	35.33	1.16	23.43	–	37.10	nr	**60.53**
610 mm × 1981 mm × 35 mm	35.33	1.16	23.43	–	37.10	nr	**60.53**

24 DOORS, SHUTTERS AND HATCHES

Item	PC £	Labour hours	Labour £	Plant £	Material £	Unit	Total rate £
686 mm × 1981 mm × 35 mm	35.33	1.16	23.43	–	37.10	nr	**60.53**
762 mm × 1981 mm × 35 mm	35.33	1.16	23.43	–	37.10	nr	**60.53**
838 mm × 1981 mm × 35 mm	37.42	1.16	23.43	–	39.29	nr	**62.72**
526 mm × 2040 mm × 40 mm	36.37	1.16	23.43	–	38.19	nr	**61.62**
626 mm × 2040 mm × 40 mm	36.37	1.16	23.43	–	38.19	nr	**61.62**
726 mm × 2040 mm × 40 mm	37.42	1.16	23.43	–	39.29	nr	**62.72**
826 mm × 2040 mm × 40 mm	43.66	1.16	23.43	–	45.84	nr	**69.27**
Flush door; internal quality; skeleton or cellular core; chipboard veneered; faced both sides; lipped on two long edges; Jeld-Wen Sapele veneered or other equal and approved							
457 mm × 1981 mm × 35 mm	58.20	1.25	25.24	–	61.11	nr	**86.35**
533 mm × 1981 mm × 35 mm	58.20	1.25	25.24	–	61.11	nr	**86.35**
610 mm × 1981 mm × 35 mm	58.20	1.25	25.24	–	61.11	nr	**86.35**
686 mm × 1981 mm × 35 mm	58.20	1.25	25.24	–	61.11	nr	**86.35**
762 mm × 1981 mm × 35 mm	58.20	1.25	25.24	–	61.11	nr	**86.35**
838 mm × 1981 mm × 35 mm	64.43	1.25	25.24	–	67.65	nr	**92.89**
526 mm × 2040 mm × 40 mm	64.43	1.25	25.24	–	67.65	nr	**92.89**
626 mm × 2040 mm × 40 mm	64.43	1.25	25.24	–	67.65	nr	**92.89**
726 mm × 2040 mm × 40 mm	64.43	1.25	25.24	–	67.65	nr	**92.89**
826 mm × 2040 mm × 40 mm	68.59	1.25	25.24	–	72.02	nr	**97.26**
926 mm × 2040 mm × 40 mm	70.67	1.25	25.24	–	74.20	nr	**99.44**
Doors; purpose-made panelled; wrought softwood							
Panelled doors; one open panel for glass; including glazing beads							
686 mm × 1981 mm × 44 mm	82.35	1.62	32.72	–	86.47	nr	**119.19**
762 mm × 1981 mm × 44 mm	83.03	1.62	32.72	–	87.18	nr	**119.90**
838 mm × 1981 mm × 44 mm	83.72	1.62	32.72	–	87.91	nr	**120.63**
Panelled doors; two open panel for glass; including glazing beads							
686 mm × 1981 mm × 44 mm	115.13	1.62	32.72	–	120.89	nr	**153.61**
762 mm × 1981 mm × 44 mm	116.13	1.62	32.72	–	121.94	nr	**154.66**
838 mm × 1981 mm × 44 mm	117.12	1.62	32.72	–	122.98	nr	**155.70**
Panelled doors; four 19 mm thick plywood panels; mouldings worked on solid both sides							
686 mm × 1981 mm × 44 mm	174.84	1.62	32.72	–	183.58	nr	**216.30**
762 mm × 1981 mm × 44 mm	177.10	1.62	32.72	–	185.96	nr	**218.68**
838 mm × 1981 mm × 44 mm	179.36	1.62	32.72	–	188.33	nr	**221.05**
Panelled doors; six 25 mm thick panels raised and fielded; mouldings worked on solid both sides							
686 mm × 1981 mm × 44 mm	321.70	1.94	39.18	–	337.79	nr	**376.97**
762 mm × 1981 mm × 44 mm	324.91	1.94	39.18	–	341.16	nr	**380.34**
838 mm × 1981 mm × 44 mm	328.12	1.94	39.18	–	344.53	nr	**383.71**
rebated edges beaded	–	–	–	–	2.04	m	**2.04**
rounded edges or heels	–	–	–	–	0.46	m	**0.46**
weatherboard fixed to bottom rail	–	0.23	4.64	–	7.21	m	**11.85**
stopped groove for weatherboard	–	–	–	–	2.31	m	**2.31**

24 DOORS, SHUTTERS AND HATCHES

Item	PC £	Labour hours	Labour £	Plant £	Material £	Unit	Total rate £
INTERNAL DOORS AND FRAMES – cont							
Doors; purpose-made panelled; selected Sapele							
Panelled doors; one open panel for glass; including glazing beads							
686 mm × 1981 mm × 44 mm	111.49	2.31	46.65	–	117.06	nr	**163.71**
762 mm × 1981 mm × 44 mm	112.99	2.31	46.65	–	118.64	nr	**165.29**
838 mm × 1981 mm × 44 mm	114.51	2.31	46.65	–	120.24	nr	**166.89**
686 mm × 1981 mm × 57 mm	119.16	2.54	51.29	–	125.12	nr	**176.41**
762 mm × 1981 mm × 57 mm	120.95	2.54	51.29	–	127.00	nr	**178.29**
838 mm × 1981 mm × 57 mm	122.74	2.54	51.29	–	128.88	nr	**180.17**
Panelled doors; 250 mm wide cross tongued intermediate rail; two open panels for glass; mouldings worked on the solid one side; 19 mm × 13 mm beads one side; fixing with brass cups and screws							
686 mm × 1981 mm × 44 mm	170.43	2.31	46.65	–	178.95	nr	**225.60**
762 mm × 1981 mm × 44 mm	173.31	2.31	46.65	–	181.98	nr	**228.63**
838 mm × 1981 mm × 44 mm	181.74	2.31	46.65	–	190.83	nr	**237.48**
686 mm × 1981 mm × 57 mm	181.74	2.54	51.29	–	190.83	nr	**242.12**
762 mm × 1981 mm × 57 mm	185.14	2.54	51.29	–	194.40	nr	**245.69**
838 mm × 1981 mm × 57 mm	188.62	2.54	51.29	–	198.05	nr	**249.34**
Panelled doors; four panels (19 mm thick for 44 mm doors, 25 mm thick for 57 mm doors); mouldings worked on solid both sides							
686 mm × 1981 mm × 44 mm	239.49	2.31	46.65	–	251.46	nr	**298.11**
762 mm × 1981 mm × 44 mm	258.16	2.31	46.65	–	271.07	nr	**317.72**
838 mm × 1981 mm × 44 mm	270.93	2.31	46.65	–	284.48	nr	**331.13**
686 mm × 1981 mm × 57 mm	244.33	2.54	51.29	–	256.55	nr	**307.84**
762 mm × 1981 mm × 57 mm	264.55	2.54	51.29	–	277.78	nr	**329.07**
838 mm × 1981 mm × 57 mm	249.18	2.54	51.29	–	261.64	nr	**312.93**
Panelled doors; 150 mm wide stiles in one width; 430 mm wide cross tongued bottom rail; six panels raised and fielded one side (19 mm thick for 44 mm doors, 25 mm thick for 57 mm doors); mouldings worked on solid both sides							
686 mm × 1981 mm × 44 mm	408.82	2.31	46.65	–	429.26	nr	**475.91**
762 mm × 1981 mm × 44 mm	451.16	2.31	46.65	–	473.72	nr	**520.37**
838 mm × 1981 mm × 44 mm	459.73	2.31	46.65	–	482.72	nr	**529.37**
686 mm × 1981 mm × 57 mm	435.34	2.54	51.29	–	457.11	nr	**508.40**
762 mm × 1981 mm × 57 mm	480.04	2.54	51.29	–	504.04	nr	**555.33**
838 mm × 1981 mm × 57 mm	490.96	2.54	51.29	–	515.51	nr	**566.80**
rebated edges beaded	–	–	–	–	2.58	m	**2.58**
rounded edges or heels	–	–	–	–	0.68	m	**0.68**
weatherboard fixed to bottom rail	–	0.31	6.26	–	9.75	m	**16.01**
stopped groove for weatherboard	–	–	–	–	2.40	m	**2.40**

24 DOORS, SHUTTERS AND HATCHES

Item	PC £	Labour hours	Labour £	Plant £	Material £	Unit	Total rate £
Internal white foiled moisture-resistant MDF door lining composite standard joinery set							
22 mm × 77 mm wide (finished) set; with loose stops; for internal doors							
610 mm × 1981 mm × 35 mm	8.46	0.70	14.13	–	9.06	nr	**23.19**
686 mm × 1981 mm × 35 mm	8.41	0.70	14.13	–	9.01	nr	**23.14**
762 mm × 1981 mm × 35 mm	8.41	0.70	14.13	–	9.01	nr	**23.14**
838 mm × 1981 mm × 35 mm	9.27	0.70	14.13	–	9.90	nr	**24.03**
864 mm × 1981 mm × 35 mm	8.41	0.70	14.13	–	9.01	nr	**23.14**
22 mm × 150 mm wide (finished) set; with loose stops; for internal doors							
610 mm × 1981 mm × 35 mm	9.31	0.70	14.13	–	9.95	nr	**24.08**
686 mm × 1981 mm × 35 mm	9.27	0.70	14.13	–	9.90	nr	**24.03**
762 mm × 1981 mm × 35 mm	9.27	0.70	14.13	–	9.90	nr	**24.03**
838 mm × 1981 mm × 35 mm	9.27	0.70	14.13	–	9.90	nr	**24.03**
864 mm × 1981 mm × 35 mm	9.27	0.70	14.13	–	9.90	nr	**24.03**
Internal softwood door lining set; with loose stops							
32 mm × 115 mm wide (finished) set; with loose stops; for internal doors							
686/762/838 wide doors	12.65	0.70	14.13	–	13.46	nr	**27.59**
32 mm × 138 mm wide (finished) set; with loose stops; for internal doors							
686/762/838 wide doors	14.45	0.70	14.13	–	15.35	nr	**29.48**
Internal softwood fire door door lining set; with loose stops							
38 mm × 115 mm wide (finished) set; with loose stops; for internal doors							
686/762/838 wide doors	19.05	0.70	14.13	–	20.18	nr	**34.31**
38 mm × 138 mm wide (finished) set; with loose stops; for internal doors							
686/762/838 wide doors	22.61	0.70	14.13	–	23.92	nr	**38.05**
Door frames and door linings, sets; purpose-made; wrought softwood							
Jambs and heads; as linings							
32 mm × 63 mm	–	0.16	3.23	–	4.98	m	**8.21**
32 mm × 100 mm	–	0.16	3.23	–	5.61	m	**8.84**
32 mm × 140 mm	–	0.16	3.23	–	5.97	m	**9.20**
Jambs and heads; as frames; rebated, rounded and grooved							
44 mm × 75 mm	–	0.16	3.23	–	8.00	m	**11.23**
44 mm × 100 mm	–	0.16	3.23	–	8.61	m	**11.84**
44 mm × 115 mm	–	0.16	3.23	–	8.65	m	**11.88**
44 mm × 140 mm	–	0.19	3.83	–	9.07	m	**12.90**
57 mm × 100 mm	–	0.19	3.83	–	9.21	m	**13.04**
57 mm × 125 mm	–	0.19	3.83	–	9.72	m	**13.55**
69 mm × 88 mm	–	0.19	3.83	–	9.37	m	**13.20**
69 mm × 100 mm	–	0.19	3.83	–	10.04	m	**13.87**
69 mm × 125 mm	–	0.20	4.04	–	10.66	m	**14.70**
69 mm × 150 mm	–	0.20	4.04	–	11.28	m	**15.32**

24 DOORS, SHUTTERS AND HATCHES

Item	PC £	Labour hours	Labour £	Plant £	Material £	Unit	Total rate £
INTERNAL DOORS AND FRAMES – cont							
Door frames and door linings, sets – cont							
Jambs and heads – cont							
94 mm × 100 mm	–	0.23	4.64	–	15.21	m	**19.85**
94 mm × 150 mm	–	0.23	4.64	–	17.74	m	**22.38**
Mullions and transoms; in linings							
32 mm × 63 mm	–	0.11	2.23	–	6.64	m	**8.87**
32 mm × 100 mm	–	0.11	2.23	–	7.27	m	**9.50**
32 mm × 140 mm	–	0.11	2.23	–	7.58	m	**9.81**
Mullions and transoms; in frames; twice rebated, rounded and grooved							
44 mm × 75 mm	–	0.11	2.23	–	10.05	m	**12.28**
44 mm × 100 mm	–	0.11	2.23	–	10.47	m	**12.70**
44 mm × 115 mm	–	0.11	2.23	–	10.47	m	**12.70**
44 mm × 140 mm	–	0.13	2.63	–	10.88	m	**13.51**
57 mm × 100 mm	–	0.13	2.63	–	11.01	m	**13.64**
57 mm × 125 mm	–	0.13	2.63	–	11.53	m	**14.16**
69 mm × 88 mm	–	0.13	2.63	–	10.88	m	**13.51**
69 mm × 100 mm	–	0.13	2.63	–	11.51	m	**14.14**
Add 5% to the above material prices for selected softwood for staining							
Door frames and door linings, sets; purpose-made; medium density fireboard							
Jambs and heads; as linings							
18 mm × 126 mm	–	0.16	3.23	–	5.99	m	**9.22**
22 mm × 126 mm	–	0.16	3.23	–	6.21	m	**9.44**
25 mm × 126 mm	–	0.16	3.23	–	6.33	m	**9.56**
Door frames and door linings, sets; purpose-made; selected Sapele							
Jambs and heads; as linings							
32 mm × 63 mm	7.32	0.21	4.24	–	7.73	m	**11.97**
32 mm × 100 mm	9.04	0.21	4.24	–	9.54	m	**13.78**
32 mm × 140 mm	9.89	0.21	4.24	–	10.48	m	**14.72**
Jambs and heads; as frames; rebated, rounded and grooved							
44 mm × 75 mm	11.99	0.21	4.24	–	12.64	m	**16.88**
44 mm × 100 mm	13.66	0.21	4.24	–	14.38	m	**18.62**
44 mm × 115 mm	14.12	0.21	4.24	–	14.91	m	**19.15**
44 mm × 140 mm	14.80	0.25	5.05	–	15.62	m	**20.67**
57 mm × 100 mm	15.13	0.25	5.05	–	15.97	m	**21.02**
57 mm × 125 mm	16.56	0.25	5.05	–	17.47	m	**22.52**
69 mm × 88 mm	15.11	0.25	5.05	–	15.92	m	**20.97**
69 mm × 100 mm	16.81	0.25	5.05	–	17.74	m	**22.79**
69 mm × 125 mm	18.51	0.28	5.66	–	19.53	m	**25.19**
69 mm × 150 mm	20.20	0.28	5.66	–	21.30	m	**26.96**
94 mm × 100 mm	23.60	0.28	5.66	–	24.87	m	**30.53**
94 mm × 150 mm	28.95	0.28	5.66	–	30.49	m	**36.15**

24 DOORS, SHUTTERS AND HATCHES

Item	PC £	Labour hours	Labour £	Plant £	Material £	Unit	Total rate £
Mullions and transoms; in linings							
32 mm × 63 mm	9.19	0.15	3.03	–	9.65	m	**12.68**
32 mm × 100 mm	10.90	0.15	3.03	–	11.45	m	**14.48**
32 mm × 140 mm	11.76	0.15	3.03	–	12.35	m	**15.38**
Mullions and transoms; in frames; twice rebated, rounded and grooved							
44 mm × 75 mm	14.76	0.15	3.03	–	15.50	m	**18.53**
44 mm × 100 mm	15.89	0.15	3.03	–	16.68	m	**19.71**
44 mm × 115 mm	16.35	0.15	3.03	–	17.17	m	**20.20**
44 mm × 140 mm	17.03	0.17	3.43	–	17.88	m	**21.31**
57 mm × 100 mm	17.37	0.17	3.43	–	18.24	m	**21.67**
57 mm × 125 mm	18.79	0.17	3.43	–	19.73	m	**23.16**
69 mm × 88 mm	17.03	0.17	3.43	–	17.88	m	**21.31**
69 mm × 100 mm	19.16	0.17	3.43	–	20.12	m	**23.55**
Sills; once sunk weathered; once rebated, three times grooved							
63 mm × 175 mm	42.41	0.31	6.26	–	44.53	m	**50.79**
75 mm × 125 mm	40.85	0.31	6.26	–	42.89	m	**49.15**
75 mm × 150 mm	42.77	0.31	6.26	–	44.91	m	**51.17**
Door frames and door linings, sets; European Oak							
Sills; once sunk weathered; once rebated, three times grooved							
63 mm × 175 mm	69.71	0.31	6.26	–	73.20	m	**79.46**
75 mm × 125 mm	68.82	0.31	6.26	–	72.26	m	**78.52**
75 mm × 150 mm	75.24	0.31	6.26	–	79.00	m	**85.26**
FIRE DOORS							
Flush door; half-hour fire-resisting (FD30); hardboard faced both sides; Jeld-Wen or other equal and approved							
762 mm × 1981 mm × 44 mm	56.12	1.62	32.72	–	58.93	nr	**91.65**
838 mm × 1981 mm × 44 mm	60.28	1.62	32.72	–	63.29	nr	**96.01**
726 mm × 2040 mm × 44 mm	58.20	1.62	32.72	–	61.11	nr	**93.83**
826 mm × 2040 mm × 44 mm	58.20	1.62	32.72	–	61.11	nr	**93.83**
926 mm × 2040 mm × 44 mm	58.20	1.62	32.72	–	61.11	nr	**93.83**
Flush door; half-hour fire-resisting (FD30); chipboard veneered; faced both sides; lipped on two long edges; Jeld-Wen paint grade veneer or other equal and approved							
610 mm × 1981 mm × 44 mm	48.85	1.62	32.72	–	51.29	nr	**84.01**
686 mm × 1981 mm × 44 mm	48.85	1.62	32.72	–	51.29	nr	**84.01**
762 mm × 1981 mm × 44 mm	48.85	1.62	32.72	–	51.29	nr	**84.01**
838 mm × 1981 mm × 44 mm	48.85	1.62	32.72	–	51.29	nr	**84.01**
526 mm × 2040 mm × 44 mm	49.88	1.62	32.72	–	52.37	nr	**85.09**
626 mm × 2040 mm × 44 mm	49.88	1.62	32.72	–	52.37	nr	**85.09**
726 mm × 2040 mm × 44 mm	50.92	1.62	32.72	–	53.47	nr	**86.19**
826 mm × 2040 mm × 44 mm	57.16	1.62	32.72	–	60.02	nr	**92.74**

24 DOORS, SHUTTERS AND HATCHES

Item	PC £	Labour hours	Labour £	Plant £	Material £	Unit	Total rate £
INTERNAL DOORS AND FRAMES – cont							
FIRE DOORS – cont							
Moulded panel doors; half-hour fire-resisting (FD30); white based coated facings suitable for paint finish only; two, four or six panel options; Premdor Fireshield or other equal and approved							
526 mm × 2040 mm × 44 mm	95.99	1.10	22.22	–	100.79	nr	**123.01**
626 mm × 2040 mm × 44 mm	95.99	1.10	22.22	–	100.79	nr	**123.01**
726 mm × 2040 mm × 44 mm	95.99	1.10	22.22	–	100.79	nr	**123.01**
826 mm × 2040 mm × 44 mm	101.48	1.10	22.22	–	106.55	nr	**128.77**
926 mm × 2040 mm × 44 mm	106.05	1.10	22.22	–	111.35	nr	**133.57**
Flush door; half-hour fire-resisting (FD30); faced both sides; lipped on two long edges; Jeld-Wen Sapele veneered or other equal and approved							
610 mm × 1981 mm × 44 mm	89.38	1.71	34.53	–	93.85	nr	**128.38**
686 mm × 1981 mm × 44 mm	89.38	1.71	34.53	–	93.85	nr	**128.38**
762 mm × 1981 mm × 44 mm	89.38	1.71	34.53	–	93.85	nr	**128.38**
838 mm × 1981 mm × 44 mm	95.61	1.71	34.53	–	100.39	nr	**134.92**
726 mm × 2040 mm × 44 mm	95.61	1.71	34.53	–	100.39	nr	**134.92**
826 mm × 2040 mm × 44 mm	99.77	1.71	34.53	–	104.76	nr	**139.29**
926 mm × 2040 mm × 44 mm	101.85	1.71	34.53	–	106.94	nr	**141.47**
Flush door; half-hour fire-resisting (FD30); chipboard for painting; hardwood lipping two long edges; Premdor Fireshield or other equal and approved;							
526 mm × 2040 mm × 44 mm	50.29	1.62	32.72	–	52.80	nr	**85.52**
626 mm × 2040 mm × 44 mm	50.29	1.62	32.72	–	52.80	nr	**85.52**
726 mm × 2040 mm × 44 mm	50.29	1.62	32.72	–	52.80	nr	**85.52**
826 mm × 2040 mm × 44 mm	51.20	1.62	32.72	–	53.76	nr	**86.48**
926 mm × 2040 mm × 44 mm	64.91	1.62	32.72	–	68.16	nr	**100.88**
826 mm × 2040 mm × 44 mm; single side vision panel 150 mm × 700 mm; factory fitted clear fire-rated glass	202.05	1.71	34.53	–	212.15	nr	**246.68**
826 mm × 2040 mm × 44 mm; two side vision panels 150 mm × 700 mm; factory fitted clear fire-rated glass	261.48	1.71	34.53	–	274.55	nr	**309.08**
926 mm × 2040 mm × 44 mm; single side vision panel 150 mm × 700 mm; factory fitted clear fire-rated glass	207.54	1.71	34.53	–	217.92	nr	**252.45**
926 mm × 2040 mm × 44 mm; two side vision panels 150 mm × 700 mm; factory fitted clear fire-rated glass	266.96	1.71	34.53	–	280.31	nr	**314.84**
Flush door; half-hour fire-resisting (FD30); White Oak veneer; hardwood lipping all edges; Premdor Fireshield or other equal and approved;							
526 mm × 2040 mm × 44 mm	94.17	1.62	32.72	–	98.88	nr	**131.60**
626 mm × 2040 mm × 44 mm	94.17	1.62	32.72	–	98.88	nr	**131.60**
726 mm × 2040 mm × 44 mm	94.17	1.62	32.72	–	98.88	nr	**131.60**

24 DOORS, SHUTTERS AND HATCHES

Item	PC £	Labour hours	Labour £	Plant £	Material £	Unit	Total rate £
826 mm × 2040 mm × 44 mm	97.82	1.62	32.72	–	102.71	nr	**135.43**
926 mm × 2040 mm × 44 mm	118.86	1.62	32.72	–	124.80	nr	**157.52**
826 mm × 2040 mm × 44 mm; single side vision panel 508 mm × 1649 mm; factory fitted clear fire-rated glass	285.25	1.71	34.53	–	299.51	nr	**334.04**
826 mm × 2040 mm × 44 mm; two side vision panels 150 mm × 775 mm and 150 mm × 700 mm; factory fitted clear fire-rated glass	320.90	1.71	34.53	–	336.94	nr	**371.47**
926 mm × 2040 mm × 44 mm; single side vision panel 508 mm × 1649 mm; factory fitted clear fire-rated glass	294.38	1.71	34.53	–	309.10	nr	**343.63**
926 mm × 2040 mm × 44 mm; two side vision panels 150 mm × 775 mm and 150 mm × 700 mm; factory fitted clear fire-rated glass	328.22	1.71	34.53	–	344.63	nr	**379.16**
Flush door; one-hour fire-resisting (FD60); chipboard for painting; hardwood lipping two long edges; Premdor Firemaster or other equal and approved							
626 mm × 2040 mm × 54 mm	178.28	1.71	34.53	–	187.19	nr	**221.72**
726 mm × 2040 mm × 54 mm	178.28	1.71	34.53	–	187.19	nr	**221.72**
826 mm × 2040 mm × 54 mm	178.28	1.71	34.53	–	187.19	nr	**221.72**
926 mm × 2040 mm × 54 mm	196.57	1.71	34.53	–	206.40	nr	**240.93**
Moulded panel doors; half-hour fire-resisting (FD60); white based coated facings suitable for paint finish only; two, four or six panel options; Premdor Firemaster or other equal and approved							
626 mm × 2040 mm × 54 mm	278.85	1.71	34.53	–	292.79	nr	**327.32**
726 mm × 2040 mm × 54 mm	278.85	1.71	34.53	–	292.79	nr	**327.32**
826 mm × 2040 mm × 54 mm	280.67	1.71	34.53	–	294.70	nr	**329.23**
926 mm × 2040 mm × 54 mm	286.17	1.71	34.53	–	300.48	nr	**335.01**
Flush door; one-hour fire-resisting (FD60); White Oak veneer; hardwood lipping all edges; Premdor Firemaster or other equal and approved							
626 mm × 2040 mm × 54 mm	216.68	1.94	39.18	–	227.51	nr	**266.69**
726 mm × 2040 mm × 54 mm	216.68	1.94	39.18	–	227.51	nr	**266.69**
826 mm × 2040 mm × 54 mm	228.56	1.94	39.18	–	239.99	nr	**279.17**
926 mm × 2040 mm × 54 mm	246.85	1.94	39.18	–	259.19	nr	**298.37**
Flush door; one-hour fire-resisting (FD60); Steamed Beech veneer; hardwood lipping all edges; Premdor Firemaster or other equal and approved							
626 mm × 2040 mm × 54 mm	246.85	1.94	39.18	–	259.19	nr	**298.37**
726 mmx 2040 mm × 54 mm	246.85	1.94	39.18	–	259.19	nr	**298.37**
826 mm × 2040 mm × 54 mm	262.39	1.94	39.18	–	275.51	nr	**314.69**
926 mm × 2040 mm × 54 mm	277.02	1.94	39.18	–	290.87	nr	**330.05**

24 DOORS, SHUTTERS AND HATCHES

Item	PC £	Labour hours	Labour £	Plant £	Material £	Unit	Total rate £
INTERNAL DOORS AND FRAMES – cont							
Fire-resisting door frame; internal and external; fitted with 15 mm × 4 mm intumescent strips; 12 mm deep rebates; screwed to masonry/ concrete							
Softwood frames; no sill – open in or out							
Door size							
762 mm × 1981 mm × 44 mm; FD30; intemescent strip only	67.93	0.75	15.15	–	71.50	nr	**86.65**
838 mm × 1981 mm × 44 mm; FD30; intemescent strip only	69.79	0.75	15.15	–	73.46	nr	**88.61**
826 mm × 2040 mm × 44 mm; FD30; intemescent strip only	80.09	0.75	15.15	–	84.27	nr	**99.42**
762 mm × 1981 mm × 44 mm; FD30; intemescent strip/smoke seal	72.28	0.75	15.15	–	76.07	nr	**91.22**
838 mm × 1981 mm × 44 mm; FD30; intemescent strip/smoke seal	73.86	0.75	15.15	–	77.73	nr	**92.88**
826 mm × 2040 mm × 44 mm; FD30; intemescent strip/smoke seal	73.86	0.75	15.15	–	77.73	nr	**92.88**
Hardwood frames; no sill – open in or out							
762 mm × 1981 mm × 44 mm; FD60; intemescent strip/smoke seal	137.02	0.75	15.15	–	144.05	nr	**159.20**
838 mm × 1981 mm × 44 mm; FD60; intemescent strip/smoke seal	140.10	0.75	15.15	–	147.28	nr	**162.43**
826 mm × 2040 mm × 44 mm; FD60; intemescent strip/smoke seal	142.22	0.75	15.15	–	149.51	nr	**164.66**
HYGENIC DOORSETS							
Altro Whiterock flush doors complete with frame and stops, leaf and integral hinges. NOTE EXCLUDING IRONMONGERY							
946 mm × 2040 mm single leaf doorset; non-fire-rated	1005.96	3.00	60.59	–	1056.44	nr	**1117.03**
946 mm × 2040 mm single leaf doorset; FD30	1069.59	3.00	60.59	–	1123.25	nr	**1183.84**
946 mm × 2040 mm single leaf doorset; FD60	1281.69	3.00	60.59	–	1345.95	nr	**1406.54**
1350 mm × 2040 mm leaf and & half doorset; non-fire-rated	1445.31	4.00	80.78	–	1517.80	nr	**1598.58**
1350 mm × 2040 mm leaf & half doorset; FD30	1538.23	4.00	80.78	–	1615.36	nr	**1696.14**
1350 mm × 2040 mm leaf and half doorset; FD60	1864.46	4.00	80.78	–	1957.90	nr	**2038.68**
1800 mm × 2040 mm double leaf doorset; non-fire-rated	1566.51	6.00	121.17	–	1645.10	nr	**1766.27**
1800 mm × 2040 mm double leaf doorset; FD30	1693.77	6.00	121.17	–	1778.72	nr	**1899.89**
1800 mm × 2040 mm double leaf doorset; FD60	2121.00	6.00	121.17	–	2227.31	nr	**2348.48**
Intumescent strips							
Factory installed intumescent smoke seals to fire doors; 3 edges, jambs and head; site fixed intumescent strips	–	–	–	–	25.35	door	**25.35**
fire and smoke intumescent strips							
15 mm × 4 mm – FD30 doors	–	0.15	3.03	–	0.70	m	**3.73**
fire and smoke intumescent strips							
20 mm × 4 mm – FD60 doors	–	0.15	3.03	–	0.99	m	**4.02**

24 DOORS, SHUTTERS AND HATCHES

Item	PC £	Labour hours	Labour £	Plant £	Material £	Unit	Total rate £
Bedding and pointing frames							
Pointing wood frames or sills with mastic							
one side	–	0.09	1.47	–	0.95	m	**2.42**
both sides	–	0.19	3.10	–	1.89	m	**4.99**
Pointing wood frames or sills with polysulphide sealant							
one side	–	0.09	1.47	–	2.40	m	**3.87**
both sides	–	0.19	3.10	–	4.81	m	**7.91**
Bedding wood frames in cement mortar (1:3) and point							
one side	–	0.07	1.51	–	0.08	m	**1.59**
both sides	–	0.09	1.94	–	0.10	m	**2.04**
one side in mortar; other side in mastic	–	0.19	3.62	–	1.02	m	**4.64**
SHUTTER DOORS AND PANEL DOORS							
ASSA ABBLOY Entrance Systems Ltd Sectional Insulated Overhead Doors							
Insulated overhead sectional doors; Thermal transmittance (EN12428) = 0.43 W/m^2 K (Installed Door 1.1 W/m^2 K); Water penetration (EN12425) – Class 3; Air permeability (EN12426) Class 3; Wind-Load (EN12424) Class 3 reinforced steel panel; spring break device; versatile track system; Single phase electrically operated; manual override, deadman control, pre-coated panel in a standard colour; 1 row DAOP vision panels.							
Crawford 1042P, Standard Lift; 2500 mm × 3000 mm; 42 mm thick insulated sandwich panels	–	–	–	–	–	nr	**1650.00**
Crawford 1042P, Standard Lift; 3000 mm × 3000 mm; 42 mm thick insulated sandwich panels	–	–	–	–	–	nr	**1750.00**
Crawford 1042P, Standard Lift; 4000 mm × 5000 mm; 42 mm thick insulated sandwich panels	–	–	–	–	–	nr	**2475.00**
Crawford 1042P, Standard Lift; 6000 mm × 6000 mm; 42 mm thick insulated sandwich panels	–	–	–	–	–	nr	**5000.00**
Crawford 1042P, Standard Lift; 8000 mm × 6000 mm; 42 mm thick insulated sandwich panels	–	–	–	–	–	nr	**7595.00**

24 DOORS, SHUTTERS AND HATCHES

Item	PC £	Labour hours	Labour £	Plant £	Material £	Unit	Total rate £
SHUTTER DOORS AND PANEL DOORS – cont							
ASSA ABBLOY Entrance Systems Ltd Sectional Insulated Overhead Doors – cont							
High speed internal rapid action fabric door; breakaway facility with auto repair system; electrically operated frequency inverter motor with opening speeds to 2.4 m/s; lifetime expectations 1,000,000 cycles; galvanized steel guide bodies; soft bottom edge for safety; high tensile fabric curtain; stop and return safety-edge, safety photo cells, vision strip, push button operation to both sides							
Crawford HS9010P internal; one row windows 1000 mm × 2800 mm	–	–	–	–	–	nr	**5825.00**
Crawford HS9010P internal; one row windows 2000 mm × 3000 mm	–	–	–	–	–	nr	**6040.00**
Crawford HS9010P internal; one row windows 4000 mm × 4000 mm	–	–	–	–	–	nr	**6665.00**
Crawford HS9010P internal; one row windows 4000 mm × 5500 mm	–	–	–	–	–	nr	**7395.00**
High speed rapid action fabric door for external applications; ASSA ABLOY Entrance Systems Ltd; breakaway facility with auto repair system; air permeability(EN12426) Class 2; Wind load resistance Class 4; lifetime expectations 1,000,000 cycles; electrically operated frequency inverter motor, opening speeds up to 2.4 m/s; galvanized steel guide bodies; soft bottom edge for safety; high tensile fabric curtain; stop and return safety-edge, safety photo cells, push button operation to both sides, vision strip							
Crawford HS8010P external; one row windows 1000 mm × 2800 mm	–	–	–	–	–	nr	**6325.00**
Crawford HS8010P external; one row windows 2000 mm × 3000 mm	–	–	–	–	–	nr	**6520.00**
Crawford HS8010P external; one row windows 3000 mm × 3000 mm	–	–	–	–	–	nr	**6700.00**
Crawford HS8010P external; one row windows 4000 mm × 5500 mm	–	–	–	–	–	nr	**7975.00**

24 DOORS, SHUTTERS AND HATCHES

Item	PC £	Labour hours	Labour £	Plant £	Material £	Unit	Total rate £
DOCK LEVELLERS							
Swingdock leveller from ASSA ABLOY Entrance Systems; electro-hydraulic operation with two lifting cylinders; integrated frame to suit various pit applications; Control for door and leveller from one communal control panel, including interlocking of leveller/door; hold-to-run operation; mains power switch; load bearing capacity 60 kN in line with EN1398; colour RAL5010; toe guards, warning signs							
Crawford 6010S 500 mm swing lip; 2500 mm length × 2000 mm wide × 600 mm height	–	–	–	–	–	nr	2285.00
Crawford 6010S 500 mm swing lip; 4000 mm length × 2000 mm wide × 1100 mm height	–	–	–	–	–	nr	2835.00
Teledock leveller from ASSA ABLOY Entrance Systems; electro-hydraulic operation with two lifting cylinders; one lip ram cylinder; movable telescopic lip; integrated frame to suit various pit applications; Control for door and leveller from one communal control panel, including interlocking of leveller/door; hold-to-run operation; mains power switch; load bearing capacity 60 kN in line with EN1398; colour RAL5010; toe guards, warning signs							
Crawford 6020T 500 mm telescopic lip; 2500 mm length × 2000 mm wide	–	–		–	–	nr	2400.00
Crawford 6020T 1000 mm telescopic lip; 4000 mm length × 2000 mm wide	–	–	–	–	–	nr	3255.00
DOCK SHELTERS							
Curtain mechanical shelter; ASSA ABLOY Entrance Systems; extruded aluminium frame; spring loaded-frame connected with parallel bracing arms; 2 side curtains, one top curtain; double-layered high quality polyester, coated on both sides, longitudinal strength: approx. 750 N, transverse strength: approx. 900 N; 4500 mm above outside yard level, 600 mm projection, 1000 mm top curtain, top curtain to have one slit at the contact point on each side, integrated rain-channels							
Crawford DS6060 A; 3200 mm height × 3250 mm width	–	–	–	–	–	nr	865.00
Crawford DS6060 A; 600 mm projection, 1000 mm top curtain; 3400 mm height × 3250 width	–	–	–	–	–	nr	880.00
Crawford DS6060 A; 600 mm projection, 1000 mm top curtain; 3400 mm height × 3450 width	–	–	–	–	–	nr	890.00

24 DOORS, SHUTTERS AND HATCHES

Item	PC £	Labour hours	Labour £	Plant £	Material £	Unit	Total rate £
SHUTTER DOORS AND PANEL DOORS – cont							
DOCK SHELTERS – cont							
Inflatable mechanical shelter; ASSA ABLOY Entrance Systems; hot dipped galvanized surface treatment, polyester painted, corrosion protection class 3; top bag with polyester fabric panels, 1000 mm extension; side bags with polyester fabric panels, 650 mm extension, strong blower for straight movement of top and side bags; side section with steel guards; colour from standard range							
Crawford DS6070B; 3755 mm height × 3600 width × 770 mm depth	–	–	–	–	–	nr	**3275.00**
Crawford DS6070B; 4055 mm height × 3600 width × 770 mm depth	–	–	–	–	–	nr	**3320.00**
Crawford DS6070B; 4555 mm height × 3600 width × 770 mm depth	–	–	–	–	–	nr	**3420.00**
LOAD HOUSES							
Stand-alone Load House; ASSA ABLOY Entrance Systems; a complete loading unit attached externally to building; modular steel skin/insulated cladding; two wall elements, one roof element; hot-dipped galvanized steel frame; installed at 90°/45° to building; NL 2000/3000 mm; NW3300/3600 mm; DH 950/1300 mm; 4° roof slope; drainpipe, gutter; standard colour range							
Crawford LH6081L Load house,6010SA Swingdock c/w Autodock platform; 2000 mm length × 3300 mm width × 3495 mm height	–	–	–	–	–	nr	**5030.00**
Crawford LH6081L Load house, 6010SA Swingdock c/w Autodock platform; 3000 mm length × 3600 mm width × 3575 mm height	–	–	–	–	–	nr	**6290.00**
Rolling shutters and collapsible gates; steel counter shutters; push-up, self-coiling; polyester power coated; fixing by bolting							
Shutter doors							
3000 mm × 1000 mm	–	–	–	–	–	nr	**1309.00**
4000 mm × 1000 mm; in two panels	–	–	–	–	–	nr	**2266.00**
Rolling shutters with collapsible gates; galvanized steel; one hour fire-resisting; self-coiling; activated by fusible link; fixing with bolts							
Shutter doors							
1000 mm × 2750 mm	–	–	–	–	–	nr	**1562.00**
1500 mm × 2750 mm	–	–	–	–	–	nr	**1650.00**
2400 mm × 2750 mm	–	–	–	–	–	nr	**1925.00**

24 DOORS, SHUTTERS AND HATCHES

Item	PC £	Labour hours	Labour £	Plant £	Material £	Unit	Total rate £
Translucent GRP stacking door (78% translucency); U-value = 2.6 W/m² K; electrically operated; Envirodoor Ltd; fuly enclosed aluminium track system with SAA finish; manual over-ride, lock interlock, stop and return safety cage, deadmans down button, anti-flip device, photo-electric cell and beam deflectors; fixing by bolting; standard panel finishes							
Stacking doors							
HT40 stacking door 2500 mm × 3000 mm; 40 mm thick × 500 mm high twin walled GRP translucent panels	–	–	–	–	–	nr	**5000.00**
HT40 stacking door 4500 mm × 6000 mm; 40 mm thick × 500 mm high twin walled translucent panels	–	–	–	–	–	nr	**9200.00**
HT60-N stacking door 6000 mm × 6000 mm: 60 mm thick × 500 mm high twinn walled translucent panels	–	–	–	–	–	nr	**13000.00**
HT60-H stacking door 7000 mm × 8000 mm: 60 mm thick × 1000 mm high twin walled translucent panels	–	–	–	–	–	nr	**20000.00**
HT80–400 stacking door 10000 mm × 10000 mm; 80 mm thick × 1000 mm high twin walled translucent panels	–	–	–	–	–	nr	**47500.00**
Doors; galvanized steel up and over type garage doors; Catnic Horizon 90 or other equal and approved; spring counter balanced; fixed to timber frame (not included)							
Garage door							
2135 mm × 1980 mm	337.50	3.70	74.72	–	354.53	nr	**429.25**
2135 mm × 2135 mm	369.00	3.70	74.72	–	387.61	nr	**462.33**
2400 mm × 2135 mm	467.25	3.70	74.72	–	490.79	nr	**565.51**
3965 mm × 2135 mm	1040.25	5.55	112.09	–	1092.62	nr	**1204.71**
Grilles; Galaxy nylon rolling counter grille or other equal and approved; Bolton Brady Ltd; colour, off-white; self-coiling; fixing by bolting							
Grilles							
3000 mm × 1000 mm	–	–	–	–	–	nr	**950.00**
4000 mm × 1000 mm	–	–	–	–	–	nr	**1550.00**
Sliding/folding partitions; Alco Beldan Ltd or equal and approved							
Sliding/folding partitions							
ref. NW100 Moveable Wall; 5000 mm (wide) × 2495 mm (high) comprising 4 nr 954 mm (wide) standard panels and 1 nr 954 mm (wide) telescopic panel; sealing; fixing	–	–	–	–	–	nr	**10000.00**

24 DOORS, SHUTTERS AND HATCHES

Item	PC £	Labour hours	Labour £	Plant £	Material £	Unit	Total rate £
ASSOCIATED IRONMONGERY							
Associated Ironmongery							
NOTE: Ironmongery is largely a matter of selection and specification and prices vary considerably; indicative prices for reasonable quantities of good quality ironmongery are given below.							
Allgood Ltd or other equal; to softwood							
Bolts							
75 × 35 mm Modric anodized aluminium straight barrel bolt	8.50	0.30	6.06	–	8.92	nr	**14.98**
150 × 35 mm Modric anodized aluminium straight barrel bolt	9.70	0.30	6.06	–	10.18	nr	**16.24**
75 × 35 mm Modric anodized aluminium necked barrel bolt	9.52	0.30	6.06	–	10.00	nr	**16.06**
150 × 35 mm Modric anodized aluminium necked barrel bolt	12.17	0.30	6.06	–	12.78	nr	**18.84**
11 mm Easiclean socket for wood or stone	4.15	0.10	2.02	–	4.36	nr	**6.38**
Security hinge bolt chubb WS12	8.61	0.50	10.10	–	9.04	nr	**19.14**
203 × 19 × 11 mm lever action flush bolt set, with coil spring and intumescent pack for FD30 and FD60 fire doors	29.20	0.60	12.12	–	30.66	nr	**42.78**
609 × 19 mm lever action flush bolt set, with coil sprin and intumescent pack for FD30 and FD60 fire doors	85.13	0.60	12.12	–	89.39	nr	**101.51**
Stainless steel indicating bolt complete with outside indicator and emergency release	61.01	0.60	12.12	–	64.06	nr	**76.18**
Catches							
Magnetic catch	0.43	0.20	4.04	–	0.45	nr	**4.49**
Door closers and furniture							
13 mm stainless steel rebate component for 7104/08/78/79/86	25.31	0.60	12.12	–	26.58	nr	**38.70**
70 × 70 mm Modric anodized aluminium electrically powered hold open wall magnet (excluding power supply and connection)	119.07	0.40	8.07	–	125.02	nr	**133.09**
Modric anodized aluminium bathroom configuration with quadaxial assembly, turn, release and optional indicator	46.40	0.80	16.16	–	48.72	nr	**64.88**

24 DOORS, SHUTTERS AND HATCHES

Item	PC £	Labour hours	Labour £	Plant £	Material £	Unit	Total rate £
Concealed jamb door closer check action 75 × 57 × 170 mm Modric anodized aluminium	138.89	1.00	20.19	–	145.83	nr	**166.02**
door coordinator for pairs of rebated leaves, CE Marked to BS EN1158 3–5-3/5–1–1–0	30.73	0.80	16.16	–	32.27	nr	**48.43**
290 × 48 × 50 mm Modric anodized aluminium rectangular overhead door closer with adjustable power and adjustable backcheck intumescent protected arm heavy duty U.L. & certifire listed & CE Marked to BS EN1154 4–8-2/4–1–1–3 and Kitemarked.	89.22	1.00	20.19	–	93.68	nr	**113.87**
Stainless steel overhead door closer. Projecting armset, Power EN 2–5, CE marked, c/w Backcheck, Latch action and Speed control. Max door width 1100 mm, Max door weight 100 kg	88.98	1.00	20.19	–	93.43	nr	**113.62**
288 × 45 × 32 mm fully concealed overhead door closer complete with track and arm for single action doors, adjustable power, latch action and backcheck. Certifire approved	149.07	0.80	16.16	–	156.52	nr	**172.68**
75 × 45 mm heavy duty floor pivot set with thrust roller bearing 200 kg load capacity. Complete with forged steel intumescent protected double action strap with 10 mm height adjustment and matching cover plate	227.17	2.30	46.45	–	238.53	nr	**284.98**
Cavalier floor spring unit with cover plate and loose box for concrete, adjustable power 25 mm offset strap & top centre, intumescent pack. Certfire listed	240.92	2.30	46.45	–	252.97	nr	**299.42**
Double action pivot set for door maximum width 1100 mm and maximum weight 80 kg	76.06	2.30	46.45	–	79.86	nr	**126.31**
Surface vertical rod push bar panic bolt, reversible, to suit doors 2500x1100 mm maximum, silver finish, CE marked to EN1125 class 3–7-5-1–1-3-2–2-A	127.10	1.50	30.29	–	133.45	nr	**163.74**
Rim push bar panic latch, reversible, to suit doors 1100 mm wide maximum, silver finish, CE marked to EN1125 class 3–7-5-1–1-3-2–2-A	90.62	1.30	26.25	–	95.15	nr	**121.40**
76 × 51 × 13 mm adjustable heavy roller catch, stainless steel forend and strike complete with satin nickel plate roller bolt	7.16	0.60	12.12	–	7.52	nr	**19.64**
External access device for use with XX10280/2 panic hardware to suit door thickness 45–55 mm, complete with SS3006N lever, SS755 rose, SS796 profile escutcheon and spindle.For use with MA7420 A51 or MA7420 A55 profile cylinders	30.11	1.30	26.25	–	31.62	nr	**57.87**
142 × 22 mm Ø Concealed jamb door closer light duty	16.12	0.80	16.16	–	16.93	nr	**33.09**

24 DOORS, SHUTTERS AND HATCHES

Item	PC £	Labour hours	Labour £	Plant £	Material £	Unit	Total rate £
ASSOCIATED IRONMONGERY – cont							
Allgood Ltd or other equal – cont							
Door closers and furniture – cont							
80 × 40 × 45 mm Emergency release door stop with holdback facility	76.06	1.00	20.19	–	79.86	nr	**100.05**
Modric anodized aluminium quadaxial lever MA3503 assembly tested to BS EN1906 4/7/-/1/1/4/0/U	28.20	0.80	16.16	–	29.61	pair	**45.77**
Modric anodized aluminium quadaxial lever MA3502 assembly Tested to BS EN1906 4/7/-/1/1/4/0/U	30.06	0.80	16.16	–	31.56	pair	**47.72**
Modric anodized aluminium quadaxial lever assembly Tested to BS EN1906 4/7/-/1/1/4/0/U with Biocote® anti-bacterial protection	54.20	0.80	16.16	–	56.91	pair	**73.07**
Modric stainless steel quadaxial lever assembly Tested to BS EN1906 4/7/-/1/1/4/0/U	51.60	0.80	16.16	–	54.18	pair	**70.34**
152 × 38 × 13 mm Modric anodized aluminium security door chain leather covered	40.20	0.40	8.07	–	42.21	nr	**50.28**
50 Ø × 3 mm Modric anodized aluminium circular covered rose for profile cylinder	3.95	0.10	2.02	–	4.15	nr	**6.17**
50 Ø × 3 mm Modric anodized aluminium circular covered rose with indicator and emergency release	8.15	0.15	3.03	–	8.56	nr	**11.59**
50 Ø × 3 mm Modric anodized aluminium circular covered rose with heavy turn, 5–8 mm spindle	14.30	0.15	3.03	–	15.02	nr	**18.05**
Budget lock escutcheon – satin stainless steel 316	7.79	0.10	2.02	–	8.18	nr	**10.20**
50 Ø × 3 mm Stainless steel circular covered rose for profile cylinder	6.33	0.10	2.02	–	6.65	nr	**8.67**
50 Ø × 3 mm Stainless steel circular covered rose with indicator and emergency release	8.54	0.15	3.03	–	8.97	nr	**12.00**
50 Ø × 3 mm Stainless steel circular covered rose with heavy turn, 5–8 mm spindle	17.59	0.15	3.03	–	18.47	nr	**21.50**
330 × 76 × 1.6 mm Modric anodized aluminium push plate	3.12	0.15	3.03	–	3.28	nr	**6.31**
330 × 76 × 1.6 mm Stainless steel push plate	7.21	0.15	3.03	–	7.57	nr	**10.60**
800 × 150 × 1.5 mm Modric anodized aluminium kicking plate, drilled and countersunk with screws.	7.68	0.25	5.05	–	8.06	nr	**13.11**
900 × 150 × 1.5 mm Modric anodized aluminium kicking plate, drilled and countersunk with screws.	8.63	0.25	5.05	–	9.06	nr	**14.11**
1000 × 150 × 1.5 mm Modric anodized aluminium kicking plate, drilled & countersunk with screws.	9.58	0.25	5.05	–	10.06	nr	**15.11**
800 × 150 × 1.5 mm Stainless steel kicking plate, drilled and countersunk with screws.	13.58	0.25	5.05	–	14.26	nr	**19.31**
900 × 150 × 1.5 mm Stainless steel kicking plate, drilled and countersunk with screws.	15.28	0.25	5.05	–	16.04	nr	**21.09**
1000 × 150 × 1.5 mm Stainless steel kicking plate, drilled & countersunk with screws.	16.97	0.25	5.05	–	17.82	nr	**22.87**
610 × 70 × 19 mm Ø Modric anodized aluminium grab handle bolt through fixing for doors 10 to 55 mm thick	27.74	0.40	8.07	–	29.13	nr	**37.20**

24 DOORS, SHUTTERS AND HATCHES

Item	PC £	Labour hours	Labour £	Plant £	Material £	Unit	**Total rate £**
400 × 19 mm Ø Stainless steel D line straight pull handle with M8 threaded holes, fixing centres 300 mm	42.92	0.33	6.67	–	49.85	nr	**56.52**
Hinges							
100 × 75 × 3 mm Stainless steel triple knuckle concealed twin bearings, button tipped butt hinges, jig drilled for metal doors/frames, complete with M6x12MT 'undercut' machine screws, stainless steel 316 CE marked to EN1935	17.04	0.25	5.05	–	17.89	pair	**22.94**
100 × 100 × 3 mm Stainless steel triple knuckle concealed twin Newton bearings, button tipped hinges, jig drilled, stainless steel grade 316 CE marked to EN1935	30.25	0.25	5.05	–	31.76	pair	**36.81**
Latches							
Modric anodized aluminium round cylinder for rim night latch, 2 keyed satin nickel plated	23.95	0.40	8.07	–	25.15	nr	**33.22**
93 × 75 mm Cylinder rim non-deadlocking night latch case only 60 mm backset	20.52	0.40	8.07	–	21.55	nr	**29.62**
71 series mortice latch, case only, low friction latchbolt, griptight follower, heavy spring for levers. Radius forend and sq strike. CE marked to BS EN12209 3/X/8/1/0G/-/B/02/0	15.51	0.80	16.16	–	16.29	nr	**32.45**
Modric anodized aluminium latch configuration with quadaxial assembly	28.02	0.80	16.16	–	29.42	nr	**45.58**
Modric anodized aluminium Nightlatch configuration with quadaxial assembly and single cylinder	51.97	0.80	16.16	–	54.57	nr	**70.73**
Locks							
44 mm case Bright zinc plated steel mortice budget lock with slotted strike plate 33 mm backset	26.33	0.80	16.16	–	27.65	nr	**43.81**
76 × 58 mm b/s Stainless steel cubicle mortice deadlock with 8 mm follower	12.47	0.80	16.16	–	13.09	nr	**29.25**
'A' length European profile double cylinder lock, 2 keyed satin nickel plated	24.56	0.80	16.16	–	25.79	nr	**41.95**
'A' length European profile cylinder and large turn, 2 keyed satin nickel plated	27.91	0.80	16.16	–	29.31	nr	**45.47**
'A' length European profile cylinder and large turn, 2 keyed under master key, satin nickel plated	25.68	0.80	16.16	–	26.96	nr	**43.12**
'A' length European profile single cylinder, 2 keyed satin nickel plated	18.99	0.80	16.16	–	19.94	nr	**36.10**
'A' length European profile single cylinder, 2 keyed under master key, satin nickel plated	18.99	0.80	16.16	–	19.94	nr	**36.10**

24 DOORS, SHUTTERS AND HATCHES

Item	PC £	Labour hours	Labour £	Plant £	Material £	Unit	Total rate £
ASSOCIATED IRONMONGERY – cont							
Allgood Ltd or other equal – cont							
Locks – cont							
93 × 60 mm b/s 71 series profile cylinder mortice deadlock, case only. Single throw 22 mm deadbolt. Radius forend and square strike. CE marked to BS EN12209 3/X/8/1/0/G/4/B/A/0/0	15.51	0.80	16.16	–	16.29	nr	**32.45**
92 × 60 mm b/s 71 series bathroom lock, case only, low friction latchbolt, griptight follower, heavy spring for levers, twin 8 mm followers at 78 mm centres. Radius forend and square strike. CE marked to BS EN12209 3/X/8/0/0/G-/B/0/2/0	18.38	0.80	16.16	–	19.30	nr	**35.46**
93 × 60 mm b/s 71 series profile cylinder mortice lock, case only, low friction latchbolt, griptight follower. Heavy spring for levers, 22 mm throw deadbolt, cylinder withdraws bolt bolts. Radius forend and square strike. CE marked to BS EN12209 3/X/8/1/0G/4/B/A2/0	18.38	0.80	16.16	–	19.30	nr	**35.46**
92 × 60 mm b/s71 series profile cylinder emergency lock, case only. Low friction latchbolt, griptight follower, heavy spring for lever, single throw 22 mm deadbolt, lever can withdraw both bolts. Radius forend and strike	60.48	0.80	16.16	–	63.50	nr	**79.66**
Modric anodized aluminium lock configuration with quadaxial assembly and cylinder with turn	80.06	0.80	16.16	–	84.06	nr	**100.22**
Sundries							
76 mm Ø Modric anodized aluminium circular sex symbol male	4.35	0.08	1.62	–	4.57	nr	**6.19**
76 mm Ø Modric anodized aluminium circular symbol fire door keep locked	4.35	0.08	1.62	–	4.57	nr	**6.19**
76 mm Ø Modric anodized aluminium circular symbol fire door keep shut	4.35	0.10	2.02	–	4.57	nr	**6.59**
38 × 47 mm Ø Modric anodized aluminium heavy circular floor door stop with cover	8.79	0.10	2.02	–	9.23	nr	**11.25**
38 × 47 mm Ø Stainless steel heavy circular floor door stop with cover	15.69	0.10	2.02	–	16.47	nr	**18.49**
63 × 19 mm Ø Modric ancdised aluminium Circular heavy duty skirting buffer with thief-resistant insert	5.75	0.10	2.02	–	6.04	nr	**8.06**
102 × 25 mm Ø Stainless steel circular heavy duty skirting buffer with thief-resistant insert	9.96	0.10	2.02	–	10.46	nr	**12.48**
152 mm Cabin hook satin chrome on brass	20.81	0.15	3.03	–	21.85	nr	**24.88**
14 mm Ø × 145 × 94 mm Toilet roll holder, length 145 mm, colour white, satin stainless steel 316	57.95	0.15	3.03	–	62.44	nr	**65.47**
Towel rail with bushes, fixing centres 450 mm, satin stainless steel 316	78.75	0.25	5.05	–	87.48	nr	**92.53**
Toilet brush holder with toilet brush, with bushes, satin stainless steel 316	122.28	0.20	4.04	–	131.59	nr	**135.63**

24 DOORS, SHUTTERS AND HATCHES

Item	PC £	Labour hours	Labour £	Plant £	Material £	Unit	Total rate £
Bathline 850 mm lift up support rail	185.23	0.75	15.15	–	194.49	set	**209.64**
Bathline 600 × 95 × 35 mm support rail with concealed fixing roses	94.94	0.50	10.10	–	99.69	set	**109.79**
Bathline 400 × 250 × 35 mm backrest rail with concealed fixing roses	94.94	0.50	10.10	–	99.69	set	**109.79**
Allgood Ltd or other equal; to hardwood							
Bolts							
75 × 35 mm Modric anodized aluminium straight barrel bolt	8.50	0.40	8.07	–	8.92	nr	**16.99**
150 × 35 mm Modric anodized aluminium straight barrel bolt	9.70	0.40	8.07	–	10.18	nr	**18.25**
75 × 35 mm Modric anodized aluminium necked barrel bolt	9.52	0.40	8.07	–	10.00	nr	**18.07**
150 × 35 mm Modric anodized aluminium necked barrel bolt	12.17	0.40	8.07	–	12.78	nr	**20.85**
11 mm Easiclean socket for wood or stone	4.15	0.15	3.03	–	4.36	nr	**7.39**
Security hinge bolt chubb WS12	8.61	0.65	13.13	–	9.04	nr	**22.17**
203 × 19 × 11 mm lever action flush bolt set with coil spring and intumescent pack for FD30 and FD60 fire doors	29.20	0.80	16.16	–	30.66	nr	**46.82**
609 × 19 mm lever action flush bolt set with coil spring and intumescent pack for FD30 and FD60 fire doors	85.13	0.80	16.16	–	89.39	nr	**105.55**
Stainless steel indicating bolt complete with outside indicator and emergency release	61.01	0.80	16.16	–	64.06	nr	**80.22**
Catches							
Magnetic catch	0.43	0.25	5.05	–	0.45	nr	**5.50**
Door closers and furniture							
13 mm stainless steel rebate component for 7104/08/78/79/86	25.31	0.80	16.16	–	26.58	nr	**42.74**
70 × 70 mm Modric anodized aluminium electrically powered hold open wall magnet. CE marked to BS EN1155:1997 & A1:2002 3–5-6/3–1-1–3	119.07	0.55	11.11	–	125.02	nr	**136.13**
Modric anodized aluminium bathroom configuration with quadaxial assembly, turn, release and optional indicator	46.40	1.05	21.21	–	48.72	nr	**69.93**

24 DOORS, SHUTTERS AND HATCHES

Item	PC £	Labour hours	Labour £	Plant £	Material £	Unit	Total rate £
ASSOCIATED IRONMONGERY – cont							
Allgood Ltd or other equal – cont							
Door closers and furniture – cont							
495 × 17.5 × 16 mm concealed overhead door restraining stay with aluminium channel. Stainless steel plated arm and bracket with adjustable friction slide. Block and spring buffer to cussion door opening. Not for use with door closing devices	23.87	1.35	27.27	–	25.06	nr	**52.33**
Concealed jamb door closer check action	138.89	1.35	27.27	–	145.83	nr	**173.10**
75 × 57 × 170 mm Modric anodized aluminium door co-ordinator for pairs of rebated leaves, CE Marked to BS EN1158 3–5–3/5–1–1–0	30.73	1.05	21.21	–	32.27	nr	**53.48**
290 × 48 × 50 mm Modric anodized aluminium rectangular overhead door closer with adjustable power and adjustable backcheck intumescent protected arm heavy duty U.L. & certifire listed & CE Marked to BS EN1154 4–8–2/4–1–1–3 and Kitemarked	89.22	1.35	27.27	–	93.68	nr	**120.95**
Stainless steel overhead door closer. Projecting armset, Power EN 2–5, CE marked, c/w Backcheck, Latch action and Speed control. Max door width 1100 mm, Max door weight 100 kg	88.98	1.35	27.27	–	93.43	nr	**120.70**
288 × 45 × 32 mm fully concealed overhead door closer complete with track and arm for single action doors, adjustable power, latch action and backcheck. Certifire approved	149.07	1.05	21.21	–	156.52	nr	**177.73**
75 × 45 mm heavy duty floor pivot set with thrust roller bearing 200 kg load capacity. Complete with forged steel intumescent protected double action strap with 10 mm height adjustment and matching cover plate	227.17	3.05	61.59	–	238.53	nr	**300.12**
Cavalier floor spring unit with cover plate and loose box for concrete, adjustable power 25 mm offset strap & top centre with intumescent pack. Certfire listed	240.92	2.30	46.45	–	252.97	nr	**299.42**
Double action pivot set for door maximum width 1100 mm and maximum weight 80 kg	76.06	3.05	61.59	–	79.86	nr	**141.45**

24 DOORS, SHUTTERS AND HATCHES

Item	PC £	Labour hours	Labour £	Plant £	Material £	Unit	Total rate £
Surface vertical rod push bar panic bolt, reversible, to suit doors 2500 × 1100 mm maximum, silver finish, CE marked to EN1125 class 3–7-5-1–1-3-2–2-A	127.10	2.00	40.39	–	133.45	nr	**173.84**
Rim push bar panic latch, reversible, to suit doors 1100 mm wide maximum, silver finish, CE marked to EN1125 class 3–7-5-1–1-3-2–2-A	90.62	1.75	35.34	–	95.15	nr	**130.49**
76 × 51 × 13 mm adjustable heavy roller catch, stainless steel forend and strike complete with satin nickel roller bolt satin chrome	7.16	0.80	16.16	–	7.52	nr	**23.68**
External access device for use with XX10280/2 panic hardware to suit door thickness 45–55 mm, complete with SS3006N lever, SS755 rose, SS796 profile escutcheon and spindle.For use with MA7420 A51 or MA7420 A55 profile cylinders	30.11	1.75	35.34	–	31.62	nr	**66.96**
142 × 22 mm Ø Concealed jamb door closer light duty	16.12	1.05	21.21	–	16.93	nr	**38.14**
80 × 40 × 45 mm Emergency release door stop with holdback facility	76.06	1.35	27.27	–	79.86	nr	**107.13**
Modric anodized aluminium quadaxial lever MA3503 assembly tested to BS EN1906 4/7/-/1/1/4/0/U	28.20	1.05	21.21	–	29.61	pair	**50.82**
Modric anodized aluminium quadaxial lever MA3502 assembly Tested to BS EN1906 4/7/-/1/1/4/0/U	30.06	1.05	21.21	–	31.56	pair	**52.77**
Modric anodized aluminium quadaxial lever assembly Tested to BS EN1906 4/7/-/1/1/4/0/U with Biocote® anti-bacterial protection	54.20	1.05	21.21	–	56.91	pair	**78.12**
Modric stainless steel quadaxial lever assembly Tested to BS EN1906 4/7/-/1/1/4/0/U	51.60	1.05	21.21	–	54.18	pair	**75.39**
152 × 38 × 13 mm Modric anodized aluminium security door chain leather covered	40.20	0.55	11.11	–	42.21	nr	**53.32**
50 Ø × 3 mm Modric anodized aluminium circular covered rose for profile cylinder	3.95	0.15	3.03	–	4.15	nr	**7.18**
50 Ø × 3 mm Modric anodized aluminium circular covered rose with indicator and emergency release	8.15	0.20	4.04	–	8.56	nr	**12.60**
50 Ø × 3 mm Modric anodized aluminium circular covered rose with heavy turn, 5–8 mm spindle	14.30	0.20	4.04	–	15.02	nr	**19.06**
Budget lock escutcheon – satin stainless steel 316	7.79	0.15	3.03	–	8.18	nr	**11.21**
50 Ø × 3 mm Stainless steel circular covered rose for profile cylinder	6.33	0.15	3.03	–	6.65	nr	**9.68**
50 Ø × 3 mm Stainless steel circular covered rose with indicator and emergency release	8.54	0.20	4.04	–	8.97	nr	**13.01**
50 Ø × 3 mm Stainless steel circular covered rose with heavy turn, 5–8 mm spindle	17.59	0.20	4.04	–	18.47	nr	**22.51**
330 × 76 × 1.6 mm Modric anodized aluminium push plate	3.12	0.20	4.04	–	3.28	nr	**7.32**
330 × 76 × 1.6 mm Stainless steel push plate	7.21	0.20	4.04	–	7.57	nr	**11.61**
800 × 150 × 1.5 mm Modric anodized aluminium kicking plate, drilled and countersunk with screws.	7.68	0.35	7.07	–	8.06	nr	**15.13**

24 DOORS, SHUTTERS AND HATCHES

Item	PC £	Labour hours	Labour £	Plant £	Material £	Unit	Total rate £
ASSOCIATED IRONMONGERY – cont							
Allgood Ltd or other equal – cont							
Door closers and furniture – cont							
900 × 150 × 1.5 mm Modric anodized aluminium kicking plate, drilled and countersunk with screws.	8.63	0.35	7.07	–	9.06	nr	**16.13**
1000 × 150 × 1.5 mm Modric anodized aluminium kicking plate, drilled & countersunk with screws.	9.58	0.35	7.07	–	10.06	nr	**17.13**
800 × 150 × 1.5 mm Stainless steel kicking plate, drilled and countersunk with screws.	13.58	0.35	7.07	–	14.26	nr	**21.33**
900 × 150 × 1.5 mm Stainless steel kicking plate, drilled and countersunk with screws.	15.28	0.35	7.07	–	16.04	nr	**23.11**
1000 × 150 × 1.5 mm Stainless steel kicking plate, drilled & countersunk with screws.	16.97	0.35	7.07	–	17.82	nr	**24.89**
610 × 70 × 19 mm Ø Modric anodized aluminium grab handle bolt through fixing for doors 10 to 55 mm thick	27.74	0.55	11.11	–	29.13	nr	**40.24**
400 × 19 mm Ø Stainless steel D line straight pull handle with M8 threaded holes, fixing centres 300 mm	42.92	0.45	9.09	–	49.85	nr	**58.94**
Hinges							
100 × 75 × 3 mm Stainless steel triple knuckle concealed twin Newtonbearings, button tipped butt hinges, jig drilled for metal doors/frames, complete with M6x12MT 'undercut' machine screws, stainless steel 316 CE marked to EN1935	17.04	0.35	7.07	–	17.89	pair	**24.96**
100 × 100 × 3 mm Stainless steel triple knuckle concealed twin Newton bearings, button tipped hinges, jig drilled, stainless steel grade 316 CE marked to EN1935	30.25	0.35	7.07	–	31.76	pair	**38.83**
Latches							
Modric anodized aluminium round cylinder for rim night latch, 2 keyed satin nickel plated	23.95	0.55	11.11	–	25.15	nr	**36.26**
93 × 75 mm Cylinder rim non-deadlocking night latch case only 60 mm backset	20.52	0.55	11.11	–	21.55	nr	**32.66**
71 series mortice latch, case only, low friction latchbolt, griptight follower, heavy spring for levers. Radius forend and sq strike. CE marked to BS EN12209 3/X/8/1/0G/-/B/02/0	15.51	1.05	21.21	–	16.29	nr	**37.50**
Modric anodized aluminium latch configuration with quadaxial assembly	28.02	1.05	21.21	–	29.42	nr	**50.63**
Modric anodized aluminium Nightlatch configuration with quadaxial assembly and single cylinder	51.97	1.05	21.21	–	54.57	nr	**75.78**

24 DOORS, SHUTTERS AND HATCHES

Item	PC £	Labour hours	Labour £	Plant £	Material £	Unit	Total rate £
Locks							
44 mm case Bright zinc plated steel mortice budget lock with slotted strike plate 33 mm backset	26.33	1.05	21.21	–	27.65	nr	**48.86**
76 × 58 mm b/s Stainless steel cubicle mortice deadlock with 8 mm follower	12.47	1.05	21.21	–	13.09	nr	**34.30**
'A' length European profile double cylinder lock, 2 keyed satin nickel plated	24.56	1.05	21.21	–	25.79	nr	**47.00**
'A' length European profile cylinder and large turn, 2 keyed satin nickel plated	27.91	1.05	21.21	–	29.31	nr	**50.52**
'A' length European profile cylinder and large turn, 2 keyed under master key, satin nickel plated	25.68	1.05	21.21	–	26.96	nr	**48.17**
'A' length European profile single cylinder, 2 keyed satin nickel plated	18.99	1.05	21.21	–	19.94	nr	**41.15**
'A' length European profile single cylinder, 2 keyed under master key, satin nickel plated	18.99	1.05	21.21	–	19.94	nr	**41.15**
93 × 60 mm b/s 71 series profile cylinder mortice deadlock, case only. Single throw 22 mm deadbolt. Radius forend and square strike. CE marked to BS EN12209 3/X/8/1/0/G/4/B/A/0/0	15.51	1.05	21.21	–	16.29	nr	**37.50**
92 × 60 mm b/s 71 series bathroom lock, case only, low friction latchbolt, griptight follower, heavy spring for levers, twin 8 mm followers at 78 mm centres. Radius forend and square strike. CE marked to BS EN12209 3/X/8/0/0/G-/B/0/2/0	18.38	1.05	21.21	–	19.30	nr	**40.51**
93 × 60 mm b/s 71 series profile cylinder mortice lock, case only, low friction latchbolt, griptight follower. Heavy spring for levers, 22 mm throw deadbolt, cylinder withdraws bolt bolts. Radius forend and square strike. CE marked to BS EN12209 3/X/8/1/0G/4/B/A2/0	18.38	1.05	21.21	–	19.30	nr	**40.51**
92 × 60 mm b/s 71 series profile cylinder emergency lock, case only. Low friction latchbolt, griptight follower, heavy spring for lever, single throw 22 mm deadbolt, lever can withdraw both bolts. Radius forend and strike	60.48	1.05	21.21	–	63.50	nr	**84.71**
Modric anodized aluminium lock configuration with quadaxial assembly and cylinder with turn	80.06	1.05	21.21	–	84.06	nr	**105.27**
Sundries							
76 mm Ø Modric anodized aluminium circular sex symbol male	4.35	0.10	2.02	–	4.57	nr	**6.59**
76 mm Ø Modric anodized aluminium circular symbol fire door keep locked	4.35	0.10	2.02	–	4.57	nr	**6.59**
76 mm Ø Modric anodized aluminium circular symbol fire door keep shut	4.35	0.15	3.03	–	4.57	nr	**7.60**

24 DOORS, SHUTTERS AND HATCHES

Item	PC £	Labour hours	Labour £	Plant £	Material £	Unit	Total rate £
ASSOCIATED IRONMONGERY – cont							
Allgood Ltd or other equal – cont							
Sundries – cont							
38 × 47 mm Ø Modric anodized aluminium heavy circular floor door stop with cover	8.79	0.15	3.03	–	9.23	nr	**12.26**
38 × 47 mm Ø Stainless steel heavy circular floor door stop with cover	15.69	0.15	3.03	–	16.47	nr	**19.50**
63 × 19 mm Ø Modric anodised aluminium Circular heavy duty skirting buffer with thief-resistant insert	5.75	0.15	3.03	–	6.04	nr	**9.07**
102 × 25 mm Ø Stainless steel circular heavy duty skirting buffer with thief-resistant insert	9.96	0.15	3.03	–	10.46	nr	**13.49**
152 mm Cabin hook satin chrome on brass	20.81	0.20	4.04	–	21.85	nr	**25.89**
14 mm × 145 × 94 mm Ø Toilet roll holder, length 145 mm, colour white, satin stainless steel 316	57.95	0.20	4.04	–	62.44	nr	**66.48**
Towel rail with bushes, fixing centres 450 mm, satin stainless steel 316	78.75	0.35	7.07	–	87.48	nr	**94.55**
Toilet brush holder with toilet brush, with bushes, satin stainless steel 316	122.28	0.25	5.05	–	131.59	nr	**136.64**
Bathline 850 mm lift up support rail	185.23	0.75	15.15	–	194.49	set	**209.64**
Bathline 600 × 95 × 35 mm support rail with concealed fixing roses	94.94	0.50	10.10	–	99.69	set	**109.79**
Bathline 400 × 250 × 35 mm backrest rail with concealed fixing roses	94.94	0.50	10.10	–	99.69	set	**109.79**
ASSA ABBLOY Ltd typical door ironmongery sets							
These are standard doorsets compilations which ASSA ABBLOY most commonly supply							
Classroom doorset (fire-rated)	–	2.00	40.39	–	376.29	nr	**416.68**
Ironmongery schedule and cost							
UNION PowerLOAD 603 bushed bearing butt hinges (1.5 pair) with a fixed in and radius corners, 100 mm × 88 mm, satin stainless steel, grade 13 BS EN 1935	8.35	–	–	–	–	nr	–
ASSA ABBLOY DC700 A overhead cam motion backcheck closure with slide arm size 2–6. Full cover standard silver	142.55	–	–	–	–	nr	–
UNION keyULTRA euro profile key and turn cylinder, turn/key, 35 mm/35 mm , satin chrome, BS EN 1303 Security Grade 2 when fitted with security escutcheon (sold seperately)	34.02	–	–	–	–	nr	–
Union 2S21 euro-profile sashlock, 72 mm centresm, 55 mm backset, satin stainless steel, Grade 3 category of use	10.34	–	–	–	–	nr	–
Union 1000 01 style, 19 mm dia. lever on 8 mm round rose, sprung, satin stainless steel grade 304, bolt through connections	7.14	–	–	–	–	pair	–
7012F ASSA finger guard 2.05 m push side, silver	84.58	–	–	–	–	nr	–
UNION 1000EE Euro Escutcheon, 54 mm dia. × 8 mm projection, satin stainless steel	0.57	–	–	–	–	nr	–

24 DOORS, SHUTTERS AND HATCHES

Item	PC £	Labour hours	Labour £	Plant £	Material £	Unit	Total rate £
Kick Plate 900 mm × 150 mm, satin stainless steel softened & finished edges with radius corners	15.78	–	–	–	–	nr	–
'Fire Door Keep Shut' 75 mm dia. sign, satin stainless steel	4.39	–	–	–	–	nr	–
UNION DS100 Floor Mounted Door Stop, half moon design, 45 mm dia. × 24 mm high, satin stainless steel	2.14	–	–	–	–	nr	–
Non-locking office doorset (fire-rated)	–	2.00	40.39	–	238.99	nr	**279.38**
Ironmongery schedule and cost							
UNION PowerLOAD 603 bushed bearing butt hinges (1.5 pair) with a fixed pin and radius corners, 100 mm × 88 mm, satin stainless steel, grade 13 BS EN 1935	25.06	–	–	–	–	nr	–
ASSA ABBLOY DC700 A overhead cam motion backcheck closure with slide arm size 2–6. Full cover standard silver	142.55	–	–	–	–	nr	–
UNION keyULTRA euro profile key and turn cylinder, turn/key, 35 mm/35 mm , satin chrome, BS EN 1303 Security Grade 2 when fitted with security escutcheon (sold seperately)	34.02	–	–	–	–	nr	–
UNION 2C23 DIN mortice latch, 55 mm backset, satin stainless steel, radius forend, Grade 3 Category of Use, BS EN 12209, CE marked	10.39	–	–	–	–	nr	–
Union 1000 01 style, 19 mm dia. lever on 8 mm round rose, sprung, satin stainless steel grade 304, bolt through connections	7.14	–	–	–	–	pair	–
Kick Plate 900 mm × 150 mm, satin stainless steel softened & finished edges with radius corners	15.78	–	–	–	–	nr	–
Fire Door Keep Shut' 75 mm dia. sign, satin stainless steel	4.39	–	–	–	–	nr	–
UNION DS100 Floor Mounted Door Stop, half moon design, 45 mm dia. × 24 mm high, satin stainless steel	2.14	–	–	–	–	nr	–
Office/Store locking doorset (fire-rated)	–	2.00	40.39	–	275.86	nr	**316.25**
Ironmongery schedule and cost							
UNION PowerLOAD 603 bushed bearing butt hinges (1.5 pair) with a fixed pin and radius corners, 100 mm × 88 mm, satin stainless steel, grade 13 BS EN 1935	25.06	–	–	–	–	nr	–
ASSA ABBLOY DC700 A overhead cam motion backcheck closure with slide arm size 2–6. Full cover standard silver	142.55	–	–	–	–	nr	–
UNION keyULTRA euro profile key and turn cylinder, turn/key, 35 mm/35 mm , satin chrome, BS EN 1303 Security Grade 2 when fitted with security escutcheon (sold seperately)	34.02	–	–	–	–	nr	–

24 DOORS, SHUTTERS AND HATCHES

Item	PC £	Labour hours	Labour £	Plant £	Material £	Unit	Total rate £
ASSOCIATED IRONMONGERY – cont							
ASSA ABBLOY Ltd typical door ironmongery sets – cont							
Ironmongery schedule and cost – cont							
Union 2S21 euro-profile sashlock, 72 mm centresm, 55 mm backset, satin stainless steel, Grade 3 category of use	10.34	–	–	–	–	nr	–
Union 1000 01 style, 19 mm dia. lever on 8 mm round rose, sprung, satin stainless steel grade 304, bolt through connections	7.14	–	–	–	–	pair	–
Kick Plate 200 mm × 1.5 mm, satin stainless steel softened & finished edges with radius corners	15.78	–	–	–	–	nr	–
'Fire Door Keep Shut' 75 mm dia. sign, satin stainless steel	4.39	–	–	–	–	nr	–
UNION DS100 Floor Mounted Door Stop, half moon design, 45 mm dia. × 24 mm high, satin stainless steel	2.14	–	–	–	–	nr	–
Common room locking doorset (fire-rated)	–	2.50	50.48	–	384.56	nr	**435.04**
Ironmongery schedule and cost							
UNION PowerLOAD 603 bushed bearing butt hinges (1.5 pair) with a fixed pin and radius corners, 100 mm × 88 mm, satin stainless steel, Grade 13 BS EN 1935	25.06	–	–	–	–	nr	–
ASSA ABBLOY DC700 A overhead cam motion backcheck closure with slide arm size 2–6. Full cover standard silver	142.55	–	–	–	–	nr	–
UNION keyULTRA euro profile key and turn cylinder, turn/key (Large), 35 mm/35 mm , 6pin, satin chrome, BS EN 1303 Security Grade 2 when fitted with security escutcheon (sold seperately)	45.09	–	–	–	–	nr	–
UNION 2C22 DIN euro-profile deadlock, 55 mm backset, satin stainless steel, radius forend, Grade 3 Category of Use, BS EN 12209, CE marked	15.12	–	–	–	–	nr	-
UNION 01 style pull handle 19 mm dia. × 425 mm centres, satin stainless steel, Grade 304, bolt through fixing	4.54	–	–	–	–	nr	-
Kick Plate 200 mm × 1.5 mm, satin stainless steel softened & finished edges with radius corners	15.78	–	–	–	–	nr	-
Push Plate 450 mm × 100 mm × 1.5 mm satin stainless steel, softened & finished edges with radius corners	1.78	–	–	–	–	nr	-

24 DOORS, SHUTTERS AND HATCHES

Item	PC £	Labour hours	Labour £	Plant £	Material £	Unit	Total rate £
7012F ASSA finger guard 2.05 m push side, silver	84.58	–	–	–	–	nr	–
UNION 1000EE Euro Escutcheon, 54 mm dia. × 8 mm projection, satin stainless steel	0.57	–	–	–	–	nr	–
UNION DS100 Floor Mounted Door Stop, Half Moon Design, 45 mm dia. × 24 mm high, Satin Stainless Steel	2.14	–	–	–	–	nr	–
'Fire Door Keep Shut' 75 mm dia. sign, satin stainless steel	4.39	–	–	–	–	nr	–
PULL or PUSH signage, 75 mm dia., satin stainless steel	1.95	–	–	–	–	nr	–
Maintenance/Plant room doorset (fire-rated)	–	1.75	35.34	–	263.79	nr	**299.13**
Ironmongery schedule and cost							
1 ½ pairs Union Powerload 603 brushed bearing butt hinge with a fixed pin; 100 mm × 88 mm; satin stainless steel grade 13 BS EN 1935	25.06	–	–	–	–	nr	–
ASSA ABBLOY DC700A overhead cam motion backcheck closure with slide arm size 2–6. Full cover standard silver	142.55	–	–	–	–	nr	–
UNION 2C22 DIN euro-profile deadlock, 55 mm backset, satin stainless steel, radius forend, Grade 3 Category of Use, BS EN 12209, CE marked	15.12	–	–	–	–	nr	–
UNION keyULTRA euro profile single cylinder, 35 mm/35 mm, satin chrome, BS EN 1303 Securily Grade 2 when fitted with security escutcheon	45.09	–	–	–	–	nr	–
Kick Plate 900 mm × 150 mm, satin stainless steel softened & finished edges With Radius Corner	15.78	–	–	–	–	nr	–
'Fire Door Keep Locked' 75 mm dia. Sign, Satin Stainless Steel	4.39	–	–	–	–	nr	–
Euro Rim Cylinder Pull, stainless steel	0.54	–	–	–	–	nr	–
UNION 1000EE Euro Escutcheon, 54 mm dia. × 8 mm projection, satin stainless steel	0.57	–	–	–	–	nr	–
UNION DS100 Floor Mounted Door Stop, half moon design, 45 mm Dia. × 24 mm high, satin stainless steel	2.14	–	–	–	–	nr	–

24 DOORS, SHUTTERS AND HATCHES

Item	PC £	Labour hours	Labour £	Plant £	Material £	Unit	Total rate £
ASSOCIATED IRONMONGERY – cont							
Standard bathroom doorset (Unisex)	–	1.78	36.05	–	210.02	nr	**246.07**
Ironmongery schedule and cost							
1 ½ pairs Union Powerload 603 brushed bearing butt hinge with a fixed pin; 100 mm × 88 mm; satin stainless steel Grade 13 BS EN 1935	25.06	–	–	–	–	nr	–
ASSA ABBLOY DC700 A overhead cam motion backcheck closure with slide arm size 2–6. Full cover standard silver	142.55	–	–	–	–	nr	–
UNION 2C27 DIN bathroom lock, 78 mm centres, 55 mm backset, satin stainless steel, radius forend, Grade 3 Category of Use, BS EN 12209, CE marked	10.75	–	–	–	–	nr	–
Union 1000 01 style, 19 mm dia. lever on 8 mm round rose, sprung, satin stainless steel grade 304, bolt through connections	7.14	–	–	–	–	pair	–
UNION 1000ER emergency release, 54 mm dia. × 8 mm projection, satin stainless steel	3.42	–	–	–	–	nr	–
UNION 1000T standard turn, 54 mm dia. × 8 mm projection, satin stainless steel	5.09	–	–	–	–	nr	–
UNION DS100 Floor Mounted Door Stop, half moon design, 45 mm dia. × 24 mm high, satin stainless steel	2.14	–	–	–	–	nr	–
UNION US200 Unisex Symbol, 75 mm dia., satin stainless steel, Grade 304	1.95	–	–	–	–	nr	–
UNION HC100 hat & coat hook with a buffer, satin stainless steel	1.93	–	–	–	–	nr	–
Accessible toilet doorset	–	2.55	51.50	–	90.99	nr	**142.49**
Ironmongery schedule and cost							
1 ½ pairs Union Powerload 603 brushed bearing butt hinge with a fixed pin; 100 mm × 88 mm; satin stainless steel Grade 13 BS EN 1935	25.06	–	–	–	–	nr	–
UNION 2C27 DIN bathroom lock, 78 mm centres, 55 mm backset, satin stainless steel, radius forend, Grade 3 Category of Use, BS EN 12209, CE marked	10.75	–	–	–	–	nr	–
Union 1000 01 style, 19 mm dia. lever on 8 mm round rose, sprung, satin stainless steel grade 304, bolt through connections	7.14	–	–	–	–	pair	–
Kick plates 900 mm × 150 mm, stainless steel	15.78	–	–	–	–	nr	–
UNION WS200 Disabled Symbol, 75 mm dia., satin stainless steel, Grade 304	1.95	–	–	–	–	nr	–
UNION 1000ER emergency release, 54 mm dia. × 8 mm projection, satin stainless steel	3.42	–	–	–	–	nr	–
UNION 1000LT Large Turn, 54 mm dia. × 8 mm projection, satin stainless steel	2.73	–	–	–	–	nr	–
UNION DS100 Floor Mounted Door Stop, half moon design, 45 mm dia. × 24 mm high, satin stainless steel	2.14	–	–	–	–	nr	–
UNION HC100 hat & coat hook with a buffer, satin stainless steel	1.93	–	–	–	–	nr	–

24 DOORS, SHUTTERS AND HATCHES

Item	PC £	Labour hours	Labour £	Plant £	Material £	Unit	Total rate £
Single Lobby/Corridor doorset (fire-rated)	–	2.25	45.44	–	320.14	nr	**365.58**
Ironmongery schedule and cost							
UNION PowerLOAD 603 bushed bearing butt hinges (1.5 pair) with a fixed pin and radius corners, 100 mm × 88 mm, satin stainless steel, Grade 13 BS EN 1935	25.06	–	–	–	–	nr	-
UNION 01 style pull handle 19 mm dia. × 425 mm centres, satin stainless steel, Grade 304, bolt through fixing	4.54	–	–	–	–	nr	-
ASSA ABBLOY DC700 A overhead cam motion backcheck closure with slide arm size 2–6. Full cover standard silver	142.55	–	–	–	–	nr	-
7012F ASSA finger guard 2.05 m push side, silver	84.58	–	–	–	–	nr	-
Kick Plate 900 mm × 150 mm, satin stainless steel softened & finished edges with radius corners	15.78	–	–	–	–	nr	-
'Fire Door Keep Shut' 75 mm dia. sign, satin stainless steel	4.39	–	–	–	–	nr	-
PULL or PUSH signage, 75 mm dia., satin stainless steel	1.95	–	–	–	–	nr	-
UNION DS100 Floor Mounted Door Stop, half moon design, 45 mm dia. × 24 mm high, satin stainless steel	2.14	–	–	–	–	nr	-
Double Lobby/Corridor doorset (fire-rated)	–	4.00	80.78	–	661.26	nr	**742.04**
Ironmongery schedule and cost							
UNION PowerLOAD 603 bushed bearing butt hinges (3 pair) with a fixed pin and radius corners, 100 mm × 88 mm, satin stainless steel, Grade 13 BS EN 1935	50.11	–	–	–	–	nr	-
UNION 8899 Electro Mag closer hold open/swing free, fixed strength size 4, painted silver, app 1, 61 and 66, BS EN 1155, CE Marked, Certifire approved.Supplied with a curve cover c/w matching arms.	152.53	–	–	–	–	nr	-
UNION 01 style pull handle 19 mm dia. × 425 mm centres, satin stainless steel, Grade 304, bolt through fixing	4.54	–	–	–	–	nr	-
Push Plate 450 mm × 100 mm × 1.5 mm satin stainless steel, softened & finished edges with radius corners	1.78	–	–	–	–	nr	-
7012F ASSA finger guard 2.05 m push side, silver	84.58	–	–	–	–	nr	-
Kick Plate 900 mm × 150 mm, satin stainless steel softened & finished edges with radius corners	15.78	–	–	–	–	nr	-
'Fire Door Keep Shut' 75 mm dia. sign, satin stainless steel	4.39	–	–	–	–	nr	-
PULL or PUSH signage, 75 mm dia., satin stainless steel	1.95	–	–	–	–	nr	-
UNION DS100 Floor Mounted Door Stop, half moon design, 45 mm dia. × 24 mm high, satin stainless steel	2.14	–	–	–	–	nr	-

24 DOORS, SHUTTERS AND HATCHES

Item	PC £	Labour hours	Labour £	Plant £	Material £	Unit	Total rate £
ASSOCIATED IRONMONGERY – cont							
External Rebated Double Fire Escape doorset	–	4.00	80.78	–	1449.15	nr	**1529.93**
Ironmongery schedule and cost							
UNION PowerLOAD 603 bushed bearing butt hinges (1.5 pair) with a fixed pin and radius corners, 100 mm × 88 mm, satin stainless steel, Grade 13 BS EN 1935	25.06	–	–	–	–	nr	–
UNION satin stainless steel, Ball Bearing Dog Bolt Hinge	126.83	–	–	–	–	nr	–
ASSA ABBLOY DC700 A overhead cam motion backcheck closure with slide arm size 2–6. Full cover standard silver	142.55	–	–	–	–	nr	–
UNION 883T panic exit double rebated doorset, 3 point locking with pullman latches, touch bar design, 900 mm maximum door width, silver, EN1125, CE marked	179.13	–	–	–	–	nr	–
UNION 885 outside access device (OAD), lever handle to suit 880 series only, silver, uses a UNION 2X50 5 pin euro profile cylinder to be ordered separately	43.68	–	–	–	–	nr	–
UNION keyULTRA euro profile single cylinder, 35 mm/35 mm, satin chrome, BS EN 1303 Security Grade 2 when fitted with security escutcheon	45.09	–	–	–	–	nr	–
Generic door door limiting stays	38.55	–	–	–	–	nr	–
Kick Plate 900 mm × 150 mm, satin stainless steel softened & finished edges with radius corners	15.78	–	–	–	–	nr	–
'Fire Door Keep Shut' 75 mm dia. sign, satin stainless steel	4.39	–	–	–	–	nr	–
SP(102)FA1201 – Aluminium	25.90	–	–	–	–	nr	–
Double Aluminium Entrance doorset	–	5.00	100.98	–	665.67	nr	**766.65**
Ironmongery schedule and cost							
Adams Rite SENTINAL Radius Weatherstrip Faceplate, satin anodized aluminium	5.19	–	–	–	–	nr	–
Adams Rite 4700 Series Faceplate Radius w/s, satin anodized aluminium	4.34	–	–	–	–	nr	–
Adams Rite SENTINEL 6 Deadlock, 30 mm Backset BS EN 12209	14.13	–	–	–	–	nr	–
ADAMS RITE SENTINEL Armoured Strike Plate	4.95	–	–	–	–	nr	–
ADAMS RITE SENTINEL Radius Armoured Trim Strike, satin anodized aluminium	9.55	–	–	–	–	nr	–
ADAMS RITE 4596 Euro Paddle Handle, Pull to Left 45 mm Door Satin Anodized Aluminium	72.67	–	–	–	–	nr	–
ADAMS RITE 4750 Deadlatch EuroProfile 28 mm backset body only – left hand	38.04	–	–	–	–	nr	–

24 DOORS, SHUTTERS AND HATCHES

Item	PC £	Labour hours	Labour £	Plant £	Material £	Unit	Total rate £
ADAMS RITE 2 Point Latch Push to Left, satin anodized aluminium	141.19	–	–	–	–	nr	–
ADAMS RITE Transom ARC–51N EN3 Spring Closer HO side load 70 mm pivot	69.98	–	–	–	–	nr	–
ADAMS RITE 7108 12vDC Fail Safe Escape Electric Release, Satin Anodized Aluminium. Specifer to confirm depth of section as could clash with mechanism of 4781.	165.66	–	–	–	–	nr	–
ADAMS RITE Sentinal Euro Profile Security Escutcheon	38.27	–	–	–	–	nr	–
External Toilet Lobby doorset (fire-rated)	–	2.50	50.48	–	386.01	nr	**436.49**
Ironmongery schedule and cost							
1 ½ pairs Union Powerload 603 brushed bearing butt hinge with a fixed pin; 100 mm × 88 mm; satin stainless steel Grade 13 BS EN 1935	25.06	–	–	–	–	nr	–
ASSA ABBLOY DC700 A overhead cam motion backcheck closure with slide arm size 2–6. Full cover standard silver	142.55	–	–	–	–	nr	–
UNION 2C22 DIN euro-profile deadlock, 55 mm backset, satin stainless steel, radius forend, Grade 3 Category of Use, BS EN 12209, CE marked	15.12	–	–	–	–	nr	–
UNION 01 style pull handle 19 mm dia. × 425 mm centres, satin stainless steel, Grade 304, bolt through fixing	4.54	–	–	–	–	nr	–
UNION keyULTRA euro profile single cylinder, 35 mm/35 mm, satin chrome, BS EN 1303 Security Grade 2 when fitted with security escutcheon	45.09	–	–	–	–	nr	–
Push Plate 450 mm × 100 mm × 1.5 mm satin stainless steel, softened & finished edges with radius corners	1.78	–	–	–	–	nr	–
Kick Plate 900 mm × 150 mm, satin stainless steel softened & finished edges with radius corner	15.78	–	–	–	–	nr	–
7012F ASSA finger guard 2.05 m push side, silver	84.58	–	–	–	–	nr	–
'Fire Door Keep Locked' 75 mm dia. sign, satin stainless steel	4.39	–	–	–	–	nr	–
UNION 1000EE Euro Escutcheon, 54 mm dia. × 8 mm projection, satin stainless steel	0.57	–	–	–	–	nr	–
UNION DS100 Floor Mounted Door Stop, half moon design, 45 mm dia. × 24 mm high, satin stainless steel	2.14	–	–	–	–	nr	–
PULL or PUSH signage, 75 mm dia., satin stainless steel	1.95	–	–	–	–	nr	–
UNION US200 Unisex Symbol, 75 mm dia., satin stainless steel, Grade 304	1.95	–	–	–	–	nr	–

24 DOORS, SHUTTERS AND HATCHES

Item	PC £	Labour hours	Labour £	Plant £	Material £	Unit	Total rate £
ASSOCIATED IRONMONGERY – cont							
Internal Toilet Lobby doorset (non-fire-rated)	–	2.00	40.39	–	222.12	nr	**262.51**
Ironmongery schedule and cost							
1 ½ pairs Union Powerload 603 brushed bearing butt hinge with a fixed pin; 100 mm × 88 mm; satin stainless steel Grade 13 BS EN 1935	25.06	–	–	–	–	nr	–
ASSA ABBLOY DC700 A overhead cam motion backcheck closure with slide arm size 2–6. Full cover standard silver	142.55	–	–	–	–	nr	–
UNION 01 style pull handle 19 mm dia. × 425 mm centres, satin stainless steel, Grade 304, bolt through fixing	4.54	–	–	–	–	nr	–
Push Plate 450 mm × 100 mm × 1.5 mm satin stainless steel, softened & finished edges with radius corners	1.78	–	–	–	–	nr	–
Kick Plate 900 mm × 150 mm, satin stainless steel softened & finished edges with radius corners	15.78	–	–	–	–	nr	–
UNION DS100 Floor Mounted Door Stop, half moon design, 45 mm Dia. × 24 mm high, satin stainless steel	2.14	–	–	–	–	nr	–
PULL or PUSH signage, 75 mm dia., satin stainless steel	1.95	–	–	–	–	nr	-
Double Classroom Store doorset (non-fire-rated)	–	1.50	30.29	–	323.52	nr	**353.81**
Ironmongery schedule and cost							
UNION PowerLOAD 603 bushed bearing butt hinges (3 pair) with a fixed pin and radius corners, 100 mm × 88 mm, satin stainless steel, Grade 13 BS EN 1935	25.06	–	–	–	–	nr	–
Union 2S21 euro-profile sashlock, 72 mm centresm, 55 mm backset, satin stainless steel, Grade 3 category of use	10.34	–	–	–	–	nr	–
UNION keyULTRA euro profile key and turn cylinder, turn/key (Large), 35 mm/35 mm, 6pin, satin chrome, BS EN 1303 Security Grade 2 when fitted with security escutcheon (sold seperately)	45.09	–	–	–	–	nr	–
Union 1000 01 style, 19 mm dia. lever on 8 mm round rose, sprung, satin stainless steel grade 304, bolt through connections	7.14	–	–	–	–	pair	–
UNION 1000EE Euro Escutcheon, 54 mm dia. × 8 mm projection, Satin Stainless Steel	0.57	–	–	–	–	nr	–
Kick Plate 900 mm × 150 mm, satin stainless steel softened & finished edges with radius corners	15.78	–	–	–	–	nr	–
UNION 8060 aluminium lever action flush bolt 200 mm satin stainless steel satin chrome	6.78	–	–	–	–	pair	–

24 DOORS, SHUTTERS AND HATCHES

Item	PC £	Labour hours	Labour £	Plant £	Material £	Unit	Total rate £
Sliding door gear; Hillaldam Coburn Ltd or other equal and approved; Commercial/Light industrial; for top hung timber/metal doors, weight not exceeding 365 kg							
Sliding door gear							
bottom guide; fixed to concrete in groove	24.28	0.46	9.29	–	25.49	m	**34.78**
top track	33.00	0.23	4.64	–	34.65	m	**39.29**
detachable locking bar	41.28	0.31	6.26	–	43.34	nr	**49.60**
hangers; timber doors	66.60	0.46	9.29	–	69.93	nr	**79.22**
hangers; metal doors	42.56	0.46	9.29	–	44.69	nr	**53.98**
head brackets; open, soffit fixing; screwing to timber	8.49	0.32	6.46	–	8.94	nr	**15.40**
head brackets; open, side fixing; bolting to masonry	8.90	0.46	9.29	–	10.25	nr	**19.54**
door guide to timber door	7.46	0.23	4.64	–	7.83	nr	**12.47**
door stop; rubber buffers; to masonry	32.90	0.69	13.93	–	34.55	nr	**48.48**
drop bolt; screwing to timber	28.78	0.46	9.29	–	30.22	nr	**39.51**
bow handle; to timber	11.41	0.23	4.64	–	11.98	nr	**16.62**
Sundries							
rubber door stop; plugged and screwed to concrete	5.55	0.09	1.82	–	5.83	nr	**7.65**

25 STAIRS, WALKWAYS AND BALUSTRADES

Item	PC £	Labour hours	Labour £	Plant £	Material £	Unit	Total rate £
STAIRS							
Standard staircases; wrought softwood (Parana pine)							
Stairs; 25 mm thick treads with rounded nosings; 9 mm thick plywood risers; 32 mm thick strings; bullnose bottom tread; 50 mm × 75 mm hardwood handrail; 32 mm square plain balusters; 100 mm square plain newel posts							
straight flight; 838 mm wide; 2676 mm going; 2600 mm rise; with two newel posts	–	6.48	130.86	–	371.25	nr	**502.11**
straight flight with turn; 838 mm wide; 2676 mm going; 2600 mm rise; with two newel posts; three top treads winding	–	6.48	130.86	–	484.12	nr	**614.98**
dogleg staircase; 838 mm wide; 2676 mm going; 2600 mm rise; with two newel posts; quarter space landing third riser from top	–	6.48	130.86	–	453.84	nr	**584.70**
dogleg staircase; 838 mm wide; 2676 mm going; 2600 mm rise; with two newel posts; half space landing third riser from top	–	7.40	149.45	–	558.45	nr	**707.90**
Hardwood staircases; purpose-made; assembled at works							
Fixing only complete staircase including landings, balustrades, etc.							
plugging and screwing to brickwork or blockwork	–	13.88	280.31	–	2.77	nr	**283.08**
The following are supply only prices for purpose-made staircase components in selected Sapele supplied as part of an assembled staircase and may be used to arrive at a guide price for a complete hardwood staircase							
Board landings; cross-tongued joints; 100 mm × 50 mm sawn softwood bearers							
25 mm thick	–	–	–	–	87.35	m²	**87.35**
32 mm thick	–	–	–	–	98.19	m²	**98.19**
Treads; cross-tongued joints and risers; rounded nosings; tongued, grooved, glued and blocked together; one 175 mm × 50 mm sawn softwood carriage							
25 mm treads; 19 mm risers	–	–	–	–	175.55	m²	**175.55**
ends; quadrant	–	–	–	–	53.48	nr	**53.48**
ends; housed to hardwood	–	–	–	–	0.99	nr	**0.99**
32 mm treads; 25 mm risers	–	–	–	–	181.87	m²	**181.87**
ends; quadrant	–	–	–	–	68.73	nr	**68.73**
ends; housed to hardwood	–	–	–	–	0.99	nr	**0.99**

25 STAIRS, WALKWAYS AND BALUSTRADES

Item	PC £	Labour hours	Labour £	Plant £	Material £	Unit	Total rate £
Winders; cross-tongued joints and risers in one width; rounded nosings; tongued, grooved glued and blocked together; one 175 mm × 50 mm sawn softwood carriage							
25 mm treads; 19 mm risers	–	–	–	–	244.04	m²	**244.04**
32 mm treads; 25 mm risers	–	–	–	–	249.77	m²	**249.77**
wide ends; housed to hardwood	–	–	–	–	1.96	nr	**1.96**
narrow ends; housed to hardwood	–	–	–	–	1.48	nr	**1.48**
Closed strings; in one width; 230 mm wide; rounded twice							
32 mm thick	–	–	–	–	32.15	m	**32.15**
38 mm thick	–	–	–	–	35.01	m	**35.01**
50 mm thick	–	–	–	–	39.00	m	**39.00**
Closed strings; cross-tongued joints; 280 mm wide; once rounded							
32 mm thick	–	–	–	–	41.73	m	**41.73**
extra for short ramp	–	–	–	–	21.44	nr	**21.44**
38 mm thick	–	–	–	–	45.51	m	**45.51**
extra for short ramp	–	–	–	–	24.37	nr	**24.37**
50 mm thick	–	–	–	–	50.79	m	**50.79**
extra for short ramp	–	–	–	–	30.17	nr	**30.17**
The following labours are irrespective of timber width							
ends; fitted	–	–	–	–	1.28	nr	**1.28**
ends; framed	–	–	–	–	7.47	nr	**7.47**
extra for tongued heading joint	–	–	–	–	3.69	nr	**3.69**
Closed strings; ramped; crossed tongued joints 280 mm wide; once rounded							
32 mm thick	–	–	–	–	41.73	m	**41.73**
44 mm thick	–	–	–	–	45.51	m	**45.51**
57 mm thick	–	–	–	–	50.79	m	**50.79**
Apron linings; in one width 230 mm wide							
19 mm thick	–	–	–	–	11.10	m	**11.10**
25 mm thick	–	–	–	–	13.07	m	**13.07**
The following are supply only prices for purpose-made staircase components in selected American Oak; supplied as part of an assembled staircase							
Board landings; cross-tongued joints; 100 mm × 50 mm sawn softwood bearers							
25 mm thick	–	–	–	–	128.87	m²	**128.87**
32 mm thick	–	–	–	–	155.61	m²	**155.61**
Treads; cross-tongued joints and risers; rounded nosings; tongued, grooved, glued and blocked together; one 175 mm × 50 mm sawn softwood carriage							
25 mm treads; 19 mm risers	–	–	–	–	213.93	m²	**213.93**
ends; quadrant	–	–	–	–	106.87	nr	**106.87**
ends; housed to hardwood	–	–	–	–	1.30	nr	**1.30**
32 mm treads; 25 mm risers	–	–	–	–	245.12	m²	**245.12**
ends; quadrant	–	–	–	–	131.54	nr	**131.54**
ends; housed to hardwood	–	–	–	–	1.30	nr	**1.30**

25 STAIRS, WALKWAYS AND BALUSTRADES

Item	PC £	Labour hours	Labour £	Plant £	Material £	Unit	Total rate £
STAIRS – cont							
The following are supply only prices for purpose-made staircase components in selected American Oak – cont							
Winders; cross-tongued joints and risers in one width; rounded nosings; tongued, grooved glued and blocked together; one 175 mm × 50 mm sawn softwood carriage							
25 mm treads; 19 mm risers	–	–	–	–	271.17	m²	**271.17**
32 mm treads; 25 mm risers	–	–	–	–	295.88	m²	**295.88**
wide ends; housed to hardwood	–	–	–	–	2.64	nr	**2.64**
narrow ends; housed to hardwood	–	–	–	–	1.97	nr	**1.97**
Closed strings; in one width; 230 mm wide; rounded twice							
32 mm thick	–	–	–	–	50.54	m	**50.54**
44 mm thick	–	–	–	–	58.34	m	**58.34**
57 mm thick	–	–	–	–	80.08	m	**80.08**
Closed strings; cross-tongued joints; 280 mm wide; once rounded							
32 mm thick	–	–	–	–	64.21	m	**64.21**
extra for short ramp	–	–	–	–	36.72	nr	**36.72**
38 mm thick	–	–	–	–	74.41	m	**74.41**
extra for short ramp	–	–	–	–	41.82	nr	**41.82**
50 mm thick	–	–	–	–	102.04	m	**102.04**
extra for short ramp	–	–	–	–	55.63	nr	**55.63**
Closed strings; ramped; crossed tongued joints 280 mm wide; once rounded							
32 mm thick	–	–	–	–	73.86	m	**73.86**
44 mm thick	–	–	–	–	85.60	m	**85.60**
57 mm thick	–	–	–	–	117.34	m	**117.34**
Apron linings; in one width 230 mm wide							
19 mm thick	–	–	–	–	17.21	m	**17.21**
25 mm thick	–	–	–	–	21.09	m	**21.09**
Handrails; rounded							
40 mm × 50 mm	–	–	–	–	13.94	m	**13.94**
50 mm × 75 mm	–	–	–	–	17.88	m	**17.88**
57 mm × 87 mm	–	–	–	–	26.84	m	**26.84**
69 mm × 100 mm	–	–	–	–	36.13	m	**36.13**
Handrails; moulded							
40 mm × 50 mm	–	–	–	–	15.31	m	**15.31**
50 mm × 75 mm	–	–	–	–	19.23	m	**19.23**
57 mm × 87 mm	–	–	–	–	28.18	m	**28.18**
69 mm × 100 mm	–	–	–	–	37.46	m	**37.46**
Add to above for							
grooved once	–	–	–	–	0.69	m	**0.69**
ends; framed	–	–	–	–	6.91	nr	**6.91**
ends; framed on rake	–	–	–	–	8.76	nr	**8.76**

25 STAIRS, WALKWAYS AND BALUSTRADES

Item	PC £	Labour hours	Labour £	Plant £	Material £	Unit	Total rate £
Heading joints to handrail; mitred or raked							
overall size not exceeding 50 mm × 75 mm	–	–	–	–	36.88	nr	**36.88**
overall size not exceeding 69 mm × 100 mm	–	–	–	–	43.78	nr	**43.78**
Knee piece to handrail; mitred or raked							
overall size not exceeding 69 mm × 100 mm	–	–	–	–	78.36	nr	**78.36**
Balusters; stiffeners							
25 mm × 25 mm	–	–	–	–	3.18	m	**3.18**
32 mm × 32 mm	–	–	–	–	4.02	m	**4.02**
44 mm × 44 mm	–	–	–	–	6.34	m	**6.34**
ends; housed	–	–	–	–	1.61	nr	**1.61**
Sub-rails							
32 mm × 63 mm	–	–	–	–	8.59	m	**8.59**
ends; framed joint to newel	–	–	–	–	6.91	nr	**6.91**
Knee rails							
32 mm × 140 mm	–	–	–	–	14.91	m	**14.91**
ends; framed joint to newel	–	–	–	–	6.91	nr	**6.91**
Newel posts							
44 mm × 94 mm; half newel	–	–	–	–	11.19	m	**11.19**
69 mm × 69 mm	–	–	–	–	19.07	m	**19.07**
94 mm × 94 mm	–	–	–	–	47.58	m	**47.58**
Newel caps; splayed on four sides							
62.50 mm × 125 mm × 50 mm	–	–	–	–	9.61	nr	**9.61**
100 mm × 100 mm × 50 mm	–	–	–	–	10.14	nr	**10.14**
125 mm × 125 mm × 50 mm	–	–	–	–	11.15	nr	**11.15**
Spiral staircases, balustrades and handrails; mild steel; galvanized and polyester powder coated							
Staircase							
2080 mm dia. × 3695 mm high; 18 nr treads; 16 mm dia. intermediate balusters; 1040 mm × 1350 mm landing unit with matching balustrade both sides; fixing with 16 mm dia. resin anchors to masonry at landing and with 12 mm dia. expanding bolts to concrete at base	–	–	–	–	–	nr	**7000.00**
BALUSTRADES							
Standard balustrades; wrought softwood							
Landing balustrade; 50 mm × 75 mm hardwood handrail; 32 mm square plain balusters; one end of handrail jointed to newel post; other end built into wall; balusters housed in at bottom (newel post and mortices both not included)							
3.00 m long	–	3.70	74.72	–	93.61	nr	**168.33**

25 STAIRS, WALKWAYS AND BALUSTRADES

Item	PC £	Labour hours	Labour £	Plant £	Material £	Unit	Total rate £
BALUSTRADES – cont							
Steel							
Balustrades; galvanized mild steel CHS posts and top rail, with one infill rail							
1100 mm high	–	–	–	–	–	m	**210.00**
Balustrades; painted mild steel flat bar posts and CHS top rail, with 3 nr stainless steel infills							
1100 mm high	–	–	–	–	–	m	**300.00**
Stainless steel							
Balustrades; stainless steel flat bar posts and circular handrail, with 3 nr stainless steel infills							
1100 mm high	–	–	–	–	–	m	**360.00**
Balustrades; stainless steel 50 mm Ø posts and circular handrail, with 10 mm thick toughened glass infill panels							
1100 mm high	–	–	–	–	–	m	**600.00**
Glass							
Balustrades; laminated glass; with stainless steel cap channel to top and including all necessary support fixings							
1100 mm high	–	–	–	–	–	m	**600.00**
Softwood handrails							
Handrails; rounded							
40 mm × 50 mm	–	–	–	–	12.04	m	**12.04**
50 mm × 75 mm	–	–	–	–	14.53	m	**14.53**
57 mm × 87 mm	–	–	–	–	17.04	m	**17.04**
69 mm × 100 mm	–	–	–	–	21.19	m	**21.19**
Handrails; moulded							
40 mm × 50 mm	–	–	–	–	13.40	m	**13.40**
50 mm × 75 mm	–	–	–	–	15.88	m	**15.88**
57 mm × 87 mm	–	–	–	–	18.41	m	**18.41**
69 mm × 100 mm	–	–	–	–	22.53	m	**22.53**
Add to above for							
grooved once	–	–	–	–	0.60	m	**0.60**
ends; framed	–	–	–	–	5.63	nr	**5.63**
ends; framed on rake	–	–	–	–	6.91	nr	**6.91**
Heading joints to handrail; mitred or raked							
overall size not exceeding 50 mm × 75 mm	–	–	–	–	27.66	nr	**27.66**
overall size not exceeding 69 mm × 100 mm	–	–	–	–	34.58	nr	**34.58**
Knee piece to handrail; mitred or raked							
overall size not exceeding 69 mm × 100 mm	–	–	–	–	73.75	nr	**73.75**
Balusters; stiffeners							
25 mm × 25 mm	–	–	–	–	3.09	m	**3.09**
32 mm × 32 mm	–	–	–	–	3.51	m	**3.51**
44 mm × 44 mm	–	–	–	–	4.61	m	**4.61**
ends; housed	–	–	–	–	1.39	nr	**1.39**

25 STAIRS, WALKWAYS AND BALUSTRADES

Item	PC £	Labour hours	Labour £	Plant £	Material £	Unit	Total rate £
Sub-rails							
32 mm × 63 mm	–	–	–	–	7.11	m	**7.11**
ends; framed joint to newel	–	–	–	–	5.99	nr	**5.99**
Knee rails							
32 mm × 140 mm	–	–	–	–	11.78	m	**11.78**
ends; framed joint to newel	–	–	–	–	5.99	nr	**5.99**
Newel posts							
44 mm × 94 mm; half newel	–	–	–	–	8.18	m	**8.18**
69 mm × 69 mm	–	–	–	–	8.87	m	**8.87**
94 mm × 94 mm	–	–	–	–	18.14	m	**18.14**
Newel caps; splayed on four sides							
62.50 mm × 125 mm × 50 mm	–	–	–	–	8.66	nr	**8.66**
100 mm × 100 mm × 50 mm	–	–	–	–	8.83	nr	**8.83**
125 mm × 125 mm × 50 mm	–	–	–	–	9.28	nr	**9.28**
Metal handrails							
Wallrails; painted mild steel CHS wall rail; with wall rose bracket							
42 mm dia.	–	–	–	–	–	m	**85.00**
Wallrails; stainless steel circular wall rail; with wall rose bracket							
42 mm dia.	–	–	–	–	–	m	**110.00**
WALKWAYS							
Flooring; metalwork							
Chequer plate flooring; galvanized mild steel; over 300 mm wide; bolted to steel supports							
6 mm thick	–	–	–	–	–	m²	**260.00**
8 mm thick	–	–	–	–	–	m²	**270.00**
Open mesh flooring; galvanized; over 300 mm wide; bolted to steel supports							
8 mm thick	–	–	–	–	–	m²	**260.00**
Surface treatment							
At works							
galvanizing	–	–	–	–	–	tonne	**345.00**
shotblasting	–	–	–	–	–	m²	**4.50**
touch up primer and one coat of two pack epoxy zinc phosphate or chromate primer	–	–	–	–	–	m²	**9.00**
LADDERS							
Loft ladders; fixing with screws to timber lining (not included)							
Loft ladders							
Youngman Easiway 3 section aluminium ladder; 2.3 m to 3.0 m ceiling height	–	0.93	18.78	–	112.30	nr	**131.08**
Youngman Eco folding 3 section timber ladder; 2.8 m ceiling height	–	2.00	40.39	–	141.82	nr	**182.21**
Youngman Spacemeaker aluminium sliding ladder; 2.6 m ceiling height	–	1.25	25.24	–	61.50	nr	**86.74**

25 STAIRS, WALKWAYS AND BALUSTRADES

Item	PC £	Labour hours	Labour £	Plant £	Material £	Unit	Total rate £
LADDERS – cont							
Loft ladders – cont							
Loft ladders – cont							
Youngman Deluxe 2 section aluminium ladder; 3.25 m ceiling height; spring assisted	–	2.50	50.48	–	330.82	nr	**381.30**
Access ladders; mild steel							
Ladders							
400 mm wide; 3850 mm long (overall); 12 mm dia. rungs; 65 mm × 15 mm strings; 50 mm × 5 mm safety hoops; fixing with expanded bolts; to masonry; mortices; welded fabrication	–	–	–	–	–	nr	**1250.00**

27 GLAZING

Item	PC £	Labour hours	Labour £	Plant £	Material £	Unit	Total rate £
SUPPLY ONLY RATES FOR GLASS							
BASIC GLASS PRICES SUPPLY ONLY							
Ordinary transluscent/patterned glass							
3 mm	–	–	–	–	30.74	m²	**30.74**
4 mm	–	–	–	–	32.62	m²	**32.62**
5 mm	–	–	–	–	39.67	m²	**39.67**
6 mm	–	–	–	–	43.55	m²	**43.55**
Obscured ground sheet glass – patterned							
4 mm white	–	–	–	–	46.13	m²	**46.13**
6 mm white	–	–	–	–	50.77	m²	**50.77**
Rough cast							
6 mm	–	–	–	–	43.23	m²	**43.23**
Ordinary Georgian wired							
7 mm cast	–	–	–	–	44.09	m²	**44.09**
6 mm polish	–	–	–	–	67.81	m²	**67.81**
Cetuff toughened; float							
4 mm	–	–	–	–	36.11	m²	**36.11**
5 mm	–	–	–	–	47.87	m²	**47.87**
6 mm	–	–	–	–	52.71	m²	**52.71**
10 mm	–	–	–	–	86.49	m²	**86.49**
Clear laminated; safety							
4.40 mm	–	–	–	–	43.04	m²	**43.04**
6.40 mm	–	–	–	–	51.39	m²	**51.39**
GENERAL GLAZING							
SUPPLY AND FIX PRICES							
NOTE: The following measured rates are provided by a glazing contractor and assume in excess of 500 m², within 20 miles of the suppliers branch.							
Standard plain glass; BS EN 14449; clear float; panes area 0.15 m²–4.00 m²							
3 mm thick; glazed with							
screwed beads	–	–	–	–	–	m²	**53.12**
4 mm thick; glazed with							
screwed beads	–	–	–	–	–	m²	**56.35**
5 mm thick; glazed with							
screwed beads	–	–	–	–	–	m²	**68.62**
6 mm thick; glazed with							
screwed beads	–	–	–	–	–	m²	**75.17**
Standard plain glass; BS EN 14449; obscure patterned; panes area 0.15 m²–4.00 m²							
4 mm thick; glazed with							
screwed beads	–	–	–	–	–	m²	**79.68**
6 mm thick; glazed with							
screwed beads	–	–	–	–	–	m²	**87.67**

27 GLAZING

Item	PC £	Labour hours	Labour £	Plant £	Material £	Unit	Total rate £
GENERAL GLAZING – cont							
Standard plain glass; BS EN 14449; rough cast; panes area 0.15 m²–4.00 m²							
6 mm thick; glazed with							
screwed beads	–	–	–	–	–	m²	72.09
Standard plain glass; BS EN 14449; Georgian wired cast; panes area 0.15 m²–4.00 m²							
7 mm thick; glazed with							
screwed beads	–	–	–	–	–	m²	64.06
extra for lining up wired glass	–	–	–	–	–	m²	3.80
Standard plain glass; BS EN 14449; Georgian wired polished; panes area 0.15 m²–4.00 m²							
6 mm thick; glazed with							
screwed beads	–	–	–	–	–	m²	99.39
extra for lining up wired glass	–	–	–	–	–	m²	3.80
Factory-made double hermetically sealed units; to wood or metal with screwed or clipped beads							
Two panes; BS EN 14449; clear float glass; 4 mm thick; 6 mm air space							
0.35 m²–2.00 m²	–	–	–	–	–	m²	142.46
Two panes; BS 952; clear float glass; 6 mm thick; 6 mm air space							
0.35 m²–2.0 m²	–	–	–	–	–	m²	165.90
2.00 m²–4.00 m²	–	–	–	–	–	m²	249.46
Factory-made double hermetically sealed units; with inner pane of Pilkington's K low emissivity coated glass; to wood or metal with screwed or clipped beads							
Two panes; BS EN 14449; clear float glass; 4 mm thick; 6 mm air space							
0.35 m²–2.00 m²	–	–	–	–	–	m²	173.11
Two panes; BS EN 14449; clear float glass; 6 mm thick; 6 mm air space							
0.35 m²–2.0 m²	–	–	–	–	–	m²	201.76
2.00 m²–4.00 m²	–	–	–	–	–	m²	303.44
Factory-made triple hermetically sealed units; with inner pane of Pilkington's K low emissivity coated glass; to wood or metal with screwed or clipped beads							
Three panes; BS EN 14449; clear float glass; 4 mm thick; 6 mm air spaces							
0.35 m²–2.00 m²	–	–	–	–	–	m²	279.00

27 GLAZING

Item	PC £	Labour hours	Labour £	Plant £	Material £	Unit	Total rate £
Three panes; BS EN 14449; clear float glass; 6 mm thick; 6 mm air spaces							
0.35 m²–2.0 m²	–	–	–	–	–	m²	324.79
2.00 m²–4.00 m²	–	–	–	–	–	m²	488.57
FIRE-RESISTING GLAZING							
Special glass; BS EN 14449; Pyran half-hour fire-resisting glass or other equal or approved							
6.50 mm thick rectangular panes; glazed with screwed hardwood beads and Sealmaster Fireglaze intumescent compound or other equal and approved to rebated frame							
300 mm × 400 mm pane	–	0.37	12.80	–	57.80	nr	70.60
400 mm × 800 mm pane	–	0.46	15.92	–	148.35	nr	164.27
500 mm × 1400 mm pane	–	0.74	25.61	–	314.59	nr	340.20
600 mm × 1800 mm pane	–	0.93	32.18	–	505.46	nr	537.64
Special glass; BS EN 14449; Pyrostop one-hour fire-resisting glass or other equal and approved							
15 mm thick regular panes; glazed with screwed hardwood beads and Sealmaster Fireglaze intumescent liner and compound or other equal and approved both sides							
300 mm × 400 mm pane	–	1.11	38.41	–	107.02	nr	145.43
400 mm × 800 mm pane	–	1.39	48.10	–	217.60	nr	265.70
500 mm × 1400 mm pane	–	1.85	64.02	–	452.65	nr	516.67
600 mm × 1800 mm pane	–	2.31	79.93	–	678.75	nr	758.68
TOUGHENED AND LAMINATED GLAZING							
Special glass; BS EN 14449; toughened clear float; panes area 0.15 m²–4.00 m²							
4 mm thick; glazed with screwed beads	–	–	–	–	–	m²	56.49
5 mm thick; glazed with screwed beads	–	–	–	–	–	m²	75.04
6 mm thick; glazed with screwed beads	–	–	–	–	–	m²	82.54
10 mm thick; glazed with screwed beads	–	–	–	–	–	m²	136.98
Special glass; BS EN 14449; clear laminated safety glass; panes area 0.15 m²–4.00 m²							
4.40 mm thick; glazed with screwed beads	–	–	–	–	–	m²	75.78
6.40 mm thick; glazed with screwed beads	–	–	–	–	–	m²	90.45

27 GLAZING

Item	PC £	Labour hours	Labour £	Plant £	Material £	Unit	Total rate £
TOUGHENED AND LAMINATED GLAZING – cont							
Special glass; BS EN 14449; clear laminated security glass							
7.50 mm thick regular panes; glazed with screwed hardwood beads and Intergens intumescent strip							
300 mm × 400 mm pane	–	0.37	12.80	–	34.84	nr	**47.64**
400 mm × 800 mm pane	–	0.46	15.92	–	87.80	nr	**103.72**
500 mm × 1400 mm pane	–	0.74	25.61	–	186.93	nr	**212.54**
600 mm × 1800 mm pane	–	0.93	32.18	–	294.29	nr	**326.47**
MIRRORS AND LOUVRES							
Mirror panels; BS EN 14449; silvered; insulation backing							
4 mm thick float; fixing with adhesive							
1000 mm × 1000 mm	–	–	–	–	–	nr	**56.66**
1000 mm × 2000 mm	–	–	–	–	–	nr	**113.42**
1000 mm × 4000 mm	–	–	–	–	–	nr	**413.39**
Glass louvres; BS EN 14449; with long edges ground or smooth							
6 mm thick float							
150 mm wide	–	–	–	–	–	m	**28.24**
7 mm thick Georgian wired cast							
150 mm wide	–	–	–	–	–	m	**44.45**
6 mm thick Georgian wire polished							
150 mm wide	–	–	–	–	–	m	**62.45**
ADDITIONAL LABOURS							
Drill holes in glass							
Drill holes 6 mm to 15 mm dia. to glass thickness:							
not exceeding 6 mm thick	–	–	–	–	3.49	nr	**3.49**
not exceeding 10 mm thick	–	–	–	–	4.48	nr	**4.48**
not exceeding 12 mm thick	–	–	–	–	5.55	nr	**5.55**
not exceeding 19 mm thick	–	–	–	–	7.00	nr	**7.00**
not exceeding 25 mm thick	–	–	–	–	8.68	nr	**8.68**
Drill holes 16 mm to 38 mm dia. to glass thickness:							
not exceeding 6 mm thick	–	–	–	–	5.01	nr	**5.01**
not exceeding 10 mm thick	–	–	–	–	6.62	nr	**6.62**
not exceeding 12 mm thick	–	–	–	–	7.90	nr	**7.90**
not exceeding 19 mm thick	–	–	–	–	9.95	nr	**9.95**
not exceeding 25 mm thick	–	–	–	–	12.32	nr	**12.32**
Drill holes over 38 mm dia. to glass thickness:							
not exceeding 6 mm thick	–	–	–	–	9.95	nr	**9.95**
not exceeding 10 mm thick	–	–	–	–	12.01	nr	**12.01**
not exceeding 12 mm thick	–	–	–	–	14.22	nr	**14.22**
not exceeding 19 mm thick	–	–	–	–	17.49	nr	**17.49**
not exceeding 25 mm thick	–	–	–	–	21.83	nr	**21.83**

27 GLAZING

Item	PC £	Labour hours	Labour £	Plant £	Material £	Unit	**Total rate £**
Other works to glass							
Curved cutting to glass							
to 4 mm thick panes	–	–	–	–	4.79	m	**4.79**
to 6 mm thick panes	–	–	–	–	4.79	m	**4.79**
to 6 mm thick wired panes	–	–	–	–	7.30	m	**7.30**
Intumescant paste to glazed panels for die doors; per side treated	–	–	–	–	10.11	m	**10.11**
Imitation washleather/black velvet bedding to edge of glass	–	–	–	–	1.74	m	**1.74**

28 FLOOR, WALL, CEILING AND ROOF FINISHES

Item	PC £	Labour hours	Labour £	Plant £	Material £	Unit	Total rate £
FLOORS; SCREEDS (CEMENT: SAND; CONCRETE; GRANOLITHIC)							
Cement and sand (1:3) screeds; steel trowelled							
Work to floors; one coat level; to concrete base; screeded; over 600 mm wide							
25 mm thick	–	0.41	9.39	0.38	0.49	m²	**10.26**
50 mm thick	–	0.52	11.91	0.38	1.52	m²	**13.81**
75 mm thick	–	0.66	15.11	0.38	2.29	m²	**17.78**
100 mm thick	–	0.70	16.02	0.38	3.06	m²	**19.46**
Add to the above for work to falls and crossfalls and to slopes							
not exceeding 15° from horizontal	–	0.02	0.46	–	–	m²	**0.46**
over 15° from horizontal	–	0.09	2.06	–	–	m²	**2.06**
water repellent additive incorporated in the mix	–	0.02	0.46	–	4.00	m²	**4.46**
oil repellent additive incorporated in the mix	–	0.07	1.61	–	3.60	m²	**5.21**
Fine concrete (1: 4–5) levelling screeds; steel trowelled							
Work to floors; one coat; level; to concrete base; over 600 mm wide							
50 mm thick	–	0.52	11.91	–	4.82	m²	**16.73**
75 mm thick	–	0.66	15.11	–	7.23	m²	**22.34**
extra over last for isolation joint to perimeter	–	0.20	4.58	–	3.32	m	**7.90**
Early drying floor screed; RMC Mortars Readyscreed; or other equal and approved; steel trowelled							
Work to floors; one coat; level; to concrete base; over 600 mm wide							
100 mm thick	–	0.70	16.02	–	11.52	m²	**27.54**
extra over last for galvanized chicken wire anticrack reinforcement	–	0.01	0.23	–	1.09	m²	**1.32**
Granolithic paving; cement and granite chippings 5 to dust (1:1:2); steel trowelled							
Work to floors; one coat; level; laid on concrete while green; bonded; over 600 mm wide							
25 mm thick	–	0.75	17.17	0.38	3.94	m²	**21.49**
38 mm thick	–	0.80	18.31	0.38	6.29	m²	**24.98**
Work to floors; two coat; laid on hacked concrete with slurry; over 600 mm wide							
50 mm thick	–	1.20	27.47	0.38	7.85	m²	**35.70**
75 mm thick	–	1.50	34.34	0.38	12.57	m²	**47.29**
Work to landings; one coat; level; laid on concrete while green; bonded; over 600 mm wide							
25 mm thick	–	0.80	18.31	0.38	3.94	m²	**22.63**
38 mm thick	–	1.00	22.89	0.38	6.29	m²	**29.56**

28 FLOOR, WALL, CEILING AND ROOF FINISHES

Item	PC £	Labour hours	Labour £	Plant £	Material £	Unit	Total rate £
Work to landings; two coat; laid on hacked concrete with slurry; over 600 mm wide							
50 mm thick	–	2.00	45.78	0.38	7.85	m²	**54.01**
75 mm thick	–	2.50	57.24	0.38	12.57	m²	**70.19**
Add to the above over 600 mm wide for							
liquid hardening additive incorporated in the mix	–	0.04	0.91	–	0.46	m²	**1.37**
oil-repellent additive incorporated in the mix	–	0.07	1.61	–	3.60	m²	**5.21**
25 mm work to treads; one coat; to concrete base							
225 mm wide	–	0.83	19.01	–	8.74	m	**27.75**
275 mm wide	–	0.83	19.01	–	9.79	m	**28.80**
returned end	–	0.17	3.90	–	–	nr	**3.90**
13 mm skirtings; rounded top edge and coved bottom junction; to brickwork or blockwork base							
75 mm wide on face	–	0.51	11.68	–	0.42	m	**12.10**
150 mm wide on face	–	0.69	15.79	–	7.70	m	**23.49**
ends; fair	–	0.04	0.91	–	–	nr	**0.91**
angles	–	0.06	1.38	–	–	nr	**1.38**
13 mm outer margin to stairs; to follow profile of and with rounded nosing to treads and risers; fair edge and arris at bottom, to concrete base							
75 mm wide	–	0.83	19.01	–	4.20	m	**23.21**
angles	–	0.06	1.38	–	–	nr	**1.38**
13 mm wall string to stairs; fair edge and arris on top; coved bottom junction with treads and risers; to brickwork or blockwork base							
275 mm (extreme) wide	–	0.74	16.94	–	7.34	m	**24.28**
ends	–	0.04	0.91	–	–	nr	**0.91**
angles	–	0.06	1.38	–	–	nr	**1.38**
ramps	–	0.07	1.61	–	–	nr	**1.61**
ramped and wreathed corners	–	0.09	2.06	–	–	nr	**2.06**
13 mm outer string to stairs; rounded nosing on top at junction with treads and risers; fair edge and arris at bottom; to concrete base							
300 mm (extreme) wide	–	0.74	16.94	–	9.09	m	**26.03**
ends	–	0.04	0.91	–	–	nr	**0.91**
angles	–	0.06	1.38	–	–	nr	**1.38**
ramps	–	0.07	1.61	–	–	nr	**1.61**
ramps and wreathed corners	–	0.09	2.06	–	–	nr	**2.06**
19 mm thick skirtings; rounded top edge and coved bottom junction; to brickwork or blockwork base							
75 mm wide on face	–	0.51	11.68	–	7.70	m	**19.38**
150 mm wide on face	–	0.69	15.79	–	11.89	m	**27.68**
ends; fair	–	0.04	0.91	–	–	nr	**0.91**
angles	–	0.06	1.38	–	–	nr	**1.38**
19 mm riser; one rounded nosing; to concrete base							
150 mm high; plain	–	0.83	19.01	–	6.65	m	**25.66**
150 mm high; undercut	–	0.83	19.01	–	6.65	m	**25.66**
180 mm high; plain	–	0.83	19.01	–	9.09	m	**28.10**
180 mm high; undercut	–	0.83	19.01	–	9.09	m	**28.10**

28 FLOOR, WALL, CEILING AND ROOF FINISHES

Item	PC £	Labour hours	Labour £	Plant £	Material £	Unit	Total rate £
FLOORS; SCREEDS (RESIN)							
Resins can be difficult to price as there may be site-specific variables, such as substrate condition and moisture content. There are numerous variables of thicknesses and surface textures, but we have provided cost guides for the most commonly used products. The installed cost will reduce by amounts increasing to approx. 10% as the overall area increases above 200 m² toward 4,000 m² and rise significantly as the area reduces toward m²							
Latex self-levelling floor screeds; steel trowelled							
Work to floors; level; to concrete base; over 600 mm wide							
3 mm thick; one coat	–	–	–	–	–	m²	**5.00**
5 mm thick; two coats	–	–	–	–	–	m²	**7.00**
Resin flooring; Altro epoxy and polyurethane resin flooring system, level, to concrete; sand cement screed or securely bonded 25 mm plywood							
Work to floors; level; to concrete base; over 600 mm wide and over 300 m² total area							
Altrocoat water-based epoxy coating; 140 micron thick for two coats	–	0.15	6.49	–	7.15	m²	**13.64**
AltroTect solvent free high build epoxy coating; 350 micron thick for two coats	–	0.15	6.49	–	8.69	m²	**15.18**
Altroseal UVR chemically resistant polyurethane coating; 140 micron thick for two coats	–	0.15	6.49	–	8.25	m²	**14.74**
AltroFlow EP, epoxy self-smoothing resin screed; 2 mm thick including primer	–	0.30	12.98	–	19.74	m²	**32.72**
Altroflow PU, polyurethane self-smoothing resin screed; 2 mm thickness including primer	–	0.30	12.98	–	13.05	m²	**26.03**
Altroflow PU, polyurethane self-smoothing resin screed; 4 mm thickness including primer	–	0.30	12.98	–	21.91	m²	**34.89**
Altro Flexiflow, flexible polyurethame self-smoothing resin screed; 2 mm thickness including primer	–	0.30	12.98	–	22.73	m²	**35.71**
Altro TB screed, trowel applied multi-coloured epoxy screed; 3 mm thickness including primer	–	0.40	17.30	–	23.66	m²	**40.96**
AltroScreed Quartz EP, trowel applied multi-coloured epoxy screed; 4 mm thickness including primer	–	0.40	17.30	–	25.52	m²	**42.82**
Altro MultiScreed, trowel applied flecked epoxy screed; 4 mm thickness including primer	–	0.40	17.30	–	23.46	m²	**40.76**
AltroGrip EP, multilayer, textured, slip-resistant, plain coloured; 4 mm thickness including primer	–	0.40	17.30	–	24.19	m²	**41.49**

28 FLOOR, WALL, CEILING AND ROOF FINISHES

Item	PC £	Labour hours	Labour £	Plant £	Material £	Unit	Total rate £
AltroGrip PU, multilayer, textured, slip-resistant, plain coloured; 4 mm thickness including primer	–	0.40	17.30	–	19.48	m²	**36.78**
Altrocrete PU Excel, heavy duty polyurethane screed; 6 mm thickness including primer	–	0.40	17.30	–	26.32	m²	**43.62**
Altrocrete PU Excel, heavy duty polyurethane screed; 8 mm thickness including primer	–	0.40	17.30	–	30.26	m²	**47.56**
Perimeter site-formed resin cove, 100 mm high, 40 mm radius	–	–	–	–	26.25	m	**26.25**
Isocrete K screeds or other equal; steel trowelled							
Work to floors; level; to concrete base; over 600 mm wide							
35 mm thick; plus polymer bonder coat	–	–	–	–	–	m²	**15.53**
40 mm thick	–	–	–	–	–	m²	**14.34**
45 mm thick	–	–	–	–	–	m²	**15.16**
50 mm thick	–	–	–	–	–	m²	**15.97**
Work to floors; to falls or cross-falls; to concrete base; over 600 mm wide							
55 mm (average) thick	–	–	–	–	–	m²	**16.77**
60 mm (average) thick	–	–	–	–	–	m²	**17.61**
65 mm (average) thick	–	–	–	–	–	m²	**18.41**
75 mm (average) thick	–	–	–	–	–	m²	**20.05**
90 mm (average) thick	–	–	–	–	–	m²	**22.51**
Isocrete K screeds; quick drying; or other equal and approved; steel trowelled							
Work to floors; level or to floors n.e. 15° frojm the horizontal; to concrete base; over 600 mm wide							
55 mm thick	–	–	–	–	–	m²	**21.74**
75 mm thick	–	–	–	–	–	m²	**27.16**
Isocrete pumpable Self-Level Plus screeds; or other equal and approved; protected with Corex type polythene; knifed off prior to laying floor finish; flat smooth finish							
Work to floors; level or to floors n.e. 15° frojm the horizontal; to concrete base; over 600 mm wide							
20 mm thick	–	–	–	–	–	m²	**26.48**
50 mm thick	–	–	–	–	–	m²	**35.32**
FLOORS; MASTIC ASPHALT							
Mastic asphalt flooring to BS 6925 Type F 1076; black							
20 mm thick; one coat coverings; felt isolating membrane; to concrete base; flat							
over 300 mm wide	–	–	–	–	–	m²	**18.92**
225 mm–300 mm wide	–	–	–	–	–	m²	**35.16**
150 mm–225 mm wide	–	–	–	–	–	m²	**38.63**
not exceeding 150 mm wide	–	–	–	–	–	m²	**47.28**

28 FLOOR, WALL, CEILING AND ROOF FINISHES

Item	PC £	Labour hours	Labour £	Plant £	Material £	Unit	Total rate £
FLOORS; MASTIC ASPHALT – cont							
Mastic asphalt flooring to BS 6925 Type F 1076 – cont							
25 mm thick; one coat coverings; felt isolating membrane; to concrete base; flat							
over 300 mm wide	–	–	–	–	–	m²	**21.97**
225 mm–300 mm wide	–	–	–	–	–	m²	**37.51**
150 mm–225 mm wide	–	–	–	–	–	m²	**40.88**
not exceeding 150 mm wide	–	–	–	–	–	m²	**49.56**
20 mm three coat skirtings to brickwork base							
not exceeding 150 mm girth	–	–	–	–	–	m	**19.34**
150 mm–225 mm girth	–	–	–	–	–	m	**23.64**
225 mm–300 mm girth	–	–	–	–	–	m	**27.94**
Mastic asphalt flooring; acid-resisting; black							
20 mm thick; one coat coverings; felt isolating membrane; to concrete base flat							
over 300 mm wide	–	–	–	–	–	m²	**22.16**
225 mm–300 mm wide	–	–	–	–	–	m²	**40.55**
150 mm–225 mm wide	–	–	–	–	–	m²	**41.88**
not exceeding 150 mm wide	–	–	–	–	–	m²	**50.52**
25 mm thick; one coat coverings; felt isolating membrane; to concrete base; flat							
over 300 mm wide	–	–	–	–	–	m²	**26.19**
225 mm–300 mm wide	–	–	–	–	–	m²	**41.68**
150 mm–225 mm wide	–	–	–	–	–	m²	**45.12**
not exceeding 150 mm wide	–	–	–	–	–	m²	**53.79**
20 mm thick; three coat skirtings to brickwork base							
not exceeding 150 mm girth	–	–	–	–	–	m	**19.54**
150 mm–225 mm girth	–	–	–	–	–	m	**22.76**
225 mm–300 mm girth	–	–	–	–	–	m	**25.83**
Mastic asphalt flooring to BS 6925 Type F 1451; red							
20 mm thick; one coat coverings; felt isolating membrane; to concrete base; flat							
over 300 mm wide	–	–	–	–	–	m²	**31.03**
225 mm–300 mm wide	–	–	–	–	–	m²	**51.24**
150 mm–225 mm wide	–	–	–	–	–	m²	**55.36**
not exceeding 150 mm wide	–	–	–	–	–	m²	**66.22**
20 mm thick; three coat skirtings to brickwork base							
not exceeding 150 mm girth	–	–	–	–	–	m	**24.35**
150 mm–225 mm girth	–	–	–	–	–	m	**31.03**

28 FLOOR, WALL, CEILING AND ROOF FINISHES

Item	PC £	Labour hours	Labour £	Plant £	Material £	Unit	Total rate £
FLOORS; EDGE FIXED CARPETING; CARPET TILES							
Fitted carpeting; Wilton wool/nylon or other equal and approved; 80/20 velvet pile; heavy domestic plain							
Work to floors							
over 600 mm wide	40.00	0.37	5.74	–	46.20	m²	**51.94**
Work to treads and risers							
not exceeding 600 mm wide	24.00	0.74	11.49	–	27.72	m	**39.21**
Fitted carpet; Forbo Flooring; Flocked flooring							
Work to floors							
Flotex classic textile; 2.0 m wide roll	22.00	0.45	9.09	–	25.00	m²	**34.09**
Flotex HD textile; 2.0 m wide roll	24.63	0.45	9.09	–	27.97	m²	**37.06**
Carpet tiles; Forbo Flooring; 500 mm × 500 mm carpet tiles; adhesive applied to subfloor							
Work to floors:							
Teviot; low level loop pile; 2.50 mm thick	10.56	0.45	9.09	–	12.09	m²	**21.18**
Barcode; low level loop pile; 2.50 mm thick	15.84	0.45	9.09	–	18.04	m²	**27.13**
Circulate; random lay, batchless	20.23	0.45	9.09	–	23.01	m²	**32.10**
Arran; textured loop; 4.00 mm thick	24.63	0.45	9.09	–	27.97	m²	**37.06**
Alignment; textured cut and loop; 4.00 mm thick	24.63	0.45	9.09	–	27.97	m²	**37.06**
Fitted carpeting; Gradus broadloom polypropylene or other equal and approved							
Work to floors over 600 mm wide							
Bodega/Pacific/Predator ranges, loop pile	15.87	0.37	5.74	–	18.33	m²	**24.07**
Genus/Volnay ranges, cut pile	26.81	0.33	5.18	–	30.96	m²	**36.14**
Work to treads and risers							
not exceeding 600 mm wide	16.09	0.74	11.49	–	18.57	m	**30.06**
Underlay to carpeting							
Work to floors							
over 600 mm wide	3.04	0.07	1.08	–	3.35	m²	**4.43**
raking cutting	–	0.07	0.92	–	–	m	**0.92**
Carpet tiles to cement and sand base							
Work to floors over 600 mm wide							
Heuga 530 heavy duty loop pile	21.51	0.28	6.40	–	23.72	m²	**30.12**
Gradus Adventure range; bitumen backing; tufted pattern loop pile	30.90	0.28	6.40	–	34.07	m²	**40.47**
Latour/Predator ranges; bitumen backing; loop pile	16.23	0.28	6.40	–	17.89	m²	**24.29**
Sundries							
Carpet gripper fixed to floor; standard edging							
22 mm wide	–	0.04	0.53	–	0.38	m	**0.91**

28 FLOOR, WALL, CEILING AND ROOF FINISHES

Item	PC £	Labour hours	Labour £	Plant £	Material £	Unit	Total rate £
FLOORS; EDGE FIXED CARPETING; CARPET TILES – cont							
Stair nosings; aluminium; Gradus or equivalent							
Medium duty hard aluminium alloy stair tread nosings; plugged and screwed in concrete							
56 mm × 32 mm; ref AS11	8.52	0.23	3.76	–	9.24	m	**13.00**
84 mm × 32 mm; ref AS12	11.81	0.28	4.57	–	12.78	m	**17.35**
Heavy duty aluminium alloy stair tread nosings; plugged and screwed to concrete							
48 mm × 38 mm; ref HE1	9.95	0.28	4.57	–	10.77	m	**15.34**
82 mm × 38 mm; ref HE2	13.81	0.32	5.22	–	14.93	m	**20.15**
Door entrance mats							
Entrance mat systems; aluminium wiper bar; laying in position; 18 mm thick							
Gradus Topguard; 900 mm × 550 mm	137.28	0.46	6.05	–	144.14	nr	**150.19**
Gradus Topguard; 1200 mm × 750 mm	247.09	0.46	6.05	–	259.44	nr	**265.49**
Gradus Topguard; 2400 mm × 1200 mm	790.70	0.93	12.22	–	830.24	nr	**842.46**
Nuway Tuftiguard 500 mm × 1300 mm	–	–	–	–	–	nr	**238.76**
Nuway Tuftiguard 1000 mm × 1600 mm	–	–	–	–	–	nr	**501.03**
Nuway Tuftiguard 1800 mm × 3000 mm	–	–	–	–	–	nr	**1983.55**
extra for brass wipers	–	–	–	–	200.81	m²	**200.81**
Nuway Tuftiguard; single side aluminium scraper bar	–	–	–	–	–	m²	**367.32**
Nuway Tuftiguard; double side aluminium rigidsctaper bars	–	–	–	–	–	m²	**524.76**
Coral Classic textile secondary and circulation matting system; 2.0 m wide roll	–	–	–	–	–	m²	**62.97**
Coral Brush Activ secondary and circulation matting system; 2.0 m wide roll	–	–	–	–	–	m²	**62.97**
Coral Duo secondary matting system; 2.0 m wide roll	–	–	–	–	–	m²	**62.97**
Matwells							
Polished stainless steel matwell; angle rim with lugs; bedding in screed							
Stainless steel frame bed in screed 500 × 1300 mm	–	–	–	–	–	nr	**47.75**
Stainless steel frame bed in screed 1000 × 1600 mm	–	–	–	–	–	nr	**95.51**
Stainless steel frame bed in screed 1100 × 1400 mm	–	–	–	–	–	nr	**91.83**
Stainless steel frame bed in screed 1800 × 3000 mm	–	–	–	–	–	nr	**176.32**

28 FLOOR, WALL, CEILING AND ROOF FINISHES

Item	PC £	Labour hours	Labour £	Plant £	Material £	Unit	Total rate £
Polished aluminium matwell; angle rim with lugs brazed on; bedding in screed							
900 mm × 550 mm; constructed with							
25 × 25 × 3 mm angle	26.18	0.93	12.22	–	27.49	nr	**39.71**
1200 mm × 750 mm; constructed with							
34 × 26 × 6 mm angle	43.79	1.00	13.14	–	45.98	nr	**59.12**
2400 mm × 1200 mm; constructed with							
50 × 50 × 6 mm angle	167.75	1.50	19.71	–	176.14	nr	**195.85**
Polished brass matwell; comprising angle rim with lugs brazed on; bedding in screed							
900 mm × 550 mm; constructed with							
25 × 25 × 5 mm angle	66.11	0.93	12.22	–	69.42	nr	**81.64**
1200 mm × 750 mm; constructed with							
38 × 38 × 6 mm angle	179.61	1.00	13.14	–	188.59	nr	**201.73**
2400 mm × 1200 mm; constructed with							
38 × 38 × 6 mm angle	331.58	1.50	19.71	–	348.16	nr	**367.87**
FLOORS; GRANITE; MARBLE; SLATE; TERRAZZO							
ALTERNATIVE TILE MATERIALS							
Prime cost rates for Dennis Ruabon clay floor quarries or equivalent (£/1000)							
194 mm × 194 mm × 12.5 mm; square; red	–	–	–	–	588.00	1000	**588.00**
194 mm × 194 mm × 12.5 mm; polygon; red	–	–	–	–	43.76	m²	**43.76**
150 mm × 150 mm × 12.5 mm; square; heatherbrown	–	–	–	–	672.00	1000	**672.00**
150 mm × 150 mm × 12.5 mm; studded square; heatherbrown or red	–	–	–	–	974.40	1000	**974.40**
150 mm × 150 mm × 12.50 mm; polygon; red	–	–	–	–	62.53	m²	**62.53**
SUPPLY AND FIX PRICES							
Clay floor quarries; BS EN 10545; class 1; Dennis Ruabon tiles or other equal; level bedding 10 mm thick and jointing in cement and sand (1:3); butt joints; straight both ways; flush pointing with grout; to cement and sand base							
Work to floors; over 600 mm wide							
150 mm × 150 mm × 12.50 mm thick; heatherbrown	26.88	0.74	16.94	–	37.11	m²	**54.05**
150 mm × 150 mm × 12.50 mm thick; red	23.52	0.74	16.94	–	33.40	m²	**50.34**
194 mm × 194 mm × 12.50 mm thick; heatherbrown	20.00	0.60	13.73	–	29.52	m²	**43.25**
Works to floors; in staircase areas or plant rooms							
150 mm × 150 mm × 12.50 mm thick; heatherbrown	26.88	0.83	19.01	–	37.11	m²	**56.12**
150 mm × 150 mm × 12.50 mm thick; red	24.70	0.83	19.01	–	33.40	m²	**52.41**
194 mm × 194 mm × 12.50 mm thick; heatherbrown	20.00	0.69	15.79	–	29.52	m²	**45.31**
Work to floors; not exceeding 600 mm wide							
150 mm × 150 mm × 12.50 mm thick; heatherbrown	–	0.37	8.47	–	9.85	m	**18.32**
150 mm × 150 mm × 12.50 mm thick; red	–	0.37	8.47	–	8.72	m	**17.19**
194 mm × 194 mm × 12.50 mm thick; heatherbrown	–	0.31	7.10	–	7.36	m	**14.46**

28 FLOOR, WALL, CEILING AND ROOF FINISHES

Item	PC £	Labour hours	Labour £	Plant £	Material £	Unit	Total rate £
FLOORS; GRANITE; MARBLE; SLATE; TERRAZZO – cont							
Clay floor quarries – cont							
Work to floors – cont							
fair square cutting against flush edges of existing finishes	–	0.11	1.70	–	2.43	m	4.13
raking cutting	–	0.19	2.99	–	2.73	m	5.72
cutting around pipes; not exceeding 0.30 m girth	–	0.14	2.29	–	–	nr	2.29
extra for cutting and fitting into recessed manhole cover 600 mm × 600 mm	–	0.93	15.18	–	–	nr	15.18
Work to sills; 150 mm wide; rounded edge tiles							
200 mm × 150 mm × 22 mm thick; interior; heatherbrown or red	–	0.31	7.10	–	9.22	m	16.32
150 mm × 173 mm × 58 mm thick; exterior; heatherbrown or red	–	0.32	7.33	–	38.35	m	45.68
fitted end	–	0.14	2.29	–	–	nr	2.29
Coved skirtings; 150 mm high; rounded top edge							
150 mm × 150 mm × 12.50 mm thick; ref. CBTR; heatherbrown or red	–	0.23	5.26	–	10.61	m	15.87
ends	–	0.04	0.65	–	–	nr	0.65
angles	–	0.14	2.29	–	2.81	nr	5.10
50 mm × 50 mm × 5.50 mm thick slip-resistant mosaic floor tiles, fixing with adhesive; butt joints; straight both ways; flush pointing with white grout; to cement and sand base							
Work to floors							
over 600 mm wide	89.25	1.76	40.29	–	95.74	m²	136.03
not exceeding 600 mm wide	–	1.38	31.59	–	58.05	m	89.64
Dakota mahogany granite cladding; polished finish; jointed and pointed in coloured mortar (1:2:8)							
20 mm work to floors; level; to cement and sand base							
over 600 mm wide	–	–	–	–	–	m²	350.00
20 mm × 300 mm treads; plain nosings	–	–	–	–	–	m	200.00
raking, cutting	–	–	–	–	–	m	35.00
polished edges	–	–	–	–	–	m	450.00
birdsmouth	–	–	–	–	–	m	45.00
Riven Welsh slate floor tiles; level; bedding 10 mm thick and jointing in cement and sand (1:3); butt joints; straight both ways; flush pointing with coloured mortar; to cement and sand base							
Work to floors; over 600 mm wide							
regular size and patter; 12 mm–15 mm thick	28.50	0.56	12.82	–	42.88	m²	55.70

28 FLOOR, WALL, CEILING AND ROOF FINISHES

Item	PC £	Labour hours	Labour £	Plant £	Material £	Unit	Total rate £
Work to floors; not exceeding 600 mm wide							
regular size and patter; 12 mm–15 mm thick	–	0.56	12.82	–	25.89	m	**38.71**
Riven Chinese or Spanish slate floor tiles; level; bedding 10 mm thick and jointing in cement and sand (1:3); butt joints; straight both ways; flush pointing with coloured mortar; to cement and sand base							
Work to floors; over 600 mm wide							
regular size and patter; 12 mm–15 mm thick	13.30	0.56	12.82	–	25.33	m²	**38.15**
Work to floors; not exceeding 600 mm wide							
regular size and patter; 12 mm–15 mm thick	–	0.56	12.82	–	25.89	m	**38.71**
Roman Travertine marble cladding; polished finish; jointed and pointed in coloured mortar (1:2:8)							
20 mm thick work to floors; level; to cement and sand base							
over 600 mm wide	–	–	–	–	–	m²	**230.00**
20 mm × 300 mm treads; plain nosings	–	–	–	–	–	m	**138.00**
raking cutting	–	–	–	–	–	m	**27.00**
polished edges	–	–	–	–	–	m	**25.00**
birdsmouth	–	–	–	–	–	m	**50.00**
Terrazzo tiles; BS EN 13748; aggregate size random ground grouted and polished to 80's grit finish; standard colour range; 3 mm joints symmetrical layout; bedding in 42 mm cement semi-dry mix (1:4); grouting with neat matching cement							
300 mm × 300 mm × 28 mm (nominal) Terrazzo tile units; hydraulically pressed, mechanically vibrated, steam cured; to floors on concrete base (not included); sealed with penetrating case hardener or other equal and approved; 2 coats applied immediately after final polishing							
plain; laid level	–	–	–	–	–	m²	**46.95**
plain; to slopes exceeding 15° from horizontal	–	–	–	–	–	m²	**57.23**
to small areas/toilets	–	–	–	–	–	m²	**107.46**
Accessories							
plastic division strips; 6 mm × 38 mm; set into floor tiling above crack inducing joints, to the nearest full tile module	–	–	–	–	–	m	**3.24**

28 FLOOR, WALL, CEILING AND ROOF FINISHES

Item	PC £	Labour hours	Labour £	Plant £	Material £	Unit	Total rate £
FLOORS; GRANITE; MARBLE; SLATE; TERRAZZO – cont							
Specially made terrazzo precast units; BS EN 13748–1; aggregate size random; standard colour range; 3 mm joints; grouting with neat matching cement							
Standard tread and riser square combined terrazzo units (with riser cast down) or other equal and approved; 280 mm wide; 150 mm high; 40 mm thick; machine-made; vibrated and fully machine polished; incorporating 1 nr. Ferodo anti-slip insert ref. OT40D or other equal and approved cast-in during manufacture; one end polished only							
fixed with cement: sand (1:4) mortar on prepared backgrounds (not included); grouted in neat tinted cement; wiped clean on completion of fixing	–	–	–	–	–	m	**245.85**
Standard tread square terrazzo units or other equal and approved; 40 mm thick; 280 mm wide; factory polished; incorporating 1 nr. Ferodo anti-slip insert ref. OT40D or other equal and approved							
fixed with cement: sand (1:4) mortar on prepared backgrounds (not included); grouted in neat tinted cement; wiped clean on completion of fixing	–	–	–	–	–	m	**147.92**
extra over for 55 × 55 mm contrasting colour to step nosing	–	–	–	–	–	m	**51.70**
Standard riser square terrazzo units or other equal and approved; 40 mm thick; 150 mm high; factory polished							
fixed with cemnt: sand (1:4) mortar on prepared backgrounds (not included); grouted in neat tinted cement; wiped clean on completion of fixing	–	–	–	–	–	m	**91.79**
Standard coved terrazzo skirting units or other equal and approved; 904 mm long; 150 mm high; nominal finish; 23 mm thick; with square top edge							
fixed with cement: sand (1:4) mortar on prepared backgrounds (by others); grouted in neat tinted cement; wiped clean on completion of fixing	–	–	–	–	–	m	**86.23**
extra over for special internal/external angle pieces to match	–	–	–	–	–	m	**24.49**
extra over for special polished ends	–	–	–	–	–	nr	**8.07**
FLOORS; CORK; RUBBER; VINYL							
Linoleum sheet; Forbo Flooring; laid level; fixing with adhesive; welded joints; to cement and sand base							

28 FLOOR, WALL, CEILING AND ROOF FINISHES

Item	PC £	Labour hours	Labour £	Plant £	Material £	Unit	Total rate £
Work to floors; over 600 mm wide							
Marmoleum Real marbled sheet, 2.0 m wide × 2.50 mm thick	15.84	0.37	8.47	–	17.63	m²	**26.10**
Marmoleum Dual marbled tile 333 mm × 333 mm × 3.20 mm thick	19.35	0.45	10.30	–	21.50	m²	**31.80**
Decibel 17db marbled acoustic sheet, 2.0 m wide; 3.50 mm thick	19.35	0.45	10.30	–	21.50	m²	**31.80**
High performance vinyl sheet; Forbo Flooring Eternal; laid level with welded seams; fixing with adhesive; to cement and sand base							
Work to floors; over 600 mm wide							
2.0 m wide roll × 2.00 mm thick	13.20	0.35	8.01	–	14.72	m²	**22.73**
Slip-resistant vinyl sheet; Forbo Flooring Surestep; laid level with welded seams; fixing with adhesive; to cement and sand base							
Work to floors; over 600 mm wide							
2.0 m wide roll × 2.00 mm thick	13.20	0.37	8.47	–	14.72	m²	**23.19**
Anti-static vinyl tiles; Forbo Flooring Colorex SD; laid level; fixing with adhesive; to cement and sand base							
Work to floors; over 600 mm wide							
615 mm × 615 mm tiles × 2.00 mm thick	19.35	0.50	11.45	–	21.50	m²	**32.95**
Acoustic vinyl sheet; Forbo Flooring Sarlon Traffic 19 db; level with welded seams; fixing with adhesive; to cement and sand base							
Work to floors; over 600 mm wide							
2.0 m wide roll × 3.40 mm thick	15.84	0.37	8.47	–	17.63	m²	**26.10**
Vinyl sheet; Altro range; with welded seams; level; fixing with adhesive where appropriate; to cement and sand base							
Work to floors; over 600 mm wide and over 200 m² total area							
2.00 mm thick; Walkway 20 and Walkway Plus 20	10.35	0.56	12.82	–	12.89	m²	**25.71**
2.00 mm thick; Marine 20	17.00	0.55	12.59	–	20.64	m²	**33.23**
2.00 mm thick; Aquarius	17.00	0.55	12.59	–	20.64	m²	**33.23**
2.00 mm thick; Suprema II	18.25	0.55	12.59	–	21.81	m²	**34.40**
2.00 mm thick; Wood Safety	13.70	0.55	12.59	–	17.01	m²	**29.60**
2.20 mm thick; Xpresslay and Xpresslay Plus(NB No adhesive required)	12.50	0.44	10.07	–	13.79	m²	**23.86**
2.50 mm thick; Classic 25	18.70	0.60	13.73	–	22.10	m²	**35.83**
2.50 mm thick; Designer 25	18.70	0.60	13.73	–	22.10	m²	**35.83**
2.50 mm thick; Unity 25	20.30	0.60	13.73	–	23.87	m²	**37.60**
3.50 mm thick; Stronghold 30	22.20	0.65	14.88	–	26.40	m²	**41.28**

28 FLOOR, WALL, CEILING AND ROOF FINISHES

Item	PC £	Labour hours	Labour £	Plant £	Material £	Unit	Total rate £
FLOORS; CORK; RUBBER; VINYL – cont							
Homogeneous Vinyl sheet; Marleyflor Plus or other equal; level; with welded seams; fixing with adhesive; level; to cement and sand base							
Work to floors; over 600 mm wide							
2.00 mm thick	6.16	0.42	9.62	–	7.18	m²	**16.80**
100 mm high skirtings	–	0.11	2.52	–	1.75	m	**4.27**
Safety sheet; Marleyflor Granite Multisafe or other equal; level; with welded seams; fixing with adhesive; level; to cement and sand base							
Work to floors; over 600 mm wide							
2.00 mm thick	12.22	0.42	9.62	–	13.87	m²	**23.49**
Vinyl sheet; Marleyflor Omnisports or other equal; level; with welded seams; fixing with adhesive; level; to cement and sand base							
Work to floors; over 600 mm wide							
7.65 mm thick; Pro	24.36	0.90	20.60	–	27.25	m²	**47.85**
8.75 mm thick; Competition	28.55	1.00	22.89	–	31.87	m²	**54.76**
Vinyl semi-flexible tiles; Marleyflor Homogeneous tiles range; level; fixing with adhesive; butt joints; straight both ways; to cement and sand base							
Work to floors; over 600 mm wide							
300 mm × 300 mm × 2.00 mm thick; Vylon Plus	5.21	0.23	5.26	–	6.13	m²	**11.39**
500 mm × 500 mm × 2.00 mm thick; Marleyflor Plus	6.40	0.20	4.58	–	7.44	m²	**12.02**
Vinyl tiles; Polyflex Plus; level; fixing with adhesive; butt joints; straight both ways; to cement and sand base							
Work to floors; over 600 mm wide							
300 mm × 300 mm × 2.00 mm thick	5.66	0.23	5.26	–	6.63	m²	**11.89**
Vinyl tiles; Polyflex Camaro; level; fixing with adhesive; butt joints; straight both ways; to cement and sand base							
Work to floors; over 600 mm wide							
300 mm × 300 mm × 2.00 mm thick	12.73	0.30	6.87	–	14.43	m²	**21.30**
Vinyl tiles; Polyflor XL; level; fixing with adhesive; butt joints; straight both ways; to cement and sand base							
Work to floors; over 600 mm wide							
300 mm × 300 mm × 2.00 mm thick	7.06	0.32	7.33	–	8.17	m²	**15.50**
Vinyl sheet; Polyflor XL; level; fixing with adhesive; butt joints; straight both ways; to cement and sand base							

28 FLOOR, WALL, CEILING AND ROOF FINISHES

Item	PC £	Labour hours	Labour £	Plant £	Material £	Unit	Total rate £
Work to floors; over 600 mm wide							
2.00 mm thick	4.84	0.28	6.30	–	5.73	m²	**12.03**
Vinyl sheet; Polysafe Standard; level; fixing with adhesive; welded seams; to cement and sand base							
Work to floors; over 600 mm wide							
2.00 mm thick	8.89	0.28	6.30	–	10.20	m²	**16.50**
2.50 mm thick	13.12	0.28	6.30	–	14.85	m²	**21.15**
Vinyl sheet; Polysafe hydro; level; fixing with adhesive; welded seams; to cement and sand base							
Work to floors; over 600 mm wide							
2.00 mm thick	11.50	0.28	6.30	–	13.06	m²	**19.36**
Luxury mineral vinyl tiles; Marley I D Naturelle; level; fixing with adhesive; butt joints; straight both ways; to cement and sand base							
Work to floors; over 600 mm wide							
330 mm × 330 mm × 2.00 mm thick	8.54	0.23	5.26	–	9.81	m²	**15.07**
Acoustic vinyl tiles; Marley Tapiflex 243; level; fixing with adhesive; butt joints; straight both ways; to cement and sand base							
Work to floors; over 600 mm wide							
500 mm × 500 mm × 2.00 mm thick	11.54	0.20	4.58	–	13.10	m²	**17.68**
Linoleum tiles; Marley Veneto XF; level; fixing with adhesive; butt joints; straight both ways; to cement and sand base							
Work to floors; over 600 mm wide							
500 mm × 500 mm × 2.50 mm thick	13.67	0.20	4.58	–	15.47	m²	**20.05**
Cork tiles; natural finish, unsealed; level; fixing with adhesive; butt joints; straight both ways; to cement and sand base							
Work to floors; over 600 mm wide							
300 mm × 300 mm × 8.00 mm thick; contract quality	20.09	0.40	9.16	–	25.98	m²	**35.14**
300 mm × 300 mm × 6.00 mm thick; high density	16.25	0.40	9.16	–	21.54	m²	**30.70**
300 mm × 300 mm × 4.00 mm thick; domestic	14.15	0.37	8.47	–	19.12	m²	**27.59**
300 mm × 300 mm × 3.20 mm thick; domestic	11.51	0.37	8.47	–	16.07	m²	**24.54**
Cork tiles; PVC surfaced for heaviest wear areas; level; fixing with adhesive; butt joints; straight both ways; to cement and sand base							
600 mm × 300 mm × 3.20 mm thick	32.40	0.37	8.47	–	36.78	m²	**45.25**

28 FLOOR, WALL, CEILING AND ROOF FINISHES

Item	PC £	Labour hours	Labour £	Plant £	Material £	Unit	Total rate £
FLOORS; CORK; RUBBER; VINYL – cont							
Rubber studded tiles; Altro Mondopave; level; fixing with adhesive; butt joints; straight to cement and sand base							
Work to floors; over 600 mm wide							
500 mm × 500 mm × 2.50 mm thick; type MRB; black	21.44	0.60	13.73	–	25.69	m²	**39.42**
500 mm × 500 mm × 4.00 mm thick; type MRB; black	25.16	0.60	13.73	–	29.80	m²	**43.53**
Work to landings; over 600 mm wide							
500 mm × 500 mm × 4.00 mm thick; type MRB; black	25.16	0.74	16.94	–	32.55	m²	**49.49**
Work to stairs							
tread; 275 mm wide	–	0.46	10.53	–	9.69	m	**20.22**
riser; 180 mm wide	–	0.56	12.82	–	6.78	m	**19.60**
Sundry floor sheeting underlays							
For floor finishings; over 600 mm wide							
building paper to BS 1521; class A; 75 mm lap (laying only)	–	0.05	0.66	–	–	m²	**0.66**
3.20 mm thick hardboard	–	0.19	4.35	–	1.60	m²	**5.95**
6.00 mm thick plywood	–	0.28	6.40	–	6.13	m²	**12.53**
Skirtings; plastic; Gradus or equivalent							
Set-in skirtings							
100 mm high; ref. SI100	–	0.11	2.52	–	2.24	m	**4.76**
150 mm high; ref. SI150	–	0.22	5.04	–	3.65	m	**8.69**
Set-on skirtings							
100 mm high; ref. SO100	–	0.22	5.04	–	1.83	m	**6.87**
150 mm high; ref. SO150	–	0.40	9.16	–	3.42	m	**12.58**
Stair nosings; aluminium; Gradus or equivalent							
Medium duty hard aluminium alloy stair tread nosings; plugged and screwed in concrete							
56 mm × 32 mm; ref AS11	8.73	0.23	3.76	–	9.24	m	**13.00**
84 mm × 32 mm; ref AS12	12.10	0.28	4.57	–	12.78	m	**17.35**
Heavy duty aluminium alloy stair tread nosings; plugged and screwed to concrete							
48 mm × 38 mm; ref HE1	10.20	0.28	4.57	–	10.77	m	**15.34**
82 mm × 38 mm; ref HE2	14.15	0.32	5.22	–	14.93	m	**20.15**

28 FLOOR, WALL, CEILING AND ROOF FINISHES

Item	PC £	Labour hours	Labour £	Plant £	Material £	Unit	Total rate £
FLOORS; WOOD BLOCK							
Wood blocks; Havwoods or other equal and approved; 25 mm thick; level; laid to herringbone pattern with 2 block borderl; fixing with adhesive; to cement: sand base; sanded and sealed							
Work to floors; over 300 mm wide							
Merbau	–	–	–	–	–	m²	**111.36**
Iroko	–	–	–	–	–	m²	**111.36**
American Oak	–	–	–	–	–	m²	**116.72**
European Oak	–	–	–	–	–	m²	**122.06**
Add to wood block flooring over 300 mm wide for							
buff; one coat seal	–	–	–	–	–	m²	**4.28**
buff; two coats seal	–	–	–	–	–	m²	**6.96**
sand; three coats for seal or oil	–	–	–	–	–	m²	**19.27**
FLOORS; RAISED ACCESS FLOORS							
Raised flooring system; laid on or fixed to concrete floor							
Full access system; 150 mm high overall; pedestal supports							
PSA light grade; steel finish	–	–	–	–	–	m²	**29.95**
PSA medium grade; steel finish	–	–	–	–	–	m²	**29.95**
PSA heavy grade; steel finish	–	–	–	–	–	m²	**43.06**
Extra for							
factory applied needlepunch carpet	–	–	–	–	–	m²	**14.04**
factory applied anti-static vinyl	–	–	–	–	–	m²	**23.40**
factory applied black PVC edge strips	–	–	–	–	–	m	**4.25**
ramps; 3.00 m × 1.40 m (no finish)	–	–	–	–	–	nr	**655.20**
steps (no finish)	–	–	–	–	–	m	**37.44**
forming cut-out for electrical boxes	–	–	–	–	–	nr	**3.74**
supply and lay protection to raised floor; 2440 × 1220 mm polypropylene sheets with taped joints	–	–	–	–	–	m²	**1.64**
WALLS; PLASTERED; RENDERED; ROUGHCAST							
Cement and sand (1:3) beds and backings							
10 mm thick work to walls; one coat; to brickwork or blockwork base							
over 600 mm wide	–	0.60	13.73	0.38	0.98	m²	**15.09**
not exceeding 600 mm wide	–	0.35	8.01	0.19	0.56	m	**8.76**
13 mm thick; work to walls; two coats; to brickwork or blockwork base							
over 600 mm wide	–	0.80	18.31	0.38	1.01	m²	**19.70**
not exceeding 600 mm wide	–	0.40	9.16	0.19	0.56	m	**9.91**

28 FLOOR, WALL, CEILING AND ROOF FINISHES

Item	PC £	Labour hours	Labour £	Plant £	Material £	Unit	Total rate £
WALLS; PLASTERED; RENDERED; ROUGHCAST – cont							
Cement and sand (1:3) beds and backings – cont							
15 mm thick work to walls; two coats; to brickwork or blockwork base							
over 600 mm wide	–	0.90	20.60	0.38	1.07	m²	**22.05**
not exceeding 600 mm wide	–	0.40	9.16	0.19	0.64	m	**9.99**
Cement and sand (1:3); steel trowelled							
13 mm thick work to walls; two coats; to brickwork or blockwork base							
over 600 mm wide	–	0.80	18.31	0.38	1.01	m²	**19.70**
not exceeding 600 mm wide	–	0.40	9.16	0.19	0.56	m	**9.91**
16 mm thick work to walls; two coats; to brickwork or blockwork base							
over 600 mm wide	–	0.90	20.60	0.38	1.07	m²	**22.05**
not exceeding 600 mm wide	–	0.40	9.16	0.19	0.67	m	**10.02**
19 mm thick work to walls; two coats; to brickwork or blockwork base							
over 600 mm wide	–	0.95	21.75	0.38	1.20	m²	**23.33**
not exceeding 600 mm wide	–	0.40	9.16	0.19	0.88	m	**10.23**
ADD to above							
over 600 mm wide in water repellent cement	–	0.10	2.29	–	0.95	m²	**3.24**
finishing coat in colour cement	–	0.35	8.01	0.19	0.89	m²	**9.09**
Cement-lime-sand (1:2:9); steel trowelled							
19 mm thick work to walls; two coats; to brickwork or blockwork base							
over 600 mm wide	–	0.95	21.75	0.38	0.95	m²	**23.08**
not exceeding 600 mm wide	–	0.40	9.16	0.19	3.95	m	**13.30**
Cement-lime-sand (1:1:6); steel trowelled							
13 mm thick work to walls; two coats; to brickwork or blockwork base							
over 600 mm wide	–	0.95	21.75	0.38	1.10	m²	**23.23**
not exceeding 600 mm wide	–	0.40	9.16	0.19	3.95	m	**13.30**
Add to the above over 600 mm wide for							
waterr epellent cement	–	0.10	2.29	–	0.95	m²	**3.24**
Plaster; first 11 mm coat of Thistle Hardwall plaster; second 2 mm finishing coat of Thistle Multi-Finish plaster; steel trowelled							
13 mm thick work to walls; two coats; to brickwork or blockwork base							
over 600 mm wide	–	0.50	11.45	–	1.80	m²	**13.25**
over 600 mm wide; in staircase areas, plant rooms and similar restricted areas	–	0.60	13.73	–	1.80	m²	**15.53**
not exceeding 600 mm wide	–	0.25	5.72	–	1.07	m	**6.79**

28 FLOOR, WALL, CEILING AND ROOF FINISHES

Item	PC £	Labour hours	Labour £	Plant £	Material £	Unit	Total rate £
13 mm thick work to isolated brickwork or blockwork columns; two coats							
over 600 mm wide	–	1.00	22.89	–	1.80	m²	**24.69**
not exceeding 600 mm wide	–	0.50	11.45	–	1.07	m	**12.52**
Plaster; first 8 mm or 11 mm coat of Thistle Bonding plaster; second 2 mm finishing coat of Thistle Multi-Finish plaster; steel trowelled finish							
13 mm thick work to walls; two coats; to concrete base							
over 600 mm wide	–	0.50	11.45	–	1.80	m²	**13.25**
over 600 mm wide; in staircase areas, plant rooms and similar restricted areas	–	0.60	13.73	–	1.80	m²	**15.53**
not exceeding 600 mm wide	–	0.25	5.72	–	1.07	m	**6.79**
13 mm thick work to isolated piers or columns; two coats; to concrete base							
over 600 mm wide	–	1.00	22.89	–	1.80	m²	**24.69**
not exceeding 600 mm wide	–	0.50	11.45	–	1.08	m	**12.53**
13 mm thick work to vertical face of metal lathing arch former							
not exceeding 0.50 m² per side	–	0.75	17.17	–	0.03	nr	**17.20**
0.50 m²–1 m² per side	–	1.00	22.89	–	0.04	nr	**22.93**
Plaster; first coat of Limelite renovating plaster; finishing coat of Limelite finishing plaster; or other equal							
13 mm thick work to walls; two coats; to brickwork or blockwork base							
over 600 mm wide	–	0.90	20.60	–	2.19	m²	**22.79**
over 600 mm wide; in staircase areas, plant rooms and similar restricted areas	–	1.00	22.89	–	2.19	m²	**25.08**
not exceeding 600 mm wide	–	0.45	10.30	–	1.27	m	**11.57**
Dubbing out existing walls with plaster; average 6 mm thick							
over 600 mm wide	–	0.25	5.72	–	0.06	m²	**5.78**
not exceeding 600 mm wide	–	0.15	3.43	–	0.04	m	**3.47**
Dubbing out existing walls with undercoat plaster; average 12 mm thick							
over 600 mm wide	–	0.50	11.45	–	0.11	m²	**11.56**
not exceeding 600 mm wide	–	0.25	5.72	–	0.07	m	**5.79**
Plaster; first coat of Thistle X-ray plaster or other equal; finishing coat of Thistle X-ray finishing plaster or other equal and approved; steel trowelled							
17 mm thick work to walls; two coats; to brickwork or blockwork base							
over 600 mm wide	–	–	–	–	–	m²	**65.13**
over 600 mm wide; in staircase areas or plant rooms	–	–	–	–	–	m²	**69.97**
not exceeding 600 mm wide	–	–	–	–	–	m	**26.04**
17 mm thick work to isolated columns; two coats							
over 600 mm wide	–	–	–	–	–	m²	**105.65**
not exceeding 600 mm wide	–	–	–	–	–	m	**42.23**

28 FLOOR, WALL, CEILING AND ROOF FINISHES

Item	PC £	Labour hours	Labour £	Plant £	Material £	Unit	Total rate £
WALLS; PLASTERED; RENDERED; ROUGHCAST – cont							
Plaster; one coat Thistle projection plaster or other equal; steel trowelled							
13 mm thick work to walls; one coat; to brickwork or blockwork base							
over 600 mm wide	–	–	–	–	–	m²	13.96
over 600 mm wide; in staircase areas or plant rooms	–	–	–	–	–	m²	15.97
not exceeding 600 mm wide	–	–	–	–	–	m	6.97
10 mm thick work to isolated columns; one coat							
over 600 mm wide	–	–	–	–	–	m²	17.00
not exceeding 600 mm wide	–	–	–	–	–	m	8.48
Cemrend self-coloured render or other equal; one coat; to brickwork or blockwork base							
20 mm thick work to walls; to brickwork or blockwork base							
over 600 mm wide	–	–	–	–	–	m²	34.00
not exceeding 600 mm wide	–	–	–	–	–	m	20.00
Tyrolean decorative rendering or similar; 13 mm thick first coat of cement-lime-sand (1:1:6); finishing three coats of Cullamix or other equal and approved; applied with approved hand operated machine external							
To walls; four coats; to brickwork or blockwork base							
over 600 mm wide	–	–	–	–	–	m²	33.37
not exceeding 600 mm wide	–	–	–	–	–	m	16.67
Drydash (pebbledash) finish of Derbyshire Spar chippings or other equal and approved on and including cement-lime-sand (1:2:9) backing							
18 mm thick work to walls; two coats; to brickwork or blockwork base							
over 600 mm wide	–	–	–	–	–	m²	28.93
not exceeding 600 mm wide	–	–	–	–	–	m	14.47
Plaster; one coat Thistle board finish or other equal; steel trowelled							
3 mm thick work to walls or ceilings; one coat; to plasterboard base							
over 600 mm wide	–	0.25	5.72	–	0.06	m²	5.78
over 600 mm wide; in staircase areas or plant rooms	–	0.35	8.01	–	0.06	m²	8.07
not exceeding 600 mm wide	–	0.15	3.43	–	0.03	m	3.46

28 FLOOR, WALL, CEILING AND ROOF FINISHES

Item	PC £	Labour hours	Labour £	Plant £	Material £	Unit	Total rate £
Plaster; one coat Thistle board finish, 3 mm work to walls or ceilings; one coat on and including gypsum plasterboard; fixing with nails; 3 mm joints filled with plaster scrim; to softwood base; plain grade baseboard or lath with rounded edges							
9.50 mm thick boards to walls							
over 600 mm wide	–	0.97	15.04	–	4.40	m²	**19.44**
not exceeding 600 mm wide	–	0.37	6.04	–	1.27	m	**7.31**
9.50 mm thick boards to walls; in staircase areas or plant rooms							
over 600 mm wide	–	1.06	16.51	–	4.40	m²	**20.91**
not exceeding 600 mm wide	–	0.46	7.51	–	1.27	m	**8.78**
9.50 mm thick boards to isolated columns							
over 600 mm wide	–	1.06	16.51	–	4.40	m²	**20.91**
not exceeding 600 mm wide	–	0.56	9.15	–	1.27	m	**10.42**
12.50 mm thick boards to walls; in staircase areas or plant rooms							
over 600 mm wide	–	1.12	17.48	–	4.40	m²	**21.88**
not exceeding 600 mm wide	–	0.50	8.16	–	1.27	m	**9.43**
12.50 mm thick boards to isolated columns							
over 600 mm wide	–	1.12	17.48	–	4.40	m²	**21.88**
not exceeding 600 mm wide	–	0.59	9.63	–	1.27	m	**10.90**
Accessories							
Expamet render beads or other equal and approved; white PVC nosings; to brickwork or blockwork base							
external stop bead; ref 573	–	0.07	1.14	–	4.05	m	**5.19**
Expamet render beads or other equal and approved; stainless steel; to brickwork or blockwork base							
stop bead; ref 546	–	0.07	1.14	–	3.32	m	**4.46**
stop bead; ref 547	–	0.07	1.14	–	3.32	m	**4.46**
Expamet plaster beads or other equal and approved; to brickwork or blockwork base							
angle bead; ref 550	–	0.08	1.30	–	0.63	m	**1.93**
architrave bead; ref 579	–	0.10	1.63	–	2.92	m	**4.55**
stop bead; ref 562	–	0.07	1.14	–	0.84	m	**1.98**
stop beads; ref 563	–	0.07	1.14	–	0.84	m	**1.98**
movement bead; ref 588	–	0.09	1.47	–	10.30	m	**11.77**
Expamet plaster beads or other equal and approved; stainless steel; to brickwork or blockwork base							
angle bead; ref 545	–	0.08	1.30	–	1.42	m	**2.72**
stop bead; ref 534	–	0.07	1.14	–	3.32	m	**4.46**
stop bead; ref 533	–	0.07	1.14	–	3.32	m	**4.46**
Expamet thin coat plaster beads or other equal and approved; galvanized steel; to timber base							
angle bead; ref 553	–	0.07	1.14	–	0.63	m	**1.77**
stop bead; ref 560	–	0.06	0.98	–	0.58	m	**1.56**
stop bead; ref 561	–	0.06	0.98	–	0.58	m	**1.56**

Prices for Measured Works

28 FLOOR, WALL, CEILING AND ROOF FINISHES

Item	PC £	Labour hours	Labour £	Plant £	Material £	Unit	Total rate £
WALLS; INSULATED RENDER							
Sto Therm Classic M-system insulation render							
70 mm EPS insulation fixed with adhesive to SFS structure (measured separately) with horizontal PVC intermediate track and vertical T-spines; with glassfibre mesh reinforcement embedded in Sto Armat Classic Basecoat Render and Stolit K 1.5 Decorative Topcoat Render (white)							
over 300 mm wide	–	–	–	–	–	m²	**62.34**
70 mm EPS insulation mechanically fixed to SFS structure (measured separately) with horizontal PVC intermediate track and vertical T-spines; with glassfibre mesh reinforcement embedded in Sto Armat Classic Basecoat Render and Stolit K 1.5 Decorative Topcoat Render (white)							
over 300 mm wide	–	–	–	–	–	m²	**68.59**
rendered heads and reveals not exceeding 100 mm wide; including angle beads	–	–	–	–	–	m	**17.22**
Extra for							
aluminium starter track at base of insulated render system	–	–	–	–	–	m	**10.48**
external angle with PVC mesh angle bead	–	–	–	–	–	m	**4.61**
internal angle with Sto Armor angle	–	–	–	–	–	m	**4.61**
render stop bead	–	–	–	–	–	m	**4.61**
Sto seal tape to all vertical abutments	–	–	–	–	–	m	**4.29**
Sto Armor mat HD mesh reinforcement to areas prone to physical damage (e.g. 1800 mm high adjoining floor level)							
over 300 mm wide	–	–	–	–	–	m²	**13.79**
WALLS; CORK; CERAMIC TILING; MARBLE							
Glazed ceramic wall tiles; BS EN 10545; fixing with adhesive; butt joints; straight both ways; flush pointing with white grout; to plaster base							
Work to walls; over 600 mm wide							
152 mm × 152 mm × 5.50 mm thick; white	11.88	0.56	12.82	–	14.12	m²	**26.94**
152 mm × 152 mm × 5.50 mm thick; light colours	12.73	0.56	12.82	–	15.02	m²	**27.84**
152 mm × 152 mm × 5.50 mm thick; dark colours	14.85	0.56	12.82	–	17.25	m²	**30.07**
extra for RE or REX tile	–	–	–	–	5.80	m²	**5.80**
200 mm × 100 mm × 6.50 mm thick; white and light colours	10.61	0.56	12.82	–	12.80	m²	**25.62**
250 mm × 200 mm × 7.00 mm thick; white and light colours	12.73	0.56	12.82	–	15.02	m²	**27.84**
Work to walls; in staircase areas or plant rooms							
152 mm × 152 mm × 5.50 mm thick; white	–	0.62	14.20	–	14.12	m²	**28.32**

28 FLOOR, WALL, CEILING AND ROOF FINISHES

Item	PC £	Labour hours	Labour £	Plant £	Material £	Unit	Total rate £
Work to walls; not exceeding 600 mm wide							
152 mm × 152 mm × 5.50 mm thick; white	–	0.56	12.82	–	8.43	m	**21.25**
152 mm × 152 mm × 5.50 mm thick; light colours	–	0.56	12.82	–	9.49	m	**22.31**
152 mm × 152 mm × 5.50 mm thick; dark colours	–	0.56	12.82	–	10.84	m	**23.66**
200 mm × 100 mm × 6.50 mm thick; white and light colours	–	0.56	12.82	–	7.63	m	**20.45**
250 mm × 200 mm × 7.00 mm thick; white and light colours	–	0.46	10.53	–	8.97	m	**19.50**
cutting around pipes; not exceeding 0.30 m girth	–	0.09	1.47	–	–	nr	**1.47**
Work to sills; 150 mm wide; rounded edge tiles							
152 mm × 152 mm × 5.50 mm thick; white	–	0.23	5.26	–	2.11	m	**7.37**
fitted end	–	0.09	1.47	–	–	nr	**1.47**
198 mm × 64.50 mm × 6 mm thick wall tiles; fixing with adhesive; butt joints; straight both ways; flush pointing with white grout; to plaster base							
Work to walls							
over 600 mm wide	24.00	1.67	38.23	–	29.38	m²	**67.61**
not exceeding 600 mm wide	14.40	1.30	29.76	–	17.58	m	**47.34**
20 mm × 20 mm × 5.50 mm thick glazed mosaic wall tiles; fixing with adhesive; butt joints; straight both ways; flush pointing with white grout; to plaster base							
Work to walls							
over 600 mm wide	28.00	1.76	40.29	–	34.09	m²	**74.38**
not exceeding 600 mm wide	16.80	1.38	31.59	–	21.23	m	**52.82**
Dakota mahogany granite cladding; polished finish; jointed and pointed in coloured mortar (1:2:8)							
20 mm thick work to walls; to cement and sand base							
over 600 mm wide	–	–	–	–	–	m²	**360.00**
not exceeding 600 mm wide	–	–	–	–	–	m	**160.00**
40 mm thick work to walls; to cement and sand base							
over 600 mm wide	–	–	–	–	–	m²	**600.00**
not exceeding 600 mm wide	–	–	–	–	–	m	**270.00**
Roman Travertine marble cladding; polished finish; jointed and pointed in coloured mortar (1:2:8)							
20 mm thick work to walls; to cement and sand base							
over 600 mm wide	–	–	–	–	–	m²	**273.00**
not exceeding 600 mm wide	–	–	–	–	–	m	**125.00**
40 mm thick work to walls; to cement and sand base							
over 600 mm wide	–	–	–	–	–	m²	**376.00**
not exceeding 600 mm wide	–	–	–	–	–	m	**170.00**

Prices for Measured Works

28 FLOOR, WALL, CEILING AND ROOF FINISHES

Item	PC £	Labour hours	Labour £	Plant £	Material £	Unit	Total rate £
WALLS; CORK; CERAMIC TILING; MARBLE – cont							
Cork tiles; fixing with adhesive; butt joints; straight both ways; toplasterd walls							
600 mm × 300 mm × 3.00 mm thick; standard natural finish	9.26	0.40	9.16	–	13.47	m²	**22.63**
600 mm × 300 mm × 4.00 mm thick; character natural finish	9.31	0.40	9.16	–	13.52	m²	**22.68**
300 mm × 300 mm × 3.00 mm thick; pinboard 2 tone	13.44	0.45	10.30	–	18.29	m²	**28.59**
WALLS; WATERPROOF PANELLING							
Waterproof panelling system with moisture-resistant MDF core and high pressure laminate backing.and decorative laminate face							
Supply and fit Showerwall 10.5 mm thick panels complete with aluminium trims, sealants and adhesive							
White gloss finish fixed to substrate with adhesive	–	0.25	5.05	–	60.72	m²	**65.77**
Perganmon Marble gloss finish fixed to substrate with adhesive	–	0.25	5.05	–	66.35	m²	**71.40**
CEILINGS; PLASTERED; RENDERED							
Plaster; first 8 mm or 11 mm coat of Thistle Bonding plaster; second 2 mm finishing coat of Thistle Multi-Finish plaster; steel trowelled finish							
10 mm thick work to ceilings; two coats; to concrete base							
over 600 mm wide	–	0.50	11.45	–	0.13	m²	**11.58**
over 600 wide; 3.50 m–5.00 m high	–	0.75	17.17	–	0.13	m²	**17.30**
over 600 mm wide; in staircase areas, plant rooms and similar restricted areas	–	0.75	17.17	–	0.13	m²	**17.30**
not exceeding 600 mm wide	–	0.33	7.62	–	0.07	m	**7.69**
10 mm thick work to isolated beams; two coats; to concrete base							
over 600 mm wide	–	1.10	25.18	–	0.13	m²	**25.31**
over 600 mm wide; 3.50 m–5.00 m high	–	1.20	27.47	–	0.13	m²	**27.60**
not exceeding 600 mm wide	–	0.50	11.45	–	0.07	m	**11.52**
Plaster; first 11 mm coat of Thistle Bonding plaster; second coat 2 mm finishing coat of Thistle Multi-Finish plaster; steel trowelled							
13 mm thick work to ceilings; to metal lathing base							
over 600 mm wide	–	0.50	11.45	–	0.13	m²	**11.58**
over 600 mm wide; in staircase areas, plant rooms and similar restricted areas	–	0.75	17.17	–	0.13	m²	**17.30**
not exceeding 600 mm wide	–	0.33	7.62	–	0.07	m	**7.69**

28 FLOOR, WALL, CEILING AND ROOF FINISHES

Item	PC £	Labour hours	Labour £	Plant £	Material £	Unit	Total rate £
Plaster; one coat Thistle board finish or other and approved; steel trowelled 3 mm work to ceilings; one coat on and including gypsum plasterboard; fixing with nails; 3 mm joints filled with plaster and jute scrim cloth; to softwood base; plain grade baseboard or lath with rounded edges							
9.50 mm thick boards to ceilings							
over 600 mm wide	–	0.89	13.73	–	4.40	m²	**18.13**
over 600 mm wide; 3.50 m–5.00 m high	–	1.03	16.01	–	4.40	m²	**20.41**
not exceeding 600 mm wide	–	0.43	7.01	–	1.27	m	**8.28**
9.50 mm thick boards to ceilings; in staircase areas or plant rooms							
over 600 mm wide	–	0.98	15.20	–	4.40	m²	**19.60**
not exceeding 600 mm wide	–	0.47	7.68	–	1.27	m	**8.95**
9.50 mm thick boards to isolated beams							
over 600 mm wide	–	1.05	16.34	–	4.40	m²	**20.74**
not exceeding 600 mm wide	–	0.50	8.16	–	1.27	m	**9.43**
12.50 mm thick boards to ceilings							
over 600 mm wide	–	0.95	14.71	–	4.40	m²	**19.11**
over 600 mm wide; 3.50 m–5.00 m high	–	1.06	16.51	–	4.40	m²	**20.91**
not exceeding 600 mm wide	–	0.45	7.35	–	1.27	m	**8.62**
12.50 mm thick boards to ceilings; in staircase areas or plant rooms							
over 600 mm wide	–	1.06	16.51	–	4.40	m²	**20.91**
not exceeding 600 mm wide	–	0.51	8.33	–	1.27	m	**9.60**
12.50 mm thick boards to isolated beams							
over 600 mm wide	–	1.15	17.98	–	4.40	m²	**22.38**
not exceeding 600 mm wide	–	0.56	9.15	–	1.27	m	**10.42**
Plaster; one coat Thistle board finish or other equal; steel trowelled							
3 mm thick work to walls or ceilings; one coat; to plasterboard base							
over 600 mm wide	–	0.25	5.72	–	0.06	m²	**5.78**
over 600 mm wide; in staircase areas, plant rooms and similar restricted areas	–	0.35	8.01	–	0.06	m²	**8.07**
not exceeding 600 mm wide	–	0.15	3.43	–	0.03	m	**3.46**
Cement-lime-sand (1:1:6); steel trowelled							
19 mm thick work to ceilings; three coats; to metal lathing base							
over 600 mm wide	–	0.95	21.75	0.38	0.95	m²	**23.08**
not exceeding 600 mm wide	–	0.40	9.16	0.19	3.95	m	**13.30**
Gyproc cove or other equal and approved; fixing with adhesive; filling and pointing joints with plaster							
Cove							
125 mm girth	–	0.19	3.10	–	1.77	m	**4.87**
angles	–	0.03	0.49	–	1.11	nr	**1.60**

28 FLOOR, WALL, CEILING AND ROOF FINISHES

Item	PC £	Labour hours	Labour £	Plant £	Material £	Unit	Total rate £
METAL MESH LATHING FOR PLASTERED COATINGS							
Accessories							
Preformed galvanized expanded steel semi-circular arch-frames; Expamet or other equal; to suit walls up to 230 mm thick							
for 760 mm opening; ref ESC 30	34.07	0.46	6.77	–	35.77	nr	**42.54**
for 920 mm opening; ref ESC 36	42.84	0.46	6.77	–	44.98	nr	**51.75**
for 1220 mm opening; ref ESC 48	48.68	0.46	6.77	–	51.11	nr	**57.88**
Lathing; Expamet BB expanded metal lathing or other equal; BS EN 13658; 50 mm laps							
6 mm thick mesh linings to ceilings; fixing with staples; to softwood base; over 300 mm wide							
ref BB263; 0.500 mm thick	8.07	0.56	8.25	–	8.47	m²	**16.72**
ref BB264; 0.675 mm thick	12.59	0.56	8.25	–	13.22	m²	**21.47**
6 mm thick mesh linings to ceilings; fixing with wire; to steelwork; over 300 mm wide							
ref BB263; 0.500 mm thick	–	0.59	8.70	–	8.47	m²	**17.17**
ref BB264; 0.675 mm thick	–	0.59	8.70	–	13.22	m²	**21.92**
6 mm thick mesh linings to ceilings; fixing with wire; to steelwork; not exceeding 300 mm wide							
ref BB263; 0.500 mm thick	–	0.37	5.44	–	8.47	m²	**13.91**
ref BB264; 0.675 mm thick	–	0.37	5.44	–	13.22	m²	**18.66**
raking cutting	–	0.19	3.10	–	–	m	**3.10**
cutting and fitting around pipes; not exceeding 0.30 m girth	–	0.28	4.57	–	–	nr	**4.57**
Lathing; Expamet Riblath or Spreedpro or other equal stiffened expanded metal lathing; 50 mm laps; 2.50 m × 0.60 m sheets							
rigid rib mesh lining to walls; fixing with nails; to softwood base; over 300 mm wide							
Riblath ref 269; galvanized steel	6.96	0.46	6.77	–	7.43	m²	**14.20**
Riblath ref 267; stainless steel	16.29	0.46	6.77	–	17.23	m²	**24.00**
rigid rib mesh lining to walls; fixing with nails; to softwood base; not exceeding 300 mm wide							
Riblath ref 269; galvanized steel	–	0.28	4.13	–	2.28	m	**6.41**
Riblath ref 267; stainless steel	–	0.28	4.13	–	5.22	m	**9.35**
rigid rib mesh lining to ceilings; fixing with wire; to steelwork; over 300 mm wide							
Riblath ref 269; galvanized steel	–	0.59	8.70	–	7.85	m²	**16.55**
Riblath ref 267; stainless steel	–	0.59	8.70	–	17.64	m²	**26.34**

28 FLOOR, WALL, CEILING AND ROOF FINISHES

Item	PC £	Labour hours	Labour £	Plant £	Material £	Unit	Total rate £
SPRAYED MINERAL FIBRE COATINGS							
Prepare and apply by spray Mandolite CP2 fire protection or other equal and approved on structural steel/metalwork							
16 mm thick (one hour) fire protection							
to walls and columns	–	–	–	–	–	m²	**10.57**
to ceilings and beams	–	–	–	–	–	m²	**11.66**
to isolated metalwork	–	–	–	–	–	m²	**23.24**
22 mm thick (one and a half hour) fire protection							
to walls and columns	–	–	–	–	–	m²	**12.29**
to ceilings and beams	–	–	–	–	–	m²	**13.63**
to isolated metalwork	–	–	–	–	–	m²	**27.27**
28 mm thick (two hour) fire protection							
to walls and columns	–	–	–	–	–	m²	**14.41**
to ceilings and beams	–	–	–	–	–	m²	**15.74**
to isolated metalwork	–	–	–	–	–	m²	**31.47**
52 mm thick (four hour) fire protection							
to walls and columns	–	–	–	–	–	m²	**21.80**
to ceilings and beams	–	–	–	–	–	m²	**24.27**
to isolated metalwork	–	–	–	–	–	m²	**48.29**
Prepare and apply by spray; cementitious Pyrok WF26 render or other equal and approved; on expanded metal lathing (not included)							
15 mm thick							
to ceilings and beams	–	–	–	–	–	m²	**33.45**

29 DECORATION

Item	PC £	Labour hours	Labour £	Plant £	Material £	Unit	Total rate £
SUPPLY ONLY RATES FOR PAINTS							
BASIC PAINT PRICES – MATERIAL ONLY							
Paints							
matt emulsion	–	–	–	–	14.94	5 litre	**14.94**
gloss	–	–	–	–	25.96	5 litre	**25.96**
Weathershield gloss	–	–	–	–	49.76	5 litre	**49.76**
Weathershield undercoat	–	–	–	–	61.89	5 litre	**61.89**
Sandtex masonry paint							
brilliant white	–	–	–	–	14.27	5 litre	**14.27**
coloured	–	–	–	–	25.38	5 litre	**25.38**
Primer/undercoats							
oil-based undercoat	–	–	–	–	23.35	5 litre	**23.35**
acrylic	–	–	–	–	26.92	5 litre	**26.92**
red oxide	–	–	–	–	29.36	5 litre	**29.36**
water-based	–	–	–	–	24.00	5 litre	**24.00**
zinc phosphate	–	–	–	–	45.54	5 litre	**45.54**
mdf primer	–	–	–	–	55.20	5 litre	**55.20**
knotting solution	–	–	–	–	87.31	5 litre	**87.31**
Special paints							
solar reflective aluminium	–	–	–	–	47.39	5 litre	**47.39**
anti-graffiti	–	–	–	–	151.87	5 litre	**151.87**
bituminous emulsion	–	–	–	–	16.98	5 litre	**16.98**
Hammerite	–	–	–	–	59.39	5 litre	**59.39**
fire retardant							
undercoat	–	–	–	–	74.51	5 litre	**74.51**
top coat	–	–	–	–	64.26	5 litre	**64.26**
Stains and Preservatives							
Cuprinol							
Clear	–	–	–	–	37.38	5 litre	**37.38**
Boiled linseed oil	–	–	–	–	41.57	5 litre	**41.57**
Sadolin							
Extra	–	–	–	–	72.44	5 litre	**72.44**
New Base	–	–	–	–	25.04	5 litre	**25.04**
Sikkens							
Cetol HLS	–	–	–	–	59.76	5 litre	**59.76**
Cetol TS	–	–	–	–	85.96	5 litre	**85.96**
Cetol Filter 7	–	–	–	–	90.90	5 litre	**90.90**
Protim Solignum							
Architectural	–	–	–	–	113.70	5 litre	**113.70**
Green	–	–	–	–	45.49	5 litre	**45.49**
Cedar	–	–	–	–	18.50	5 litre	**18.50**
Varnishes							
polyurethane	–	–	–	–	43.39	5 litre	**43.39**

29 DECORATION

Item	PC £	Labour hours	Labour £	Plant £	Material £	Unit	Total rate £
PAINTING AND CLEAR FINISHES – EXTERNAL							
SUPPLY AND FIX PRICES							
NOTE: The following prices include for preparing surfaces. Painting woodwork also includes for knotting prior to applying the priming coat and for all stopping of nail holes etc.							
Two coats of cement paint, Sandtex Matt or other equal and approved							
Brick or block walls							
over 300 mm girth	–	0.26	4.24	–	1.83	m²	**6.07**
Cement render or concrete walls							
over 300 mm girth	–	0.23	3.76	–	1.21	m²	**4.97**
Roughcast walls							
over 300 mm girth	–	0.40	6.53	–	1.21	m²	**7.74**
One coat sealer and two coats of external grade emulsion paint, Dulux Weathershield or other equal and approved							
Brick or block walls							
over 300 mm girth	–	0.43	7.01	–	8.66	m²	**15.67**
Cement render or concrete walls							
over 300 mm girth	–	0.35	5.71	–	5.78	m²	**11.49**
Concrete soffits							
over 300 mm girth	–	0.40	6.53	–	5.78	m²	**12.31**
One coat sealer (applied by brush) and two coats of external grade emulsion paint, Dulux Weathershield or other equal and approved (spray applied)							
Roughcast							
over 300 mm girth	–	0.29	4.74	–	11.75	m²	**16.49**
One coat sealer and two coats of anti-graffiti paint (spray applied)							
Brick or block walls							
over 300 mm girth	–	0.01	0.09	0.45	4.61	m²	**5.15**
Cement render or concrete walls							
over 300 mm girth	–	0.01	0.09	0.45	5.57	m²	**6.11**
2.5 mm of Vandalene or similar anti-climb paint (spray applied)							
General surfaces							
over 300 mm girth	–	0.01	0.09	0.45	4.56	m²	**5.10**
Two coats solar reflective aluminium paint; on bituminous roofing							
General surfaces							
over 300 mm girth	–	0.44	7.18	–	16.03	m²	**23.21**

29 DECORATION

Item	PC £	Labour hours	Labour £	Plant £	Material £	Unit	Total rate £
PAINTING AND CLEAR FINISHES – EXTERNAL – cont							
Touch up primer; two undercoats and one finishing coat of gloss oil paint; on wood surfaces							
General surfaces							
over 300 mm girth	–	0.35	5.71	–	2.34	m²	**8.05**
isolated surfaces not exceeding 300 mm girth	–	0.15	2.45	–	0.63	m	**3.08**
isolated areas not exceeding 1.00 m²; irrespective of girth	–	0.27	4.41	–	1.27	nr	**5.68**
Glazed windows and screens							
panes; area not exceeding 0.10 m²	–	0.59	9.63	–	2.08	m²	**11.71**
panes; area 0.10 m²–0.50 m²	–	0.59	9.63	–	1.75	m²	**11.38**
panes; area 0.50 m²–1.00 m²	–	0.47	7.68	–	1.53	m²	**9.21**
panes; area over 1.00 m²	–	0.35	5.71	–	1.27	m²	**6.98**
Glazed windows and screens; multi-coloured work							
panes; area not exceeding 0.10 m²	–	0.68	11.10	–	2.08	m²	**13.18**
panes; area 0.10 m²–0.50 m²	–	0.55	8.98	–	1.81	m²	**10.79**
panes; area 0.50 m²–1.00 m²	–	0.47	7.68	–	1.53	m²	**9.21**
panes; area over 1.00 m²	–	0.41	6.69	–	1.27	m²	**7.96**
Knot; one coat primer; two undercoats and one finishing coat of gloss oil paint; on wood surfaces							
General surfaces							
over 300 mm girth	–	0.46	7.51	–	2.54	m²	**10.05**
isolated surfaces not exceeding 300 mm girth	–	0.19	3.10	–	4.14	m	**7.24**
isolated areas not exceeding 1.00 m²; irrespective of girth	–	0.35	5.71	–	1.71	nr	**7.42**
Glazed windows and screens							
panes; area not exceeding 0.10 m²	–	0.78	12.74	–	2.84	m²	**15.58**
panes; area 0.10 m²–0.50 m²	–	0.62	10.12	–	2.52	m²	**12.64**
panes; area 0.50 m²–1.00 m²	–	0.55	8.98	–	1.93	m²	**10.91**
panes; area over 1.00 m²	–	0.46	7.51	–	1.35	m²	**8.86**
Glazed windows and screens; multi-coloured work							
panes; area not exceeding 0.10 m²	–	0.89	14.53	–	2.84	m²	**17.37**
panes; area 0.10 m²–0.50 m²	–	0.72	11.75	–	2.54	m²	**14.29**
panes; area 0.50 m²–1.00 m²	–	0.64	10.45	–	1.93	m²	**12.38**
panes; area over 1.00 m²	–	0.54	8.81	–	1.35	m²	**10.16**
Touch up primer; two undercoats and one finishing coat of gloss oil paint; on iron or steel surfaces							
General surfaces							
over 300 mm girth	–	0.35	5.71	–	1.89	m²	**7.60**
isolated surfaces not exceeding 300 mm girth	–	0.14	2.29	–	0.53	m	**2.82**
isolated areas not exceeding 1.00 m²; irrespective of girth	–	0.26	4.24	–	1.07	nr	**5.31**

29 DECORATION

Item	PC £	Labour hours	Labour £	Plant £	Material £	Unit	Total rate £
Glazed windows and screens							
panes; area not exceeding 0.10 m²	–	0.59	9.63	–	1.95	m²	**11.58**
panes; area 0.10 m²–0.50 m²	–	0.47	7.68	–	1.69	m²	**9.37**
panes; area 0.50 m²–1.00 m²	–	0.41	6.69	–	1.43	m²	**8.12**
panes; area over 1.00 m²	–	0.35	5.71	–	1.17	m²	**6.88**
Structural steelwork							
over 300 mm girth	–	0.40	6.53	–	1.98	m²	**8.51**
Members of roof trusses							
over 300 mm girth	–	0.54	8.81	–	2.25	m²	**11.06**
Ornamental railings and the like; each side measured overall							
over 300 mm girth	–	0.60	9.80	–	2.32	m²	**12.12**
Eaves gutters							
over 300 mm girth	–	0.64	10.45	–	2.57	m²	**13.02**
not exceeding 300 mm girth	–	0.25	4.08	–	1.07	m	**5.15**
Pipes or conduits							
over 300 mm girth	–	0.54	8.81	–	2.57	m²	**11.38**
not exceeding 300 mm girth	–	0.21	3.42	–	0.86	m	**4.28**
One coat primer; two undercoats and one finishing coat of gloss oil paint; on iron or steel surfaces							
General surfaces							
over 300 mm girth	–	0.43	7.01	–	2.18	m²	**9.19**
isolated surfaces not exceeding 300 mm girth	–	0.18	2.94	–	0.57	m	**3.51**
isolated areas not exceeding 1.00 m²; irrespective of girth	–	0.32	5.22	–	1.14	nr	**6.36**
Glazed windows and screens							
panes; area not exceeding 0.10 m²	–	0.71	11.59	–	2.03	m²	**13.62**
panes; area 0.10 m²–0.50 m²	–	0.56	9.15	–	1.76	m²	**10.91**
panes; area 0.50 m²–1.00 m²	–	0.50	8.16	–	1.50	m²	**9.66**
panes; area over 1.00 m²	–	0.43	7.01	–	1.14	m²	**8.15**
Structural steelwork							
over 300 mm girth	–	0.48	7.83	–	2.29	m²	**10.12**
Members of roof trusses							
over 300 mm girth	–	0.64	10.45	–	2.55	m²	**13.00**
Ornamental railings and the like; each side measured overall							
over 300 mm girth	–	0.72	11.75	–	2.55	m²	**14.30**
Eaves gutters							
over 300 mm girth	–	0.76	12.40	–	2.92	m²	**15.32**
not exceeding 300 mm girth	–	0.31	5.06	–	1.01	m	**6.07**
Pipes or conduits							
over 300 mm girth	–	0.64	10.45	–	2.92	m²	**13.37**
not exceeding 300 mm girth	–	0.25	4.08	–	4.20	m	**8.28**

29 DECORATION

Item	PC £	Labour hours	Labour £	Plant £	Material £	Unit	Total rate £
PAINTING AND CLEAR FINISHES – EXTERNAL – cont							
One coat of Andrews Hammerite paint or other equal and approved; on iron or steel surfaces							
General surfaces							
over 300 mm girth	–	0.15	2.45	–	1.87	m²	**4.32**
isolated surfaces not exceeding 300 mm girth	–	0.08	1.30	–	0.59	m	**1.89**
isolated areas not exceeding 1.00 m²; irrespective of girth	–	0.11	1.80	–	1.07	nr	**2.87**
Glazed windows and screens							
panes; area not exceeding 0.10 m²	–	0.25	4.08	–	1.39	m²	**5.47**
panes; area 0.10 m²–0.50 m²	–	0.19	3.10	–	1.56	m²	**4.66**
panes; area 0.50 m²–1.00 m²	–	0.18	2.94	–	1.41	m²	**4.35**
panes; area over 1.00 m²	–	0.15	2.45	–	1.41	m²	**3.86**
Structural steelwork							
over 300 mm girth	–	0.17	2.77	–	1.71	m²	**4.48**
Members of roof trusses							
over 300 mm girth	–	0.23	3.76	–	1.87	m²	**5.63**
Ornamental railings and the like; each side measured overall							
over 300 mm girth	–	0.26	4.24	–	1.87	m²	**6.11**
Eaves gutters							
over 300 mm girth	–	0.27	4.41	–	2.02	m²	**6.43**
not exceeding 300 mm girth	–	0.08	1.30	–	0.99	m	**2.29**
Pipes or conduits							
over 300 mm girth	–	0.26	4.24	–	1.71	m²	**5.95**
not exceeding 300 mm girth	–	0.08	1.30	–	0.80	m	**2.10**
Two coats of creosote; on wood surfaces							
General surfaces							
over 300 mm girth	–	0.16	2.61	–	0.48	m²	**3.09**
isolated surfaces not exceeding 300 mm girth	–	0.01	0.09	–	0.29	m	**0.38**
Two coats of Solignum wood preservative or other equal and approved; on wood surfaces							
General surfaces							
over 300 mm girth	–	0.14	2.29	–	1.03	m²	**3.32**
isolated surfaces not exceeding 300 mm girth	–	0.05	0.82	–	0.35	m	**1.17**
Three coats of polyurethane; on wood surfaces							
General surfaces							
over 300 mm girth	–	0.29	4.74	–	2.71	m²	**7.45**
isolated surfaces not exceeding 300 mm girth	–	0.11	1.80	–	1.35	m	**3.15**
isolated areas not exceeding 1.00 m²; irrespective of girth	–	0.21	3.42	–	1.56	nr	**4.98**

29 DECORATION

Item	PC £	Labour hours	Labour £	Plant £	Material £	Unit	Total rate £
Two coats of New Base primer or other equal and approved; and two coats of Extra or other equal and approved; Sadolin Ltd; pigmented; on wood surfaces							
General surfaces							
over 300 mm girth	–	0.43	7.01	–	4.78	m²	**11.79**
isolated surfaces not exceeding 300 mm girth	–	0.13	2.12	–	0.83	m	**2.95**
Glazed windows and screens							
panes; area not exceeding 0.10 m²	–	0.71	11.59	–	3.41	m²	**15.00**
panes; area 0.10 m²–0.50 m²	–	0.57	9.30	–	3.21	m²	**12.51**
panes; area 0.50 m²–1.00 m²	–	0.50	8.16	–	3.02	m²	**11.18**
panes; area over 1.00 m²	–	0.43	7.01	–	2.44	m²	**9.45**
Two coats Sikkens Cetol Filter 7 exterior stain or other equal and approved; on wood surfaces							
General surfaces							
over 300 mm girth	–	0.20	3.27	–	4.57	m²	**7.84**
isolated surfaces not exceeding 300 mm girth	–	0.09	1.47	–	1.59	m	**3.06**
isolated areas not exceeding 1.00 m²; irrespective of girth	–	0.14	2.29	–	2.33	nr	**4.62**
PAINTING AND CLEAR FINISHES – INTERNAL							
SUPPLY AND FIX PRICES							
NOTE: The following prices include for preparing surfaces. Painting woodwork also includes for knotting prior to applying the priming coat and for all stopping of nail holes etc.							
Touch up primer; two undercoats and one finishing coat of gloss oil paint; on wood surfaces							
General surfaces							
over 300 mm girth	–	0.35	5.71	–	2.34	m²	**8.05**
isolated surfaces not exceeding 300 mm girth	–	0.15	2.45	–	0.63	m	**3.08**
isolated areas not exceeding 1.00 m²; irrespective of girth	–	0.27	4.41	–	1.27	nr	**5.68**
One coat primer; on wood surfaces before fixing							
General surfaces							
over 300 mm girth	–	0.08	1.30	–	1.05	m²	**2.35**
isolated surfaces not exceeding 300 mm girth	–	0.02	0.33	–	0.38	m	**0.71**
isolated areas not exceeding 1.00 m²; irrespective of girth	–	0.06	0.98	–	0.40	nr	**1.38**
One coat polyurethane sealer; on wood surfaces before fixing							
General surfaces							
over 300 mm girth	–	0.10	1.63	–	0.93	m²	**2.56**
isolated surfaces not exceeding 300 mm girth	–	0.03	0.49	–	0.34	m	**0.83**
isolated areas not exceeding 1.00 m²; irrespective of girth	–	0.08	1.30	–	0.44	nr	**1.74**

29 DECORATION

Item	PC £	Labour hours	Labour £	Plant £	Material £	Unit	Total rate £
PAINTING AND CLEAR FINISHES – INTERNAL – cont							
One coat of Sikkens Cetol HLS stain or other equal and approved; on wood surfaces before fixing							
General surfaces							
over 300 mm girth	–	0.11	1.80	–	1.12	m²	**2.92**
isolated surfaces not exceeding 300 mm girth	–	0.03	0.49	–	0.81	m	**1.30**
isolated areas not exceeding 1.00 m²; irrespective of girth	–	0.08	1.30	–	0.54	nr	**1.84**
One coat of Sikkens Cetol TS interior stain or other equal and approved; on wood surfaces before fixing							
General surfaces							
over 300 mm girth	–	0.11	1.80	–	1.55	m²	**3.35**
isolated surfaces not exceeding 300 mm girth	–	0.03	0.49	–	0.74	m	**1.23**
isolated areas not exceeding 1.00 m²; irrespective of girth	–	0.08	1.30	–	0.75	nr	**2.05**
One coat Cuprinol clear wood preservative or other equal and approved; on wood surfaces before fixing							
General surfaces							
over 300 mm girth	–	0.08	1.30	–	0.92	m²	**2.22**
isolated surfaces not exceeding 300 mm girth	–	0.02	0.33	–	0.34	m	**0.67**
isolated areas not exceeding 1.00 m²; irrespective of girth	–	0.05	0.82	–	0.43	nr	**1.25**
One coat HCC Protective Coatings Ltd Permacor urethane alkyd gloss finishing coat or other equal and approved; on previously primed steelwork							
Members of roof trusses							
over 300 mm girth	–	0.15	2.45	–	0.50	m²	**2.95**
Two coats emulsion paint							
Brick or block walls							
over 300 mm girth	–	0.21	3.42	–	0.86	m²	**4.28**
Cement render or concrete							
over 300 mm girth	–	0.20	3.27	–	0.77	m²	**4.04**
isolated surfaces not exceeding 300 mm girth	–	0.10	1.63	–	0.26	m	**1.89**
Plaster walls or plaster/plasterboard ceilings							
over 300 mm girth	–	0.18	2.94	–	0.75	m²	**3.69**
over 300 mm girth; in multi colours	–	0.24	3.92	–	0.88	m²	**4.80**
over 300 mm girth; in staircase areas	–	0.21	3.42	–	0.84	m²	**4.26**
cutting in edges on flush surfaces	–	0.08	1.30	–	–	m	**1.30**
Plaster/plasterboard ceilings							
over 300 mm girth; 3.50 m–5.00 m high	–	0.21	3.42	–	0.76	m²	**4.18**

29 DECORATION

Item	PC £	Labour hours	Labour £	Plant £	Material £	Unit	Total rate £
One mist and two coats emulsion paint							
Brick or block walls							
over 300 mm girth	–	0.19	3.10	–	1.11	m²	**4.21**
Cement render or concrete							
over 300 mm girth	–	0.19	3.10	–	1.04	m²	**4.14**
Plaster walls or plaster/plasterboard ceilings							
over 300 mm girth	–	0.18	2.94	–	1.04	m²	**3.98**
over 300 mm girth; in multi colours	–	0.25	4.08	–	1.05	m²	**5.13**
over 300 mm girth; in staircase areas	–	0.21	3.42	–	1.04	m²	**4.46**
cutting in edges on flush surfaces	–	0.09	1.47	–	–	m	**1.47**
Plaster/plasterboard ceilings							
over 300 mm girth; 3.50 m–5.00 m high	–	0.21	3.42	–	1.04	m²	**4.46**
One mist Supermatt; one full Supermatt and one full coat of quick drying Acrylic Eggshell							
Brick or block walls							
over 300 mm girth	–	0.19	3.10	–	1.76	m²	**4.86**
Cement render or concrete							
over 300 mm girth	–	0.19	3.10	–	1.64	m²	**4.74**
Plaster walls or plaster/plasterboard ceilings							
over 300 mm girth	–	0.18	2.94	–	1.64	m²	**4.58**
over 300 mm girth; in multi colours	–	0.25	4.08	–	1.64	m²	**5.72**
over 300 mm girth; in staircase areas	–	0.21	3.42	–	1.64	m²	**5.06**
cutting in edges on flush surfaces	–	0.09	1.47	–	–	m	**1.47**
Plaster/plasterboard ceilings							
over 300 mm girth; 3.50 m–5.00 m high	–	0.21	3.42	–	1.64	m²	**5.06**
Clean and prepare surfaces, one coat Keim dilution, one coat primer and two coats of Keim Ecosil paint							
Brick or block walls							
over 300 mm girth	–	0.30	4.89	–	5.03	m²	**9.92**
Cement render or concrete							
over 300 mm girth	–	0.30	4.89	–	4.41	m²	**9.30**
Plaster walls or plaster/plasterboard ceilings							
over 300 mm girth	–	0.30	4.89	–	4.41	m²	**9.30**
over 300 mm girth; in staircase areas	–	0.30	4.89	–	4.41	m²	**9.30**
cutting in edges on flush surfaces	–	0.09	1.47	–	–	m	**1.47**
Plaster/plasterboard ceilings							
over 300 mm girth; 3.50 m–5.00 m high	–	0.25	4.08	–	4.58	m²	**8.66**
One coat Tretol No 10 Sealer or other equal and approved; two coats Tretol sprayed Supercover Spraytone emulsion paint or other equal and approved							
Plaster walls or plaster/plasterboard ceilings							
over 300 mm girth	–	0.25	5.72	–	0.22	m²	**5.94**

29 DECORATION

Item	PC £	Labour hours	Labour £	Plant £	Material £	Unit	Total rate £
PAINTING AND CLEAR FINISHES – INTERNAL – cont							
Textured plastic; Artex or other equal and approved finish							
Plasterboard ceilings							
over 300 mm girth	–	0.19	3.10	–	3.18	m²	**6.28**
Concrete walls or ceilings							
over 300 mm girth	–	0.23	3.76	–	2.96	m²	**6.72**
Touch up primer; one undercoat and one finishing coat of gloss oil paint; on wood surfaces							
General surfaces							
over 300 mm girth	–	0.20	3.27	–	2.28	m²	**5.55**
isolated surfaces not exceeding 300 mm girth	–	0.08	1.30	–	0.80	m	**2.10**
isolated areas not exceeding 1.00 m²; irrespective of girth	–	0.18	2.94	–	1.24	nr	**4.18**
Glazed windows and screens							
panes; area not exceeding 0.10 m²	–	0.38	6.21	–	1.84	m²	**8.05**
panes; area 0.10 m²–0.50 m²	–	0.31	5.06	–	1.41	m²	**6.47**
panes; area 0.50 m²–1.00 m²	–	0.26	4.24	–	1.13	m²	**5.37**
panes; area over 1.00 m²	–	0.23	3.76	–	0.97	m²	**4.73**
Knot; one coat primer; stop; one undercoat and one finishing coat of gloss oil paint; on wood surfaces							
General surfaces							
over 300 mm girth	–	0.33	5.39	–	2.34	m²	**7.73**
isolated surfaces not exceeding 300 mm girth	–	0.13	2.12	–	0.80	m	**2.92**
isolated areas not exceeding 0.50 m²; irrespective of girth	–	0.25	4.08	–	1.54	nr	**5.62**
Glazed windows and screens							
panes; area not exceeding 0.10 m²	–	0.56	9.15	–	2.34	m²	**11.49**
panes; area 0.10 m²–0.50 m²	–	0.45	7.35	–	1.95	m²	**9.30**
panes; area 0.50 m²–1.00 m²	–	0.40	6.53	–	1.95	m²	**8.48**
panes; area over 1.00 m²	–	0.33	5.39	–	1.45	m²	**6.84**
One coat primer; one undercoat and one finishing coat of gloss oil paint							
Plaster surfaces							
over 300 mm girth	–	0.30	4.89	–	2.98	m²	**7.87**
One coat primer; two undercoats and one finishing coat of gloss oil paint							
Plaster surfaces							
over 300 mm girth	–	0.40	6.53	–	3.94	m²	**10.47**

29 DECORATION

Item	PC £	Labour hours	Labour £	Plant £	Material £	Unit	Total rate £
One coat primer; two undercoats and one finishing coat of eggshell paint							
Plaster surfaces							
over 300 mm girth	–	0.40	6.53	–	4.25	m²	**10.78**
Touch up primer; one undercoat and one finishing coat of gloss paint; on iron or steel surfaces							
General surfaces							
over 300 mm girth	–	0.23	3.76	–	1.74	m²	**5.50**
isolated surfaces not exceeding 300 mm girth	–	0.09	1.47	–	0.60	m	**2.07**
isolated areas not exceeding 0.50 m²; irrespective of girth	–	0.18	2.94	–	0.97	nr	**3.91**
Glazed windows and screens							
panes; area not exceeding 0.10 m²	–	0.38	6.21	–	1.83	m²	**8.04**
panes; area 0.10 m²–0.50 m²	–	0.31	5.06	–	1.42	m²	**6.48**
panes; area 0.50 m²–1.00 m²	–	0.26	4.24	–	1.10	m²	**5.34**
panes; area over 1.00 m²	–	0.23	3.76	–	0.93	m²	**4.69**
Structural steelwork							
over 300 mm girth	–	0.25	4.08	–	1.84	m²	**5.92**
Members of roof trusses							
over 300 mm girth	–	0.34	5.55	–	2.10	m²	**7.65**
Ornamental railings and the like; each side measured overall							
over 300 mm girth	–	0.40	6.53	–	2.31	m²	**8.84**
Iron or steel radiators							
over 300 mm girth	–	0.23	3.76	–	1.92	m²	**5.68**
Pipes or conduits							
over 300 mm girth	–	0.34	5.55	–	2.02	m²	**7.57**
not exceeding 300 mm girth	–	0.13	2.12	–	0.67	m	**2.79**
One coat primer; one undercoat and one finishing coat of gloss oil paint; on iron or steel surfaces							
General surfaces							
over 300 mm girth	–	0.30	4.89	–	1.76	m²	**6.65**
isolated surfaces not exceeding 300 mm girth	–	0.12	1.96	–	1.02	m	**2.98**
isolated areas not exceeding 0.50 m²; irrespective of girth	–	0.23	3.76	–	1.71	nr	**5.47**
Glazed windows and screens							
panes; area not exceeding 0.10 m²	–	0.50	8.16	–	2.78	m²	**10.94**
panes; area 0.10 m²–0.50 m²	–	0.40	6.53	–	2.23	m²	**8.76**
panes; area 0.50 m²–1.00 m²	–	0.34	5.55	–	1.91	m²	**7.46**
panes; area over 1.00 m²	–	0.30	4.89	–	1.71	m²	**6.60**
Structural steelwork							
over 300 mm girth	–	0.33	5.39	–	2.72	m²	**8.11**
Members of roof trusses							
over 300 mm girth	–	0.45	7.35	–	2.88	m²	**10.23**

29 DECORATION

Item	PC £	Labour hours	Labour £	Plant £	Material £	Unit	Total rate £
PAINTING AND CLEAR FINISHES – INTERNAL – cont							
One coat primer – cont							
Ornamental railings and the like; each side measured overall							
over 300 mm girth	–	0.51	8.33	–	3.42	m²	**11.75**
Iron or steel radiators							
over 300 mm girth	–	0.30	4.89	–	2.88	m²	**7.77**
Pipes or conduits							
over 300 mm girth	–	0.45	7.35	–	2.88	m²	**10.23**
not exceeding 300 mm girth	–	0.18	2.94	–	0.96	m	**3.90**
Two coats of bituminous paint; on iron or steel surfaces							
General surfaces							
over 300 mm girth	–	0.23	3.76	–	0.89	m²	**4.65**
Inside of galvanized steel cistern							
over 300 mm girth	–	0.34	5.55	–	1.06	m²	**6.61**
Two coats bituminous paint; first coat blinded with clean sand prior to second coat; on concrete surfaces							
General surfaces							
over 300 mm girth	–	0.79	12.89	–	2.64	m²	**15.53**
Mordant solution; one coat HCC Protective Coatings Ltd Permacor Alkyd MIO or other equal and approved; one coat Permatex Epoxy Gloss finishing coat or other equal and approved on galvanized steelwork							
Structural steelwork							
over 300 mm girth	–	0.44	7.18	–	3.36	m²	**10.54**
One coat HCC Protective Coatings Ltd Epoxy Zinc Primer or other equal and approved; two coats Permacor Alkyd MIO or other equal and approved; one coat Permacor Epoxy Gloss finishing coat or other equal and approved on steelwork							
Structural steelwork							
over 300 mm girth	–	0.63	10.28	–	6.33	m²	**16.61**
Steel protection; HCC Protective Coatings Ltd Unitherm or other equal and approved; two coats to steelwork							
Structural steelwork							
over 300 mm girth	–	0.99	16.16	–	2.26	m²	**18.42**

29 DECORATION

Item	PC £	Labour hours	Labour £	Plant £	Material £	Unit	Total rate £
Two coats of epoxy anti-slip floor paint; on screeded concrete surfaces							
General surfaces							
over 300 mm girth	–	0.25	4.08	–	14.85	m²	**18.93**
Nitoflor Lithurin floor hardener and dust proofer or other equal and approved; Fosroc Expandite Ltd; two coats; on concrete surfaces							
General surfaces							
over 300 mm girth	–	0.24	3.15	–	0.47	m²	**3.62**
Two coats of boiled linseed oil; on hardwood surfaces							
General surfaces							
over 300 mm girth	–	0.18	2.94	–	3.29	m²	**6.23**
isolated surfaces not exceeding 300 mm girth	–	0.07	1.14	–	1.06	m	**2.20**
isolated areas not exceeding 0.50 m²; irrespective of girth	–	0.13	2.12	–	1.91	nr	**4.03**
Two coats polyurethane varnish; on wood surfaces							
General surfaces							
over 300 mm girth	–	0.18	2.94	–	1.66	m²	**4.60**
isolated surfaces not exceeding 300 mm girth	–	0.07	1.14	–	0.62	m	**1.76**
isolated areas not exceeding 0.50 m²; irrespective of girth	–	0.13	2.12	–	0.22	nr	**2.34**
Three coats polyurethane varnish; on wood surfaces							
General surfaces							
over 300 mm girth	–	0.26	4.24	–	2.49	m²	**6.73**
isolated surfaces not exceeding 300 mm girth	–	0.10	1.63	–	0.80	m	**2.43**
isolated areas not exceeding 0.50 m²; irrespective of girth	–	0.19	3.10	–	1.40	nr	**4.50**
One undercoat; and one finishing coat; of Albi clear flame retardant surface coating or other equal and approved; on wood surfaces							
General surfaces							
over 300 mm girth	–	0.34	5.55	–	5.52	m²	**11.07**
isolated surfaces not exceeding 300 mm girth	–	0.14	2.29	–	1.92	m	**4.21**
isolated areas not exceeding 0.50 m²; irrespective of girth	–	0.19	3.10	–	4.21	nr	**7.31**
Two undercoats; and one finishing coat; of Albi clear flame retardant surface coating or other equal and approved; on wood surfaces							
General surfaces							
over 300 mm girth	–	0.40	6.53	–	6.99	m²	**13.52**
isolated surfaces not exceeding 300 mm girth	–	0.20	3.27	–	2.81	m	**6.08**
isolated areas not exceeding 0.50 m²; irrespective of girth	–	0.33	5.39	–	3.74	nr	**9.13**

29 DECORATION

Item	PC £	Labour hours	Labour £	Plant £	Material £	Unit	Total rate £
PAINTING AND CLEAR FINISHES – INTERNAL – cont							
Seal and wax polish; dull gloss finish on wood surfaces							
General surfaces							
over 300 mm girth	–	0.50	11.45	–	2.14	m²	**13.59**
isolated surfaces not exceeding 300 mm girth	–	0.25	5.72	–	0.41	m	**6.13**
isolated areas not exceeding 0.50 m²; irrespective of girth	–	0.35	8.01	–	1.02	nr	**9.03**
One coat of Sadolin Extra or other equal and approved; clear or pigmented; one further coat of Holdex clear interior silk matt lacquer or similar							
General surfaces							
over 300 mm girth	–	0.25	4.08	–	5.91	m²	**9.99**
isolated surfaces not exceeding 300 mm girth	–	0.10	1.63	–	2.76	m	**4.39**
isolated areas not exceeding 0.50 m²; irrespective of girth	–	0.20	3.27	–	2.88	nr	**6.15**
Glazed windows and screens							
panes; area not exceeding 0.10 m²	–	0.42	6.86	–	3.38	m²	**10.24**
panes; area 0.10 m²–0.50 m²	–	0.33	5.39	–	3.15	m²	**8.54**
panes; area 0.50 m²–1.00 m²	–	0.29	4.74	–	2.92	m²	**7.66**
panes; area over 1.00 m²	–	0.25	4.08	–	2.76	m²	**6.84**
Two coats of Sadolin Extra or other equal and approved; clear or pigmented; two further coats of PV67 clear interior silk matt lacquer or similar							
General surfaces							
over 300 mm girth	–	0.40	6.53	–	10.87	m²	**17.40**
isolated surfaces not exceeding 300 mm girth	–	0.16	2.61	–	5.44	m	**8.05**
isolated areas not exceeding 0.50 m²; irrespective of girth	–	0.30	4.89	–	6.20	nr	**11.09**
Glazed windows and screens							
panes; area not exceeding 0.10 m²	–	0.66	10.77	–	6.66	m²	**17.43**
panes; area 0.10 m²–0.50 m²	–	0.52	8.48	–	6.20	m²	**14.68**
panes; area 0.50 m²–1.00 m²	–	0.45	7.35	–	5.74	m²	**13.09**
panes; area over 1.00 m²	–	0.40	6.53	–	5.44	m²	**11.97**
Two coats of Sikkens Cetol TS interior stain or other equal and approved; on wood surfaces							
General surfaces							
over 300 mm girth	–	0.19	3.10	–	2.74	m²	**5.84**
isolated surfaces not exceeding 300 mm girth	–	0.08	1.30	–	0.98	m	**2.28**
isolated areas not exceeding 0.50 m²; irrespective of girth	–	0.13	2.12	–	1.51	nr	**3.63**

29 DECORATION

Item	PC £	Labour hours	Labour £	Plant £	Material £	Unit	Total rate £
Body in and wax polish; dull gloss finish; on hardwood surfaces							
General surfaces							
over 300 mm girth	–	0.60	13.73	–	1.02	m²	**14.75**
isolated surfaces not exceeding 300 mm girth	–	0.30	6.87	–	0.48	m	**7.35**
isolated areas not exceeding 0.50 m²; irrespective of girth	–	0.40	9.16	–	0.48	nr	**9.64**
Stain; body in and wax polish; dull gloss finish; on hardwood surfaces							
General surfaces							
over 300 mm girth	–	0.75	17.17	–	1.02	m²	**18.19**
isolated surfaces not exceeding 300 mm girth	–	0.30	6.87	–	1.02	m	**7.89**
isolated areas not exceeding 0.50 m²; irrespective of girth	–	0.50	11.45	–	1.02	nr	**12.47**
Seal; two coats of synthetic resin lacquer; decorative flatted finish; wire down, wax and burnish; on wood surfaces							
General surfaces							
over 300 mm girth	–	1.00	22.89	–	0.99	m²	**23.88**
isolated surfaces not exceeding 300 mm girth	–	0.40	9.16	–	0.82	m	**9.98**
isolated areas not exceeding 0.50 m²; irrespective of girth	–	0.65	14.88	–	0.82	nr	**15.70**
Stain; body in and fully French polish; full gloss finish; on hardwood surfaces							
General surfaces							
over 300 mm girth	–	1.10	25.18	–	1.73	m²	**26.91**
isolated surfaces not exceeding 300 mm girth	–	0.50	11.45	–	0.98	m	**12.43**
isolated areas not exceeding 0.50 m²; irrespective of girth	–	0.75	17.17	–	1.68	nr	**18.85**
Stain; fill grain and fully French polish; full gloss finish; on hardwood surfaces							
General surfaces							
over 300 mm girth	–	1.60	36.62	–	2.93	m²	**39.55**
isolated surfaces not exceeding 300 mm girth	–	0.60	13.73	–	0.98	m	**14.71**
isolated areas not exceeding 0.50 m²; irrespective of girth	–	1.00	22.89	–	1.68	nr	**24.57**
Stain black; body in and fully French polish; ebonized finish; on hardwood surfaces							
General surfaces							
over 300 mm girth	–	1.75	40.06	–	3.28	m²	**43.34**
isolated surfaces not exceeding 300 mm girth	–	0.80	18.31	–	0.98	m	**19.29**
isolated areas not exceeding 0.50 m²; irrespective of girth	–	1.30	29.76	–	1.68	nr	**31.44**

29 DECORATION

Item	PC £	Labour hours	Labour £	Plant £	Material £	Unit	Total rate £
DECORATIVE PAPERS OR FABRICS							
Lining paper; and hanging							
Plaster walls or columns							
over 1.00 m² (PC £ per roll)	2.25	0.19	3.10	–	0.33	m²	**3.43**
Plaster ceilings or beams							
over 1.00 m² girth (PC £ per roll)	2.25	0.23	3.76	–	0.33	m²	**4.09**
Decorative paper-backed vinyl wallpaper; and hanging. Because of the enormous range of papers available we have simply allowed a prime cost amount to purchase the paper							
Plaster walls or columns							
over 1.00 m² girth (PC £ per roll)	25.00	0.23	3.76	–	4.46	m²	**8.22**
PVC Wall lining; Altro Whiterock; or other equal and approved; fixed directly to plastered brick or blockwork							
Work to walls over 300 mm wide; over 100 m² total area							
Standard white	32.14	1.25	20.40	–	35.44	m²	**55.84**
Satins	40.66	1.25	20.40	–	44.82	m²	**65.22**
Chameleon	62.87	1.25	20.40	–	69.31	m²	**89.71**

30 SUSPENDED CEILINGS

Item	PC £	Labour hours	Labour £	Plant £	Material £	Unit	Total rate £
SUSPENDED CEILINGS							
Suspended Ceilings							
Gyproc M/F suspended ceiling system or other equal and approved; hangers plugged and screwed to concrete soffit, 900 mm × 1800 mm × 12.50 mm tapered edge wallboard infill; joints filled with joint filler and taped to receive direct decoration							
Lining to ceilings; hangers 150 mm–500 mm long							
over 600 mm wide	–	–	–	–	–	m²	32.64
not exceeding 600 mm wide in isolated strips	–	–	–	–	–	m	33.08
Edge treatments							
20 × 20 mm SAS perimeter shadow gap; screwed to plasterboard	–	–	–	–	–	m	5.94
20 × 20 mm SAS shadow gap around 450 mm dia. column; including 15 × 44 mm batten plugged and screwed to concrete	–	–	–	–	–	nr	70.55
Vertical bulkhead; including additional hangers							
over 600 mm wide	–	–	–	–	–	m²	41.20
not exceeding 600 mm wide in isolated strips	–	–	–	–	–	m	40.64
Rockfon, or other equal and approved; Z demountable suspended concealed ceiling system; 400 mm long hangers plugged and screwed to concrete soffit.							
Lining to ceilings; 600 mm × 600 mm × 20 mm Sonar suspended ceiling tiles							
over 600 mm wide	–	–	–	–	–	m²	41.78
not exceeding 600 mm wide in isolated strips	–	–	–	–	–	m	24.63
edge trim; shadow-line trim	–	–	–	–	–	m	4.88
Vertical bulkhead, as upstand to rooflight well; including additional hangers; perimeter trim							
300 mm × 600 mm wide	–	–	–	–	–	m	45.73
Ecophon, or other equal and approved; Z demountable suspended concealed ceiling system; 400 mm long hangers plugged and screwed to concrete soffit.							
Lining to ceilings; 600 mm × 600 mm × 20 mm Gedina ET15 suspended ceiling tiles							
over 600 mm wide	–	–	–	–	–	m²	36.25
not exceeding 600 mm wide in isolated strips	–	–	–	–	–	m	22.85
edge trim; shadow-line trim	–	–	–	–	–	m	4.51
Vertical bulkhead, as upstand to rooflight well; including additional hangers; perimeter trim							
300 mm × 600 mm wide	–	–	–	–	–	m	42.08

30 SUSPENDED CEILINGS

Item	PC £	Labour hours	Labour £	Plant £	Material £	Unit	Total rate £
SUSPENDED CEILINGS – cont							
Ecophon, or other equal and approved – cont							
Lining to ceilings; 600 mm × 600 mm × 20 mm Hygiene Performance washable suspended ceiling tiles							
over 600 mm wide	–	–	–	–	–	m²	**48.41**
not exceeding 600 mm wide in isolated strips	–	–	–	–	–	m	**39.68**
edge trim; shadow-line trim	–	–	–	–	–	m	**6.38**
Vertical bulkhead, as upstand to rooflight well; including additional hangers; perimeter trim							
not exceeding 600 mm wide in isolated strips	–	–	–	–	–	m	**43.69**
Lining to ceilings; 1200 mm × 1200 mm × 20 mm Focus DG suspended ceiling tiles							
over 600 mm wide	–	–	–	–	–	m²	**43.91**
not exceeding 600 mm wide in isolated strips	–	–	–	–	–	m	**26.03**
edge trim; shadow-line trim	–	–	–	–	–	m	**4.51**
Vertical bulkhead, as upstand to rooflight well; including additional hangers; perimeter trim							
300 mm × 600 mm wide	–	–	–	–	–	m	**44.34**
Z demountable suspended ceiling system or other equal and approved; hangers plugged and screwed to concrete soffit, 600 mm × 600 mm × 19 mm Echostop glass reinforced fibrous plaster lightweight plain bevelled edge tiles							
Lining to ceilings; hangers 150 mm–500 mm long							
over 600 mm wide	–	–	–	–	–	m²	**89.73**
not exceeding 600 mm wide in isolated strips	–	–	–	–	–	m	**63.98**
Concealed galvanized steel suspension system; hangers plugged and screwed to concrete soffit, Burgess white stove enamelled perforated mild steel tiles 600 mm × 600 mm							
Lining to ceilings; hangers average 150 mm–500 mm long							
over 600 mm wide	–	–	–	–	–	m²	**43.14**
not exceeding 600 mm wide in isolated strips	–	–	–	–	–	m	**37.83**
Semi-concealed galvanized steel Trulok suspension system or other equal; hangers plugged and screwed to concrete; Armstrong Microlook Ultima Plain 600 mm × 600 mm × 18 mm mineral ceiling tiles							
Linings to ceilings; hangers 500 mm–1000 mm long							
over 600 mm wide	–	–	–	–	–	m²	**35.00**
over 600 mm wide; 3.50 m–5.00 m high	–	–	–	–	–	m²	**40.00**
over 600 mm wide; in staircase areas or plant rooms	–	–	–	–	–	m²	**40.00**

30 SUSPENDED CEILINGS

Item	PC £	Labour hours	Labour £	Plant £	Material £	Unit	Total rate £
not exceeding 600 mm wide in isolated strips	–	–	–	–	–	m	35.00
cutting and fitting around modular downlighter including yoke	–	–	–	–	–	nr	16.31
24 mm × 19 mm white finished angle edge trim	–	–	–	–	–	m	4.04
Vertical bulkhead; including additional hangers							
over 600 mm wide	–	–	–	–	–	m²	50.04
not exceeding 600 mm wide in isolated strips	–	–	–	–	–	m	45.13
Metal; SAS system 330; EMAC suspension system; 100 mm Omega C profiles at 1500 mm centres filled in with 1400 mm × 250 mm perforated metal tiles with 18 mm thick × 80 kg/ m³ density foil wrapped tissue-faced acoustic pad adhered above; ceiling to achieve 40dB with 0.7 absorption coefficient							
Linings to ceilings; hangers 500 mm–700 mm long							
over 600 mm wide	–	–	–	–	–	m²	45.18
not exceeding 600 mm wide in isolated strips	–	–	–	–	–	m	23.05
cutting and reinforcing to receive a recessed light maximum 1300 mm × 500 mm.	–	–	–	–	–	nr	12.91
edge trim; to perimeter	–	–	–	–	–	m	11.53
edge trim around 450 mm dia. column;	–	–	–	–	–	nr	44.25
Galvanized steel suspension system; hangers plugged and screwed to concrete soffit, Luxalon stove enamelled aluminium linear panel ceiling, type 80B or other equal and approved, complete with mineral insulation							
Linings to ceilings; hangers 500 mm–700 mm long							
over 600 mm wide	–	–	–	–	–	m²	81.43
not exceeding 600 mm wide in isolated strips	–	–	–	–	–	m	39.48
SOLID SUSPENDED CEILINGS							
Gypsum plasterboard; BS EN 520; plain grade tapered edge wallboard; fixing on dabs or with nails; joints left open to receive Artex finish or other equal and approved; to softwood base							
9.50 mm board to ceilings							
over 600 mm wide	–	0.23	4.64	–	2.88	m²	7.52
9.50 mm board to beams							
girth not exceeding 600 mm	–	0.28	5.66	–	1.75	m²	7.41
girth 600 mm–1200 mm	–	0.37	7.48	–	3.48	m²	10.96
12.50 mm board to ceilings							
over 300 mm wide	–	0.31	6.26	–	2.96	m²	9.22
12.50 mm board to beams							
girth not exceeding 600 mm	–	0.28	5.66	–	1.82	m²	7.48
girth 600 mm–1200 mm	–	0.37	7.48	–	3.56	m²	11.04

30 SUSPENDED CEILINGS

Item	PC £	Labour hours	Labour £	Plant £	Material £	Unit	Total rate £
SOLID SUSPENDED CEILINGS – cont							
Gypsum plasterboard to BS EN 520; fixing on dabs or with nails; joints filled with joint filler and joint tape to receive direct decoration; to softwood base (measured elsewhere)							
Plain grade tapered edge wallboard							
9.50 mm board to ceilings							
over 600 mm wide	–	0.29	6.12	–	3.70	m²	**9.82**
9.50 mm board to faces of beams – 3 nr							
not exceeding 600 mm total girth	–	0.46	9.83	–	4.00	m	**13.83**
600 mm–1200 mm total girth	–	0.67	14.21	–	5.75	m	**19.96**
1200 mm–1800 mm total girth	–	0.75	15.96	–	7.51	m	**23.47**
12.50 mm board to ceilings							
over 600 mm wide	–	0.31	6.53	–	5.64	m²	**12.17**
12.50 mm board to faces of beams – 3 nr							
not exceeding 600 mm total girth	–	0.46	9.83	–	5.20	m	**15.03**
600 mm–1200 mm total girth	–	0.67	14.21	–	8.12	m	**22.33**
1200 mm–1800 mm total girth	–	0.75	15.96	–	11.03	m	**26.99**
external angle; with joint tape bedded and covered with Jointex or other equal and approved	–	0.11	2.52	–	0.54	m	**3.06**
Tapered edge wallboard TEN							
12.50 mm board to ceilings							
over 600 mm wide	–	0.41	8.82	–	4.22	m²	**13.04**
12.50 mm board to faces of beams – 3 nr							
not exceeding 600 mm total girth	–	0.56	12.12	–	4.35	m	**16.47**
600 mm–1200 mm total girth	–	1.02	22.22	–	6.42	m	**28.64**
1200 mm–1800 mm total girth	–	1.30	28.55	–	8.48	m	**37.03**
external angle; with joint tape bedded and covered with Jointex or other equal and approved	–	0.11	2.52	–	0.54	m	**3.06**
Tapered edge plank							
19 mm plank to ceilings							
over 600 mm wide	–	0.43	9.22	–	6.80	m²	**16.02**
19 mm plank to faces of beams – 3 nr							
not exceeding 600 mm total girth	–	0.60	12.93	–	5.89	m	**18.82**
600 mm–1200 mm total girth	–	1.06	23.03	–	9.50	m	**32.53**
1200 mm–1800 mm total girth	–	1.34	29.36	–	13.13	m	**42.49**
ThermaLine plus board							
27 mm board to ceilings							
over 600 mm wide	–	0.46	9.29	–	10.83	m²	**20.12**
27 mm board to faces of beams – 3 nr							
not exceeding 600 mm total girth	–	0.56	11.31	–	8.31	m	**19.62**
600 mm–1200 mm total girth	–	1.06	21.41	–	14.33	m	**35.74**
1200 mm–1800 mm total girth	–	1.43	28.88	–	20.36	m	**49.24**

30 SUSPENDED CEILINGS

Item	PC £	Labour hours	Labour £	Plant £	Material £	Unit	Total rate £
48 mm board to ceilings							
over 600 mm wide	–	0.49	9.89	–	14.66	m²	**24.55**
48 mm board to faces of beams – 3 nr							
not exceeding 600 mm total girth	–	0.58	11.72	–	10.76	m	**22.48**
600 mm–1200 mm total girth	–	1.17	23.63	–	19.12	m	**42.75**
1200 mm–1800 mm total girth	–	1.57	31.71	–	27.48	m	**59.19**
ThermaLine super boards							
50 mm board to ceilings							
over 600 mm wide	–	0.49	9.89	–	19.82	m²	**29.71**
50 mm board to faces of beams – 3 nr							
not exceeding 600 mm total girth	–	0.58	11.72	–	13.86	m	**25.58**
600 mm–1200 mm total girth	–	1.17	23.63	–	25.32	m	**48.95**
1200 mm–1800 mm total girth	–	1.57	31.71	–	36.77	m	**68.48**
60 mm board to ceilings							
over 600 mm wide	–	0.49	9.89	–	23.38	m²	**33.27**
60 mm board to faces of beams – 3 nr							
not exceeding 600 mm total girth	–	0.58	11.72	–	16.00	m	**27.72**
600 mm–1200 mm total girth	–	1.17	23.63	–	29.59	m	**53.22**
1200 mm–1800 mm total girth	–	1.57	31.71	–	43.19	m	**74.90**

31 INSULATION, FIRE STOPPING AND FIRE PROTECTION

Item	PC £	Labour hours	Labour £	Plant £	Material £	Unit	Total rate £
INSULATION							
Altrernative insulation supply only prices, delivered in full loads							
Crown FrameTherm Roll 40 (Thermal conductivity 0.040 W/mK)							
90 mm thick	–	–	–	–	1.29	m²	**1.29**
140 mm thick	–	–	–	–	1.86	m²	**1.86**
Crown FrameTherm Roll 35 (Thermal conductivity 0.035 W/mK)							
90 mm thick	–	–	–	–	3.07	m²	**3.07**
140 mm thick	–	–	–	–	4.35	m²	**4.35**
Crown Factoryclad 40 (Thermal conductivity 0.040 W/mK)							
80 mm thick	–	–	–	–	0.87	m²	**0.87**
100 mm thick	–	–	–	–	1.05	m²	**1.05**
Crown Factoryclad 37 (Thermal conductivity 0.037 W/mK)							
100 mm thick	–	–	–	–	2.01	m²	**2.01**
120 mm thick	–	–	–	–	2.36	m²	**2.36**
Crown Factoryclad 32 (Thermal conductivity 0.032 W/mK)							
100 mm thick	–	–	–	–	4.67	m²	**4.67**
Thermafleece EcoRoll (0.039W/mK); 75% sheep's wool, 15% recycled polyester and 10% polyester binder with a high recycled content							
50 mm thick	–	–	–	–	3.32	m²	**3.32**
75 mm thick	–	–	–	–	4.96	m²	**4.96**
100 mm thick	–	–	–	–	6.60	m²	**6.60**
140 mm thick	–	–	–	–	9.25	m²	**9.25**
Thermafleece TF35 high density wool insulating batts (0.035 W/mK); 60% British wool, 30% recycled polyester and 10% polyester binder with a high recycled content							
50 mm thick	–	–	–	–	7.18	m²	**7.18**
70 mm thick	–	–	–	–	10.05	m²	**10.05**
SUPPLY AND FIX PRICES							
Sisalkraft building papers/vapour barriers or other equal and approved							
Building paper; 150 mm laps; fixed to softwood							
Moistop grade 728 (class A1F)	–	0.08	1.24	–	1.34	m²	**2.58**
Vapour barrier/reflective insulation 150 mm laps; fixed to softwood							
Insulex grade 714; single sided	–	0.08	1.24	–	1.60	m²	**2.84**

31 INSULATION, FIRE STOPPING AND FIRE PROTECTION

Item	PC £	Labour hours	Labour £	Plant £	Material £	Unit	Total rate £
Mat or quilt insulation							
Glass fibre roll; Crown Loft Roll 44 (Thermal conductivity 0.044 W/mK) or other equal; laid loose							
100 mm thick	1.11	0.09	1.40	–	1.17	m²	**2.57**
150 mm thick	1.65	0.10	1.55	–	1.73	m²	**3.28**
200 mm thick	2.21	0.11	1.71	–	2.32	m²	**4.03**
Glass fibre quilt; Isover Modular roll (Thermal conductivity 0.043 W/mK) or other equal and approved; laid loose							
100 mm thick	1.84	0.09	1.40	–	1.93	m²	**3.33**
150 mm thick	2.70	0.10	1.55	–	2.84	m²	**4.39**
170 mm thick	3.35	0.12	1.86	–	3.52	m²	**5.38**
200 mm thick	4.02	0.15	2.33	–	4.22	m²	**6.55**
Mineral fibre quilt; Isover Acoustic Partition Roll (APR 1200) or other equal; pinned vertically to softwood							
25 mm thick	1.05	0.08	1.24	–	1.37	m²	**2.61**
50 mm thick	1.73	0.09	1.40	–	2.08	m²	**3.48**
75 mm thick	3.16	0.10	1.55	–	3.58	m²	**5.13**
100 mm thick	4.64	0.15	2.33	–	5.13	m²	**7.46**
Crown Dritherm Cavity Slab 37 (Thermal conductivity 0.037 W/mK) glass fibre batt or other equal; as full or partial cavity fill; including cutting and fitting around wall ties and retaining discs							
50 mm thick	3.15	0.12	1.86	–	3.57	m²	**5.43**
75 mm thick	3.50	0.13	2.02	–	3.94	m²	**5.96**
100 mm thick	3.34	0.14	2.17	–	3.77	m²	**5.94**
Crown Dritherm Cavity Slab 34 (Thermal conductivity 0.034 W/mK) glass fibre batt or other equal; as full or partial cavity fill; including cutting and fitting around wall ties and retaining discs							
65 mm thick	2.28	0.12	1.86	–	2.66	m²	**4.52**
75 mm thick	2.65	0.13	2.02	–	3.04	m²	**5.06**
85 mm thick	4.02	0.13	2.02	–	4.48	m²	**6.50**
100 mm thick	4.02	0.14	2.17	–	4.48	m²	**6.65**
Crown Dritherm Cavity Slab 32 (Thermal conductivity 0.032 W/mK) glass fibre batt or other equal; as full or partial cavity fill; including cutting and fitting around wall ties and retaining discs							
65 mm thick	2.97	0.12	1.86	–	3.38	m²	**5.24**
75 mm thick	3.46	0.13	2.02	–	3.90	m²	**5.92**
85 mm thick	3.92	0.13	2.02	–	4.38	m²	**6.40**
100 mm thick	4.52	0.14	2.17	–	5.01	m²	**7.18**
Crown Frametherm Roll 40 (Thermal conductivity 0.040 W/mK) glass fibre semi-rigid or rigid batt or other equal; pinned vertically in timber frame construction							
90 mm thick	3.69	0.14	2.17	–	3.87	m²	**6.04**
140 mm thick	5.36	0.16	2.49	–	5.63	m²	**8.12**

31 INSULATION, FIRE STOPPING AND FIRE PROTECTION

Item	PC £	Labour hours	Labour £	Plant £	Material £	Unit	Total rate £
INSULATION – cont							
Mat or quilt insulation – cont							
Crown Rafter Roll 32 (Thermal conductivity 0.032 W/mK) glass fibre flanged building roll; pinned vertically or to slope between timber framing							
50 mm thick	3.45	0.13	2.02	–	3.62	m²	**5.64**
75 mm thick	4.93	0.14	2.17	–	5.18	m²	**7.35**
100 mm thick	6.35	0.15	2.33	–	6.67	m²	**9.00**
Board or slab insulation							
Kay-Cel (Thermal conductivity 0.033 W/mK) expanded polystyrene board standard grade SD/N or other equal; fixed with adhesive							
20 mm thick	–	0.14	2.82	–	1.48	m²	**4.30**
25 mm thick	–	0.14	2.82	–	1.61	m²	**4.43**
30 mm thick	–	0.14	2.82	–	1.73	m²	**4.55**
40 mm thick	–	0.15	3.03	–	2.02	m²	**5.05**
50 mm thick	–	0.16	3.23	–	2.26	m²	**5.49**
60 mm thick	–	0.17	3.43	–	2.53	m²	**5.96**
75 mm thick	–	0.18	3.63	–	2.92	m²	**6.55**
100 mm thick	–	0.19	3.83	–	3.58	m²	**7.41**
KIngspan Kooltherm K8 (Thermal conductivity 0.022 W/mK) Premium performance rigid thermoset modified resin insulant manufactured with a blowing agent that has zero Ozone Depletion Potential (ODP) and low Global Warming Potential (GWP) or other equal; as partial cavity fill; including cutting and fitting around wall ties and retaining discs							
40 mm thick	–	0.15	3.03	–	7.21	m²	**10.24**
50 mm thick	–	0.15	3.03	–	9.01	m²	**12.04**
60 mm thick	–	0.15	3.03	–	10.71	m²	**13.74**
75 mm thick	–	0.15	3.03	–	13.28	m²	**16.31**
Kingspan Thermawall TW50 (Thermal conductivity 0.022 W/mK) High performance rigid thermoset polyisocyanurate (PIR) insulant or other equal and approved; as partial cavity fill; including cutting and fitting around wall ties and retaining discs							
25 mm thick	–	0.17	3.43	–	6.73	m²	**10.16**
50 mm thick	–	0.17	3.43	–	11.40	m²	**14.83**
75 mm thick	–	0.18	3.63	–	17.18	m²	**20.81**
100 mm thick	–	0.19	3.83	–	22.25	m²	**26.08**
Kingspan Thermafloor TF70 (Thermal conductivity 0.022 W/mK) high performance rigid urethane floor insulation for solid concrete and suspended ground floors							
50 mm thick	–	0.17	3.54	–	12.39	m²	**15.93**
75 mm thick	–	0.17	3.54	–	18.53	m²	**22.07**
100 mm thick	–	0.17	3.54	–	24.09	m²	**27.63**
115 mm thick	–	0.20	4.04	–	10.75	m²	**14.79**
125 mm thick	–	0.17	3.54	–	29.13	m²	**32.67**
150 mm thick	–	0.20	4.04	–	35.68	m²	**39.72**

31 INSULATION, FIRE STOPPING AND FIRE PROTECTION

Item	PC £	Labour hours	Labour £	Plant £	Material £	Unit	Total rate £
Kingspan Thermafloor TF73 (Thermal conductivity 0.029 W/mK) high performance rigid extruded polystyrene insulation for floating and suspended ground floors; upper face 18 mm moisture-resistant chipboard							
40 mm thick	–	0.17	3.54	–	16.28	m²	**19.82**
50 mm thick	–	0.17	3.54	–	18.12	m²	**21.66**
60 mm thick	–	0.20	4.04	–	21.99	m²	**26.03**
80 mm thick	–	0.17	3.54	–	26.12	m²	**29.66**
Styrofoam Floormate 500 (Thermal conductivity 0.033 W/mK) extruded polystyrene foam or other equal and approved							
50 mm thick	–	0.46	9.29	–	10.38	m²	**19.67**
80 mm thick	–	0.46	9.29	–	15.86	m²	**25.15**
120 mm thick	–	0.46	9.29	–	22.47	m²	**31.76**
Dow Chemicals Styrofoam SP or other equal and approved; cold bridging insulation fixed with adhesive to brick, block or concrete base							
Insulation to walls							
50 mm thick	–	0.33	6.67	–	8.68	m²	**15.35**
75 mm thick	–	0.35	7.07	–	10.20	m²	**17.27**
Insulation to isolated columns							
50 mm thick	–	0.41	8.28	–	8.68	m²	**16.96**
75 mm thick	–	0.43	8.68	–	10.20	m²	**18.88**
Insulation to ceilings							
50 mm thick	–	0.36	7.27	–	8.68	m²	**15.95**
75 mm thick	–	0.39	7.88	–	10.20	m²	**18.08**
Insulation to isolated beams							
50 mm thick	–	0.43	8.68	–	8.68	m²	**17.36**
75 mm thick	–	0.46	9.29	–	10.20	m²	**19.49**
Pittsburgh Company Foamglas cellular glass insulation with PC 56 component adhesive as required							
Flat roof insulation; Foamglas T4+; thermal conductivity 0.041 W/mK; size 600 mm × 450 mm slabs							
80 mm thick slab with adhered joints	29.36	1.20	25.87	–	37.34	m²	**63.21**
100 mm thick slab with adhered joints	36.71	1.20	25.87	–	45.06	m²	**70.93**
150 mm thick slab with adhered joints	55.06	1.30	28.02	–	64.32	m²	**92.34**
180 mm thick slab with adhered joints	66.08	1.40	30.19	–	75.89	m²	**106.08**
Cavity wall insulation slabs; Foamglas W+F; thermal conductivity 0038 W/mK; 600 mm × 450 mm slabs							
80 mm thick slab with adhered joints	–	1.25	26.95	–	37.34	m²	**64.29**
100 mm thick slab with adhered joints	–	1.25	26.95	–	45.06	m²	**72.01**
150 mm thick slab with adhered joints	–	1.25	26.95	–	64.32	m²	**91.27**

31 INSULATION, FIRE STOPPING AND FIRE PROTECTION

Item	PC £	Labour hours	Labour £	Plant £	Material £	Unit	Total rate £
INSULATION – cont							
Pittsburgh Company Foamglas cellular glass insulation with PC 56 component adhesive as required – cont							
Cavity wall insulation boards; Foamglas W+F; thermal conductivity 0038 W/mK 1200 mm × 600 mm boards							
80 mm thick board with adhered joints	31.75	1.10	23.72	–	39.85	m²	**63.57**
100 mm thick board with adhered joints	–	1.20	25.87	–	48.19	m²	**74.06**
150 mm thick board with adhered joints	–	1.25	26.95	–	69.03	m²	**95.98**
Sheepswool mix							
Thermafleece TF35 high density wool insulating batts (0.035 W/mK); 60% British wool, 30% recycled polyester and 10% polyester binder with a high recycled content							
50 mm thick	–	0.35	7.07	–	7.18	m²	**14.25**
70 mm thick	–	0.38	7.57	–	10.05	m²	**17.62**
Thermafleece EcoRoll (0.039 W/mK); 75% sheep's wool, 15% recycled polyester and 10% polyester binder with a high recycled content							
50 mm thick	–	0.15	2.33	–	3.32	m²	**5.65**
75 mm thick	–	0.14	2.17	–	4.96	m²	**7.13**
100 mm thick	–	0.15	2.33	–	6.60	m²	**8.93**
140 mm thick	–	0.17	2.72	–	9.25	m²	**11.97**
FIRE STOPPING							
Fire stopping							
Cape Firecheck channel; intumescent coatings on cut mitres; fixing with brass cups and screws							
19 mm × 44 mm or 19 mm × 50 mm	–	0.56	11.31	–	17.71	m	**29.02**
Sealmaster intumescent fire and smoke seals or other equal and approved; pinned into groove in timber							
type N30; for single leaf half hour door	–	0.28	5.66	–	2.53	m	**8.19**
type N60; for single leaf one hour door	–	0.31	6.26	–	3.31	m	**9.57**
type IMN or IMP; for meeting or pivot stiles of pair of one hour doors; per stile	–	0.31	6.26	–	3.31	m	**9.57**
intumescent plugs in timber; including boring	–	0.09	1.82	–	0.37	nr	**2.19**
Rockwool fire stops or other equal and approved; between top of brick/block wall and concrete soffit							
30 mm deep × 100 mm wide	–	0.07	1.42	–	4.46	m	**5.88**
30 mm deep × 150 mm wide	–	0.09	1.82	–	6.76	m	**8.58**
30 mm deep × 200 mm wide	–	0.11	2.23	–	9.03	m	**11.26**
60 mm deep × 100 mm wide	–	0.08	1.62	–	5.87	m	**7.49**
60 mm deep × 150 mm wide	–	0.10	2.02	–	8.78	m	**10.80**
60 mm deep × 200 mm wide	–	0.12	2.43	–	11.81	m	**14.24**
90 mm deep × 100 mm wide	–	0.10	2.02	–	9.38	m	**11.40**
90 mm deep × 150 mm wide	–	0.12	2.43	–	14.03	m	**16.46**
90 mm deep × 200 mm wide	–	0.14	2.82	–	18.74	m	**21.56**

31 INSULATION, FIRE STOPPING AND FIRE PROTECTION

Item	PC £	Labour hours	Labour £	Plant £	Material £	Unit	Total rate £
Fire protection compound							
Quelfire QF4, fire protection compound or other equal and approved; filling around pipes, ducts and the like; including all necessary formwork							
300 mm × 300 mm × 250 mm; pipes – 2	–	0.93	16.36	–	14.32	nr	**30.68**
500 mm × 500 mm × 250 mm; pipes – 2	–	1.16	19.88	–	42.98	nr	**62.86**
Fire barriers							
Rockwool fire barrier or other equal and approved; between top of suspended ceiling and concrete soffit							
one 50 mm layer × 900 mm wide; half hour	–	0.56	11.31	–	25.69	m²	**37.00**
two 50 mm layers × 900 mm wide; one hour	–	0.83	16.76	–	51.11	m²	**67.87**
three 50 mm layers × 900 mm wide; two hour	–	1.10	22.22	–	73.28	m²	**95.50**
Corofil C144 fire barrier to edge of slab; fixed with non-flammable contact adhesive							
to suit void 30 mm wide × 100 mm deep; one hour	–	–	–	–	–	m	**13.01**
Lamatherm fire barrier or other equal and approved; to void below raised access floors							
75 mm thick × 300 mm high; half hour	–	0.17	3.43	–	9.38	m	**12.81**
75 mm thick × 600 mm high; half hour	–	0.17	3.43	–	20.55	m	**23.98**
90 mm thick × 300 mm high; half hour	–	0.17	3.43	–	13.17	m	**16.60**
90 mm thick × 600 mm high; half hour	–	0.17	3.43	–	27.43	m	**30.86**

32 FURNITURE, FITTINGS AND EQUIPMENT

Item	PC £	Labour hours	Labour £	Plant £	Material £	Unit	Total rate £
GENERAL FIXTURES, FURNISHES AND EQUIPMENT							
SUPPLY ONLY PRICES							
NOTE: The fixing of general fixtures will vary considerably dependent upon the size of the fixture and the method of fixing employed. Prices for fixing like sized kitchen fittings may be suitable for certain fixtures, although adjustment to those rates will almost invariably be necessary							
The following supply only prices are for purpose-made fittings components in various materials supplied as part of an assembled fitting and therefore may be used to arrive at a guide price for a complete fitting							
Fitting components; medium density fibreboard							
Backs, fronts, sides or divisions; over 300 mm wide							
12 mm thick	–	–	–	–	21.39	m²	**21.39**
18 mm thick	–	–	–	–	22.67	m²	**22.67**
25 mm thick	–	–	–	–	25.25	m²	**25.25**
Shelves or worktops; over 300 mm wide							
18 mm thick	–	–	–	–	22.67	m²	**22.67**
25 mm thick	–	–	–	–	25.25	m²	**25.25**
Flush doors; lipped on four edges							
450 mm × 750 mm × 18 mm	–	–	–	–	32.83	nr	**32.83**
450 mm × 750 mm × 25 mm	–	–	–	–	33.47	nr	**33.47**
600 mm × 900 mm × 18 mm	–	–	–	–	38.77	nr	**38.77**
600 mm × 900 mm × 25 mm	–	–	–	–	39.77	nr	**39.77**
Fitting components; moisture-resistant medium density fibreboard							
Backs, fronts, sides or divisions; over 300 mm wide							
12 mm thick	–	–	–	–	23.95	m²	**23.95**
18 mm thick	–	–	–	–	26.51	m²	**26.51**
25 mm thick	–	–	–	–	29.06	m²	**29.06**
Shelves or worktops; over 300 mm wide							
18 mm thick	–	–	–	–	26.51	m²	**26.51**
25 mm thick	–	–	–	–	29.06	m²	**29.06**
Flush doors; lipped on four edges							
450 mm × 750 mm × 18 mm	PC	–	–	–	33.47	nr	**33.47**
450 mm × 750 mm × 25 mm	–	–	–	–	34.41	nr	**34.41**
600 mm × 900 mm × 18 mm	–	–	–	–	39.77	nr	**39.77**
600 mm × 900 mm × 25 mm	–	–	–	–	41.32	nr	**41.32**
Fitting components; medium density fibreboard; melamine faced both sides							
Backs, fronts, sides or divisions; over 300 mm wide							
12 mm thick	–	–	–	–	28.23	m²	**28.23**
18 mm thick	–	–	–	–	31.44	m²	**31.44**

32 FURNITURE, FITTINGS AND EQUIPMENT

Item	PC £	Labour hours	Labour £	Plant £	Material £	Unit	Total rate £
Shelves or worktops; over 300 mm wide							
18 mm thick	–	–	–	–	31.44	m^2	**31.44**
Flush doors; lipped on four edges							
450 mm × 750 mm × 18 mm	–	–	–	–	23.48	nr	**23.48**
600 mm × 900 mm × 25 mm	–	–	–	–	29.79	nr	**29.79**
Fitting components; medium density fibreboard; formica faced both sides							
Backs, fronts, sides or divisions; over 300 mm wide							
12 mm thick	–	–	–	–	85.95	m^2	**85.95**
18 mm thick	–	–	–	–	89.39	m^2	**89.39**
Shelves or worktops; over 300 mm wide							
18 mm thick	–	–	–	–	89.39	m^2	**89.39**
Flush doors; lipped on four edges							
450 mm × 750 mm × 18 mm	–	–	–	–	47.33	nr	**47.33**
600 mm × 900 mm × 25 mm	–	–	–	–	48.35	nr	**48.35**
Fitting components; wrought softwood							
Backs, fronts, sides or divisions; cross-tongued joints; over 300 mm wide							
25 mm thick	–	–	–	–	41.09	m^2	**41.09**
Shelves or worktops; cross-tongued joints; over 300 mm wide							
25 mm thick	–	–	–	–	41.09	m^2	**41.09**
Bearers							
19 mm × 38 mm	–	–	–	–	2.15	m	**2.15**
25 mm × 50 mm	–	–	–	–	2.38	m	**2.38**
44 mm × 44 mm	–	–	–	–	2.54	m	**2.54**
44 mm × 75 mm	–	–	–	–	2.93	m	**2.93**
Bearers; framed; to backs, fronts or sides							
19 mm × 38 mm	–	–	–	–	4.91	m	**4.91**
25 mm × 50 mm	–	–	–	–	5.32	m	**5.32**
50 mm × 50 mm	–	–	–	–	6.87	m	**6.87**
50 mm × 75 mm	–	–	–	–	7.89	m	**7.89**
Add 5% to the above material prices for selected softwood staining							
Fitting components; selected Sapele							
Backs, fronts, sides or divisions; cross-tongued joints; over 300 mm wide							
25 mm thick	–	–	–	–	64.66	m^2	**64.66**
Shelves or worktops; cross-tongued joints; over 300 mm wide							
25 mm thick	–	–	–	–	64.66	m^2	**64.66**
Bearers							
19 mm × 38 mm	–	–	–	–	3.24	m	**3.24**
25 mm × 50 mm	–	–	–	–	4.03	m	**4.03**
50 mm × 50 mm	–	–	–	–	4.54	m	**4.54**
50 mm × 75 mm	–	–	–	–	5.90	m	**5.90**

32 FURNITURE, FITTINGS AND EQUIPMENT

Item	PC £	Labour hours	Labour £	Plant £	Material £	Unit	Total rate £
GENERAL FIXTURES, FURNISHES AND EQUIPMENT – cont							
Fitting components – cont							
Bearers; framed; to backs, fronts or sides							
19 mm × 38 mm	–	–	–	–	6.27	m	**6.27**
25 mm × 50 mm	–	–	–	–	6.94	m	**6.94**
50 mm × 50 mm	–	–	–	–	9.31	m	**9.31**
50 mm × 75 mm	–	–	–	–	11.68	m	**11.68**
Fitting components; Iroko							
Backs, fronts, sides or divisions; cross-tongued joints; over 300 mm wide							
25 mm thick	–	–	–	–	71.27	m²	**71.27**
Shelves or worktops; cross-tongued joints; over 300 mm wide							
25 mm thick	–	–	–	–	71.27	m²	**71.27**
Draining boards; cross-tongued joints; over 300 mm wide							
25 mm thick	–	–	–	–	89.29	m²	**89.29**
stopped flutes	–	–	–	–	4.61	m	**4.61**
grooves; cross-grain	–	–	–	–	0.68	m	**0.68**
Bearers							
19 mm × 38 mm	–	–	–	–	3.54	m	**3.54**
25 mm × 50 mm	–	–	–	–	4.53	m	**4.53**
50 mm × 50 mm	–	–	–	–	5.15	m	**5.15**
50 mm × 75 mm	–	–	–	–	6.78	m	**6.78**
Bearers; framed; to backs, fronts or sides							
19 mm × 38 mm	–	–	–	–	6.48	m	**6.48**
25 mm × 50 mm	–	–	–	–	7.21	m	**7.21**
50 mm × 50 mm	–	–	–	–	9.73	m	**9.73**
50 mm × 75 mm	–	–	–	–	12.64	m	**12.64**
Lockers and cupboards; Welconstruct Distribution or other equal and approved							
Standard clothes lockers; steel body and door within reinforced 19 G frame, powder coated finish, cam locks							
1 compartment; placing in position							
300 mm × 300 mm × 1800 mm	–	0.23	3.02	–	51.63	nr	**54.65**
380 mm × 380 mm × 1800 mm	–	0.23	3.02	–	73.89	nr	**76.91**
450 mm × 450 mm × 1800 mm	–	0.28	3.67	–	75.43	nr	**79.10**
Compartment lockers; steel body and door within reinforced 19 G frame, powder coated finish, cam locks							
2 compartments; placing in position							
300 mm × 300 mm × 1800 mm	–	0.23	3.02	–	59.70	nr	**62.72**
380 mm × 380 mm × 1800 mm	–	0.23	3.02	–	71.99	nr	**75.01**
450 mm × 450 mm × 1800 mm	–	0.28	3.67	–	78.08	nr	**81.75**

32 FURNITURE, FITTINGS AND EQUIPMENT

Item	PC £	Labour hours	Labour £	Plant £	Material £	Unit	Total rate £
4 compartments; placing in position							
300 mm × 300 mm × 1800 mm	–	0.23	3.02	–	70.20	nr	**73.22**
380 mm × 380 mm × 1800 mm	–	0.23	3.02	–	84.28	nr	**87.30**
450 mm × 450 mm × 1800 mm	–	0.28	3.67	–	84.28	nr	**87.95**
Timber clothes lockers; veneered MDF finish, routed door, cam locks							
1 compartment; placing in position							
380 mm × 380 mm × 1830 mm	–	0.28	3.67	–	183.89	nr	**187.56**
4 compartments; placing in position							
380 mm × 380 mm × 1830 mm	–	0.28	3.67	–	274.08	nr	**277.75**
Vandal-resistant lockers							
1030 high mm × 370 mm × 560 mm; one compartment	–	0.23	3.02	–	194.94	nr	**197.96**
1930 mm × 370 mm × 560 mm; two compartments	–	0.23	3.02	–	302.39	nr	**305.41**
850 mm × 740 mm × 560 mm; 2 high × 2 wide	–	0.23	3.02	–	396.96	nr	**399.98**
1930 mm × 740 mm × 560 mm; 5 high × 2 wide	–	0.23	3.02	–	898.73	nr	**901.75**
Shelving support systems; The Welconstruct Company or other equal and approved standard duty; maximum bayload of 2000 kg							
Shelving support systems; steel body; stove enamelled finish; assembling							
open initial bay; 5 shelves; placing in position							
1000 mm × 300 mm × 1850 mm	–	0.69	10.71	–	143.08	nr	**153.79**
1000 mm × 600 mm × 1850 mm	–	0.69	10.71	–	181.41	nr	**192.12**
open extension bay; 5 shelves; placing in position							
1000 mm × 300 mm × 1850 mm	–	0.83	12.88	–	92.91	nr	**105.79**
1000 mm × 600 mm × 1850 mm	–	0.83	12.88	–	127.60	nr	**140.48**
closed initial bay; 5 shelves; placing in position							
1000 mm × 300 mm × 1850 mm	–	0.69	10.71	–	200.59	nr	**211.30**
1000 mm × 600 mm × 1850 mm	–	0.69	10.71	–	259.77	nr	**270.48**
closed extension bay; 5 shelves; placing in position							
1000 mm × 300 mm × 1850 mm	–	0.83	12.88	–	149.04	nr	**161.92**
1000 mm × 600 mm × 1850 mm	–	0.83	12.88	–	194.19	nr	**207.07**
extra for pair of doors; fixing in position							
1000 mm × 1850 mm	–	0.75	11.64	–	281.91	nr	**293.55**
Cloakroom racks; The Welconstruct Company or other equal and approved							
Cloakroom racks; 40 mm × 40 mm square tube framing, polyester powder coated finish; beech slatted seats and rails to one side only; placing in position							
1675 mm × 325 mm × 1500 mm; 5 nr coat hooks	–	0.30	4.66	–	321.09	nr	**325.75**
1825 mm × 325 mm × 1500 mm; 15 nr coat hangers	–	0.30	4.66	–	367.93	nr	**372.59**
Extra for							
shoe baskets	–	–	–	–	72.30	nr	**72.30**
mesh bottom shelf	–	–	–	–	50.49	nr	**50.49**

32 FURNITURE, FITTINGS AND EQUIPMENT

Item	PC £	Labour hours	Labour £	Plant £	Material £	Unit	Total rate £
GENERAL FIXTURES, FURNISHES AND EQUIPMENT – cont							
Cloakroom racks – cont							
Cloakroom racks; 40 mm × 40 mm square tube framing, polyester powder coated finish; beech slatted seats and rails to both sides; placing in position							
1675 mm × 600 mm × 1500 mm; 10 nr coat hooks	–	0.40	6.21	–	440.33	nr	**446.54**
1825 mm × 600 mm × 1500 mm; 30 nr coat hangers	–	0.40	6.21	–	459.06	nr	**465.27**
Extra for							
shoe baskets	–	–	–	–	90.36	nr	**90.36**
mesh bottom shelf	–	–	–	–	61.40	nr	**61.40**
6 mm thick rectangular glass mirrors; silver backed; fixed with chromium plated domed headed screws; to background requiring plugging							
Mirror with polished edges							
365 mm × 254 mm	10.32	0.74	12.07	–	11.13	nr	**23.20**
400 mm × 300 mm	13.46	0.74	12.07	–	14.43	nr	**26.50**
560 mm × 380 mm	23.32	0.83	13.55	–	24.78	nr	**38.33**
640 mm × 460 mm	30.48	0.93	15.18	–	32.30	nr	**47.48**
Mirror with bevelled edges							
365 mm × 254 mm	18.39	0.74	12.07	–	19.59	nr	**31.66**
400 mm × 300 mm	21.53	0.74	12.07	–	22.90	nr	**34.97**
560 mm × 380 mm	35.89	0.83	13.55	–	37.97	nr	**51.52**
640 mm × 460 mm	44.85	0.93	15.18	–	47.38	nr	**62.56**
Internal blinds							
Roller blinds; Luxaflex EOS type 10 roller; Compact Fabric; plain type material; 1219 mm drop; fixing with screws							
1016 mm wide	45.53	0.93	12.22	–	47.81	nr	**60.03**
2031 mm wide	67.22	1.45	19.05	–	70.58	nr	**89.63**
2843 mm wide	83.50	1.97	25.88	–	87.67	nr	**113.55**
Roller blinds; Luxaflex EOS type 10 roller; Compact Fabric; fire-resisting material; 1219 mm drop; fixing with screws							
1016 mm wide	59.64	0.93	12.22	–	62.62	nr	**74.84**
2031 mm wide	88.91	1.45	19.05	–	93.36	nr	**112.41**
2843 mm wide	112.77	1.97	25.88	–	118.41	nr	**144.29**
Roller blinds; Luxaflex EOS type 10 roller; Light-resistant; blackout material; 1219 mm drop; fixing with screws							
1016 mm wide	76.99	0.93	12.22	–	80.84	nr	**93.06**
2031 mm wide	129.03	1.45	19.05	–	135.48	nr	**154.53**
2843 mm wide	174.57	1.97	25.88	–	183.30	nr	**209.18**

32 FURNITURE, FITTINGS AND EQUIPMENT

Item	PC £	Labour hours	Labour £	Plant £	Material £	Unit	Total rate £
Roller blinds; Luxaflex Lite-master Crank Op; 100% blackout; 1219 mm drop; fixing with screws							
1016 mm wide	214.69	1.96	25.76	–	225.42	nr	**251.18**
2031 mm wide	287.34	2.75	36.13	–	301.71	nr	**337.84**
2843 mm wide	370.83	3.53	46.38	–	389.37	nr	**435.75**
Vertical louvre blinds; 89 mm wide louvres; Luxaflex EOS type; Florida Fabric; 1219 mm drop; fixing with screws							
1016 mm wide	61.81	0.82	10.77	–	64.90	nr	**75.67**
2031 mm wide	94.33	1.30	17.08	–	99.05	nr	**116.13**
3046 mm wide	129.03	1.77	23.26	–	135.48	nr	**158.74**
Vertical louvre blinds; 127 mm wide louvres; Luxaflex EOS type; Florida Fabric; 1219 mm drop; fixing with screws							
1016 mm wide	52.04	0.88	11.56	–	54.64	nr	**66.20**
2031 mm wide	79.16	1.35	17.73	–	83.12	nr	**100.85**
3046 mm wide	106.27	1.81	23.78	–	111.58	nr	**135.36**
Venetian blinds; 80 mm wide louvres; Levolux 480 type; solid slat; stock colour; manual hand crank; 2700 mm drop; standard brackets fixed to suitable grounds							
875 mm wide	–	–	–	–	–	nr	**190.00**
1150 mm wide	–	–	–	–	–	nr	**220.00**
1500 mm wide	–	–	–	–	–	nr	**260.00**
2500 mm wide	–	–	–	–	–	nr	**350.00**
3000 mm wide	–	–	–	–	–	nr	**390.00**
3500 mm wide	–	–	–	–	–	nr	**440.00**
4000 mm wide	–	–	–	–	–	nr	**495.00**
Door entrance mats							
Entrance mats; aluminium wiper bar; laying in position; 18 mm thick							
Gradus Topguard; 900 mm × 550 mm	–	0.46	6.05	–	144.14	nr	**150.19**
Gradus Topguard; 1200 mm × 750 mm	–	0.46	6.05	–	259.44	nr	**265.49**
Gradus Topguard; 2400 mm × 1200 mm	–	0.93	12.22	–	830.24	nr	**842.46**
Nuway Tuftiguard; 500 mm × 1300 mm	–	–	–	–	–	nr	**236.40**
Nuway Tuftiguard; 1000 mm × 1600 mm	–	–	–	–	–	nr	**496.07**
Nuway Tuftiguard; 1800 mm × 3000 mm	–	–	–	–	–	nr	**1963.91**
extra for brass wipers	–	–	–	–	200.81	m²	**200.81**
Nuway Tuftiguard; single side aluminium scraper bar	–	–	–	–	–	m²	**363.68**
Nuway Tuftiguard; double side aluminium rigidsctaper bars	–	–	–	–	–	m²	**519.56**
Coral Classic textile secondary and circulation matting system for moisture removal; 2.0 m wide roll	–	–	–	–	–	m²	**62.35**
Coral Brush Activ secondary and circulation matting system for soil and moisture removal; 2.0 m wide roll	–	–	–	–	–	m²	**62.35**
Coral Duo secondary matting system ribbed; 2.0 m wide roll	–	–	–	–	–	m²	**62.35**

32 FURNITURE, FITTINGS AND EQUIPMENT

Item	PC £	Labour hours	Labour £	Plant £	Material £	Unit	Total rate £
GENERAL FIXTURES, FURNISHES AND EQUIPMENT – cont							
Matwells							
Polished stainless steel matwell; angle rim with lugs; bedding in screed							
Stainless steel frame bed in screed							
500 mm × 1300 mm	–	–	–	–	–	nr	**47.28**
Stainless steel frame bed in screed							
1000 mm × 1600 mm	–	–	–	–	–	nr	**94.56**
Stainless steel frame bed in screed							
1100 mm × 1400 mm	–	–	–	–	–	nr	**90.92**
Stainless steel frame bed in screed							
1800 mm × 3000 mm	–	–	–	–	–	nr	**174.57**
Polished aluminium matwell; angle rim with lugs brazed on; bedding in screed							
900 mm × 550 mm; constructed with							
25 × 25 × 3 mm angle	26.18	0.93	12.22	–	27.49	nr	**39.71**
1200 mm × 750 mm; constructed with							
34 × 26 × 6 mm angle	43.79	1.00	13.14	–	45.98	nr	**59.12**
2400 mm × 1200 mm; constructed with							
50 × 50 × 6 mm angle	167.75	1.50	19.71	–	176.14	nr	**195.85**
Polished brass matwell; comprising angle rim with lugs brazed on; bedding in screed							
900 mm × 550 mm; constructed with							
25 × 25 × 5 mm angle	66.11	0.93	12.22	–	69.42	nr	**81.64**
1200 mm × 750 mm; constructed with							
38 × 38 × 6 mm angle	179.61	1.00	13.14	–	188.59	nr	**201.73**
2400 mm x1200 mm; constructed with							
38 × 38 × 6 mm angle	331.58	1.50	19.71	–	348.16	nr	**367.87**
KITCHEN FITTINGS							
SUPPLY AND FIX PRICES							
NOTE: Kitchen fittings vary considerably. PC supply prices for reasonable quantities for a moderately priced range of kitchen fittings have been shown.							
Kitchen Units							
Supplying and fixing to backgrounds requiring plugging; including any pre-assembly							
Wall units							
300 mm × 300 mm × 720 mm	100.00	1.11	18.12	–	105.15	nr	**123.27**
500 mm × 300 mm × 720 mm	120.00	1.16	18.93	–	126.15	nr	**145.08**
600 mm × 300 mm × 720 mm	130.00	1.30	21.22	–	136.65	nr	**157.87**
800 mm × 300 mm × 720 mm	200.00	1.48	24.16	–	210.15	nr	**234.31**

32 FURNITURE, FITTINGS AND EQUIPMENT

Item	PC £	Labour hours	Labour £	Plant £	Material £	Unit	**Total rate £**
Floor units with drawers							
500 mm × 600 mm × 870 mm	175.00	1.16	18.93	–	183.90	nr	**202.83**
600 mm × 600 mm × 870 mm	200.00	1.30	21.22	–	210.15	nr	**231.37**
1000 mm × 600 mm × 870 mm	300.00	1.57	25.63	–	315.15	nr	**340.78**
Sink units (excluding sink top)							
1000 mm × 600 mm × 870 mm	275.00	1.48	24.16	–	288.90	nr	**313.06**
Kitchen Worktop							
Laminated plastics worktops; single rolled edge; prices include for fixing							
38 mm thick; 600 mm wide	50.75	0.37	6.04	–	53.34	m	**59.38**
extra for forming hole for inset sink	–	0.69	11.27	–	–	nr	**11.27**
extra for jointing strip at corner intersection of worktops	–	0.14	2.29	–	10.50	nr	**12.79**
extra for butt and scribe joint at corner intersection of worktops	–	4.16	67.90	–	–	nr	**67.90**
SANITARY APPLIANCES AND FITTINGS							
Sinks; Armitage Shanks or equal							
Sinks; white glazed fireclay; BS 6465; pointing all round with silicone sealant							
Belfast sink; 46 cm × 38 cm × 21 cm; pair of Nuastyle 21 basin taps with dual indices, chrome handle; wall mounts 38 mm slotted waste, chain and plug, screw stay; pair of 40.5 cm aluminium alloy build-in brackets with 35.5 cm studs	269.72	2.78	48.10	–	361.06	nr	**409.16**
Belfast sink; 61 cm × 38 cm × 21 cm; pair of Nuastyle 21 basin taps with dual indices, chrome handle; wall mounts; 38 mm slotted waste, chain and plug, screw stay; pair of 40.5 cm aluminium alloy build-in brackets with 35.5 cm studs	321.01	2.78	48.10	–	415.18	nr	**463.28**
Belfast sink; 76 cm × 38 cm × 21 cm; pair of Nuastyle 21 basin taps with dual indices, chrome handle; wall mounts; 38 mm slotted waste, chain and plug, screw stay; pair of 40.5 cm aluminium alloy build-in brackets with 35.5 cm studs	450.84	2.78	48.10	–	551.75	nr	**599.85**

32 FURNITURE, FITTINGS AND EQUIPMENT

Item	PC £	Labour hours	Labour £	Plant £	Material £	Unit	Total rate £
SANITARY APPLIANCES AND FITTINGS – cont							
Lavatory basins; Armitage Shanks or equal							
Basins; white vitreous china; BS 6465 Part 3;							
pointing all round with silcone sealant							
Portman 21 40 cm basin; with overflow, chain hole and two tapholes; pair of Nuastyle 21 basin taps with dual indices; slotted basin waste with plastic plug, chain waste and plug; 32 × 75 mm seal plastic standard bottle trap; pair of Portman concealed brackets with waste support; Isovalve 15 mm plastic servicing valve with outlet for copper	126.65	2.13	36.86	–	183.43	nr	**220.29**
Portman 21 50 cm basin; with overflow, chain hole and two tapholes; pair of Nuastyle 21 basin taps with dual indices; slotted basin waste with plastic plug, chain waste and plug; 32 × 75 mm seal plastic standard bottle trap; pair of Portman concealed brackets with waste support; Isovalve 15 mm plastic servicing valve with outlet for copper	154.60	2.13	36.86	–	212.95	nr	**249.81**
Portman 21 60 cm basin; with overflow, chain hole and two tapholes; pair of Nuastyle 21 basin taps with dual indices; slotted basin waste with plastic plug, chain waste and plug; 32 × 75 mm seal plastic standard bottle trap; pair of Portman concealed brackets with waste support; Isovalve 15 mm plastic servicing valve with outlet for copper	203.38	2.13	36.86	–	264.34	nr	**301.20**
Tiffany 50 cm pedestal basin; with two tapholes; Millenia STD dual control one taphole standard basin mixer with pop-up waste; pair of Millenia STD handles; Full pedestal; Isovalve 15 mm plastic servicing valve with outlet for copper	232.93	2.31	39.96	–	268.66	nr	**308.62**
Tiffany or similar 55 cm pedestal basin; with two tapholes; Millenia STD dual control one taphole standard basin mixer with pop-up waste; pair of Millenia STD handles; Full pedestal; Isovalve 15 mm plastic servicing valve with outlet for copper	228.05	2.31	39.96	–	263.54	nr	**303.50**
Tiffany 60 cm pedestal basin; with two tapholes; Millenia STD dual control one taphole standard basin mixer with pop-up waste; pair of Millenia STD handles; Full pedestal; Isovalve 15 mm plastic servicing valve with outlet for copper	237.85	2.31	39.96	–	276.87	nr	**316.83**
Montana 51 cm pedestal basin; with one taphole; Millenia STD dual control one taphole standard basin mixer with pop-up waste; pair of Millenia STD handles; Full pedestal; Isovalve 15 mm plastic servicing valve with outlet for copper	220.40	2.31	39.96	–	255.42	nr	**295.38**
Montana 58 cm pedestal basin; with one taphole ; Millenia STD dual control one taphole standard basin mixer with pop-up waste; pair of Millenia STD handles; Full pedestal; Isovalve 15 mm plastic servicing valve with outlet for copper	225.55	2.31	39.96	–	260.95	nr	**300.91**

32 FURNITURE, FITTINGS AND EQUIPMENT

Item	PC £	Labour hours	Labour £	Plant £	Material £	Unit	Total rate £
Drinking fountains; Armitage Shanks or equal							
White vitreous china fountains; pointing all round with silicone selant							
Aqualon wall mounted drinking fountain; Aqualon self-closing valve with fittings and plastic waste; 32 × 75 mm seal plastic standard bottle trap	379.28	2.31	39.96	–	409.87	nr	**449.83**
Polished stainless steel fountains; pointing all round with silicone selant							
Purita wall mounted drinking fountain with self-closing valve and fittings; 32 mm unslotted basin strainer waste	300.31	2.31	39.96	–	315.92	nr	**355.88**
Purita pedestal mounted drinking fountain 90 cm high with self-closing valve and fittings; 32 mm unslotted basin strainer waste	773.45	2.78	48.10	–	814.04	nr	**862.14**
Baths; Armitage Shanks or equal							
Pointing all round with silicone selant							
Sandringham acrylic rectangular bath with chrome plated grips and two tapholes; standard pair of standard bath taps with chrome handles; bath chain waste with plastic plug and overflow; cast brass P trap with plain outlet and overflow connection; 170 cm long × 70 cm wide; white or coloured	187.84	3.50	60.55	–	197.23	nr	**257.78**
Nisa lowline heavy gauge steel rectangular bath with chrome plated grips and two tapholes; standard pair of standard bath taps with chrome handles; bath chain waste with plastic plug and overflow; cast brass P trap with plain outlet and overflow connection; 170 cm long × 70 cm wide; white or coloured	476.85	3.50	60.55	–	500.69	nr	**561.24**
Water closets; Armitage Shanks or equal							
White vitreous china pans and cisterns; pointing all round base with silicone sealant							
Wentworth close coupled washdown closet pan with horizontal outlet; Orion 3 plastic toilet seat and cover; Panketa pan connector 14° finned; Universal close coupled bottom inlet cistern with syphon	216.27	3.05	52.77	–	236.65	nr	**289.42**
Tiffany back to wall washdown closet pan with horizontal outlet; Saturn plastic toilet seat and cover; Panketa pan connector 14° finned; Conceala 2 6 litre low level side inlet cistern with syphon and lever	284.57	3.05	52.77	–	308.36	nr	**361.13**
Extra for; Panketa pan connector 90° finned	–	–	–	–	2.24	nr	**2.24**

32 FURNITURE, FITTINGS AND EQUIPMENT

Item	PC £	Labour hours	Labour £	Plant £	Material £	Unit	Total rate £
SANITARY APPLIANCES AND FITTINGS – cont							
Water closets – cont							
White vitreous china pans and cisterns – cont							
Tiffany close coupled washdown closet pan with horizontal outlet; Saturn plastic toilet seat and cover; Panketa pan connector 14° finned; Tiffany 7½ litre close coupled cistern with dual flush valve	293.97	3.05	52.77	–	318.23	nr	**371.00**
Extra for; Panketa pan connector 90° finned	–	–	–	–	2.24	nr	**2.24**
Cameo close coupled washdown closet pan with horizontal outlet; Accolade/Cameo plastic toilet seat and cover; Panketa pan connector 14° finned; Cameo 6 litre close coupled cistern with dual flush valve	366.61	3.05	52.77	–	394.51	nr	**447.28**
Extr for; Panketa pan connector 90° finned	–	–	–	–	2.24	nr	**2.24**
Wall urinals; Armitage Shanks or equal							
White vitreous china bowls and cisterns; pointing all round with silicone sealant							
Single Sanura 40 cm urinal bowl; Sanura top inlet spreader; pair of wall hangers for urinal bowl; 38 mm plastic domed waste; 38 × 75 mm seal plastic standard bottle trap; Conceala 4½ litres capacity auto cistern and cover; Sanura concealed flushpipe for single urinal bowl; screwing	254.49	3.70	64.02	–	296.57	nr	**360.59**
Single Sanura 40 cm urinal bowl; Sanura top inlet spreader; pair of wall hangers for urinal bowl; 38 mm plastic domed waste; 38 × 75 mm seal plastic standard bottle trap; Mura 4½ litres capacity auto cistern and cover; Sanura/Mura exposed flushpipe for single urinal bowl	289.94	3.70	64.02	–	333.79	nr	**397.81**
Single Sanura 50 cm urinal bowl; Sanura top inlet spreader; pair of wall hangers for urinal bowl; 38 mm plastic domed waste; 38 × 75 mm seal plastic standard bottle trap; Conceala 4½ litres capacity auto cistern and cover; Sanura concealed flushpipe for single urinal bowl	347.50	3.70	64.02	–	394.36	nr	**458.38**

32 FURNITURE, FITTINGS AND EQUIPMENT

Item	PC £	Labour hours	Labour £	Plant £	Material £	Unit	Total rate £
Single Sanura 50 cm urinal bowl; Sanura top inlet spreader; pair of wall hangers for urinal bowl; 38 mm plastic domed waste; 38 × 75 mm seal plastic standard bottle trap; Mura 4½ litres capacity auto cistern and cover; Sanura/Mura exposed flushpipe for single urinal bowl	382.95	3.70	64.02	–	431.58	nr	**495.60**
Range of 2 nr Sanura 40 cm urinal bowls; Sanura top inlet spreader; pairs of wall hangers for urinal bowls; 38 mm plastic domed wastes; 38 × 75 mm seal plastic standard bottle traps; Conceala 9 litres capacity auto cistern and cover; Sanura concealed flushpipe for range of 2 nr urinal bowls	428.47	6.95	120.25	–	508.61	nr	**628.86**
Range of 2 nr Sanura 50 cm urinal bowls; Sanura top inlet spreader; pairs of wall hangers for urinal bowls; 38 mm plastic domed wastes; 38 × 75 mm seal plastic standard bottle traps; Conceala 9 litres capacity auto cistern and cover; Sanura concealed flushpipe for range of 2 nr urinal bowls	614.49	6.95	120.25	–	704.18	nr	**824.43**
Range of 3 nr Sanura 40 cm urinal bowls; Sanura top inlet spreader; pairs of wall hangers for urinal bowls; 38 mm plastic domed wastes; 38 × 75 mm seal plastic standard bottle traps; Conceala 9 litres capacity auto cistern and cover; Sanura concealed flushpipe for range of 3 nr urinal bowls	596.52	10.15	175.60	–	714.41	nr	**890.01**
Range of 3 nr Sanura 50 cm urinal bowls; Sanura top inlet spreader; pairs of wall hangers for urinal bowls; 38 mm plastic domed wastes; 38 × 75 mm seal plastic standard bottle traps; Conceala 9 litres capacity auto cistern and cover; Sanura concealed flushpipe for range of 3 nr urinal bowls	875.55	10.15	175.60	–	1007.39	nr	**1182.99**
Range of 4 nr Sanura 40 cm urinal bowls; Sanura top inlet spreader; pairs of wall hangers for urinal bowls; 38 mm plastic domed wastes; 38 × 75 mm seal plastic standard bottle traps; Conceala 9 litres capacity auto cistern and cover; Sanura concealed flushpipe for range of 4 nr urinal bowls; screwing	770.70	13.40	231.83	–	926.66	nr	**1158.49**

32 FURNITURE, FITTINGS AND EQUIPMENT

Item	PC £	Labour hours	Labour £	Plant £	Material £	Unit	Total rate £
SANITARY APPLIANCES AND FITTINGS – cont							
Wall urinals – cont							
White vitreous china bowls and cisterns – cont							
Range of 4 nr Sanura 50 cm urinal bowls; Sanura top inlet spreader; pairs of wall hangers for urinal bowls; 38 mm plastic domed wastes; 38 × 75 mm seal plastic standard bottle traps; Conceala 9 litres capacity auto cistern and cover; Sanura concealed flushpipe for range of 4 nr urinal bowls	1142.74	13.40	231.83	–	1317.81	nr	**1549.64**
Range of 5 nr Sanura 40 cm urinal bowls; Sanura top inlet spreader; pairs of wall hangers for urinal bowls; 38 mm plastic domed wastes; 38 × 75 mm seal plastic standard bottle traps; Conceala 9 litres capacity auto cistern and cover; Sanura concealed flushpipe for range of 5 nr urinal bowls	938.82	16.65	288.07	–	1132.54	nr	**1420.61**
Range of 5 nr Sanura 50 cm urinal bowls; Sanura top inlet spreader; pairs of wall hangers for urinal bowls; 38 mm plastic domed wastes; 38 × 75 mm seal plastic standard bottle traps; Conceala 9 litres capacity auto cistern and cover; Sanura concealed flushpipe for range of 5 nr urinal bowls	1403.87	16.65	288.07	–	1621.48	nr	**1909.55**
White vitreous china division panels; pointing all round with silicone sealant							
Urinal division with screw and hanger	79.77	0.70	12.11	–	85.17	nr	**97.28**
Bidets; Armitage Shanks or equal							
Tiffany back to wall bidet with one taphole; vitreous china; chromium plated pop-up waste and mixer tap with hand wheels; 58 cm × 39 cm; white or coloured	423.38	3.50	60.55	–	444.55	nr	**505.10**
Shower tray and fittings							
Simplicity shower tray; acrylic; with outlet and grated waste; chain and plug; bedding and pointing in waterproof cement mortar							
760 mm × 760 mm; white or coloured	67.18	3.00	51.90	–	70.54	nr	**122.44**
Shower fitting; riser pipe with mixing valve and shower rose; chromium plated; plugging and screwing mixing valve and pipe bracket							
15 mm dia. riser pipe; 127 mm dia. shower rose	361.51	5.00	86.51	–	379.59	nr	**466.10**
Corner fitting shower enclosure							
Bliss flat top hinged door with front panel and clear glass side panel	711.74	3.00	39.42	–	747.33	nr	**786.75**
Miscellaneous fittings; Magrini Ltd or equal							
Vertical nappy changing unit							
ref KBCS; screwing	–	0.60	9.80	–	215.25	nr	**225.05**

32 FURNITURE, FITTINGS AND EQUIPMENT

Item	PC £	Labour hours	Labour £	Plant £	Material £	Unit	Total rate £
Horizontal nappy changing unit							
ref KBHS; screwing	–	0.60	9.80	–	215.25	nr	**225.05**
Stay Safe baby seat							
ref KBPS; screwing	–	0.55	8.98	–	74.26	nr	**83.24**
Miscellaneous fittings; Pressalit Ltd or equal							
Grab rails							
300 mm long ref RT100000; screw fix to wall	–	0.50	8.16	–	45.85	nr	**54.01**
450 mm long ref RT101000; screw fix to wall	–	0.50	8.16	–	53.01	nr	**61.17**
600 mm long ref RT102000; screw fix to wall	–	0.50	8.16	–	60.81	nr	**68.97**
800 mm long ref RT103000; screw fix to wall	–	0.50	8.16	–	68.38	nr	**76.54**
1000 mm long ref RT104000; screw fix to wall	–	0.50	8.16	–	78.83	nr	**86.99**
Angled grab rails							
900 mm long, angled 135° ref RT110000; screw fix to wall	–	0.50	8.16	–	99.10	nr	**107.26**
1300 mm long, angled 90° ref RT119000; screw fix to wall	–	0.75	12.24	–	155.21	nr	**167.45**
Hinged grab rails							
600 mm long ref R3016000; screw fix to wall	–	0.35	5.71	–	160.95	nr	**166.66**
600 mm long with spring counter balance ref RF016000; screw fix to wall	–	0.35	5.71	–	224.62	nr	**230.33**
850 mm long ref R3010000; screw fix to wall	–	0.35	5.71	–	195.45	nr	**201.16**
850 mm long with spring counter balance ref RF010000; screw fix to wall	–	0.35	5.71	–	241.08	nr	**246.79**
Shower seat; wall mounted; with padded seat and back	–	1.50	24.49	–	238.56	nr	**263.05**
NOTICES AND SIGNS							
Plain script; in gloss oil paint; on painted or varnished surfaces							
Capital letters; lower case letters or numerals							
per coat; per 25 mm high	–	0.09	1.47	–	–	nr	**1.47**

33 DRAINAGE ABOVE GROUND

Item	PC £	Labour hours	Labour £	Plant £	Material £	Unit	Total rate £
RAINWATER INSTALLATIONS							
Aluminium							
Pipes and fittings; BS EN 612; ears cast on;							
polyester powder coated finish							
63 mm dia. pipes; plugged and screwed	19.04	0.34	4.97	–	21.37	m	**26.34**
extra for							
fittings with one end	–	0.20	2.92	–	11.05	nr	**13.97**
fittings with two ends	–	0.39	5.70	–	11.17	nr	**16.87**
fittings with three ends	–	0.56	8.18	–	17.35	nr	**25.53**
shoe	11.38	0.20	2.92	–	12.64	nr	**15.56**
bend	12.13	0.39	5.70	–	12.78	nr	**18.48**
single branch	15.82	0.56	8.18	–	17.35	nr	**25.53**
offset 228 projection	27.98	0.39	7.88	–	30.08	nr	**37.96**
offset 304 projection	31.20	0.39	5.70	–	33.45	nr	**39.15**
access pipe	34.72	0.39	5.70	–	36.48	nr	**42.18**
connection to clay pipes; cement and sand							
(1:2) joint	–	0.14	2.05	–	0.13	nr	**2.18**
76.50 mm dia. pipes; plugged and screwed	22.18	0.37	5.41	–	24.66	m	**30.07**
extra for							
shoe	15.61	0.23	3.36	–	17.09	nr	**20.45**
bend	15.33	0.42	6.13	–	16.14	nr	**22.27**
single branch	19.05	0.60	8.77	–	20.75	nr	**29.52**
offset 228 projection	30.94	0.42	6.13	–	33.19	nr	**39.32**
offset 304 projection	34.22	0.42	6.13	–	36.63	nr	**42.76**
access pipe	47.54	0.42	6.13	–	49.95	nr	**56.08**
connection to clay pipes; cement and sand							
(1:2) joint	–	0.16	2.34	–	0.13	nr	**2.47**
100 mm dia. pipes; plugged and screwed	37.85	0.42	6.13	–	41.13	m	**47.26**
extra for							
shoe	18.81	0.26	3.80	–	20.46	nr	**24.26**
bend	21.35	0.46	6.72	–	22.48	nr	**29.20**
single branch	25.51	0.69	10.08	–	27.56	nr	**37.64**
offset 228 projection	35.80	0.46	6.72	–	37.62	nr	**44.34**
offset 304 projection	39.74	0.46	6.72	–	41.76	nr	**48.48**
access pipe	44.29	0.46	6.72	–	46.54	nr	**53.26**
connection to clay pipes; cement and sand							
(1:2) joint	–	0.19	2.77	–	0.13	nr	**2.90**
Roof outlets; circular aluminium; with flat or domed							
grating; joint to pipe							
50 mm dia.	68.92	0.56	12.82	–	72.37	nr	**85.19**
75 mm dia.	90.32	0.60	13.73	–	94.84	nr	**108.57**
100 mm dia.	117.62	0.65	14.88	–	123.50	nr	**138.38**
150 mm dia.	150.69	0.69	15.79	–	158.22	nr	**174.01**
Roof outlets; d-shaped; balcony; with flat or domed							
grating; joint to pipe							
50 mm dia.	68.92	0.56	12.82	–	72.37	nr	**85.19**
75 mm dia.	91.46	0.60	13.73	–	96.03	nr	**109.76**
100 mm dia.	128.29	0.65	14.88	–	134.70	nr	**149.58**

33 DRAINAGE ABOVE GROUND

Item	PC £	Labour hours	Labour £	Plant £	Material £	Unit	Total rate £
PVC balloon grating							
110 mm dia.	4.97	0.06	1.38	–	5.22	nr	**6.60**
63 mm dia.	3.11	0.06	1.38	–	3.27	nr	**4.65**
Aluminium gutters and fittings; BS EN 612; polyester powder coated finish							
100 mm half round gutters; on brackets; screwed to timber	20.72	0.32	6.46	–	27.23	m	**33.69**
extra for							
stop end	5.48	0.15	3.03	–	10.92	nr	**13.95**
running outlet	12.14	0.31	6.26	–	14.16	nr	**20.42**
stop end outlet	10.82	0.15	3.03	–	16.52	nr	**19.55**
angle	1.11	0.31	6.26	–	2.60	nr	**8.86**
113 mm half round gutters; on brackets; screwed to timber	21.70	0.32	6.46	–	28.31	m	**34.77**
extra for							
stop end	5.77	0.15	3.03	–	11.26	nr	**14.29**
running outlet	13.25	0.31	6.26	–	15.33	nr	**21.59**
stop end outlet	12.39	0.15	3.03	–	18.18	nr	**21.21**
angle	12.65	0.31	6.26	–	10.84	nr	**17.10**
125 mm half round gutters; on brackets; screwed to timber	24.37	0.37	7.48	–	33.99	m	**41.47**
extra for							
stop end	7.03	0.17	3.43	–	15.20	nr	**18.63**
running outlet	14.33	0.32	6.46	–	16.47	nr	**22.93**
stop end outlet	13.15	0.17	3.43	–	21.63	nr	**25.06**
angle	14.03	0.32	6.46	–	16.16	nr	**22.62**
100 mm ogee gutters; on brackets; screwed to timber	25.88	0.34	6.87	–	35.02	m	**41.89**
extra for							
stop end	5.79	0.16	3.23	–	7.23	nr	**10.46**
running outlet	14.27	0.32	6.46	–	15.47	nr	**21.93**
stop end outlet	11.05	0.16	3.23	–	18.85	nr	**22.08**
angle	12.03	0.32	6.46	–	8.52	nr	**14.98**
112 mm ogee gutters; on brackets; screwed to timber	28.78	0.39	7.88	–	38.54	m	**46.42**
extra for							
stop end	6.19	0.16	3.23	–	7.66	nr	**10.89**
running outlet	14.43	0.32	6.46	–	15.63	nr	**22.09**
stop end outlet	12.38	0.16	3.23	–	20.66	nr	**23.89**
angle	14.33	0.32	6.46	–	15.55	nr	**22.01**
125 mm ogee gutters; on brackets; screwed to timber	31.78	0.39	7.88	–	42.36	m	**50.24**
extra for							
stop end	6.77	0.18	3.63	–	8.27	nr	**11.90**
running outlet	15.77	0.34	6.87	–	17.05	nr	**23.92**
stop end outlet	14.06	0.18	3.63	–	23.04	nr	**26.67**
angle	16.70	0.34	6.87	–	12.37	nr	**19.24**

33 DRAINAGE ABOVE GROUND

Item	PC £	Labour hours	Labour £	Plant £	Material £	Unit	Total rate £
RAINWATER INSTALLATIONS – cont							
Cast Iron							
Pipes and fittings; BS 416; ears cast on; joints							
65 mm pipes; primed; nailed to masonry	40.22	0.48	7.01	–	42.70	m	**49.71**
extra for							
shoe	34.46	0.30	4.38	–	34.24	nr	**38.62**
bend	21.09	0.53	7.74	–	20.21	nr	**27.95**
single branch	41.45	0.67	9.79	–	41.14	nr	**50.93**
offset 225 mm projection	37.59	0.53	7.74	–	35.00	nr	**42.74**
offset 305 mm projection	44.03	0.53	7.74	–	40.92	nr	**48.66**
connection to clay pipes; cement and sand (1:2) joint	–	0.14	2.05	–	0.15	nr	**2.20**
75 mm pipes; primed; nailed to masonry	40.22	0.51	7.45	–	42.97	m	**50.42**
extra for							
shoe	34.46	0.32	4.67	–	34.69	nr	**39.36**
bend	25.60	1.11	19.57	–	25.39	nr	**44.96**
single branch	45.71	0.69	10.08	–	46.50	nr	**56.58**
offset 225 mm projection	37.59	0.56	8.18	–	35.45	nr	**43.63**
offset 305 mm projection	46.20	0.56	8.18	–	43.64	nr	**51.82**
connection to clay pipes; cement and sand (1:2) joint	–	0.16	2.34	–	0.15	nr	**2.49**
100 mm pipes; primed; nailed to masonry	54.00	0.56	8.18	–	57.73	m	**65.91**
extra for							
shoe	45.74	0.37	5.41	–	46.21	nr	**51.62**
bend	36.15	0.60	8.77	–	36.14	nr	**44.91**
single branch	53.26	0.74	10.82	–	54.35	nr	**65.17**
offset 225 mm projection	73.75	0.60	8.77	–	72.22	nr	**80.99**
offset 305 mm projection	75.21	0.60	8.77	–	72.62	nr	**81.39**
connection to clay pipes; cement and sand (1:2) joint	–	0.19	2.77	–	0.13	nr	**2.90**
100 mm × 75 mm rectangular pipes; primed; nailing to masonry	108.61	0.56	8.18	–	115.07	m	**123.25**
extra for							
shoe	129.05	0.37	5.41	–	129.10	nr	**134.51**
bend	122.89	0.60	8.77	–	122.62	nr	**131.39**
offset 225 mm projection	173.04	0.37	5.41	–	168.45	nr	**173.86**
offset 305 mm projection	184.94	0.37	5.41	–	178.66	nr	**184.07**
connection to clay pipes; cement and sand (1:2) joint	–	0.19	2.77	–	0.13	nr	**2.90**
Rainwater head; rectangular; for pipes							
65 mm dia.	105.89	0.53	7.74	–	112.62	nr	**120.36**
75 mm dia.	105.89	0.56	8.18	–	113.07	nr	**121.25**
100 mm dia.	146.21	0.60	8.77	–	156.24	nr	**165.01**
Rainwater head; octagonal; for pipes							
65 mm dia.	76.16	0.53	7.74	–	81.42	nr	**89.16**
75 mm dia.	76.16	0.56	8.18	–	81.86	nr	**90.04**
100 mm dia.	90.28	0.60	8.77	–	97.51	nr	**106.28**
Gutters and fittings; BS EN 877							
100 mm half round gutters; primed; on brackets; screwed to timber	19.95	0.37	7.48	–	26.72	m	**34.20**

33 DRAINAGE ABOVE GROUND

Item	PC £	Labour hours	Labour £	Plant £	Material £	Unit	Total rate £
extra for							
stop end	4.77	0.16	3.23	–	7.94	nr	**11.17**
running outlet	13.90	0.32	6.46	–	13.59	nr	**20.05**
angle	14.26	0.32	6.46	–	16.55	nr	**23.01**
115 mm half round gutters; primed; on brackets;							
screwed to timber	20.79	0.37	7.48	–	27.71	m	**35.19**
extra for							
stop end	6.20	0.16	3.23	–	9.44	nr	**12.67**
running outlet	15.14	0.32	6.46	–	14.88	nr	**21.34**
angle	14.66	0.32	6.46	–	16.88	nr	**23.34**
125 mm half round gutters; primed; on brackets;							
screwed to timber	24.34	0.42	8.48	–	31.50	m	**39.98**
extra for							
stop end	6.20	0.19	3.83	–	9.57	nr	**13.40**
running outlet	17.29	0.37	7.48	–	16.91	nr	**24.39**
angle	17.29	0.37	7.48	–	19.07	nr	**26.55**
150 mm half round gutters; primed; on brackets;							
screwed to timber	39.60	0.46	9.29	–	48.92	m	**58.21**
extra for							
stop end	8.62	0.20	4.04	–	15.55	nr	**19.59**
running outlet	29.94	0.42	8.48	–	28.98	nr	**37.46**
angle	31.60	0.42	8.48	–	32.61	nr	**41.09**
100 mm ogee gutters; primed; on brackets;							
screwed to timber	22.24	0.39	7.88	–	29.27	m	**37.15**
extra for							
stop end	4.90	0.17	3.43	–	10.77	nr	**14.20**
running outlet	15.15	0.34	6.87	–	14.84	nr	**21.71**
angle	14.87	0.34	6.87	–	17.27	nr	**24.14**
115 mm ogee gutters; primed; on brackets;							
screwed to timber	24.48	0.39	7.88	–	31.67	m	**39.55**
extra for							
stop end	6.34	0.17	3.43	–	12.35	nr	**15.78**
running outlet	16.13	0.34	6.87	–	15.68	nr	**22.55**
angle	16.13	0.34	6.87	–	18.20	nr	**25.07**
125 mm ogee gutters; primed; on brackets;							
screwed to timber	25.67	0.43	8.68	–	33.63	m	**42.31**
extra for							
stop end	6.34	0.19	3.83	–	13.05	nr	**16.88**
running outlet	17.59	0.39	7.88	–	17.24	nr	**25.12**
angle	17.59	0.39	7.88	–	20.29	nr	**28.17**
Steel							
3 mm thick zincalume coated pressed steel gutters and fittings; joggle joints; including bracket and stiffeners							
200 mm × 100 mm (400 mm girth) box gutter;							
screwed to timber	–	0.60	8.77	–	14.01	m	**22.78**
extra for							
stop end	–	0.32	4.67	–	4.45	nr	**9.12**
running outlet	–	0.65	9.49	–	25.07	nr	**34.56**
stop end outlet	–	0.32	4.67	–	31.82	nr	**36.49**

33 DRAINAGE ABOVE GROUND

Item	PC £	Labour hours	Labour £	Plant £	Material £	Unit	Total rate £
RAINWATER INSTALLATIONS – cont							
Steel – cont							
3 mm thick zincalume coated pressed steel gutters and fittings – cont							
angle	–	0.65	9.49	–	27.35	nr	**36.84**
381 mm boundary wall gutters (900 mm girth); bent twice; screwed to timber	–	0.60	8.77	–	15.67	m	**24.44**
extra for							
stop end	–	0.37	5.41	–	4.94	nr	**10.35**
running outlet	–	0.65	9.49	–	32.78	nr	**42.27**
stop end outlet	–	0.32	4.67	–	59.44	nr	**64.11**
angle	–	0.65	9.49	–	37.99	nr	**47.48**
457 mm boundary wall gutters (1200 mm girth); bent twice; screwed to timber	–	0.69	10.08	–	64.88	m	**74.96**
extra for							
stop end	–	0.37	5.41	–	22.80	nr	**28.21**
running outlet	–	0.74	10.82	–	46.68	nr	**57.50**
stop end outlet	–	0.37	5.41	–	47.41	nr	**52.82**
angle	–	0.74	10.82	–	50.18	nr	**61.00**
750 mm girth; valley gutters; Kalzip Membrane lined composite gutter system	–	–	–	–	–	m	**93.48**
extra for							
stop end	–	–	–	–	–	nr	**70.11**
running outlet	–	–	–	–	–	nr	**88.80**
uPVC							
External rainwater pipes and fittings; BS EN 12200; slip-in joints							
50 mm pipes; fixing with pipe or socket brackets; plugged and screwed	8.95	0.28	4.10	–	12.60	m	**16.70**
extra for							
shoe	5.25	0.19	2.77	–	6.77	nr	**9.54**
bend	6.16	0.28	4.10	–	7.73	nr	**11.83**
two bends to form offset 229 mm projection	12.32	0.28	4.10	–	12.59	nr	**16.69**
connection to clay pipes; cement and sand (1:2) joint	–	0.12	1.75	–	0.15	nr	**1.90**
68 mm pipes; fixing with pipe or socket brackets; plugged and screwed	6.92	0.31	4.53	–	11.11	m	**15.64**
extra for							
shoe	5.25	0.20	2.92	–	7.35	nr	**10.27**
bend	6.74	0.31	4.53	–	8.91	nr	**13.44**
single branch	13.55	0.41	5.99	–	16.07	nr	**22.06**
two bends to form offset 229 mm projection	13.48	0.31	4.53	–	14.76	nr	**19.29**
loose drain connector; cement and sand (1:2) joint	–	0.14	2.05	–	16.65	nr	**18.70**
110 mm pipes; fixing with pipe or socket brackets; plugged and screwed	14.99	0.33	4.82	–	23.97	m	**28.79**

33 DRAINAGE ABOVE GROUND

Item	PC £	Labour hours	Labour £	Plant £	Material £	Unit	**Total rate £**
extra for							
shoe	17.29	0.22	3.21	–	20.39	nr	**23.60**
bend	25.53	0.33	4.82	–	29.04	nr	**33.86**
single branch	38.56	0.44	6.43	–	42.72	nr	**49.15**
two bends to form offset 229 mm projection	51.07	0.33	4.82	–	53.66	nr	**58.48**
loose drain connector; cement and sand (1:2) joint	–	0.32	4.67	–	17.20	nr	**21.87**
65 mm square pipes; fixing with pipe or socket brackets; plugged and screwed	6.02	0.31	4.53	–	9.53	m	**14.06**
extra for							
shoe	4.40	0.20	2.92	–	6.16	nr	**9.08**
bend	6.74	0.31	4.53	–	8.62	nr	**13.15**
single branch	13.55	0.41	5.99	–	15.76	nr	**21.75**
two bends to form offset 229 mm projection	13.48	0.31	4.53	–	14.94	nr	**19.47**
drain connector; square to round; cement and sand (1:2) joint	–	0.32	4.67	–	7.08	nr	**11.75**
Rainwater head; rectangular; for pipes							
50 mm dia.	28.63	0.42	6.13	–	32.62	nr	**38.75**
68 mm dia.	23.12	0.43	6.28	–	28.00	nr	**34.28**
110 mm dia.	49.53	0.51	7.45	–	56.52	nr	**63.97**
65 mm square	19.34	0.43	6.28	–	23.44	nr	**29.72**
Gutters and fittings; BS EN 12200							
76 mm half round gutters; on brackets screwed to timber	6.80	0.28	5.66	–	10.35	m	**16.01**
extra for							
stop end	2.42	0.12	2.43	–	3.26	nr	**5.69**
running outlet	6.80	0.23	4.64	–	6.56	nr	**11.20**
stop end outlet	6.79	0.12	2.43	–	7.27	nr	**9.70**
angle	6.80	0.23	4.64	–	7.96	nr	**12.60**
112 mm half round gutters; on brackets screwed to timber	6.88	0.31	6.26	–	12.18	m	**18.44**
extra for							
stop end	4.00	0.12	2.43	–	5.44	nr	**7.87**
running outlet	7.83	0.26	5.25	–	7.64	nr	**12.89**
stop end outlet	7.83	0.12	2.43	–	8.88	nr	**11.31**
angle	8.75	0.26	5.25	–	11.08	nr	**16.33**
170 mm half round gutters; on brackets; screwed to timber	14.95	0.31	6.26	–	23.71	m	**29.97**
extra for							
stop end	6.53	0.15	3.03	–	9.24	nr	**12.27**
running outlet	14.61	0.29	5.86	–	14.09	nr	**19.95**
stop end outlet	13.89	0.15	3.03	–	15.72	nr	**18.75**
angle	15.52	0.29	5.86	–	19.82	nr	**25.68**
114 mm rectangular gutters; on brackets; screwed to timber	7.08	0.31	6.26	–	13.64	m	**19.90**
extra for							
stop end	4.00	0.12	2.43	–	5.44	nr	**7.87**
running outlet	7.83	0.29	5.86	–	7.62	nr	**13.48**
stop end outlet	7.83	0.12	2.43	–	8.86	nr	**11.29**
angle	8.75	0.26	5.25	–	11.07	nr	**16.32**

33 DRAINAGE ABOVE GROUND

Item	PC £	Labour hours	Labour £	Plant £	Material £	Unit	Total rate £
RAINWATER INSTALLATIONS – cont							
Kingspan Insulated Gutter							
Pre-laminated insulated membrane gutters							
Eaves; 1250 mm girth	–	–	–	–	–	m	**103.19**
Valley; 1250 mm girth	–	–	–	–	–	m	**103.19**
Weir overflows	–	–	–	–	–	nr	**171.99**
Stop ends	–	–	–	–	–	nr	**212.12**
T-sections	–	–	–	–	–	nr	**343.98**
FOUL DRAINAGE INSTALLATIONS							
Cast Iron							
Cast iron Timesaver pipes and fittings or other equal and approved; BS 416							
50 mm pipes; primed; 3 m lengths; fixing with expanding bolts; to masonry	21.24	0.51	7.45	–	31.91	m	**39.36**
extra for							
fittings with two ends	–	0.51	7.47	–	27.79	nr	**35.26**
fittings with three ends	–	0.69	10.08	–	47.07	nr	**57.15**
bends; short radius	18.91	0.51	7.45	–	27.79	nr	**35.24**
access bends; short radius	46.59	0.51	7.45	–	56.86	nr	**64.31**
boss; 38 BSP	39.14	0.51	7.45	–	48.47	nr	**55.92**
single branch	28.44	0.69	10.08	–	48.07	nr	**58.15**
isolated Timesaver coupling joint	10.74	0.28	4.10	–	11.28	nr	**15.38**
connection to clay pipes; cement and sand (1:2) joint	–	0.12	1.75	–	0.13	nr	**1.88**
75 mm pipes; primed; 3 m lengths; fixing with standard brackets; plugged and screwed to masonry	23.75	0.51	7.45	–	39.47	m	**46.92**
extra for							
bends; short radius	21.39	0.55	8.03	–	31.16	nr	**39.19**
access bends; short radius	50.52	0.51	7.45	–	61.75	nr	**69.20**
boss; 38 BSP	39.14	0.55	8.03	–	49.79	nr	**57.82**
single branch	32.20	0.79	11.54	–	53.57	nr	**65.11**
double branch	47.82	1.02	14.90	–	82.41	nr	**97.31**
offset 115 mm projection	30.67	0.55	8.03	–	38.04	nr	**46.07**
offset 150 mm projection	36.05	0.55	8.03	–	42.94	nr	**50.97**
access pipe	45.49	0.55	8.03	–	53.21	nr	**61.24**
isolated Timesaver coupling joint	11.85	0.32	4.67	–	12.44	nr	**17.11**
connection to clay pipes; cement and sand (1:2) joint	–	0.14	2.05	–	0.13	nr	**2.18**
100 mm pipes; primed; 3 m lengths; fixing with standard brackets; plugged and screwed to masonry	28.71	0.55	8.03	–	54.91	m	**62.94**
extra for							
WC bent connector; 450 mm long tail	41.90	0.55	8.03	–	49.54	nr	**57.57**
bends; short radius	26.16	0.62	9.06	–	39.20	nr	**48.26**
access bends; short radius	55.37	0.62	9.06	–	69.86	nr	**78.92**

33 DRAINAGE ABOVE GROUND

Item	PC £	Labour hours	Labour £	Plant £	Material £	Unit	Total rate £
boss; 38 BSP	46.76	0.62	9.06	–	60.83	nr	**69.89**
single branch	40.45	0.93	13.59	–	67.88	nr	**81.47**
double branch	50.03	1.20	17.52	–	94.20	nr	**111.72**
offset 225 mm projection	39.38	0.62	9.06	–	49.16	nr	**58.22**
offset 300 mm projection	42.39	0.62	9.06	–	51.41	nr	**60.47**
access pipe	47.82	0.62	9.06	–	57.41	nr	**66.47**
roof connector; for asphalt	45.19	0.62	9.06	–	58.12	nr	**67.18**
isolated Timesaver coupling joint	15.47	0.39	5.70	–	16.24	nr	**21.94**
transitional clayware socket; cement and sand (1:2) joint	33.55	0.37	5.41	–	51.60	nr	**57.01**
150 mm pipes; primed; 3 m lengths; fixing with standard brackets; plugged and screwed to masonry	59.94	0.69	10.08	–	111.11	m	**121.19**
extra for							
bends; short radius	46.76	0.77	11.25	–	72.08	nr	**83.33**
access bends; short radius	78.60	0.77	11.25	–	105.51	nr	**116.76**
boss; 38 BSP	76.28	0.77	11.25	–	101.50	nr	**112.75**
single branch	100.28	1.11	16.21	–	153.14	nr	**169.35**
double branch	140.90	1.48	21.62	–	225.40	nr	**247.02**
access pipe	79.53	0.77	11.25	–	93.65	nr	**104.90**
isolated Timesaver coupling joint	–	0.46	6.72	–	32.42	nr	**39.14**
transitional clayware socket; cement and sand (1:2) joint	58.77	0.48	7.01	–	94.26	nr	**101.27**
Cast iron Ensign lightweight pipes and fittings or other equal and approved; BS EN 877							
50 mm pipes; primed; 3 m lengths; fixing with standard brackets; plugged and screwed to masonry	14.65	0.31	4.78	–	22.80	m	**27.58**
extra for							
bends; short radius	11.40	0.27	4.17	–	18.90	nr	**23.07**
single branch	18.29	0.33	5.08	–	33.06	nr	**38.14**
access pipe	30.37	0.27	4.04	–	38.82	nr	**42.86**
70 mm pipes; primed; 3 m lengths; fixing with standard brackets; plugged and screwed to masonry	16.95	0.34	5.26	–	25.43	m	**30.69**
extra for							
bends; short radius	12.82	0.30	4.61	–	21.08	nr	**25.69**
single branch	19.30	0.37	5.69	–	35.50	nr	**41.19**
access pipe	32.12	0.30	4.61	–	41.35	nr	**45.96**
100 mm pipes; primed; 3 m lengths; fixing with standard brackets; plugged and screwed to masonry	20.17	0.37	5.69	–	30.24	m	**35.93**
extra for							
bends; short radius	15.19	0.32	4.96	–	25.86	nr	**30.82**
single branch	26.47	0.39	6.00	–	47.62	nr	**53.62**
double branch	35.38	0.46	7.09	–	66.89	nr	**73.98**
access pipe	35.31	0.32	4.96	–	46.99	nr	**51.95**
connector	32.15	0.21	3.21	–	43.67	nr	**46.88**
reducer	20.62	0.32	4.96	–	31.56	nr	**36.52**

33 DRAINAGE ABOVE GROUND

Item	PC £	Labour hours	Labour £	Plant £	Material £	Unit	Total rate £
FOUL DRAINAGE INSTALLATIONS – cont							
uPVC							
muPVC waste pipes and fittings; BS EN 1329;							
solvent welded joints							
32 mm pipes; fixing with pipe clips; plugged and							
screwed	1.60	0.23	3.36	–	2.47	m	**5.83**
extra for							
fittings with one end	–	0.16	2.34	–	1.26	nr	**3.60**
fittings with two ends	–	0.23	3.36	–	1.35	nr	**4.71**
fittings with three ends	–	0.31	4.53	–	1.78	nr	**6.31**
access plug	0.90	0.16	2.34	–	1.26	nr	**3.60**
straight coupling	0.97	0.16	2.34	–	1.33	nr	**3.67**
expansion coupling	1.72	0.23	3.36	–	2.11	nr	**5.47**
male iron to muPVC coupling	1.73	0.35	5.11	–	1.97	nr	**7.08**
sweep bend	0.99	0.23	3.36	–	1.35	nr	**4.71**
spigot/socket bend	–	0.23	3.36	–	2.02	nr	**5.38**
sweep tee	1.33	0.31	4.53	–	1.78	nr	**6.31**
40 mm pipes; fixing with pipe clips; plugged and							
screwed	1.98	0.28	4.10	–	2.90	m	**7.00**
extra for							
fittings with one end	–	0.18	2.63	–	1.26	nr	**3.89**
fittings with two ends	–	0.28	4.10	–	1.47	nr	**5.57**
fittings with three ends	–	0.37	5.41	–	2.16	nr	**7.57**
fittings with four ends	4.09	0.49	7.16	–	4.84	nr	**12.00**
access plug	0.90	0.18	2.63	–	1.26	nr	**3.89**
straight coupling	0.97	0.19	2.77	–	1.32	nr	**4.09**
expansion coupling	2.06	0.28	4.10	–	2.48	nr	**6.58**
male iron to muPVC coupling	1.73	0.35	5.11	–	1.97	nr	**7.08**
level invert taper	1.22	0.28	4.10	–	1.59	nr	**5.69**
sweep bend	1.10	0.28	4.10	–	1.47	nr	**5.57**
spigot/socket bend	1.86	0.28	4.10	–	2.26	nr	**6.36**
sweep tee	1.69	0.37	5.41	–	2.16	nr	**7.57**
sweep cross	4.09	0.49	7.16	–	4.84	nr	**12.00**
50 mm pipes; fixing with pipe clips; plugged and							
screwed	2.98	0.32	4.67	–	4.68	m	**9.35**
extra for							
fittings with one end	–	0.19	2.77	–	1.68	nr	**4.45**
fittings with two ends	–	0.32	4.67	–	2.34	nr	**7.01**
fittings with three ends	–	0.43	6.28	–	3.83	nr	**10.11**
fittings with four ends	–	0.57	8.33	–	5.04	nr	**13.37**
access plug	1.30	0.19	2.77	–	1.68	nr	**4.45**
straight coupling	1.78	0.21	3.07	–	2.17	nr	**5.24**
expansion coupling	2.79	0.32	4.67	–	3.24	nr	**7.91**
male iron to muPVC coupling	2.50	0.42	6.13	–	2.78	nr	**8.91**
level invert taper	1.51	0.32	4.67	–	1.89	nr	**6.56**
sweep bend	1.94	0.32	4.67	–	2.34	nr	**7.01**
spigot/socket bend	2.64	0.32	4.67	–	3.09	nr	**7.76**
sweep tee	1.69	0.37	5.41	–	2.16	nr	**7.57**
sweep cross	4.29	0.57	8.33	–	5.04	nr	**13.37**

33 DRAINAGE ABOVE GROUND

Item	PC £	Labour hours	Labour £	Plant £	Material £	Unit	Total rate £
uPVC overflow pipes and fittings; solvent welded joints							
19 mm pipes; fixing with pipe clips; plugged and screwed	1.24	0.20	2.92	–	1.93	m	**4.85**
extra for							
splay cut end	–	0.01	0.15	–	–	nr	**0.15**
fittings with one end	–	0.16	2.34	–	1.14	nr	**3.48**
fittings with two ends	–	0.16	2.34	–	1.32	nr	**3.66**
fittings with three ends	–	0.20	2.92	–	1.50	nr	**4.42**
straight connector	0.94	0.16	2.34	–	1.14	nr	**3.48**
female iron to uPVC coupling	–	0.19	2.77	–	1.76	nr	**4.53**
bend	1.11	0.16	2.34	–	1.32	nr	**3.66**
bent tank connector	1.74	0.19	2.77	–	1.90	nr	**4.67**
uPVC pipes and fittings; with solvent welded joints (unless otherwise described)							
82 mm pipes; fixing with holderbats; plugged and screwed	9.70	0.37	5.41	–	13.89	m	**19.30**
extra for							
socket plug	5.60	0.19	2.77	–	6.99	nr	**9.76**
slip coupling; push fit	9.59	0.34	4.97	–	10.07	nr	**15.04**
expansion coupling	7.84	0.37	5.41	–	9.35	nr	**14.76**
sweep bend	13.05	0.37	5.41	–	14.82	nr	**20.23**
boss connector	5.41	0.25	3.65	–	6.79	nr	**10.44**
single branch	13.79	0.49	7.16	–	16.26	nr	**23.42**
access door	13.15	0.56	8.18	–	14.36	nr	**22.54**
110 mm pipes; fixing with holderbats; plugged and screwed	8.43	0.41	5.99	–	12.56	m	**18.55**
extra for							
socket plug	6.79	0.20	2.92	–	8.58	nr	**11.50**
slip coupling; push fit	8.97	0.37	5.41	–	9.42	nr	**14.83**
expansion coupling	8.02	0.41	5.99	–	9.87	nr	**15.86**
WC connector	10.93	0.27	3.95	–	12.25	nr	**16.20**
sweep bend	11.86	0.41	5.99	–	13.90	nr	**19.89**
WC connecting bend	17.93	0.27	3.95	–	19.61	nr	**23.56**
access bend	32.05	0.43	6.28	–	35.10	nr	**41.38**
boss connector	5.41	0.27	3.95	–	7.12	nr	**11.07**
single branch	15.28	0.54	7.89	–	18.27	nr	**26.16**
single branch with access	26.16	0.56	8.18	–	29.69	nr	**37.87**
double branch	37.77	0.68	9.93	–	42.67	nr	**52.60**
WC manifold	15.01	0.27	3.95	–	17.99	nr	**21.94**
access door	–	0.56	8.18	–	14.36	nr	**22.54**
access pipe connector	24.56	0.46	6.72	–	27.24	nr	**33.96**
connection to clay pipes; caulking ring and cement and sand (1:2) joint	–	0.39	5.70	–	10.59	nr	**16.29**
160 mm pipes; fixing with holderbats; plugged and screwed	26.22	0.46	6.72	–	36.49	m	**43.21**
extra for							
socket plug	12.47	0.23	3.36	–	16.32	nr	**19.68**
slip coupling; push fit	23.57	0.42	6.13	–	24.75	nr	**30.88**
expansion coupling	24.14	0.46	6.72	–	28.57	nr	**35.29**

33 DRAINAGE ABOVE GROUND

Item	PC £	Labour hours	Labour £	Plant £	Material £	Unit	Total rate £
FOUL DRAINAGE INSTALLATIONS – cont							
uPVC – cont							
uPVC pipes and fittings – cont							
extra for – cont							
sweep bend	31.61	0.46	6.72	–	36.42	nr	**43.14**
boss connector	7.63	0.31	4.53	–	11.25	nr	**15.78**
single branch	32.45	0.61	8.91	–	38.86	nr	**47.77**
double branch	68.25	0.77	11.25	–	78.00	nr	**89.25**
access door	23.48	0.56	8.18	–	25.21	nr	**33.39**
access pipe connector	24.56	0.46	6.72	–	27.24	nr	**33.96**
Weathering apron; for pipe							
82 mm dia.	2.77	0.31	4.53	–	3.46	nr	**7.99**
110 mm dia.	3.19	0.35	5.11	–	4.13	nr	**9.24**
160 mm dia.	9.61	0.39	5.70	–	11.64	nr	**17.34**
Weathering slate; for pipe							
110 mm dia.	33.94	0.83	12.13	–	36.41	nr	**48.54**
Vent cowl; for pipe							
82 mm dia.	2.77	0.31	4.53	–	3.46	nr	**7.99**
110 mm dia.	2.82	0.31	4.53	–	3.74	nr	**8.27**
160 mm dia.	7.37	0.31	4.53	–	9.29	nr	**13.82**
Polypropylene							
Polypropylene (PP) waste pipes and fittings; BS EN 1451; push fit O – ring joints							
32 mm pipes; fixing with pipe clips; plugged and screwed	1.58	0.20	2.92	–	2.38	m	**5.30**
extra for							
fittings with one end	–	0.15	2.19	–	1.35	nr	**3.54**
fittings with two ends	–	0.20	2.92	–	1.38	nr	**4.30**
fittings with three ends	–	0.28	4.10	–	2.37	nr	**6.47**
access plug	1.29	0.15	2.19	–	1.35	nr	**3.54**
double socket	0.99	0.14	2.05	–	1.04	nr	**3.09**
male iron to PP coupling	1.78	0.26	3.80	–	1.87	nr	**5.67**
sweep bend	1.23	0.20	2.92	–	1.29	nr	**4.21**
spigot bend	1.78	0.23	3.36	–	1.87	nr	**5.23**
40 mm pipes; fixing with pipe clips; plugged and screwed	1.95	0.20	2.92	–	2.78	m	**5.70**
extra for							
fittings with one end	–	0.18	2.63	–	1.42	nr	**4.05**
fittings with two ends	–	0.28	4.10	–	1.63	nr	**5.73**
fittings with three ends	–	0.37	5.41	–	2.51	nr	**7.92**
access plug	1.35	0.18	2.63	–	1.42	nr	**4.05**
double socket	0.99	0.19	2.77	–	1.04	nr	**3.81**
universal connector	1.91	0.23	3.36	–	2.01	nr	**5.37**
sweep bend	1.38	0.28	4.10	–	1.45	nr	**5.55**
spigot bend	1.73	0.28	4.10	–	1.82	nr	**5.92**
reducer 40 mm–32 mm	0.92	0.28	4.10	–	0.97	nr	**5.07**
50 mm pipes; fixing with pipe clips; plugged and screwed	2.50	0.32	4.67	–	4.08	m	**8.75**

33 DRAINAGE ABOVE GROUND

Item	PC £	Labour hours	Labour £	Plant £	Material £	Unit	Total rate £
extra for							
fittings with one end	–	0.19	2.77	–	2.51	nr	**5.28**
fittings with two ends	–	0.32	4.67	–	2.70	nr	**7.37**
fittings with three ends	–	0.43	6.28	–	3.73	nr	**10.01**
access plug	2.39	0.19	2.77	–	2.51	nr	**5.28**
double socket	2.02	0.21	3.07	–	2.12	nr	**5.19**
sweep bend	2.63	0.32	4.67	–	2.76	nr	**7.43**
spigot bend	2.21	0.32	4.67	–	2.32	nr	**6.99**
reducer 50 mm–40 mm	1.59	0.32	4.67	–	1.67	nr	**6.34**
Polypropylene ancillaries; screwed joint to waste fitting							
Tubular S trap; bath; shallow seal							
40 mm dia.	5.54	0.51	7.45	–	5.82	nr	**13.27**
Trap; P; two piece; 76 mm seal							
32 mm dia.	3.75	0.35	5.11	–	3.94	nr	**9.05**
40 mm dia.	4.33	0.42	6.13	–	4.55	nr	**10.68**
Trap; S; two piece; 76 mm seal							
32 mm dia.	4.74	0.35	5.11	–	4.98	nr	**10.09**
40 mm dia.	5.54	0.42	6.13	–	5.82	nr	**11.95**
Bottle trap; P; 76 mm seal							
32 dia.	4.18	0.35	5.11	–	4.39	nr	**9.50**
40 dia.	4.98	0.42	6.13	–	5.23	nr	**11.36**

34 DRAINAGE BELOW GROUND

Item	PC £	Labour hours	Labour £	Plant £	Material £	Unit	Total rate £
URBAN AND LANDSCAPE DRAINAGE							
Slot and grate drainage							
ACO RoadDrain one piece channel drainage system for medium to heavy duty highway and distribution yards							
100 F900 units, 500 mm long	–	0.46	6.05	–	204.75	m	**210.80**
200 F900 units, 500 mm long	–	0.48	6.31	–	262.60	m	**268.91**
ACO MultiDrain Monoblock PD100D one piece channel drainage system for pedestrian and medium duty vehicle applications made from high performance recycled materials							
PD100 units, 500 mm long with integral heelguard inlets	–	0.45	5.91	–	146.06	m	**151.97**
ACO MultiDrain M100D polymer concrete channel drainage system; galvanized steel edge trim; nominal bore 100 mm; type of fall constant; bedding and haunching with in situ concrete (not included)							
Slotted grating							
galvanized steel grating, load class A15 (pedestrian areas)	–	0.46	6.05	–	101.69	m	**107.74**
galvanized steel grating, load class C250 (cars and light vans)	–	0.46	6.05	–	140.07	m	**146.12**
ductile iron grating, load class D400 (driving lanes of roads)	–	0.46	6.05	–	146.06	m	**152.11**
Heelguard resin composite grating, load class C250 (cars and light vans)	–	0.46	6.05	–	138.24	m	**144.29**
extra for end caps	–	0.09	1.19	–	8.05	nr	**9.24**
extra for sump unit	–	1.39	18.26	–	139.44	nr	**157.70**
extra for ACO universal gully	–	1.50	19.71	–	832.70	nr	**852.41**
ACO MultiDrain M150D polymer concrete channel drainage system; galvanized steel edge trim; nominal bore 150 mm; type of fall constant; bedding and haunching with in situ concrete (not included)							
Slotted grating							
galvanized steel grating, load class A15 (pedestrian areas)	–	0.46	6.05	–	121.43	m	**127.48**
galvanized steel grating, load class C250 (cars and light vans)	–	0.46	6.05	–	124.48	m	**130.53**
extra for sump unit	–	1.45	19.05	–	266.65	nr	**285.70**
ACO MultiDrain M200D polymer concrete channel drainage system; galvanized steel edge trim; nominal bore 200 mm; type of fall constant; bedding and haunching with in situ concrete (not included)							
Slotted grating							
galvanized steel grating, load class A15 (pedestrian areas)	–	0.46	6.05	–	158.55	m	**164.60**
extra for sump unit	–	1.50	19.71	–	271.63	nr	**291.34**

34 DRAINAGE BELOW GROUND

Item	PC £	Labour hours	Labour £	Plant £	Material £	Unit	Total rate £
ACO S Range polymer concrete channel drainage system; bolted ductile iron grating, load class F900 (airfields); bedding and haunching with in situ concrete (not included)							
S100 F900 channel and grate	–	1.00	13.14	–	154.35	m	**167.49**
S150 F900 channel and grate	–	1.00	13.14	–	166.00	m	**179.14**
S200 F900 channel and grate	–	1.10	14.45	–	176.03	m	**190.48**
S300 F900 channel and grate	–	1.20	15.77	–	274.05	m	**289.82**
extra for end caps	–	0.09	1.19	–	22.27	nr	**23.46**
extra for sump unit	–	1.50	19.71	–	282.40	nr	**302.11**
ACO Qmax large capacity slot drainage channel with MDPE body and hot dipped galvanized steel edge rail, up to load class F900; bedding and haunching with in situ concrete (not included)							
ACO Qmax 225	–	0.75	9.85	–	95.37	m	**105.22**
ACO Qmax 350	–	1.00	13.14	–	125.53	m	**138.67**
ACO Qmax 900	–	1.50	19.71	–	239.61	m	**259.32**
shallow access chamber	–	1.50	19.71	–	270.58	nr	**290.29**
deep access chamber	–	2.00	26.28	–	645.59	nr	**671.87**
ACO Kerbdrain one-piece polymer concrete combined drainage system, load class D400; bedding and haunching in in situ concrete (not included). Manufactured from recycled and recyclable material							
KerbDrain KD305	–	0.50	6.57	–	91.67	m	**98.24**
KerbDrain KD480	–	0.65	8.54	–	99.54	m	**108.08**
KerbDrain KD305 drop kerb (left drop, one centre stone and right drop) total length 2745 mm	–	2.00	26.28	–	185.43	nr	**211.71**
KerbDrain KD305 mitre unit	–	0.25	3.29	–	99.96	nr	**103.25**
KerbDrain KD end cap	–	0.09	1.19	–	54.91	nr	**56.10**
KerbDrain KD610 shallow gully assembly	–	1.50	19.71	–	788.45	nr	**808.16**
Interconnecting drainage channel; Birco-lite ref 8012 or other equal and approved; Marshalls Plc; galvanized steel grating ref 8041; bedding and haunching in in situ concrete (not included)							
100 mm wide							
laid level or to falls	–	0.46	6.05	–	43.03	m	**49.08**
100 mm dia. trapped outlet unit	–	1.39	18.26	–	94.70	nr	**112.96**
end caps	–	0.09	1.19	–	5.36	nr	**6.55**
Oil separators (polyethylene single chamber design)							
Supply and install only oil separator complete with lockable cover. (excavations, filling etc. measured elsewhere)							
1000 litre ACO Q-ceptor by-pass oil separators NSB3 Class 1 (discharge concentrations of less than 5 mg/litre of oil) for discharges to surface water drains	–	2.00	26.28	48.52	1034.50	nr	**1109.30**
1000 litre ACO Q-ceptor full retention oil separators NS3 Class 1	–	2.25	29.56	55.45	1123.23	nr	**1208.24**

34 DRAINAGE BELOW GROUND

Item	PC £	Labour hours	Labour £	Plant £	Material £	Unit	Total rate £
TRENCHES							
NOTE: Prices for drain trenches are for excavation in firm soil and it has been assumed that earthwork support will only be required for trenches 1.00 m or more in depth.							
Excavating trenches; by machine; grading bottoms; earthwork support; filling with excavated material and compacting; disposal of surplus soil; spreading on site average 50 m from excavations							
Pipes not exceeding 200 mm nominal size							
average depth of trench 0.50 m	–	0.28	3.71	1.56	–	m	**5.27**
average depth of trench 0.75 m	–	0.37	4.90	2.31	–	m	**7.21**
average depth of trench 1.00 m	–	0.79	10.46	3.30	0.81	m	**14.57**
average depth of trench 1.25 m	–	1.16	15.36	3.46	1.10	m	**19.92**
average depth of trench 1.50 m	–	1.48	19.59	3.87	1.30	m	**24.76**
average depth of trench 1.75 m	–	1.85	24.50	4.04	1.61	m	**30.15**
average depth of trench 2.00 m	–	2.13	28.20	4.62	1.81	m	**34.63**
average depth of trench 2.25 m	–	2.64	34.95	5.61	2.41	m	**42.97**
average depth of trench 2.50 m	–	3.10	41.04	6.53	2.81	m	**50.38**
average depth of trench 2.75 m	–	3.42	45.29	7.34	3.12	m	**55.75**
average depth of trench 3.00 m	–	3.75	49.65	8.09	3.41	m	**61.15**
average depth of trench 3.25 m	–	4.07	53.90	8.49	3.72	m	**66.11**
average depth of trench 3.50 m	–	4.35	57.60	8.84	4.02	m	**70.46**
Pipes 225 mm nominal size							
average depth of trench 0.50 m	–	0.28	3.71	1.56	–	m	**5.27**
average depth of trench 0.75 m	–	0.37	4.90	2.31	–	m	**7.21**
average depth of trench 1.00 m	–	0.79	10.46	3.30	0.81	m	**14.57**
average depth of trench 1.25 m	–	1.16	15.36	3.46	1.10	m	**19.92**
average depth of trench 1.50 m	–	1.48	19.59	3.87	1.30	m	**24.76**
average depth of trench 1.75 m	–	1.85	24.50	4.04	1.61	m	**30.15**
average depth of trench 2.00 m	–	2.13	28.20	4.62	1.81	m	**34.63**
average depth of trench 2.25 m	–	2.64	34.95	5.61	2.41	m	**42.97**
average depth of trench 2.50 m	–	3.10	41.04	6.53	2.81	m	**50.38**
average depth of trench 2.75 m	–	3.42	45.29	7.34	3.12	m	**55.75**
average depth of trench 3.00 m	–	3.75	49.65	8.09	3.41	m	**61.15**
average depth of trench 3.25 m	–	4.07	53.90	8.49	3.72	m	**66.11**
average depth of trench 3.50 m	–	4.35	57.60	8.84	4.02	m	**70.46**
Pipes 300 mm nominal size							
average depth of trench 0.75 m	–	0.44	5.83	2.89	–	m	**8.72**
average depth of trench 1.00 m	–	0.93	12.32	3.30	0.81	m	**16.43**
average depth of trench 1.25 m	–	1.25	16.55	3.64	1.10	m	**21.29**
average depth of trench 1.50 m	–	1.62	21.45	4.04	1.30	m	**26.79**
average depth of trench 1.75 m	–	1.85	24.50	4.22	1.61	m	**30.33**
average depth of trench 2.00 m	–	2.13	28.20	5.20	1.81	m	**35.21**
average depth of trench 2.25 m	–	2.64	34.95	5.95	2.41	m	**43.31**
average depth of trench 2.50 m	–	3.10	41.04	6.76	2.81	m	**50.61**
average depth of trench 2.75 m	–	3.42	45.29	7.51	3.12	m	**55.92**
average depth of trench 3.00 m	–	3.75	49.65	8.26	3.41	m	**61.32**
average depth of trench 3.25 m	–	4.07	53.90	9.07	3.72	m	**66.69**
average depth of trench 3.50 m	–	4.35	57.60	9.24	4.02	m	**70.86**

34 DRAINAGE BELOW GROUND

Item	PC £	Labour hours	Labour £	Plant £	Material £	Unit	Total rate £
Pipes 375 mm nominal size							
average depth of trench 0.75 m	–	0.46	6.09	3.46	–	m	9.55
average depth of trench 1.00 m	–	0.97	12.84	3.87	0.81	m	17.52
average depth of trench 1.25 m	–	1.34	17.74	4.62	1.10	m	23.46
average depth of trench 1.50 m	–	1.71	22.65	4.80	1.30	m	28.75
average depth of trench 1.75 m	–	1.99	26.36	5.20	1.61	m	33.17
average depth of trench 2.00 m	–	2.27	30.06	5.38	1.81	m	37.25
average depth of trench 2.25 m	–	2.82	37.34	6.53	2.41	m	46.28
average depth of trench 2.50 m	–	3.38	44.76	7.51	2.81	m	55.08
average depth of trench 2.75 m	–	3.70	48.99	8.09	3.12	m	60.20
average depth of trench 3.00 m	–	4.02	53.24	8.66	3.41	m	65.31
average depth of trench 3.25 m	–	4.35	57.60	9.42	3.72	m	70.74
average depth of trench 3.50 m	–	4.67	61.83	10.00	4.02	m	75.85
Pipes 450 mm nominal size							
average depth of trench 0.75 m	–	0.51	6.75	3.46	–	m	10.21
average depth of trench 1.00 m	–	1.02	13.50	4.22	0.81	m	18.53
average depth of trench 1.25 m	–	1.48	19.59	5.03	1.10	m	25.72
average depth of trench 1.50 m	–	1.85	24.50	5.38	1.30	m	31.18
average depth of trench 1.75 m	–	2.13	28.20	5.61	1.61	m	35.42
average depth of trench 2.00 m	–	2.45	32.44	5.95	1.81	m	40.20
average depth of trench 2.25 m	–	3.05	40.38	6.93	2.41	m	49.72
average depth of trench 2.50 m	–	3.61	47.81	8.09	2.81	m	58.71
average depth of trench 2.75 m	–	3.98	52.70	8.84	3.12	m	64.66
average depth of trench 3.00 m	–	4.26	56.41	9.65	3.41	m	69.47
average depth of trench 3.25 m	–	4.63	61.31	10.57	3.72	m	75.60
average depth of trench 3.50 m	–	5.00	66.21	11.55	4.02	m	81.78
Pipes 600 mm nominal size							
average depth of trench 1.00 m	–	1.11	14.70	4.62	0.81	m	20.13
average depth of trench 1.25 m	–	1.57	20.79	5.38	1.10	m	27.27
average depth of trench 1.50 m	–	2.04	27.02	6.18	1.30	m	34.50
average depth of trench 1.75 m	–	2.31	30.59	6.18	1.61	m	38.38
average depth of trench 2.00 m	–	2.73	36.15	6.76	1.81	m	44.72
average depth of trench 2.25 m	–	3.28	43.44	8.09	2.41	m	53.94
average depth of trench 2.50 m	–	3.89	51.51	9.42	2.81	m	63.74
average depth of trench 2.75 m	–	4.30	56.94	10.57	3.12	m	70.63
average depth of trench 3.00 m	–	4.72	62.51	11.55	3.41	m	77.47
average depth of trench 3.25 m	–	5.09	67.40	12.31	3.72	m	83.43
average depth of trench 3.50 m	–	5.46	72.30	12.88	4.02	m	89.20
Pipes 900 mm nominal size							
average depth of trench 1.25 m	–	1.90	25.16	6.53	1.10	m	32.79
average depth of trench 1.50 m	–	2.41	31.91	7.34	1.30	m	40.55
average depth of trench 1.75 m	–	2.78	36.81	7.51	1.61	m	45.93
average depth of trench 2.00 m	–	3.10	41.04	8.66	1.81	m	51.51
average depth of trench 2.25 m	–	3.84	50.85	10.23	2.41	m	63.49
average depth of trench 2.50 m	–	4.53	59.99	11.73	2.81	m	74.53
average depth of trench 2.75 m	–	5.00	66.21	12.88	3.12	m	82.21
average depth of trench 3.00 m	–	5.46	72.30	14.04	3.41	m	89.75
average depth of trench 3.25 m	–	5.92	78.39	15.19	3.72	m	97.30
average depth of trench 3.50 m	–	6.38	84.48	16.17	4.02	m	104.67

34 DRAINAGE BELOW GROUND

Item	PC £	Labour hours	Labour £	Plant £	Material £	Unit	Total rate £
TRENCHES – cont							
Excavating trenches – cont							
Pipes 1200 mm nominal size							
average depth of trench 1.50 m	–	2.73	36.15	7.92	1.30	m	**45.37**
average depth of trench 1.75 m	–	3.19	42.24	9.07	1.61	m	**52.92**
average depth of trench 2.00 m	–	3.56	47.15	10.40	1.81	m	**59.36**
average depth of trench 2.25 m	–	4.35	57.60	12.31	2.41	m	**72.32**
average depth of trench 2.50 m	–	5.18	68.60	14.04	2.81	m	**85.45**
average depth of trench 2.75 m	–	5.69	75.35	15.59	3.12	m	**94.06**
average depth of trench 3.00 m	–	6.20	82.10	16.93	3.41	m	**102.44**
average depth of trench 3.25 m	–	6.75	89.39	18.31	3.72	m	**111.42**
average depth of trench 3.50 m	–	7.26	96.14	19.63	4.02	m	**119.79**
Extra over excavating trenches; irrespective of depth; breaking out existing materials							
brick	–	1.80	23.84	8.56	–	m³	**32.40**
concrete	–	2.54	33.63	11.81	–	m³	**45.44**
reinforced concrete	–	3.61	47.81	17.05	–	m³	**64.86**
Extra over excavating trenches; irrespective of depth; breaking out existing hard pavings; 75 mm thick							
tarmacadam	–	0.19	2.52	0.87	–	m²	**3.39**
Extra over excavating trenches; irrsepective of depth; breaking out existing hard pavings; 150 mm thick							
concrete	–	0.37	4.90	1.91	–	m²	**6.81**
tarmacadam and hardcore	–	0.28	3.71	1.06	–	m²	**4.77**
Excavating trenches; by hand; grading bottoms; earthwork support; filling with excavated material and compacting; disposal of surplus soil on site; spreading on site average 25 m from excavations							
Pipes not exceeding 200 mm nominal size							
average depth of trench 0.50 m	–	0.93	12.32	–	–	m	**12.32**
average depth of trench 0.75 m	–	1.39	18.41	–	–	m	**18.41**
average depth of trench 1.00 m	–	2.04	27.02	–	0.81	m	**27.83**
average depth of trench 1.25 m	–	2.87	38.00	–	1.10	m	**39.10**
average depth of trench 1.50 m	–	3.93	52.04	–	1.34	m	**53.38**
average depth of trench 1.75 m	–	5.18	68.60	–	1.61	m	**70.21**
average depth of trench 2.00 m	–	5.92	78.39	–	1.81	m	**80.20**
average depth of trench 2.25 m	–	7.40	97.99	–	2.41	m	**100.40**
average depth of trench 2.50 m	–	8.88	117.59	–	2.81	m	**120.40**
average depth of trench 2.75 m	–	9.76	129.24	–	3.12	m	**132.36**
average depth of trench 3.00 m	–	10.64	140.89	–	3.41	m	**144.30**
average depth of trench 3.25 m	–	11.52	152.54	–	3.72	m	**156.26**
average depth of trench 3.50 m	–	12.40	164.20	–	4.02	m	**168.22**
Pipes 225 mm nominal size							
average depth of trench 0.50 m	–	0.93	12.32	–	–	m	**12.32**
average depth of trench 0.75 m	–	1.39	18.41	–	–	m	**18.41**
average depth of trench 1.00 m	–	2.04	27.02	–	0.81	m	**27.83**

34 DRAINAGE BELOW GROUND

Item	PC £	Labour hours	Labour £	Plant £	Material £	Unit	Total rate £
average depth of trench 1.25 m	–	2.87	38.00	–	1.10	m	**39.10**
average depth of trench 1.50 m	–	3.93	52.04	–	1.34	m	**53.38**
average depth of trench 1.75 m	–	5.18	68.60	–	1.61	m	**70.21**
average depth of trench 2.00 m	–	5.92	78.39	–	1.81	m	**80.20**
average depth of trench 2.25 m	–	7.40	97.99	–	2.41	m	**100.40**
average depth of trench 2.50 m	–	8.88	117.59	–	2.81	m	**120.40**
average depth of trench 2.75 m	–	9.76	129.24	–	3.12	m	**132.36**
average depth of trench 3.00 m	–	10.64	140.89	–	3.41	m	**144.30**
average depth of trench 3.25 m	–	11.52	152.54	–	3.72	m	**156.26**
average depth of trench 3.50 m	–	12.40	164.20	–	4.02	m	**168.22**
Pipes 300 mm nominal size							
average depth of trench 0.75 m	–	1.62	21.45	–	–	m	**21.45**
average depth of trench 1.00 m	–	2.36	31.25	–	0.81	m	**32.06**
average depth of trench 1.25 m	–	3.33	44.10	–	1.10	m	**45.20**
average depth of trench 1.50 m	–	4.44	58.79	–	1.34	m	**60.13**
average depth of trench 1.75 m	–	5.18	68.60	–	1.61	m	**70.21**
average depth of trench 2.00 m	–	5.92	78.39	–	1.81	m	**80.20**
average depth of trench 2.25 m	–	7.40	97.99	–	2.41	m	**100.40**
average depth of trench 2.50 m	–	8.88	117.59	–	2.81	m	**120.40**
average depth of trench 2.75 m	–	9.76	129.24	–	3.12	m	**132.36**
average depth of trench 3.00 m	–	10.64	140.89	–	3.41	m	**144.30**
average depth of trench 3.25 m	–	11.52	152.54	–	3.72	m	**156.26**
average depth of trench 3.50 m	–	12.40	164.20	–	4.02	m	**168.22**
Pipes 375 mm nominal size							
average depth of trench 0.75 m	–	1.80	23.84	–	–	m	**23.84**
average depth of trench 1.00 m	–	2.64	34.95	–	0.81	m	**35.76**
average depth of trench 1.25 m	–	3.70	48.99	–	1.10	m	**50.09**
average depth of trench 1.50 m	–	4.93	65.28	–	1.34	m	**66.62**
average depth of trench 1.75 m	–	5.74	76.01	–	1.61	m	**77.62**
average depth of trench 2.00 m	–	6.57	87.00	–	1.81	m	**88.81**
average depth of trench 2.25 m	–	8.23	108.98	–	2.41	m	**111.39**
average depth of trench 2.50 m	–	9.90	131.09	–	2.81	m	**133.90**
average depth of trench 2.75 m	–	10.87	143.93	–	3.12	m	**147.05**
average depth of trench 3.00 m	–	11.84	156.79	–	3.41	m	**160.20**
average depth of trench 3.25 m	–	12.86	170.29	–	3.72	m	**174.01**
average depth of trench 3.50 m	–	13.88	183.79	–	4.02	m	**187.81**
Pipes 450 mm nominal size							
average depth of trench 0.75 m	–	2.04	27.02	–	–	m	**27.02**
average depth of trench 1.00 m	–	2.94	38.93	–	0.81	m	**39.74**
average depth of trench 1.25 m	–	4.13	54.68	–	1.10	m	**55.78**
average depth of trench 1.50 m	–	5.41	71.64	–	1.34	m	**72.98**
average depth of trench 1.75 m	–	6.31	83.56	–	1.61	m	**85.17**
average depth of trench 2.00 m	–	7.22	95.60	–	1.81	m	**97.41**
average depth of trench 2.25 m	–	9.05	119.84	–	2.41	m	**122.25**
average depth of trench 2.50 m	–	10.87	143.93	–	2.81	m	**146.74**
average depth of trench 2.75 m	–	11.96	158.37	–	3.12	m	**161.49**
average depth of trench 3.00 m	–	13.04	172.67	–	3.41	m	**176.08**
average depth of trench 3.25 m	–	14.11	186.85	–	3.72	m	**190.57**
average depth of trench 3.50 m	–	15.17	200.88	–	4.02	m	**204.90**

34 DRAINAGE BELOW GROUND

Item	PC £	Labour hours	Labour £	Plant £	Material £	Unit	Total rate £
TRENCHES – cont							
Excavating trenches – cont							
Pipes 600 mm nominal size							
average depth of trench 1.00 m	–	3.24	42.90	–	0.81	m	**43.71**
average depth of trench 1.25 m	–	4.63	61.31	–	1.10	m	**62.41**
average depth of trench 1.50 m	–	6.20	82.10	–	1.34	m	**83.44**
average depth of trench 1.75 m	–	7.17	94.94	–	1.61	m	**96.55**
average depth of trench 2.00 m	–	8.19	108.45	–	1.81	m	**110.26**
average depth of trench 2.25 m	–	9.20	121.82	–	2.41	m	**124.23**
average depth of trench 2.50 m	–	11.56	153.08	–	2.81	m	**155.89**
average depth of trench 2.75 m	–	12.35	163.54	–	3.12	m	**166.66**
average depth of trench 3.00 m	–	14.80	195.98	–	3.41	m	**199.39**
average depth of trench 3.25 m	–	16.03	212.27	–	3.72	m	**215.99**
average depth of trench 3.50 m	–	17.25	228.42	–	4.02	m	**232.44**
Pipes 900 mm nominal size							
average depth of trench 1.25 m	–	5.78	76.53	–	1.10	m	**77.63**
average depth of trench 1.50 m	–	7.63	101.03	–	1.34	m	**102.37**
average depth of trench 1.75 m	–	8.88	117.59	–	1.61	m	**119.20**
average depth of trench 2.00 m	–	10.13	134.14	–	1.81	m	**135.95**
average depth of trench 2.25 m	–	12.72	168.44	–	2.41	m	**170.85**
average depth of trench 2.50 m	–	15.31	202.73	–	2.81	m	**205.54**
average depth of trench 2.75 m	–	16.84	222.99	–	3.12	m	**226.11**
average depth of trench 3.00 m	–	18.32	242.59	–	3.41	m	**246.00**
average depth of trench 3.25 m	–	19.84	262.72	–	3.72	m	**266.44**
average depth of trench 3.50 m	–	21.37	282.97	–	4.02	m	**286.99**
Pipes 1200 mm nominal size							
average depth of trench 1.50 m	–	9.11	120.63	–	1.34	m	**121.97**
average depth of trench 1.75 m	–	10.59	140.23	–	1.61	m	**141.84**
average depth of trench 2.00 m	–	12.12	160.49	–	1.81	m	**162.30**
average depth of trench 2.25 m	–	15.20	201.27	–	2.41	m	**203.68**
average depth of trench 2.50 m	–	18.27	241.93	–	2.81	m	**244.74**
average depth of trench 2.75 m	–	20.07	265.77	–	3.12	m	**268.89**
average depth of trench 3.00 m	–	21.88	289.74	–	3.41	m	**293.15**
average depth of trench 3.25 m	–	23.66	313.30	–	3.72	m	**317.02**
average depth of trench 3.50 m	–	25.44	336.87	–	4.02	m	**340.89**
Extra over excavating trenches irrespective of depth; breaking out existing materials							
brick	–	2.78	36.81	7.00	–	m³	**43.81**
concrete	–	4.16	55.08	11.68	–	m³	**66.76**
reinforced concrete	–	5.55	73.49	16.36	–	m³	**89.85**
concrete; 150 mm thick	–	0.65	8.61	1.64	–	m²	**10.25**
tarmacadam and hardcore; 150 mm thick	–	0.46	6.09	1.17	–	m²	**7.26**
Extra over excavating trenches irrespective of depth; breaking out existing hard pavings, 75 mm thick							
tarmacadam	–	0.37	4.90	0.93	–	m²	**5.83**
Extra over excavating trenches irrespective of depth; breaking out existing hard pavings, 150 mm thick							
concrete	–	0.65	8.61	1.64	–	m²	**10.25**
tarmacadam and hardcore	–	0.46	6.09	1.17	–	m²	**7.26**

34 DRAINAGE BELOW GROUND

Item	PC £	Labour hours	Labour £	Plant £	Material £	Unit	Total rate £
BEDS AND FILLINGS							
Sand filling							
Beds; to receive pitch fibre pipes							
600 mm × 50 mm thick	–	0.07	0.92	0.53	1.07	m	**2.52**
700 mm × 50 mm thick	–	0.09	1.20	0.53	1.25	m	**2.98**
800 mm × 50 mm thick	–	0.11	1.46	0.53	1.43	m	**3.42**
Granular (shingle) filling							
Beds; 100 mm thick; to pipes							
100 mm nominal size	–	0.09	1.20	0.53	2.18	m	**3.91**
150 mm nominal size	–	0.09	1.20	0.53	2.55	m	**4.28**
225 mm nominal size	–	0.11	1.46	0.53	2.91	m	**4.90**
300 mm nominal size	–	0.13	1.72	0.53	3.28	m	**5.53**
375 mm nominal size	–	0.15	1.98	0.53	3.64	m	**6.15**
450 mm nominal size	–	0.17	2.25	0.53	4.00	m	**6.78**
600 mm nominal size	–	0.19	2.52	0.53	4.37	m	**7.42**
Beds; 150 mm thick; to pipes							
100 mm nominal size	–	0.13	1.72	0.53	3.28	m	**5.53**
150 mm nominal size	–	0.15	1.98	0.53	3.64	m	**6.15**
225 mm nominal size	–	0.17	2.25	0.53	4.00	m	**6.78**
300 mm nominal size	–	0.19	2.52	0.53	4.37	m	**7.42**
375 mm nominal size	–	0.22	2.91	0.53	5.46	m	**8.90**
450 mm nominal size	–	0.24	3.18	0.53	5.82	m	**9.53**
600 mm nominal size	–	0.28	3.71	0.53	6.92	m	**11.16**
Beds and benchings; beds 100 mm thick; to pipes							
100 nominal size	–	0.21	2.78	0.53	4.00	m	**7.31**
150 nominal size	–	0.23	3.04	0.53	4.00	m	**7.57**
225 nominal size	–	0.28	3.71	0.53	5.46	m	**9.70**
300 nominal size	–	0.32	4.24	0.53	6.18	m	**10.95**
375 nominal size	–	0.42	5.57	0.53	8.37	m	**14.47**
450 nominal size	–	0.48	6.35	0.53	9.46	m	**16.34**
600 nominal size	–	0.62	8.21	0.53	12.37	m	**21.11**
Beds and benchings; beds 150 mm thick; to pipes							
100 nominal size	–	0.23	3.04	0.53	4.37	m	**7.94**
150 nominal size	–	0.26	3.44	0.53	4.74	m	**8.71**
225 nominal size	–	0.32	4.24	0.53	6.55	m	**11.32**
300 nominal size	–	0.42	5.57	0.53	8.00	m	**14.10**
375 nominal size	–	0.48	6.35	0.53	9.46	m	**16.34**
450 nominal size	–	0.57	7.55	0.53	11.28	m	**19.36**
600 nominal size	–	0.68	9.01	0.53	14.55	m	**24.09**
Beds and coverings; 100 mm thick; to pipes							
100 nominal size	–	0.33	4.37	0.53	5.46	m	**10.36**
150 nominal size	–	0.42	5.57	0.53	6.55	m	**12.65**
225 nominal size	–	0.56	7.41	0.53	9.09	m	**17.03**
300 nominal size	–	0.67	8.87	0.53	10.92	m	**20.32**
375 nominal size	–	0.80	10.59	0.53	13.09	m	**24.21**
450 nominal size	–	0.94	12.44	0.53	15.65	m	**28.62**
600 nominal size	–	1.22	16.16	0.53	20.01	m	**36.70**

34 DRAINAGE BELOW GROUND

Item	PC £	Labour hours	Labour £	Plant £	Material £	Unit	Total rate £
BEDS AND FILLINGS – cont							
Granular (shingle) filling – cont							
Beds and coverings; 150 mm thick; to pipes							
100 nominal size	–	0.50	6.63	0.53	8.00	m	**15.16**
150 nominal size	–	0.56	7.41	0.53	9.09	m	**17.03**
225 nominal size	–	0.72	9.53	0.53	11.64	m	**21.70**
300 nominal size	–	0.86	11.39	0.53	13.83	m	**25.75**
375 nominal size	–	1.00	13.24	0.53	16.38	m	**30.15**
450 nominal size	–	1.19	15.76	0.53	19.65	m	**35.94**
600 nominal size	–	1.44	19.07	0.53	23.66	m	**43.26**
Plain in situ ready mixed designated concrete; C10–40 mm aggregate							
Beds; 100 mm thick; to pipes							
100 mm nominal size	–	0.17	2.64	0.53	4.83	m	**8.00**
150 mm nominal size	–	0.17	2.64	0.53	4.83	m	**8.00**
225 mm nominal size	–	0.20	3.11	0.53	5.79	m	**9.43**
300 mm nominal size	–	0.23	3.57	0.53	6.76	m	**10.86**
375 mm nominal size	–	0.27	4.19	0.53	7.72	m	**12.44**
450 mm nominal size	–	0.30	4.66	0.53	8.69	m	**13.88**
600 mm nominal size	–	0.33	5.12	0.53	9.65	m	**15.30**
900 mm nominal size	–	0.40	6.21	0.53	11.58	m	**18.32**
1200 mm nominal size	–	0.54	8.38	0.53	15.45	m	**24.36**
Beds; 150 mm thick; to pipes							
100 mm nominal size	–	0.23	3.57	0.53	6.76	m	**10.86**
150 mm nominal size	–	0.27	4.19	0.53	7.72	m	**12.44**
225 mm nominal size	–	0.30	4.66	0.53	8.69	m	**13.88**
300 mm nominal size	–	0.33	5.12	0.53	9.65	m	**15.30**
375 mm nominal size	–	0.40	6.21	0.53	11.58	m	**18.32**
450 mm nominal size	–	0.43	6.68	0.53	12.55	m	**19.76**
600 mm nominal size	–	0.50	7.76	0.53	14.48	m	**22.77**
900 mm nominal size	–	0.63	9.78	0.53	18.34	m	**28.65**
1200 mm nominal size	–	0.77	11.95	0.53	22.20	m	**34.68**
Beds and benchings; beds 100 mm thick; to pipes							
100 mm nominal size	–	0.33	5.12	0.53	8.69	m	**14.34**
150 mm nominal size	–	0.38	5.90	0.53	9.65	m	**16.08**
225 mm nominal size	–	0.45	6.98	0.53	11.58	m	**19.09**
300 mm nominal size	–	0.53	8.23	0.53	13.51	m	**22.27**
375 mm nominal size	–	0.68	10.55	0.53	17.37	m	**28.45**
450 mm nominal size	–	0.80	12.42	0.53	20.27	m	**33.22**
600 mm nominal size	–	1.02	15.83	0.53	26.06	m	**42.42**
900 mm nominal size	–	1.65	25.61	0.53	42.46	m	**68.60**
1200 mm nominal size	–	2.44	37.87	0.53	62.73	m	**101.13**
Beds and benchings; beds 150 mm thick; to pipes							
100 mm nominal size	–	0.38	5.90	0.53	9.65	m	**16.08**
150 mm nominal size	–	0.42	6.52	0.53	10.62	m	**17.67**
225 mm nominal size	–	0.53	8.23	0.53	13.51	m	**22.27**
300 mm nominal size	–	0.68	10.55	0.53	17.37	m	**28.45**
375 mm nominal size	–	0.80	12.42	0.53	20.27	m	**33.22**
450 mm nominal size	–	0.94	14.60	0.53	24.12	m	**39.25**
600 mm nominal size	–	1.20	18.63	0.53	30.88	m	**50.04**

34 DRAINAGE BELOW GROUND

Item	PC £	Labour hours	Labour £	Plant £	Material £	Unit	Total rate £
900 mm nominal size	–	1.91	29.65	0.53	49.22	m	**79.40**
1200 mm nominal size	–	2.70	41.92	0.53	69.49	m	**111.94**
Beds and coverings; 100 mm thick; to pipes							
100 mm nominal size	–	0.50	7.76	0.53	11.58	m	**19.87**
150 mm nominal size	–	0.58	9.00	0.53	13.51	m	**23.04**
225 mm nominal size	–	0.83	12.88	0.53	19.30	m	**32.71**
300 mm nominal size	–	1.00	15.52	0.53	23.16	m	**39.21**
375 mm nominal size	–	1.21	18.78	0.53	27.98	m	**47.29**
450 mm nominal size	–	1.42	22.04	0.53	32.81	m	**55.38**
600 mm nominal size	–	1.83	28.40	0.53	42.46	m	**71.39**
900 mm nominal size	–	2.79	43.31	0.53	64.67	m	**108.51**
1200 mm nominal size	–	3.83	59.45	0.53	88.79	m	**148.77**
Beds and coverings; 150 mm thick; to pipes							
100 mm nominal size	–	0.75	11.64	0.53	17.37	m	**29.54**
150 mm nominal size	–	0.83	12.88	0.53	19.30	m	**32.71**
225 mm nominal size	–	1.08	16.77	0.53	25.09	m	**42.39**
300 mm nominal size	–	1.30	20.18	0.53	29.91	m	**50.62**
375 mm nominal size	–	1.50	23.29	0.53	34.74	m	**58.56**
450 mm nominal size	–	1.79	27.78	0.53	41.50	m	**69.81**
600 mm nominal size	–	2.16	33.53	0.53	50.18	m	**84.24**
900 mm nominal size	–	3.54	54.96	0.53	82.03	m	**137.52**
1200 mm nominal size	–	5.00	77.62	0.53	115.81	m	**193.96**
Plain in situ ready mixed designated concrete; C20–40 mm aggregate							
Beds; 100 mm thick; to pipes							
100 mm nominal size	–	0.17	2.64	0.53	5.02	m	**8.19**
150 mm nominal size	–	0.17	2.64	0.53	5.02	m	**8.19**
225 mm nominal size	–	0.20	3.11	0.53	6.02	m	**9.66**
300 mm nominal size	–	0.23	3.57	0.53	7.02	m	**11.12**
375 mm nominal size	–	0.27	4.19	0.53	8.02	m	**12.74**
450 mm nominal size	–	0.30	4.66	0.53	9.03	m	**14.22**
600 mm nominal size	–	0.33	5.12	0.53	10.03	m	**15.68**
900 mm nominal size	–	0.40	6.21	0.53	12.03	m	**18.77**
1200 mm nominal size	–	0.54	8.38	0.53	16.04	m	**24.95**
Beds; 150 mm thick; to pipes							
100 mm nominal size	–	0.23	3.57	0.53	7.02	m	**11.12**
150 mm nominal size	–	0.27	4.19	0.53	8.02	m	**12.74**
225 mm nominal size	–	0.30	4.66	0.53	9.03	m	**14.22**
300 mm nominal size	–	0.33	5.12	0.53	10.03	m	**15.68**
375 mm nominal size	–	0.40	6.21	0.53	12.03	m	**18.77**
450 mm nominal size	–	0.43	6.68	0.53	13.04	m	**20.25**
600 mm nominal size	–	0.50	7.76	0.53	15.05	m	**23.34**
900 mm nominal size	–	0.63	9.78	0.53	19.06	m	**29.37**
1200 mm nominal size	–	0.77	11.95	0.53	23.07	m	**35.55**
Beds and benchings; beds 100 mm thick; to pipes							
100 mm nominal size	–	0.33	5.12	0.53	9.03	m	**14.68**
150 mm nominal size	–	0.38	5.90	0.53	10.03	m	**16.46**
225 mm nominal size	–	0.45	6.98	0.53	12.03	m	**19.54**

34 DRAINAGE BELOW GROUND

Item	PC £	Labour hours	Labour £	Plant £	Material £	Unit	Total rate £
BEDS AND FILLINGS – cont							
Plain in situ ready mixed designated concrete – cont							
Beds and benchings – cont							
300 mm nominal size	–	0.53	8.23	0.53	14.04	m	**22.80**
375 mm nominal size	–	0.68	10.55	0.53	18.05	m	**29.13**
450 mm nominal size	–	0.80	12.42	0.53	21.05	m	**34.00**
600 mm nominal size	–	1.02	15.83	0.53	27.08	m	**43.44**
900 mm nominal size	–	1.65	25.61	0.53	44.11	m	**70.25**
1200 mm nominal size	–	2.44	37.87	0.53	65.16	m	**103.56**
Beds and benchings; beds 150 mm thick; to pipes							
100 mm nominal size	–	0.38	5.90	0.53	10.03	m	**16.46**
150 mm nominal size	–	0.42	6.52	0.53	11.03	m	**18.08**
225 mm nominal size	–	0.53	8.23	0.53	14.04	m	**22.80**
300 mm nominal size	–	0.68	10.55	0.53	18.05	m	**29.13**
375 mm nominal size	–	0.80	12.42	0.53	21.05	m	**34.00**
450 mm nominal size	–	0.94	14.60	0.53	25.06	m	**40.19**
600 mm nominal size	–	1.20	18.63	0.53	32.09	m	**51.25**
900 mm nominal size	–	1.91	29.65	0.53	51.14	m	**81.32**
1200 mm nominal size	–	2.70	41.92	0.53	72.19	m	**114.64**
Beds and coverings; 100 mm thick; to pipes							
100 mm nominal size	–	0.50	7.76	0.53	12.03	m	**20.32**
150 mm nominal size	–	0.58	9.00	0.53	14.04	m	**23.57**
225 mm nominal size	–	0.83	12.88	0.53	20.06	m	**33.47**
300 mm nominal size	–	1.00	15.52	0.53	24.07	m	**40.12**
375 mm nominal size	–	1.21	18.78	0.53	29.07	m	**48.38**
450 mm nominal size	–	1.42	22.04	0.53	34.09	m	**56.66**
600 mm nominal size	–	1.83	28.40	0.53	44.11	m	**73.04**
900 mm nominal size	–	2.79	43.31	0.53	67.18	m	**111.02**
1200 mm nominal size	–	3.83	59.45	0.53	92.24	m	**152.22**
Beds and coverings; 150 mm thick; to pipes							
100 mm nominal size	–	0.75	11.64	0.53	18.05	m	**30.22**
150 mm nominal size	–	0.83	12.88	0.53	20.06	m	**33.47**
225 mm nominal size	–	1.08	16.77	0.53	26.07	m	**43.37**
300 mm nominal size	–	1.30	20.18	0.53	31.08	m	**51.79**
375 mm nominal size	–	1.50	23.29	0.53	36.10	m	**59.92**
450 mm nominal size	–	1.79	27.78	0.53	43.10	m	**71.41**
600 mm nominal size	–	2.16	33.53	0.53	52.13	m	**86.19**
900 mm nominal size	–	3.54	54.96	0.53	85.22	m	**140.71**
1200 mm nominal size	–	5.00	77.62	0.53	120.31	m	**198.46**
PIPES, FITTINGS AND ACCESSORIES							
NOTE: The following items unless otherwise described include for all appropriate joints/couplings in the running length. The prices for gullies and rainwater shoes, etc. include for appropriate joints to pipes and for setting on and surrounding accessory with site mixed in situ concrete 10.00 N/mm²–40 mm aggregate (1:3:6).							

34 DRAINAGE BELOW GROUND

Item	PC £	Labour hours	Labour £	Plant £	Material £	Unit	Total rate £
Cast Iron							
Timesaver drain pipes and fittings or other equal; BS 437; coated; with mechanical coupling joints							
100 mm dia. pipes							
laid straight	39.57	0.46	6.05	–	52.91	m	**58.96**
in runs not exceeding 3 m long	39.57	0.63	8.27	–	82.56	m	**90.83**
extra for							
bend; medium radius	48.63	0.56	7.36	–	71.81	nr	**79.17**
bend; medium radius with access	135.12	0.56	7.36	–	162.62	nr	**169.98**
bend; long radius	80.34	0.56	7.36	–	102.92	nr	**110.28**
rest bend	55.78	0.56	7.36	–	77.14	nr	**84.50**
single branch	64.51	0.69	9.06	–	111.64	nr	**120.70**
single branch; with access	148.80	0.79	10.38	–	200.14	nr	**210.52**
double branch	109.67	0.88	11.56	–	182.85	nr	**194.41**
isolated Timesaver joint	25.99	0.32	4.20	–	27.29	nr	**31.49**
transitional pipe; for WC	38.08	0.46	6.05	–	67.27	nr	**73.32**
150 mm dia. pipes							
laid straight	73.26	0.56	7.36	–	92.01	m	**99.37**
in runs not exceeding 3 m long	73.26	0.76	9.99	–	135.37	m	**145.36**
extra for							
bend; medium radius	111.87	0.65	8.54	–	138.41	nr	**146.95**
bend; medium radius with access	237.23	0.65	8.54	–	270.04	nr	**278.58**
bend; long radius	149.81	0.65	8.54	–	174.22	nr	**182.76**
diminishing pipe	63.39	0.65	8.54	–	83.46	nr	**92.00**
single branch	139.29	0.79	10.38	–	149.43	nr	**159.81**
isolated Timesaver joint	31.49	0.39	5.12	–	33.06	nr	**38.18**
Accessories in Timesaver cast iron or other equal and approved; with mechanical coupling joints							
Gully fittings; comprising low invert gully trap and round hopper							
100 mm outlet	56.63	0.88	11.56	–	93.25	nr	**104.81**
150 mm outlet	140.91	1.20	15.77	–	190.73	nr	**206.50**
Add to above for bellmouth 300 mm high; circular plain grating							
100 mm nominal size; 200 mm grating	58.98	0.42	5.52	–	98.55	nr	**104.07**
100 mm nominal size; 100 mm horizontal inlet; 200 mm grating	72.11	0.42	5.52	–	112.68	nr	**118.20**
100 mm nominal size; 100 mm vertical inlet; 200 mm grating	73.94	0.42	5.52	–	114.66	nr	**120.18**
Yard gully (Deans); trapped; galvanized sediment pan; 267 mm round heavy grating							
100 mm outlet	382.71	2.68	35.22	–	474.95	nr	**510.17**
Yard gully (garage); trapless; galvanized sediment pan; 267 mm round heavy grating							
100 mm outlet	390.39	2.50	32.84	–	453.46	nr	**486.30**

34 DRAINAGE BELOW GROUND

Item	PC £	Labour hours	Labour £	Plant £	Material £	Unit	Total rate £
PIPES, FITTINGS AND ACCESSORIES – cont							
Accessories in Timesaver cast iron or other equal and approved – cont							
Yard gully (garage); trapped; with rodding eye, galvanized perforated sediment pan; stopper; 267 mm round heavy grating							
100 mm outlet	744.19	2.50	32.84	–	890.57	nr	**923.41**
Grease trap; internal access; galvanized perforated bucket; lid and frame							
100 mm outlet; 20 gallon capacity	809.76	3.70	48.61	–	919.88	nr	**968.49**
Ensign lightweight drain pipes and fittings or other equal and approved; BS EN 877; ductile iron couplings							
100 mm dia. pipes							
laid straight	28.80	0.19	2.91	–	35.26	m	**38.17**
extra for							
bend; long radius	44.58	0.19	2.91	–	61.89	nr	**64.80**
single branch	30.81	0.23	3.56	–	62.49	nr	**66.05**
150 mm dia. pipes							
laid straight	55.97	0.22	3.39	–	69.03	m	**72.42**
extra for							
bend; medium radius	133.64	0.22	3.39	–	171.10	nr	**174.49**
single branch	72.42	0.28	4.30	–	137.59	nr	**141.89**
Clay							
Extra strength vitrified clay pipes and fittings; Hepworth Supersleve or other equal; plain ends with push fit polypropylene flexible couplings							
100 mm dia. pipes							
laid straight	5.03	0.19	2.50	–	9.19	m	**11.69**
extra for							
bend	6.23	0.19	2.50	–	12.76	nr	**15.26**
rest bend	12.02	0.19	2.50	–	18.83	nr	**21.33**
rodding point	30.17	0.19	2.50	–	37.36	nr	**39.86**
socket adaptor	8.14	0.16	2.10	–	11.92	nr	**14.02**
saddle	14.40	0.69	9.06	–	19.02	nr	**28.08**
single junction	13.46	0.23	3.02	–	23.71	nr	**26.73**
single access junction	43.87	0.23	3.02	–	55.65	nr	**58.67**
150 mm dia. pipes							
laid straight	9.34	0.23	3.02	–	16.31	m	**19.33**
extra for							
bend	12.83	0.22	2.89	–	23.53	nr	**26.42**
access bend	7.42	0.22	2.89	–	59.90	nr	**62.79**
90° bend	12.83	0.22	2.89	–	23.53	nr	**26.42**
taper pipe	15.89	0.22	2.89	–	24.14	nr	**27.03**
rodding point	46.29	0.22	2.89	–	57.68	nr	**60.57**
socket adaptor	13.01	0.19	2.50	–	19.18	nr	**21.68**
adaptor to HepSeal pipe	8.56	0.19	2.50	–	14.50	nr	**17.00**

34 DRAINAGE BELOW GROUND

Item	PC £	Labour hours	Labour £	Plant £	Material £	Unit	Total rate £
saddle	19.66	0.83	10.91	–	27.14	nr	**38.05**
single junction	17.45	0.28	3.67	–	33.89	nr	**37.56**
single access junction	65.22	0.28	3.67	–	84.05	nr	**87.72**
225 mm dia. pipes							
laid straight	32.83	0.23	3.02	–	48.05	m	**51.07**
extra for							
bend	71.25	0.22	2.89	–	91.62	nr	**94.51**
90° bend	71.25	0.22	2.89	–	91.62	nr	**94.51**
taper pipe	17.30	0.22	2.89	–	21.41	nr	**24.30**
socket adaptor	30.46	0.19	2.50	–	42.10	nr	**44.60**
300 mm dia. pipes includiing EPDM coupling							
laid straight	50.26	0.23	3.02	–	80.65	m	**83.67**
extra for							
bend	135.29	0.24	3.15	–	181.98	nr	**185.13**
90° bend	135.29	0.24	3.15	–	181.98	nr	**185.13**
taper pipe	17.30	0.24	3.15	–	30.21	nr	**33.36**
socket adaptor	48.36	0.20	2.63	–	73.38	nr	**76.01**
Extra strength vitrified clay pipes and fittings; Hepworth SuperSeal/Hepseal or equivalent; socketted; with push-fit flexible joints							
150 mm dia. pipes							
SuperSeal pipes; laid straight	14.94	0.30	3.94	–	15.69	m	**19.63**
extra for							
bend	28.71	0.23	3.02	–	25.44	nr	**28.46**
rest bend	15.41	0.20	2.63	–	11.48	nr	**14.11**
stopper	8.66	0.15	1.97	–	9.09	nr	**11.06**
taper reducer	14.86	0.23	3.02	–	10.90	nr	**13.92**
saddle	18.38	0.75	9.85	–	19.30	nr	**29.15**
single junction	37.52	0.30	3.94	–	33.13	nr	**37.07**
225 mm dia. pipes							
SuperSeal pipes; laid straight	31.00	0.38	4.99	–	32.55	m	**37.54**
extra for							
bend	67.27	0.30	3.94	–	60.87	nr	**64.81**
rest bend	82.18	0.30	3.94	–	76.52	nr	**80.46**
stopper	14.57	0.19	2.50	–	15.30	nr	**17.80**
taper reducer	46.35	0.30	3.94	–	38.90	nr	**42.84**
saddle	68.37	1.00	13.14	–	71.79	nr	**84.93**
single junction	119.50	0.38	4.99	–	112.45	nr	**117.44**
300 mm Superseal pipes							
SuperSeal pipes; laid straight	47.54	0.50	6.57	–	49.92	m	**56.49**
extra for							
bend	127.77	0.40	5.26	–	119.19	nr	**124.45**
rest bend	182.08	0.40	5.26	–	176.20	nr	**181.46**
stopper	31.12	0.25	3.29	–	32.68	nr	**35.97**
taper reducer	127.91	0.40	5.26	–	119.32	nr	**124.58**
saddle	119.05	1.33	17.47	–	125.00	nr	**142.47**
single junction	226.35	0.50	6.57	–	217.71	nr	**224.28**

34 DRAINAGE BELOW GROUND

Item	PC £	Labour hours	Labour £	Plant £	Material £	Unit	Total rate £
PIPES, FITTINGS AND ACCESSORIES – cont							
Extra strength vitrified clay pipes and fittings – cont							
400 mm dia. pipes							
Hepseal pipes; laid straight	116.93	0.67	8.80	–	122.78	m	**131.58**
extra for							
bend	439.40	0.54	7.10	–	424.54	nr	**431.64**
single unequal junction	411.72	0.67	8.80	–	383.20	nr	**392.00**
450 mm pipes							
Hepseal pipes; laid straight	151.89	0.83	10.91	–	159.48	m	**170.39**
extra for							
bend	578.63	0.67	8.80	–	559.71	nr	**568.51**
single unequal junction	492.12	0.83	10.91	–	452.94	nr	**463.85**
British Standard quality vitrified clay pipes and fittings; socketted; cement and sand (1:2) joints							
100 mm dia. pipes							
laid straight	9.44	0.37	4.86	–	10.05	m	**14.91**
extra for							
bend (short/medium/knuckle)	6.61	0.30	3.94	–	7.07	nr	**11.01**
bend (long/rest/elbow)	15.52	0.30	3.94	–	13.45	nr	**17.39**
single junction	17.34	0.37	4.86	–	14.40	nr	**19.26**
double collar	11.38	0.25	3.29	–	12.07	nr	**15.36**
150 mm dia. pipes							
laid straight	14.53	0.42	5.52	–	15.38	m	**20.90**
extra for							
bend (short/medium/knuckle)	14.40	0.33	4.34	–	10.67	nr	**15.01**
bend (long/rest/elbow)	25.99	0.33	4.34	–	22.84	nr	**27.18**
taper	34.41	0.33	4.34	–	31.22	nr	**35.56**
single junction	28.44	0.42	5.52	–	23.92	nr	**29.44**
double collar	18.96	0.28	3.67	–	20.03	nr	**23.70**
225 mm dia. pipes							
laid straight	28.78	0.51	6.70	–	30.52	m	**37.22**
extra for							
double collar	44.39	0.33	4.34	–	46.74	nr	**51.08**
300 mm pipes							
laid straight	48.25	0.69	9.06	–	50.97	m	**60.03**
Accessories in vitrified clay; set in concrete; with polypropylene coupling joints to pipes							
Rodding point; with square aluminium plate							
100 mm nominal size	32.87	0.46	6.05	–	40.34	nr	**46.39**
Gully fittings; comprising low back trap and square hopper; 150 mm × 150 mm square gully grid							
100 mm nominal size	31.09	0.79	10.38	–	42.37	nr	**52.75**
Gully fittings; comprising low back trap and square hopper with back inlet; 150 mm × 150 mm square gully grid							
100 mm nominal size	53.35	0.85	11.17	–	65.74	nr	**76.91**

34 DRAINAGE BELOW GROUND

Item	PC £	Labour hours	Labour £	Plant £	Material £	Unit	Total rate £
Accessories in vitrified clay; set in concrete; with cement and sand (1:2) joints to pipes							
Yard gully; 225 mm dia.; including domestic duty grating and frame (up to 1 tonne) and combined filter and silk bucket							
100 mm outlet	141.82	2.50	32.84	–	149.50	nr	**182.34**
100 mm outlet; 100 mm back inlet	199.59	2.70	35.48	–	210.16	nr	**245.64**
150 mm outlet	141.82	3.50	45.99	–	149.50	nr	**195.49**
150 mm outlet; 150 mm back inlet	203.76	3.70	48.61	–	214.54	nr	**263.15**
Yard gully; 225 mm dia.; including medium duty grating and frame (up to 5 tonnes) and combined filter and silk bucket							
100 mm outlet	186.60	2.50	32.84	–	196.51	nr	**229.35**
100 mm outlet; 100 mm back inlet	248.44	2.70	35.48	–	261.44	nr	**296.92**
150 mm outle	201.23	3.50	45.99	–	211.87	nr	**257.86**
150 mm outlet; 150 mm back inlet	252.61	3.70	48.61	–	265.83	nr	**314.44**
Road gully; trapped with rodding eye and stopper (grate not included)							
300 mm × 600 mm × 100 mm outlet	104.10	3.05	40.07	–	131.63	nr	**171.70**
300 mm × 600 mm × 150 mm outlet	106.60	3.05	40.07	–	134.25	nr	**174.32**
400 mm × 750 mm × 150 mm outlet	123.62	3.70	48.61	–	163.71	nr	**212.32**
450 mm × 900 mm × 150 mm outlet	167.25	4.65	61.10	–	216.28	nr	**277.38**
Grease trap; with internal access; galvanized perforated bucket; lid and frame							
600 mm × 450 mm × 600 mm deep; 100 mm outlet	852.21	3.89	51.11	–	928.13	nr	**979.24**
Interceptor; trapped with inspection arm; lever locking stopper; chain and staple; cement and sand (1:2) joints to pipes; building in, and cutting and fitting brickwork around							
100 mm outlet; 100 mm inlet	114.38	3.70	48.61	–	120.65	nr	**169.26**
150 mm outlet; 150 mm inlet	162.33	4.16	54.65	–	170.99	nr	**225.64**
225 mm outlet; 225 mm inlet	442.58	4.63	60.84	–	465.32	nr	**526.16**
Accessories in polypropylene; cover set in concrete; with coupling joints to pipes							
Inspection chamber; 5 nr 100 mm inlets; cast iron cover and frame							
475 mm dia. × 595 mm deep	217.47	2.13	27.98	–	237.03	nr	**265.01**
475 mm dia. × 940 mm deep	264.92	2.31	30.36	–	286.86	nr	**317.22**
Accessories; grates and covers							
Aluminium alloy gully grids; set in position							
120 mm × 120 mm	3.58	0.09	1.19	–	3.76	nr	**4.95**
150 mm × 150 mm	3.43	0.09	1.19	–	3.60	nr	**4.79**
225 mm × 225 mm	10.67	0.09	1.19	–	11.20	nr	**12.39**
100 mm dia.	3.58	0.09	1.19	–	3.76	nr	**4.95**
150 mm dia.	5.49	0.09	1.19	–	5.76	nr	**6.95**
225 mm dia.	11.94	0.09	1.19	–	12.54	nr	**13.73**

34 DRAINAGE BELOW GROUND

Item	PC £	Labour hours	Labour £	Plant £	Material £	Unit	Total rate £
PIPES, FITTINGS AND ACCESSORIES – cont							
Accessories – cont							
Aluminium alloy sealing plates and frames; set in cement and sand (1:3)							
150 mm × 150 mm	13.78	0.23	3.02	–	14.57	nr	**17.59**
225 mm × 225 mm	25.22	0.23	3.02	–	26.59	nr	**29.61**
140 mm dia. (for 100 mm)	11.22	0.23	3.02	–	11.90	nr	**14.92**
197 mm dia. (for 150 mm)	16.15	0.23	3.02	–	17.07	nr	**20.09**
273 mm dia. (for 225 mm)	25.86	0.23	3.02	–	27.26	nr	**30.28**
Polypropylene access covers and frames; supplied by Manhole Covers Ltd or other equal and approved; to suit PPIC inspection chambers; bedding and pointing in frame.							
450 mm dia.; class A15	42.64	1.30	17.08	–	46.48	nr	**63.56**
450 mm dia.; class B125; kitemarked	32.41	1.30	17.08	–	35.74	nr	**52.82**
Ductile iron heavy duty road gratings and frame; supplied by Manhole Covers Ltd or other equal and approved; bedding and pointing in cement and sand (1:3); one course half brick thick wall in semi-engineering bricks in cement mortar (1:3)							
225 mm × 225 mm × 80 mm hinged and dished road grating and frame; class C250	28.44	2.25	29.56	–	34.04	nr	**63.60**
300 mm × 300 mm × 80 mm hinged and dished road grating and frame; class C250	46.99	2.25	29.56	–	53.52	nr	**83.08**
420 mm × 420 mm × 75 mm hinged road grating and frame; class C250; kitemarked	58.12	2.25	29.56	–	65.20	nr	**94.76**
445 mm × 445 mm × 75 mm double triangular road grating and frame; class C250; kitemarked	61.83	2.25	29.56	–	69.10	nr	**98.66**
435 mm × 435 mm × 100 mm pedestrian mesh road grating and frame; class D400	63.07	2.25	29.56	–	70.40	nr	**99.96**
440 mm × 400 mm × 150 mm hinged road grating and frame; class D400; kitemarked	81.61	2.25	29.56	–	89.87	nr	**119.43**
Concrete							
Vibrated concrete pipes and fittings; with flexible joints; BS 5911 Part 1; trench 2.00 m deep							
300 mm dia. pipes							
Class M; laid straight	17.08	0.65	8.54	6.93	17.93	m	**33.40**
extra for							
bend; <= 45°	–	0.65	8.54	–	179.31	nr	**187.85**
bend; > 45°	–	0.65	8.54	–	268.97	nr	**277.51**
junction; 300 mm × 100 mm	–	0.46	6.05	–	94.14	nr	**100.19**
450 mm dia. pipes							
Class H; laid straight	24.96	1.02	13.40	6.93	26.21	m	**46.54**
extra for							
bend; <= 45°	–	1.02	13.40	–	262.13	nr	**275.53**
bend; > 45°	–	1.02	13.40	–	393.19	nr	**406.59**
junction; 450 mm × 150 mm	–	0.65	8.54	–	137.62	nr	**146.16**

34 DRAINAGE BELOW GROUND

Item	PC £	Labour hours	Labour £	Plant £	Material £	Unit	Total rate £
525 mm dia. pipes							
Class H; laid straight	32.76	1.48	19.45	6.93	34.40	m	**60.78**
extra for							
bend; <= 45°	–	1.48	19.45	6.93	343.98	nr	**370.36**
bend; > 45°	–	1.48	19.45	6.93	515.97	nr	**542.35**
junction; 600 mm × 150 mm	–	0.83	10.91	6.93	240.79	nr	**258.63**
900 mm dia. pipes							
Class H; laid straight	101.91	2.59	34.03	6.93	107.01	m	**147.97**
extra for							
bend; <= 45°	–	2.59	34.03	6.93	1070.05	nr	**1111.01**
bend; > 45°	–	2.59	34.03	6.93	1605.09	nr	**1646.05**
junction; 900 mm × 150 mm	–	1.02	13.40	6.93	481.53	nr	**501.86**
1200 mm dia. pipes							
Class H; laid straight	174.89	3.70	48.61	6.93	183.63	m	**239.17**
extra for							
bend; <= 45°	–	3.70	48.61	6.93	1836.32	nr	**1891.86**
bend; > 45°	–	3.70	48.61	6.93	2754.49	nr	**2810.03**
junction; 1200 mm × 150 mm	–	1.48	19.45	6.93	826.35	nr	**852.73**
Accessories in precast concrete; top set in with rodding eye and stopper; cement and sand (1:2) joint to pipe							
Concrete road gully; BS 5911; trapped with rodding eye and stopper; cement and sand (1:2) joint to pipe							
450 mm dia. × 900 mm deep; 100 mm or 150 mm outlet	47.43	4.39	57.68	–	75.53	nr	**133.21**
450 mm dia. × 1050 mm deep; 100 mm or 150 mm outlet	51.71	4.39	57.68	–	80.02	nr	**137.70**
Gully adapter type 1501	–	–	–	–	4.68	nr	**4.68**
uPVC							
Osmadrain uPVC pipes and fittings or other equal and approved; BS 4660; with ring seal joints							
82 mm dia. pipes							
laid straight	9.43	0.15	1.97	–	9.90	m	**11.87**
extra for							
bend; short radius	18.56	0.13	1.71	–	19.49	nr	**21.20**
spigot/socket bend	15.60	0.13	1.71	–	16.38	nr	**18.09**
adaptor	8.15	0.07	0.92	–	8.56	nr	**9.48**
single junction	24.14	0.18	2.36	–	25.35	nr	**27.71**

34 DRAINAGE BELOW GROUND

Item	PC £	Labour hours	Labour £	Plant £	Material £	Unit	Total rate £
PIPES, FITTINGS AND ACCESSORIES – cont							
Osmadrain uPVC pipes and fittings or other equal and approved – cont							
110 mm dia. pipes							
laid straight	5.77	0.17	2.24	–	7.23	m	**9.47**
extra for							
bend; short radius	17.54	0.15	1.97	–	18.05	nr	**20.02**
bend; long radius	33.23	0.15	1.97	–	33.07	nr	**35.04**
spigot/socket bend	14.82	0.15	1.97	–	21.29	nr	**23.26**
socket plug	7.68	0.04	0.53	–	8.06	nr	**8.59**
adjustable double socket bend	20.98	0.15	1.97	–	27.87	nr	**29.84**
adaptor to clay	22.74	0.09	1.19	–	23.65	nr	**24.84**
single junction	20.92	0.21	2.76	–	20.15	nr	**22.91**
sealed access junction	54.12	0.19	2.50	–	55.01	nr	**57.51**
slip coupler	10.15	0.09	1.19	–	10.66	nr	**11.85**
160 mm dia. pipes							
laid straight	13.24	0.21	2.76	–	16.07	m	**18.83**
extra for							
bend; short radius	41.79	0.18	2.36	–	43.05	nr	**45.41**
spigot/socket bend	37.82	0.18	2.36	–	50.76	nr	**53.12**
socket plug	16.47	0.07	0.92	–	17.29	nr	**18.21**
adaptor to clay	49.42	0.12	1.58	–	51.19	nr	**52.77**
level invert taper	20.23	0.18	2.36	–	31.45	nr	**33.81**
single junction	68.30	0.24	3.15	–	71.72	nr	**74.87**
slip coupler	24.65	0.11	1.45	–	25.88	nr	**27.33**
uPVC Osma Ultra-Rib ribbed pipes and fittings or other equal and approved; WIS approval; with sealed ring push-fit joints							
150 mm dia. pipes							
laid straight	7.49	0.19	2.50	–	7.86	m	**10.36**
extra for							
bend; short radius	21.64	0.17	2.24	–	22.25	nr	**24.49**
adaptor to 160 mm dia. uPVC	28.78	0.10	1.31	–	29.27	nr	**30.58**
adaptor to clay	70.34	0.10	1.31	–	73.38	nr	**74.69**
level invert taper	10.57	0.18	2.36	–	9.69	nr	**12.05**
single junction	43.55	0.22	2.89	–	43.38	nr	**46.27**
225 mm dia. pipes							
laid straight	19.40	0.22	2.89	–	20.37	m	**23.26**
extra for							
bend; short radius	101.78	0.20	2.63	–	105.64	nr	**108.27**
adaptor to clay	91.60	0.13	1.71	–	93.73	nr	**95.44**
level invert taper	17.72	0.20	2.63	–	14.93	nr	**17.56**
single junction	151.09	0.27	3.55	–	152.53	nr	**156.08**
300 mm dia. pipes							
laid straight	27.53	0.32	4.20	–	28.91	m	**33.11**
extra for							
bend; short radius	160.31	0.29	3.81	–	166.59	nr	**170.40**
adaptor to clay	240.96	0.14	1.84	–	249.54	nr	**251.38**
level invert taper	57.50	0.29	3.81	–	55.18	nr	**58.99**
single junction	349.09	0.37	4.86	–	357.88	nr	**362.74**

34 DRAINAGE BELOW GROUND

Item	PC £	Labour hours	Labour £	Plant £	Material £	Unit	Total rate £
Accessories in uPVC; with ring seal joints to pipes (unless otherwise described)							
Rodding eye							
Square top, sealed with coupling	27.88	0.43	5.65	–	34.10	nr	**39.75**
Universal gully fitting; comprising gully trap, plain hopper							
150 mm × 150 mm grate	31.23	0.93	12.22	–	39.54	nr	**51.76**
Bottle gully; square comprising gully with rotating							
217 mm × 217 mm grate	57.66	0.78	10.25	–	67.30	nr	**77.55**
Shallow inspection chamber; 250 mm dia.; 600 mm deep; sealed cover and frame (1.0 tonne loading)							
4 nr 110 mm outlets/inlets	112.55	1.28	16.82	–	142.31	nr	**159.13**
Universal inspection chamber; 450 mm dia.; 4 nr 110 mm branch outlet; single seal cast iron cover and frame (12.5 tonnes loading)							
500 mm deep	228.99	1.35	17.73	–	264.56	nr	**282.29**
730 mm deep	264.07	1.60	21.02	–	306.23	nr	**327.25**
960 mm deep	299.16	1.85	24.31	–	347.89	nr	**372.20**
Equal manhole base; 500 mm dia.							
6 nr 110 mm branch outlets	131.56	1.21	15.90	–	152.62	nr	**168.52**
Unequal manhole base; 500 mm dia.							
2 nr 160 mm, 4nr 110 mm branch outlets	131.45	1.21	15.90	–	152.50	nr	**168.40**
Kerb to gullies; class B engineering bricks on edge to three sides in cement mortar (1:3) rendering in cement mortar (1:3) to top and two sides and skirting to brickwork 230 mm high; dishing in cement mortar (1:3) to gully; steel trowelled							
230 mm × 230 mm internally	–	1.39	18.26	–	1.34	nr	**19.60**
LAND DRAINAGE							
Excavating; by hand; grading bottoms; earthwork support; filling to within 150 mm of surface with gravel rejects; remainder filled with excavated material and compacting; disposal of surplus soil on site; spreading on site average 50 m							
Pipes not exceeding 200 nominal size							
average depth of trench 0.75 m	–	1.57	20.79	–	10.30	m	**31.09**
average depth of trench 1.00 m	–	2.08	27.54	–	17.05	m	**44.59**
average depth of trench 1.25 m	–	2.91	38.54	–	21.62	m	**60.16**
average depth of trench 1.50 m	–	5.00	66.21	–	26.55	m	**92.76**
average depth of trench 1.75 m	–	5.92	78.39	–	31.12	m	**109.51**
average depth of trench 2.00 m	–	6.85	90.71	–	36.05	m	**126.76**
Disposal; load lorry by machine							
Excavated material							
off site; to tip not exceeding 13 km (using lorries);							
including Landfill Tax based on inactive waste	–	–	–	1.23	25.00	m^3	**26.23**

34 DRAINAGE BELOW GROUND

Item	PC £	Labour hours	Labour £	Plant £	Material £	Unit	Total rate £
LAND DRAINAGE – cont							
Disposal; load lorry by hand							
Excavated material							
off site; to tip not exceeding 13 km (using lorries); including Landfill Tax based on inactive waste	–	0.75	9.93	–	25.00	m³	**34.93**
Vitrified clay perforated subsoil pipes; BS 65; Hepworth Hepline or other equal and approved							
Pipes; laid straight							
100 mm dia.	8.44	0.20	2.63	–	8.86	m	**11.49**
150 mm dia.	14.50	0.25	3.29	–	15.22	m	**18.51**
225 mm dia.	32.44	0.33	4.34	–	34.06	m	**38.40**
Concrete Canvas cement impregnated cloth lining to form ditch linings; holding ponds; slope protection and similar							
Type CC8, 8 mm thick and sprayed with water	–	0.10	1.55	0.09	37.79	m²	**39.43**
MANHOLES AND SOAKAWAYS							
Excavating; by machine							
Manholes							
maximum depth not exceeding 1.00 m	–	0.19	2.52	4.04	–	m³	**6.56**
maximum depth not exceeding 2.00 m	–	0.21	2.78	4.45	–	m³	**7.23**
maximum depth not exceeding 4.00 m	–	0.25	3.31	5.20	–	m³	**8.51**
Excavating; by hand							
Manholes							
maximum depth not exceeding 1.00 m	–	3.05	40.38	–	–	m³	**40.38**
maximum depth not exceeding 2.00 m	–	3.61	47.81	–	–	m³	**47.81**
maximum depth not exceeding 4.00 m	–	4.63	61.31	–	–	m³	**61.31**
Earthwork support (average risk prices)							
Maximum depth not exceeding 1.00 m							
distance between opposing faces not exceeding 2.00 m	–	0.14	1.86	–	1.89	m²	**3.75**
Maximum depth not exceeding 2.00 m							
distance between opposing faces not exceeding 2.00 m	–	0.18	2.38	–	3.57	m²	**5.95**
Maximum depth not exceeding 4.00 m							
distance between opposing faces not exceeding 2.00 m	–	0.22	2.91	–	5.26	m²	**8.17**
Disposal; by machine							
Excavated material							
off site; to tip not exceeding 13 km (using lorries) including Landfill Tax based on inactive waste	–	–	–	1.23	25.00	m³	**26.23**
on site; depositing on site in spoil heaps; average 50 m distance	–	0.14	1.86	3.61	–	m³	**5.47**

34 DRAINAGE BELOW GROUND

Item	PC £	Labour hours	Labour £	Plant £	Material £	Unit	Total rate £
Disposal; by hand							
Excavated material							
off site; to tip not exceeding 13 km (using lorries) including Landfill Tax based on inactive waste	–	0.75	9.93	–	25.00	m³	**34.93**
on site; depositing on site in spoil heaps; average 50 m distance	–	1.20	15.89	–	–	m³	**15.89**
Filling to excavations; by machine							
Average thickness not exceeding 0.25 m							
arising excavations	–	0.14	1.86	1.91	–	m³	**3.77**
Filling to excavations; by hand							
Average thickness not exceeding 0.25 m							
arising from excavations	–	0.93	12.32	–	–	m³	**12.32**
Plain in situ ready mixed designated concrete; C10–40 mm aggregate							
Beds							
thickness not exceeding 150 mm	96.51	2.78	43.16	–	101.34	m³	**144.50**
thickness 150 mm–450 mm	–	2.08	32.29	–	101.34	m³	**133.63**
thickness exceeding 450 mm	–	1.76	27.32	–	101.34	m³	**128.66**
Plain in situ ready mixed designated concrete; C20–20 mm aggregate							
Beds							
thickness not exceeding 150 mm	100.27	2.78	43.16	–	105.28	m³	**148.44**
thickness 150 mm–450 mm	–	2.08	32.29	–	105.28	m³	**137.57**
thickness exceeding 450 mm	–	1.76	27.32	–	105.28	m³	**132.60**
Plain in situ ready mixed designated concrete; C25–20 mm aggregate; (small quantities)							
Benching in bottoms							
150 mm–450 mm average thickness	102.13	8.33	141.38	–	107.24	m³	**248.62**
Reinforced in situ ready mixed designated concrete; C20–20 mm aggregate; (small quantities)							
Isolated cover slabs							
thickness not exceeding 150 mm	95.49	6.48	100.59	–	100.26	m³	**200.85**
Reinforcement; fabric to BS 4449; lapped; in beds or suspended slabs							
Ref D98 (1.54 kg/m²)							
400 mm minimum laps	0.97	0.11	2.02	–	1.02	m²	**3.04**
Ref A142 (2.22 kg/m²)							
400 mm minimum laps	1.10	0.11	2.02	–	1.16	m²	**3.18**
Ref A193 (3.02 kg/m²)							
400 mm minimum laps	1.49	0.11	2.02	–	1.56	m²	**3.58**

34 DRAINAGE BELOW GROUND

Item	PC £	Labour hours	Labour £	Plant £	Material £	Unit	Total rate £
MANHOLES AND SOAKAWAYS – cont							
Formwork; basic finish							
Soffits of isolated cover slabs							
horizontal	–	2.64	48.45	–	4.20	m²	**52.65**
Edges of isolated cover slabs							
height not exceeding 250 mm	–	0.78	14.31	–	1.60	m	**15.91**
Precast concrete circular manhole rings; bedding, jointing and pointing in cement mortar (1:3) on prepared bed							
Chamber or shaft rings; plain							
900 mm dia.	53.28	5.09	66.87	–	56.69	m	**123.56**
1050 mm dia.	56.53	6.01	78.96	–	60.85	m	**139.81**
1200 mm dia.	68.88	6.94	91.18	–	74.56	m	**165.74**
Chamber or shaft rings; reinforced							
1350 mm dia.	108.11	7.86	103.27	34.66	116.51	m	**254.44**
1500 mm dia.	119.92	8.79	115.49	34.66	130.40	m	**280.55**
1800 mm dia.	182.80	11.10	145.85	62.38	198.66	m	**406.89**
2100 mm dia.	368.08	13.88	182.36	69.31	395.45	m	**647.12**
extra for step irons built in	6.08	0.14	1.84	–	6.38	nr	**8.22**
extra for integrated ladder system 150 mm projection polypropylene encapusulated steps and rails	52.16	1.00	13.14	–	54.77	m	**67.91**
Reducing slabs							
1200 mm dia.	97.94	5.55	72.92	34.66	104.34	nr	**211.92**
1350 mm dia.	157.02	8.79	115.49	34.66	167.86	nr	**318.01**
1500 mm dia.	178.96	10.18	133.75	34.66	191.64	nr	**360.05**
1800 mm dia.	281.68	12.95	170.14	62.38	301.74	nr	**534.26**
Heavy duty cover slabs; to suit rings							
900 mm dia.	58.50	2.78	36.53	13.86	62.17	nr	**112.56**
1050 mm dia.	62.05	3.24	42.57	13.86	66.06	nr	**122.49**
1200 mm dia.	75.34	3.70	48.61	15.94	80.61	nr	**145.16**
1350 mm dia.	120.78	4.16	54.65	17.32	129.06	nr	**201.03**
1500 mm dia.	137.67	4.63	60.84	29.12	147.55	nr	**237.51**
1800 mm dia.	216.68	5.55	72.92	33.27	231.65	nr	**337.84**
2100 mm dia.	459.79	6.48	85.13	34.66	488.38	nr	**608.17**
Precast concrete circular manhole; CPM Perfect Manhole or similar; complete with preformed benching and outlets to base; elastomeric seal to joints to rings; single steps as required							
1200 mm with up to four outlets 100 mm or 150 mm outlets; effective internal depth:							
1250 mm deep	–	5.00	68.88	27.72	845.65	nr	**942.25**
1500 mm deep	–	5.00	68.88	27.72	890.58	nr	**987.18**
1800 mm deep	–	5.00	68.88	27.72	898.30	nr	**994.90**
2000 mm deep	–	5.00	68.88	27.72	951.49	nr	**1048.09**
2500 mm deep	–	4.50	59.13	34.66	1103.36	nr	**1197.15**

34 DRAINAGE BELOW GROUND

Item	PC £	Labour hours	Labour £	Plant £	Material £	Unit	Total rate £
1500 mm dia. with up to four outlets; effective internal depth:							
2000 mm deep; upto 450 mm pipe	–	4.50	59.13	34.66	1580.86	nr	**1674.65**
2500 mm deep; up to 450 mm pipe	–	4.75	62.41	38.82	1834.02	nr	**1935.25**
3000 mm deep; up to 450 mm pipe	–	4.75	62.41	38.82	1928.70	nr	**2029.93**
3500 mm deep; up to 450 mm pipe	–	5.00	65.70	41.59	2026.16	nr	**2133.45**
4000 mm deep; up to 450 mm pipe	–	5.00	65.70	41.59	2279.33	nr	**2386.62**
4500 mm deep; up to 450 mm pipe	–	5.00	65.70	41.59	2374.01	nr	**2481.30**
4000 mm deep; 600 mm pipe	–	5.00	65.70	41.59	2390.82	nr	**2498.11**
4500 mm deep; 600 mm pipe	–	5.00	65.70	41.59	2485.50	nr	**2592.79**
Common bricks; in cement mortar (1:3)							
Walls to manholes							
one brick thick	380.00	2.22	47.86	–	66.24	m²	**114.10**
one and a half brick thick	–	3.24	69.85	–	99.37	m²	**169.22**
Projections of footings							
two brick thick	–	4.53	97.66	–	132.50	m²	**230.16**
Class A engineering bricks; in cement mortar (1:3)							
Walls to manholes							
one brick thick (PC £ per 1000)	750.00	2.50	53.90	–	85.55	m²	**139.45**
one and a half brick thick	–	3.61	77.83	–	87.80	m²	**165.63**
Projections of footings							
two brick thick	–	5.09	109.74	–	171.12	m²	**280.86**
Class B engineering bricks; in cement mortar (1:3)							
Walls to manholes							
one brick thick (PC £ per 1000)	590.00	2.50	53.90	–	72.71	m²	**126.61**
one and a half brick thick	–	3.61	77.83	–	109.07	m²	**186.90**
Projections of footings							
two brick thick	–	5.09	109.74	–	145.44	m²	**255.18**
Brickwork sundries							
Extra over for fair face; flush smooth pointing							
manhole walls	–	0.19	4.10	–	–	m²	**4.10**
Building ends of pipes into brickwork; making good fair face or rendering							
not exceeding 55 mm nominal size	–	0.09	1.94	–	–	nr	**1.94**
55 mm–110 mm nominal size	–	0.14	3.01	–	–	nr	**3.01**
over 110 mm nominal size	–	0.19	4.10	–	–	nr	**4.10**
Step irons; BS 1247; malleable; galvanized; building into joints							
general purpose pattern	–	0.14	3.01	–	6.38	nr	**9.39**

34 DRAINAGE BELOW GROUND

Item	PC £	Labour hours	Labour £	Plant £	Material £	Unit	Total rate £
MANHOLES AND SOAKAWAYS – cont							
Cement and sand (1:3) in situ finishings; steel trowelled							
13 mm work to manhole walls; one coat; to							
brickwork base over 300 wide	–	0.65	14.02	–	1.49	m²	**15.51**
Cast iron inspection chambers; with bolted flat covers; BS 437; bedded in cement mortar (1:3); with mechanical coupling joints							
100 mm × 100 mm							
one branch either side	282.50	1.40	18.40	–	297.37	nr	**315.77**
two branches either side	532.60	2.00	26.28	–	559.98	nr	**586.26**
150 mm × 100 mm							
one branch either side	348.46	1.55	20.37	–	366.64	nr	**387.01**
two branches either side	673.67	2.15	28.24	–	708.86	nr	**737.10**
150 mm × 150 mm							
one branch either side	430.74	1.80	23.65	–	453.77	nr	**477.42**
two branches either side	828.91	2.60	34.16	–	871.86	nr	**906.02**
Coated cast or ductile iron access covers and frames; to BS EN124; supplied by Manhole Covers Ltd or other equal; bedding frame in cement and sand (1:3); cover in grease and sand							
Light duty; cast iron; rectangular single seal solid top							
450 mm × 450 mm; class A15	45.10	1.50	19.71	–	49.07	nr	**68.78**
600 mm × 450 mm; class A15	45.10	1.50	19.71	–	49.22	nr	**68.93**
600 mm × 600 mm; class A15	69.70	1.50	19.71	–	75.22	nr	**94.93**
750 mm × 600 mm; class A15	136.94	1.50	19.71	–	145.82	nr	**165.53**
Light duty; cast iron; rectangular double seal solid top							
Medium duty; ductile iron; rectangular single seal solid top							
450 mm × 450 mm × 40 mm; class C250; kitemarked	83.64	2.00	26.28	–	89.86	nr	**116.14**
600 mm × 450 mm × 40 mm; slide-out; class C250; kitemarked	86.10	2.00	26.28	–	92.44	nr	**118.72**
600 mm × 600 mm × 40 mm; slide-out; class C250; kitemarked	104.14	2.00	26.28	–	111.38	nr	**137.66**
760 mm × 600 mm × 40 mm; slide-out; class C250; kitemarked	154.57	2.00	26.28	–	164.34	nr	**190.62**
Heavy duty; ductile iron; solid top							
450 mm × 450 mm × 75 mm; single seal; class C250; kitemarked	122.42	2.50	32.84	–	130.57	nr	**163.41**
600 mm × 450 mm × 75 mm; single seal; class C250; kitemarked	136.02	2.50	32.84	–	144.86	nr	**177.70**
600 mm × 600 mm × 75 mm; single seal; class C250; kitemarked	155.81	2.50	32.84	–	165.63	nr	**198.47**

34 DRAINAGE BELOW GROUND

Item	PC £	Labour hours	Labour £	Plant £	Material £	Unit	Total rate £
450 mm × 450 mm × 100 mm; double triangular; class D400; kitemarked	111.29	2.50	32.84	–	118.89	nr	**151.73**
600 mm × 450 mm × 100 mm; double triangular; class D400; kitemarked	145.91	2.50	32.84	–	155.24	nr	**188.08**
600 mm × 600 mm × 100 mm; double triangular; class D400; kitemarked	89.04	2.50	32.84	–	95.52	nr	**128.36**
750 mm × 600 mm × 100 mm; double triangular; class D400; kitemarked	227.53	2.50	32.84	–	240.93	nr	**273.77**
1220 mm × 675 mm × 100 mm; double triangular; class D400; kitemarked	253.49	3.50	45.99	–	268.20	nr	**314.19**
British Standard best quality vitrified clay channels; bedding and jointing in cement and sand (1:2)							
Half section straight							
100 mm dia. × 1 m long	8.69	0.74	9.72	–	9.12	nr	**18.84**
150 mm dia. × 1 m long	15.30	0.93	12.22	–	16.07	nr	**28.29**
225 mm dia. × 1 m long	36.63	1.20	15.77	–	38.46	nr	**54.23**
300 mm dia. × 1 m long	77.17	1.48	19.45	–	81.03	nr	**100.48**
Half section bend							
100 mm dia.	8.82	0.56	7.36	–	9.26	nr	**16.62**
150 mm dia.	15.24	0.69	9.06	–	16.00	nr	**25.06**
225 mm dia.	59.06	0.93	12.22	–	62.01	nr	**74.23**
Taper straight							
150 mm–100 mm dia.	22.38	0.65	8.54	–	23.50	nr	**32.04**
300 mm–225 mm dia.	195.02	0.83	10.91	–	204.77	nr	**215.68**
Taper bend							
150 mm–100 mm dia.	69.69	0.83	10.91	–	73.17	nr	**84.08**
225 mm–150 mm dia.	199.47	1.06	13.92	–	209.44	nr	**223.36**
Three quarter section branch bend							
100 mm dia.	24.82	0.46	6.05	–	26.06	nr	**32.11**
150 mm dia.	41.66	0.69	9.06	–	43.74	nr	**52.80**
225 mm dia.	152.01	0.93	12.22	–	159.61	nr	**171.83**
Glass fibre septic tank; Klargester or other equal and approved; fixing lockable manhole cover and frame; placing in position (excavations, fill etc. measured elsewhere)							
2800 litre capacity; depth to invert							
1000 mm deep	–	2.45	32.19	55.45	853.64	nr	**941.28**
1500 mm deep	–	2.73	35.87	55.45	904.51	nr	**995.83**
3800 litre capacity; depth to invert							
1000 mm deep	–	2.64	34.68	69.31	1090.71	nr	**1194.70**
1500 mm deep	–	2.91	38.23	76.24	1141.58	nr	**1256.05**
4600 litre capacity; depth to invert							
1000 mm deep	–	2.75	36.13	69.31	1250.45	nr	**1355.89**
1500 mm deep	–	3.20	42.04	76.24	1301.32	nr	**1419.60**

34 DRAINAGE BELOW GROUND

Item	PC £	Labour hours	Labour £	Plant £	Material £	Unit	Total rate £
MANHOLES AND SOAKAWAYS – cont							
SOAKAWAYS							
Soakaway crates, lightweight modular water storage cells for stormwater attenuation with geotextile membrane wrapping							
Polystorm Heavy Duty (60 tonne) crates							
1000 mm × 500 mm × 400 mm deep	–	1.25	16.42	–	137.20	m³	**153.62**
Polystorm Lite Duty (20 tonne) crates							
1000 mm × 500 mm × 400 mm deep	–	1.25	16.42	–	136.27	m³	**152.69**
Brett Martin Heavy Duty (60 tonne) crates							
1200 mm × 600 mm × 420 mm deep	–	1.25	16.42	–	84.30	m³	**100.72**
Brett Martin Light Duty (20 tonne) crates							
1200 mm × 600 mm × 420 mm deep	–	1.25	16.42	–	77.72	m³	**94.14**

35 SITE WORK

Item	PC £	Labour hours	Labour £	Plant £	Material £	Unit	Total rate £
KERBS, EDGINGS AND CHANNELS							
Excavating; by machine							
Excavating trenches; to receive kerb foundations; average size							
300 mm × 100 mm	–	0.02	0.26	0.29	–	m	**0.55**
450 mm × 150 mm	–	0.02	0.26	0.58	–	m	**0.84**
600 mm × 200 mm	–	0.03	0.42	0.81	–	m	**1.23**
Excavating curved trenches; to receive kerb foundations; average size							
300 mm × 100 mm	–	0.01	0.14	0.46	–	m	**0.60**
450 mm × 150 mm	–	0.03	0.40	0.69	–	m	**1.09**
600 mm × 200 mm	–	0.04	0.53	0.87	–	m	**1.40**
Excavating; by hand							
Excavating trenches; to receive kerb foundations; average size							
150 mm × 50 mm	–	0.02	0.26	–	–	m	**0.26**
200 mm × 75 mm	–	0.06	0.80	–	–	m	**0.80**
250 mm × 100 mm	–	0.10	1.32	–	–	m	**1.32**
300 mm × 100 mm	–	0.13	1.72	–	–	m	**1.72**
Excavating curved trenches; to receive kerb foundations; average size							
150 mm × 50 mm	–	0.03	0.40	–	–	m	**0.40**
200 mm × 75 mm		0.07	0.92	–	–	rn	**0.92**
250 mm × 100 mm	–	0.11	1.46	–	–	m	**1.46**
300 mm × 100 mm	–	0.14	1.86	–	–	m	**1.86**
Plain in situ ready mixed designated concrete; C7.5–40 mm aggregate; poured on or against earth or unblinded hardcore							
Foundations	95.44	1.16	18.01	–	100.21	m³	**118.22**
Blinding beds							
thickness not exceeding 150 mm	95.44	1.71	26.54	–	100.21	m³	**126.75**
Plain in situ ready mixed designated concrete; C10–40 mm aggregate; poured on or against earth or unblinded hardcore							
Foundations	96.51	1.16	18.01	–	101.34	m³	**119.35**
Blinding beds							
thickness not exceeding 150 mm	96.51	1.71	26.54	–	101.34	m³	**127.88**
Plain in situ ready mixed designated concrete; C20–20 mm aggregate; poured on or against earth or unblinded hardcore							
Foundations	100.27	1.16	18.01	–	105.28	m³	**123.29**
Blinding beds							
thickness not exceeding 150 mm	100.27	1.71	26.54	–	105.28	m³	**131.82**

35 SITE WORK

Item	PC £	Labour hours	Labour £	Plant £	Material £	Unit	Total rate £
KERBS, EDGINGS AND CHANNELS – cont							
Filling to make up levels; by machine							
Average thickness not exceeding 0.25 m							
obtained off site; hardcore	–	0.28	3.71	0.96	36.87	m^3	**41.54**
obtained off site; granular fill type one	–	0.28	3.71	0.96	57.55	m^3	**62.22**
obtained off site; granular fill type two	–	0.28	3.71	0.96	44.48	m^3	**49.15**
Average thickness exceeding 0.25 m							
obtained off site; hardcore	–	0.24	3.18	0.87	31.59	m^3	**35.64**
obtained off site; granular fill type one	–	0.24	3.18	0.87	57.55	m^3	**61.60**
obtained off site; granular fill type two	–	0.24	3.18	0.87	44.48	m^3	**48.53**
Filling to make up levels; by hand							
Average thickness not exceeding 0.25 m							
obtained off site; hardcore	–	0.61	8.07	1.65	36.87	m^3	**46.59**
obtained off site; sand	–	0.71	9.40	1.93	48.72	m^3	**60.05**
Average thickness exceeding 0.25 m							
obtained off site; hardcore	–	0.51	6.75	1.38	31.59	m^3	**39.72**
obtained off site; sand	–	0.60	7.95	1.63	48.72	m^3	**58.30**
Surface treatments							
Compacting							
filling; blinding with sand	–	0.04	0.53	0.04	2.28	m^2	**2.85**
Precast concrete kerbs, channels, edgings, etc.; BS 340; bedded, jointed and pointed in cement mortar (1:3); including haunching up one side with in situ ready mix designated concrete C10– 40 mm aggregate; to concrete base							
Edgings; straight; square edge							
50 mm × 150 mm	–	0.23	5.26	–	4.75	m	**10.01**
50 mm × 200 mm	–	0.23	5.26	–	8.23	m	**13.49**
50 mm × 255 mm	–	0.23	5.26	–	10.93	m	**16.19**
Kerbs; straight							
125 mm × 255 mm; half battered	–	0.31	7.10	–	11.16	m	**18.26**
125 mm × 255 mm; half battered drop kerb	–	0.31	7.10	–	16.76	m	**23.86**
150 mm × 305 mm; half battered	–	0.31	7.10	–	28.20	m	**35.30**
150 mm × 305 mm; half battered drop kerb	–	0.31	7.10	–	62.96	m	**70.06**
Kerbs; curved							
125 mm × 255 mm; half battered	–	0.46	10.53	–	17.45	m	**27.98**
150 mm × 305 mm; half battered	–	0.46	10.53	–	31.16	m	**41.69**
Channels; 255 × 125 mm							
straight	–	0.31	7.10	–	19.51	m	**26.61**
Quadrants; half battered							
305 mm × 305 mm × 150 mm	–	0.32	7.33	–	28.98	nr	**36.31**
305 mm × 305 mm × 255 mm	–	0.32	7.33	–	31.80	nr	**39.13**
455 mm × 455 mm × 255 mm	–	0.37	8.47	–	41.71	nr	**50.18**

35 SITE WORK

Item	PC £	Labour hours	Labour £	Plant £	Material £	Unit	Total rate £
Natural stone kerbs; Silver Grey Granite bedded, jointed and pointed in cement mortar (1:3); including haunching up one side with in situ ready mix designated concrete C10–40 mm aggregate; to concrete base							
Kerbs; straight							
Kerbs (type C) with granite setts							
100 mm × 100 mm × 200 mm, split all sides	9.37	0.35	8.01	–	15.96	m	**23.97**
Kerbs (type A) with granite setts							
300 mm × 50 mm × 600 mm, sawn all sides, flame textured top and front face	6.21	0.35	8.01	–	11.65	m	**19.66**
Kerbs (type A) with granite setts							
250 mm × 50 mm × 600 mm, sawn all sides, flame textured top and front face	5.64	0.35	8.01	–	11.65	m	**19.66**
Bullnose kerbs (type B) with granite setts							
150 mm × 150 mm × 600 mm, 10 mm rounding to top one long side, flame textured top and front face	11.09	0.50	11.45	–	17.96	m	**29.41**
Bullnose kerbs (type B) with granite setts							
300 mm × 150 mm × 600 mm, 10 mm rounding to top one long side, flame textured top and front face	21.85	0.50	11.45	–	30.38	m	**41.83**
Transition Bullnose kerbs (type B) with granite setts 300 mm to 150 mm width change across length × 150 mm × 600 mm, 10 mm rounding to top one long side, flame textured top and front face	9.98	0.03	0.68	–	16.66	nr	**17.34**
Kerbs; curved							
Bullnose kerbs (type B) with granite setts 150 mm × 175 mm × 600 mm, sawn all sides, 10 mm rounding to top one long side, flame textured top and front face, curved on plan 1–12 m	19.00	0.50	11.45	–	27.09	m	**38.54**
Bullnose kerbs (type B) with granite setts 300 mm × 150 mm × 600 mm, 10 mm rounding to top one long side, flame textured top and front face, curved on plan 1–12 m	26.12	0.50	11.45	–	35.31	m	**46.76**
IN SITU CONCRETE ROADS AND PAVINGS							
Reinforced in situ ready mixed designated concrete; C10–40 mm aggregate							
Roads; to hardcore base							
thickness not exceeding 150 mm	91.91	1.85	28.72	–	96.51	m³	**125.23**
thickness 150 mm–450 mm	91.91	1.30	20.18	–	96.51	m³	**116.69**

35 SITE WORK

Item	PC £	Labour hours	Labour £	Plant £	Material £	Unit	Total rate £
IN SITU CONCRETE ROADS AND PAVINGS – cont							
Reinforced in situ ready mixed designated concrete; C20–20 mm aggregate							
Roads; to hardcore base							
thickness not exceeding 150 mm	95.49	1.85	28.72	–	100.26	m³	**128.98**
thickness 150 mm–450 mm	95.49	1.30	20.18	–	100.26	m³	**120.44**
Reinforced in situ ready mixed designated concrete; C25–20 mm aggregate							
Roads; to hardcore base							
thickness ot exceeding 150 mm	102.13	1.85	28.72	–	107.24	m³	**135.96**
thickness 150 mm–450 mm	102.13	1.30	20.18	–	107.24	m³	**127.42**
Formwork; sides of foundations; basic finish							
Plain vertical							
height not exceeding 250 mm	–	0.39	7.16	–	1.90	m	**9.06**
height 250 mm–500 mm	–	0.57	10.46	–	3.09	m	**13.55**
height 500 mm–1.00 m	–	0.83	15.24	–	4.45	m	**19.69**
add to above for curved radius 6 m	–	0.03	0.55	–	0.18	m	**0.73**
Reinforcement; fabric; BS 4449; lapped; in roads, footpaths or pavings							
Ref A142 (2.22 kg/m²)							
400 mm minimum laps	1.10	0.14	2.57	–	1.16	m²	**3.73**
Ref A193 (3.02 kg/m²)							
400 mm minimum laps	–	0.14	2.57	–	1.56	m²	**4.13**
Formed joints; Fosroc Expandite Flexcell impregnated joint filler or other equal and approved							
Width not exceeding 150 mm							
12.50 mm thick	–	0.14	2.57	–	1.95	m	**4.52**
25 mm thick	–	0.19	3.49	–	5.34	m	**8.83**
Width 150–300 mm							
12.50 mm thick	–	0.19	3.49	–	2.94	m	**6.43**
25 mm thick	–	0.19	3.49	–	5.29	m	**8.78**
Width 300–450 mm							
12.50 mm thick	–	0.23	4.22	–	4.41	m	**8.63**
25 mm thick	–	0.23	4.22	–	7.94	m	**12.16**
Sealants; Fosroc Expandite Pliastic N2 hot poured rubberized bituminous compound or other equal and approved							
Width 25 mm							
25 mm depth	–	0.20	3.67	–	1.67	m	**5.34**

35 SITE WORK

Item	PC £	Labour hours	Labour £	Plant £	Material £	Unit	Total rate £
Concrete sundries							
Treating surfaces							
unset concrete; grading to cambers; tamping with a 75 mm thick steel shod tamper	–	0.23	3.57	–	–	m²	**3.57**
Sundries							
Line marking							
width not exceeding 300 mm; NB minimum charge usually applies	–	0.04	0.65	–	0.23	m	**0.88**
COATED MACADAM, ASPHALT ROADS AND PAVINGS							
NOTE: The prices for all bitumen macadam and hot rolled asphalt materials are for individual courses to roads and footpaths and need combining to arrive at complete specifications and costs for full construction. Costs include for work to falls, crossfalls or slopes not exceeding 15° from horizontal; for laying on prepared bases (prices not included) and for rolling with an appropriate roller. The following rates are based on black bitumen macadam. Red bitumen macadam rates are approximately 50% dearer. PSV is Polished Stone Value.							
Dense bitumen macadam base course; BS EN 13108; bitumen penetration 100/125							
Carriageway, hardshoulder and hardstrip							
100 mm thick; one coat; with 0/32 mm aggregate size	–	–	–	–	–	m²	**18.53**
200 mm thick; one coat; with 0/32 mm aggregate size	–	–	–	–	–	m²	**32.63**
extra for increase thickness in 10 mm increments	–	–	–	–	–	m²	**1.33**
Hot rolled asphalt base course; BS EN 13108							
Carriageway, hardshoulder and hardstrip							
150 mm thick; one coat; 60% 0/32 mm aggregate size; to column 2/5	–	–	–	–	–	m²	**30.81**
200 mm thick; one coat; 60% 0/32 mm aggregate size; to column 2/5	–	–	–	–	–	m²	**40.99**
extra for increase thickness in 10 mm increments	–	–	–	–	–	m²	**1.70**
Dense bitumen macadam binder course; BS EN 13108; bitumen penetration 100/125							
Carriageway, hardshoulder and hardstrip							
60 mm thick; one coat; with 0/32 mm aggregate size	–	–	–	–	–	m²	**11.33**
60 mm thick; one coat; with 0/32 mm aggregate size; to clause 6.5	–	–	–	–	–	m²	**11.43**
extra for increase thickness in 10 mm increments	–	–	–	–	–	m²	**1.51**

35 SITE WORK

Item	PC £	Labour hours	Labour £	Plant £	Material £	Unit	Total rate £
COATED MACADAM, ASPHALT ROADS AND PAVINGS – cont							
Hot rolled asphalt binder course; BS EN 13108							
Carriageway, hardshoulder and hardstrip							
40 mm thick; one coat; 50% 0/14 mm aggregate size; to column 2/2; 55 PSV	–	–	–	–	–	m²	10.81
60 mm thick; one coat; 50% 0/14 mm aggregate size; to column 2/2	–	–	–	–	–	m²	12.45
60 mm thick; one coat; 50% 0/20 mm aggregate size; to column 2/3	–	–	–	–	–	m²	12.24
60 mm thick; one coat; 60% 0/32 mm aggregate size; to column 2/5	–	–	–	–	–	m²	11.62
100 mm thick; one coat; 60% 0/32 mm aggregate size; to column 2/5	–	–	–	–	–	m²	18.76
extra for increase thickness in 10 mm increments	–	–	–	–	–	m²	2.24
Macadam surface course; BS EN 13108; bitumen penetration 100/125							
Carriageway, hardshoulder and hardstrip							
30 mm thick; one coat; medium graded with 0/6 mm nominal aggregate binder	–	–	–	–	–	m²	8.75
40 mm thick; one coat; close graded with 0/14 mm nominal aggregate binder; to clause 7.3	–	–	–	–	–	m²	8.02
40 mm thick; one coat; close graded with 0/10 mm nominal aggregate binder; to clause 7.4	–	–	–	–	–	m²	8.75
extra over above items for increase/reduction in 10 mm thick increments	–	–	–	–	–	m²	1.58
extra over above items for coarse aggregate 60–64 PSV	–	–	–	–	–	m²	1.59
extra over above items for coarse aggregate 65–67 PSV	–	–	–	–	–	m²	1.74
extra over above items for coarse aggregate 68 PSV	–	–	–	–	–	m²	2.31
Hot rolled asphalt surface course; BS EN 13108; bitumen penetration 40/60							
Carriageway, hardshoulder and hardstrip							
40 mm thick; one coat; 30% mix 0/10 mm aggregate size; to column 3/2; with 20 mm pre-coated chippings 60–64 PSV	–	–	–	–	–	m²	10.80
40 mm thick; one coat; 30% mix 0/10 mm aggregate size; to column 3/2; with 14 mm pre-coated chippings 60–64 PSV	–	–	–	–	–	m²	10.89
extra over above items for increase/reduction in 10 mm increments	–	–	–	–	–	m²	1.80
extra over above items for chippings with 65–67 PSV	–	–	–	–	–	m²	0.10
extra over above items for chippings with 68 PSV	–	–	–	–	–	m²	0.16
extra over above items for 6–10 KN High Traffic Flows	–	–	–	–	–	m²	0.78

35 SITE WORK

Item	PC £	Labour hours	Labour £	Plant £	Material £	Unit	Total rate £
Stone mastic asphalt surface course; BS EN 13108							
Carriageway, hardshoulder and hardstrip							
35 mm thick; one coat; with 0/14 mm nominal aggregate size; 55 PSV	–	–	–	–	–	m²	**9.93**
35 mm thick; one coat; with 0/10 mm nominal aggregate size; 55 PSV	–	–	–	–	–	m²	**9.93**
extra for increase thickness in 10 mm increments	–	–	–	–	–	m²	**2.22**
Thin surface course with 60 PSV							
Carriageway, hardshoulder and hardstrip							
35 mm thick; one coat; with 0/10 mm nominal aggregate size	–	–	–	–	–	m²	**9.93**
extra for increase thickness in 10 mm increments	–	–	–	–	–	m²	**1.12**
extra over above items for coarse aggregate 60–64 PSV	–	–	–	–	–	m²	**0.27**
extra over above items for coarse aggregate 65–67 PSV	–	–	–	–	–	m²	**0.27**
extra over above items for coarse aggregate 68 PSV	–	–	–	–	–	m²	**0.51**
Regulating courses							
Carriageway, hardshoulder and hardstrip							
Dense Bitumen Macadam; bitumen penetration 100/125; with 0/20 mm nominal aggregate regulating course	–	–	–	–	–	tonne	**84.97**
Hot rolled asphalt; 50% 0/20 mm aggregate size	–	–	–	–	–	tonne	**93.96**
Stone mastic asphalt; 0/6 mm aggregate	–	–	–	–	–	tonne	**119.95**
Bitumen emulsion tack coats							
Carriageway, hardshoulder and hardstrip							
K1–40; applied 0.35–0.45l/m²	–	–	–	–	–	m²	**0.16**
K1–70; applied 0.35–0.45l/m²	–	–	–	–	–	m²	**0.26**
Sundries							
Line marking							
width not exceeding 300 mm; NB minimum charge usually applies	–	0.04	0.65	–	0.23	m	**0.88**
GRAVEL PAVINGS							
Two coat gravel paving; level and to falls; first layer course clinker aggregate and wearing layer fine gravel aggregate							
Pavings; over 300 mm wide							
50 mm thick	–	0.07	1.61	0.06	2.47	m²	**4.14**
63 mm thick	–	0.09	2.06	0.08	2.99	m²	**5.13**

35 SITE WORK

Item	PC £	Labour hours	Labour £	Plant £	Material £	Unit	Total rate £
GRAVEL PAVINGS – cont							
Resin bonded gravel paving; level and to falls							
Pavings; over 300 mm wide							
50 mm thick	–	–	–	–	–	m²	**43.98**
SLAB, BRICK/BLOCK SETTS AND COBBLE PAVINGS							
Artificial stone paving; Charcon's Moordale Textured or other equal and approved; to falls or crossfalls; bedding 25 mm thick in cement mortar (1:3); staggered joints; jointing in coloured cement mortar (1:3), brushed in; to sand base							
Pavings; over 300 mm wide							
600 mm × 600 mm × 50 mm thick; natural	14.91	0.39	8.92	–	18.73	m²	**27.65**
Brick paviors; 215 mm × 103 mm × 65 mm rough stock bricks; to falls or crossfalls; bedding on 50 mm sharp sand; kiln dried sand to joints							
Pavings; over 300 mm wide; straight joints both ways							
bricks laid flat (PC £ per 1000)	450.00	1.50	32.34	–	21.90	m²	**54.24**
bricks laid on edge	–	2.50	53.90	–	32.06	m²	**85.96**
Pavings; over 300 mm wide; laid to herringbone pattern							
bricks laid flat bricks laid flat (PC £ per 1000)	450.00	0.93	20.04	–	21.90	m²	**41.94**
bricks laid on edge	–	1.30	28.02	–	32.06	m²	**60.08**
Add or deduct for variation of £10.00/1000 in PC of brick paviors							
bricks laid flat bricks laid flat (PC £ per 1000)	–	–	–	–	0.47	m²	**0.47**
bricks laid on edge	–	–	–	–	0.71	m²	**0.71**
River washed cobble paving; 50 mm–75 mm; to falls or crossfalls; bedding 13 mm thick in cement mortar (1:3); jointing to a height of two thirds of cobbles in dry mortar (1:3); tightly butted, washed and brushed; to concrete							
Pavings; over 300 mm wide							
regular (PC £ per tonne)	97.52	3.70	84.70	–	23.43	m²	**108.13**
laid to pattern	–	4.63	105.99	–	23.43	m²	**129.42**
Concrete paving flags; BS EN 1339; to falls or crossfalls; bedding 25 mm thick in cement and sand mortar (1:4); butt joints straight both ways; jointing in cement and sand (1:3); brushed in; to sand base							
Pavings; over 300 mm wide							
450 mm × 600 mm × 50 mm thick; grey	8.79	0.42	9.62	–	11.18	m²	**20.80**
450 mm × 600 mm × 60 mm thick; coloured	9.82	0.42	9.62	–	12.33	m²	**21.95**
600 mm × 600 mm × 50 mm thick; grey	6.82	0.39	8.92	–	9.03	m²	**17.95**

35 SITE WORK

Item	PC £	Labour hours	Labour £	Plant £	Material £	Unit	Total rate £
600 mm × 600 mm × 50 mm thick; coloured	8.20	0.39	8.92	–	10.55	m²	**19.47**
750 mm × 600 mm × 50 mm thick; grey	6.14	0.36	8.24	–	8.26	m²	**16.50**
750 mm × 600 mm × 50 mm thick; coloured	8.16	0.36	8.24	–	10.50	m²	**18.74**
900 mm × 600 mm × 50 mm thick; grey	5.47	0.33	7.55	–	7.52	m²	**15.07**
900 mm × 600 mm × 50 mm thick; coloured	7.48	0.33	7.55	–	9.74	m²	**17.29**
Blister Tactile paving flags; to falls or crossfalls; bedding 25 mm thick in cement and sand mortar (1:4); butt joints straight both ways; jointing in cement and sand (1:3); brushed in; to sand base							
Pavings; over 300 mm wide							
400 mm × 400 mm × 50 mm thick; buff	21.10	0.45	10.30	–	24.75	m²	**35.05**
400 mm × 400 mm × 60 mm thick; buff	22.91	0.45	10.30	–	26.75	m²	**37.05**
450 mm × 450 mm × 50 mm thick; buff	23.61	0.45	10.30	–	27.52	m²	**37.82**
450 mm × 450 mm × 70 mm thick; buff	26.24	0.45	10.30	–	30.43	m²	**40.73**
Concrete rectangular paving blocks; to falls or crossfalls; herringbone pattern; on prepared base (not included here); bedding on 50 mm thick dry sharp sand; filling joints with sharp sand brushed in							
Pavings; Keyblock or other equal and approved; over 300 mm wide; straight joints both ways							
200 mm × 100 mm × 60 mm thick; grey	7.71	1.25	28.61	–	11.93	m²	**40.54**
200 mm × 100 mm × 60 mm thick; coloured	8.31	1.25	28.61	–	12.59	m²	**41.20**
200 mm × 100 mm × 60 mm thick; brindle	8.43	1.25	28.61	–	12.73	m²	**41.34**
200 mm × 100 mm × 80 mm thick; grey	8.60	1.35	30.90	–	13.19	m²	**44.09**
200 mm × 100 mm × 80 mm thick; coloured	9.70	1.35	30.90	–	14.42	m²	**45.32**
200 mm × 100 mm × 80 mm thick; brindle	9.93	1.35	30.90	–	14.66	m²	**45.56**
Pavings; Cemex or other equal; over 300 mm wide; laid to herringbone pattern							
200 mm × 100 mm × 60 mm thick; grey	7.71	1.35	30.90	–	11.93	m²	**42.83**
200 mm × 100 mm × 60 mm thick; coloured	8.31	1.35	30.90	–	12.59	m²	**43.49**
200 mm × 100 mm × 80 mm thick; grey	8.60	1.35	30.90	–	13.19	m²	**44.09**
200 mm × 100 mm × 80 mm thick; coloured	9.70	1.35	30.90	–	14.42	m²	**45.32**
Extra for two row boundary edging to herringbone pavings; 200 mm wide; including a 150 mm high in situ concrete mix C10–40 mm aggregate haunching to one side; blocks laid breaking joint							
200 mm × 100 mm × 60 mm; coloured	–	0.28	6.40	–	2.57	m	**8.97**
200 mm × 100 mm × 80 mm; coloured	–	0.28	6.40	–	2.68	m	**9.08**
Pavings; Europa or other equal; over 300 mm wide; straight joints both ways							
200 mm × 100 mm × 60 mm thick; grey	8.04	1.25	28.61	–	12.28	m²	**40.89**
200 mm × 100 mm × 60 mm thick; coloured	8.92	1.25	28.61	–	13.26	m²	**41.87**
200 mm × 100 mm × 80 mm thick; grey	9.57	1.35	30.90	–	14.26	m²	**45.16**
200 mm × 100 mm × 80 mm thick; coloured	10.48	1.35	30.90	–	15.28	m²	**46.18**
Pavings; Metropolitan or other equal; over 300 mm wide; straight joints both ways							
200 mm × 100 mm × 80 mm thick; grey	20.13	1.35	30.90	–	25.91	m²	**56.81**
200 mm × 100 mm × 80 mm thick; coloured	20.13	1.35	30.90	–	25.91	m²	**56.81**

35 SITE WORK

Item	PC £	Labour hours	Labour £	Plant £	Material £	Unit	Total rate £
SLAB, BRICK/BLOCK SETTS AND COBBLE PAVINGS – cont							
Concrete rectangular paving blocks – cont							
Pavings; Intersett or other equal and approved; over 300 mm wide; straight joints both ways							
200 mm × 100 mm × 60 mm thick; grey	9.50	1.25	28.61	–	13.90	m²	**42.51**
200 mm × 100 mm × 60 mm thick; coloured	10.55	1.25	28.61	–	15.06	m²	**43.67**
200 mm × 100 mm × 80 mm thick; grey	11.37	1.35	30.90	–	16.25	m²	**47.15**
200 mm × 100 mm × 80 mm thick; coloured	12.62	1.35	30.90	–	17.63	m²	**48.53**
Concrete rectangular paving blocks; to falls or crossfalls; 6 mm wide joints; symmetrical layout; bedding in 15 mm semi-dry cement mortar (1:4); jointing and pointing in cement and sand (1:4); on concrete base							
Pavings; Trafica or other equal and approved; over 300 mm wide							
400 mm × 400 mm × 65 mm; Saxon textured; natural	42.03	0.44	10.07	–	23.90	m²	**33.97**
400 mm × 400 mm × 65 mm; Saxon textured; buff	23.88	0.44	10.07	–	27.69	m²	**37.76**
400 mm × 400 mm × 65 mm; Perfecta; natural	27.58	0.44	10.07	–	30.31	m²	**40.38**
400 mm × 400 mm × 65 mm; Perfecta; buff	30.31	0.44	10.07	–	34.78	m²	**44.85**
450 mm × 450 mm × 70 mm; Saxon textured; natural	22.45	0.43	9.84	–	26.11	m²	**35.95**
450 mm × 450 mm × 70 mm; Saxon textured; buff	25.79	0.43	9.84	–	29.79	m²	**39.63**
450 mm × 450 mm × 70 mm; Perfecta; natural	24.25	0.43	9.84	–	28.09	m²	**37.93**
450 mm × 450 mm × 70 mm; Perfecta; buff	28.38	0.43	9.84	–	32.63	m²	**42.47**
Bllue Grey granite setts tactile pavings (blister or corduroy); 400 mm × 80 mm × 400 mm; standard B dressing; sandblasted top finish, tightly butted to falls or crossfalls; bedding 25 mm thick in cement mortar (1:3); filling joints with dry mortar (1:6); washed and brushed; on concrete base							
Pavings; over 300 mm wide							
straight joints	67.54	1.48	33.88	–	77.83	m²	**111.71**
York stone slab pavings; to falls or crossfalls; bedding 25 mm thick in cement: sand mortar (1:4); 5 mm wide joints; jointing in coloured cement mortar (1:3); brushed in; to sand base							
Pavings; over 300 mm wide							
50 mm thick; random rectangular pattern	74.29	0.69	14.88	–	81.76	m²	**96.64**
600 mm × 600 mm × 50 mm thick	70.74	0.39	8.41	–	77.95	m²	**86.36**
600 mm × 900 mm × 50 mm thick	70.74	0.33	7.12	–	77.95	m²	**85.07**

35 SITE WORK

Item	PC £	Labour hours	Labour £	Plant £	Material £	Unit	Total rate £
Granite setts; 200 mm × 100 mm × 100 mm; standard C dressing; tightly butted to falls or crossfalls; bedding 25 mm thick in cement mortar (1:3); filling joints with dry mortar (1:6); washed and brushed; on concrete base							
Pavings; over 300 mm wide							
straight joints (PC £ per tonne)	162.55	1.48	33.88	–	51.32	m²	**85.20**
laid to pattern	162.55	1.85	42.35	–	51.32	m²	**93.67**
Boundary edging							
two rows of granite setts as boundary edging; 200 mm wide; including a 150 mm high ready mixed designated concrete C10–40 mm aggregate; haunching to one side; blocks laid breaking joint	–	0.65	14.88	–	12.21	m	**27.09**
Blue Grey granite setts paving bands; 100 mm × 100 mm × 100 mm; standard C dressing; flame textured top, tightly butted to falls or crossfalls; bedding 25 mm thick in cement mortar (1:3); filling joints with dry mortar (1:6); washed and brushed; on concrete base							
Pavings; over 300 mm wide							
straight joints	46.97	1.48	33.88	–	55.69	m²	**89.57**
laid to pattern	46.97	1.85	42.35	–	55.69	m²	**98.04**
Boundary edging							
two rows of granite setts as boundary edging; 200 mm wide; including a 150 mm high ready mixed designated concrete C10–40 mm aggregate; haunching to one side; blocks laid breaking joint	–	0.65	14.88	–	13.34	m	**28.22**
Black Basalt setts paving bands; 100 mm × 100 mm × 100 mm; standard C dressing; flame textured top, tightly butted to falls or crossfalls; bedding 25 mm thick in cement mortar (1:3); filling joints with dry mortar (1:6); washed and brushed; on concrete base							
Pavings; over 300 mm wide							
straight joints	69.50	1.48	33.88	–	79.94	m²	**113.82**
laid to pattern	69.50	1.85	42.35	–	79.94	m²	**122.29**
Boundary edging							
two rows of granite setts as boundary edging; 200 mm wide; including a 150 mm high ready mixed designated concrete C10–40 mm aggregate; haunching to one side; blocks laid breaking joint	–	0.65	14.88	–	18.30	m	**33.18**

35 SITE WORK

Item	PC £	Labour hours	Labour £	Plant £	Material £	Unit	Total rate £
EXTERNAL UNDERGROUND SERVICE RUNS							
Excavating trenches; by machine; grading bottoms; earthwork support; filling with excavated material and compacting; disposal of surplus soil on site; spreading on site average 50 m from excavations							
Services not exceeding 200 mm nominal size							
average depth of run not exceeding 0.50 m	–	0.28	3.71	1.16	–	m	**4.87**
average depth of run not exceeding 0.75 m	–	0.37	4.90	1.91	–	m	**6.81**
average depth of run not exceeding 1.00 m	–	0.79	10.46	2.49	0.75	m	**13.70**
average depth of run not exceeding 1.25 m	–	1.16	15.36	3.30	1.10	m	**19.76**
average depth of run not exceeding 1.50 m	–	1.48	19.59	4.45	1.34	m	**25.38**
average depth of run not exceeding 1.75 m	–	1.85	24.50	5.78	1.65	m	**31.93**
average depth of run not exceeding 2.00 m	–	2.13	28.20	6.76	1.80	m	**36.76**
Excavating trenches; by hand; grading bottoms; earthwork support; filling with excavated material and compacting; disposal; of surplus soil on site; spreading on site average 50 m from excavations							
Services not exceeding 200 mm nominal size							
average depth of run not exceeding 0.50 m	–	0.93	12.32	–	–	m	**12.32**
average depth of run not exceeding 0.75 m	–	1.39	18.41	–	–	m	**18.41**
average depth of run not exceeding 1.00 m	–	2.04	27.02	–	0.81	m	**27.83**
average depth of run not exceeding 1.25 m	–	2.87	38.00	–	1.10	m	**39.10**
average depth of run not exceeding 1.50 m	–	3.93	52.04	–	1.34	m	**53.38**
average depth of run not exceeding 1.75 m	–	5.18	68.60	–	1.63	m	**70.23**
average depth of run not exceeding 2.00 m	–	5.92	78.39	–	1.78	m	**80.17**
Stop cock pits, valve chambers and the like; excavating; half brick thick walls in common bricks in cement mortar (1:3); on in situ concrete designated mix C20–20 mm aggregate bed; 100 mm thick							
Pits							
100 mm × 100 mm × 750 mm deep; internal holes for one small pipe; polypropylene hinged box cover; bedding in cement mortar (1:3)	–	3.89	83.86	–	46.05	nr	**129.91**

36 FENCING

Item	PC £	Labour hours	Labour £	Plant £	Material £	Unit	Total rate £
FENCING							
NOTE: The prices for all fencing include for setting posts in position, to a depth of 0.60 m for fences not exceeding 1.40 m high and of 0.76 m for fences over 1.40 m high. The prices allow for excavating post holes; filling to within 150 mm of ground level with concrete and all necessary backfilling.							
For more examples please consult Spon's External Works and Landscaping Book							
Fencing; strained wire							
Strained wire fencing; BS 1722 Part 3; 4 mm dia. galvanized mild steel plain wire threaded through posts and strained with eye bolts							
900 mm high; three line; concrete posts at 2.75 m centres	–	–	–	–	–	m	18.61
end concrete straining post; one strut	–	–	–	–	–	nr	45.24
angle concrete straining post; two struts	–	–	–	–	–	nr	52.54
1.07 m high; six line; concrete posts at 2.75 m centres	–	–	–	–	–	m	19.36
end concrete straining post; one strut	–	–	–	–	–	nr	50.88
angle concrete straining post; two struts	–	–	–	–	–	nr	58.18
1.20 m high; six line; concrete posts at 2.75 m centres	–	–	–	–	–	m	19.47
end concrete straining post; one strut	–	–	–	–	–	nr	52.32
angle concrete straining post; two struts	–	–	–	–	–	nr	59.60
1.40 m high; eight line; concrete posts at 2.75 m centres	–	–	–	–	–	m	20.00
end concrete straining post; one strut	–	–	–	–	–	nr	53.46
angle concrete straining post; two struts	–	–	–	–	–	nr	60.73
Chain link fencing; BS 1722 Part 1; 3 mm dia. galvanized mild steel wire; 50 mm mesh; galvanized mild steel tying and line wire; three line wires threaded through posts and strained with eye bolts and winding brackets							
900 mm high; galvanized mild steel angle posts at 3.00 m centres	–	–	–	–	–	m	26.18
end steel straining post; one strut	–	–	–	–	–	nr	75.42
angle steel straining post; two struts	–	–	–	–	–	nr	86.99
900 mm high; concrete posts at 3.00 m centres	–	–	–	–	–	m	18.96
end concrete straining post; one strut	–	–	–	–	–	nr	40.51
angle concrete straining post; two struts	–	–	–	–	–	nr	47.82
1.20 m high; galvanized mild steel angle posts at 3.00 m centres	–	–	–	–	–	m	19.30
end steel straining post; one strut	–	–	–	–	–	nr	80.48
angle steel straining post; two struts	–	–	–	–	–	nr	103.08
1.20 m high; concrete posts at 3.00 m centres	–	–	–	–	–	m	18.49
end concrete straining post; one strut	–	–	–	–	–	nr	46.38

36 FENCING

Item	PC £	Labour hours	Labour £	Plant £	Material £	Unit	Total rate £
FENCING – cont							
Fencing – cont							
Chain link fencing – cont							
angle concrete straining post; two struts	–	–	–	–	–	nr	54.79
1.80 m high; galvanized mild steel angle posts at 3.00 m centres	–	–	–	–	–	m	21.69
end steel straining post; one strut	–	–	–	–	–	nr	81.86
angle steel straining post; two struts	–	–	–	–	–	nr	101.83
1.80 m high; concrete posts at 3.00 m centres	–	–	–	–	–	m	25.06
end concrete straining post; one strut	–	–	–	–	–	nr	64.86
angle concrete straining post; two struts	–	–	–	–	–	nr	76.50
Gates							
Pair of gates and gate posts; gates to match galvanized chain link fencing, with angle framing, braces, etc., complete with hinges, locking bar, lock and bolts; two 100 mm × 100 mm angle section gate posts; each with one strut							
2.44 m × 0.90 m	–	–	–	–	–	nr	636.03
2.44 m × 1.20 m	–	–	–	–	–	nr	656.39
2.44 m × 1.80 m	–	–	–	–	–	nr	708.08
Fencing; chain link							
Chain link fencing; BS 1722 Part 1; 3 mm dia. plastic coated mild steel wire; 50 mm mesh; plastic coated mild steel tying and line wire; three line wires threaded through posts and strained with eye bolts and winding brackets							
900 mm high; galvanized mild steel angle posts at 3.00 m centres	–	–	–	–	–	m	24.00
end steel straining post; one strut	–	–	–	–	–	nr	66.51
angle steel straining post; two struts	–	–	–	–	–	nr	74.04
900 mm high; concrete posts at 3.00 m centres	–	–	–	–	–	m	17.90
end concrete straining post; one strut	–	–	–	–	–	nr	40.51
angle concrete straining post; two struts	–	–	–	–	–	nr	47.82
1.20 m high; galvanized mild steel angle posts at 3.00 m centres	–	–	–	–	–	m	17.72
end steel straining post; one strut	–	–	–	–	–	nr	69.77
angle steel straining post; two struts	–	–	–	–	–	nr	74.53
1.20 m high; concrete posts at 3.00 m centres	–	–	–	–	–	m	18.13
end concrete straining post; one strut	–	–	–	–	–	nr	46.38
angle concrete straining post; two struts	–	–	–	–	–	nr	54.79
1.80 m high; galvanized mild steel angle posts at 3.00 m centres	–	–	–	–	–	m	20.11
end steel straining post; one strut	–	–	–	–	–	nr	69.05
angle steel straining post; two struts	–	–	–	–	–	nr	82.47
1.80 m high; concrete posts at 3.00 mm centres	–	–	–	–	–	m	23.12
end concrete straining post; one strut	–	–	–	–	–	nr	64.86
angle concrete straining post; two struts	–	–	–	–	–	nr	76.50

36 FENCING

Item	PC £	Labour hours	Labour £	Plant £	Material £	Unit	Total rate £
Chain link fencing for tennis courts; BS 1722 Part 13; 2.5 dia. galvanized mild wire; 45 mm mesh; line and tying wires threaded through 45 mm × 45 mm × 5 mm galvanized mild steel angle standards, posts and struts; 60 mm × 60 mm × 6 mm straining posts and gate posts							
fencing to tennis court							
36.00 m × 18.00 m × 2.745 m high; standards at 3.00 m centres	–	–	–	–	–	nr	**2541.53**
fencing to tennis court 36.00 m × 18.00 m × 3.66 m high; standards at 2.50 mm centres	–	–	–	–	–	nr	**3893.40**
Gates							
Pair of gates and gate posts; gates to match plastic chain link fencing; with angle framing, braces, etc. complete with hinges, locking bar, lock and bolts; two 100 mm × 100 mm angle section gate posts; each with one strut							
2.44 m × 0.90 m	–	–	–	–	–	nr	**556.14**
2.44 m × 1.20 m	–	–	–	–	–	nr	**570.64**
2.44 m × 1.80 m	–	–	–	–	–	nr	**614.60**
Fencing; timber							
Cleft chestnut pale fencing; BS 1722 Part 4; pales spaced 51 mm apart; on two lines of galvanized wire; 64 mm dia. posts; 76 mm × 51 mm struts							
900 mm high; posts at 2.50 m centres	–	–	–	–	–	m	**10.81**
straining post; one strut	–	–	–	–	–	nr	**28.66**
corner straining post; two struts	–	–	–	–	–	nr	**28.66**
1.05 m high; posts at 2.50 m centres	–	–	–	–	–	m	**12.12**
straining post; one strut	–	–	–	–	–	nr	**28.88**
corner straining post; two struts	–	–	–	–	–	nr	**28.88**
Closeboarded fencing; BS 1722 Part 5; 76 mm × 38 mm softwood rails; 89 mm × 19 mm softwood pales lapped 13 mm; 152 mm × 25 mm softwood gravel boards; all softwood treated; posts at 3.00 m centres							
Fencing; two rail; concrete posts							
height 1.00 m	–	–	–	–	–	m	**32.88**
height 1.20 m	–	–	–	–	–	m	**33.25**
Fencing; three rail; concrete posts							
height 1.40 m	–	–	–	–	–	m	**36.50**
height 1.60 m	–	–	–	–	–	m	**36.50**
height 1.80 m	–	–	–	–	–	m	**37.85**
Fencing; concrete							
Precast concrete slab fencing; 305 mm × 38 mm × 1753 mm slabs; fitted into twice grooved concrete posts at 1.83 m centres							
height 1.50 m	–	–	–	–	–	m	**64.89**
height 1.80 m	–	–	–	–	–	m	**70.30**

Prices for Measured Works

36 FENCING

Item	PC £	Labour hours	Labour £	Plant £	Material £	Unit	Total rate £
FENCING – cont							
Fencing; security							
Mild steel unclimbable fencing; in rivetted panels 2440 mm long; 44 mm × 13 mm flat section top and bottom rails; two 44 mm × 19 mm flat section standards; one with foot plate; and 38 mm × 13 mm raking stay with foot plate; 20 mm dia. pointed verticals at 120 mm centres; two 44 mm × 19 mm supports 760 mm long with ragged ends to bottom rail; the whole bolted together; coated with red oxide primer; setting standards and stays in ground at 2440 mm centres and supports at 815 mm centres							
height 1.67 m	–	–	–	–	–	m	**121.13**
height 2.13 m	–	–	–	–	–	m	**140.59**
Pair of gates and gate posts, to match mild steel unclimbable fencing; with flat section framing, braces, etc., complete with locking bar, lock, handles, drop bolt, gate stop and holding back catches; two 102 mm × 102 mm hollow section gate posts with cap and foot plates							
2.44 m × 1.67 m	–	–	–	–	–	nr	**1043.65**
2.44 m × 2.13 m	–	–	–	–	–	nr	**1211.28**
4.88 m × 1.67 m	–	–	–	–	–	nr	**1622.25**
4.88 m × 2.13 m	–	–	–	–	–	nr	**2044.04**
PVC coated, galvanized mild steel high security fencing; Sentinal Sterling fencing or other equal and approved; 50 mm × 50 mm mesh; 3/3.50 mm gauge wire; barbed edge – 1; Sentinal Bi-steel colour coated posts or other equal and approved at 2.44 m centres							
1.80 m	–	0.93	12.32	–	34.50	m	**46.82**
2.10 m	–	1.16	15.36	–	38.10	m	**53.46**

37 SOFT LANDSCAPING

Item	PC £	Labour hours	Labour £	Plant £	Material £	Unit	Total rate £
SEEDING AND TURFING							
For more examples and depth of information please consult *Spon's External Works and Landscape Price Book*							
Top soil							
By machine							
Selected from spoil heaps not exceeding 50 m; grading; prepared for turfing or seeding; to general surfaces							
average 100 mm thick	–	–	0.06	1.18	–	m²	**1.24**
average 150 mm thick	–	0.01	0.07	1.56	–	m²	**1.63**
average 200 mm thick	–	0.01	0.10	2.73	–	m²	**2.83**
Selected from spoil heaps; grading; prepared for turfing or seeding; to cuttings or embankments							
average 100 mm thick	–	–	0.07	1.32	–	m²	**1.39**
average 150 mm thick	–	0.01	0.08	1.80	–	m²	**1.88**
average 200 mm thick	–	0.01	0.11	3.12	–	m²	**3.23**
By hand							
Selected from spoil heaps not exceeding 20 m; grading; prepared for turfing or seeding; to general surfaces							
average 100 mm thick	–	0.40	5.29	–	–	m²	**5.29**
average 150 mm thick	–	0.50	6.63	–	–	m²	**6.63**
average 200 mm thick	–	0.60	7.95	–	–	m²	**7.95**
Selected from spoil heaps; grading; prepared for turfing or seeding; to cuttings or embankments							
average 100 mm thick	–	0.51	6.70	–	–	m²	**6.70**
average 150 mm thick	–	0.63	8.38	–	–	m²	**8.38**
average 200 mm thick	–	0.76	10.05	–	–	m²	**10.05**
Imported top soil, planting quality							
By machine							
Grading; prepared for turfing or seeding; to general surfaces							
average 100 mm thick	–	–	0.06	1.18	2.84	m²	**4.08**
average 150 mm thick	–	0.01	0.07	1.56	4.25	m²	**5.88**
average 200 mm thick	–	0.01	0.10	2.73	5.20	m²	**8.03**
Grading; preparing for turfing or seeding; to cuttings or embankments							
average 100 mm thick	–	–	0.07	1.32	3.46	m²	**4.85**
average 150 mm thick	–	0.01	0.08	1.80	5.67	m²	**7.55**
average 200 mm thick	–	0.01	0.11	3.12	6.93	m²	**10.16**
By hand							
Grading; prepared for turfing or seeding; to general surfaces							
average 100 mm thick	–	0.51	6.70	–	3.46	m²	**10.16**
average 150 mm thick	–	0.63	8.38	–	5.67	m²	**14.05**
average 200 mm thick	–	0.76	10.05	–	6.93	m²	**16.98**

37 SOFT LANDSCAPING

Item	PC £	Labour hours	Labour £	Plant £	Material £	Unit	Total rate £
SEEDING AND TURFING – cont							
Fertilizer							
Fertilizer 0.07 kg/m²; raking in							
general surfaces (PC £ per 25 kg)	25.00	0.03	0.40	–	0.08	m²	**0.48**
Selected grass seed							
Grass seed; sowing at a rate of 40 g/m² two applications; raking in							
general surfaces (PC £ per kg)	4.75	0.17	2.25	–	0.25	m²	**2.50**
cuttings or embankments	–	0.20	2.65	–	0.25	m²	**2.90**
Turfing							
Imported turf; cultivated							
general surfaces	2.39	0.10	1.32	–	2.51	m²	**3.83**
cuttings or embankments; shallow	2.39	0.15	1.98	–	2.51	m²	**4.49**
cuttings or embankments; steep; pegged	2.39	0.28	3.71	–	2.51	m²	**6.22**
Preserved turf from stack on site; lay only							
general surfaces	–	0.19	2.52	–	–	m²	**2.52**
cuttings or embankments; shallow	–	0.20	2.65	–	–	m²	**2.65**
cuttings or embankments; steep; pegged	–	0.28	3.71	–	–	m²	**3.71**
PLANTING							
Hedge plants							
height not exceeding 750 mm	–	0.23	3.04	–	–	nr	**3.04**
height 750 mm–1.50 m	–	0.56	7.41	–	–	nr	**7.41**
Saplings							
height not exceeding 3.00 m	–	1.57	20.79	–	–	nr	**20.79**

39 ELECTRICAL SERVICES

Item	PC £	Labour hours	Labour £	Plant £	Material £	Unit	Total rate £
LIGHTNING PROTECTION							
For more examples and depth of information please consult Spon's Mechanical and Electrical Services Price Book							
Lightning protection equipment							
Copper strip roof or down conductors fixed with bracket or saddle clips							
20 mm × 3 mm flat section	–	–	–	–	–	m	**21.28**
25 mm × 3 mm flat section	–	–	–	–	–	m	**24.83**
Aluminium strip roof or down conductors fixed with bracket or saddle clips							
20 mm × 3 mm flat section	–	–	–	–	–	m	**15.62**
25 mm × 3 mm flat section	–	–	–	–	–	m	**17.04**
Joints in tapes	–	–	–	–	–	nr	**12.08**
Bonding connections to roof and structural metalwork	–	–	–	–	–	nr	**70.96**
Testing points	–	–	–	–	–	nr	**58.31**
Earth electrodes							
16 mm dia. driven copper electrodes in 1220 mm long sectional lengths; 2440 mm long overall first 2440 mm length driven and tested	–	–	–	–	–	nr	**184.49**
25 mm × 3 mm copper strip electrode in 457 mm deep prepared trench	–	–	–	–	–	m	**14.19**

41 BUILDER'S WORK IN CONNECTION WITH SERVICES

Item	PC £	Labour hours	Labour £	Plant £	Material £	Unit	Total rate £
HOLES, CHASES FOR SERVICES							
Electrical installations; cutting away for and making good after electrician; including cutting or leaving all holes, notches, mortices, sinkings and chases, in both the structure and its coverings, for the following electrical points							
Exposed installation							
lighting points	–	0.28	4.65	–	–	nr	**4.65**
socket outlet points	–	0.46	7.99	–	–	nr	**7.99**
fitting outlet points	–	0.46	7.99	–	–	nr	**7.99**
equipment points or control gear points	–	0.65	11.47	–	–	nr	**11.47**
Concealed installation							
lighting points	–	0.37	6.32	–	–	nr	**6.32**
socket outlet points	–	0.65	11.47	–	–	nr	**11.47**
fitting outlet points	–	0.65	11.47	–	–	nr	**11.47**
equipment points or control gear points	–	0.93	16.12	–	–	nr	**16.12**
Builders' work for other services installations							
Cutting chases in brickwork							
for one pipe; not exceeding 55 mm nominal size; vertical	–	0.37	4.86	–	–	m	**4.86**
for one pipe; 55 mm–110 mm nominal size; vertical	–	0.65	8.54	–	–	m	**8.54**
Cutting and pinning to brickwork or blockwork; ends of supports							
for pipes not exceeding 55 mm nominal size	–	0.19	4.10	–	–	nr	**4.10**
for cast iron pipes 55 mm–110 mm nominal size	–	0.31	6.69	–	–	nr	**6.69**
Cutting or forming holes for pipes or the like; not exceeding 55 mm nominal size; making good							
reinforced concrete; not exceeding 100 mm deep	–	0.75	11.64	0.71	–	nr	**12.35**
reinforced concrete; 100 mm–200 mm deep	–	1.15	17.85	1.09	–	nr	**18.94**
reinforced concrete; 200 mm–300 mm deep	–	1.50	23.29	1.43	–	nr	**24.72**
half brick thick	–	0.31	4.81	–	–	nr	**4.81**
one brick thick	–	0.51	7.92	–	–	nr	**7.92**
one and a half brick thick	–	0.83	12.88	–	–	nr	**12.88**
100 mm blockwork	–	0.28	4.35	–	–	nr	**4.35**
140 mm blockwork	–	0.37	5.74	–	–	nr	**5.74**
215 mm blockwork	–	0.46	7.14	–	–	nr	**7.14**
plasterboard partition or suspended ceiling	–	0.35	5.43	–	–	nr	**5.43**
Cutting or forming holes for pipes or the like; 55 mm–110 mm nominal size; making good							
reinforced concrete; not exceeding 100 mm deep	–	1.15	17.85	1.09	–	nr	**18.94**
reinforced concrete; 100 mm–200 mm deep	–	1.75	27.16	1.67	–	nr	**28.83**
reinforced concrete; 200 mm–300 mm deep	–	2.25	34.92	2.14	–	nr	**37.06**
half brick thick	–	0.37	5.74	–	–	nr	**5.74**
one brick thick	–	0.65	10.09	–	–	nr	**10.09**
one and a half brick thick	–	1.02	15.83	–	–	nr	**15.83**
100 mm blockwork	–	0.32	4.97	–	–	nr	**4.97**
140 mm blockwork	–	0.46	7.14	–	–	nr	**7.14**
215 mm blockwork	–	0.56	8.69	–	–	nr	**8.69**
plasterboard partition or suspended ceiling	–	0.40	6.21	–	–	nr	**6.21**

41 BUILDER'S WORK IN CONNECTION WITH SERVICES

Item	PC £	Labour hours	Labour £	Plant £	Material £	Unit	Total rate £
Cutting or forming holes for pipes or the like; over 110 mm nominal size; making good							
reinforced concrete; not exceeding 100 mm deep	–	1.15	17.85	1.09	–	nr	**18.94**
reinforced concrete; 100 mm–200 mm deep	–	1.75	27.16	1.67	–	nr	**28.83**
reinforced concrete; 200 mm–300 mm deep	–	2.25	34.92	2.14	–	nr	**37.06**
half brick thick	–	0.46	7.14	–	–	nr	**7.14**
one brick thick	–	0.79	12.26	–	–	nr	**12.26**
one and a half brick thick	–	1.25	19.40	–	–	nr	**19.40**
100 mm blockwork	–	0.42	6.52	–	–	nr	**6.52**
140 mm blockwork	–	0.56	8.69	–	–	nr	**8.69**
215 mm blockwork	–	0.69	10.71	–	–	nr	**10.71**
plasterboard partition or suspended ceiling	–	0.45	6.98	–	–	nr	**6.98**
Add for making good fair face or facings one side							
pipe; not exceeding 55 mm nominal size	–	0.07	1.51	–	–	nr	**1.51**
pipe; 55 mm–110 mm nominal size	–	0.09	1.94	–	–	nr	**1.94**
pipe; over 110 mm nominal size	–	0.11	2.37	–	–	nr	**2.37**
Add for fixing sleeve (supply not included)							
for pipe; small	–	0.14	3.01	–	–	nr	**3.01**
for pipe; large	–	0.19	4.10	–	–	nr	**4.10**
for pipe; extra large	–	0.28	6.04	–	–	nr	**6.04**
Add for supplying and fixing two hour intumescent sleeve							
for 55 mm uPVC pipe	–	0.25	3.89	–	6.25	nr	**10.14**
for 110 mm uPVC pipe	–	0.28	4.35	–	6.79	nr	**11.14**
for 200 mm uPVC pipe	–	0.30	4.66	–	44.46	nr	**49.12**
Cutting or forming holes for ducts; girth not exceeding 1.00 m; making good							
half brick thick	–	0.56	8.69	–	–	nr	**8.69**
one brick thick	–	0.93	14.44	–	–	nr	**14.44**
one and a half brick thick	–	1.48	22.97	–	–	nr	**22.97**
100 mm blockwork	–	0.46	7.14	–	–	nr	**7.14**
140 mm blockwork	–	0.65	10.09	–	–	nr	**10.09**
215 mm blockwork	–	0.83	12.88	–	–	nr	**12.88**
plasterboard partition or suspended ceiling	–	0.65	10.09	–	–	nr	**10.09**
Cutting or forming holes for ducts; girth 1.00 m–2.00 m; making good							
half brick thick	–	0.65	10.09	–	–	nr	**10.09**
one brick thick	–	1.11	17.23	–	–	nr	**17.23**
one and a half brick thick	–	1.76	27.32	–	–	nr	**27.32**
100 mm blockwork	–	0.56	8.69	–	–	nr	**8.69**
140 mm blockwork	–	0.74	11.49	–	–	nr	**11.49**
215 mm blockwork	–	0.93	14.44	–	–	nr	**14.44**
plasterboard partition or suspended ceiling	–	0.75	11.64	–	–	nr	**11.64**
Cutting or forming holes for ducts; girth 2.00 m–3.00 m; making good							
half brick thick	–	1.02	15.83	–	–	nr	**15.83**
one brick thick	–	1.76	27.32	–	–	nr	**27.32**
one and a half brick thick	–	2.78	43.16	–	–	nr	**43.16**
100 mm blockwork	–	0.88	13.66	–	–	nr	**13.66**
140 mm blockwork	–	1.20	18.63	–	–	nr	**18.63**
215 mm blockwork	–	1.53	23.75	–	–	nr	**23.75**
plasterboard partition or suspended ceiling	–	1.00	15.52	–	–	nr	**15.52**

41 BUILDER'S WORK IN CONNECTION WITH SERVICES

Item	PC £	Labour hours	Labour £	Plant £	Material £	Unit	Total rate £
HOLES, CHASES FOR SERVICES – cont							
Builders' work for other services installations – cont							
Cutting or forming holes for ducts; girth 3.00 m– 4.00 m; making good							
half brick thick	–	1.39	21.58	–	–	nr	**21.58**
one brick thick	–	2.31	35.86	–	–	nr	**35.86**
one and a half brick thick	–	3.70	57.44	–	–	nr	**57.44**
100 mm blockwork	–	1.02	15.83	–	–	nr	**15.83**
140 mm blockwork	–	1.39	21.58	–	–	nr	**21.58**
215 mm blockwork	–	1.76	27.32	–	–	nr	**27.32**
plasterboard partition or suspended ceiling	–	1.25	19.40	–	–	nr	**19.40**
Mortices in brickwork							
for expansion bolt	–	0.19	2.95	–	–	nr	**2.95**
for 20 mm dia. bolt; 75 mm deep	–	0.14	2.17	–	–	nr	**2.17**
for 20 mm dia. bolt; 150 mm deep	–	0.23	3.57	–	–	nr	**3.57**
Mortices in brickwork; grouting with cement mortar (1:1)							
75 mm × 75 mm × 200 mm deep	–	0.28	4.35	–	0.14	nr	**4.49**
75 mm × 75 mm × 300 mm deep	–	0.37	5.74	–	0.21	nr	**5.95**
Holes in softwood for pipes, bars, cables and the like							
12 mm thick	–	0.03	0.61	–	–	nr	**0.61**
25 mm thick	–	0.05	1.01	–	–	nr	**1.01**
50 mm thick	–	0.09	1.82	–	–	nr	**1.82**
100 mm thick	–	0.14	2.82	–	–	nr	**2.82**
Holes in hardwood for pipes, bars, cables and the like							
12 mm thick	–	0.05	1.01	–	–	nr	**1.01**
25 mm thick	–	0.08	1.62	–	–	nr	**1.62**
50 mm thick	–	0.14	2.82	–	–	nr	**2.82**
100 mm thick	–	0.20	4.04	–	–	nr	**4.04**
Diamond drilling for cutting holes and mortices in masonry or concrete							
Cutting holes and mortices in brickwork; per 25 mm depth							
25 mm dia.	–	–	–	–	–	nr	**1.76**
32 mm dia.	–	–	–	–	–	nr	**1.41**
52 mm dia.	–	–	–	–	–	nr	**1.70**
78 mm dia.	–	–	–	–	–	nr	**1.87**
107 mm dia.	–	–	–	–	–	nr	**1.96**
127 mm dia.	–	–	–	–	–	nr	**2.42**
152 mm dia.	–	–	–	–	–	nr	**2.86**
200 mm dia.	–	–	–	–	–	nr	**3.67**
250 mm dia.	–	–	–	–	–	nr	**5.54**
300 mm dia.	–	–	–	–	–	nr	**7.35**
Diamond chasing; per 25 × 25 mm section							
in facing or common brickwork	–	–	–	–	–	m	**3.34**
in semi-engineering brickwork	–	–	–	–	–	m	**6.68**
in engineering brickwork	–	–	–	–	–	m	**9.32**

41 BUILDER'S WORK IN CONNECTION WITH SERVICES

Item	PC £	Labour hours	Labour £	Plant £	Material £	Unit	Total rate £
in lightweight blockwork	–	–	–	–	–	m	2.62
in heavyweight blockwork	–	–	–	–	–	m	5.26
in render/screed	–	–	–	–	–	m	10.35
Forming boxes; 100 × 100 mm; per 25 mm depth							
in facing or common brickwork	–	–	–	–	–	nr	1.34
in semi-engineering brickwork	–	–	–	–	–	nr	2.67
in engineering brickwork	–	–	–	–	–	nr	3.72
in lightweight blockwork	–	–	–	–	–	nr	1.06
in heavyweight blockwork	–	–	–	–	–	nr	2.11
in render/screed	–	–	–	–	–	nr	4.14
Other items							
diamond track mount or ring sawing brickwork	–	–	–	–	–	m	6.58
diamond floor sawing asphalte	–	–	–	–	–	m	1.10
stitch drilling 107 mm dia. hole in brickwork	–	–	–	–	–	nr	1.41
INTERNAL FLOOR DUCTS FOR SERVICES							
Screed floor ducting; with side flanges; laid within floor screed; galvanized mild steel							
Floor ducting							
100 mm wide × 50 mm deep	10.88	0.19	3.83	–	11.42	m	15.25
extra for							
bend	–	0.09	1.82	–	17.68	nr	19.50
tee section	–	0.09	1.82	–	17.68	nr	19.50
connector/stop end	–	0.09	1.82	–	2.04	nr	3.86
ply cover 15 mm/16 mm thick WBP exterior grade	–	0.09	1.82	–	2.53	m	4.35
100 mm wide × 70 mm deep	11.95	0.20	4.04	–	12.55	m	16.59
extra for							
bend	–	0.09	1.82	–	17.68	nr	19.50
tee section	–	0.09	1.82	–	17.68	nr	19.50
connector/stop end	–	0.09	1.82	–	2.04	nr	3.86
ply cover 15 mm/16 mm thick WBP exterior grade	–	0.09	1.82	–	2.53	m	4.35
200 mm wide × 50 mm deep	15.04	0.19	3.83	–	15.79	m	19.62
extra for							
bend	–	0.09	1.82	–	19.69	nr	21.51
tee section	–	0.09	1.82	–	19.69	nr	21.51
connector/stop end	–	0.09	1.82	–	2.04	nr	3.86
ply cover 15 mm/16 mm thick WBP exterior grade	–	0.09	1.82	–	4.60	m	6.42
EXTERNAL SERVICES							
Service runs							
Water main; all laid in trenches including excavation and backfill with type 1 hardcore. Surface finish not included							
MDPE pipe; 20 mm dia.	–	2.85	35.94	7.50	57.33	m	100.77
MDPE pipe; 25 mm dia.	–	2.85	35.94	7.50	57.45	m	100.89
MDPE pipe; 32 mm dia.	–	2.85	35.94	7.50	57.74	m	101.18
MDPE pipe; 50 mm dia.	–	2.85	35.94	7.50	59.25	m	102.69

41 BUILDER'S WORK IN CONNECTION WITH SERVICES

Item	PC £	Labour hours	Labour £	Plant £	Material £	Unit	Total rate £
EXTERNAL SERVICES – cont							
Service runs – cont							
Electric mains 25 mm 2 core armoured cable laid in service duct, including excavation and backfill with type 1 hardcore. Surface finsih not included							
25 mm 2 core armoured cable	–	3.00	37.83	7.50	75.25	m	**120.58**
25 mm 3 core armoured cable	–	3.00	37.83	7.50	76.66	m	**121.99**
25 mm 5 core armoured cable	–	3.00	37.83	7.50	82.90	m	**128.23**

Spon's Asia Pacific Construction Costs Handbook, Fifth Edition

LANGDON & SEAH

In the last few years, the global economic outlook has continued to be shrouded in uncertainty and volatility following the financial crisis in the Euro zone. While the US and Europe are going through a difficult period, investors are focusing more keenly on Asia. This fifth edition provides overarching construction cost data for 16 countries: Brunei, Cambodia, China, Hong Kong, India, Indonesia, Japan, Malaysia, Myanmar, Philippines, Singapore, South Korea, Sri Lanka, Taiwan, Thailand and Vietnam.

May 2015: 234X156 mm: 452 pp
Hb: 978-1-4822-4358-1: £160.00

To Order: Tel: +44 (0) 1235 400524 Fax: +44 (0) 1235 400525
or Post: Taylor and Francis Customer Services,
Bookpoint Ltd, Unit T1, 200 Milton Park, Abingdon, Oxon, OX14 4TA UK
Email: book.orders@tandf.co.uk

For a complete listing of all our titles visit:
www.tandf.co.uk

BIM and Quantity Surveying
PITTARD & SELL

The sudden arrival of Building Information Modelling (BIM) as a key part of the building industry is redefining the roles and working practices of its stakeholders. Many clients, designers, contractors, quantity surveyors, and building managers are still finding their feet in an industry where BIM compliance can bring great rewards.

This guide is designed to help quantity surveying practitioners and students understand what BIM means for them, and how they should prepare to work successfully on BIM compliant projects. The case studies show how firms at the forefront of this technology have integrated core quantity surveying responsibilities like cost estimating, tendering, and development appraisal into high profile BIM projects. In addition to this, the implications for project management, facilities management, contract administration and dispute resolution are also explored through case studies, making this a highly valuable guide for those in a range of construction project management roles.

Featuring a chapter describing how the role of the quantity surveyor is likely to permanently shift as a result of this development, as well as descriptions of tools used, this covers both the organisational and practical aspects of a crucial topic.

December 2015: 234 x 156 mm: 258 pp
Pb: 978-0-415-87043-6; £24.99

To Order: Tel: +44 (0) 1235 400524 Fax: +44 (0) 1235 400525
or Post: Taylor and Francis Customer Services,
Bookpoint Ltd, Unit T1, 200 Milton Park, Abingdon, Oxon, OX14 4TA UK
Email: book.orders@tandf.co.uk

For a complete listing of all our titles visit:
www.tandf.co.uk

PART 5

Fees for Professional Services

This part contains the following sections:

Estimator's Pocket Book

Duncan Cartlidge

The Estimator's Pocket Book is a concise and practical reference covering the main pricing approaches, as well as useful information such as how to process sub-contractor quotations, tender settlement and adjudication. It is fully up-to-date with NRM2 throughout, features a look ahead to NRM3 and describes the implications of BIM for estimators.

It includes instructions on how to handle:

*   the NRM order of cost estimate;
*   unit-rate pricing for different trades;
*   pro rata pricing and dayworks
*   builders' quantities;
*   approximate quantities.

Worked examples show how each of these techniques should be carried out in clear, easy-to-follow steps. This is the indispensible estimating reference for all quantity surveyors, cost managers, project managers and anybody else with estimating responsibilities. Particular attention is given to NRM2, but the overall focus is on the core estimating skills needed in practice.

May 2013 186x123: 310pp
Pb: 978-0-415-52711-8: £21.99

To Order: Tel: +44 (0) 1235 400524 Fax: +44 (0) 1235 400525
or Post: Taylor and Francis Customer Services,
Bookpoint Ltd, Unit T1, 200 Milton Park, Abingdon, Oxon, OX14 4TA UK
Email: book.orders@tandf.co.uk

For a complete listing of all our titles visit:
www.tandf.co.uk

Taylor & Francis
Taylor & Francis Group

QUANTITY SURVEYORS' FEES

Guidance on basic quantity surveying services is set out in the RICS Standard Form of Consultant's Appointment for Quantity Surveyors. Services are separated into core services for a variety of contracts and supplementary services and it is advisable to refer to this guidance. Copies can be obtained from the Royal Institution of Chartered Surveyors (RICS) at www.ricsbooks.com.

Quantity surveying services

Preparation:

- liaising with clients and the professional team
- advice on cost
- preparation of initial budget/cost plan/cash flow forecasts

Design:

- prepare and maintain cost plan
- advise design team on impact of design development on cost

Pre-construction:

- liaise with professional team
- advise on procurement strategy
- liaise with client's legal advisors on contract matters
- prepare tender documents
- define prospective tenderers
- obtain tenders/check tenders/prepare recommendation for client
- maintain and develop cost plan

Construction:

- visit the site
- prepare interim valuations
- advise on the cost of variations
- agree the cost of claims
- advise on contractual matters

Supplementary services may include:

- preparation of mechanical and electrical tender documentation
- preparation of cost analyses
- advice on insurance claims
- facilitate value management exercises
- prepare life cycle calculations/sustainability
- capital allowance advice/VAT
- attend adjudication/mediation proceedings

Fee Guide

Quantity Surveying Services Benchmark	Mean	Lower Quartile	Upper Quartile
	2%	1.9%	2.8%

The level of fees above, are expressed as a percentage of the contract value of £3,500,000 for a new build project and do not include VAT.

Fee levels vary depending on many factors, including, but not limited to the following; type of project; complexity; procurement route.

ARCHITECTS' FEES

RIBA Agreements are designed to be:

- in line with current working practices, legislative changes and procurement methods
- attractive to clients, architects and other consultants, with robust but fair terms
- a flexible system of components that can be assembled and customized to create tailored and bespoke contracts
- suitable for a wide range of projects and services
- based upon the updated RIBA Outline Plan of Work 2013
- available in electronic and printed formats

RIBA Agreements 2013	Suitable for
Standard Agreement	• a commission where detailed contract terms are necessary for a wide range of projects using most procurement methods • where the client is acting for business or commercial purposes • where the commission is for work to the client's home where the size or value of the Project merits use of the JCT Standard or Intermediate forms of building contract or similar and the terms have been negotiated with the client as a consumer
Concise Agreement	• for a commission where the concise contract terms are compatible with the complexity of the Project and the risks to each party • where the Client is acting for business or commercial purposes • where the commission is for work to the Client's home and the terms have been negotiated with the Client as a 'consumer'. A consumer is 'a natural person acting for purposes outside his trade, business or profession' • where the building works, including extensions and alterations, will be carried out using forms of building contract, such as JCT Agreement for Minor Works or JCT Intermediate Form of Building Contract
Domestic Project Agreement	• the commission relates to work to the client's home, provided that they have elected to use these conditions in their own name, i.e. not as a limited company or other legal entity • the contract terms are compatible with the complexity of the project and the risks to each party and have been negotiated with the client as a consumer • the building works, including extensions and alterations, will be carried out using forms of building contract, such as the JCT Building Contract for a homeowner/occupier, JCT Agreement for Minor Works or JCT Intermediate Form of Building Contract
Sub-Consultant's Agreement	• a consultant wishes or perhaps is required by the client to appoint another consultant (thus, a sub-consultant) to perform part of the consultant's services • the contract terms are compatible with the (head) agreement between the consultant and the client, with the complexity of the project and the risks to each party
Electronic and Print Formats	• All the RIBA Agreements 2013 and their components are available as electronic files. A limited number of the conditions and core components are published in print. The electronic and printed versions may be used in combination
Electronic Components	• Conditions, notes and guides are available as locked PDFs and all other components, e.g. schedules and model letters, are available in Rich Text Format (RTF), which can be customized using most commonly used word processing software to meet project requirements or modified to match the house style of the practice

Each agreement comprises the selected Conditions of Appointment (i.e. Standard, Concise or Domestic), related components, and a schedule or schedules of Services.

ARCHITECTS' FEES

Notes on use and completion and model letters for business clients and domestic clients are included with each pack.

For further information, readers are advised to log onto the RIBA Publications website at *www.ribabookshops.com/agreements*

RIBA Plan of Work 2013

The table below shows how the new Plan of Work maps to the old RIBA Plan of Work 2007 and the CIC's work stages.

RIBA Plan of Work 2013	RIBA Plan of Work 2007	CIC/BIM Task Group Coordinated work stages
0 Strategic definition	A Appraisal	0 Strategy
1 Preparation and brief	B Design brief	1 Brief
2 Concept design	C Concept	2 Concept
3 Developed design	D Design development	3 Definition
4 Technical design	E Technical design	4 Design
	F Production information	
	G Tender documentation	
	H Tender action	
5 Construction	J Mobilization	5 Build and commission
	K Construction to practical competition	
6 Handover and close out		6 Handover and close out
7 In use	L Post practical completion	7 Operation and end of life

More details can be found at the RIBA website.

Appointment Guidance

A guide, 'A Client's Guide to Engaging an Architect', is available from RIBA Bookshops at www.ribabookshops.com.

This guide includes an introduction to the services an Architect can be expected to provide, advice on the forms to use, linking the RIBA Plan of Work Stages with fees (which are a matter of negotiation) and classifying buildings according to three levels of complexity.

Generally, the more complex the building the higher the level of fee.

Example categories include:

- Simple: for buildings such as car parks, warehouses, factories and speculative retail schemes
- Average: for buildings such as offices, most retail outlets, general housing, schools etc.
- Complex: for multi-purpose developments, specialist buildings e.g. hospitals, research laboratories etc.

Procurement option can also influence the architects' fees.

ARCHITECTS' FEES

Fee Guide

Architectural Services Benchmark	Mean	Lower Quartile	Upper Quartile
	4.4%	3.6%	5.0%

The level of fees above, are expressed as a percentage of the contract value of £3,500,000 for a new build project and do not include VAT.

Projects carried out under the traditional form of contracts tend to attract higher fees than for design and build contract led jobs.

Sectors project have less of an impact on fee levels, however health and education projects attract slightly higher fees than private housing and industrial.

CONSULTING ENGINEERS' FEES

CONDITIONS OF APPOINTMENT

A scale of professional charges for consulting engineering services is published by the Association for Consultancy and Engineering (ACE).

Copies of the document can be obtained direct from:

Association for Consultancy and Engineering
Alliance House
12 Caxton Street
London SW1H OQL
Tel 020 7222 6557
Fax: 020 7222 0750

Comparisons

Instead of the previous arrangement of having different agreements designed for each major discipline of engineering, the current agreements have been developed primarily to suit the different roles that Consulting Engineers may be required to perform, with variants of some of them for different disciplines. The agreements have been standardized as far as possible whilst retaining essential differences.

Greater attention is required than with previous agreements to ensure the documents are completed properly. This is because of the perceived need to allow for a wider choice of arrangements, particularly of methods of payment.

The agreements are not intended to be used as unsigned reference material with the details of an engagement being covered in an exchange of letters, although much of their content could be used as a basis for drafting such correspondence.

For 2009 the ACE has published a new suite of Agreements with a broader set of services, these are listed below.

Forms of Agreement

- ACE Agreement 1: Design
- ACE Agreement 2: Advise and Report
- ACE Agreement 3: Design and Construct
- ACE Agreement 4: Sub-Consultancy
- ACE Agreement 5: Homeowner
- ACE Agreement 6: Expert Witness (Sole Practitioner)
- ACE Agreement 7: Expert Witness (Firm)
- ACE Agreement 8: Adjudicator

To a number of the the above ACE Agreements Schedules of Services are appended and currently these are:

For use with ACE Agreement 1: Design

- ACE Schedule of Services – Part G(a):
  Civil and Structural Engineer – single consultant or non-lead consultant
- ACE Schedule of Services – Part G(b):
  Mechanical and Electrical Engineering (detailed design in buildings)
- ACE Schedule of Services – Part G(c):
  Mechanical and Electrical Engineering (performance design in buildings)
- ACE Schedule of Services – Part G(d):
  Civil and Structural Engineer – Lead consultant
- ACE Schedule of Services – Part G(e):
  Mechanical and Electrical Engineering Design in buildings – Lead consultant

CONSULTING ENGINEERS' FEES

For use with ACE Agreement 3: Design and Construct

- ACE Schedule of Services – Part G(f):
  Civil and Structural Engineer
- ACE Schedule of Services – Part G(g):
  Mechanical and Electrical Engineering (detailed design in buildings)
- ACE Schedule of Services – Part G(h):
  Mechanical and Electrical Engineering (performance design in buildings)

ACE Agreement 1: Design

Design for the appointment of a consultant by a client to undertake detailed design and/or specification of permanent works to be undertaken or installed by a contractor including any studies, appraisals, investigations, contract administration or construction monitoring leading to or resulting from such detailed design and/or specification.

ACE Agreement 2: Advise and Report

Advise and report for the appointment of a consultant by a client to provide any type of advisory, research, checking, reviewing, investigatory, monitoring, reporting or technical services in the built and natural environments where such services do not consist of detailed design or specification of permanent works to be constructed or installed by a contractor.

ACE Agreement 3: Design and Construct

For the appointment of a consultant by a contractor in circumstances where the contractor is to construct permanent works designed by the consultant.

ACE Agreement 4: Sub-Consultancy

For the appointment of a sub-consultant by a consultant in circumstances where the consultant is appointed on the terms of an ACE Agreement by its client.

ACE Agreement 5: Homeowner

Model letter for the appointment of a consultant by a homeowner.

ACE Agreement 6: Expert Witness (Sole Practitioner)

For the appointment of an individual to act as an expert witness.

ACE Agreement 7: Expert Witness (Firm)

For the appointment of a firm to provide an expert witness.

ACE Agreement 8: Adjudicator

For the appointment of an adjudicator.

Fee Guide

Profession

Structural Engineering Services Benchmark	Mean	Lower Quartile	Upper Quartile
	2.5%	1.7%	2.9%

Services Engineering Benchmark	Mean	Lower Quartile	Upper Quartile
	2.0%	1.5%	2.6%

THE TOWN AND COUNTRY PLANNING FEES AND BUILDING REGULATION FEES

THE TOWN AND COUNTRY PLANNING APPLICATION FEES 2014

Author's Note

This is only a small extract of typical the fees chargeable. Users should always obtain actual fees from the local authority concerned with the particular planning application.

All Outline Applications		
£385 per 0.1 hectare for sites up to and including 2.5 hectares	Not more than 2.5 hectares	£385 per 0.1 hectare
£9,527 + £115 for each 0.1 in excess of 2.5 hectares to a maximum of £125,000	More than 2.5 hectares	£9,527 + £115 per 0.1 hectare

There is a very useful online calculator which will give prices for planning applications at
http://www.planningportal.gov.uk/pins/FeeCalculatorStandalone

THE BUILDING (LOCAL AUTHORITY CHARGES) REGULATIONS

CHARGE SCHEDULES

Fees vary from one authority to another, so always check with your Local Authority if you decide to use their officers to provide building control services.

The Construction Industry Council (CIC) has been designated by the government as a body for approving inspectors (AI). Individual and Corporate Approved Inspectors registered with CIC are qualified to undertake building control work in accordance with section 49 of the Building Act 1984 and regulation 4 of the Building (Approved Inspectors etc.) Regulations 1985, and the Building (Approved Inspectors etc.) Regulations 2010.

Approved Inspectors provide building control services on all types of construction projects. The Construction Industry Council (CIC) maintains a list of approved inspectors, see www.cic.org.uk.

Fees vary. Each Local Authority publishes a schedule of rates applicable to project values commonly up to £250,000. The fee for a project of £250,000 in value is approximately £1,500.

For higher value projects the fees are individually determined but are typically approximately 0.1% of the estimated project cost.

Metric Handbook
Planning and Design
Data, 5th Edition

Pamela Buxton

The Metric Handbook is the major handbook of planning and design information for architects and architecture students. Covering basic design data for all the major building types, it is the ideal starting point for any project. For each building type, the book gives the basic design requirements and all the principal dimensional data, and succinct guidance on how to use the information and what regulations the designer needs to be aware of.

As well as building types, the Metric Handbook deals with broader aspects of design such as materials, acoustics and lighting, and general design data on human dimensions and space requirements. The Metric Handbook provides an invaluable resource for solving everyday design and planning problems.

Mar 2015: 203 x 292: 858 pp
Pb: 978-0-415-72542-2 £39.99

To Order: Tel: +44 (0) 1235 400524 Fax: +44 (0) 1235 400525
or Post: Taylor and Francis Customer Services,
Bookpoint Ltd, Unit T1, 200 Milton Park, Abingdon, Oxon, OX14 4TA UK
Email: book.orders@tandf.co.uk

For a complete listing of all our titles visit:
www.tandf.co.uk

PART 6

Daywork and Prime Cost

Architect's Legal Pocket Book
Second Edition

Matthew Cousins

A little book that's big on information. The Architect's Legal Pocket Book is the definite reference on legal issues for architects and architectural students. This handy pocket guide covers key legal principles which will help you to quickly understand the law and where to go for further information.

Now in a fully updated new edition, this bestselling book covers a wide range of subjects focused on the UK including building legislation and the Localism Bill, negligence, liability, planning policy and development, listed buildings, party wall legislation, and rights of light. This edition also contains greater coverage of contracts including the RIBA contracts, dispute resolution and legal issues in professional practice.

Illustrated with clear diagrams and featuring key cases, this is an invaluable source of practical information and a comprehensive guide of the current law for architects. It is a book no architect should be without.

August 2015 186x123: 402pp
Pb: 978-1-138-82144-6: £24.99

To Order: Tel: +44 (0) 1235 400524 Fax: +44 (0) 1235 400525
or Post: Taylor and Francis Customer Services,
Bookpoint Ltd, Unit T1, 200 Milton Park, Abingdon, Oxon, OX14 4TA UK
Email: book.orders@tandf.co.uk

For a complete listing of all our titles visit:
www.tandf.co.uk

BUILDING INDUSTRY

When work is carried out which cannot be valued in any other way it is customary to assess the value on a cost basis with an allowance to cover overheads and profit. The basis of costing is a matter for agreement between the parties concerned, but definitions of prime cost for the building industry have been prepared and published jointly by the Royal Institution of Chartered Surveyors and the National Federation of Building Trades Employers (now the Construction Confederation) for the convenience of those who wish to use them. These documents are reproduced with the permission of the Royal Institution of Chartered Surveyors, which owns the copyright.

The daywork schedule published by the Civil Engineering Contractors Association is included in the *Architects' & Builders'* companion title, *Spon's Civil Engineering and Highway Works Price Book*.

For larger Prime Cost contracts the reader is referred to the form of contract issued by the Royal Institute of British Architects.

DEFINITION OF PRIME COST OF BUILDING WORKS OF A JOBBING OR MAINTENANCE CHARACTER (2010 EDITION)

This definition of Prime Cost is published by the Royal Institution of Chartered Surveyors and the National Federation of Building Trades Employers, for convenience and for use by people who choose to use it. Members of the National Federation of Building Trades Employers are not in any way debarred from defining Prime Cost and rendering their accounts for work carried out on that basis in any way they choose. Building owners are advised to reach agreement with contractors on the Definition of Prime Cost to be used prior to issuing instructions.

SECTION 1 – APPLICATION

1.1. This definition provides a basis for the valuation of work of a jobbing or maintenance character executed under such building contracts as provide for its use.

1.2. It is not applicable in any other circumstances, such as daywork executed under or incidental to a building contract.

SECTION 2 – COMPOSITION OF TOTAL CHARGES

2.1. The prime cost of jobbing work comprises the sum of the following costs:
   (a) Labour as defined in Section 3.
   (b) Materials and goods as defined in Section 4.
   (c) Plant, consumable stores and services as defined in Section 5.
   (d) Subcontracts as defined in Section 6.

2.2. Incidental costs, overhead and profit as defined in Section 7 and expressed as percentage adjustments are applicable to each of 2.1 (a)–(d).

SECTION 3 – LABOUR

3.1. Labour costs comprise all payments made to or in respect of all persons directly engaged upon the work, whether on or off the site, except those included in Section 7.

3.2. Such payments are based upon the standard wage rates, emoluments and expenses as laid down for the time being in the rules or decisions of the National Joint Council for the Building Industry and the terms of the Building and Civil Engineering Annual and Public Holiday Agreements applying to the works, or the rules of decisions or agreements of such other body as may relate to the class of labour concerned, at the time when and in the area where the work is executed, together with the Contractor's statutory obligations, including:
   (a) Guaranteed minimum weekly earnings (e.g. Standard Basic Rate of Wages and Guaranteed Minimum Bonus Payment in the case of NJCBI rules).
   (b) All other guaranteed minimum payments (unless included in Section 7).
   (c) Payments in respect of incentive schemes or productivity agreements applicable to the works.
   (d) Payments in respect of overtime normally worked; or necessitated by the particular circumstances of the work; or as otherwise agreed between the parties.
   (e) Differential or extra payments in respect of skill, responsibility, discomfort or inconvenience.
   (f) Tool allowance.
   (g) Subsistence and periodic allowances.
   (h) Fares, travelling and lodging allowances.

BUILDING INDUSTRY

(i)   Employer's contributions to annual holiday credits.
(j)   Employer's contributions to death benefit schemes.
(k)   Any amounts which may become payable by the Contractor to or in respect of operatives arising from the operation of the rules referred to in 3.2 which are not provided for in 3.2 (a)–(k) or in Section 7.
(l)   Employer's National Insurance contributions and any contribution, levy or tax imposed by statute, payable by the Contractor in his capacity as employer.

Note:

Any payments normally made by the Contractor which are of a similar character to those described in 3.2 (a)–(c) but which are not within the terms of the rules and decisions referred to above are applicable subject to the prior agreement of the parties, as an alternative to 3.2 (a)–(c).

3.3.   The wages or salaries of supervisory staff, timekeepers, storekeepers, and the like, employed on or regularly visiting site, where the standard wage rates, etc., are not applicable, are those normally paid by the Contractor together with any incidental payments of a similar character to 3.2 (c)–(k).
3.4.   Where principals are working manually their time is chargeable, in respect of the trades practised, in accordance with 3.2.

SECTION 4 – MATERIALS AND GOODS

4.1.   The prime cost of materials and goods obtained by the Contractor from stockists or manufacturers is the invoice cost after deduction of all trade discounts but including cash discounts not exceeding 5%, and includes the cost of delivery to site.
4.2.   The prime cost of materials and goods supplied from the Contractor's stock is based upon the current market prices plus any appropriate handling charges.
4.3.   The prime cost under 4.1 and 4.2 also includes any costs of:
(a)   Non-returnable crates or other packaging.
(b)   Returning crates and other packaging less any credit obtainable.
4.4.   Any value added tax which is treated, or is capable of being treated, as input tax (as defined in the Finance Act, 1972 or any re-enactment thereof) by the Contractor is excluded.

SECTION 5 – PLANT, CONSUMABLE STORES AND SERVICES

5.1.   The prime cost of plant and consumable stores as listed below is the cost at hire rates agreed between the parties or in the absence of prior agreement at rates not exceeding those normally applied in the locality at the time when the works are carried out, or on a use and waste basis where applicable:
(a)   Machinery in workshops.
(b)   Mechanical plant and power-operated tools.
(c)   Scaffolding and scaffold boards.
(d)   Non-mechanical plant excluding hand tools.
(e)   Transport including collection and disposal of rubbish.
(f)   Tarpaulins and dust sheets.
(g)   Temporary roadways, shoring, planking and strutting, hoarding, centring, formwork, temporary fans, partitions or the like.
(h)   Fuel and consumable stores for plant and power-operated tools unless included in 5.1 (a), (b), (d) or (e) above.
(i)   Fuel and equipment for drying out the works and fuel for testing mechanical services.
5.2.   The prime cost also includes the net cost incurred by the Contractor of the following services, excluding any such cost included under Sections 3, 4 or 7:
(a)   Charges for temporary water supply including the use of temporary plumbing and storage.
(b)   Charges for temporary electricity or other power and lighting including the use of temporary installations.
(c)   Charges arising from work carried out by local authorities or public undertakings.
(d)   Fees, royalties and similar charges.
(e)   Testing of materials.
(f)   The use of temporary buildings including rates and telephone and including heating and lighting not charged under (b) above.

BUILDING INDUSTRY

(g)   The use of canteens, sanitary accommodation, protective clothing and other provision for the welfare of persons engaged in the work in accordance with the current Working Rule Agreement and any Act of Parliament, statutory instrument, rule, order, regulation or bye-law.

(h)   The provision of safety measures necessary to comply with any Act of Parliament.

(i)   Premiums or charges for any performance bonds or insurances which are required by the Building Owner and which are not referred to elsewhere in this Definition.

SECTION 6 – SUBCONTRACTS

6.1.   The prime cost of work executed by subontractors, whether nominated by the Building Owner or appointed by the Contractor, is the amount which is due from the Contractor to the subontractors in accordance with the terms of the subontracts after deduction of all discounts except any cash discount offered by any subontractor to the Contractor not exceeding 2.5%.

SECTION 7 – INCIDENTAL COSTS, OVERHEADS AND PROFIT

7.1.   The percentage adjustments provided in the building contract, which are applicable to each of the totals of Sections 3-6, provide for the following:

(a)   Head Office charges.

(b)   Off-site staff including supervisory and other administrative staff in the Contractor's workshops and yard.

(c)   Payments in respect of public holidays.

(d)   Payments in respect of apprentices' study time.

(e)   Sick pay or insurance in respect thereof.

(f)   Third party employer's liability insurance.

(g)   Liability in respect of redundancy payments made to employees.

(h)   Use, repair and sharpening of non-mechanical hand tools.

(i)   Any variations to basic rates required by the Contractor in cases where the building contract provides for the use of a specified schedule of basic plant charges (to the extent that no other provision is made for such variation).

(j)   All other liabilities and obligations whatsoever not specifically referred to in this Section nor chargeable under any other section.

(k)   Profit.

BUILDING INDUSTRY

SPECIMEN ACCOUNT FORMAT

If this Definition of Prime Cost is followed the Contractor's account could be in the following format:

	£
Labour (as defined in Section 3)	
Add __ % (see Section 7)	
Materials and goods (as defined in Section 4)	
Add __ % (see Section 7)	£
Plant, consumable stores and services (as defined in Section 5)	
Add __ % (see Section 7)	£
Subcontracts (as defined in Section 6)	
Add __ % (see Section 7)	£
Total	£

VAT to be added if applicable.

Example Calculations of Prime Cost of Labour in Daywork

Example of calculation of typical standard hourly base rate (as defined in Section 3) for CIJC Building Craft operative and General Operative based upon rates applicable June 2016.

		Craft Operative		General Operative	
		Rate £	Annual Cost £	Rate £	Annual Cost £
Basic Wages	52 weeks	457.34	23,781.44	343.83	17,879.32
CITB levy (0.5% of payroll)	0.5%		118.91		89.40
Pension and welfare benefit	52 weeks	12.86	668.72	12.86	668.72
Employer's National Insurance contribution *(13.8% after the first £155 per week)*	46.2 weeks		<u>2,287.25</u>		<u>1,472.75</u>
Annual labour cost:			**26,856.32**		**20,110.19**
Hourly base rate:			**14.91**		**11.16**

This is the prime cost of employment per person, which the employer has to meet even if there is no work for the employee to do

Note:

(1)   Calculated following Definition of Prime Cost of Daywork carried out under a Building Contract, published by the Royal Institution of Chartered Surveyors and the Construction Confederation.
(2)   Standard basic rates effective from 27 June 2016.
(3)   Standard working hours per annum calculated as follows:

BUILDING INDUSTRY

52 weeks @ 39 hours =		2028
Less		
22 days holiday @ 39 hours =	163.8	
8 days public holidays @ 7.8 hours =	<u>62.4</u>	
		= −226.2
Standard working hours per year =		1801.8

(4)   All labour costs incurred by the contractor in his capacity as an employer, other than those contained in the hourly base rate, are to be taken into account under Section 6.

(5)   The above example is for guidance only and does not form part of the Definition; all the basic costs are subject to re-examination according to the time when and in the area where the daywork is executed.

(6)   NI payments are at not-contracted out rates applicable from April 2016.

(7)   Basic rate and GMP number of weeks =
         52 Weeks − 4.2 weeks annual holiday − 1.6 weeks public holiday = 46.2 weeks

BUILDING INDUSTRY

SCHEDULE OF BASIC PLANT CHARGES (1 JULY 2010 ISSUE)

This Schedule is published by the Royal Institution of Chartered Surveyors, Parliament Square, London, and is for use in connection with Dayworks under a Building Contract.

EXPLANATORY NOTES

1. The rates in the Schedule are intended to apply solely to daywork carried out under and incidental to a Building Contract. They are NOT intended to apply to:
    (i)   Jobbing or any other work carried out as a main or separate contract; or
    (ii)   Work carried out after the date of commencement of the Defects Liability Period.
2. The rates apply only to plant and machinery already on site, whether hired or owned by the Contractor.
3. The rates, unless otherwise stated, include the cost of fuel and power of every description, lubricating oils, grease, maintenance, sharpening of tools, replacement of spare parts, all consumable stores and for licences and insurances applicable to items of plant.
4. The rates, unless otherwise stated, do not include the costs of drivers and attendants.
5. The rates do not include for any possible discounts which may be given to the Contractor if the plant or machinery was hired.
6. The rates are base costs and may be subject to the overall adjustment for price movement, overheads and profit, quoted by the Contractor prior to the placing of the Contract.
7. The rates should be applied to the time during which the plant is actually engaged in daywork.
8. Whether or not plant is chargeable on daywork depends on the daywork agreement in use and the inclusion of an item of plant in this schedule does not necessarily indicate that the item is chargeable.
9. Rates for plant not included in the Schedule or which is not already on site and is specifically provided or hired for daywork shall be settled at prices which are reasonably related to the rates in the Schedule having regard to any overall adjustment quoted by the Contractor in the Conditions of Contract.

Item of plant	Size/Rating	Unit	Rate per hour (£)
PUMPS			
Mobile Pumps			
Including pump hoses, values and strainers, etc.			
Diaphragm	50 mm diameter	Each	1.17
Diaphragm	76 mm diameter	Each	1.89
Diaphragm	102 mm diameter	Each	3.54
Submersible	50 mm diameter	Each	0.76
Submersible	76 mm diameter	Each	0.86
Submersible	102 mm diameter	Each	1.03
Induced Flow	50 mm diameter	Each	0.77
Induced Flow	76 mm diameter	Each	1.67
Centrifugal, self-priming	25 mm diameter	Each	1.30
Centrifugal, self-priming	50 mm diameter	Each	1.92
Centrifugal, self-priming	75 mm diameter	Each	2.74
Centrifugal, self-priming	102 mm diameter	Each	3.35
Centrifugal, self-priming	152 mm diameter	Each	4.27

BUILDING INDUSTRY

Item of plant	Size/Rating	Unit	Rate per hour (£)
SCAFFOLDING, SHORING, FENCING			
Complete Scaffolding			
Mobile working towers, single width	2.0 m × 0.72 m base × 7.45 m high	Each	3.36
Mobile working towers, single width	2.0 m × 0.72 m base × 8.84 m high	Each	3.79
Mobile working towers, double width	2.0 m × 1.35 m × 7.45 m high	Each	3.79
Mobile working towers, double width	2.0 m × 1.35 m × 15.8 m high	Each	7.13
Chimney scaffold, single unit		Each	1.92
Chimney scaffold, twin unit		Each	3.59
Push along access platform	1.63–3.1 m	Each	5.00
Push along access platform	1.80 m × 0.70 m	Each	1.79
Trestles			
Trestle, adjustable	Any height	Pair	0.41
Trestle, painters	1.8 m high	Pair	0.31
Trestle, painters	2.4 m high	Pair	0.36
Shoring, Planking and Strutting			
Acrow adjustable prop	Sizes up to 4.9 m (open)	Each	0.06
Strong Boy support attachment		Each	0.22
Adjustable trench strut	Sizes up to 1.67m (open)	Each	0.16
Trench sheet		Metre	0.03
Backhoe trench box	Base unit	Each	1.23
Backhoe trench box	Top unit	Each	0.87
Temporary Fencing			
Including block and coupler			
Site fencing steel grid panel	3.5 m × 2.0 m	Each	0.05
Anti-climb site steel grid fence panel	3.5 m × 2.0 m	Each	0.08
Solid panel Heras	2.0 m × 2.0 m	Each	0.09
Pedestrian gate		Each	0.36
Roadway gate		Each	0.60
LIFTING APPLIANCES AND CONVEYORS			
Cranes			
<u>Mobile Cranes</u>			
Rates are inclusive of drivers			
Lorry mounted, telescopic jib			
Two wheel drive	5 tonnes	Each	19.00
Two wheel drive	8 tonnes	Each	42.00
Two wheel drive	10 tonnes	Each	50.00

BUILDING INDUSTRY

Item of plant	Size/Rating		Unit	Rate per hour (£)
LIFTING APPLIANCES AND CONVEYORS – cont				
Two wheel drive	12 tonnes		Each	77.00
Two wheel drive	20 tonnes		Each	89.69
Four wheel drive	18 tonnes		Each	46.51
Four wheel drive	25 tonnes		Each	35.90
Four wheel drive	30 tonnes		Each	38.46
Four wheel drive	45 tonnes		Each	46.15
Four wheel drive	50 tonnes		Each	53.85
Four wheel drive	60 tonnes		Each	61.54
Four wheel drive	70 tonnes		Each	71.79
<u>Static tower crane</u>				
Rates inclusive of driver				

Note: Capacity equals maximum lift in tonnes times maximum radius at which it can be lifted

	Capacity (Metre/tonnes Up to	Height under hook above ground (m) Up to		
Tower crane	30	22	Each	22.23
Tower crane	40	22	Each	26.62
Tower crane	40	30	Each	33.33
Tower crane	50	22	Each	29.16
Tower crane	60	22	Each	35.90
Tower crane	60	36	Each	35.90
Tower crane	70	22	Each	41.03
Tower crane	80	22	Each	39.12
Tower crane	90	42	Each	37.18
Tower crane	110	36	Each	47.62
Tower crane	140	36	Each	55.77
Tower crane	170	36	Each	64.11
Tower crane	200	36	Each	71.95
Tower crane	250	36	Each	84.77
Tower crane with luffing jig	30	25	Each	22.23
Tower crane with luffing jig	40	30	Each	26.62
Tower crane with luffing jig	50	30	Each	29.16
Tower crane with luffing jig	60	36	Each	41.03
Tower crane with luffing jig	65	30	Each	33.13
Tower crane with luffing jig	80	22	Each	48.72
Tower crane with luffing jig	100	45	Each	48.72
Tower crane with luffing jig	125	30	Each	53.85

BUILDING INDUSTRY

Item of plant	Size/Rating		Unit	Rate per hour (£)
LIFTING APPLIANCES AND CONVEYORS				
Tower crane with luffing jig	160	50	Each	53.85
Tower crane with luffing jig	200	50	Each	74.36
Tower crane with luffing jig	300	60	Each	100.00
Crane Equipment				
Muck tipping skip	Up to 200 litres		Each	0.67
Muck tipping skip	500 litres		Each	0.82
Muck tipping skip	750 litres		Each	1.08
Muck tipping skip	1000 litres		Each	1.28
Muck tipping skip	1500 litres		Each	1.41
Muck tipping skip	2000 litres		Each	1.67
Mortar skip	250 litres, plastic		Each	0.41
Mortar skip	350 litres steel		Each	0.77
Boat skip	250 litres		Each	0.92
Boat skip	500 litres		Each	1.08
Boat skip	750 litres		Each	1.23
Boat skip	1000 litres		Each	1.38
Boat skip	1500 litres		Each	1.64
Boat skip	2000 litres		Each	1.90
Boat skip	3000 litres		Each	2.82
Boat skip	4000 litres		Each	3.23
Master flow skip	250 litres		Each	0.77
Master flow skip	500 litres		Each	1.03
Master flow skip	750 litres		Each	1.28
Master flow skip	1000 litres		Each	1.44
Master flow skip	1500 litres		Each	1.69
Master flow skip	2000 litres		Each	1.85
Grand master flow skip	500 litres		Each	1.28
Grand master flow skip	750 litres		Each	1.64
Grand master flow skip	1000 litres		Each	1.69
Grand master flow skip	1500 litres		Each	1.95
Grand master flow skip	2000 litres		Each	2.21
Cone flow skip	500 litres		Each	1.33
Cone flow skip	1000 litres		Each	1.69
Geared rollover skip	500 litres		Each	1.28
Geared rollover skip	750 litres		Each	1.64
Geared rollover skip	1000 litres		Each	1.69

Daywork and Prime Cost

BUILDING INDUSTRY

Item of plant	Size/Rating	Unit	Rate per hour (£)
LIFTING APPLIANCES AND CONVEYORS – cont			
Geared rollover skip	1500 litres	Each	1.95
Geared rollover skip	2000 litres	Each	2.21
Multi skip, rope operated	200 mm outlet size, 500 litres	Each	1.49
Multi skip, rope operated	200 mm outlet size, 750 litres	Each	1.64
Multi skip, rope operated	200 mm outlet size, 1000 litres	Each	1.74
Multi skip, rope operated	200 mm outlet size, 1500 litres	Each	2.00
Multi skip, rope operated	200 mm outlet size, 2000 litres	Each	2.26
Multi skip, man riding	200 mm outlet size, 1000 litres	Each	2.00
Multi skip	4 point lifting frame	Each	0.90
Multi skip	Chain brothers	Set	0.87
Crane Accessories			
Multi-purpose crane forks	1.5 and 2 tonnes S.W.L.	Each	1.13
Self-levelling crane forks		Each	1.28
Man cage	1 man, 230 kg S.W.L.	Each	1.90
Man cage	2 man, 500 kg S.W.L.	Each	1.95
Man cage	4 man, 750 kg S.W.L.	Each	2.15
Man cage	8 man, 1000 kg S.W.L.	Each	3.33
Stretcher cage	500 kg, S.W.L.	Each	2.69
Goods carrying cage	1500 kg, S.W.L.	Each	1.33
Goods carrying cage	3000 kg, S.W.L.	Each	1.85
Builders' skip lifting cradle	12 tonnes, S.W.L.	Each	2.31
Board/pallet fork	1600 kg, S.W.L.	Each	1.90
Gas bottle carrier	500 kg, S.W.L.	Each	0.92
Hoists			
Scaffold hoist	200 kg	Each	2.46
Rack and pinion (goods only)	500 kg	Each	4.56
Rack and pinion (goods only)	1100 kg	Each	5.90
Rack and pinion (goods and passenger)	8 person, 80 kg	Each	7.44
Rack and pinion (goods and passenger)	14 person, 1400 kg	Each	8.72
Wheelbarrow chain sling		Each	1.67

BUILDING INDUSTRY

Item of plant	Size/Rating		Unit	Rate per hour (£)
LIFTING APPLIANCES AND CONVEYORS				
Conveyors				
<u>Belt Conveyors</u>				
Conveyor	8 m long × 450 mm wide		Each	5.90
Miniveyor, control box and loading hopper	3 m unit		Each	4.49
<u>Other Conveying Equipment</u>				
Wheelbarrow			Each	0.62
Hydraulic superlift			Each	4.56
Pavac slab lifter (tile hoist)			Each	4.49
High lift pallet truck			Each	3.08
Lifting Trucks	Payload	Maximum Lift		
Fork lift, two wheel drive	1100 kg	up to 3.0 m	Each	5.64
Fork lift, two wheel drive	2540 kg	up to 3.7 m	Each	5.64
Fork lift, four wheel drive	1524 kg	up to 6.0 m	Each	5.64
Fork lift, four wheel drive	2600 kg	up to 5.4 m	Each	7.44
Fork life, four wheel drive	4000 kg	up to 17 m	Each	10.77
Lifting Platforms				
Hydraulic platform (Cherry picker)	9 m		Each	4.62
Hydraulic platform (Cherry picker)	12 m		Each	7.56
Hydraulic platform (Cherry picker)	15 m		Each	10.13
Hydraulic platform (Cherry picker)	17 m		Each	15.63
Hydraulic platform (Cherry picker)	20 m		Each	18.13
Hydraulic platform (Cherry picker)	25.6 m		Each	32.38
Scissor lift	7.6 m, electric		Each	3.85
Scissor lift	7.8 m, electric		Each	5.13
Scissor lift	9.7 m, electric		Each	4.23
Scissor lift	10 m, diesel		Each	6.41
Telescopic handler	7 m, 2 tonnes		Each	5.13
Telescopic handler	13 m, 3 tonnes		Each	7.18
Lifting and Jacking Gear				
Pipe winch including gantry	1 tonne		Set	1.92
Pipe winch including gantry	3 tonnes		Set	3.21
Chain block	1 tonne		Each	0.35
Chain block	2 tonnes		Each	0.58
Chain block	5 tonnes		Each	1.14
Pull lift (Tirfor winch)	1 tonne		Each	0.64
Pull lift (Tirfor winch)	1.6 tonnes		Each	0.90

BUILDING INDUSTRY

Item of plant	Size/Rating	Unit	Rate per hour (£)
LIFTING APPLIANCES AND CONVEYORS – cont			
Pull lift (Tirfor winch)	3.2 tonnes	Each	1.15
Brother or chain slings, two legs	not exceeding 3.1 tonnes	Set	0.21
Brother or chain slings, two legs	not exceeding 4.25 tonnes	Set	0.31
Brother or chain slings, four legs	not exceeding 11.2 tonnes	Set	1.09
CONSTRUCTION VEHICLES			
Lorries			
Plated lorries (Rates are inclusive of driver)			
Platform lorry	7.5 tonnes	Each	16.21
Platform lorry	17 tonnes	Each	22.90
Platform lorry	24 tonnes	Each	30.68
Extra for lorry with crane attachment	up to 2.5 tonnes	Each	3.25
Extra for lorry with crane attachment	up to 5 tonnes	Each	6.00
Extra for lorry with crane attachment	up to 7.5 tonnes	Each	9.10
Tipper Lorries			
(Rates are inclusive of driver)			
Tipper lorry	up to 11 tonnes	Each	15.78
Tipper lorry	up to 17 tonnes	Each	23.95
Tipper lorry	up to 25 tonnes	Each	31.35
Tipper lorry	up to 31 tonnes	Each	37.79
Dumpers			
Site use only (excl. tax, insurance			
and extra cost of DERV etc. when			
operating on highway)	Makers Capacity		
Two wheel drive	1 tonnes	Each	1.71
Four wheel drive	2 tonnes	Each	2.43
Four wheel drive	3 tonnes	Each	2.44
Four wheel drive	5 tonnes	Each	3.08
Four wheel drive	6 tonnes	Each	3.85
Four wheel drive	9 tonnes	Each	5.65
Tracked	0.5 tonnes	Each	3.33
Tracked	1.5 tonnes	Each	4.23
Tracked	3.0 tonnes	Each	8.33
Tracked	6.0 tonnes	Each	16.03

BUILDING INDUSTRY

Item of plant	Size/Rating	Unit	Rate per hour (£)
CONSTRUCTION VEHICLES			
Dumper Trucks (*Rates are inclusive of drivers*)			
Dumper truck	up to 15 tonnes	Each	28.56
Dumper truck	up to 17 tonnes	Each	32.82
Dumper truck	up to 23 tonnes	Each	54.64
Dumper truck	up to 30 tonnes	Each	63.50
Dumper truck	up to 35 tonnes	Each	73.02
Dumper truck	up to 40 tonnes	Each	87.84
Dumper truck	up to 50 tonnes	Each	133.44
Tractors			
Agricultural Type			
Wheeled, rubber-clad tyred	up to 40 kW	Each	8.63
Wheeled, rubber-clad tyred	up to 90 kW	Each	25.31
Wheeled, rubber-clad tyred	up to 140 kW	Each	36.49
Crawler Tractors			
With bull or angle dozer	up to 70 kW	Each	29.38
With bull or angle dozer	up to 85kW	Each	38.63
With bull or angle dozer	up to 100 kW	Each	52.59
With bull or angle dozer	up to 115kW	Each	55.85
With bull or angle dozer	up to 135kW	Each	60.43
With bull or angle dozer	up to 185kW	Each	76.44
With bull or angle dozer	up to 200 kW	Each	96.43
With bull or angle dozer	up to 250 kW	Each	117.68
With bull or angle dozer	up to 350 kW	Each	160.03
With bull or angle dozer	up to 450 kW	Each	219.86
With loading shovel	0.8 m³	Each	26.92
With loading shovel	1.0 m³	Each	32.59
With loading shovel	1.2 m³	Each	37.53
With loading shovel	1.4 m³	Each	42.89
With loading shovel	1.8 m³	Each	52.22
With loading shovel	2.0 m³	Each	57.22
With loading shovel	2.1 m³	Each	60.12
With loading shovel	3.5 m³	Each	87.26
Light Vans			
VW Caddivan or the like		Each	5.26
VW Transport transit or the like	1.0 tonnes	Each	6.03
Luton Box Van or the like	1.8 tonnes	Each	9.87

BUILDING INDUSTRY

Item of plant	Size/Rating	Unit	Rate per hour (£)
CONSTRUCTION VEHICLES – cont			
Water/Fuel Storage			
Mobile water container	110 litres	Each	0.62
Water bowser	1100 litres	Each	0.72
Water bowser	3000 litres	Each	0.87
Mobile fuel container	110 litres	Each	0.62
Fuel bowser	1100 litres	Each	1.23
Fuel bowser	3000 litres	Each	1.87
EXCAVATIONS AND LOADERS			
Excavators			
Wheeled, hydraulic	up to 11 tonnes	Each	25.86
Wheeled, hydraulic	up to 14 tonnes	Each	30.82
Wheeled, hydraulic	up to 16 tonnes	Each	34.50
Wheeled, hydraulic	up to 21 tonnes	Each	39.10
Wheeled, hydraulic	up to 25 tonnes	Each	43.81
Wheeled, hydraulic	up to 30 tonnes	Each	55.30
Crawler, hydraulic	up to 11 tonnes	Each	25.86
Crawler, hydraulic	up to 14 tonnes	Each	30.82
Crawler, hydraulic	up to 17 tonnes	Each	34.50
Crawler, hydraulic	up to 23 tonnes	Each	39.10
Crawler, hydraulic	up to 30 tonnes	Each	43.81
Crawler, hydraulic	up to 35 tonnes	Each	55.30
Crawler, hydraulic	up to 38 tonnes	Each	71.73
Crawler, hydraulic	up to 55 tonnes	Each	95.63
Mini excavator	1000/1500 kg	Each	4.87
Mini excavator	2150/2400 kg	Each	6.67
Mini excavator	2700/3500 kg	Each	7.31
Mini excavator	3500/4500 kg	Each	8.21
Mini excavator	4500/6000 kg	Each	9.23
Mini excavator	7000 kg	Each	14.10
Micro excavator	725 mm wide	Each	5.13
Loaders			
Shovel loader	0.4 m³	Each	7.69
Shovel loader	1.57 m³	Each	8.97
Shovel loader, four wheel drive	1.7 m³	Each	4.83
Shovel loader, four wheel drive	2.3 m³	Each	4.38

BUILDING INDUSTRY

Item of plant	Size/Rating	Unit	Rate per hour (£)
EXCAVATIONS AND LOADERS			
Shovel loader, four wheel drive	3.3 m³	Each	5.06
Skid steer loader wheeled	300/400 kg payload	Each	7.31
Skid steer loader wheeled	625 kg payload	Each	7.67
Tracked skip loader	650 kg	Each	4.42
Excavator Loaders			
Wheeled tractor type with black-hoe excavator			
Four wheel drive			
Four wheel drive, 2 wheel steer	6 tonnes	Each	6.41
Four wheel drive, 2 wheel steer	8 tonnes	Each	8.59
Attachments			
Breakers for excavator		Each	8.72
Breakers for mini excavator		Each	1.75
Breakers for back-hoe excavator/loader		Each	5.13
COMPACTION EQUIPMENT			
Rollers			
Vibrating roller	368–420 kg	Each	1.43
Single roller	533 kg	Each	1.94
Single roller	750 kg	Each	3.43
Twin roller	up to 650 kg	Each	6.03
Twin roller	up to 950 kg	Each	6.62
Twin roller with seat end steering wheel	up to 1400 kg	Each	7.68
Twin roller with seat end steering wheel	up to 2500 kg	Each	10.61
Pavement roller	3–4 tonnes dead weight	Each	6.00
Pavement roller	4–6 tonnes	Each	6.86
Pavement roller	6–10 tonnes	Each	7.17
Pavement roller	10–13 tonnes	Each	19.86
Rammers			
Tamper rammer 2 stroke-petrol	225 mm–275 mm	Each	1.52
Soil Compactors			
Plate compactor	75 mm–400 mm	Each	1.53
Plate compactor rubber pad	375 mm–1400 mm	Each	1.53
Plate compactor reversible plate-petrol	400 mm	Each	2.44

BUILDING INDUSTRY

Item of plant	Size/Rating	Unit	Rate per hour (£)
CONCRETE EQUIPMENT			
Concrete/Mortar Mixers			
Open drum without hopper	0.09/0.06 m³	Each	0.61
Open drum without hopper	0.12/0.09 m³	Each	1.22
Open drum without hopper	0.15/0.10 m³	Each	0.72
Concrete/Mortar Transport Equipment			
Concrete pump incl. hose, valve and couplers			
Lorry mounted concrete pump	24 m max. distance	Each	50.00
Lorry mounted concrete pump	34 m max. distance	Each	66.00
Lorry mounted concrete pump	42 m max. distance	Each	91.50
Concrete Equipment			
Vibrator, poker, petrol type	up to 75 mm dia.	Each	0.69
Air vibrator (*excluding compressor and hose*)	up to 75 mm dia.	Each	0.64
Extra poker heads	25/36/60 mm diameter	Each	0.76
Vibrating screed unit with beam	5.00 m	Each	2.48
Vibrating screed unit with adjustable beam	3.00–5.00 m	Each	3.54
Power float	725 mm–900 mm	Each	2.56
Power float finishing pan		Each	0.62
Floor grinder	660 × 1016 mm, 110 V electric	Each	4.31
Floor plane	450 × 1100 mm	Each	4.31
TESTING EQUIPMENT			
Pipe Testing Equipment			
Pressure testing pump, electric		Set	2.19
Pressure test pump		Set	0.80
SITE ACCOMODATION AND TEMPORARY SERVICES			
Heating Equipment			
Space heater – propane	80,000 Btu/hr	Each	1.03
Space heater – propane/electric	125,000 Btu/hr	Each	2.09
Space heater – propane/electric	250,000 Btu/hr	Each	2.33
Space heater – propane	125,000 Btu/hr	Each	1.54
Space heater – propane	260,000 Btu/hr	Each	1.88
Cabinet heater		Each	0.82
Cabinet heater, catalytic		Each	0.57

BUILDING INDUSTRY

Item of plant	Size/Rating	Unit	Rate per hour (£)
SITE ACCOMODATION AND TEMPORARY SERVICES			
Electric halogen heater		Each	1.27
Ceramic heater	3 kW	Each	0.99
Fan heater	3 kW	Each	0.66
Cooling fan		Each	1.92
Mobile cooling unit, small		Each	3.60
Mobile cooling unit, large		Each	4.98
Air conditioning unit		Each	2.81
Site Lighting and Equipment			
Tripod floodlight	500 W	Each	0.48
Tripod floodlight	1000 W	Each	0.62
Towable floodlight	4 × 100 W	Each	3.85
Hand held floodlight	500 W	Each	0.51
Rechargeable light		Each	0.41
Inspection light		Each	0.37
Plasterer's light		Each	0.65
Lighting mast		Each	2.87
Festoon light string	25 m	Each	0.55
Site Electrical Equipment			
Extension leads	240 V/14 m	Each	0.26
Extension leads	110 V/14 m	Each	0.36
Cable reel	25 m 110 V/240 V	Each	0.46
Cable reel	50 m 110 V/240 V	Each	0.88
4 way junction box	110 V	Each	0.56
Power Generating Units			
Generator – petrol	2 kVA	Each	1.23
Generator – silenced petrol	2 kVA	Each	2.87
Generator – petrol	3 kVA	Each	1.47
Generator – diesel	5 kVA	Each	2.44
Generator – silenced diesel	10 kVA	Each	1.90
Generator – silenced diesel	15 kVA	Each	2.26
Generator – silenced diesel	30 kVA	Each	3.33
Generator – silenced diesel	50 kVA	Each	4.10
Generator – silenced diesel	75 kVA	Each	4.62
Generator – silenced diesel	100 kVA	Each	5.64
Generator – silenced diesel	150 kVA	Each	7.18
Generator – silenced diesel	200 kVA	Each	9.74

BUILDING INDUSTRY

Item of plant	Size/Rating	Unit	Rate per hour (£)
SITE ACCOMODATION AND TEMPORARY SERVICES – cont			
Generator – silenced diesel	250 kVA	Each	11.28
Generator – silenced diesel	350 kVA	Each	14.36
Generator – silenced diesel	500 kVA	Each	15.38
Tail adaptor	240 V	Each	0.10
Transformers			
Transformer	3 kVA	Each	0.32
Transformer	5 kVA	Each	1.23
Transformer	7.5 kVA	Each	0.59
Transformer	10 kVA	Each	2.00
Rubbish Collection and Disposal			
Equipment			
Rubbish Chutes			
Standard plastic module	1 m section	Each	0.15
Steel liner insert		Each	0.30
Steel top hopper		Each	0.22
Plastic side entry hopper		Each	0.22
Plastic side entry hopper liner		Each	0.22
Dust Extraction Plant			
Dust extraction unit, light duty		Each	2.97
Dust extraction unit, heavy duty		Each	2.97
SITE EQUIPMENT – Welding Equipment			
Arc-(Electric) Complete with Leads			
Welder generator – petrol	200 amp	Each	3.53
Welder generator – diesel	300/350 amp	Each	3.78
Welder generator – diesel	4000 amp	Each	7.92
Extra welding lead sets		Each	0.69
Gas-Oxy Welder			
Welding and cutting set (including oxygen and acetylene, excluding underwater equipment and thermic boring)			
Small		Each	2.24
Large		Each	3.75
Lead burning gun		Each	0.50
Mig welder		Each	1.38
Fume extractor		Each	2.46

BUILDING INDUSTRY

Item of plant	Size/Rating	Unit	Rate per hour (£)
SITE EQUIPMENT – Welding Equipment			
Road Works Equipment			
Traffic lights, mains/generator	2-way	Set	10.94
Traffic lights, mains/generator	3-way	Set	11.56
Traffic lights, mains/generator	4-way	Set	12.19
Flashing light		Each	0.10
Road safety cone	450 mm	Each	0.08
Safety cone	750 mm	Each	0.10
Safety barrier plank	1.25 m	Each	0.13
Safety barrier plank	2 m	Each	0.15
Safety barrier plank post		Each	0.13
Safety barrier plank post base		Each	0.10
Safety four gate barrier	1 m each gate	Set	0.77
Guard barrier	2 m	Each	0.19
Road sign	750 mm	Each	0.23
Road sign	900 mm	Each	0.31
Road sign	1200 mm	Each	0.42
Speed ramp/cable protection	500 mm section	Each	0.14
Hose ramp open top	3 m section	Each	0.07
DPC Equipment			
Damp-proofing injection machine		Each	2.56
Cleaning Equipment			
Vacuum cleaner (industrial wet) single motor		Each	1.08
Vacuum cleaner (industrial wet) twin motor	30 litre capacity	Each	1.79
Vacuum cleaner (industrial wet) twin motor	70 litre capacity	Each	2.21
Steam cleaner	Diesel/electric 1 phase	Each	3.33
Steam cleaner	Diesel/electric 3 phase	Each	3.85
Pressure washer, light duty electric	1450 PSI	Each	0.72
Pressure washer, heavy duty, diesel	2500 PSI	Each	1.33
Pressure washer, heavy duty, diesel	4000 PSI	Each	2.18
Cold pressure washer, electric		Each	2.39
Hot pressure washer, petrol		Each	4.19
Hot pressure washer, electric		Each	5.13
Cold pressure washer, petrol		Each	2.92

BUILDING INDUSTRY

Item of plant	Size/Rating	Unit	Rate per hour (£)
SITE EQUIPMENT – Welding Equipment – cont			
Sandblast attachment to last washer		Each	1.23
Drain cleaning attachment to last washer		Each	1.03
Surface Preparation Equipment			
Rotavator	5 h.p.	Each	2.46
Rotavator	9 h.p.	Each	5.00
Scabbler, up to three heads		Each	1.53
Scabbler, pole		Each	2.68
Scrabbler, multi-headed floor		Each	3.89
Floor preparation machine		Each	1.05
Compressors and Equipment			
Portable Compressors			
Compressor – electric	4 cfm	Each	1.36
Compressor – electric	8 cfm lightweight	Each	1.31
Compressor – electric	8 cfm	Each	1.36
Compressor – electric	14 cfm	Each	1.56
Compressor – petrol	24 cfm	Each	2.15
Compressor – electric	25 cfm	Each	2.10
Compressor – electric	30 cfm	Each	2.36
Compressor – diesel	100 cfm	Each	2.56
Compressor – diesel	250 cfm	Each	5.54
Compressor – diesel	400 cfm	Each	8.72
Mobile Compressors			
Lorry mounted compressor			
(machine plus lorry only)	up to 3 m³	Each	41.47
(machine plus lorry only)	up to 5 m³	Each	48.94
Tractor mounted compressor			
(machine plus rubber tyred tractor)	Up to 4 m³	Each	21.03
Accessories (Pneumatic Tools) *(with and including up to 15 m of air hose)*			
Demolition pick, medium duty		Each	0.90
Demolition pick, heavy duty		Each	1.03
Breakers (with six steels) light	up to 150 kg	Each	1.19
Breakers (with six steels) medium	295 kg	Each	1.24
Breakers (with six steels) heavy	386 kg	Each	1.44
Rock drill (for use with compressor) hand held		Each	1.18

BUILDING INDUSTRY

Item of plant	Size/Rating	Unit	Rate per hour (£)
SITE EQUIPMENT – Welding Equipment			
Additional hoses	15 m	Each	0.09
Breakers			
Demolition hammer drill, heavy duty, electric		Each	1.54
Road breaker, electric		Each	2.41
Road breaker, 2 stroke, petrol		Each	4.06
Hydraulic breaker unit, light duty, petrol		Each	3.06
Hydraulic breaker unit, heavy duty, petrol		Each	3.46
Hydraulic breaker unit, heavy duty, diesel		Each	4.62
Quarrying and Tooling Equipment			
Block and stone splitter, hydraulic	600 mm × 600 mm	Each	1.90
Block and stone splitter, manual		Each	1.64
Steel Reinforcement Equipment			
Bar bending machine – manual	up to 13 mm dia. rods	Each	1.03
Bar bending machine – manual	up to 20 mm dia. rods	Each	1.41
Bar shearing machine – electric	up to 38 mm dia. rods	Each	3.08
Bar shearing machine – electric	up to 40 mm dia. rods	Each	4.62
Bar cropper machine – electric	up to 13 mm dia. rods	Each	2.05
Bar cropper machine – electric	up to 20 mm dia. rods	Each	2.56
Bar cropper machine – electric	up to 40 mm dia. rods	Each	4.62
Bar cropper machine – 3 phase	up to 40 mm dia. rods	Each	4.62
Dehumidifiers			
110/240 V Water	68 litres extraction per 24 hours	Each	2.46
110/240 V Water	90 litres extraction per 24 hours	Each	3.38
SMALL TOOLS			
Saws			
Masonry bench saw	350 mm–500 mm dia.	Each	1.13
Floor saw	125 mm max. cut	Each	1.15
Floor saw	150 mm max. cut	Each	3.83
Floor saw, reversible	350 mm max. cut	Each	3.32
Wall saw, electric		Each	2.05
Chop/cut off saw, electric	350 mm dia.	Each	1.79
Circular saw, electric	230 mm dia.	Each	0.72
Tyrannosaw		Each	1.74
Reciprocating saw		Each	0.79

Daywork and Prime Cost

BUILDING INDUSTRY

Item of plant	Size/Rating	Unit	Rate per hour (£)
SMALL TOOLS – cont			
Door trimmer		Each	1.17
Stone saw	300 mm	Each	1.44
Chainsaw, petrol	500 mm	Each	3.92
Full chainsaw safety kit		Each	0.41
Worktop jig		Each	1.08
Pipe Work Equipment			
Pipe bender	15 mm–22 mm	Each	0.92
Pipe bender, hydraulic	50 mm	Each	1.76
Pipe bender, electric	50 mm–150 mm dia.	Each	2.19
Pipe cutter, hydraulic		Each	0.46
Tripod pipe vice		Set	0.75
Ratchet threader	12 mm–32 mm	Each	0.93
Pipe threading machine, electric	12 mm–75 mm	Each	3.07
Pipe threading machine, electric	12 mm–100 mm	Each	4.93
Impact wrench, electric		Each	1.33
Hand-Held Drills and equipment			
Impact or hammer drill	up to 25 mm dia.	Each	1.03
Impact or hammer drill	35 mm diameter	Each	1.29
Dry diamond core cutter		Each	0.99
Angle head drill		Each	0.90
Stirrer, mixer drill		Each	1.13
Paint, Insulation Application Equipment			
Airless spray unit		Each	4.13
Portaspray unit		Each	1.16
HPVL turbine spray unit		Each	2.23
Compressor and spray gun		Each	1.91
Other Handtools			
Staple gun		Each	0.96
Air nail gun	110 V	Each	1.01
Cartridge hammer		Each	1.08
Tongue and groove nailer complete with mallet		Each	1.59
Diamond wall chasing machine		Each	2.63
Masonry chain saw	300 mm	Each	5.49
Floor grinder		Each	3.99
Floor plane		Each	1.79

BUILDING INDUSTRY

Item of plant	Size/Rating	Unit	Rate per hour (£)
SMALL TOOLS			
Diamond concrete planer		Each	1.93
Autofeed screwdriver, electric		Each	1.38
Laminate trimmer		Each	0.91
Biscuit jointer		Each	1.49
Random orbital sander		Each	0.97
Floor sander		Each	1.54
Palm, delta, flap or belt sander		Each	0.75
Disc cutter, electric	300 mm	Each	1.49
Disc cutter, 2 stroke petrol	300 mm	Each	1.24
Dust suppressor for petrol disc cutter		Each	0.51
Cutter cart for petrol disc cutter		Each	1.21
Grinder, angle or cutter	up to 225 mm	Each	0.50
Grinder, angle or cutter	300 mm	Each	1.41
Mortar raking tool attachment		Each	0.19
Floor/polisher scrubber	325 mm	Each	1.76
Floor tile stripper		Each	2.44
Wallpaper stripper, electric		Each	0.81
Hot air paint stripper		Each	0.50
Electric diamond tile cutter	all sizes	Each	2.42
Hand tile cutter		Each	0.82
Electric needle gun		Each	1.29
Needle chipping gun		Each	1.85
Pedestrian floor sweeper	250 mm dia.	Each	0.82
Pedestrian floor sweeper	Petrol	Each	2.20
Diamond tile saw		Each	1.84
Blow lamp equipment and glass		Set	0.50

NRM1 Cost Management Handbook

David P Benge

The definitive guide to measurement and estimating using NRM1, written by the author of NRM1

The 'RICS New rules of measurement: Order of cost estimating and cost planning of capital building works' (referred to as NRM1) is the cornerstone of good cost management of capital building works projects - enabling more effective and accurate cost advice to be given to clients and other project team members, while facilitating better cost control.

The NRM1 Cost Management Handbook is the essential guide to how to successfully interpret and apply these rules, including explanations of how to:

- quantify building works and prepare order of cost estimates and cost plans
- use the rules as a toolkit for risk management and procurement
- analyse actual costs for the purpose of collecting benchmark data and preparing cost analyses
- capture historical cost data for future order of cost estimates and elemental cost plans
- employ the rules to aid communication
- manage the complete 'cost management cycle'
- use the elemental breakdown and cost structures, together with the coding system developed for NRM1, to effectively integrate cost management with Building Information Modelling (BIM).

March 2014: 246 x 174: 640pp
Pb: 978-0-415-72077-9: £41.99

To Order: Tel: +44 (0) 1235 400524 Fax: +44 (0) 1235 400525
or Post: Taylor and Francis Customer Services,
Bookpoint Ltd, Unit T1, 200 Milton Park, Abingdon, Oxon, OX14 4TA UK
Email: book.orders@tandf.co.uk

For a complete listing of all our titles visit:
www.tandf.co.uk

PART 7

Useful Addresses for Further Information

ACOUSTICAL INVESTIGATION & RESEARCH
ORGANISATION LTD (AIRO)
Duxon's Turn
Maylands Avenue
Hemel Hempstead
Hertfordshire
HP2 4SB
Tel: 01442 247 146
Fax: 01442 256 749
Website: www.airo.co.uk

ALUMINIUM FEDERATION LTD (ALFED)
National Metalforming Centre
47 Birmingham Road
West Bromwich
West Midlands
B70 6PY
Tel: 0121 601 6361
Fax: 0870 138 9714
Website: www.alfed.org.uk

AMERICAN HARDWOOD EXPORT COUNCIL (AHEC)
3 St Michaels Alley
London
EC3 V 9DS
Tel: 020 7626 4111
Fax: 020 7626 4222
Website: www.ahec-europe.org

ANCIENT MONUMENTS SOCIETY (AMS)
Saint Ann's Vestry Hall
2 Church Entry
London
EC4 V 5HB
Tel: 020 7236 3934
Fax: 020 7329 3677
Website: www.ancientmonumentssociety.org.uk

APA – THE ENGINEERED WOOD ASSOCIATION
Claridge House
29 Barnes, High Street
London
SW13 9LW
Tel: 0845 123 3721
Fax: 020 8282 1660
Website: http://www.apa-europe.org/

ARBORICULTURAL ASSOCIATION
The Moult House
Stround Green
Standish
Stonehouse
Gloucestershire
GL10 3DL
Tel: 01242 522152
Fax: 01242 577766
Website: www.trees.org.uk

ARCHITECTURAL ADVISORY SERVICE CENTRE
(POWDER/ANODIC METAL FINISHES)
Barn One
Barn Road
Longwick
Buckinghamshire
HP27 9RW
Tel: 01844 342 425
Fax: 01844 274 781
Website: http://www.aasc.org.uk/

ARCHITECTURAL ASSOCIATION (AA)
34–36 Bedford Square
London
WC1B 3ES
Tel: 020 7887 4000
Fax: 020 7414 0782
Website: http://www.aaschool.ac.uk/

ARCHITECTURAL CLADDING ASSOCIATION (ACA)
60 Charles Street
Leicester
Leicestershire
LE1 1FB
Tel: 0116 253 6161
Fax: 0116 251 4568
Website: http://www.architectural-cladding-
        association.org.uk/

ASBESTOS INFORMATION CENTRE (AIC)
ARCA House,
237 Branston Road,
Burton upon Trent,
Staffordshire
DE14 3BT
Tel: 01283 531 126
Fax: 01283 568 228
Website: http://www.aic.org.uk

ASSOCIATION FOR CONSULTANCY AND ENGINEERING
Alliance House
12 Caxton Street
London
SW1H 0QL
Tel: 020 7222 6557
Fax: 020 7990 9202
Website: www.acenet.co.uk

ASSOCIATION OF INTERIOR SPECIALISTS
Olton Bridge
245 Warwick Road
Solihull
West Midlands
B92 7AH
Tel: 0121 707 0077
Fax: 0121 706 1949
Website: www.ais-interiors.org.uk

ASSOCIATION OF LOADING AND ELEVATING
EQUIPMENT MANUFACTURERS
Airport House
Purley Way
Croydon
Surrey
CR0 0XY
Tel: 020 8253 4501
Fax: 020 8253 4510
Website: www.alem.org.uk

ASSOCIATION OF PROJECT MANAGEMENT
IBIS House
Summerleys Road
Princes Risborough
Bucks
HP27 9LE
Tel: 0845 458 1944
Fax: 01494 528 937
Website: www.apm.org.uk

BOX CULVERT ASSOCIATION (BCA)
60 Charles Street
Leicester
Leicestershire
LE1 1FB
Tel: 0116 253 6161
Fax: 0116 251 4568
Website: www.boxculvert.org.uk

BRITISH ADHESIVES AND SEALANTS ASSOCIATION
5 Alderson Road
Worksop
Notts
S80 1UZ
Tel: 01909 480 888
Fax: 01909 473 834

BRITISH AGGREGATE CONSTRUCTION MATERIALS
INDUSTRIES LTD (BACMI)
156 Buckingham Palace Road
London
SW1W 9TR
Tel: 020 7730 8194

BRITISH APPROVALS FOR FIRE EQUIPMENT (BAFE)
Bridges 2
The Fire College
London Road
Moreton in Marsh
Gloucestershire
GL56 0RH
Tel: 0844 335 0897
Fax: 01608 653359
Website: www.bafe.org.uk

BRITISH APPROVALS SERVICE FOR CABLES (BASEC)
23 Presley Way
Crownhill
Milton Keynes
Buckinghamshire
MK8 0ES
Tel: 01908 267 300
Fax: 01908 267 255
Website: www.basec.org.uk

BRITISH ARCHITECTURAL LIBRARY (BAL)
Royal Institute of British Architects
66 Portland Place
London
W1B 1 AD
Tel: 020 7580 5533
Fax: 020 7631 1802
Website: www.architecture.com

BRITISH ASSOCIATION OF LANDSCAPE INDUSTRIES
(BALI)
Landscape House
National Agricultural Centre
Stoneleigh Park
Warwickshire
CV8 2LG
Tel: 024 7669 0333
Fax: 024 7669 0077
Website: www.bali.co.uk

BRITISH ASSOCIATION OF REINFORCEMENT
Riverside House
4 Meadows Business Park
Station Approach
Camberley
Surrey
GU17 9 AB
Tel: 07802 747031
Website: www.uk-bar.org

BRITISH BATHROOM COUNCIL
(BATHROOM MANUFACTURERS ASSOCIATION)
Federation House
Station Road
Stoke-on-Trent
Staffordshire
ST4 2RT
Tel: 01782 747 123
Fax: 01782 747 161
Website: www.bathroom-assciation.org

BRITISH BOARD OF AGREMENT (BBA)
PO Box 195
Bucknalls Lane
Garston
Watford
Hertfordshire
WD25 9BA
Tel: 01923 665 300
Fax: 01923 665 301
Website: www.bbacerts.co.uk

BRITISH CABLES ASSOCIATION (BCA)
37a Walton Road
East Molesey
Surrey
KT8 0DH
Tel: 020 8941 4079
Fax: 020 8783 0104
Website: www.bcauk.org

BRITISH CARPET MANUFACTURERS ASSOCIATION
LTD (BCMA)
PO Box 1155
MCF Complex
60 New Road
Kidderminster
Worcestershire
DY10 1 AQ
Tel: 01562 755 568
Fax: 01562 865 4055
Website: www.carpetfoundation.com

BRITISH CEMENT ASSOCIATION (BCA)
CENTRE FOR CONCRETE INFORMATION
Century House
Telford Avenue
Crowthorne
Berkshire
RG45 6YS
Tel: 01344 466 007
Fax: 01344 466 008
Website: www.cementindustry.co.uk

BRITISH CERAMIC CONFEDERATION (BCC)
Federation House
Station Road
Stoke-on-Trent
Staffordshire
ST4 2SA
Tel: 01782 744 631
Fax: 01782 744 102
Website: www.ceramfed.co.uk

BRITISH CERAMIC RESEARCH LTD (BCR)
Queens Road
Penkhull
Stoke-on-Trent
Staffordshire
ST4 7LQ
Tel: 01782 764 444
Fax: 01782 412 331
Website: www.ceram.co.uk

BRITISH CERAMIC TILE COUNCIL (BCTC TILE
ASSOCIATION)
Federation house
Station Road
Stoke-on-Trent
ST4 2RT
Tel: 01782 747 147
Fax: 01782 747 161
Website: www.tpb.org.uk/

BRITISH COMBUSTION EQUIPMENT MANUFACTURERS
ASSOCIATION (BCEMA)
58 London Road
Leicester
LE2 0QD
Tel: 0116 275 7111
Fax: 0116 275 7222
Website: bcema.co.uk

BRITISH CONSTRUCTIONAL STEELWORK ASSOCIATION
LTD (BCSA)
4 Whitehall Court
Westminster
London
SW1A 2ES
Tel: 020 7839 8566
Fax: 020 7976 1634
Website: www.steelconstruction.org

BRITISH CONTRACT FURNISHING ASSOCIATION (BCFA)
Suite 2/4
The Business Design Centre
52 Upper Street
Islington Green
London
N1 0QH
Tel: 020 7226 6641
Fax: 020 7288 6190
Website: thebcfa.com

BRITISH ELECTROTECHNICAL APPROVALS BOARD
(BEAB)
1 Station View
Guildford
Surrey
GU1 4JY
Tel: 01483 455 466
Fax: 01483 455 477
Website: www.beab.co.uk

BRITISH FIRE PROTECTION SYSTEMS ASSOCIATION
LTD (BFPSA)
Neville House
55 Eden Street
KT1 1BW
Tel: 020 8549 5855
Fax: 020 8547 1564

BRITISH FURNITURE MANUFACTURERS FEDERATION
LTD (BFM Ltd)
30 Harcourt Street
London
W1H 2 AA
Tel: 020 7724 0851
Fax: 020 7723 0622
Website: www.bfm.org.uk

BRITISH GEOLOGICAL SURVEY (BGS)
Keyworth Headquarters
Kingsley Drive
Dunham Centre
Nottingham
Nottinghamshire
NG12 5GG
Tel: 0115 936 3100
Fax: 0115 936 3200
Website: www.thebgs.co.uk

BRITISH INSTITUTE OF ARCHITECTURAL
TECHNOLOGISTS (BIAT)
397 City Road
London
EC1V 1NH
Tel: 020 7278 2206
Fax: 020 7837 3194
Website: www.biat.org.uk

BRITISH LAMINATE FABRICATORS ASSOCIATION
PO Box 775
Broseley Wood
TF7 9FG
Tel: 0845 056 8496
Website: www.blfa.co.uk

BRITISH LIBRARY BIBLIOGRAPHIC SERVICE AND
DOCUMENT SUPPLY
Boston Spa
Wetherby
West Yorkshire
LS23 7BQ
Tel: 01937 546 548
Fax: 01937 546 586
Website: www.bl.uk

BRITISH LIBRARY ENVIRONMENTAL INFORMATION
SERVICE
96 Euston Road
London
NW1 2DB
Tel: 020 7412 7000
Website: www.bl.uk/environment

BRITISH NON-FERROUS METALS FEDERATION
Broadway House
60 Calthorpe Road
Edgbaston
Birmingham
West Midlands
B15 1TN
Tel: 0121 456 6110
Fax: 0121 456 2274

BRITISH PLASTICS FEDERATION (BPF)
Plastics & Rubber Advisory Service
6 Bath Place
Rivington Street
London
EC2A 3JE
Tel: 020 7457 5000
Fax: 020 7457 5020
Website: www.bpf.co.uk

BRITISH PRECAST CONCRETE FEDERATION LTD
The Old Rectory
Main Street
Glenfield
LE3 8DG
Tel: 0116 232 5170
Website: www.britishprecast.org

BRITISH PROPERTY FEDERATION (BPF)
1 Warwick Row
London
SW1E 5ER
Tel: 020 7828 0111
Fax: 020 7824 3442
Website: www.bpf.org.uk

BRITISH RUBBER MANUFACTURERS'
ASSOCIATION LTD (BRMA)
6 Bath Place
Rivington Street
London
EC2A 3JE
Tel: 020 7457 5040
Fax: 020 7972 9008
Website: www.brma.co.uk

BRITISH STAINLESS STEEL ASSOCIATION
Broomgrove
59 Clarkehouse Road
Sheffield
S10 2LE
Tel: 0114 267 1260
Fax: 0114 266 1252
Website: www.bssa.org.uk

BRITISH STANDARDS INSTITUTION (BSI)
389 Chiswick High Road
London
W4 4 AL
Tel: 020 8996 9001
Fax: 020 8996 7001
Website: www.bsigroup.com

BRITISH WATER
1 Queen Anne's Gate
London
SW1H 9BT
Tel: 020 7957 4554
Fax: 020 7957 4565
Website: www.britishwater.co.uk

BRITISH WOOD PRESERVING & DAMP-PROOFING
ASSOCIATION (BWPDA)
6 Office Village
Romford Road
London
E15 4ED
Tel: 020 8519 2588
Fax: 020 8519 3444
Website: www.bwpda.co.uk

BRITISH WOODWORKING FEDERATION
55 Tufton Street
London
SW1 3QL
Tel: 0870 458 6939
Fax: 0870 458 6949
Website: www.bwf.org.uk

BUILDERS MERCHANTS FEDERATION
Soho Square
London
W1D 3HL
Tel: 020 7439 1753
Fax: 020 7734 2766
Website: www.bmf.org.uk

BUILDING & ENGINEERING SERVICES ASSOCIATION
ESCA House
34 Palace Court
Bayswater
London
W2 4JG
Tel: 020 7313 4900
Fax: 020 7727 9268
Website: www.hvca.org.uk

BUILDING CENTRE
The Building Centre
26 Store Street
London
WC1E 7BT
Tel: 020 7692 4000
Fax: 020 7580 9641
Website: www.buildingcentre.co.uk

BUILDING COST INFORMATION SERVICE LTD (BCIS)
Royal Institution of Chartered Surveyors
12 Great George Street
London
SW1P 3 AD
Tel: 020 7695 1500
Fax: 020 7695 1501
Website: www.bcis.co.uk

BUILDING EMPLOYERS CONFEDERATION (BEC)
55 Tufton Street
Westminster
London
SW1P 3QL
Tel: 0870 898 9090
Fax: 0870 898 9095
Website:
www.thecc.org.uk

BUILDING MAINTENANCE INFORMATION (BMI)
Royal Institution of Chartered Surveyors
12 Great George Street
London
SW1P 3 AD
Tel: 020 7695 1500
Fax: 020 7695 1501
Website: www.bcis.co.uk

BUILDING RESEARCH ESTABLISHMENT (BRE)
BRE Garston
Watford
WD5 9XX
Tel: 01923 664 000
Website: www.bre.co.uk

BUILDING RESEARCH ESTABLISHMENT: SCOTLAND (BRE)
Kelvin Road
East Kilbride
Glasgow
G75 0RZ
Tel: 01355 576 200
Fax: 01355 241 895
Website: www.bre.co.uk

BUILDING SERVICES RESEARCH AND INFORMATION ASSOCIATION LTD
Old Bracknell Lane West
Bracknell
Berkshire
RG12 7 AH
Tel: 01344 465 600
Fax: 01344 465 626
Website: www.bsria.co.uk

CASTINGS TECHNOLOGY INTERNATIONAL
7 East Bank Road,
Sheffield
S2 3PT
Tel: 0114 272 8647
Fax: 0114 273 0854
Website: www.castingsdev.com

CATERING EQUIPMENT MANUFACTURERS ASSOCIATION (CEMA)
Carlyle House
235 Vauxhall Bridge Road
London
SW1 V 1EJ
Tel: 020 7233 7724
Fax: 020 7828 0667

CEMENT ADMIXTURES ASSOCIATION
38 Tilehouse
Green Lane
Knowle
West Midlands
B93 9EY
Tel: + Fax: 01564 776 362
Website: www.admixtures.org.uk

CHARTERED INSTITUTE OF ARBITRATORS
12 Bloomsbury Square
London
WC1 A 2LP
Tel: 020 7421 7444
Fax: 020 7404 4023
Website: www.ciarb.org

CHARTERED INSTITUTE OF ARCHITECTURAL TECHNOLOGISTS
397 City Road
London
EC1 V 1NH
Tel: 020 7278 2206
Fax: 020 7837 3194
Website: www.ciat.org.uk

CHARTERED INSTITUTE OF BUILDING (CIOB)
Englemere
Kings Ride
Ascot
Berkshire
SL5 8BJ
Tel: 01344 630 700
Fax: 01344 630 777
Website: www.ciob.org.uk

CHARTERED INSTITUTE OF WASTES MANAGEMENT
9 Saxon Court
St Peter's Gardens
Northampton
NN1 1SX
Tel: 01604 620 426
Fax: 01604 621 339
Website: www.iwm.co.uk

CHARTERED INSTITUTION OF BUILDING
SERVICES ENGINEERS (CIBSE)
Delta House
222 Balham High Road
London
SW12 9BS
Tel: 020 8675 5211
Fax: 020 8675 5449
Website: www.cibse.org

CIVIL ENGINEERING CONTRACTORS ASSOCIATION
1 Birdcage Walk
London
SW1H 9JJ
Tel: 020 7340 0450
Website: www.ceca.co.uk

CLAY PIPE DEVELOPMENT ASSOCIATION (CPDA)
Copsham House
53 Broad Street
Chesham
HP5 3EA
Tel: 01494 791 456
Fax: 01494 792 378
Website: www.cpda.co.uk

CLAY ROOF TILE COUNCIL
Federation House
Station Road
Stoke-on-Trent
Staffordshire
ST4 2SA
Tel: 01782 744 631
Fax: 01782 744 102
Website: www.clayroof.co.uk

COLD ROLLED SECTIONS ASSOCIATION (CRSA)
National Metal Forming Centre
47 Birmingham Road
West Bromwich
West Midlands
B70 6PY
Tel: 0121 601 6350
Fax: 0121 601 6373
Website: www.crsauk.com

COMMONWEALTH ASSOCIATION OF ARCHITECTS
(CAA)
PO BOX 508
Edgware
HA8 9XZ
Tel: 020 8951 0550
Website: www.comarchitect.org

CONCRETE BRIDGE DEVELOPMENT GROUP
Riverside House
4 Meadows Business Park
Station Approach
Blackwater
Camberley
Surrey
GU17 9 AB
Tel: 01276 33777
Fax: 01276 38899
Website: www.concrete.org.uk

CONCRETE PIPE ASSOCIATION (CPA)
Main Street
Glenfield
LE3 8DG
Tel: 0116 232 5170
Website: www.concretepipes.co.uk

CONCRETE REPAIR ASSOCIATION (CRA)
Association House
235 Ash Road
Aldershot
Hampshire
GU12 4DD
Tel: 01252 321 302
Fax: 01252 333 901
Website: www.concreterepair.org.uk

CONCRETE SOCIETY ADVISORY SERVICE
Riverside House
4 Meadows Business Park
Station Approach
Blackwater
Camberley
Surrey
GU17 9 AB
Tel: 01276 607 140
Fax: 01276 607 141
Website: www.concrete.org.uk

CONFEDERATION OF BRITISH INDUSTRY (CBI)
Centre Point
103 New Oxford Street
London
WC1A 1DU
Tel: 020 7379 7400
Fax: 020 7240 1578
Website: www.cbi.org.uk

CONSTRUCT – CONCRETE STRUCTURES GROUP LTD
Riverside House
4 Meadows Business Park
Station Approach
Blackwater
Camberley
Surrey
GU17 9AB
Tel: 01276 38444
Fax: 01276 38899
Website: www.construct.org.uk

CONSTRUCTION EMPLOYERS FEDERATION LTD (CEF)
143 Malone Road
Belfast
Northern Ireland
BT9 6SU
Tel: 028 9087 7143
Fax: 028 9087 7155
Website: www.cefni.co.uk

CONSTRUCTION INDUSTRY JOINT COUNCIL (CIJC)
55 Tufton Street
London
SW1P 3QL
Tel: 0870 898 9090
Fax: 0870 898 9095
Website: http: //www.nscc.org.uk/Website:

CONSTRUCTION INDUSTRY RESEARCH AND
INFORMATION ASSOCIATION
Classic House
174–180 Old Street
London
EC1V 9BP
Tel: 020 7549 3300
Fax: 020 7253 0523
Website: www.ciria.org.uk

CONSTRUCTION PLANT-HIRE ASSOCIATION (CPA)
27–28 Newbury Street
London
EC1A 7HU
Tel: 020 7796 3366
Website: www.cpa.uk.net

CONTRACT FLOORING ASSOCIATION (CFA)
4c Saint Mary's Place
The Lace Market
Nottingham
Nottinghamshire
NG1 1PH
Tel: 0115 941 1126
Fax: 0115 941 2238
Website: www.cfa.org.uk

CONTRACTORS MECHANICAL PLANT ENGINEERS
(CMPE)
43 Portsmouth Road
Horndeam
Waterlooville
Hampshire
PO8 9LN
Tel: 023 925 70011
Fax: 023 925 70022
Website: www.cmpe.co.uk/

COPPER DEVELOPMENT ASSOCIATION
Verulam Industrial Estate
224 London Road
Saint Albans
Hertfordshire
AL1 1AQ
Tel: 01727 731 200
Fax: 01727 731 216
Website: www.cda.org.uk

CORUS RESEARCH DEVELOPMENT AND TECHNOLOGY
Swinden Technology Centre
Moorgate
Rotherham
South Yorkshire
S60 3AR
Tel: 01709 825 335
Fax: 01709 825 464
Website: www.corusgroup.com/en/technology/
research_and_development/

COUNCIL FOR ALUMINIUM IN BUILDING (CAB)
Bank House
Bond's Mill
Stonehouse
Gloucestershire
GL10 3RF
Tel: 01453 828851
Fax: 01453 828861
Website: www.c-a-b.org.uk

DEPARTMENT FOR BUSINESS, INNOVATIN AND SKILLS
1 Victoria Street
London
SW1H 0ET
Tel: 020 7215 5000
Website: www.gov.uk/government/organizations/department-
            for-business-innovation-skills

DEPARTMENT FOR TRANSPORT
Great Minister House
33 Horseferry Road,
London
SW1P 4DR
Tel: 030 0330 3000
Website: www.dft.gov.uk

DOORS & HARDWARE FEDERATION
42 Heath Street
Tamworth
Staffordshire
B79 7JH
Tel: 01827 52337
Fax: 01827 310 827
Website: www.abhm.org.uk

DRY STONE WALLING ASSOCIATION OF GREAT BRITAIN
(DSWA)
Westmorland County Showground
Lane Fram
Crooklands
Milnthorpe
Cumbria
LA7 7NH
Tel: 01539 567 953
Website: www.dswa.org.uk

ELECTRICAL CONTRACTORS ASSOCIATION (ECA)
ESCA House
34 Palace Court
Bayswater
London
W2 4HY
Tel: 020 7313 4800
Fax: 020 7221 7344
Website: www.eca.co.uk

ELECTRICAL CONTRACTORS ASSOCIATION OF
SCOTLAND (SELECT)
The Walled Gardens
Bush Estate
Midlothian
Scotland
EH26 0SB
Tel: 0131 445 5577
Fax: 0131 445 5548
Website: www.select.org.uk

ELECTRICAL INSTALLATION EQUIPMENT
MANUFACTURERS ASSOCIATION LTD (EIEMA)
Beama Installation Ltd
Westminster Tower
3 Albert Embankment
London
SE1 7SL
Tel: 020 7793 3000
Fax: 020 7793 3003
Website: www.eiema.org.uk

EUROPEAN LIQUID ROOFING ASSOCIATION (ELRA)
Fields House
Gower Road
Haywards Heath
West Sussex
RH16 4PL
Tel: 01444 417 458
Fax: 01444 415 616
Website: www.elra.org.uk

FEDERATION OF ENVIRONMENTAL TRADE
ASSOCIATIONS
2 Waltham Court
Milley Lane
Hare Hatch
Reading
RG10 9TH
Tel: 0118 940 3416
Fax: 0118 940 6258
Website: www.feta.co.uk

FEDERATION OF MANUFACTURERS OF
CONSTRUCTION EQUIPMENT & CRANES
Ambassador House
Brigstock Road
Thornton Heath
Surrey
CR7 7JG
Tel: 020 8665 5727
Fax: 020 8665 6447
Website: www.coneq.org.uk

FEDERATION OF MASTER BUILDERS
Gordon Fisher House
14–15 Great James Street
London
WC1N 3DP
Tel: 020 7242 7583
Fax: 020 7404 0296
Website: www.fmb.org.uk/

FEDERATION OF PILING SPECIALISTS
Forum Court
83 Coppers Cope Road
Beckenham
Kent
BR3 1NR
Tel: 020 8663 0947
Fax: 020 8663 0949
Website: www.fps.org.uk

FEDERATION OF PLASTERING & DRYWALL
CONTRACTORS
Construction House
56–64 Leonard Street
London
EC2A 4JX
Tel: 020 7608 5092
Fax: 020 7608 5081
Website: www.fpdc.org

FENCING CONTRACTORS ASSOCIATION
Warren Road
Trellech
Monmouthshire
NP5 4PQ
Tel: 07000 560 722
Fax: 01600 860 614
Website: www.fencingcontractors.org

FINNISH PLYWOOD INTERNATIONAL
PO BOX 99
Welwyn Garden City
Herts
AL6 0HS
Tel: 01438 798 746
Fax: 01438 798 305

FLAT ROOFING ALLIANCE
Fields House
Gower Road
Haywards Heath
West Sussex
RH16 4PL
Tel: 01444 440 027
Fax: 01444 415 616
Website: www.fra.org.uk

FURNITURE INDUSTRY RESEARCH ASSOCIATION (FIRA
INTERNATIONAL LTD)
Maxwell Road
Stevenage
Hertfordshire
SG1 2EW
Tel: 01438 777 700
Fax: 01438 777 800
Website: www.fira.co.uk

GLASS & GLAZING FEDERATION (GGF)
44–48 Borough High Street
London
SE1 1XB
Tel: 020 7403 7177
Website: www.ggf.org.uk

HOUSING CORPORATION HEADQUARTERS
Maple House
149 Tottenham Court Road
London
W1N 7BN
Tel: 0845 230 7000
Fax: 020 7393 2111
Website: www.housingcorp.gov.uk

ICOM Energy Association
Camden House
Warwick Road
Kenilworth
Warwickshire
CV8 1TH
Tel: 01926 513748
Fax: 01926 21 855017
Website: www.icome.org.uk

INSTITUTE OF ACOUSTICS
77 A Saint Peter' Street
Saint Albans
Hertfordshire
AL1 3BN
Tel: 01727 848 195
Fax: 01727 850 553
Website: www.ioa.org.uk

INSTITUTE OF ASPHALT TECHNOLOGY
Paper Mews Place
290 High Street
Dorking
Surrey
RH4 1QT
Tel: 01306 742 792
Fax: 01306 888 902
Website: www.instofasphalt.org

INSTITUTE OF MAINTENANCE AND BUILDING
MANAGEMENT
Keets House
30 East Street
Farnham
Surrey
GU9 7SW
Tel: 01252 710 994
Fax: 01252 737 741
Website: www.imbm.org.uk

INSTITUTE OF MATERIALS
Headquarters
1 Carlton House Terrace
London
SW1Y 5 AF
Tel: 020 7451 7300
Fax: 020 7839 1702
Website: www.materials.org.uk

INSTITUTE OF PLUMBING
64 Station Lane
Hornchurch
Essex
RM12 6NB
Tel: 01708 472 791
Fax: 01708 448 987
Website: www.plumbers.org.uk

INSTITUTE OF WOOD SCIENCE
Stocking Lane
Hughenden Valley
High Wycombe
Buckinghamshire
HP14 4NU
Tel: 01494 565 374
Fax: 01494 565 395
Website: www.iwsc.org.uk

INSTITUTION OF CIVIL ENGINEERS (ICE)
1 Great George Street
London
SW1P 3 AA
Tel: 020 7222 7722
Fax: 020 7222 7500
Website: www.ice.org.uk

INSTITUTION OF ENGINEERING AND TECHNOLOGY
The Institution of Engineering and Technology
Michael Faraday House
Stevenage
Herts
SG1 2 AY
Tel: 01438 313 311
Fax: 01438 765 526
Website: www.theiet.org

INSTITUTION OF MECHANICAL ENGINEERS
1 Birdcage Walk
London
SW1H 9JJ
Tel: 020 7222 7899
Fax: 020 7222 4557
Website: www.imeche.org

INSTITUTION OF STRUCTURAL ENGINEERS (ISE)
11 Upper Belgrave Street
London
SW1X 8BH
Tel: 020 7235 4535
Fax: 020 7235 4294
Website: www.istructe.org

INTERNATIONAL LEAD ASSOCIATION
17 A Welbeck Way
London
W1G 9YJ
Tel: 020 7499 8422
Fax: 020 7493 1555
Website: www.ldaint.org

INTERPAVE (THE PRECAST CONCRETE PAVING
& KERB ASSOCIATION)
60 Charles Street
Leicester
Leicestershire
LE1 1FB
Tel: 0116 253 6161
Fax: 0116 251 4568
Website: www.paving.org.uk/

JOINT CONTRACTS TRIBUNAL LTD
9 Cavendish Place
London
W1G 0GD
Tel: 020 7630 8650
Fax: 020 7630 8670
Website: www.jctltd.co.uk

KITCHEN SPECIALISTS ASSOCIATION
12 TopBarn Business Centre
Holt Heath
Worcester
Worcestershire
WR6 6NH
Tel: 01905 621 787
Fax: 01905 621 887
Website: www.kbsa.co.uk

LIGHTING ASSOCIATION LTD
Stafford Park 7
Telford
Shropshire
TF3 3BQ
Tel: 01952 290 905
Fax: 01952 290 906
Website: www.lightingassociation.com/

MASTIC ASPHALT COUNCIL LTD
PO BOX 77
Hastings
Kent
TN35 4WL
Tel: 01424 814 400
Fax: 01424 814 446
Website: www.masticasphaltcouncil.co.uk

METAL CLADDING & ROOFING MANUFACTURERS
ASSOCIATION
18 Mere Farm Road
Prenton
Wirral
Cheshire
CH43 9TT
Tel: 0151 652 3846
Fax: 0151 653 4080
Website: www.mcrma.co.uk

METAL GUTTER MANUFACTURERS ASSOCIATION
106 Ruskin Avenue
Rogerstone
Newport
South Wales
NP10 0BD
Tel: 01633 891584
Website: www.mgma.co.uk

MET OFFICE
Fitzroy Road
Exeter
Devon
EX1 3PB
Tel: 0870 900 0100
Fax: 0870 900 5050
Website: www.metoffice.gov.uk

NATIONAL ASSOCIATION OF STEEL STOCKHOLDERS
The Citadel
190 Corporation Street
Birmingham
B4 6QD
Tel: 0121 200 2288
Fax: 0121 236 7444
Website: www.nass.org.uk

NATIONAL HOUSE-BUILDING COUNCIL (NHBC)
NHBC
NHBC House
Davy Avenue
Knowlhill
Milton Keynes
MK5 8FP
Tel: 0844 633 1000
Website: www.nhbc.co.uk

NATURAL SLATE QUARRIES ASSOCIATION
26 Store Street
London
WC1E 7BT
Tel: 020 7323 3770
Fax: 020 7323 0307
Website: http: //slateassociation.org

NHS ESTATES
Departments of Health
1 Trevelyan Square
Boar Lane
Leeds
West Yorkshire
LS1 6 AE
Tel: 0113 254 7000
Fax: 0113 254 7299
Website: http: //www.buyingsolutions.gov.uk/healthcms/

ORDNANCE SURVEY
Romsey Road
Southampton
SO16 4GU
Tel: 08456 050 504
Fax: 02380 792 615
Website: www.ordnancesurvey.co.uk

PAINTING AND DECORATING ASSOCIATION
32 Coton Road
Nuneaton
Warwickshire
CV11 5TW
Tel: 01203 353 776
Fax: 01203 354 4513
Website: www.paintingdecoratingassociation.co.uk

PIPELINE INDUSTRIES GUILD
14–15 Belgrave Square
London
SW1X 8PS
Tel: 020 7235 7938
Fax: 020 7235 0074
Website: www.pipeguild.co.uk

PLASTIC PIPES GROUP
c/o British Plastics Federation
6 Bath Place
Rivington Street
London
EC2 A 3JE
Tel: 020 7457 5024
Website: www.plasticpipesgroup.com

PRECAST FLOORING FEDERATION
60 Charles Street
Leicester
Leicestershire
LE1 1FB
Tel: 0116 253 6161
Fax: 0116 251 4568
Website: www.pff.org.uk

PRESTRESSED CONCRETE ASSOCIATION
60 Charles Street
Leicester
Leicestershire
LE1 1FB
Tel: 0116 253 6161
Fax: 0116 251 4568
Website: www.britishprecast.org

PROPERTY CONSULTANTS SOCIETY LTD
Basement Office
1 Surrey Street
Arundel
West Sussex
BN18 9DT
Tel: 01903 883 787
Fax: 01903 889 590
Website: www.p-c-s.org.uk

QUARRY PRODUCTS ASSOCIATION
Gillingham House
38–44 Gillingham Street
London
SW1 V1HU
Tel: 020 7963 8000
Fax: 020 7963 8001
Website: www.qpa.org

READY-MIXED CONCRETE BUREAU
Century House
Telford Avenue
Crowthorne
Berkshire
RG45 6YS
Tel: 01344 725 732
Fax: 01344 774 976
Website: www.rcb.org.uk

REINFORCED CONCRETE COUNCIL
Riverside House
4 Meadows Business Park
Station Approach
Camberley
Berkshire
GU17 9AB
Tel: 01276 607140
Fax: 01276 607141
Website: www.rcc-info.org.uk

ROYAL INCORPORATION OF ARCHITECTS IN SCOTLAND
(RIAS)
15 Rutland Square
Edinburgh
Scotland
EH1 2BE
Tel: 0131 229 7545
Fax: 0131 228 2188
Website: www.rias.org.uk

ROYAL INSTITUTE OF BRITISH ARCHITECTS (RIBA)
66 Portland Place
London
W1B 1AD
Tel: 020 7580 5533
Fax: 020 7255 1541
Website: www.architecture.com

ROYAL INSTITUTION OF CHARTERED SURVEYORS
(RICS)
12 Great George Street
Parliament Square
London
SW1P 3AD
Tel: 020 7222 7000
Fax: 020 7222 9430
Website: www.rics.org

ROYAL TOWN PLANNING INSTITUTE (RTPI)
41 Botolph Lane
London
EC3R 8DL
Tel: 020 7929 9494
Fax: 020 7323 1582
Website: www.rtpi.org.uk/

RURAL DESIGN AND BUILDING ASSOCIATION
ATSS House
Station Road East
Stowmarket
Suffolk
IP14 1RQ
Tel: 01449 676 049
Fax: 01449 770 028
Website: www.rdba.org.uk

SCOTTISH BUILDING EMPLOYERS FEDERATION
Carron Grange
Carron Grange Avenue
Stenhousemuir
Scotland
FK5 3BQ
Tel: 01324 555 550
Fax: 01324 555 551
Website: www.scottish-building.co.uk

SCOTTISH HOMES – Community Scotland
Thistle House
91 Haymarket Terrace
Edinburgh
Scotland
EH12 5HE
Tel: 0131 313 0044
Fax: 0131 313 2680
Website: www.scotland.gov.uk

SCOTTISH NATURAL HERITAGE
Communications Directorate
12 Hope Terrace
Edinburgh
EH9 2AS
Tel: 0131 447 4784
Fax: 0131 446 2277
Website: www.snh.org.uk

SINGLE PLY ROOFING ASSOCIATION
177 Bagnall Road
Basford
Nottinghamshire
NG6 8SJ
Tel: 0115 914 4445
Fax: 0115 974 9827
Website: www.spra.co.uk/

SMOKE CONTROL ASSOCIATION
2 Waltham Court
Milley Lane, Hare Hatch
Reading
Berkshire
RG10 9TH
Tel: 0118 940 3416
Fax: 0118 940 6258
Website: www.feta.co.uk

SOCIETY FOR THE PROTECTION OF ANCIENT
BUILDINGS (SPAB)
37 Spital Square
London
E1 6DY
Tel: 020 7377 1644
Fax: 020 7247 5296
Website: www.spab.org.uk

SOCIETY OF GLASS TECHNOLOGY
Don Valley House
Saville Street East
Sheffield
South Yorkshire
S4 7UQ
Tel: 0114 263 4455
Fax: 0114 263 4411
Website: www.sgt.org

SOIL SURVEY AND LAND RESEARCH INSTITUTE
Cranfield University
Silsoe Campus
Bedford
Bedfordshire
MK45 4DT
Tel: 01525 863 000
Fax: 01525 863 253
Website: www.cranfield.ac.uk/sslrc

SOLAR ENERGY SOCIETY
c/o School of Engineering
Oxford Brookes University
Gipsy Lane
Headington, Oxford
OX3 0BP
Tel: 01865 741 111
Fax: 01525 863 253
Website: www.uk-ises.org/

SPON'S PRICE BOOK EDITORS
AECOM
Aldgate Tower
2 Leman Street
London
E1 8FA
Tel: 020 7061 7000
Website: www.aecom.com

SPORT ENGLAND
3rd Floor, Victoria House
Bloomsbury Square
London
WC1B 4SE
Tel: 0845 850 8508
Fax: 020 7383 5740
Website: www.sportengland.org

SPORT SCOTLAND
Caledonia House
South Gyle
Edinburgh
Scotland
EH12 9DQ
Tel: 0131 317 7200
Fax: 0131 317 7202
Website: www.sportscotland.org.uk

SPORTS COUNCIL FOR WALES
Welsh Institute of Sport
Sophia Gardens
Cardiff
CF11 9SW
Tel: 0845 045 0904
Fax: 0845 846 0014
Website: www.sports-council-wales.co.uk

SPORTS TURF RESEARCH INSTITUTE (STRI)
Saint Ives Estate
Bingley
West Yorkshire
BD16 1 AU
Tel: 01274 565 131
Fax: 01274 561 891
Website: www.stri.co.uk

SPRAYED CONCRETE ASSOCIATION
Association House
235 Ash Road
Aldershot
Hampshire
GU12 4DD
Tel: 01252 321 302
Fax: 01252 333 901
Website: www.sca.org.uk

STEEL CONSTRUCTION INSTITUTE
Silwood Park
Ascot
Berkshire
SL5 7QN
Tel: 01344 636 505
Fax: 01344 636 570
Website: www.steel-sci.org

STEEL WINDOW ASSOCIATION
The Building Centre
26 Store Street
London
WC1E 7BT
Tel: 020 7637 3571
Fax: 020 7637 3572
Website: www.steel-window-association.co.uk

STONE FEDERATION GREAT BRITAIN
Channel Business Centre
Ingles Manor
Castle Hill Avenue
Folkestone
Kent
CT20 2RD
Tel: 01303 856123
Fax: 01303 856117
Website: www.stone-federationgb.org.uk

SUSPENDED ACCESS EQUIPMENT
MANUFACTURERS ASSOCIATION
56–54 Leonard Street
London
EC2A 4JX
Tel: 020 7608 5098
Fax: 020 7636 5984
Website: www.saema.org

SWIMMING POOL & ALLIED TRADES ASSOCIATION
(SPATA)
Spata House
1a Junction Road
Andover
Hampshire
SP10 3QT
Tel: 01264 356210
Fax: 01264 332628
Website: www.spata.co.uk

THE PIPELINE INDUSTRIES GUILD
F150 First Floor
Cherwell Business Village
Southam Road
Banbury
OX16 2SP
Tel: 020 7235 7938
Fax; 020 7235 0074
Website: www.pipeline.com

THERMAL INSULATION CONTRACTORS ASSOCIATION
Tica House
Allington Way
Yarm Road Business Park
Darlington
County Durham
DL1 4QB
Tel: 01325 466 704
Fax: 01325 487 691
Website: www.tica-acad.co.uk

TIMBER RESEARCH & DEVELOPMENT ASSOCIATION
(TRADA)
Stocking Lane
Hughenden Valley
High Wycombe
Buckinghamshire
HP14 4ND
Tel: 01494 569 600
Fax: 01494 565 487
Website www.trada.co.uk

TIMBER TRADE FEDERATION
4th Floor
Clareville House
26–27 Oxenden Street
London
SW1Y 4EL
Tel: 020 7839 1891
Fax: 020 7930 0094
Website: www.ttf.co.uk

TOWN & COUNTRY PLANNING ASSOCIATION (TCPA)
17 Carlton House Terrace
London
SW1Y 5 AS
Tel: 020 7930 8903
Fax: 020 7930 3280
Website: tcpa.org.uk

TREE COUNCIL
71 Newcomen Street
London
SE1 1WT
Tel: 020 7407 9992
Fax: 020 7407 9908
Website: www.treecouncil.org.uk

TRUSSED RAFTER ASSOCIATION
31 Station Road
Sutton
Retford
Nottinghamshire
DN22 8PZ
Tel: 01777 869 281
Fax: 01777 869 281
Website: www.tra.org.uk

TWI
Granta Park
Great Abington
Cambridge
Cambridgeshire
CB1 6 AL
Tel: 01223 899 000
Fax: 01223 892 588
Website: www.twi.co.uk

UNDERFLOOR HEATING MANUFACTURERS'
ASSOCIATION
Belhaven House
67 Walton Road
East Moseley
Surrey
KT8 0DB
Tel: 020 8941 7177
Fax: 020 8941 815
Website: www.uhma.org.uk

VERMICULITE INFORMATION SERVICE
1 A Guildford Business Park
Guildford
Surrey
GU2 8XG
Tel: 01483 242 100
Fax: 01483 242 101

WALLCOVERING MANUFACTURERS ASSOCIATION
James House
Bridge Street
Leatherhead
Surrey
KT22 7EP
Tel: 01372 360 660
Fax: 01372 376 069

WATERHEATER MANUFACTURERS ASSOCIATION
c/o Andrews Waterheaters
Wednesbury One
Black Country New Road
Wednesbury
WS10 7NZ
Tel: 07775 754456
Fax: 0161 456 7106
Website: www.waterheating.fsnet.co.uk/wma.htm

WATER RESEARCH CENTRE
Henley Road
Medmenham
Marlow
Buckinghamshire
SL7 2HD
Tel: 01491 636 500
Fax: 01491 636 501
Website: www.wrcplc.co.uk

WATER UK
1 Queen Anne's Gate
London
SW1H 9BT
Tel: 020 7344 1844
Fax: 020 7344 1866
Website: www.water.org.uk/

WELDING MANUFACTURERS' ASSOCIATION
Westminster Tower
3 Albert Embankment
London
SE1 7SL
Tel: 020 7793 3041
Fax: 020 7582 8020
Website: www.wma.uk.com

WOOD PANEL INDUSTRIES FEDERATION
Grantham
Lincolnshire
NG31 6LR
Tel: 01476 563 707
Fax: 01476 579 314
Website: www.wpif.org.uk

WRAP
The Old Academy
21 Horse Fair
Banbury
OX16 0 AH
Tel: 01295 819 900
Website: www.wrap.org.uk

ZINC DEVELOPMENT ASSOCIATION
42 Weymouth Street
London
W1N 3LQ
Tel: 020 7499 6636
Fax: 020 7493 135
Website: www.zincinfocentre.org

Cost Studies of Buildings

Allan Ashworth & Srinath Perera

This practical guide to cost studies of buildings has been updated and revised throughout for the 6th edition. New developments in RICS New Rules of Measurement (NRM) are incorporated throughout the book, in addition to new material on e-business, the internet, social media, building information modelling, sustainability, building resilience and carbon estimating.

This trusted and easy to use guide to the cost management role:

Focuses on the importance of costs of constructing projects during the different phases of the construction process

Features learning outcomes and self-assessment questions for each chapter

Addresses the requirements of international readers

From introductory data on the construction industry and the history of construction economics, to recommended methods for cost analysis and post-contract cost control, Cost Studies of Buildings is an ideal companion for anyone learning about cost management.

July 2015: 246x174: 544pp Pb:
978-1-138-01735-1: £42.99

PART 8

Tables and Memoranda

This part contains the following sections:

JCT Contract Administration Pocket Book

Andy Atkinson

Successfully managing your JCT contracts is a must, and this handy reference is the swiftest way to doing just that. Making reference to best practice throughout, the JCT Standard Building Contracts SBC/Q, DB and MW are used as examples to take you through all the essential contract administration tasks, including:

- Procurement
- Payment
- Final accounts
- Progress, completion and delay
- Subcontracting
- Defects and quality control

In addition to the day to day tasks, this also gives you an overview of what to expect from common sorts of dispute resolution under the JCT, as well as a look at how to administer contracts for BIM-compliant projects. This is an essential starting point for all students of construction contract administration, as well as practitioners needing a handy reference to working with JCT contracts.

June 2015 186x123: 144pp
Pb: 978-1-138-78192-4: £24.99

To Order: Tel: +44 (0) 1235 400524 Fax: +44 (0) 1235 400525
or Post: Taylor and Francis Customer Services,
Bookpoint Ltd, Unit T1, 200 Milton Park, Abingdon, Oxon, OX14 4TA UK
Email: book.orders@tandf.co.uk

For a complete listing of all our titles visit:
www.tandf.co.uk

CONVERSION TABLES

CONVERSION TABLES

Length	Unit	Conversion factors			
Millimetre	mm	1 in	= 25.4 mm	1 mm	= 0.0394 in
Centimetre	cm	1 in	= 2.54 cm	1 cm	= 0.3937 in
Metre	m	1 ft	= 0.3048 m	1 m	= 3.2808 ft
		1 yd	= 0.9144 m		= 1.0936 yd
Kilometre	km	1 mile	= 1.6093 km	1 km	= 0.6214 mile

Note:

1 cm	= 10 mm	1 ft	= 12 in
1 m	= 1 000 mm	1 yd	= 3 ft
1 km	= 1 000 m	1 mile	= 1 760 yd

Area	Unit	Conversion factors			
Square Millimetre	mm^2	$1\ in^2$	$= 645.2\ mm^2$	$1\ mm^2$	$= 0.0016\ in^2$
Square Centimetre	cm^2	$1\ in^2$	$= 6.4516\ cm^2$	$1\ cm^2$	$= 1.1550\ in^2$
Square Metre	m^2	$1\ ft^2$	$= 0.0929\ m^2$	$1\ m^2$	$= 10.764\ ft^2$
		$1\ yd^2$	$= 0.8361\ m^2$	$1\ m^2$	$= 1.1960\ yd^2$
Square Kilometre	km^2	$1\ mile^2$	$= 2.590\ km^2$	$1\ km^2$	$= 0.3861\ mile^2$

Note:

$1\ cm^2$	$= 100\ mm^2$	$1\ ft^2$	$= 144\ in^2$
$1\ m^2$	$= 10\ 000\ cm^2$	$1\ yd^2$	$= 9\ ft^2$
$1\ km^2$	$= 100\ hectares$	$1\ acre$	$= 4\ 840\ yd^2$
		$1\ mile^2$	$= 640\ acres$

Volume	Unit	Conversion factors			
Cubic Centimetre	cm^3	$1\ cm^3$	$= 0.0610\ in^3$	$1\ in^3$	$= 16.387\ cm^3$
Cubic Decimetre	dm^3	$1\ dm^3$	$= 0.0353\ ft^3$	$1\ ft^3$	$= 28.329\ dm^3$
Cubic Metre	m^3	$1\ m^3$	$= 35.3147\ ft^3$	$1\ ft^3$	$= 0.0283\ m^3$
		$1\ m^3$	$= 1.3080\ yd^3$	$1\ yd^3$	$= 0.7646\ m^3$
Litre	l	1 l	= 1.76 pint	1 pint	= 0.5683 l
			= 2.113 US pt		= 0.4733 US l

Note:

$1\ dm^3$	$= 1\ 000\ cm^3$	$1\ ft^3$	$= 1\ 728\ in^3$	1 pint	= 20 fl oz
$1\ m^3$	$= 1\ 000\ dm^3$	$1\ yd^3$	$= 27\ ft^3$	1 gal	= 8 pints
1 l	$= 1\ dm^3$				

Neither the Centimetre nor Decimetre are SI units, and as such their use, particularly that of the Decimetre, is not widespread outside educational circles.

Mass	Unit	Conversion factors			
Milligram	mg	1 mg	= 0.0154 grain	1 grain	= 64.935 mg
Gram	g	1 g	= 0.0353 oz	1 oz	= 28.35 g
Kilogram	kg	1 kg	= 2.2046 lb	1 lb	= 0.4536 kg
Tonne	t	1 t	= 0.9842 ton	1 ton	= 1.016 t

Note:

1 g	= 1000 mg	1 oz	= 437.5 grains	1 cwt	= 112 lb
1 kg	= 1000 g	1 lb	= 16 oz	1 ton	= 20 cwt
1 t	= 1000 kg	1 stone	= 14 lb		

Force	Unit	Conversion factors			
Newton	N	1 lbf	= 4.448 N	1 kgf	= 9.807 N
Kilonewton	kN	1 lbf	= 0.004448 kN	1 ton f	= 9.964 kN
Meganewton	MN	100 tonf	= 0.9964 MN		

CONVERSION TABLES

Pressure and stress	Unit	Conversion factors	
Kilonewton per square metre	kN/m^2	1 lbf/in^2	= 6.895 kN/m^2
		1 bar	= 100 kN/m^2
Meganewton per square metre	MN/m^2	1 tonf/ft^2	= 107.3 kN/m^2 = 0.1073 MN/m^2
		1 kgf/cm^2	= 98.07 kN/m^2
		1 lbf/ft^2	= 0.04788 kN/m^2

Coefficient of consolidation (Cv) or swelling	Unit	Conversion factors	
Square metre per year	m^2/year	1 cm^2/s	= 3 154 m^2/year
		1 ft^2/year	= 0.0929 m^2/year

Coefficient of permeability	Unit	Conversion factors	
Metre per second	m/s	1 cm/s	= 0.01 m/s
Metre per year	m/year	1 ft/year	= 0.3048 m/year
			= 0.9651 $\times$ (10)^{8}m/s

Temperature	Unit	Conversion factors	
Degree Celsius	°C	°C = 5/9 $\times$ (°F − 32)	°F = (9 $\times$ °C)/ 5 + 32

CONVERSION TABLES

SPEED CONVERSION

km/h	m/min	mph	fpm
1	16.7	0.6	54.7
2	33.3	1.2	109.4
3	50.0	1.9	164.0
4	66.7	2.5	218.7
5	83.3	3.1	273.4
6	100.0	3.7	328.1
7	116.7	4.3	382.8
8	133.3	5.0	437.4
9	150.0	5.6	492.1
10	166.7	6.2	546.8
11	183.3	6.8	601.5
12	200.0	7.5	656.2
13	216.7	8.1	710.8
14	233.3	8.7	765.5
15	250.0	9.3	820.2
16	266.7	9.9	874.9
17	283.3	10.6	929.6
18	300.0	11.2	984.3
19	316.7	11.8	1038.9
20	333.3	12.4	1093.6
21	350.0	13.0	1148.3
22	366.7	13.7	1203.0
23	383.3	14.3	1257.7
24	400.0	14.9	1312.3
25	416.7	15.5	1367.0
26	433.3	16.2	1421.7
27	450.0	16.8	1476.4
28	466.7	17.4	1531.1
29	483.3	18.0	1585.7
30	500.0	18.6	1640.4
31	516.7	19.3	1695.1
32	533.3	19.9	1749.8
33	550.0	20.5	1804.5
34	566.7	21.1	1859.1
35	583.3	21.7	1913.8
36	600.0	22.4	1968.5
37	616.7	23.0	2023.2
38	633.3	23.6	2077.9
39	650.0	24.2	2132.5
40	666.7	24.9	2187.2

CONVERSION TABLES

km/h	m/min	mph	fpm
41	683.3	25.5	2241.9
42	700.0	26.1	2296.6
43	716.7	26.7	2351.3
44	733.3	27.3	2405.9
45	750.0	28.0	2460.6
46	766.7	28.6	2515.3
47	783.3	29.2	2570.0
48	800.0	29.8	2624.7
49	816.7	30.4	2679.4
50	833.3	31.1	2734.0

GEOMETRY

Two dimensional figures

Figure	Diagram of figure	Surface area	Perimeter
Square		a^2	$4a$
Rectangle		ab	$2(a+b)$
Triangle		$\frac{1}{2}ch$	$a+b+c$
Circle		πr^2 $\frac{1}{4}\pi d^2$ where $2r = d$	$2\pi r$ πd
Parallelogram		ah	$2(a+b)$
Trapezium		$\frac{1}{2}h(a+b)$	$a+b+c+d$
Ellipse		Approximately πab	$\pi(a+b)$
Hexagon		$2.6 \times a^2$	

GEOMETRY

Figure	Diagram of figure	Surface area	Perimeter
Octagon		$4.83 \times a^2$	$6a$
Sector of a circle		$\frac{1}{2}rb$ or $\frac{q}{360}\pi r^2$ note b = angle $\frac{q}{360} \times \pi 2r$	
Segment of a circle		$S - T$ where S = area of sector, T = area of triangle	
Bellmouth		$\frac{3}{14} \times r^2$	

GEOMETRY

Three dimensional figures

Figure	Diagram of figure	Surface area	Volume
Cube		$6a^2$	a^3
Cuboid/ rectangular block		$2(ab + ac + bc)$	abc
Prism/ triangular block		$bd + hc + dc + ad$	$\frac{1}{2}hcd$
Cylinder		$2\pi r^2 + 2\pi h$	$\pi r^2 h$
Sphere		$4\pi r^2$	$\frac{4}{3}\pi r^3$
Segment of sphere		$2\pi Rh$	$\frac{1}{6}\pi h(3r^2 + h^2)$ $\frac{1}{3}\pi h^2(3R - H)$
Pyramid		$(a + b)l + ab$	$\frac{1}{3}abh$

GEOMETRY

Figure	Diagram of figure	Surface area	Volume
Frustum of a pyramid		$l(a+b+c+d)+\sqrt{(ab+cd)}$ [rectangular figure only]	$\frac{h}{3}(ab+cd+\sqrt{abcd})$
Cone		πrl (excluding base) $\pi rl + \pi r^2$ (including base)	$\frac{1}{3}\pi r^2 h$ $\frac{1}{12}\pi d^2 h$
Frustum of a cone		$\pi r^2 + \pi R^2 + \pi l(R+r)$	$\frac{1}{3}\pi(R^2 + Rr + r^2)$

Formulae

Formula	Description
Pythagoras Theorem	$A^2 = B^2 + C^2$ where A is the hypotenuse of a right-angled triangle and B and C are the two adjacent sides
Simpsons Rule	The Area is divided into an even number of strips of equal width, and therefore has an odd number of ordinates at the division points $$\text{area} = \frac{S(A + 2B + 4C)}{3}$$ where S = common interval (strip width) A = sum of first and last ordinates B = sum of remaining odd ordinates C = sum of the even ordinates The Volume can be calculated by the same formula, but by substituting the area of each coordinate rather than its length
Trapezoidal Rule	A given trench is divided into two equal sections, giving three ordinates, the first, the middle and the last $$\text{volume} = \frac{S \times (A + B + 2C)}{2}$$ where S = width of the strips A = area of the first section B = area of the last section C = area of the rest of the sections
Prismoidal Rule	A given trench is divided into two equal sections, giving three ordinates, the first, the middle and the last $$\text{volume} = \frac{L \times (A + 4B + C)}{6}$$ where L = total length of trench A = area of the first section B = area of the middle section C = area of the last section

TYPICAL THERMAL CONDUCTIVITY OF BUILDING MATERIALS

(Always check manufacturer's details – variation will occur depending on product and nature of materials)

	Thermal conductivity (W/mK)		Thermal conductivity (W/mK)
Acoustic plasterboard	0.25	Oriented strand board	0.13
Aerated concrete slab (500 kg/m^3)	0.16	Outer leaf brick	0.77
Aluminium	237	Plasterboard	0.22
Asphalt (1700 kg/m^3)	0.5	Plaster dense (1300 kg/m^3)	0.5
Bitumen-impregnated fibreboard	0.05	Plaster lightweight (600 kg/m^3)	0.16
Blocks (standard grade 600 kg/m^3)	0.15	Plywood (950 kg/m^3)	0.16
Blocks (solar grade 460 kg/m^3)	0.11	Prefabricated timber wall panels (check manufacturer)	0.12
Brickwork (outer leaf 1700 kg/m^3)	0.84	Screed (1200 kg/m^3)	0.41
Brickwork (inner leaf 1700 kg/m^3)	0.62	Stone chippings (1800 kg/m^3)	0.96
Dense aggregate concrete block 1800 kg/m^3 (exposed)	1.21	Tile hanging (1900 kg/m^3)	0.84
Dense aggregate concrete block 1800 kg/m^3 (protected)	1.13	Timber (650 kg/m^3)	0.14
Calcium silicate board (600 kg/m^3)	0.17	Timber flooring (650 kg/m^3)	0.14
Concrete general	1.28	Timber rafters	0.13
Concrete (heavyweight 2300 kg/m^3)	1.63	Timber roof or floor joists	0.13
Concrete (dense 2100 kg/m^3 typical floor)	1.4	Roof tile (1900 kg/m^3)	0.84
Concrete (dense 2000 kg/m^3 typical floor)	1.13	Timber blocks (650 kg/m^3)	0.14
Concrete (medium 1400 kg/m^3)	0.51	Cellular glass	0.045
Concrete (lightweight 1200 kg/m^3)	0.38	Expanded polystyrene	0.034
Concrete (lightweight 600 kg/m^3)	0.19	Expanded polystyrene slab (25 kg/m^3)	0.035
Concrete slab (aerated 500 kg/m^3)	0.16	Extruded polystyrene	0.035
Copper	390	Glass mineral wool	0.04
External render sand/cement finish	1	Mineral quilt (12 kg/m^3)	0.04
External render (1300 kg/m^3)	0.5	Mineral wool slab (25 kg/m^3)	0.035
Felt – Bitumen layers (1700 kg/m^3)	0.5	Phenolic foam	0.022
Fibreboard (300 kg/m^3)	0.06	Polyisocyanurate	0.025
Glass	0.93	Polyurethane	0.025
Marble	3	Rigid polyurethane	0.025
Metal tray used in wriggly tin concrete floors (7800 kg/m^3)	50	Rock mineral wool	0.038
Mortar (1750 kg/m^3)	0.8		

EARTHWORK

Weights of Typical Materials Handled by Excavators

The weight of the material is that of the state in its natural bed and includes moisture. Adjustments should be made to allow for loose or compacted states

Material	Mass (kg/m³)	Mass (lb/cu yd)
Ashes, dry	610	1028
Ashes, wet	810	1365
Basalt, broken	1954	3293
Basalt, solid	2933	4943
Bauxite, crushed	1281	2159
Borax, fine	849	1431
Caliche	1440	2427
Cement, clinker	1415	2385
Chalk, fine	1221	2058
Chalk, solid	2406	4055
Cinders, coal, ash	641	1080
Cinders, furnace	913	1538
Clay, compacted	1746	2942
Clay, dry	1073	1808
Clay, wet	1602	2700
Coal, anthracite, solid	1506	2538
Coal, bituminous	1351	2277
Coke	610	1028
Dolomite, lumpy	1522	2565
Dolomite, solid	2886	4864
Earth, dense	2002	3374
Earth, dry, loam	1249	2105
Earth, Fullers, raw	673	1134
Earth, moist	1442	2430
Earth, wet	1602	2700
Felsite	2495	4205
Fieldspar, solid	2613	4404
Fluorite	3093	5213
Gabbro	3093	5213
Gneiss	2696	4544
Granite	2690	4534
Gravel, dry ¼ to 2 inch	1682	2835
Gravel, dry, loose	1522	2565
Gravel, wet ¼ to 2 inch	2002	3374
Gypsum, broken	1450	2444
Gypsum, solid	2787	4697
Hardcore (consolidated)	1928	3249
Lignite, dry	801	1350
Limestone, broken	1554	2619
Limestone, solid	2596	4375
Magnesite, magnesium ore	2993	5044
Marble	2679	4515
Marl, wet	2216	3735
Mica, broken	1602	2700
Mica, solid	2883	4859
Peat, dry	400	674
Peat, moist	700	1179

EARTHWORK

Material	Mass (kg/m^3)	Mass (lb/cu yd)
Peat, wet	1121	1889
Potash	1281	2159
Pumice, stone	640	1078
Quarry waste	1438	2423
Quartz sand	1201	2024
Quartz, solid	2584	4355
Rhyolite	2400	4045
Sand and gravel, dry	1650	2781
Sand and gravel, wet	2020	3404
Sand, dry	1602	2700
Sand, wet	1831	3086
Sandstone, solid	2412	4065
Shale, solid	2637	4444
Slag, broken	2114	3563
Slag, furnace granulated	961	1619
Slate, broken	1370	2309
Slate, solid	2667	4495
Snow, compacted	481	810
Snow, freshly fallen	160	269
Taconite	2803	4724
Trachyte	2400	4045
Trap rock, solid	2791	4704
Turf	400	674
Water	1000	1685

Transport Capacities

Type of vehicle	Capacity of vehicle	
	Payload	Heaped capacity
Wheelbarrow	150	0.10
1 tonne dumper	1250	1.00
2.5 tonne dumper	4000	2.50
Articulated dump truck (Volvo A20 6 × 4)	18500	11.00
Articulated dump truck (Volvo A35 6 × 6)	32000	19.00
Large capacity rear dumper (Euclid R35)	35000	22.00
Large capacity rear dumper (Euclid R85)	85000	50.00

EARTHWORK

Machine Volumes for Excavating and Filling

Machine type	Cycles per minute	Volume per minute (m^3)
1.5 tonne excavator	1	0.04
	2	0.08
	3	0.12
3 tonne excavator	1	0.13
	2	0.26
	3	0.39
5 tonne excavator	1	0.28
	2	0.56
	3	0.84
7 tonne excavator	1	0.28
	2	0.56
	3	0.84
21 tonne excavator	1	1.21
	2	2.42
	3	3.63
Backhoe loader JCB3CX excavator	1	0.28
Rear bucket capacity 0.28 m^3	2	0.56
	3	0.84
Backhoe loader JCB3CX loading	1	1.00
Front bucket capacity 1.00 m^3	2	2.00

Machine Volumes for Excavating and Filling

Machine type	Loads per hour	Volume per hour (m^3)
1 tonne high tip skip loader	5	2.43
Volume 0.485 m^3	7	3.40
	10	4.85
3 tonne dumper	4	7.60
Max volume 2.40 m^3	5	9.50
Available volume 1.9 m^3	7	13.30
	10	19.00
6 tonne dumper	4	15.08
Max volume 3.40 m^3	5	18.85
Available volume 3.77 m^3	7	26.39
	10	37.70

EARTHWORK

Bulkage of Soils (after excavation)

Type of soil	Approximate bulking of 1 m³ after excavation
Vegetable soil and loam	25–30%
Soft clay	30–40%
Stiff clay	10–20%
Gravel	20–25%
Sand	40–50%
Chalk	40–50%
Rock, weathered	30–40%
Rock, unweathered	50–60%

Shrinkage of Materials (on being deposited)

Type of soil	Approximate bulking of 1 m³ after excavation
Clay	10%
Gravel	8%
Gravel and sand	9%
Loam and light sandy soils	12%
Loose vegetable soils	15%

Voids in Material Used as Subbases or Beddings

Material	m³ of voids/m³
Alluvium	0.37
River grit	0.29
Quarry sand	0.24
Shingle	0.37
Gravel	0.39
Broken stone	0.45
Broken bricks	0.42

Angles of Repose

Type of soil		Degrees
Clay	– dry	30
	– damp, well drained	45
	– wet	15–20
Earth	– dry	30
	– damp	45
Gravel	– moist	48
Sand	– dry or moist	35
	– wet	25
Loam		40

EARTHWORK

Slopes and Angles

Ratio of base to height	Angle in degrees
5:1	11
4:1	14
3:1	18
2:1	27
1½:1	34
1:1	45
1:1½	56
1:2	63
1:3	72
1:4	76
1:5	79

Grades (in Degrees and Percents)

Degrees	Percent	Degrees	Percent
1	1.8	24	44.5
2	3.5	25	46.6
3	5.2	26	48.8
4	7.0	27	51.0
5	8.8	28	53.2
6	10.5	29	55.4
7	12.3	30	57.7
8	14.0	31	60.0
9	15.8	32	62.5
10	17.6	33	64.9
11	19.4	34	67.4
12	21.3	35	70.0
13	23.1	36	72.7
14	24.9	37	75.4
15	26.8	38	78.1
16	28.7	39	81.0
17	30.6	40	83.9
18	32.5	41	86.9
19	34.4	42	90.0
20	36.4	43	93.3
21	38.4	44	96.6
22	40.4	45	100.0
23	42.4		

EARTHWORK

Bearing Powers

Ground conditions		Bearing power		
		kg/m^2	lb/in^2	Metric t/m^2
Rock,	broken	483	70	50
	solid	2415	350	240
Clay,	dry or hard	380	55	40
	medium dry	190	27	20
	soft or wet	100	14	10
Gravel,	cemented	760	110	80
Sand,	compacted	380	55	40
	clean dry	190	27	20
Swamp and alluvial soils		48	7	5

Earthwork Support

Maximum depth of excavation in various soils without the use of earthwork support

Ground conditions	Feet (ft)	Metres (m)
Compact soil	12	3.66
Drained loam	6	1.83
Dry sand	1	0.3
Gravelly earth	2	0.61
Ordinary earth	3	0.91
Stiff clay	10	3.05

It is important to note that the above table should only be used as a guide. Each case must be taken on its merits and, as the limited distances given above are approached, careful watch must be kept for the slightest signs of caving in

CONCRETE WORK

Weights of Concrete and Concrete Elements

Type of material		kg/m³	lb/cu ft
Ordinary concrete (dense aggregates)			
Non-reinforced plain or mass concrete			
Nominal weight		2305	144
Aggregate	– limestone	2162 to 2407	135 to 150
	– gravel	2244 to 2407	140 to 150
	– broken brick	2000 (av)	125 (av)
	– other crushed stone	2326 to 2489	145 to 155
Reinforced concrete			
Nominal weight		2407	150
Reinforcement	– 1%	2305 to 2468	144 to 154
	– 2%	2356 to 2519	147 to 157
	– 4%	2448 to 2703	153 to 163
Special concretes			
Heavy concrete			
Aggregates	– barytes, magnetite	3210 (min)	200 (min)
	– steel shot, punchings	5280	330
Lean mixes			
Dry-lean (gravel aggregate)		2244	140
Soil-cement (normal mix)		1601	100

CONCRETE WORK

Type of material		kg/m² per mm thick	lb/sq ft per inch thick
Ordinary concrete (dense aggregates)			
Solid slabs (floors, walls etc.)			
Thickness:	75 mm or 3 in	184	37.5
	100 mm or 4 in	245	50
	150 mm or 6 in	378	75
	250 mm or 10 in	612	125
	300 mm or 12 in	734	150
Ribbed slabs			
Thickness:	125 mm or 5 in	204	42
	150 mm or 6 in	219	45
	225 mm or 9 in	281	57
	300 mm or 12 in	342	70
Special concretes			
Finishes etc.			
	Rendering, screed etc. Granolithic, terrazzo	1928 to 2401	10 to 12.5
	Glass-block (hollow) concrete	1734 (approx)	9 (approx)
Prestressed concrete		Weights as for reinforced concrete (upper limits)	
Air-entrained concrete		Weights as for plain or reinforced concrete	

CONCRETE WORK

Average Weight of Aggregates

Materials	Voids %	Weight kg/m³
Sand	39	1660
Gravel 10–20 mm	45	1440
Gravel 35–75 mm	42	1555
Crushed stone	50	1330
Crushed granite (over 15 mm)	50	1345
(n.e. 15 mm)	47	1440
'All-in' ballast	32	1800–2000

Material	kg/m³	lb/cu yd
Vermiculite (aggregate)	64–80	108–135
All-in aggregate	1999	125

Applications and Mix Design

Site mixed concrete

Recommended mix	Class of work suitable for	Cement (kg)	Sand (kg)	Coarse aggregate (kg)	Nr 25 kg bags cement per m³ of combined aggregate
1:3:6	Roughest type of mass concrete such as footings, road haunching over 300 mm thick	208	905	1509	8.30
1:2.5:5	Mass concrete of better class than 1:3:6 such as bases for machinery, walls below ground etc.	249	881	1474	10.00
1:2:4	Most ordinary uses of concrete, such as mass walls above ground, road slabs etc. and general reinforced concrete work	304	889	1431	12.20
1:1.5:3	Watertight floors, pavements and walls, tanks, pits, steps, paths, surface of 2 course roads, reinforced concrete where extra strength is required	371	801	1336	14.90
1:1:2	Works of thin section such as fence posts and small precast work	511	720	1206	20.40

CONCRETE WORK

Ready mixed concrete

Application	Designated concrete	Standardized prescribed concrete	Recommended consistence (nominal slump class)
Foundations			
Mass concrete fill or blinding	GEN 1	ST2	S3
Strip footings	GEN 1	ST2	S3
Mass concrete foundations			
Single storey buildings	GEN 1	ST2	S3
Double storey buildings	GEN 3	ST4	S3
Trench fill foundations			
Single storey buildings	GEN 1	ST2	S4
Double storey buildings	GEN 3	ST4	S4
General applications			
Kerb bedding and haunching	GEN 0	ST1	S1
Drainage works – immediate support	GEN 1	ST2	S1
Other drainage works	GEN 1	ST2	S3
Oversite below suspended slabs	GEN 1	ST2	S3
Floors			
Garage and house floors with no embedded steel	GEN 3	ST4	S2
Wearing surface: Light foot and trolley traffic	RC30	ST4	S2
Wearing surface: General industrial	RC40	N/A	S2
Wearing surface: Heavy industrial	RC50	N/A	S2
Paving			
House drives, domestic parking and external parking	PAV 1	N/A	S2
Heavy-duty external paving	PAV 2	N/A	S2

CONCRETE WORK

Prescribed Mixes for Ordinary Structural Concrete

Weights of cement and total dry aggregates in kg to produce approximately one cubic metre of fully compacted concrete together with the percentages by weight of fine aggregate in total dry aggregates

Conc. grade	Nominal max size of aggregate (mm)	40		20		14		10	
	Workability	Med.	High	Med.	High	Med.	High	Med.	High
	Limits to slump that may be expected (mm)	50–100	100–150	25–75	75–125	10–50	50–100	10–25	25–50
7	Cement (kg)	180	200	210	230	–	–	–	–
	Total aggregate (kg)	1950	1850	1900	1800	–	–	–	–
	Fine aggregate (%)	30–45	30–45	35–50	35–50	–	–	–	–
10	Cement (kg)	210	230	240	260	–	–	–	–
	Total aggregate (kg)	1900	1850	1850	1800	–	–	–	–
	Fine aggregate (%)	30–45	30–45	35–50	35–50	–	–	–	–
15	Cement (kg)	250	270	280	310	–	–	–	–
	Total aggregate (kg)	1850	1800	1800	1750	–	–	–	–
	Fine aggregate (%)	30–45	30–45	35–50	35–50	–	–	–	–
20	Cement (kg)	300	320	320	350	340	380	360	410
	Total aggregate (kg)	1850	1750	1800	1750	1750	1700	1750	1650
	Sand								
	Zone 1 (%)	35	40	40	45	45	50	50	55
	Zone 2 (%)	30	35	35	40	40	45	45	50
	Zone 3 (%)	30	30	30	35	35	40	40	45
25	Cement (kg)	340	360	360	390	380	420	400	450
	Total aggregate (kg)	1800	1750	1750	1700	1700	1650	1700	1600
	Sand								
	Zone 1 (%)	35	40	40	45	45	50	50	55
	Zone 2 (%)	30	35	35	40	40	45	45	50
	Zone 3 (%)	30	30	30	35	35	40	40	45
30	Cement (kg)	370	390	400	430	430	470	460	510
	Total aggregate (kg)	1750	1700	1700	1650	1700	1600	1650	1550
	Sand								
	Zone 1 (%)	35	40	40	45	45	50	50	55
	Zone 2 (%)	30	35	35	40	40	45	45	50
	Zone 3 (%)	30	30	30	35	35	40	40	45

REINFORCEMENT

Weights of Bar Reinforcement

Nominal sizes (mm)	Cross-sectional area (mm²)	Mass (kg/m)	Length of bar (m/tonne)
6	28.27	0.222	4505
8	50.27	0.395	2534
10	78.54	0.617	1622
12	113.10	0.888	1126
16	201.06	1.578	634
20	314.16	2.466	405
25	490.87	3.853	260
32	804.25	6.313	158
40	1265.64	9.865	101
50	1963.50	15.413	65

Weights of Bars (at specific spacings)

Weights of metric bars in kilogrammes per square metre

Size (mm)	Spacing of bars in millimetres									
	75	100	125	150	175	200	225	250	275	300
6	2.96	2.220	1.776	1.480	1.27	1.110	0.99	0.89	0.81	0.74
8	5.26	3.95	3.16	2.63	2.26	1.97	1.75	1.58	1.44	1.32
10	8.22	6.17	4.93	4.11	3.52	3.08	2.74	2.47	2.24	2.06
12	11.84	8.88	7.10	5.92	5.07	4.44	3.95	3.55	3.23	2.96
16	21.04	15.78	12.63	10.52	9.02	7.89	7.02	6.31	5.74	5.26
20	32.88	24.66	19.73	16.44	14.09	12.33	10.96	9.87	8.97	8.22
25	51.38	38.53	30.83	25.69	22.02	19.27	17.13	15.41	14.01	12.84
32	84.18	63.13	50.51	42.09	36.08	31.57	28.06	25.25	22.96	21.04
40	131.53	98.65	78.92	65.76	56.37	49.32	43.84	39.46	35.87	32.88
50	205.51	154.13	123.31	102.76	88.08	77.07	68.50	61.65	56.05	51.38

Basic weight of steelwork taken as 7850 kg/m³
Basic weight of bar reinforcement per metre run = 0.00785 kg/mm²
The value of π has been taken as 3.141592654

Fabric Reinforcement

Preferred range of designated fabric types and stock sheet sizes

Fabric reference	Longitudinal wires			Cross wires			
	Nominal wire size (mm)	Pitch (mm)	Area (mm²/m)	Nominal wire size (mm)	Pitch (mm)	Area (mm²/m)	Mass (kg/m²)
Square mesh							
A393	10	200	393	10	200	393	6.16
A252	8	200	252	8	200	252	3.95
A193	7	200	193	7	200	193	3.02
A142	6	200	142	6	200	142	2.22
A98	5	200	98	5	200	98	1.54
Structural mesh							
B1131	12	100	1131	8	200	252	10.90
B785	10	100	785	8	200	252	8.14
B503	8	100	503	8	200	252	5.93
B385	7	100	385	7	200	193	4.53
B283	6	100	283	7	200	193	3.73
B196	5	100	196	7	200	193	3.05
Long mesh							
C785	10	100	785	6	400	70.8	6.72
C636	9	100	636	6	400	70.8	5.55
C503	8	100	503	5	400	49.0	4.34
C385	7	100	385	5	400	49.0	3.41
C283	6	100	283	5	400	49.0	2.61
Wrapping mesh							
D98	5	200	98	5	200	98	1.54
D49	2.5	100	49	2.5	100	49	0.77

Stock sheet size 4.8 m × 2.4 m, Area 11.52 m²

Average weight kg/m³ of steelwork reinforcement in concrete for various building elements

Substructure	kg/m³ concrete	Substructure	kg/m³ concrete
Pile caps	110–150	Plate slab	150–220
Tie beams	130–170	Cant slab	145–210
Ground beams	230–330	Ribbed floors	130–200
Bases	125–180	Topping to block floor	30–40
Footings	100–150	Columns	210–310
Retaining walls	150–210	Beams	250–350
Raft	60–70	Stairs	130–170
Slabs – one way	120–200	Walls – normal	40–100
Slabs – two way	110–220	Walls – wind	70–125

Note: For exposed elements add the following %:
Walls 50%, Beams 100%, Columns 15%

FORMWORK

Formwork Stripping Times – Normal Curing Periods

Conditions under which concrete is maturing	Minimum periods of protection for different types of cement					
	Number of days (where the average surface temperature of the concrete exceeds 10°C during the whole period)			Equivalent maturity (degree hours) calculated as the age of the concrete in hours multiplied by the number of degrees Celsius by which the average surface temperature of the concrete exceeds 10°C		
	Other	SRPC	OPC or RHPC	Other	SRPC	OPC or RHPC
1. Hot weather or drying winds	7	4	3	3500	2000	1500
2. Conditions not covered by 1	4	3	2	2000	1500	1000

KEY
OPC – Ordinary Portland Cement
RHPC – Rapid-hardening Portland Cement
SRPC – Sulphate-resisting Portland Cement

Minimum Period before Striking Formwork

	Minimum period before striking		
	Surface temperature of concrete		
	16°C	17°C	t°C (0–25)
Vertical formwork to columns, walls and large beams	12 hours	18 hours	300 hours t+10
Soffit formwork to slabs	4 days	6 days	100 days t+10
Props to slabs	10 days	15 days	250 days t+10
Soffit formwork to beams	9 days	14 days	230 days t+10
Props to beams	14 days	21 days	360 days t+10

MASONRY

Number of Bricks Required for Various Types of Work per m² of Walling

Description	Brick size	
	215 × 102.5 × 50 mm	215 × 102.5 × 65 mm
Half brick thick		
Stretcher bond	74	59
English bond	108	86
English garden wall bond	90	72
Flemish bond	96	79
Flemish garden wall bond	83	66
One brick thick and cavity wall of two half brick skins		
Stretcher bond	148	119

Quantities of Bricks and Mortar Required per m² of Walling

	Unit	No of bricks required	Mortar required (cubic metres)		
Standard bricks			**No frogs**	**Single frogs**	**Double frogs**
Brick size 215 × 102.5 × 50 mm					
half brick wall (103 mm)	m²	72	0.022	0.027	0.032
2 × half brick cavity wall (270 mm)	m²	144	0.044	0.054	0.064
one brick wall (215 mm)	m²	144	0.052	0.064	0.076
one and a half brick wall (322 mm)	m²	216	0.073	0.091	0.108
Mass brickwork	m³	576	0.347	0.413	0.480
Brick size 215 × 102.5 × 65 mm					
half brick wall (103 mm)	m²	58	0.019	0.022	0.026
2 × half brick cavity wall (270 mm)	m²	116	0.038	0.045	0.055
one brick wall (215 mm)	m²	116	0.046	0.055	0.064
one and a half brick wall (322 mm)	m²	174	0.063	0.074	0.088
Mass brickwork	m³	464	0.307	0.360	0.413
Metric modular bricks			**Perforated**		
Brick size 200 × 100 × 75 mm					
90 mm thick	m²	67	0.016	0.019	
190 mm thick	m²	133	0.042	0.048	
290 mm thick	m²	200	0.068	0.078	
Brick size 200 × 100 × 100 mm					
90 mm thick	m²	50	0.013	0.016	
190 mm thick	m²	100	0.036	0.041	
290 mm thick	m²	150	0.059	0.067	
Brick size 300 × 100 × 75 mm					
90 mm thick	m²	33	–	0.015	
Brick size 300 × 100 × 100 mm					
90 mm thick	m²	44	0.015	0.018	

Note: Assuming 10 mm thick joints

MASONRY

Mortar Required per m² Blockwork (9.88 blocks/m²)

Wall thickness	75	90	100	125	140	190	215
Mortar m³/m²	0.005	0.006	0.007	0.008	0.009	0.013	0.014

Mortar Group	Cement: lime: sand	Masonry cement: sand	Cement: sand with plasticizer
1	1:0–0.25:3		
2	1:0.5:4–4.5	1:2.5-3.5	1:3–4
3	1:1:5–6	1:4–5	1:5–6
4	1:2:8–9	1:5.5–6.5	1:7–8
5	1:3:10–12	1:6.5–7	1:8

Group 1: strong inflexible mortar
Group 5: weak but flexible

All mixes within a group are of approximately similar strength
Frost resistance increases with the use of plasticizers
Cement: lime: sand mixes give the strongest bond and greatest resistance to rain penetration
Masonry cement equals ordinary Portland cement plus a fine neutral mineral filler and an air entraining agent

Calcium Silicate Bricks

Type	Strength	Location
Class 2 crushing strength	14.0 N/mm²	not suitable for walls
Class 3	20.5 N/mm²	walls above dpc
Class 4	27.5 N/mm²	cappings and copings
Class 5	34.5 N/mm²	retaining walls
Class 6	41.5 N/mm²	walls below ground
Class 7	48.5 N/mm²	walls below ground

The Class 7 calcium silicate bricks are therefore equal in strength to Class B bricks
Calcium silicate bricks are not suitable for DPCs

Durability of Bricks	
FL	Frost resistant with low salt content
FN	Frost resistant with normal salt content
ML	Moderately frost resistant with low salt content
MN	Moderately frost resistant with normal salt content

MASONRY

Brickwork Dimensions

No. of horizontal bricks	Dimensions (mm)	No. of vertical courses	Height of vertical courses (mm)
½	112.5	1	75
1	225.0	2	150
1½	337.5	3	225
2	450.0	4	300
2½	562.5	5	375
3	675.0	6	450
3½	787.5	7	525
4	900.0	8	600
4½	1012.5	9	675
5	1125.0	10	750
5½	1237.5	11	825
6	1350.0	12	900
6½	1462.5	13	975
7	1575.0	14	1050
7½	1687.5	15	1125
8	1800.0	16	1200
8½	1912.5	17	1275
9	2025.0	18	1350
9½	2137.5	19	1425
10	2250.0	20	1500
20	4500.0	24	1575
40	9000.0	28	2100
50	11250.0	32	2400
60	13500.0	36	2700
75	16875.0	40	3000

TIMBER

Weights of Timber

Material	kg/m³	lb/cu ft
General	806 (avg)	50 (avg)
Douglas fir	479	30
Yellow pine, spruce	479	30
Pitch pine	673	42
Larch, elm	561	35
Oak (English)	724 to 959	45 to 60
Teak	643 to 877	40 to 55
Jarrah	959	60
Greenheart	1040 to 1204	65 to 75
Quebracho	1285	80
Material	**kg/m² per mm thickness**	**lb/sq ft per inch thickness**
Wooden boarding and blocks		
Softwood	0.48	2.5
Hardwood	0.76	4
Hardboard	1.06	5.5
Chipboard	0.76	4
Plywood	0.62	3.25
Blockboard	0.48	2.5
Fibreboard	0.29	1.5
Wood-wool	0.58	3
Plasterboard	0.96	5
Weather boarding	0.35	1.8

TIMBER

Conversion Tables (for timber only)

Inches	Millimetres	Feet	Metres
1	25	1	0.300
2	50	2	0.600
3	75	3	0.900
4	100	4	1.200
5	125	5	1.500
6	150	6	1.800
7	175	7	2.100
8	200	8	2.400
9	225	9	2.700
10	250	10	3.000
11	275	11	3.300
12	300	12	3.600
13	325	13	3.900
14	350	14	4.200
15	375	15	4.500
16	400	16	4.800
17	425	17	5.100
18	450	18	5.400
19	475	19	5.700
20	500	20	6.000
21	525	21	6.300
22	550	22	6.600
23	575	23	6.900
24	600	24	7.200

Planed Softwood
The finished end section size of planed timber is usually 3/16" less than the original size from which it is produced. This however varies slightly depending upon availability of material and origin of the species used.

Standards (timber) to cubic metres and cubic metres to standards (timber)

Cubic metres	Cubic metres standards	Standards
4.672	1	0.214
9.344	2	0.428
14.017	3	0.642
18.689	4	0.856
23.361	5	1.070
28.033	6	1.284
32.706	7	1.498
37.378	8	1.712
42.050	9	1.926
46.722	10	2.140
93.445	20	4.281
140.167	30	6.421
186.890	40	8.561
233.612	50	10.702
280.335	60	12.842
327.057	70	14.982
373.779	80	17.122

TIMBER

1 cu metre = 35.3148 cu ft = 0.21403 std

1 cu ft = 0.028317 cu metres

1 std = 4.67227 cu metres

Basic sizes of sawn softwood available (cross-sectional areas)

Thickness (mm)	Width (mm)								
	75	100	125	150	175	200	225	250	300
16	X	X	X	X					
19	X	X	X	X					
22	X	X	X	X					
25	X	X	X	X	X	X	X	X	X
32	X	X	X	X	X	X	X	X	X
36	X	X	X	X					
38	X	X	X	X	X	X	X		
44	X	X	X	X	X	X	X	X	X
47*	X	X	X	X	X	X	X	X	X
50	X	X	X	X	X	X	X	X	X
63	X	X	X	X	X	X	X		
75	X	X	X	X	X	X	X	X	
100		X		X		X		X	X
150				X		X			X
200						X			
250								X	
300									X

* This range of widths for 47 mm thickness will usually be found to be available in construction quality only

Note: The smaller sizes below 100 mm thick and 250 mm width are normally but not exclusively of European origin. Sizes beyond this are usually of North and South American origin

Basic lengths of sawn softwood available (metres)

1.80	2.10	3.00	4.20	5.10	6.00	7.20
	2.40	3.30	4.50	5.40	6.30	
	2.70	3.60	4.80	5.70	6.60	
		3.90			6.90	

Note: Lengths of 6.00 m and over will generally only be available from North American species and may have to be recut from larger sizes

TIMBER

Reductions from basic size to finished size by planning of two opposed faces

Purpose		Reductions from basic sizes for timber			
		15–35 mm	36–100 mm	101–150 mm	over 150 mm
a)	Constructional timber	3 mm	3 mm	5 mm	6 mm
b)	Matching interlocking boards	4 mm	4 mm	6 mm	6 mm
c)	Wood trim not specified in BS 584	5 mm	7 mm	7 mm	9 mm
d)	Joinery and cabinet work	7 mm	9 mm	11 mm	13 mm

Note: The reduction of width or depth is overall the extreme size and is exclusive of any reduction of the face by the machining of a tongue or lap joints

Maximum Spans for Various Roof Trusses

Maximum permissible spans for rafters for Fink trussed rafters

Basic size (mm)	Actual size (mm)	Pitch (degrees)								
		15 (m)	17.5 (m)	20 (m)	22.5 (m)	25 (m)	27.5 (m)	30 (m)	32.5 (m)	35 (m)
38 × 75	35 × 72	6.03	6.16	6.29	6.41	6.51	6.60	6.70	6.80	6.90
38 × 100	35 × 97	7.48	7.67	7.83	7.97	8.10	8.22	8.34	8.47	8.61
38 × 125	35 × 120	8.80	9.00	9.20	9.37	9.54	9.68	9.82	9.98	10.16
44 × 75	41 × 72	6.45	6.59	6.71	6.83	6.93	7.03	7.14	7.24	7.35
44 × 100	41 × 97	8.05	8.23	8.40	8.55	8.68	8.81	8.93	9.09	9.22
44 × 125	41 × 120	9.38	9.60	9.81	9.99	10.15	10.31	10.45	10.64	10.81
50 × 75	47 × 72	6.87	7.01	7.13	7.25	7.35	7.45	7.53	7.67	7.78
50 × 100	47 × 97	8.62	8.80	8.97	9.12	9.25	9.38	9.50	9.66	9.80
50 × 125	47 × 120	10.01	10.24	10.44	10.62	10.77	10.94	11.00	11.00	11.00

TIMBER

Sizes of Internal and External Doorsets

Description	Internal size (mm)	Permissible deviation	External size (mm)	Permissible deviation
Coordinating dimension: height of door leaf height sets	2100		2100	
Coordinating dimension: height of ceiling height set	2300 2350 2400 2700 3000		2300 2350 2400 2700 3000	
Coordinating dimension: width of all doorsets S = Single leaf set D = Double leaf set	600 S 700 S 800 S&D 900 S&D 1000 S&D 1200 D 1500 D 1800 D 2100 D		900 S 1000 S 1200 D 1800 D 2100 D	
Work size: height of door leaf height set	2090	± 2.0	2095	± 2.0
Work size: height of ceiling height set	2285 2335 2385 2685 2985	± 2.0	2295 2345 2395 2695 2995	± 2.0
Work size: width of all doorsets S = Single leaf set D = Double leaf set	590 S 690 S 790 S&D 890 S&D 990 S&D 1190 D 1490 D 1790 D 2090 D	± 2.0	895 S 995 S 1195 D 1495 D 1795 D 2095 D	± 2.0
Width of door leaf in single leaf sets F = Flush leaf P = Panel leaf	526 F 626 F 726 F&P 826 F&P 926 F&P	± 1.5	806 F&P 906 F&P	± 1.5
Width of door leaf in double leaf sets F = Flush leaf P = Panel leaf	362 F 412 F 426 F 562 F&P 712 F&P 826 F&P 1012 F&P	± 1.5	552 F&P 702 F&P 852 F&P 1002 F&P	± 1.5
Door leaf height for all doorsets	2040	± 1.5	1994	± 1.5

ROOFING

Total Roof Loadings for Various Types of Tiles/Slates

		Roof load (slope) kg/m^2		
		Slate/Tile	Roofing underlay and battens^2	Total dead load kg/m
Asbestos cement slate (600 × 300)		21.50	3.14	24.64
Clay tile	interlocking	67.00	5.50	72.50
	plain	43.50	2.87	46.37
Concrete tile	interlocking	47.20	2.69	49.89
	plain	78.20	5.50	83.70
Natural slate (18" × 10")		35.40	3.40	38.80
		Roof load (plan) kg/m^2		
Asbestos cement slate (600 × 300)		28.45	76.50	104.95
Clay tile	interlocking	53.54	76.50	130.04
	plain	83.71	76.50	60.21
Concrete tile	interlocking	57.60	76.50	134.10
	plain	96.64	76.50	173.14

ROOFING

Tiling Data

Product		Lap (mm)	Gauge of battens	No. slates per m²	Battens (m/m²)	Weight as laid (kg/m²)
CEMENT SLATES						
Eternit slates	600 × 300 mm	100	250	13.4	4.00	19.50
(Duracem)		90	255	13.1	3.92	19.20
		80	260	12.9	3.85	19.00
		70	265	12.7	3.77	18.60
	600 × 350 mm	100	250	11.5	4.00	19.50
		90	255	11.2	3.92	19.20
	500 × 250 mm	100	200	20.0	5.00	20.00
		90	205	19.5	4.88	19.50
		80	210	19.1	4.76	19.00
		70	215	18.6	4.65	18.60
	400 × 200 mm	90	155	32.3	6.45	20.80
		80	160	31.3	6.25	20.20
		70	165	30.3	6.06	19.60
CONCRETE TILES/SLATES						
Redland Roofing						
Stonewold slate	430 × 380 mm	75	355	8.2	2.82	51.20
Double Roman tile	418 × 330 mm	75	355	8.2	2.91	45.50
Grovebury pantile	418 × 332 mm	75	343	9.7	2.91	47.90
Norfolk pantile	381 × 227 mm	75	306	16.3	3.26	44.01
		100	281	17.8	3.56	48.06
Renown interlocking tile	418 × 330 mm	75	343	9.7	2.91	46.40
'49' tile	381 × 227 mm	75	306	16.3	3.26	44.80
		100	281	17.8	3.56	48.95
Plain, vertical tiling	265 × 165 mm	35	115	52.7	8.70	62.20
Marley Roofing						
Bold roll tile	420 × 330 mm	75	344	9.7	2.90	47.00
		100	–	10.5	3.20	51.00
Modern roof tile	420 × 330 mm	75	338	10.2	3.00	54.00
		100	–	11.0	3.20	58.00
Ludlow major	420 × 330 mm	75	338	10.2	3.00	45.00
		100	–	11.0	3.20	49.00
Ludlow plus	387 × 229 mm	75	305	16.1	3.30	47.00
		100	–	17.5	3.60	51.00
Mendip tile	420 × 330 mm	75	338	10.2	3.00	47.00
		100	–	11.0	3.20	51.00
Wessex	413 × 330 mm	75	338	10.2	3.00	54.00
		100	–	11.0	3.20	58.00
Plain tile	267 × 165 mm	65	100	60.0	10.00	76.00
		75	95	64.0	10.50	81.00
		85	90	68.0	11.30	86.00
Plain vertical tiles (feature)	267 × 165 mm	35	110	53.0	8.70	67.00
		34	115	56.0	9.10	71.00

ROOFING

Slate Nails, Quantity per Kilogram

Length	Type			
	Plain wire	Galvanized wire	Copper nail	Zinc nail
28.5 mm	325	305	325	415
34.4 mm	286	256	254	292
50.8 mm	242	224	194	200

Metal Sheet Coverings

Thicknesses and weights of sheet metal coverings

Lead to BS 1178								
BS Code No	3	4	5	6	7	8		
Colour code	Green	Blue	Red	Black	White	Orange		
Thickness (mm)	1.25	1.80	2.24	2.50	3.15	3.55		
Density kg/m^2	14.18	20.41	25.40	30.05	35.72	40.26		
Copper to BS 2870								
Thickness (mm)		0.60	0.70					
Bay width								
Roll (mm)		500	650					
Seam (mm)		525	600					
Standard width to form bay	600	750						
Normal length of sheet	1.80	1.80						
Zinc to BS 849								
Zinc Gauge (Nr)	9	10	11	12	13	14	15	16
Thickness (mm)	0.43	0.48	0.56	0.64	0.71	0.79	0.91	1.04
Density (kg/m^2)	3.1	3.2	3.8	4.3	4.8	5.3	6.2	7.0
Aluminium to BS 4868								
Thickness (mm)	0.5	0.6	0.7	0.8	0.9	1.0	1.2	
Density (kg/m^2)	12.8	15.4	17.9	20.5	23.0	25.6	30.7	

ROOFING

Type of felt	Nominal mass per unit area (kg/10 m)	Nominal mass per unit area of fibre base (g/m^2)	Nominal length of roll (m)
Class 1			
1B fine granule	14	220	10 or 20
surfaced bitumen	18	330	10 or 20
	25	470	10
1E mineral surfaced bitumen	38	470	10
1F reinforced bitumen	15	160 (fibre) 110 (hessian)	15
1F reinforced bitumen, aluminium faced	13	160 (fibre) 110 (hessian)	15
Class 2			
2B fine granule surfaced bitumen asbestos	18	500	10 or 20
2E mineral surfaced bitumen asbestos	38	600	10
Class 3			
3B fine granule surfaced bitumen glass fibre	18	60	20
3E mineral surfaced bitumen glass fibre	28	60	10
3E venting base layer bitumen glass fibre	32	60*	10
3H venting base layer bitumen glass fibre	17	60*	20

* Excluding effect of perforations

GLAZING

GLAZING

Nominal thickness (mm)	Tolerance on thickness (mm)	Approximate weight (kg/m²)	Normal maximum size (mm)
Float and polished plate glass			
3	+ 0.2	7.50	2140 × 1220
4	+ 0.2	10.00	2760 × 1220
5	+ 0.2	12.50	3180 × 2100
6	+ 0.2	15.00	4600 × 3180
10	+ 0.3	25.00)	6000 × 3300
12	+ 0.3	30.00)	
15	+ 0.5	37.50	3050 × 3000
19	+ 1.0	47.50)	3000 × 2900
25	+ 1.0	63.50)	
Clear sheet glass			
2 *	+ 0.2	5.00	1920 × 1220
3	+ 0.3	7.50	2130 × 1320
4	+ 0.3	10.00	2760 × 1220
5 *	+ 0.3	12.50)	2130 × 2400
6 *	+ 0.3	15.00)	
Cast glass			
3	+ 0.4		
	− 0.2	6.00)	2140 × 1280
4	+ 0.5	7.50)	
5	+ 0.5	9.50	2140 × 1320
6	+ 0.5	11.50)	3700 × 1280
10	+ 0.8	21.50)	
Wired glass			
(Cast wired glass)			
6	+ 0.3	−)	
	− 0.7	)	3700 × 1840
7	+ 0.7	−)	
(Polished wire glass)			
6	+ 1.0	−	330 × 1830

* The 5 mm and 6 mm thickness are known as *thick drawn sheet*. Although 2 mm sheet glass is available it is not recommended for general glazing purposes

METAL

Weights of Metals

Material	kg/m³	lb/cu ft
Metals, steel construction, etc.		
Iron		
– cast	7207	450
– wrought	7687	480
– ore – general	2407	150
– (crushed) Swedish	3682	230
Steel	7854	490
Copper		
– cast	8731	545
– wrought	8945	558
Brass	8497	530
Bronze	8945	558
Aluminium	2774	173
Lead	11322	707
Zinc (rolled)	7140	446
	g/mm² per metre	**lb/sq ft per foot**
Steel bars	7.85	3.4
Structural steelwork	Net weight of member @ 7854 kg/m³	
riveted	+ 10% for cleats, rivets, bolts, etc.	
welded	+ 1.25% to 2.5% for welds, etc.	
Rolled sections		
beams	+ 2.5%	
stanchions	+ 5% (extra for caps and bases)	
Plate		
web girders	+ 10% for rivets or welds, stiffeners, etc.	
	kg/m	**lb/ft**
Steel stairs: industrial type		
1 m or 3 ft wide	84	56
Steel tubes		
50 mm or 2 in bore	5 to 6	3 to 4
Gas piping		
20 mm or ¾ in	2	1¼

METAL

Universal Beams BS 4: Part 1: 2005

Designation	Mass (kg/m)	Depth of section (mm)	Width of section (mm)	Thickness		Surface area (m²/m)
				Web (mm)	Flange (mm)	
1016 × 305 × 487	487.0	1036.1	308.5	30.0	54.1	3.20
1016 × 305 × 438	438.0	1025.9	305.4	26.9	49.0	3.17
1016 × 305 × 393	393.0	1016.0	303.0	24.4	43.9	3.15
1016 × 305 × 349	349.0	1008.1	302.0	21.1	40.0	3.13
1016 × 305 × 314	314.0	1000.0	300.0	19.1	35.9	3.11
1016 × 305 × 272	272.0	990.1	300.0	16.5	31.0	3.10
1016 × 305 × 249	249.0	980.2	300.0	16.5	26.0	3.08
1016 × 305 × 222	222.0	970.3	300.0	16.0	21.1	3.06
914 × 419 × 388	388.0	921.0	420.5	21.4	36.6	3.44
914 × 419 × 343	343.3	911.8	418.5	19.4	32.0	3.42
914 × 305 × 289	289.1	926.6	307.7	19.5	32.0	3.01
914 × 305 × 253	253.4	918.4	305.5	17.3	27.9	2.99
914 × 305 × 224	224.2	910.4	304.1	15.9	23.9	2.97
914 × 305 × 201	200.9	903.0	303.3	15.1	20.2	2.96
838 × 292 × 226	226.5	850.9	293.8	16.1	26.8	2.81
838 × 292 × 194	193.8	840.7	292.4	14.7	21.7	2.79
838 × 292 × 176	175.9	834.9	291.7	14.0	18.8	2.78
762 × 267 × 197	196.8	769.8	268.0	15.6	25.4	2.55
762 × 267 × 173	173.0	762.2	266.7	14.3	21.6	2.53
762 × 267 × 147	146.9	754.0	265.2	12.8	17.5	2.51
762 × 267 × 134	133.9	750.0	264.4	12.0	15.5	2.51
686 × 254 × 170	170.2	692.9	255.8	14.5	23.7	2.35
686 × 254 × 152	152.4	687.5	254.5	13.2	21.0	2.34
686 × 254 × 140	140.1	383.5	253.7	12.4	19.0	2.33
686 × 254 × 125	125.2	677.9	253.0	11.7	16.2	2.32
610 × 305 × 238	238.1	635.8	311.4	18.4	31.4	2.45
610 × 305 × 179	179.0	620.2	307.1	14.1	23.6	2.41
610 × 305 × 149	149.1	612.4	304.8	11.8	19.7	2.39
610 × 229 × 140	139.9	617.2	230.2	13.1	22.1	2.11
610 × 229 × 125	125.1	612.2	229.0	11.9	19.6	2.09
610 × 229 × 113	113.0	607.6	228.2	11.1	17.3	2.08
610 × 229 × 101	101.2	602.6	227.6	10.5	14.8	2.07
533 × 210 × 122	122.0	544.5	211.9	12.7	21.3	1.89
533 × 210 × 109	109.0	539.5	210.8	11.6	18.8	1.88
533 × 210 × 101	101.0	536.7	210.0	10.8	17.4	1.87
533 × 210 × 92	92.1	533.1	209.3	10.1	15.6	1.86
533 × 210 × 82	82.2	528.3	208.8	9.6	13.2	1.85
457 × 191 × 98	98.3	467.2	192.8	11.4	19.6	1.67
457 × 191 × 89	89.3	463.4	191.9	10.5	17.7	1.66
457 × 191 × 82	82.0	460.0	191.3	9.9	16.0	1.65
457 × 191 × 74	74.3	457.0	190.4	9.0	14.5	1.64
457 × 191 × 67	67.1	453.4	189.9	8.5	12.7	1.63
457 × 152 × 82	82.1	465.8	155.3	10.5	18.9	1.51
457 × 152 × 74	74.2	462.0	154.4	9.6	17.0	1.50
457 × 152 × 67	67.2	458.0	153.8	9.0	15.0	1.50
457 × 152 × 60	59.8	454.6	152.9	8.1	13.3	1.50
457 × 152 × 52	52.3	449.8	152.4	7.6	10.9	1.48
406 × 178 × 74	74.2	412.8	179.5	9.5	16.0	1.51
406 × 178 × 67	67.1	409.4	178.8	8.8	14.3	1.50
406 × 178 × 60	60.1	406.4	177.9	7.9	12.8	1.49

METAL

Designation	Mass (kg/m)	Depth of section (mm)	Width of section (mm)	Thickness		Surface area (m²/m)
				Web (mm)	Flange (mm)	
406 × 178 × 50	54.1	402.6	177.7	7.7	10.9	1.48
406 × 140 × 46	46.0	403.2	142.2	6.8	11.2	1.34
406 × 140 × 39	39.0	398.0	141.8	6.4	8.6	1.33
356 × 171 × 67	67.1	363.4	173.2	9.1	15.7	1.38
356 × 171 × 57	57.0	358.0	172.2	8.1	13.0	1.37
356 × 171 × 51	51.0	355.0	171.5	7.4	11.5	1.36
356 × 171 × 45	45.0	351.4	171.1	7.0	9.7	1.36
356 × 127 × 39	39.1	353.4	126.0	6.6	10.7	1.18
356 × 127 × 33	33.1	349.0	125.4	6.0	8.5	1.17
305 × 165 × 54	54.0	310.4	166.9	7.9	13.7	1.26
305 × 165 × 46	46.1	306.6	165.7	6.7	11.8	1.25
305 × 165 × 40	40.3	303.4	165.0	6.0	10.2	1.24
305 × 127 × 48	48.1	311.0	125.3	9.0	14.0	1.09
305 × 127 × 42	41.9	307.2	124.3	8.0	12.1	1.08
305 × 127 × 37	37.0	304.4	123.3	7.1	10.7	1.07
305 × 102 × 33	32.8	312.7	102.4	6.6	10.8	1.01
305 × 102 × 28	28.2	308.7	101.8	6.0	8.8	1.00
305 × 102 × 25	24.8	305.1	101.6	5.8	7.0	0.992
254 × 146 × 43	43.0	259.6	147.3	7.2	12.7	1.08
254 × 146 × 37	37.0	256.0	146.4	6.3	10.9	1.07
254 × 146 × 31	31.1	251.4	146.1	6.0	8.6	1.06
254 × 102 × 28	28.3	260.4	102.2	6.3	10.0	0.904
254 × 102 × 25	25.2	257.2	101.9	6.0	8.4	0.897
254 × 102 × 22	22.0	254.0	101.6	5.7	6.8	0.890
203 × 133 × 30	30.0	206.8	133.9	6.4	9.6	0.923
203 × 133 × 25	25.1	203.2	133.2	5.7	7.8	0.915
203 × 102 × 23	23.1	203.2	101.8	5.4	9.3	0.790
178 × 102 × 19	19.0	177.8	101.2	4.8	7.9	0.738
152 × 89 × 16	16.0	152.4	88.7	4.5	7.7	0.638
127 × 76 × 13	13.0	127.0	76.0	4.0	7.6	0.537

METAL

Universal Columns BS 4: Part 1: 2005

Designation	Mass (kg/m)	Depth of section (mm)	Width of section (mm)	Thickness		Surface area (m²/m)
				Web (mm)	Flange (mm)	
356 × 406 × 634	633.9	474.7	424.0	47.6	77.0	2.52
356 × 406 × 551	551.0	455.6	418.5	42.1	67.5	2.47
356 × 406 × 467	467.0	436.6	412.2	35.8	58.0	2.42
356 × 406 × 393	393.0	419.0	407.0	30.6	49.2	2.38
356 × 406 × 340	339.9	406.4	403.0	26.6	42.9	2.35
356 × 406 × 287	287.1	393.6	399.0	22.6	36.5	2.31
356 × 406 × 235	235.1	381.0	384.8	18.4	30.2	2.28
356 × 368 × 202	201.9	374.6	374.7	16.5	27.0	2.19
356 × 368 × 177	177.0	368.2	372.6	14.4	23.8	2.17
356 × 368 × 153	152.9	362.0	370.5	12.3	20.7	2.16
356 × 368 × 129	129.0	355.6	368.6	10.4	17.5	2.14
305 × 305 × 283	282.9	365.3	322.2	26.8	44.1	1.94
305 × 305 × 240	240.0	352.5	318.4	23.0	37.7	1.91
305 × 305 × 198	198.1	339.9	314.5	19.1	31.4	1.87
305 × 305 × 158	158.1	327.1	311.2	15.8	25.0	1.84
305 × 305 × 137	136.9	320.5	309.2	13.8	21.7	1.82
305 × 305 × 118	117.9	314.5	307.4	12.0	18.7	1.81
305 × 305 × 97	96.9	307.9	305.3	9.9	15.4	1.79
254 × 254 × 167	167.1	289.1	265.2	19.2	31.7	1.58
254 × 254 × 132	132.0	276.3	261.3	15.3	25.3	1.55
254 × 254 × 107	107.1	266.7	258.8	12.8	20.5	1.52
254 × 254 × 89	88.9	260.3	256.3	10.3	17.3	1.50
254 × 254 × 73	73.1	254.1	254.6	8.6	14.2	1.49
203 × 203 × 86	86.1	222.2	209.1	12.7	20.5	1.24
203 × 203 × 71	71.0	215.8	206.4	10.0	17.3	1.22
203 × 203 × 60	60.0	209.6	205.8	9.4	14.2	1.21
203 × 203 × 52	52.0	206.2	204.3	7.9	12.5	1.20
203 × 203 × 46	46.1	203.2	203.6	7.2	11.0	1.19
152 × 152 × 37	37.0	161.8	154.4	8.0	11.5	0.912
152 × 152 × 30	30.0	157.6	152.9	6.5	9.4	0.901
152 × 152 × 23	23.0	152.4	152.2	5.8	6.8	0.889

METAL

Joists BS 4: Part 1: 2005 (retained for reference, Corus have ceased manufacture in UK)

Designation	Mass (kg/m)	Depth of section (mm)	Width of section (mm)	Thickness		Surface area (m²/m)
				Web (mm)	Flange (mm)	
254 × 203 × 82	82.0	254.0	203.2	10.2	19.9	1.210
203 × 152 × 52	52.3	203.2	152.4	8.9	16.5	0.932
152 × 127 × 37	37.3	152.4	127.0	10.4	13.2	0.737
127 × 114 × 29	29.3	127.0	114.3	10.2	11.5	0.646
127 × 114 × 27	26.9	127.0	114.3	7.4	11.4	0.650
102 × 102 × 23	23.0	101.6	101.6	9.5	10.3	0.549
102 × 44 × 7	7.5	101.6	44.5	4.3	6.1	0.350
89 × 89 × 19	19.5	88.9	88.9	9.5	9.9	0.476
76 × 76 × 13	12.8	76.2	76.2	5.1	8.4	0.411

Parallel Flange Channels

Designation	Mass (kg/m)	Depth of section (mm)	Width of section (mm)	Thickness		Surface area (m²/m)
				Web (mm)	Flange (mm)	
430 × 100 × 64	64.4	430	100	11.0	19.0	1.23
380 × 100 × 54	54.0	380	100	9.5	17.5	1.13
300 × 100 × 46	45.5	300	100	9.0	16.5	0.969
300 × 90 × 41	41.4	300	90	9.0	15.5	0.932
260 × 90 × 35	34.8	260	90	8.0	14.0	0.854
260 × 75 × 28	27.6	260	75	7.0	12.0	0.79
230 × 90 × 32	32.2	230	90	7.5	14.0	0.795
230 × 75 × 26	25.7	230	75	6.5	12.5	0.737
200 × 90 × 30	29.7	200	90	7.0	14.0	0.736
200 × 75 × 23	23.4	200	75	6.0	12.5	0.678
180 × 90 × 26	26.1	180	90	6.5	12.5	0.697
180 × 75 × 20	20.3	180	75	6.0	10.5	0.638
150 × 90 × 24	23.9	150	90	6.5	12.0	0.637
150 × 75 × 18	17.9	150	75	5.5	10.0	0.579
125 × 65 × 15	14.8	125	65	5.5	9.5	0.489
100 × 50 × 10	10.2	100	50	5.0	8.5	0.382

METAL

Equal Angles BS EN 10056-1

Designation	Mass (kg/m)	Surface area (m²/m)
200 × 200 × 24	71.1	0.790
200 × 200 × 20	59.9	0.790
200 × 200 × 18	54.2	0.790
200 × 200 × 16	48.5	0.790
150 × 150 × 18	40.1	0.59
150 × 150 × 15	33.8	0.59
150 × 150 × 12	27.3	0.59
150 × 150 × 10	23.0	0.59
120 × 120 × 15	26.6	0.47
120 × 120 × 12	21.6	0.47
120 × 120 × 10	18.2	0.47
120 × 120 × 8	14.7	0.47
100 × 100 × 15	21.9	0.39
100 × 100 × 12	17.8	0.39
100 × 100 × 10	15.0	0.39
100 × 100 × 8	12.2	0.39
90 × 90 × 12	15.9	0.35
90 × 90 × 10	13.4	0.35
90 × 90 × 8	10.9	0.35
90 × 90 × 7	9.61	0.35
90 × 90 × 6	8.30	0.35

Unequal Angles BS EN 10056-1

Designation	Mass (kg/m)	Surface area (m²/m)
200 × 150 × 18	47.1	0.69
200 × 150 × 15	39.6	0.69
200 × 150 × 12	32.0	0.69
200 × 100 × 15	33.7	0.59
200 × 100 × 12	27.3	0.59
200 × 100 × 10	23.0	0.59
150 × 90 × 15	26.6	0.47
150 × 90 × 12	21.6	0.47
150 × 90 × 10	18.2	0.47
150 × 75 × 15	24.8	0.44
150 × 75 × 12	20.2	0.44
150 × 75 × 10	17.0	0.44
125 × 75 × 12	17.8	0.40
125 × 75 × 10	15.0	0.40
125 × 75 × 8	12.2	0.40
100 × 75 × 12	15.4	0.34
100 × 75 × 10	13.0	0.34
100 × 75 × 8	10.6	0.34
100 × 65 × 10	12.3	0.32
100 × 65 × 8	9.94	0.32
100 × 65 × 7	8.77	0.32

METAL

Structural Tees Split from Universal Beams BS 4: Part 1: 2005

Designation	Mass (kg/m)	Surface area (m²/m)
305 × 305 × 90	89.5	1.22
305 × 305 × 75	74.6	1.22
254 × 343 × 63	62.6	1.19
229 × 305 × 70	69.9	1.07
229 × 305 × 63	62.5	1.07
229 × 305 × 57	56.5	1.07
229 × 305 × 51	50.6	1.07
210 × 267 × 61	61.0	0.95
210 × 267 × 55	54.5	0.95
210 × 267 × 51	50.5	0.95
210 × 267 × 46	46.1	0.95
210 × 267 × 41	41.1	0.95
191 × 229 × 49	49.2	0.84
191 × 229 × 45	44.6	0.84
191 × 229 × 41	41.0	0.84
191 × 229 × 37	37.1	0.84
191 × 229 × 34	33.6	0.84
152 × 229 × 41	41.0	0.76
152 × 229 × 37	37.1	0.76
152 × 229 × 34	33.6	0.76
152 × 229 × 30	29.9	0.76
152 × 229 × 26	26.2	0.76

Universal Bearing Piles BS 4: Part 1: 2005

Designation	Mass (kg/m)	Depth of Section (mm)	Width of section (mm)	Thickness Web (mm)	Flange (mm)
356 × 368 × 174	173.9	361.4	378.5	20.3	20.4
356 × 368 × 152	152.0	356.4	376.0	17.8	17.9
356 × 368 × 133	133.0	352.0	373.8	15.6	15.7
356 × 368 × 109	108.9	346.4	371.0	12.8	12.9
305 × 305 × 223	222.9	337.9	325.7	30.3	30.4
305 × 305 × 186	186.0	328.3	320.9	25.5	25.6
305 × 305 × 149	149.1	318.5	316.0	20.6	20.7
305 × 305 × 126	126.1	312.3	312.9	17.5	17.6
305 × 305 × 110	110.0	307.9	310.7	15.3	15.4
305 × 305 × 95	94.9	303.7	308.7	13.3	13.3
305 × 305 × 88	88.0	301.7	307.8	12.4	12.3
305 × 305 × 79	78.9	299.3	306.4	11.0	11.1
254 × 254 × 85	85.1	254.3	260.4	14.4	14.3
254 × 254 × 71	71.0	249.7	258.0	12.0	12.0
254 × 254 × 63	63.0	247.1	256.6	10.6	10.7
203 × 203 × 54	53.9	204.0	207.7	11.3	11.4
203 × 203 × 45	44.9	200.2	205.9	9.5	9.5

METAL

Hot Formed Square Hollow Sections EN 10210 S275J2H & S355J2H

Size (mm)	Wall thickness (mm)	Mass (kg/m)	Superficial area (m²/m)
40 × 40	2.5	2.89	0.154
	3.0	3.41	0.152
	3.2	3.61	0.152
	3.6	4.01	0.151
	4.0	4.39	0.150
	5.0	5.28	0.147
50 × 50	2.5	3.68	0.194
	3.0	4.35	0.192
	3.2	4.62	0.192
	3.6	5.14	0.191
	4.0	5.64	0.190
	5.0	6.85	0.187
	6.0	7.99	0.185
	6.3	8.31	0.184
60 × 60	3.0	5.29	0.232
	3.2	5.62	0.232
	3.6	6.27	0.231
	4.0	6.90	0.230
	5.0	8.42	0.227
	6.0	9.87	0.225
	6.3	10.30	0.224
	8.0	12.50	0.219
70 × 70	3.0	6.24	0.272
	3.2	6.63	0.272
	3.6	7.40	0.271
	4.0	8.15	0.270
	5.0	9.99	0.267
	6.0	11.80	0.265
	6.3	12.30	0.264
	8.0	15.00	0.259
80 × 80	3.2	7.63	0.312
	3.6	8.53	0.311
	4.0	9.41	0.310
	5.0	11.60	0.307
	6.0	13.60	0.305
	6.3	14.20	0.304
	8.0	17.50	0.299
90 × 90	3.6	9.66	0.351
	4.0	10.70	0.350
	5.0	13.10	0.347
	6.0	15.50	0.345
	6.3	16.20	0.344
	8.0	20.10	0.339
100 × 100	3.6	10.80	0.391
	4.0	11.90	0.390
	5.0	14.70	0.387
	6.0	17.40	0.385
	6.3	18.20	0.384
	8.0	22.60	0.379
	10.0	27.40	0.374
120 × 120	4.0	14.40	0.470
	5.0	17.80	0.467
	6.0	21.20	0.465

Tables and Memoranda

METAL

Size (mm)	Wall thickness (mm)	Mass (kg/m)	Superficial area (m²/m)
	6.3	22.20	0.464
	8.0	27.60	0.459
	10.0	33.70	0.454
	12.0	39.50	0.449
	12.5	40.90	0.448
140 × 140	5.0	21.00	0.547
	6.0	24.90	0.545
	6.3	26.10	0.544
	8.0	32.60	0.539
	10.0	40.00	0.534
	12.0	47.00	0.529
	12.5	48.70	0.528
150 × 150	5.0	22.60	0.587
	6.0	26.80	0.585
	6.3	28.10	0.584
	8.0	35.10	0.579
	10.0	43.10	0.574
	12.0	50.80	0.569
	12.5	52.70	0.568
Hot formed from seamless hollow	16.0	65.2	0.559
160 × 160	5.0	24.10	0.627
	6.0	28.70	0.625
	6.3	30.10	0.624
	8.0	37.60	0.619
	10.0	46.30	0.614
	12.0	54.60	0.609
	12.5	56.60	0.608
	16.0	70.20	0.599
180 × 180	5.0	27.30	0.707
	6.0	32.50	0.705
	6.3	34.00	0.704
	8.0	42.70	0.699
	10.0	52.50	0.694
	12.0	62.10	0.689
	12.5	64.40	0.688
	16.0	80.20	0.679
200 × 200	5.0	30.40	0.787
	6.0	36.20	0.785
	6.3	38.00	0.784
	8.0	47.70	0.779
	10.0	58.80	0.774
	12.0	69.60	0.769
	12.5	72.30	0.768
	16.0	90.30	0.759
250 × 250	5.0	38.30	0.987
	6.0	45.70	0.985
	6.3	47.90	0.984
	8.0	60.30	0.979
	10.0	74.50	0.974
	12.0	88.50	0.969
	12.5	91.90	0.968
	16.0	115.00	0.959

METAL

Size (mm)	Wall thickness (mm)	Mass (kg/m)	Superficial area (m²/m)
300 × 300	6.0	55.10	1.18
	6.3	57.80	1.18
	8.0	72.80	1.18
	10.0	90.20	1.17
	12.0	107.00	1.17
	12.5	112.00	1.17
	16.0	141.00	1.16
350 × 350	8.0	85.40	1.38
	10.0	106.00	1.37
	12.0	126.00	1.37
	12.5	131.00	1.37
	16.0	166.00	1.36
400 × 400	8.0	97.90	1.58
	10.0	122.00	1.57
	12.0	145.00	1.57
	12.5	151.00	1.57
	16.0	191.00	1.56
(Grade S355J2H only)	20.00*	235.00	1.55

Note: * SAW process

METAL

Hot Formed Square Hollow Sections JUMBO RHS: JIS G3136

Size (mm)	Wall thickness (mm)	Mass (kg/m)	Superficial area (m²/m)
350 × 350	19.0	190.00	1.33
	22.0	217.00	1.32
	25.0	242.00	1.31
400 × 400	22.0	251.00	1.52
	25.0	282.00	1.51
450 × 450	12.0	162.00	1.76
	16.0	213.00	1.75
	19.0	250.00	1.73
	22.0	286.00	1.72
	25.0	321.00	1.71
	28.0 *	355.00	1.70
	32.0 *	399.00	1.69
500 × 500	12.0	181.00	1.96
	16.0	238.00	1.95
	19.0	280.00	1.93
	22.0	320.00	1.92
	25.0	360.00	1.91
	28.0 *	399.00	1.90
	32.0 *	450.00	1.89
	36.0 *	498.00	1.88
550 × 550	16.0	263.00	2.15
	19.0	309.00	2.13
	22.0	355.00	2.12
	25.0	399.00	2.11
	28.0 *	443.00	2.10
	32.0 *	500.00	2.09
	36.0 *	555.00	2.08
	40.0 *	608.00	2.06
600 × 600	25.0 *	439.00	2.31
	28.0 *	487.00	2.30
	32.0 *	550.00	2.29
	36.0 *	611.00	2.28
	40.0 *	671.00	2.26
700 × 700	25.0 *	517.00	2.71
	28.0 *	575.00	2.70
	32.0 *	651.00	2.69
	36.0 *	724.00	2.68
	40.0 *	797.00	2.68

Note: * SAW process

METAL

Hot Formed Rectangular Hollow Sections: EN10210 S275J2h & S355J2H

Size (mm)	Wall thickness (mm)	Mass (kg/m)	Superficial area (m²/m)
50 × 30	2.5	2.89	0.154
	3.0	3.41	0.152
	3.2	3.61	0.152
	3.6	4.01	0.151
	4.0	4.39	0.150
	5.0	5.28	0.147
60 × 40	2.5	3.68	0.194
	3.0	4.35	0.192
	3.2	4.62	0.192
	3.6	5.14	0.191
	4.0	5.64	0.190
	5.0	6.85	0.187
	6.0	7.99	0.185
	6.3	8.31	0.184
80 × 40	3.0	5.29	0.232
	3.2	5.62	0.232
	3.6	6.27	0.231
	4.0	6.90	0.230
	5.0	8.42	0.227
	6.0	9.87	0.225
	6.3	10.30	0.224
	8.0	12.50	0.219
76.2 × 50.8	3.0	5.62	0.246
	3.2	5.97	0.246
	3.6	6.66	0.245
	4.0	7.34	0.244
	5.0	8.97	0.241
	6.0	10.50	0.239
	6.3	11.00	0.238
	8.0	13.40	0.233
90 × 50	3.0	6.24	0.272
	3.2	6.63	0.272
	3.6	7.40	0.271
	4.0	8.15	0.270
	5.0	9.99	0.267
	6.0	11.80	0.265
	6.3	12.30	0.264
	8.0	15.00	0.259
100 × 50	3.0	6.71	0.292
	3.2	7.13	0.292
	3.6	7.96	0.291
	4.0	8.78	0.290
	5.0	10.80	0.287
	6.0	12.70	0.285
	6.3	13.30	0.284
	8.0	16.30	0.279

METAL

Size (mm)	Wall thickness (mm)	Mass (kg/m)	Superficial area (m²/m)
100 × 60	3.0	7.18	0.312
	3.2	7.63	0.312
	3.6	8.53	0.311
	4.0	9.41	0.310
	5.0	11.60	0.307
	6.0	13.60	0.305
	6.3	14.20	0.304
	8.0	17.50	0.299
120 × 60	3.6	9.70	0.351
	4.0	10.70	0.350
	5.0	13.10	0.347
	6.0	15.50	0.345
	6.3	16.20	0.344
	8.0	20.10	0.339
120 × 80	3.6	10.80	0.391
	4.0	11.90	0.390
	5.0	14.70	0.387
	6.0	17.40	0.385
	6.3	18.20	0.384
	8.0	22.60	0.379
	10.0	27.40	0.374
150 × 100	4.0	15.10	0.490
	5.0	18.60	0.487
	6.0	22.10	0.485
	6.3	23.10	0.484
	8.0	28.90	0.479
	10.0	35.30	0.474
	12.0	41.40	0.469
	12.5	42.80	0.468
160 × 80	4.0	14.40	0.470
	5.0	17.80	0.467
	6.0	21.20	0.465
	6.3	22.20	0.464
	8.0	27.60	0.459
	10.0	33.70	0.454
	12.0	39.50	0.449
	12.5	40.90	0.448
200 × 100	5.0	22.60	0.587
	6.0	26.80	0.585
	6.3	28.10	0.584
	8.0	35.10	0.579
	10.0	43.10	0.574
	12.0	50.80	0.569
	12.5	52.70	0.568
	16.0	65.20	0.559
250 × 150	5.0	30.40	0.787
	6.0	36.20	0.785
	6.3	38.00	0.784
	8.0	47.70	0.779
	10.0	58.80	0.774
	12.0	69.60	0.769
	12.5	72.30	0.768
	16.0	90.30	0.759

METAL

Size (mm)	Wall thickness (mm)	Mass (kg/m)	Superficial area (m²/m)
300 × 200	5.0	38.30	0.987
	6.0	45.70	0.985
	6.3	47.90	0.984
	8.0	60.30	0.979
	10.0	74.50	0.974
	12.0	88.50	0.969
	12.5	91.90	0.968
	16.0	115.00	0.959
400 × 200	6.0	55.10	1.18
	6.3	57.80	1.18
	8.0	72.80	1.18
	10.0	90.20	1.17
	12.0	107.00	1.17
	12.5	112.00	1.17
	16.0	141.00	1.16
450 × 250	8.0	85.40	1.38
	10.0	106.00	1.37
	12.0	126.00	1.37
	12.5	131.00	1.37
	16.0	166.00	1.36
500 × 300	8.0	98.00	1.58
	10.0	122.00	1.57
	12.0	145.00	1.57
	12.5	151.00	1.57
	16.0	191.00	1.56
	20.0	235.00	1.55

METAL

Hot Formed Circular Hollow Sections EN 10210 S275J2H & S355J2H

Outside diameter (mm)	Wall thickness (mm)	Mass (kg/m)	Superficial area (m²/m)
21.3	3.2	1.43	0.067
26.9	3.2	1.87	0.085
33.7	3.0	2.27	0.106
	3.2	2.41	0.106
	3.6	2.67	0.106
	4.0	2.93	0.106
42.4	3.0	2.91	0.133
	3.2	3.09	0.133
	3.6	3.44	0.133
	4.0	3.79	0.133
48.3	2.5	2.82	0.152
	3.0	3.35	0.152
	3.2	3.56	0.152
	3.6	3.97	0.152
	4.0	4.37	0.152
	5.0	5.34	0.152
60.3	2.5	3.56	0.189
	3.0	4.24	0.189
	3.2	4.51	0.189
	3.6	5.03	0.189
	4.0	5.55	0.189
	5.0	6.82	0.189
76.1	2.5	4.54	0.239
	3.0	5.41	0.239
	3.2	5.75	0.239
	3.6	6.44	0.239
	4.0	7.11	0.239
	5.0	8.77	0.239
	6.0	10.40	0.239
	6.3	10.80	0.239
88.9	2.5	5.33	0.279
	3.0	6.36	0.279
	3.2	6.76	0.27
	3.6	7.57	0.279
	4.0	8.38	0.279
	5.0	10.30	0.279
	6.0	12.30	0.279
	6.3	12.80	0.279
114.3	3.0	8.23	0.359
	3.2	8.77	0.359
	3.6	9.83	0.359
	4.0	10.09	0.359
	5.0	13.50	0.359
	6.0	16.00	0.359
	6.3	16.80	0.359

METAL

Outside diameter (mm)	Wall thickness (mm)	Mass (kg/m)	Superficial area (m²/m)
139.7	3.2	10.80	0.439
	3.6	12.10	0.439
	4.0	13.40	0.439
	5.0	16.60	0.439
	6.0	19.80	0.439
	6.3	20.70	0.439
	8.0	26.00	0.439
	10.0	32.00	0.439
168.3	3.2	13.00	0.529
	3.6	14.60	0.529
	4.0	16.20	0.529
	5.0	20.10	0.529
	6.0	24.00	0.529
	6.3	25.20	0.529
	8.0	31.60	0.529
	10.0	39.00	0.529
	12.0	46.30	0.529
	12.5	48.00	0.529
193.7	5.0	23.30	0.609
	6.0	27.80	0.609
	6.3	29.10	0.609
	8.0	36.60	0.609
	10.0	45.30	0.609
	12.0	53.80	0.609
	12.5	55.90	0.609
219.1	5.0	26.40	0.688
	6.0	31.50	0.688
	6.3	33.10	0.688
	8.0	41.60	0.688
	10.0	51.60	0.688
	12.0	61.30	0.688
	12.5	63.70	0.688
	16.0	80.10	0.688
244.5	5.0	29.50	0.768
	6.0	35.30	0.768
	6.3	37.00	0.768
	8.0	46.70	0.768
	10.0	57.80	0.768
	12.0	68.80	0.768
	12.5	71.50	0.768
	16.0	90.20	0.768
273.0	5.0	33.00	0.858
	6.0	39.50	0.858
	6.3	41.40	0.858
	8.0	52.30	0.858
	10.0	64.90	0.858
	12.0	77.20	0.858
	12.5	80.30	0.858
	16.0	101.00	0.858

METAL

Outside diameter (mm)	Wall thickness (mm)	Mass (kg/m)	Superficial area (m²/m)
323.9	5.0	39.30	1.02
	6.0	47.00	1.02
	6.3	49.30	1.02
	8.0	62.30	1.02
	10.0	77.40	1.02
	12.0	92.30	1.02
	12.5	96.00	1.02
	16.0	121.00	1.02
355.6	6.3	54.30	1.12
	8.0	68.60	1.12
	10.0	85.30	1.12
	12.0	102.00	1.12
	12.5	106.00	1.12
	16.0	134.00	1.12
406.4	6.3	62.20	1.28
	8.0	79.60	1.28
	10.0	97.80	1.28
	12.0	117.00	1.28
	12.5	121.00	1.28
	16.0	154.00	1.28
457.0	6.3	70.00	1.44
	8.0	88.60	1.44
	10.0	110.00	1.44
	12.0	132.00	1.44
	12.5	137.00	1.44
	16.0	174.00	1.44
508.0	6.3	77.90	1.60
	8.0	98.60	1.60
	10.0	123.00	1.60
	12.0	147.00	1.60
	12.5	153.00	1.60
	16.0	194.00	1.60

METAL

Spacing of Holes in Angles

Nominal leg length (mm)	Spacing of holes						Maximum diameter of bolt or rivet		
	A	B	C	D	E	F	A	B and C	D, E and F
200		75	75	55	55	55		30	20
150		55	55					20	
125		45	60					20	
120									
100	55						24		
90	50						24		
80	45						20		
75	45						20		
70	40						20		
65	35						20		
60	35						16		
50	28						12		
45	25								
40	23								
30	20								
25	15								

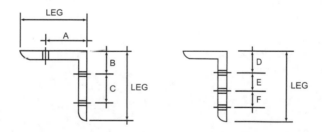

KERBS, PAVING, ETC.

KERBS/EDGINGS/CHANNELS

Precast Concrete Kerbs to BS 7263

Straight kerb units: length from 450 to 915 mm

150 mm high × 125 mm thick		
bullnosed	type BN	
half battered	type HB3	
255 mm high × 125 mm thick		
45° splayed	type SP	
half battered	type HB2	
305 mm high × 150 mm thick		
half battered	type HB1	
Quadrant kerb units		
150 mm high × 305 and 455 mm radius to match	type BN	type QBN
150 mm high × 305 and 455 mm radius to match	type HB2, HB3	type QHB
150 mm high × 305 and 455 mm radius to match	type SP	type QSP
255 mm high × 305 and 455 mm radius to match	type BN	type QBN
255 mm high × 305 and 455 mm radius to match	type HB2, HB3	type QHB
225 mm high × 305 and 455 mm radius to match	type SP	type QSP
Angle kerb units		
305 × 305 × 225 mm high × 125 mm thick		
bullnosed external angle	type XA	
splayed external angle to match type SP	type XA	
bullnosed internal angle	type IA	
splayed internal angle to match type SP	type IA	
Channels		
255 mm wide × 125 mm high flat	type CS1	
150 mm wide × 125 mm high flat type	CS2	
255 mm wide × 125 mm high dished	type CD	

KERBS, PAVING, ETC.

Transition kerb units			
from kerb type SP to HB	left handed	type TL	
	right handed	type TR	
from kerb type BN to HB	left handed	type DL1	
	right handed	type DR1	
from kerb type BN to SP	left handed	type DL2	
	right handed	type DR2	

Number of kerbs required per quarter circle (780 mm kerb lengths)

Radius (m)	Number in quarter circle
12	24
10	20
8	16
6	12
5	10
4	8
3	6
2	4
1	2

Precast Concrete Edgings

Round top type ER	Flat top type EF	Bullnosed top type EBN
150 × 50 mm	150 × 50 mm	150 × 50 mm
200 × 50 mm	200 × 50 mm	200 × 50 mm
250 × 50 mm	250 × 50 mm	250 × 50 mm

KERBS, PAVING, ETC.

BASES

Cement Bound Material for Bases and Subbases

CBM1:	very carefully graded aggregate from 37.5–75 mm, with a 7-day strength of 4.5 N/mm^2
CBM2:	same range of aggregate as CBM1 but with more tolerance in each size of aggregate with a 7-day strength of 7.0 N/mm^2
CBM3:	crushed natural aggregate or blast furnace slag, graded from 37.5–150 mm for 40 mm aggregate, and from 20–75 mm for 20 mm aggregate, with a 7-day strength of 10 N/mm^2
CBM4:	crushed natural aggregate or blast furnace slag, graded from 37.5–150 mm for 40 mm aggregate, and from 20–75 mm for 20 mm aggregate, with a 7-day strength of 15 N/mm^2

INTERLOCKING BRICK/BLOCK ROADS/PAVINGS

Sizes of Precast Concrete Paving Blocks

Type R blocks
200 × 100 × 60 mm
200 × 100 × 65 mm
200 × 100 × 80 mm
200 × 100 × 100 mm

Type S
Any shape within a 295 mm space

Sizes of clay brick pavers
200 × 100 × 50 mm
200 × 100 × 65 mm
210 × 105 × 50 mm
210 × 105 × 65 mm
215 × 102.5 × 50 mm
215 × 102.5 × 65 mm

Type PA: 3 kN
Footpaths and pedestrian areas, private driveways, car parks, light vehicle traffic and over-run

Type PB: 7 kN
Residential roads, lorry parks, factory yards, docks, petrol station forecourts, hardstandings, bus stations

KERBS, PAVING, ETC.

PAVING AND SURFACING

Weights and Sizes of Paving and Surfacing

Description of item	Size	Quantity per tonne
Paving 50 mm thick	900 × 600 mm	15
Paving 50 mm thick	750 × 600 mm	18
Paving 50 mm thick	600 × 600 mm	23
Paving 50 mm thick	450 × 600 mm	30
Paving 38 mm thick	600 × 600 mm	30
Path edging	914 × 50 × 150 mm	60
Kerb (including radius and tapers)	125 × 254 × 914 mm	15
Kerb (including radius and tapers)	125 × 150 × 914 mm	25
Square channel	125 × 254 × 914 mm	15
Dished channel	125 × 254 × 914 mm	15
Quadrants	300 × 300 × 254 mm	19
Quadrants	450 × 450 × 254 mm	12
Quadrants	300 × 300 × 150 mm	30
Internal angles	300 × 300 × 254 mm	30
Fluted pavement channel	255 × 75 × 914 mm	25
Corner stones	300 × 300 mm	80
Corner stones	360 × 360 mm	60
Cable covers	914 × 175 mm	55
Gulley kerbs	220 × 220 × 150 mm	60
Gulley kerbs	220 × 200 × 75 mm	120

KERBS, PAVING, ETC.

Weights and Sizes of Paving and Surfacing

Material	kg/m³	lb/cu yd
Tarmacadam	2306	3891
Macadam (waterbound)	2563	4325
Vermiculite (aggregate)	64–80	108–135
Terracotta	2114	3568
Cork – compressed	388	24
	kg/m²	**lb/sq ft**
Clay floor tiles, 12.7 mm	27.3	5.6
Pavement lights	122	25
Damp-proof course	5	1
	kg/m² per mm thickness	**lb/sq ft per inch thickness**
Paving slabs (stone)	2.3	12
Granite setts	2.88	15
Asphalt	2.30	12
Rubber flooring	1.68	9
Polyvinyl chloride	1.94 (avg)	10 (avg)

Coverage (m²) Per Cubic Metre of Materials Used as Subbases or Capping Layers

Consolidated thickness laid in (mm)	Square metre coverage		
	Gravel	Sand	Hardcore
50	15.80	16.50	–
75	10.50	11.00	–
100	7.92	8.20	7.42
125	6.34	6.60	5.90
150	5.28	5.50	4.95
175	–	–	4.23
200	–	–	3.71
225	–	–	3.30
300	–	–	2.47

KERBS, PAVING, ETC.

Approximate Rate of Spreads

Average thickness of course (mm)	Description	Approximate rate of spread			
		Open Textured		Dense, Medium & Fine Textured	
		(kg/m²)	(m²/t)	(kg/m²)	(m²/t)
35	14 mm open textured or dense wearing course	60–75	13–17	70–85	12–14
40	20 mm open textured or dense base course	70–85	12–14	80–100	10–12
45	20 mm open textured or dense base course	80–100	10–12	95–100	9–10
50	20 mm open textured or dense, or 28 mm dense base course	85–110	9–12	110–120	8–9
60	28 mm dense base course, 40 mm open textured of dense base course or 40 mm single course as base course		8–10	130–150	7–8
65	28 mm dense base course, 40 mm open textured or dense base course or 40 mm single course	100–135	7–10	140–160	6–7
75	40 mm single course, 40 mm open textured or dense base course, 40 mm dense roadbase	120–150	7–8	165–185	5–6
100	40 mm dense base course or roadbase	–	–	220–240	4–4.5

KERBS, PAVING, ETC.

Surface Dressing Roads: Coverage (m²) per Tonne of Material

Size in mm	Sand	Granite chips	Gravel	Limestone chips
Sand	168	–	–	–
3	–	148	152	165
6	–	130	133	144
9	–	111	114	123
13	–	85	87	95
19	–	68	71	78

Sizes of Flags

Reference	Nominal size (mm)	Thickness (mm)
A	600 × 450	50 and 63
B	600 × 600	50 and 63
C	600 × 750	50 and 63
D	600 × 900	50 and 63
E	450 × 450	50 and 70 chamfered top surface
F	400 × 400	50 and 65 chamfered top surface
G	300 × 300	50 and 60 chamfered top surface

Sizes of Natural Stone Setts

Width (mm)		Length (mm)		Depth (mm)
100	×	100	×	100
75	×	150 to 250	×	125
75	×	150 to 250	×	150
100	×	150 to 250	×	100
100	×	150 to 250	×	150

SEEDING/TURFING AND PLANTING

Topsoil Quality

Topsoil grade	Properties
Premium	Natural topsoil, high fertility, loamy texture, good soil structure, suitable for intensive cultivation.
General purpose	Natural or manufactured topsoil of lesser quality than Premium, suitable for agriculture or amenity landscape, may need fertilizer or soil structure improvement.
Economy	Selected subsoil, natural mineral deposit such as river silt or greensand. The grade comprises two subgrades; 'Low clay' and 'High clay' which is more liable to compaction in handling. This grade is suitable for low-production agricultural land and amenity woodland or conservation planting areas.

Forms of Trees

Standards:	Shall be clear with substantially straight stems. Grafted and budded trees shall have no more than a slight bend at the union. Standards shall be designated as Half, Extra light, Light, Standard, Selected standard, Heavy, and Extra heavy.
Sizes of Standards	
Heavy standard	12–14 cm girth × 3.50 to 5.00 m high
Extra Heavy standard	14–16 cm girth × 4.25 to 5.00 m high
Extra Heavy standard	16–18 cm girth × 4.25 to 6.00 m high
Extra Heavy standard	18–20 cm girth × 5.00 to 6.00 m high
Semi-mature trees:	Between 6.0 m and 12.0 m tall with a girth of 20 to 75 cm at 1.0 m above ground.
Feathered trees:	Shall have a defined upright central leader, with stem furnished with evenly spread and balanced lateral shoots down to or near the ground.
Whips:	Shall be without significant feather growth as determined by visual inspection.
Multi-stemmed trees:	Shall have two or more main stems at, near, above or below ground.

Seedlings grown from seed and not transplanted shall be specified when ordered for sale as:

1+0	one year old seedling
2+0	two year old seedling
1+1	one year seed bed, one year transplanted = two year old seedling
1+2	one year seed bed, two years transplanted = three year old seedling
2+1	two years seed bed, one year transplanted = three year old seedling
1u1	two years seed bed, undercut after 1 year = two year old seedling
2u2	four years seed bed, undercut after 2 years = four year old seedling

SEEDING/TURFING AND PLANTING

Cuttings

The age of cuttings (plants grown from shoots, stems, or roots of the mother plant) shall be specified when ordered for sale. The height of transplants and undercut seedlings/cuttings (which have been transplanted or undercut at least once) shall be stated in centimetres. The number of growing seasons before and after transplanting or undercutting shall be stated.

0 + 1	one year cutting
0 + 2	two year cutting
0 + 1 + 1	one year cutting bed, one year transplanted = two year old seedling
0 + 1 + 2	one year cutting bed, two years transplanted = three year old seedling

Grass Cutting Capacities in m² per hour

Speed mph	Width of cut in metres												
	0.5	0.7	1.0	1.2	1.5	1.7	2.0	2.0	2.1	2.5	2.8	3.0	3.4
1.0	724	1127	1529	1931	2334	2736	3138	3219	3380	4023	4506	4828	5472
1.5	1086	1690	2293	2897	3500	4104	4707	4828	5069	6035	6759	7242	8208
2.0	1448	2253	3058	3862	4667	5472	6276	6437	6759	8047	9012	9656	10944
2.5	1811	2816	3822	4828	5834	6840	7846	8047	8449	10058	11265	12070	13679
3.0	2173	3380	4587	5794	7001	8208	9415	9656	10139	12070	13518	14484	16415
3.5	2535	3943	5351	6759	8167	9576	10984	11265	11829	14082	15772	16898	19151
4.0	2897	4506	6115	7725	9334	10944	12553	12875	13518	16093	18025	19312	21887
4.5	3259	5069	6880	8690	10501	12311	14122	14484	15208	18105	20278	21726	24623
5.0	3621	5633	7644	9656	11668	13679	15691	16093	16898	20117	22531	24140	27359
5.5	3983	6196	8409	10622	12834	15047	17260	17703	18588	22128	24784	26554	30095
6.0	4345	6759	9173	11587	14001	16415	18829	19312	20278	24140	27037	28968	32831
6.5	4707	7322	9938	12553	15168	17783	20398	20921	21967	26152	29290	31382	35566
7.0	5069	7886	10702	13518	16335	19151	21967	22531	23657	28163	31543	33796	38302

Number of Plants per m²: For Plants Planted on an Evenly Spaced Grid

Planting distances

mm	0.10	0.15	0.20	0.25	0.35	0.40	0.45	0.50	0.60	0.75	0.90	1.00	1.20	1.50
0.10	100.00	66.67	50.00	40.00	28.57	25.00	22.22	20.00	16.67	13.33	11.11	10.00	8.33	6.67
0.15	66.67	44.44	33.33	26.67	19.05	16.67	14.81	13.33	11.11	8.89	7.41	6.67	5.56	4.44
0.20	50.00	33.33	25.00	20.00	14.29	12.50	11.11	10.00	8.33	6.67	5.56	5.00	4.17	3.33
0.25	40.00	26.67	20.00	16.00	11.43	10.00	8.89	8.00	6.67	5.33	4.44	4.00	3.33	2.67
0.35	28.57	19.05	14.29	11.43	8.16	7.14	6.35	5.71	4.76	3.81	3.17	2.86	2.38	1.90
0.40	25.00	16.67	12.50	10.00	7.14	6.25	5.56	5.00	4.17	3.33	2.78	2.50	2.08	1.67
0.45	22.22	14.81	11.11	8.89	6.35	5.56	4.94	4.44	3.70	2.96	2.47	2.22	1.85	1.48
0.50	20.00	13.33	10.00	8.00	5.71	5.00	4.44	4.00	3.33	2.67	2.22	2.00	1.67	1.33
0.60	16.67	11.11	8.33	6.67	4.76	4.17	3.70	3.33	2.78	2.22	1.85	1.67	1.39	1.11
0.75	13.33	8.89	6.67	5.33	3.81	3.33	2.96	2.67	2.22	1.78	1.48	1.33	1.11	0.89
0.90	11.11	7.41	5.56	4.44	3.17	2.78	2.47	2.22	1.85	1.48	1.23	1.11	0.93	0.74
1.00	10.00	6.67	5.00	4.00	2.86	2.50	2.22	2.00	1.67	1.33	1.11	1.00	0.83	0.67
1.20	8.33	5.56	4.17	3.33	2.38	2.08	1.85	1.67	1.39	1.11	0.93	0.83	0.69	0.56
1.50	6.67	4.44	3.33	2.67	1.90	1.67	1.48	1.33	1.11	0.89	0.74	0.67	0.56	0.44

SEEDING/TURFING AND PLANTING

Grass Clippings Wet: Based on 3.5 m³/tonne

Annual kg/100 m²	Average 20 cuts kg/100 m²	m²/tonne	m²/m³
32.0	1.6	61162.1	214067.3

Nr of cuts	22	20	18	16	12	4
kg/cut	1.45	1.60	1.78	2.00	2.67	8.00
Area capacity of 3 tonne vehicle per load						
m²	206250	187500	168750	150000	112500	37500
Load m³	**100 m² units/m³ of vehicle space**					
1	196.4	178.6	160.7	142.9	107.1	35.7
2	392.9	357.1	321.4	285.7	214.3	71.4
3	589.3	535.7	482.1	428.6	321.4	107.1
4	785.7	714.3	642.9	571.4	428.6	142.9
5	982.1	892.9	803.6	714.3	535.7	178.6

Transportation of Trees

To unload large trees a machine with the necessary lifting strength is required. The weight of the trees must therefore be known in advance. The following table gives a rough overview. The additional columns with root ball dimensions and the number of plants per trailer provide additional information, for example about preparing planting holes and calculating unloading times.

Girth in cm	Rootball diameter in cm	Ball height in cm	Weight in kg	Numbers of trees per trailer
16–18	50–60	40	150	100–120
18–20	60–70	40–50	200	80–100
20–25	60–70	40–50	270	50–70
25–30	80	50–60	350	50
30–35	90–100	60–70	500	12–18
35–40	100–110	60–70	650	10–15
40–45	110–120	60–70	850	8–12
45–50	110–120	60–70	1100	5–7
50–60	130–140	60–70	1600	1–3
60–70	150–160	60–70	2500	1
70–80	180–200	70	4000	1
80–90	200–220	70–80	5500	1
90–100	230–250	80–90	7500	1
100–120	250–270	80–90	9500	1

Data supplied by Lorenz von Ehren GmbH
The information in the table is approximate; deviations depend on soil type, genus and weather

FENCING AND GATES

Types of Preservative

Creosote (tar oil) can be 'factory' applied	by pressure to BS 144: pts 1&2 by immersion to BS 144: pt 1 by hot and cold open tank to BS 144: pts 1&2
Copper/chromium/arsenic (CCA)	by full cell process to BS 4072 pts 1&2
Organic solvent (OS)	by double vacuum (vacvac) to BS 5707 pts 1&3 by immersion to BS 5057 pts 1&3
Pentachlorophenol (PCP)	by heavy oil double vacuum to BS 5705 pts 2&3
Boron diffusion process (treated with disodium octaborate to BWPA Manual 1986)	

Note: Boron is used on green timber at source and the timber is supplied dry

Cleft Chestnut Pale Fences

Pales	Pale spacing	Wire lines	
900 mm	75 mm	2	temporary protection
1050 mm	75 or 100 mm	2	light protective fences
1200 mm	75 mm	3	perimeter fences
1350 mm	75 mm	3	perimeter fences
1500 mm	50 mm	3	narrow perimeter fences
1800 mm	50 mm	3	light security fences

Close-Boarded Fences

Close-boarded fences 1.05 to 1.8 m high
Type BCR (recessed) or BCM (morticed) with concrete posts 140 × 115 mm tapered and Type BW with timber posts

Palisade Fences

Wooden palisade fences
Type WPC with concrete posts 140 × 115 mm tapered and Type WPW with timber posts

For both types of fence:
Height of fence 1050 mm: two rails
Height of fence 1200 mm: two rails
Height of fence 1500 mm: three rails
Height of fence 1650 mm: three rails
Height of fence 1800 mm: three rails

FENCING AND GATES

Post and Rail Fences

Wooden post and rail fences
Type MPR 11/3 morticed rails and Type SPR 11/3 nailed rails
Height to top of rail 1100 mm
Rails: three rails 87 mm, 38 mm

Type MPR 11/4 morticed rails and Type SPR 11/4 nailed rails
Height to top of rail 1100 mm
Rails: four rails 87 mm, 38 mm

Type MPR 13/4 morticed rails and Type SPR 13/4 nailed rails
Height to top of rail 1300 mm
Rail spacing 250 mm, 250 mm, and 225 mm from top
Rails: four rails 87 mm, 38 mm

Steel Posts

Rolled steel angle iron posts for chain link fencing

Posts	Fence height	Strut	Straining post
1500 × 40 × 40 × 5 mm	900 mm	1500 × 40 × 40 × 5 mm	1500 × 50 × 50 × 6 mm
1800 × 40 × 40 × 5 mm	1200 mm	1800 × 40 × 40 × 5 mm	1800 × 50 × 50 × 6 mm
2000 × 45 × 45 × 5 mm	1400 mm	2000 × 45 × 45 × 5 mm	2000 × 60 × 60 × 6 mm
2600 × 45 × 45 × 5 mm	1800 mm	2600 × 45 × 45 × 5 mm	2600 × 60 × 60 × 6 mm
3000 × 50 × 50 × 6 mm with arms	1800 mm	2600 × 45 × 45 × 5 mm	3000 × 60 × 60 × 6 mm

Concrete Posts

Concrete posts for chain link fencing

Posts and straining posts	Fence height	Strut
1570 mm 100 × 100 mm	900 mm	1500 mm × 75 × 75 mm
1870 mm 125 × 125 mm	1200 mm	1830 mm × 100 × 75 mm
2070 mm 125 × 125 mm	1400 mm	1980 mm × 100 × 75 mm
2620 mm 125 × 125 mm	1800 mm	2590 mm × 100 × 85 mm
3040 mm 125 × 125 mm	1800 mm	2590 mm × 100 × 85 mm (with arms)

FENCING AND GATES

Rolled Steel Angle Posts

Rolled steel angle posts for rectangular wire mesh (field) fencing

Posts	Fence height	Strut	Straining post
1200 × 40 × 40 × 5 mm	600 mm	1200 × 75 × 75 mm	1350 × 100 × 100 mm
1400 × 40 × 40 × 5 mm	800 mm	1400 × 75 × 75 mm	1550 × 100 × 100 mm
1500 × 40 × 40 × 5 mm	900 mm	1500 × 75 × 75 mm	1650 × 100 × 100 mm
1600 × 40 × 40 × 5 mm	1000 mm	1600 × 75 × 75 mm	1750 × 100 × 100 mm
1750 × 40 × 40 × 5 mm	1150 mm	1750 × 75 × 100 mm	1900 × 125 × 125 mm

Concrete Posts

Concrete posts for rectangular wire mesh (field) fencing

Posts	Fence height	Strut	Straining post
1270 × 100 × 100 mm	600 mm	1200 × 75 × 75 mm	1420 × 100 × 100 mm
1470 × 100 × 100 mm	800 mm	1350 × 75 × 75 mm	1620 × 100 × 100 mm
1570 × 100 × 100 mm	900 mm	1500 × 75 × 75 mm	1720 × 100 × 100 mm
1670 × 100 × 100 mm	600 mm	1650 × 75 × 75 mm	1820 × 100 × 100 mm
1820 × 125 × 125 mm	1150 mm	1830 × 75 × 100 mm	1970 × 125 × 125 mm

Cleft Chestnut Pale Fences

Timber Posts

Timber posts for wire mesh and hexagonal wire netting fences

Round timber for general fences

Posts	Fence height	Strut	Straining post
1300 × 65 mm dia.	600 mm	1200 × 80 mm dia.	1450 × 100 mm dia.
1500 × 65 mm dia.	800 mm	1400 × 80 mm dia.	1650 × 100 mm dia.
1600 × 65 mm dia.	900 mm	1500 × 80 mm dia.	1750 × 100 mm dia.
1700 × 65 mm dia.	1050 mm	1600 × 80 mm dia.	1850 × 100 mm dia.
1800 × 65 mm dia.	1150 mm	1750 × 80 mm dia.	2000 × 120 mm dia.

Squared timber for general fences

Posts	Fence height	Strut	Straining post
1300 × 75 × 75 mm	600 mm	1200 × 75 × 75 mm	1450 × 100 × 100 mm
1500 × 75 × 75 mm	800 mm	1400 × 75 × 75 mm	1650 × 100 × 100 mm
1600 × 75 × 75 mm	900 mm	1500 × 75 × 75 mm	1750 × 100 × 100 mm
1700 × 75 × 75 mm	1050 mm	1600 × 75 × 75 mm	1850 × 100 × 100 mm
1800 × 75 × 75 mm	1150 mm	1750 × 75 × 75 mm	2000 × 125 × 100 mm

FENCING AND GATES

Steel Fences to BS 1722: Part 9: 1992

	Fence height	Top/bottom rails and flat posts	Vertical bars
Light	1000 mm	40 × 10 mm 450 mm in ground	12 mm dia. at 115 mm cs
	1200 mm	40 × 10 mm 550 mm in ground	12 mm dia. at 115 mm cs
	1400 mm	40 × 10 mm 550 mm in ground	12 mm dia. at 115 mm cs
Light	1000 mm	40 × 10 mm 450 mm in ground	16 mm dia. at 120 mm cs
	1200 mm	40 × 10 mm 550 mm in ground	16 mm dia. at 120 mm cs
	1400 mm	40 × 10 mm 550 mm in ground	16 mm dia. at 120 mm cs
Medium	1200 mm	50 × 10 mm 550 mm in ground	20 mm dia. at 125 mm cs
	1400 mm	50 × 10 mm 550 mm in ground	20 mm dia. at 125 mm cs
	1600 mm	50 × 10 mm 600 mm in ground	22 mm dia. at 145 mm cs
	1800 mm	50 × 10 mm 600 mm in ground	22 mm dia. at 145 mm cs
Heavy	1600 mm	50 × 10 mm 600 mm in ground	22 mm dia. at 145 mm cs
	1800 mm	50 × 10 mm 600 mm in ground	22 mm dia. at 145 mm cs
	2000 mm	50 × 10 mm 600 mm in ground	22 mm dia. at 145 mm cs
	2200 mm	50 × 10 mm 600 mm in ground	22 mm dia. at 145 mm cs

Notes: Mild steel fences: round or square verticals; flat standards and horizontals. Tops of vertical bars may be bow-top, blunt, or pointed. Round or square bar railings

Timber Field Gates to BS 3470: 1975

Gates made to this standard are designed to open one way only
All timber gates are 1100 mm high
Width over stiles 2400, 2700, 3000, 3300, 3600, and 4200 mm
Gates over 4200 mm should be made in two leaves

Steel Field Gates to BS 3470: 1975

All steel gates are 1100 mm high
Heavy duty: width over stiles 2400, 3000, 3600 and 4500 mm
Light duty: width over stiles 2400, 3000, and 3600 mm

FENCING AND GATES

Domestic Front Entrance Gates to BS 4092: Part 1: 1966

Metal gates:	Single gates are 900 mm high minimum, 900 mm, 1000 mm and 1100 mm wide

Domestic Front Entrance Gates to BS 4092: Part 2: 1966

Wooden gates:	All rails shall be tenoned into the stiles
	Single gates are 840 mm high minimum, 801 mm and 1020 mm wide
	Double gates are 840 mm high minimum, 2130, 2340 and 2640 mm wide

Timber Bridle Gates to BS 5709:1979 (Horse or Hunting Gates)

Gates open one way only	
Minimum width between posts	1525 mm
Minimum height	1100 mm

Timber Kissing Gates to BS 5709:1979

Minimum width	700 mm
Minimum height	1000 mm
Minimum distance between shutting posts	600 mm
Minimum clearance at mid-point	600 mm

Metal Kissing Gates to BS 5709:1979

Sizes are the same as those for timber kissing gates
Maximum gaps between rails 120 mm

Categories of Pedestrian Guard Rail to BS 3049:1976

Class A for normal use
Class B where vandalism is expected
Class C where crowd pressure is likely

DRAINAGE

Width Required for Trenches for Various Diameters of Pipes

Pipe diameter (mm)	Trench n.e. 1.50 m deep	Trench over 1.50 m deep
n.e. 100 mm	450 mm	600 mm
100–150 mm	500 mm	650 mm
150–225 mm	600 mm	750 mm
225–300 mm	650 mm	800 mm
300–400 mm	750 mm	900 mm
400–450 mm	900 mm	1050 mm
450–600 mm	1100 mm	1300 mm

Weights and Dimensions – Vitrified Clay Pipes

Product	Nominal diameter (mm)	Effective length (mm)	BS 65 limits of tolerance		Crushing strength (kN/m)	Weight	
			min (mm)	max (mm)		(kg/pipe)	(kg/m)
Supersleve	100	1600	96	105	35.00	14.71	9.19
	150	1750	146	158	35.00	29.24	16.71
Hepsleve	225	1850	221	236	28.00	84.03	45.42
	300	2500	295	313	34.00	193.05	77.22
	150	1500	146	158	22.00	37.04	24.69
Hepseal	225	1750	221	236	28.00	85.47	48.84
	300	2500	295	313	34.00	204.08	81.63
	400	2500	394	414	44.00	357.14	143.90
	450	2500	444	464	44.00	454.55	181.63
	500	2500	494	514	48.00	555.56	222.22
	600	2500	591	615	57.00	796.23	307.69
	700	3000	689	719	67.00	1111.11	370.45
	800	3000	788	822	72.00	1351.35	450.45
Hepline	100	1600	95	107	22.00	14.71	9.19
	150	1750	145	160	22.00	29.24	16.71
	225	1850	219	239	28.00	84.03	45.42
	300	1850	292	317	34.00	142.86	77.22
Hepduct (conduit)	90	1500	–	–	28.00	12.05	8.03
	100	1600	–	–	28.00	14.71	9.19
	125	1750	–	–	28.00	20.73	11.84
	150	1750	–	–	28.00	29.24	16.71
	225	1850	–	–	28.00	84.03	45.42
	300	1850	–	–	34.00	142.86	77.22

DRAINAGE

Weights and Dimensions – Vitrified Clay Pipes

Nominal internal diameter (mm)	Nominal wall thickness (mm)	Approximate weight (kg/m)
150	25	45
225	29	71
300	32	122
375	35	162
450	38	191
600	48	317
750	54	454
900	60	616
1200	76	912
1500	89	1458
1800	102	1884
2100	127	2619

Wall thickness, weights and pipe lengths vary, depending on type of pipe required
The particulars shown above represent a selection of available diameters and are applicable to strength class 1 pipes with flexible rubber ring joints
Tubes with Ogee joints are also available

DRAINAGE

Weights and Dimensions – PVC-u Pipes

	Nominal size	Mean outside diameter (mm)		Wall thickness	Weight
		min	max	(mm)	(kg/m)
Standard pipes	82.4	82.4	82.7	3.2	1.2
	110.0	110.0	110.4	3.2	1.6
	160.0	160.0	160.6	4.1	3.0
	200.0	200.0	200.6	4.9	4.6
	250.0	250.0	250.7	6.1	7.2
Perforated pipes heavy grade	As above	As above	As above	As above	As above
thin wall	82.4	82.4	82.7	1.7	–
	110.0	110.0	110.4	2.2	–
	160.0	160.0	160.6	3.2	–

Width of Trenches Required for Various Diameters of Pipes

Pipe diameter (mm)	Trench n.e. 1.5 m deep (mm)	Trench over 1.5 m deep (mm)
n.e. 100	450	600
100–150	500	650
150–225	600	750
225–300	650	800
300–400	750	900
400–450	900	1050
450–600	1100	1300

DRAINAGE

DRAINAGE BELOW GROUND AND LAND DRAINAGE

Flow of Water Which Can Be Carried by Various Sizes of Pipe

Clay or concrete pipes

	Gradient of pipeline							
	1:10	1:20	1:30	1:40	1:50	1:60	1:80	1:100
Pipe size	Flow in litres per second							
DN 100 15.0	8.5	6.8	5.8	5.2	4.7	4.0	3.5	
DN 150 28.0	19.0	16.0	14.0	12.0	11.0	9.1	8.0	
DN 225 140.0	95.0	76.0	66.0	58.0	53.0	46.0	40.0	

Plastic pipes

	Gradient of pipeline							
	1:10	1:20	1:30	1:40	1:50	1:60	1:80	1:100
Pipe size	Flow in litres per second							
82.4 mm i/dia.	12.0	8.5	6.8	5.8	5.2	4.7	4.0	3.5
110 mm i/dia.	28.0	19.0	16.0	14.0	12.0	11.0	9.1	8.0
160 mm i/dia.	76.0	53.0	43.0	37.0	33.0	29.0	25.0	22.0
200 mm i/dia.	140.0	95.0	76.0	66.0	58.0	53.0	46.0	40.0

Vitrified (Perforated) Clay Pipes and Fittings to BS En 295-5 1994

Length not specified		
75 mm bore	250 mm bore	600 mm bore
100	300	700
125	350	800
150	400	1000
200	450	1200
225	500	

Precast Concrete Pipes: Prestressed Non-pressure Pipes and Fittings: Flexible Joints to BS 5911: Pt. 103: 1994

Rationalized metric nominal sizes: 450, 500	
Length:	500–1000 by 100 increments
	1000–2200 by 200 increments
	2200–2800 by 300 increments
Angles: length:	450–600 angles 45, 22.5,11.25°
	600 or more angles 22.5, 11.25°

DRAINAGE

Precast Concrete Pipes: Unreinforced and Circular Manholes and Soakaways to BS 5911: Pt. 200: 1994

Nominal sizes:	
Shafts:	675, 900 mm
Chambers:	900, 1050, 1200, 1350, 1500, 1800, 2100, 2400, 2700, 3000 mm
Large chambers:	To have either tapered reducing rings or a flat reducing slab in order to accept the standard cover
Ring depths:	1. 300–1200 mm by 300 mm increments except for bottom slab and rings below cover slab, these are by 150 mm increments
	2. 250–1000 mm by 250 mm increments except for bottom slab and rings below cover slab, these are by 125 mm increments
Access hole:	750 × 750 mm for DN 1050 chamber 1200 × 675 mm for DN 1350 chamber

Calculation of Soakaway Depth

The following formula determines the depth of concrete ring soakaway that would be required for draining given amounts of water.

$$h = \frac{4ar}{3\pi D^2}$$

h = depth of the chamber below the invert pipe
a = the area to be drained
r = the hourly rate of rainfall (50 mm per hour)
π = pi
D = internal diameter of the soakaway

This table shows the depth of chambers in each ring size which would be required to contain the volume of water specified. These allow a recommended storage capacity of ⅓ (one third of the hourly rainfall figure).

Table Showing Required Depth of Concrete Ring Chambers in Metres

Area m²	50	100	150	200	300	400	500
Ring size							
0.9	1.31	2.62	3.93	5.24	7.86	10.48	13.10
1.1	0.96	1.92	2.89	3.85	5.77	7.70	9.62
1.2	0.74	1.47	2.21	2.95	4.42	5.89	7.37
1.4	0.58	1.16	1.75	2.33	3.49	4.66	5.82
1.5	0.47	0.94	1.41	1.89	2.83	3.77	4.72
1.8	0.33	0.65	0.98	1.31	1.96	2.62	3.27
2.1	0.24	0.48	0.72	0.96	1.44	1.92	2.41
2.4	0.18	0.37	0.55	0.74	1.11	1.47	1.84
2.7	0.15	0.29	0.44	0.58	0.87	1.16	1.46
3.0	0.12	0.24	0.35	0.47	0.71	0.94	1.18

DRAINAGE

Precast Concrete Inspection Chambers and Gullies to BS 5911: Part 230: 1994

Nominal sizes:	375 diameter, 750, 900 mm deep 450 diameter, 750, 900, 1050, 1200 mm deep
Depths:	from the top for trapped or untrapped units: centre of outlet 300 mm invert (bottom) of the outlet pipe 400 mm
Depth of water seal for trapped gullies:	85 mm, rodding eye int. dia. 100 mm
Cover slab:	65 mm min

Bedding Flexible Pipes: PVC-u Or Ductile Iron

Type 1 =	100 mm fill below pipe, 300 mm above pipe: single size material
Type 2 =	100 mm fill below pipe, 300 mm above pipe: single size or graded material
Type 3 =	100 mm fill below pipe, 75 mm above pipe with concrete protective slab over
Type 4 =	100 mm fill below pipe, fill laid level with top of pipe
Type 5 =	200 mm fill below pipe, fill laid level with top of pipe
Concrete =	25 mm sand blinding to bottom of trench, pipe supported on chocks, 100 mm concrete under the pipe, 150 mm concrete over the pipe

DRAINAGE

Bedding Rigid Pipes: Clay or Concrete
(for vitrified clay pipes the manufacturer should be consulted)

Class D:	Pipe laid on natural ground with cut-outs for joints, soil screened to remove stones over 40 mm and returned over pipe to 150 m min depth. Suitable for firm ground with trenches trimmed by hand.
Class N:	Pipe laid on 50 mm granular material of graded aggregate to Table 4 of BS 882, or 10 mm aggregate to Table 6 of BS 882, or as dug light soil (not clay) screened to remove stones over 10 mm. Suitable for machine dug trenches.
Class B:	As Class N, but with granular bedding extending half way up the pipe diameter.
Class F:	Pipe laid on 100 mm granular fill to BS 882 below pipe, minimum 150 mm granular fill above pipe: single size material. Suitable for machine dug trenches.
Class A:	Concrete 100 mm thick under the pipe extending half way up the pipe, backfilled with the appropriate class of fill. Used where there is only a very shallow fall to the drain. Class A bedding allows the pipes to be laid to an exact gradient.
Concrete surround:	25 mm sand blinding to bottom of trench, pipe supported on chocks, 100 mm concrete under the pipe, 150 mm concrete over the pipe. It is preferable to bed pipes under slabs or wall in granular material.

PIPED SUPPLY SYSTEMS

Identification of Service Tubes From Utility to Dwellings

Utility	Colour	Size	Depth
British Telecom	grey	54 mm od	450 mm
Electricity	black	38 mm od	450 mm
Gas	yellow	42 mm od rigid 60 mm od convoluted	450 mm
Water	may be blue	(normally untubed)	750 mm

ELECTRICAL SUPPLY/POWER/LIGHTING SYSTEMS

Electrical Insulation Class En 60.598 BS 4533

Class 1:	luminaires comply with class 1 (I) earthed electrical requirements
Class 2:	luminaires comply with class 2 (II) double insulated electrical requirements
Class 3:	luminaires comply with class 3 (III) electrical requirements

Protection to Light Fittings

BS EN 60529:1992 Classification for degrees of protection provided by enclosures.
(IP Code – International or ingress Protection)

1st characteristic: against ingress of solid foreign objects

The figure	2	indicates that fingers cannot enter
	3	that a 2.5 mm diameter probe cannot enter
	4	that a 1.0 mm diameter probe cannot enter
	5	the fitting is dust proof (no dust around live parts)
	6	the fitting is dust tight (no dust entry)

2nd characteristic: ingress of water with harmful effects

The figure	0	indicates unprotected
	1	vertically dripping water cannot enter
	2	water dripping 15° (tilt) cannot enter
	3	spraying water cannot enter
	4	splashing water cannot enter
	5	jetting water cannot enter
	6	powerful jetting water cannot enter
	7	proof against temporary immersion
	8	proof against continuous immersion

Optional additional codes:		A–D protects against access to hazardous parts
	H	high voltage apparatus
	M	fitting was in motion during water test
	S	fitting was static during water test
	W	protects against weather

Marking code arrangement:	(example) IPX5S = IP (International or Ingress Protection)
	X (denotes omission of first characteristic)
	5 = jetting
	S = static during water test

RAIL TRACKS

	kg/m of track	lb/ft of track
Standard gauge Bull-head rails, chairs, transverse timber (softwood) sleepers etc.	245	165
Main lines Flat-bottom rails, transverse prestressed concrete sleepers, etc.	418	280
Add for electric third rail	51	35
Add for crushed stone ballast	2600	1750
	kg/m²	**lb/sq ft**
Overall average weight – rails connections, sleepers, ballast, etc.	733	150
	kg/m of track	**lb/ft of track**
Bridge rails, longitudinal timber sleepers, etc.	112	75

RAIL TRACKS

Heavy Rails

British Standard Section No.	Rail height (mm)	Foot width (mm)	Head width (mm)	Min web thickness (mm)	Section weight (kg/m)
Flat Bottom Rails					
60 A	114.30	109.54	57.15	11.11	30.62
70 A	123.82	111.12	60.32	12.30	34.81
75 A	128.59	114.30	61.91	12.70	37.45
80 A	133.35	117.47	63.50	13.10	39.76
90 A	142.88	127.00	66.67	13.89	45.10
95 A	147.64	130.17	69.85	14.68	47.31
100 A	152.40	133.35	69.85	15.08	50.18
110 A	158.75	139.70	69.85	15.87	54.52
113 A	158.75	139.70	69.85	20.00	56.22
50 'O'	100.01	100.01	52.39	10.32	24.82
80 'O'	127.00	127.00	63.50	13.89	39.74
60R	114.30	109.54	57.15	11.11	29.85
75R	128.59	122.24	61.91	13.10	37.09
80R	133.35	127.00	63.50	13.49	39.72
90R	142.88	136.53	66.67	13.89	44.58
95R	147.64	141.29	68.26	14.29	47.21
100R	152.40	146.05	69.85	14.29	49.60
95N	147.64	139.70	69.85	13.89	47.27
Bull Head Rails					
95R BH	145.26	69.85	69.85	19.05	47.07

Light Rails

British Standard Section No.	Rail height (mm)	Foot width (mm)	Head width (mm)	Min web thickness (mm)	Section weight (kg/m)
Flat Bottom Rails					
20M	65.09	55.56	30.96	6.75	9.88
30M	75.41	69.85	38.10	9.13	14.79
35M	80.96	76.20	42.86	9.13	17.39
35R	85.73	82.55	44.45	8.33	17.40
40	88.11	80.57	45.64	12.3	19.89
Bridge Rails					
13	48.00	92	36.00	18.0	13.31
16	54.00	108	44.50	16.0	16.06
20	55.50	127	50.00	20.5	19.86
28	67.00	152	50.00	31.0	28.62
35	76.00	160	58.00	34.5	35.38
50	76.00	165	58.50	–	50.18
Crane Rails					
A65	75.00	175.00	65.00	38.0	43.10
A75	85.00	200.00	75.00	45.0	56.20
A100	95.00	200.00	100.00	60.0	74.30
A120	105.00	220.00	120.00	72.0	100.00
175CR	152.40	152.40	107.95	38.1	86.92

RAIL TRACKS

Fish Plates

British Standard Section No.	Overall plate length		Hole diameter	Finished weight per pair	
	4 Hole (mm)	6 Hole (mm)	(mm)	4 Hole (kg/pair)	6 Hole (kg/pair)
For British Standard Heavy Rails: Flat Bottom Rails					
60 A	406.40	609.60	20.64	9.87	14.76
70 A	406.40	609.60	22.22	11.15	16.65
75 A	406.40	–	23.81	11.82	17.73
80 A	406.40	609.60	23.81	13.15	19.72
90 A	457.20	685.80	25.40	17.49	26.23
100 A	508.00	–	pear	25.02	–
110 A (shallow)	507.00	–	27.00	30.11	54.64
113 A (heavy)	507.00	–	27.00	30.11	54.64
50 'O' (shallow)	406.40	–	–	6.68	10.14
80 'O' (shallow)	495.30	–	23.81	14.72	22.69
60R (shallow)	406.40	609.60	20.64	8.76	13.13
60R (angled)	406.40	609.60	20.64	11.27	16.90
75R (shallow)	406.40	–	23.81	10.94	16.42
75R (angled)	406.40	–	23.81	13.67	–
80R (shallow)	406.40	609.60	23.81	11.93	17.89
80R (angled)	406.40	609.60	23.81	14.90	22.33
For British Standard Heavy Rails: Bull head rails					
95R BH (shallow)	–	457.20	27.00	14.59	14.61
For British Standard Light Rails: Flat Bottom Rails					
30M	355.6	–	–	–	2.72
35M	355.6	–	–	–	2.83
40	355.6	–	–	3.76	–

FRACTIONS, DECIMALS AND MILLIMETRE EQUIVALENTS

FRACTIONS, DECIMALS AND MILLIMETRE EQUIVALENTS

Fractions	Decimals	(mm)		Fractions	Decimals	(mm)
1/64	0.015625	0.396875		33/64	0.515625	13.096875
1/32	0.03125	0.79375		17/32	0.53125	13.49375
3/64	0.046875	1.190625		35/64	0.546875	13.890625
1/16	0.0625	1.5875		9/16	0.5625	14.2875
5/64	0.078125	1.984375		37/64	0.578125	14.684375
3/32	0.09375	2.38125		19/32	0.59375	15.08125
7/64	0.109375	2.778125		39/64	0.609375	15.478125
1/8	0.125	3.175		5/8	0.625	15.875
9/64	0.140625	3.571875		41/64	0.640625	16.271875
5/32	0.15625	3.96875		21/32	0.65625	16.66875
11/64	0.171875	4.365625		43/64	0.671875	17.065625
3/16	0.1875	4.7625		11/16	0.6875	17.4625
13/64	0.203125	5.159375		45/64	0.703125	17.859375
7/32	0.21875	5.55625		23/32	0.71875	18.25625
15/64	0.234375	5.953125		47/64	0.734375	18.653125
1/4	0.25	6.35		3/4	0.75	19.05
17/64	0.265625	6.746875		49/64	0.765625	19.446875
9/32	0.28125	7.14375		25/32	0.78125	19.84375
19/64	0.296875	7.540625		51/64	0.796875	20.240625
5/16	0.3125	7.9375		13/16	0.8125	20.6375
21/64	0.328125	8.334375		53/64	0.828125	21.034375
11/32	0.34375	8.73125		27/32	0.84375	21.43125
23/64	0.359375	9.128125		55/64	0.859375	21.828125
3/8	0.375	9.525		7/8	0.875	22.225
25/64	0.390625	9.921875		57/64	0.890625	22.621875
13/32	0.40625	10.31875		29/32	0.90625	23.01875
27/64	0.421875	10.71563		59/64	0.921875	23.415625
7/16	0.4375	11.1125		15/16	0.9375	23.8125
29/64	0.453125	11.50938		61/64	0.953125	24.209375
15/32	0.46875	11.90625		31/32	0.96875	24.60625
31/64	0.484375	12.30313		63/64	0.984375	25.003125
1/2	0.5	12.7		1.0	1	25.4

IMPERIAL STANDARD WIRE GAUGE (SWG)

SWG	Diameter		SWG	Diameter	
No.	(inches)	(mm)	No.	(inches)	(mm)
7/0	0.5	12.7	23	0.024	0.61
6/0	0.464	11.79	24	0.022	0.559
5/0	0.432	10.97	25	0.02	0.508
4/0	0.4	10.16	26	0.018	0.457
3/0	0.372	9.45	27	0.0164	0.417
2/0	0.348	8.84	28	0.0148	0.376
1/0	0.324	8.23	29	0.0136	0.345
1	0.3	7.62	30	0.0124	0.315
2	0.276	7.01	31	0.0116	0.295
3	0.252	6.4	32	0.0108	0.274
4	0.232	5.89	33	0.01	0.254
5	0.212	5.38	34	0.009	0.234
6	0.192	4.88	35	0.008	0.213
7	0.176	4.47	36	0.008	0.193
8	0.16	4.06	37	0.007	0.173
9	0.144	3.66	38	0.006	0.152
10	0.128	3.25	39	0.005	0.132
11	0.116	2.95	40	0.005	0.122
12	0.104	2.64	41	0.004	0.112
13	0.092	2.34	42	0.004	0.102
14	0.08	2.03	43	0.004	0.091
15	0.072	1.83	44	0.003	0.081
16	0.064	1.63	45	0.003	0.071
17	0.056	1.42	46	0.002	0.061
18	0.048	1.22	47	0.002	0.051
19	0.04	1.016	48	0.002	0.041
20	0.036	0.914	49	0.001	0.031
21	0.032	0.813	50	0.001	0.025
22	0.028	0.711			

PIPES, WATER, STORAGE, INSULATION

WATER PRESSURE DUE TO HEIGHT

Imperial

Head (Feet)	Pressure (lb/in²)		Head (Feet)	Pressure (lb/in²)
1	0.43		70	30.35
5	2.17		75	32.51
10	4.34		80	34.68
15	6.5		85	36.85
20	8.67		90	39.02
25	10.84		95	41.18
30	13.01		100	43.35
35	15.17		105	45.52
40	17.34		110	47.69
45	19.51		120	52.02
50	21.68		130	56.36
55	23.84		140	60.69
60	26.01		150	65.03
65	28.18			

Metric

Head (m)	Pressure (bar)		Head (m)	Pressure (bar)
0.5	0.049		18.0	1.766
1.0	0.098		19.0	1.864
1.5	0.147		20.0	1.962
2.0	0.196		21.0	2.06
3.0	0.294		22.0	2.158
4.0	0.392		23.0	2.256
5.0	0.491		24.0	2.354
6.0	0.589		25.0	2.453
7.0	0.687		26.0	2.551
8.0	0.785		27.0	2.649
9.0	0.883		28.0	2.747
10.0	0.981		29.0	2.845
11.0	1.079		30.0	2.943
12.0	1.177		32.5	3.188
13.0	1.275		35.0	3.434
14.0	1.373		37.5	3.679
15.0	1.472		40.0	3.924
16.0	1.57		42.5	4.169
17.0	1.668		45.0	4.415

1 bar	=	14.5038 lbf/in²	
1 lbf/in²	=	0.06895 bar	
1 metre	=	3.2808 ft or 39.3701 in	
1 foot	=	0.3048 metres	
1 in wg	=	2.5 mbar (249.1 N/m²)	

PIPES, WATER, STORAGE, INSULATION

Dimensions and Weights of Copper Pipes to BSEN 1057, BSEN 12499, BSEN 14251

Outside diameter (mm)	Internal diameter (mm)	Weight per metre (kg)	Internal diameter (mm)	Weight per metre (kg)	Internal siameter (mm)	Weight per metre (kg)
	Formerly Table X		Formerly Table Y		Formerly Table Z	
6	4.80	0.0911	4.40	0.1170	5.00	0.0774
8	6.80	0.1246	6.40	0.1617	7.00	0.1054
10	8.80	0.1580	8.40	0.2064	9.00	0.1334
12	10.80	0.1914	10.40	0.2511	11.00	0.1612
15	13.60	0.2796	13.00	0.3923	14.00	0.2031
18	16.40	0.3852	16.00	0.4760	16.80	0.2918
22	20.22	0.5308	19.62	0.6974	20.82	0.3589
28	26.22	0.6814	25.62	0.8985	26.82	0.4594
35	32.63	1.1334	32.03	1.4085	33.63	0.6701
42	39.63	1.3675	39.03	1.6996	40.43	0.9216
54	51.63	1.7691	50.03	2.9052	52.23	1.3343
76.1	73.22	3.1287	72.22	4.1437	73.82	2.5131
108	105.12	4.4666	103.12	7.3745	105.72	3.5834
133	130.38	5.5151	–	–	130.38	5.5151
159	155.38	8.7795	–	–	156.38	6.6056

Dimensions of Stainless Steel Pipes to BS 4127

Outside siameter (mm)	Maximum outside siameter (mm)	Minimum outside diameter (mm)	Wall thickness (mm)	Working pressure (bar)
6	6.045	5.940	0.6	330
8	8.045	7.940	0.6	260
10	10.045	9.940	0.6	210
12	12.045	11.940	0.6	170
15	15.045	14.940	0.6	140
18	18.045	17.940	0.7	135
22	22.055	21.950	0.7	110
28	28.055	27.950	0.8	121
35	35.070	34.965	1.0	100
42	42.070	41.965	1.1	91
54	54.090	53.940	1.2	77

Dimensions of Steel Pipes to BS 1387

Nominal Size (mm)	Approx. Outside Diameter (mm)	Outside diameter				Thickness		
		Light		Medium & Heavy		Light	Medium	Heavy
		Max (mm)	Min (mm)	Max (mm)	Min (mm)	(mm)	(mm)	(mm)
6	10.20	10.10	9.70	10.40	9.80	1.80	2.00	2.65
8	13.50	13.60	13.20	13.90	13.30	1.80	2.35	2.90
10	17.20	17.10	16.70	17.40	16.80	1.80	2.35	2.90
15	21.30	21.40	21.00	21.70	21.10	2.00	2.65	3.25
20	26.90	26.90	26.40	27.20	26.60	2.35	2.65	3.25
25	33.70	33.80	33.20	34.20	33.40	2.65	3.25	4.05
32	42.40	42.50	41.90	42.90	42.10	2.65	3.25	4.05
40	48.30	48.40	47.80	48.80	48.00	2.90	3.25	4.05
50	60.30	60.20	59.60	60.80	59.80	2.90	3.65	4.50
65	76.10	76.00	75.20	76.60	75.40	3.25	3.65	4.50
80	88.90	88.70	87.90	89.50	88.10	3.25	4.05	4.85
100	114.30	113.90	113.00	114.90	113.30	3.65	4.50	5.40
125	139.70	–	–	140.60	138.70	–	4.85	5.40
150	165.1*	–	–	166.10	164.10	–	4.85	5.40

* 165.1 mm (6.5in) outside diameter is not generally recommended except where screwing to BS 21 is necessary
All dimensions are in accordance with ISO R65 except approximate outside diameters which are in accordance with ISO R64
Light quality is equivalent to ISO R65 Light Series II

Approximate Metres Per Tonne of Tubes to BS 1387

Nom. size (mm)	BLACK						GALVANIZED					
	Plain/screwed ends			Screwed & socketed			Plain/screwed ends			Screwed & socketed		
	L (m)	M (m)	H (m)	L (m)	M (m)	H (m)	L (m)	M (m)	H (m)	L (m)	M (m)	H (m)
6	2765	2461	2030	2743	2443	2018	2604	2333	1948	2584	2317	1937
8	1936	1538	1300	1920	1527	1292	1826	1467	1254	1811	1458	1247
10	1483	1173	979	1471	1165	974	1400	1120	944	1386	1113	939
15	1050	817	688	1040	811	684	996	785	665	987	779	661
20	712	634	529	704	628	525	679	609	512	673	603	508
25	498	410	336	494	407	334	478	396	327	474	394	325
32	388	319	260	384	316	259	373	308	254	369	305	252
40	307	277	226	303	273	223	296	268	220	292	264	217
50	244	196	162	239	194	160	235	191	158	231	188	157
65	172	153	127	169	151	125	167	149	124	163	146	122
80	147	118	99	143	116	98	142	115	97	139	113	96
100	101	82	69	98	81	68	98	81	68	95	79	67
125	–	62	56	–	60	55	–	60	55	–	59	54
150	–	52	47	–	50	46	–	51	46	–	49	45

The figures for 'plain or screwed ends' apply also to tubes to BS 1775 of equivalent size and thickness
Key:
L – Light
M – Medium
H – Heavy

PIPES, WATER, STORAGE, INSULATION

Flange Dimension Chart to BS 4504 & BS 10

Normal Pressure Rating (PN 6) 6 Bar

Nom. size	Flange outside dia.	Table 6/2 Forged Welding Neck	Table 6/3 Plate Slip on	Table 6/4 Forged Bossed Screwed	Table 6/5 Forged Bossed Slip on	Table 6/8 Plate Blank	Raised face		Nr. bolt hole	Size of bolt
							Dia.	T'ness		
15	80	12	12	12	12	12	40	2	4	M10 × 40
20	90	14	14	14	14	14	50	2	4	M10 × 45
25	100	14	14	14	14	14	60	2	4	M10 × 45
32	120	14	16	14	14	14	70	2	4	M12 × 45
40	130	14	16	14	14	14	80	3	4	M12 × 45
50	140	14	16	14	14	14	90	3	4	M12 × 45
65	160	14	16	14	14	14	110	3	4	M12 × 45
80	190	16	18	16	16	16	128	3	4	M16 × 55
100	210	16	18	16	16	16	148	3	4	M16 × 55
125	240	18	20	18	18	18	178	3	8	M16 × 60
150	265	18	20	18	18	18	202	3	8	M16 × 60
200	320	20	22	–	20	20	258	3	8	M16 × 60
250	375	22	24	–	22	22	312	3	12	M16 × 65
300	440	22	24	–	22	22	365	4	12	M20 × 70

Normal Pressure Rating (PN 16) 16 Bar

Nom. size	Flange outside dia.	Table 6/2 Forged Welding Neck	Table 6/3 Plate Slip on	Table 6/4 Forged Bossed Screwed	Table 6/5 Forged Bossed Slip on	Table 6/8 Plate Blank	Raised face		Nr. bolt hole	Size of bolt
							Dia.	T'ness		
15	95	14	14	14	14	14	45	2	4	M12 × 45
20	105	16	16	16	16	16	58	2	4	M12 × 50
25	115	16	16	16	16	16	68	2	4	M12 × 50
32	140	16	16	16	16	16	78	2	4	M16 × 55
40	150	16	16	16	16	16	88	3	4	M16 × 55
50	165	18	18	18	18	18	102	3	4	M16 × 60
65	185	18	18	18	18	18	122	3	4	M16 × 60
80	200	20	20	20	20	20	138	3	8	M16 × 60
100	220	20	20	20	20	20	158	3	8	M16 × 65
125	250	22	22	22	22	22	188	3	8	M16 × 70
150	285	22	22	22	22	22	212	3	8	M20 × 70
200	340	24	24	–	24	24	268	3	12	M20 × 75
250	405	26	26	–	26	26	320	3	12	M24 × 90
300	460	28	28	–	28	28	378	4	12	M24 × 90

PIPES, WATER, STORAGE, INSULATION

Minimum Distances Between Supports/Fixings

Material	BS Nominal pipe size		Pipes – Vertical	Pipes – Horizontal on to low gradients
	(inch)	(mm)	Support distance in metres	Support distance in metres
Copper	0.50	15.00	1.90	1.30
	0.75	22.00	2.50	1.90
	1.00	28.00	2.50	1.90
	1.25	35.00	2.80	2.50
	1.50	42.00	2.80	2.50
	2.00	54.00	3.90	2.50
	2.50	67.00	3.90	2.80
	3.00	76.10	3.90	2.80
	4.00	108.00	3.90	2.80
	5.00	133.00	3.90	2.80
	6.00	159.00	3.90	2.80
muPVC	1.25	32.00	1.20	0.50
	1.50	40.00	1.20	0.50
	2.00	50.00	1.20	0.60
Polypropylene	1.25	32.00	1.20	0.50
	1.50	40.00	1.20	0.50
uPVC	–	82.40	1.20	0.50
	–	110.00	1.80	0.90
	–	160.00	1.80	1.20
Steel	0.50	15.00	2.40	1.80
	0.75	20.00	3.00	2.40
	1.00	25.00	3.00	2.40
	1.25	32.00	3.00	2.40
	1.50	40.00	3.70	2.40
	2.00	50.00	3.70	2.40
	2.50	65.00	4.60	3.00
	3.00	80.40	4.60	3.00
	4.00	100.00	4.60	3.00
	5.00	125.00	5.50	3.70
	6.00	150.00	5.50	4.50
	8.00	200.00	8.50	6.00
	10.00	250.00	9.00	6.50
	12.00	300.00	10.00	7.00
	16.00	400.00	10.00	8.25

PIPES, WATER, STORAGE, INSULATION

Litres of Water Storage Required Per Person Per Building Type

Type of building	Storage (litres)
Houses and flats (up to 4 bedrooms)	120/bedroom
Houses and flats (more than 4 bedrooms)	100/bedroom
Hostels	90/bed
Hotels	200/bed
Nurses homes and medical quarters	120/bed
Offices with canteen	45/person
Offices without canteen	40/person
Restaurants	7/meal
Boarding schools	90/person
Day schools – Primary	15/person
Day schools – Secondary	20/person

Recommended Air Conditioning Design Loads

Building type	Design loading
Computer rooms	500 W/m² of floor area
Restaurants	150 W/m² of floor area
Banks (main area)	100 W/m² of floor area
Supermarkets	25 W/m² of floor area
Large office block (exterior zone)	100 W/m² of floor area
Large office block (interior zone)	80 W/m² of floor area
Small office block (interior zone)	80 W/m² of floor area

PIPES, WATER, STORAGE, INSULATION

Capacity and Dimensions of Galvanized Mild Steel Cisterns – BS 417

Capacity (litres)	BS type (SCM)	Dimensions		
		Length (mm)	Width (mm)	Depth (mm)
18	45	457	305	305
36	70	610	305	371
54	90	610	406	371
68	110	610	432	432
86	135	610	457	482
114	180	686	508	508
159	230	736	559	559
191	270	762	584	610
227	320	914	610	584
264	360	914	660	610
327	450/1	1220	610	610
336	450/2	965	686	686
423	570	965	762	787
491	680	1090	864	736
709	910	1070	889	889

Capacity of Cold Water Polypropylene Storage Cisterns – BS 4213

Capacity (litres)	BS type (PC)	Maximum height (mm)
18	4	310
36	8	380
68	15	430
91	20	510
114	25	530
182	40	610
227	50	660
273	60	660
318	70	660
455	100	760

PIPES, WATER, STORAGE, INSULATION

Minimum Insulation Thickness to Protect Against Freezing for Domestic Cold Water Systems (8 Hour Evaluation Period)

Pipe size (mm)	Insulation thickness (mm)					
	Condition 1			Condition 2		
	λ = 0.020	λ = 0.030	λ = 0.040	λ = 0.020	λ = 0.030	λ = 0.040
Copper pipes						
15	11	20	34	12	23	41
22	6	9	13	6	10	15
28	4	6	9	4	7	10
35	3	5	7	4	5	7
42	3	4	5	8	4	6
54	2	3	4	2	3	4
76	2	2	3	2	2	3
Steel pipes						
15	9	15	24	10	18	29
20	6	9	13	6	10	15
25	4	7	9	5	7	10
32	3	5	6	3	5	7
40	3	4	5	3	4	6
50	2	3	4	2	3	4
65	2	2	3	2	3	3

Condition 1: water temperature 7°C; ambient temperature –6°C; evaluation period 8 h; permitted ice formation 50%; normal installation, i.e. inside the building and inside the envelope of the structural insulation
Condition 2: water temperature 2°C; ambient temperature –6°C; evaluation period 8 h; permitted ice formation 50%; extreme installation, i.e. inside the building but outside the envelope of the structural insulation
λ = thermal conductivity [W/(mK)]

Insulation Thickness for Chilled And Cold Water Supplies to Prevent Condensation

On a Low Emissivity Outer Surface (0.05, i.e. Bright Reinforced Aluminium Foil) with an Ambient Temperature of +25°C and a Relative Humidity of 80%

Steel pipe size (mm)	t = +10			t = +5			t = 0		
	Insulation thickness (mm)			Insulation thickness (mm)			Insulation thickness (mm)		
	λ = 0.030	λ = 0.040	λ = 0.050	λ = 0.030	λ = 0.040	λ = 0.050	λ = 0.030	λ = 0.040	λ = 0.050
15	16	20	25	22	28	34	28	36	43
25	18	24	29	25	32	39	32	41	50
50	22	28	34	30	39	47	38	49	60
100	26	34	41	36	47	57	46	60	73
150	29	38	46	40	52	64	51	67	82
250	33	43	53	46	60	74	59	77	94
Flat surfaces	39	52	65	56	75	93	73	97	122

t = temperature of contents (°C)
λ = thermal conductivity at mean temperature of insulation [W/(mK)]

PIPES, WATER, STORAGE, INSULATION

Insulation Thickness for Non-domestic Heating Installations to Control Heat Loss

Steel pipe size (mm)	t = 75			t = 100			t = 150		
	Insulation thickness (mm)			Insulation thickness (mm)			Insulation thickness (mm)		
	$\lambda = 0.030$	$\lambda = 0.040$	$\lambda = 0.050$	$\lambda = 0.030$	$\lambda = 0.040$	$\lambda = 0.050$	$\lambda = 0.030$	$\lambda = 0.040$	$\lambda = 0.050$
10	18	32	55	20	36	62	23	44	77
15	19	34	56	21	38	64	26	47	80
20	21	36	57	23	40	65	28	50	83
25	23	38	58	26	43	68	31	53	85
32	24	39	59	28	45	69	33	55	87
40	25	40	60	29	47	70	35	57	88
50	27	42	61	31	49	72	37	59	90
65	29	43	62	33	51	74	40	63	92
80	30	44	62	35	52	75	42	65	94
100	31	46	63	37	54	76	45	68	96
150	33	48	64	40	57	77	50	73	100
200	35	49	65	42	59	79	53	76	103
250	36	50	66	43	61	80	55	78	105

t = hot face temperature (°C)
λ = thermal conductivity at mean temperature of insulation [W/(mK)]

Index

telephone 189
water 189

Valley gutters, steel 548
Vapour barriers 524
Variations, regional 42
VAT 9–15
Veneered plywood 394
Ventilating systems, all-in rates
   distribution centres 177–178
   hospitals 177–178
   hotels 177–178
   offices 177–178
   performing arts buildings 177–178
   residential 177–178
   schools 177–178
   shopping malls 177–178
   sports halls 177–178
   supermarkets 177–178
Vermiculite, boards 385–386
Vinyl flooring
   safety 489–491
   sports 490
Vinyl floors 489–491
Visqueen, *see* Membranes

Wage rates 45–47
   BATJIC 46
   CIJC 45
   JIB-Scotland and Northern Ireland 47
   plumbing and mechanical engineering 46
Walkways, metal
   chequer plate 471
   open mesh 471
Wallboard, supply only 374
Wallpaper, *see* Decoration
Walls
   composite panels
      flat profile 401
      profiled 400–402
   concrete 233–236
   external
      block 160
      cavity 160
      common bricks 160
      concrete panels 161
      curtain walling 162, 403–405
      facing bricks 160
      flat panel 162
      glazed 162

      metal 161
      micro-rib 162
      sheet cladding 161
   finishes 170
      cement and sand render 493–494
      Cemrend 496
      Expamet beads 497
      granite 499
      Hardwall 494
      insulated render 498
      Limelite renovating 495
      marble 499
      mosaic 499
      paint 170
      panels 171
      pebbledash 496
      plaster 170, 494–496
      render 170
      Thistle 96–497
      tiles 171
      Tyrolean 496
      vinyl paper 170
      x-ray plaster 495
   Georgian wired glazing 399
   insulation 163–164
   internal 167–169
   laminated clear 399
   panelling 500
   partitions 386–390
   profiled pcv-ue strips 400
   rainscreen
      aluminium 403
      brick tiles 163
      cellulose fibre 402
      clay 403
      Terracotta 163
      timber 403
      Trespa 163
      Western Red Cedar 162
   retaining
      concrete 183
      crib 183
      embedded 230
      gabion basket 231
   screen, masonry 184
   sheet cladding plastic 402
   soil
      modular 183
      reinforced 183
   steel cladding

Spon's Asia Pacific Construction Costs Handbook, Fifth Edition

LANGDON & SEAH

In the last few years, the global economic outlook has continued to be shrouded in uncertainty and volatility following the financial crisis in the Euro zone. While the US and Europe are going through a difficult period, investors are focusing more keenly on Asia. This fifth edition provides overarching construction cost data for 16 countries: Brunei, Cambodia, China, Hong Kong, India, Indonesia, Japan, Malaysia, Myanmar, Philippines, Singapore, South Korea, Sri Lanka, Taiwan, Thailand and Vietnam.

May 2015: 234X156 mm: 452 pp
Hb: 978-1-4822-4358-1: £160.00

To Order: Tel: +44 (0) 1235 400524 Fax: +44 (0) 1235 400525
or Post: Taylor and Francis Customer Services,
Bookpoint Ltd, Unit T1, 200 Milton Park, Abingdon, Oxon, OX14 4TA UK
Email: book.orders@tandf.co.uk

For a complete listing of all our titles visit:
www.tandf.co.uk

It is one thing to
imagine a better world.
It's another to deliver it.

**Understanding change,
unlocking potential, creating
brilliant new communities.**

The Tate Modern extension
takes an iconic building and
adds to it. Cost management
provided by AECOM.

Built to deliver a better world

aecom.com

BIM and Quantity Surveying
PITTARD & SELL

The sudden arrival of Building Information Modelling (BIM) as a key part of the building industry is redefining the roles and working practices of its stakeholders. Many clients, designers, contractors, quantity surveyors, and building managers are still finding their feet in an industry where BIM compliance can bring great rewards.

This guide is designed to help quantity surveying practitioners and students understand what BIM means for them, and how they should prepare to work successfully on BIM compliant projects. The case studies show how firms at the forefront of this technology have integrated core quantity surveying responsibilities like cost estimating, tendering, and development appraisal into high profile BIM projects. In addition to this, the implications for project management, facilities management, contract administration and dispute resolution are also explored through case studies, making this a highly valuable guide for those in a range of construction project management roles.

Featuring a chapter describing how the role of the quantity surveyor is likely to permanently shift as a result of this development, as well as descriptions of tools used, this covers both the organisational and practical aspects of a crucial topic.

December 2015: 234 x 156 mm: 258 pp
Pb: 978-0-415-87043-6; £24.99

To Order: Tel: +44 (0) 1235 400524 Fax: +44 (0) 1235 400525
or Post: Taylor and Francis Customer Services,
Bookpoint Ltd, Unit T1, 200 Milton Park, Abingdon, Oxon, OX14 4TA UK
Email: book.orders@tandf.co.uk

For a complete listing of all our titles visit:
www.tandf.co.uk

Estimator's Pocket Book

Duncan Cartlidge

The Estimator's Pocket Book is a concise and practical reference covering the main pricing approaches, as well as useful information such as how to process sub-contractor quotations, tender settlement and adjudication. It is fully up-to-date with NRM2 throughout, features a look ahead to NRM3 and describes the implications of BIM for estimators.

It includes instructions on how to handle:

- the NRM order of cost estimate;
- unit-rate pricing for different trades;
- pro rata pricing and dayworks
- builders' quantities;
- approximate quantities.

Worked examples show how each of these techniques should be carried out in clear, easy-to-follow steps. This is the indispensible estimating reference for all quantity surveyors, cost managers, project managers and anybody else with estimating responsibilities. Particular attention is given to NRM2, but the overall focus is on the core estimating skills needed in practice.

May 2013 186x123: 310pp
Pb: 978-0-415-52711-8: £21.99

To Order: Tel: +44 (0) 1235 400524 Fax: +44 (0) 1235 400525
or Post: Taylor and Francis Customer Services,
Bookpoint Ltd, Unit T1, 200 Milton Park, Abingdon, Oxon, OX14 4TA UK
Email: book.orders@tandf.co.uk

For a complete listing of all our titles visit:
www.tandf.co.uk

Ebook Single-User Licence Agreement

We welcome you as a user of this Spon Price Book ebook and hope that you find it a useful and valuable tool. Please read this document carefully. **This is a legal agreement** between you (hereinafter referred to as the "Licensee") and Taylor and Francis Books Ltd. (the "Publisher"), which defines the terms under which you may use the Product. **By accessing and retrieving the access code on the label inside the front cover of this book you agree to these terms and conditions outlined herein. If you do not agree to these terms you must return the Product to your supplier intact, with the seal on the label unbroken and with the access code not accessed.**

1. **Definition of the Product**

   The product which is the subject of this Agreement, (the "Product") consists of online and offline access to the VitalSource ebook edition of *Spon's Architects' & Builders' Price Book 2017*.

2. **Commencement and licence**

   2.1 This Agreement commences upon the breaking open of the document containing the access code by the Licensee (the "Commencement Date").

   2.2 This is a licence agreement (the "Agreement") for the use of the Product by the Licensee, and not an agreement for sale.

   2.3 The Publisher licenses the Licensee on a non-exclusive and non-transferable basis to use the Product on condition that the Licensee complies with this Agreement. The Licensee acknowledges that it is only permitted to use the Product in accordance with this Agreement.

3. **Multiple use**

   Use of the Product is not provided or allowed for more than one user or for a wide area network or consortium.

4. **Installation and Use**

   4.1 The Licensee may provide access to the Product for individual study in the following manner: The Licensee may install the Product on a secure local area network on a single site for use by one user.

   4.2 The Licensee shall be responsible for installing the Product and for the effectiveness of such installation.

   4.3 Text from the Product may be incorporated in a coursepack. Such use is only permissible with the express permission of the Publisher in writing and requires the payment of the appropriate fee as specified by the Publisher and signature of a separate licence agreement.

   4.4 The Product is a free addition to the book and the Publisher is under no obligation to provide any technical support.

5. **Permitted Activities**

   5.1 The Licensee shall be entitled to use the Product for its own internal purposes;

   5.2 The Licensee acknowledges that its rights to use the Product are strictly set out in this Agreement, and all other uses (whether expressly mentioned in Clause 6 below or not) are prohibited.

6. **Prohibited Activities**

   The following are prohibited without the express permission of the Publisher:

   6.1 The commercial exploitation of any part of the Product.

   6.2 The rental, loan, (free or for money or money's worth) or hire purchase of this product, save with the express consent of the Publisher.

   6.3 Any activity which raises the reasonable prospect of impeding the Publisher's ability or opportunities to market the Product.

   6.4 Any networking, physical or electronic distribution or dissemination of the product save as expressly permitted by this Agreement.

   6.5 Any reverse engineering, decompilation, disassembly or other alteration of the Product save in accordance with applicable national laws.

   6.6 The right to create any derivative product or service from the Product save as expressly provided for in this Agreement.

   6.7 Any alteration, amendment, modification or deletion from the Product, whether for the purposes of error correction or otherwise.

7.   General Responsibilities of the License

7.1   The Licensee will take all reasonable steps to ensure that the Product is used in accordance with the terms and conditions of this Agreement.

7.2   The Licensee acknowledges that damages may not be a sufficient remedy for the Publisher in the event of breach of this Agreement by the Licensee, and that an injunction may be appropriate.

7.3   The Licensee undertakes to keep the Product safe and to use its best endeavours to ensure that the product does not fall into the hands of third parties, whether as a result of theft or otherwise.

7.4   Where information of a confidential nature relating to the product of the business affairs of the Publisher comes into the possession of the Licensee pursuant to this Agreement (or otherwise), the Licensee agrees to use such information solely for the purposes of this Agreement, and under no circumstances to disclose any element of the information to any third party save strictly as permitted under this Agreement. For the avoidance of doubt, the Licensee's obligations under this sub-clause 7.4 shall survive the termination of this Agreement.

8.   Warrant and Liability

8.1   The Publisher warrants that it has the authority to enter into this agreement and that it has secured all rights and permissions necessary to enable the Licensee to use the Product in accordance with this Agreement.

8.2   The Publisher warrants that the Product as supplied on the Commencement Date shall be free of defects in materials and workmanship, and undertakes to replace any defective Product within 28 days of notice of such defect being received provided such notice is received within 30 days of such supply. As an alternative to replacement, the Publisher agrees fully to refund the Licensee in such circumstances, if the Licensee so requests, provided that the Licensee returns this copy of *Spon's Architects' & Builders' Price Book 2017* to the Publisher. The provisions of this sub-clause 8.2 do not apply where the defect results from an accident or from misuse of the product by the Licensee.

8.3   Sub-clause 8.2 sets out the sole and exclusive remedy of the Licensee in relation to defects in the Product.

8.4   The Publisher and the Licensee acknowledge that the Publisher supplies the Product on an "as is" basis. The Publisher gives no warranties:

   8.4.1   that the Product satisfies the individual requirements of the Licensee; or
   8.4.2   that the Product is otherwise fit for the Licensee's purpose; or
   8.4.3   that the Product is compatible with the Licensee's hardware equipment and software operating environment.

8.5   The Publisher hereby disclaims all warranties and conditions, express or implied, which are not stated above.

8.6   Nothing in this Clause 8 limits the Publisher's liability to the Licensee in the event of death or personal injury resulting from the Publisher's negligence.

8.7   The Publisher hereby excludes liability for loss of revenue, reputation, business, profits, or for indirect or consequential losses, irrespective of whether the Publisher was advised by the Licensee of the potential of such losses.

8.8   The Licensee acknowledges the merit of independently verifying the price book data prior to taking any decisions of material significance (commercial or otherwise) based on such data. It is agreed that the Publisher shall not be liable for any losses which result from the Licensee placing reliance on the data under any circumstances.

8.9   Subject to sub-clause 8.6 above, the Publisher's liability under this Agreement shall be limited to the purchase price.

9.   Intellectual Property Rights

9.1   Nothing in this Agreement affects the ownership of copyright or other intellectual property rights in the Product.

9.2   The Licensee agrees to display the Publishers' copyright notice in the manner described in the Product.

9.3   The Licensee hereby agrees to abide by copyright and similar notice requirements required by the Publisher, details of which are as follows:

"© 2017 Taylor & Francis. All rights reserved. All materials in *Spon's Architects' & Builders' Price Book 2017* are copyright protected. All rights reserved. No such materials may be used, displayed, modified, adapted, distributed, transmitted, transferred, published or otherwise reproduced in any form or by any means now or hereafter developed other than strictly in accordance with the terms of the licence agreement enclosed with *Spon's Architects' & Builders' Price Book 2017*. However, text and images may be printed and copied for research and private study within the preset program limitations. Please note the copyright notice above, and that any text or images printed or copied must credit the source."

9.4   This Product contains material proprietary to and copyedited by the Publisher and others. Except for the licence granted herein, all rights, title and interest in the Product, in all languages, formats and media throughout the world, including copyrights therein, are and remain the property of the Publisher or other copyright holders identified in the Product.

10. Non-assignment

This Agreement and the licence contained within it may not be assigned to any other person or entity without the written consent of the Publisher.

11. Termination and Consequences of Termination.

11.1 The Publisher shall have the right to terminate this Agreement if:

    11.1.1 the Licensee is in material breach of this Agreement and fails to remedy such breach (where capable of remedy) within 14 days of a written notice from the Publisher requiring it to do so; or

    11.1.2 the Licensee becomes insolvent, becomes subject to receivership, liquidation or similar external administration; or

    11.1.3 the Licensee ceases to operate in business.

11.2 The Licensee shall have the right to terminate this Agreement for any reason upon two month's written notice. The Licensee shall not be entitled to any refund for payments made under this Agreement prior to termination under this sub-clause 11.2.

11.3 Termination by either of the parties is without prejudice to any other rights or remedies under the general law to which they may be entitled, or which survive such termination (including rights of the Publisher under sub-clause 7.4 above).

11.4 Upon termination of this Agreement, or expiry of its terms, the Licensee must destroy all copies and any back up copies of the product or part thereof.

12. General

12.1 *Compliance with export provisions*

The Publisher hereby agrees to comply fully with all relevant export laws and regulations of the United Kingdom to ensure that the Product is not exported, directly or indirectly, in violation of English law.

12.2 *Force majeure*

The parties accept no responsibility for breaches of this Agreement occurring as a result of circumstances beyond their control.

12.3 *No waiver*

Any failure or delay by either party to exercise or enforce any right conferred by this Agreement shall not be deemed to be a waiver of such right.

12.4 *Entire agreement*

This Agreement represents the entire agreement between the Publisher and the Licensee concerning the Product. The terms of this Agreement supersede all prior purchase orders, written terms and conditions, written or verbal representations, advertising or statements relating in any way to the Product.

12.5 *Severability*

If any provision of this Agreement is found to be invalid or unenforceable by a court of law of competent jurisdiction, such a finding shall not affect the other provisions of this Agreement and all provisions of this Agreement unaffected by such a finding shall remain in full force and effect.

12.6 *Variations*

This agreement may only be varied in writing by means of variation signed in writing by both parties.

12.7 *Notices*

All notices to be delivered to: Spon's Price Books, Taylor & Francis Books Ltd., 3 Park Square, Milton Park, Abingdon, Oxfordshire, OX14 4RN, UK.

12.8 *Governing law*

This Agreement is governed by English law and the parties hereby agree that any dispute arising under this Agreement shall be subject to the jurisdiction of the English courts.

If you have any queries about the terms of this licence, please contact:

Spon's Price Books
Taylor & Francis Books Ltd.
3 Park Square, Milton Park, Abingdon, Oxfordshire, OX14 4RN
Tel: +44 (0) 20 7017 6000
www.sponpress.com

Spon Press
an imprint of Taylor & Francis

Ebook options

Your print copy comes with a free ebook on the VitalSource®
Bookshelf platform. Further copies of the ebook are available from
https://www.crcpress.com/search/results?kw=spon+2017
Pick your price book and select the ebook under the 'Select Format'
option.

To buy Spon's ebooks for five or more users of in your organisation
please contact:

Spon's Price Books
eBooks & Online Sales
Taylor & Francis Books Ltd.
3 Park Square, Milton Park, Oxfordshire, OX14 4RN
Tel: (020) 337 73480
onlinesales@informa.com